Preface

This *Student's Solutions Manual* contains solutions to selected exercises in the text *Introductory Algebra, Eighth Edition* by Margaret L. Lial, John Hornsby, and Terry McGinnis. It contains solutions to all margin exercises, the odd-numbered exercises in each section, all Relating Concepts exercises, as well as solutions to all the exercises in the review sections, the chapter tests, and the cumulative review sections.

This manual is a text supplement and should be read along *with* the text. You should read all exercise solutions in this manual because many concept explanations are given and then used in subsequent solutions. All concepts necessary to solve a particular problem are not reviewed for every exercise. If you are having difficulty with a previously covered concept, refer back to the section where it was covered for more complete help.

A significant number of today's students are involved in various outside activities, and find it difficult, if not impossible, to attend all class sessions; this manual should help meet the needs of these students. In addition, it is my hope that this manual's solutions will enhance the understanding of all readers of the material and provide insights to solving other exercises.

I appreciate feedback concerning errors, solution correctness or style, and manual style. Any comments may be sent directly to me at the address below, at jeff.cole@anokaramsey.edu, or in care of the publisher, Pearson Addison-Wesley.

I would like to thank Ken Grace, of Anoka-Ramsey Community College, and Jeannine Grace, for typesetting the manuscript and providing assistance with many features of the manual; Mary Johnson, of Inver Hills Community College, and Marv Riedesel, for their careful accuracy checking and valuable suggestions; Jim McLaughlin, for his help with the entire art package; and the authors and Maureen O'Connor and Lauren Morse, of Pearson Addison-Wesley, for entrusting me with this project.

Jeffery A. Cole
Anoka-Ramsey Community College
11200 Mississippi Blvd. NW
Coon Rapids, MN 55433

STUDENT'S SOLUTIONS MANUAL

JEFFERY A. COLE
Anoka-Ramsey Community College

INTRODUCTORY ALGEBRA

EIGHTH EDITION

Margaret L. Lial
American River College

John Hornsby
University of New Orleans

Terry McGinnis

Boston San Francisco New York
London Toronto Sydney Tokyo Singapore Madrid
Mexico City Munich Paris Cape Town Hong Kong Montreal

Reproduced by Pearson Addison-Wesley from electronic files supplied by the author.

Copyright © 2006 Pearson Education, Inc.
Publishing as Pearson Addison-Wesley, 75 Arlington Street, Boston, MA 02116.

All rights reserved. This manual may be reproduced for classroom use only. Printed in the United States of America.

ISBN   0-321-28580-8

3 4 5 6 BB 08 07 06

Table of Contents

R Prealgebra Review 1
    R.1   Fractions • 1
    R.2   Decimals and Percents • 6

1 The Real Number System 11
    1.1   Exponents, Order of Operations, and Inequality • 11
    1.2   Variables, Expressions, and Equations • 13
    1.3   Real Numbers and the Number Line • 17
    1.4   Adding Real Numbers • 19
    1.5   Subtracting Real Numbers • 21
    1.6   Multiplying and Dividing Real Numbers • 24
           Summary Exercises on Operations with Real Numbers • 28
    1.7   Properties of Real Numbers • 29
    1.8   Simplifying Expressions • 31
           Chapter 1 Review Exercises • 33
           Chapter 1 Test • 39

2 Equations, Inequalities, and Applications 41
    2.1   The Addition Property of Equality • 41
    2.2   The Multiplication Property of Equality • 44
    2.3   More on Solving Linear Equations • 48
           Summary Exercises on Solving Linear Equations • 53
    2.4   An Introduction to Applications of Linear Equations • 54
    2.5   Formulas and Applications from Geometry • 60
    2.6   Ratio, Proportion, and Percent • 64
           Summary Exercises on Solving Applied Problems • 69
    2.7   Solving Linear Inequalities • 71
           Chapter 2 Review Exercises • 75
           Chapter 2 Test • 83
           Cumulative Review Exercises (Chapters R–2) • 85

3 Graphs of Linear Equations and Inequalities in Two Variables 91
    3.1   Reading Graphs; Linear Equations in Two Variables • 91
    3.2   Graphing Linear Equations in Two Variables • 97
    3.3   Slope of a Line • 104
    3.4   Equations of Lines • 108
           Summary Exercises on Graphing Linear Equations • 115
    3.5   Graphing Linear Inequalities in Two Variables • 117
           Chapter 3 Review Exercises • 122
           Chapter 3 Test • 130
           Cumulative Review Exercises (Chapters R–3) • 132

4 Systems of Linear Equations and Inequalities 135
    4.1   Solving Systems of Linear Equations by Graphing • 135
    4.2   Solving Systems of Linear Equations by Substitution • 140
    4.3   Solving Systems of Linear Equations by Elimination • 145
          Summary Exercises on Solving Systems of Linear Equations • 150
    4.4   Applications of Linear Systems • 153
    4.5   Solving Systems of Linear Inequalities • 160
          Chapter 4 Review Exercises • 163
          Chapter 4 Test • 172
          Cumulative Review Exercises (Chapters R–4) • 175

5 Exponents and Polynomials 179
    5.1   Adding and Subtracting Polynomials • 179
    5.2   The Product Rule and Power Rules for Exponents • 183
    5.3   Multiplying Polynomials • 185
    5.4   Special Products • 189
    5.5   Integer Exponents and the Quotient Rule • 191
          Summary Exercises on the Rules for Exponents • 194
    5.6   Dividing a Polynomial by a Monomial • 196
    5.7   Dividing a Polynomial by a Polynomial • 197
    5.8   An Application of Exponents: Scientific Notation • 201
          Chapter 5 Review Exercises • 203
          Chapter 5 Test • 208
          Cumulative Review Exercises (Chapters R–5) • 210

6 Factoring and Applications 215
    6.1   Factors; The Greatest Common Factor • 215
    6.2   Factoring Trinomials • 219
    6.3   Factoring Trinomials by Grouping • 223
    6.4   Factoring Trinomials Using FOIL • 225
    6.5   Special Factoring Techniques • 230
          Summary Exercises on Factoring • 232
    6.6   Solving Quadratic Equations by Factoring • 234
    6.7   Applications of Quadratic Equations • 238
          Chapter 6 Review Exercises • 244
          Chapter 6 Test • 250
          Cumulative Review Exercises (Chapters R–6) • 251

7 Rational Expressions and Applications..255
- 7.1 The Fundamental Property of Rational Expressions • 255
- 7.2 Multiplying and Dividing Rational Expressions • 259
- 7.3 Least Common Denominators • 262
- 7.4 Adding and Subtracting Rational Expressions • 265
- 7.5 Complex Fractions • 272
- 7.6 Solving Equations with Rational Expressions • 276
  Summary Exercises on Rational Expressions and Equations • 283
- 7.7 Applications of Rational Expressions • 285
- 7.8 Variation • 290
  Chapter 7 Review Exercises • 292
  Chapter 7 Test • 300
  Cumulative Review Exercises (Chapters R–7) • 303

8 Roots and Radicals..309
- 8.1 Evaluating Roots • 309
- 8.2 Multiplying, Dividing, and Simplifying Radicals • 313
- 8.3 Adding and Subtracting Radicals • 316
- 8.4 Rationalizing the Denominator • 319
- 8.5 More Simplifying and Operations with Radicals • 321
  Summary Exercises on Operations with Radicals • 325
- 8.6 Solving Equations with Radicals • 327
  Chapter 8 Review Exercises • 334
  Chapter 8 Test • 340
  Cumulative Review Exercises (Chapters R–8) • 342

9 Quadratic Equations..347
- 9.1 Solving Quadratic Equations by the Square Root Property • 347
- 9.2 Solving Quadratic Equations by Completing the Square • 349
- 9.3 Solving Quadratic Equations by the Quadratic Formula • 355
  Summary Exercises on Quadratic Equations • 360
- 9.4 Graphing Quadratic Equations • 363
- 9.5 Introduction to Functions • 368
  Chapter 9 Review Exercises • 370
  Chapter 9 Test • 378
  Cumulative Review Exercises (Chapters R–9) • 381

Appendices..387
- Appendix A   Strategies for Problem Solving • 387
- Appendix B   Sets • 393
- Appendix C   Mean, Median, and Mode • 397
- Appendix D   Factoring Sums and Differences of Cubes • 401

CHAPTER R PREALGEBRA REVIEW

R.1 Fractions

R.1 Margin Exercises

1. **(a)** Since 12 can be divided by 2, it has more than two different factors, so it is a *composite* number. Note: There are always two factors of any number–the number and 1.

   (b) 13 has exactly two different factors, 1 and 13, so it is a *prime* number.

   (c) Since 27 can be divided by 3, it has more than two different factors, so it is *composite*.

   (d) 59 has exactly two different factors, 1 and 59, so it is *prime*.

   (e) 1806 can be divided by 2, so it is *composite*.

2. **(a)** To write 70 in prime factored form, first divide by the smallest prime, 2, to get
   $$70 = 2 \cdot 35.$$
   Since 35 can be factored as $5 \cdot 7$, we have
   $$70 = 2 \cdot 5 \cdot 7.$$

   (b)
   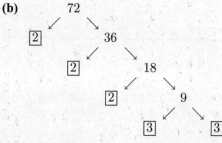
   $72 = 2 \cdot 2 \cdot 2 \cdot 3 \cdot 3$

   (c)
   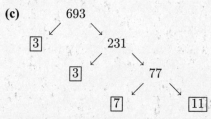
   $693 = 3 \cdot 3 \cdot 7 \cdot 11$

   (d) Since 97 is a prime number, its prime factored form is just 97.

3. **(a)** $\dfrac{8}{14} = \dfrac{4 \cdot 2}{7 \cdot 2} = \dfrac{4}{7} \cdot \dfrac{2}{2} = \dfrac{4}{7} \cdot 1 = \dfrac{4}{7}$

   (b) $\dfrac{35}{42} = \dfrac{5 \cdot 7}{6 \cdot 7} = \dfrac{5}{6} \cdot \dfrac{7}{7} = \dfrac{5}{6} \cdot 1 = \dfrac{5}{6}$

   (c) $\dfrac{120}{72} = \dfrac{5 \cdot 24}{3 \cdot 24} = \dfrac{5}{3} \cdot \dfrac{24}{24} = \dfrac{5}{3} \cdot 1 = \dfrac{5}{3}$

4. **(a)** $\dfrac{5}{8} \cdot \dfrac{2}{10} = \dfrac{5 \cdot 2}{8 \cdot 10}$ *Multiply numerators.* *Multiply denominators.*

   $= \dfrac{5 \cdot 2}{2 \cdot 4 \cdot 2 \cdot 5}$ *Factor.*

   $= \dfrac{1}{2 \cdot 4} = \dfrac{1}{8}$ *Write in lowest terms.*

   (b) $\dfrac{1}{10} \cdot \dfrac{12}{5} = \dfrac{1 \cdot 12}{10 \cdot 5}$ *Multiply numerators.* *Multiply denominators.*

   $= \dfrac{1 \cdot 2 \cdot 6}{2 \cdot 5 \cdot 5}$ *Factor.*

   $= \dfrac{6}{5 \cdot 5} = \dfrac{6}{25}$ *Write in lowest terms.*

   (c) $\dfrac{7}{9} \cdot \dfrac{12}{14} = \dfrac{7 \cdot 12}{9 \cdot 14}$ *Multiply numerators.* *Multiply denominators.*

   $= \dfrac{7 \cdot 2 \cdot 2 \cdot 3}{3 \cdot 3 \cdot 2 \cdot 7}$ *Factor.*

   $= \dfrac{2}{3}$ *Write in lowest terms.*

   (d) $3\dfrac{1}{3} \cdot 1\dfrac{3}{4} = \dfrac{10}{3} \cdot \dfrac{7}{4}$ *Write as improper fractions.*

   $= \dfrac{10 \cdot 7}{3 \cdot 4}$ *Multiply numerators.* *Multiply denominators.*

   $= \dfrac{2 \cdot 5 \cdot 7}{3 \cdot 2 \cdot 2}$ *Factor.*

   $= \dfrac{35}{6}$ or $5\dfrac{5}{6}$ *Write as a mixed number.*

5. **(a)** $\dfrac{3}{10} \div \dfrac{2}{7} = \dfrac{3}{10} \cdot \dfrac{7}{2}$ *Multiply by the reciprocal of the second fraction.*

   $= \dfrac{21}{20}$ or $1\dfrac{1}{20}$

   (b) $\dfrac{3}{4} \div \dfrac{7}{16} = \dfrac{3}{4} \cdot \dfrac{16}{7}$ *Multiply by the reciprocal of the second fraction.*

   $= \dfrac{3 \cdot 4 \cdot 4}{7 \cdot 4}$

   $= \dfrac{12}{7}$ or $1\dfrac{5}{7}$

   (c) $\dfrac{4}{3} \div 6 = \dfrac{4}{3} \div \dfrac{6}{1}$

   $= \dfrac{4}{3} \cdot \dfrac{1}{6}$ *Multiply by the reciprocal of the second fraction.*

   $= \dfrac{2 \cdot 2}{3 \cdot 2 \cdot 3}$

   $= \dfrac{2}{3 \cdot 3} = \dfrac{2}{9}$

2   Chapter R   Prealgebra Review

(d) $3\frac{1}{4} \div 1\frac{2}{5} = \frac{13}{4} \div \frac{7}{5}$ *Change both mixed numbers to improper fractions. Multiply by the reciprocal of the second fraction.*

$= \frac{13}{4} \cdot \frac{5}{7}$

$= \frac{65}{28}$ or $2\frac{9}{28}$

6. (a) $\frac{3}{5} + \frac{4}{5} = \frac{3+4}{5}$ Add numerators; denominator does not change.

$= \frac{7}{5}$ or $1\frac{2}{5}$

(b) $\frac{5}{14} + \frac{3}{14} = \frac{5+3}{14}$ Add numerators; denominator does not change.

$= \frac{8}{14}$

$= \frac{2 \cdot 4}{2 \cdot 7}$ Factor.

$= \frac{4}{7}$

7. (a) $\frac{7}{30} + \frac{2}{45}$

Since $30 = 2 \cdot 3 \cdot 5$ and $45 = 3 \cdot 3 \cdot 5$, the least common denominator must have one factor of 2 (from 30), two factors of 3 (from 45), and one factor of 5 (from either 30 or 45), so it is $2 \cdot 3 \cdot 3 \cdot 5 = 90$.

Write each fraction with a denominator of 90.

$\frac{7}{30} = \frac{7 \cdot 3}{30 \cdot 3} = \frac{21}{90}$ and $\frac{2}{45} = \frac{2 \cdot 2}{45 \cdot 2} = \frac{4}{90}$

Now add.

$\frac{7}{30} + \frac{2}{45} = \frac{21}{90} + \frac{4}{90} = \frac{21+4}{90} = \frac{25}{90}$

$\frac{25}{90}$ can be simplified.

$\frac{25}{90} = \frac{5 \cdot 5}{5 \cdot 18} = \frac{5}{18}$

(b) $\frac{17}{10} + \frac{8}{27}$

Since $10 = 2 \cdot 5$ and $27 = 3 \cdot 3 \cdot 3$, the least common denominator is $2 \cdot 5 \cdot 3 \cdot 3 \cdot 3 = 270$.

Write each fraction with a denominator of 270.

$\frac{17}{10} = \frac{17 \cdot 27}{10 \cdot 27} = \frac{459}{270}$ and $\frac{8}{27} = \frac{8 \cdot 10}{27 \cdot 10} = \frac{80}{270}$

Now add.

$\frac{17}{10} + \frac{8}{27} = \frac{459}{270} + \frac{80}{270} = \frac{539}{270}$ or $1\frac{269}{270}$

(c) $2\frac{1}{8} + 1\frac{2}{3} = \frac{17}{8} + \frac{5}{3}$ *Change both mixed numbers to improper fractions.*

The least common denominator is 24, so write each fraction with a denominator of 24.

$\frac{17}{8} = \frac{17 \cdot 3}{8 \cdot 3} = \frac{51}{24}$ and $\frac{5}{3} = \frac{5 \cdot 8}{3 \cdot 8} = \frac{40}{24}$

Now add.

$\frac{17}{8} + \frac{5}{3} = \frac{51}{24} + \frac{40}{24} = \frac{51+40}{24}$

$= \frac{91}{24}$ or $3\frac{19}{24}$

(d) $132\frac{4}{5} + 28\frac{3}{4}$

We will use a vertical method. LCD = 20

$\begin{aligned} 132\frac{4}{5} &= 132\frac{16}{20} \\ + 28\frac{3}{4} &= 28\frac{15}{20} \\ \hline &\ 160\frac{31}{20} \end{aligned}$ Add the whole numbers and the fractions separately.

$160\frac{31}{20} = 160 + \left(1 + \frac{11}{20}\right) = 161\frac{11}{20}$

8. (a) $\frac{9}{11} - \frac{3}{11} = \frac{9-3}{11}$ Subtract numerators; denominator does not change.

$= \frac{6}{11}$

(b) $\frac{13}{15} - \frac{5}{6}$

Since $15 = 3 \cdot 5$ and $6 = 2 \cdot 3$, the least common denominator is $3 \cdot 5 \cdot 2 = 30$. Write each fraction with a denominator of 30.

$\frac{13}{15} = \frac{13 \cdot 2}{15 \cdot 2} = \frac{26}{30}$ and $\frac{5}{6} = \frac{5 \cdot 5}{6 \cdot 5} = \frac{25}{30}$

Now subtract.

$\frac{13}{15} - \frac{5}{6} = \frac{26}{30} - \frac{25}{30} = \frac{1}{30}$

(c) $2\frac{3}{8} - 1\frac{1}{2} = \frac{19}{8} - \frac{3}{2}$ *Change each mixed number into an improper fraction.*

The least common denominator is 8. Write each fraction with a denominator of 8. $\frac{19}{8}$ remains unchanged, and

$\frac{3}{2} = \frac{3 \cdot 4}{2 \cdot 4} = \frac{12}{8}$.

Now subtract.

$\frac{19}{8} - \frac{3}{2} = \frac{19}{8} - \frac{12}{8} = \frac{19-12}{8} = \frac{7}{8}$

(d) $50\frac{1}{4} - 32\frac{2}{3} = \frac{201}{4} - \frac{98}{3}$ *Write as improper fractions.*

The least common denominator is 12.

$\frac{201}{4} = \frac{201 \cdot 3}{4 \cdot 3} = \frac{603}{12}$ and $\frac{98}{3} = \frac{98 \cdot 4}{3 \cdot 4} = \frac{392}{12}$

Now subtract.

$\frac{603}{12} - \frac{392}{12} = \frac{211}{12}$ or $17\frac{7}{12}$

9. To find out how much fabric Wei Jen should buy, add the amounts of fabric needed for each article of clothing.

   Write each mixed number as an improper fraction.

   $1\frac{1}{4} + 1\frac{2}{3} + 2\frac{1}{2} = \frac{5}{4} + \frac{5}{3} + \frac{5}{2}$

   The LCD is 12. Write each fraction with a denominator of 12, then add and simplify the answer.

   $\frac{5}{4} + \frac{5}{3} + \frac{5}{2} = \frac{15}{12} + \frac{20}{12} + \frac{30}{12} = \frac{65}{12}$ or $5\frac{5}{12}$

   Wei Jen should purchase $5\frac{5}{12}$ yd of fabric.

10. To find out how many gallons of paint Tram should buy, divide the total area to be painted by the area that one gallon of paint covers.

    $\frac{4200}{500} = \frac{42}{5}$ or $8\frac{2}{5}$

    $8\frac{2}{5}$ gal are needed, so he must buy 9 gal.

R.1 Section Exercises

1. True; the number above the fraction bar is called the numerator and the number below the fraction bar is called the denominator.

3. False; the fraction $\frac{17}{51}$ can be reduced to $\frac{1}{3}$ since $\frac{17}{51} = \frac{17 \cdot 1}{17 \cdot 3} = \frac{1}{3}$.

5. False; *product* refers to multiplication, so the product of 8 and 2 is 16. The *sum* of 8 and 2 is 10.

7. Since 19 has only itself and 1 as factors, it is a prime number.

9. The number 52 is composite since it has factors other than 1 and itself (2, for example).

11. 2468 can be divided by 2. It has more than two different factors, so it is composite.

13. As stated in the text, the number 1 is neither prime nor composite, by agreement.

15. $30 = 2 \cdot 15$
    $= 2 \cdot 3 \cdot 5$

17. $252 = 2 \cdot 126$
    $= 2 \cdot 2 \cdot 63$
    $= 2 \cdot 2 \cdot 3 \cdot 21$
    $= 2 \cdot 2 \cdot 3 \cdot 3 \cdot 7$

19. $124 = 2 \cdot 62$
    $= 2 \cdot 2 \cdot 31$

21. Since 29 has only itself and 1 as factors, it is a prime number. Its prime factored form is just 29.

23. $\frac{8}{16} = \frac{1 \cdot 8}{2 \cdot 8} = \frac{1}{2}$

25. $\frac{15}{18} = \frac{3 \cdot 5}{3 \cdot 6} = \frac{5}{6}$

27. $\frac{15}{75} = \frac{1 \cdot 15}{5 \cdot 15} = \frac{1}{5}$

29. $\frac{144}{120} = \frac{6 \cdot 24}{5 \cdot 24} = \frac{6}{5}$

31. A common denominator for $\frac{p}{q}$ and $\frac{r}{s}$ must be a multiple of both denominators, q and s. Such a number is $q \cdot s$. Therefore, **A** is correct.

33. $\frac{4}{5} \cdot \frac{6}{7} = \frac{4 \cdot 6}{5 \cdot 7} = \frac{24}{35}$

35. $\frac{1}{10} \cdot \frac{12}{5} = \frac{1 \cdot 12}{10 \cdot 5} = \frac{1 \cdot 2 \cdot 6}{2 \cdot 5 \cdot 5} = \frac{6}{25}$

37. $\frac{15}{4} \cdot \frac{8}{25} = \frac{15 \cdot 8}{4 \cdot 25}$

    $= \frac{3 \cdot 5 \cdot 4 \cdot 2}{4 \cdot 5 \cdot 5}$

    $= \frac{3 \cdot 2}{5}$

    $= \frac{6}{5}$ or $1\frac{1}{5}$

4 Chapter R Prealgebra Review

39. $2\frac{2}{3} \cdot 5\frac{4}{5}$

Change both mixed numbers to improper fractions.

$$2\frac{2}{3} = 2 + \frac{2}{3} = \frac{6}{3} + \frac{2}{3} = \frac{8}{3}$$

$$5\frac{4}{5} = 5 + \frac{4}{5} = \frac{25}{5} + \frac{4}{5} = \frac{29}{5}$$

$$2\frac{2}{3} \cdot 5\frac{4}{5} = \frac{8}{3} \cdot \frac{29}{5}$$

$$= \frac{8 \cdot 29}{3 \cdot 5}$$

$$= \frac{232}{15} \text{ or } 15\frac{7}{15}$$

41. $\frac{5}{4} \div \frac{3}{8} = \frac{5}{4} \cdot \frac{8}{3}$ *Multiply by the reciprocal of the second fraction.*

$$= \frac{5 \cdot 8}{4 \cdot 3}$$

$$= \frac{5 \cdot 4 \cdot 2}{4 \cdot 3}$$

$$= \frac{5 \cdot 2}{3}$$

$$= \frac{10}{3} \text{ or } 3\frac{1}{3}$$

43. $\frac{32}{5} \div \frac{8}{15} = \frac{32}{5} \cdot \frac{15}{8}$ *Multiply by the reciprocal of the second fraction.*

$$= \frac{32 \cdot 15}{5 \cdot 8}$$

$$= \frac{8 \cdot 4 \cdot 3 \cdot 5}{1 \cdot 5 \cdot 8}$$

$$= \frac{4 \cdot 3}{1} = 12$$

45. $\frac{3}{4} \div 12 = \frac{3}{4} \cdot \frac{1}{12}$ *Multiply by the reciprocal of 12.*

$$= \frac{3 \cdot 1}{4 \cdot 12}$$

$$= \frac{3 \cdot 1}{4 \cdot 3 \cdot 4}$$

$$= \frac{1}{4 \cdot 4} = \frac{1}{16}$$

47. $2\frac{5}{8} \div 1\frac{15}{32}$

Change both mixed numbers to improper fractions.

$$2\frac{5}{8} = 2 + \frac{5}{8} = \frac{16}{8} + \frac{5}{8} = \frac{21}{8}$$

$$1\frac{15}{32} = 1 + \frac{15}{32} = \frac{32}{32} + \frac{15}{32} = \frac{47}{32}$$

$$2\frac{5}{8} \div 1\frac{15}{32} = \frac{21}{8} \div \frac{47}{32}$$

$$= \frac{21}{8} \cdot \frac{32}{47}$$

$$= \frac{21 \cdot 32}{8 \cdot 47}$$

$$= \frac{21 \cdot 8 \cdot 4}{8 \cdot 47}$$

$$= \frac{21 \cdot 4}{47}$$

$$= \frac{84}{47} \text{ or } 1\frac{37}{47}$$

49. Multiply the first fraction (the dividend) by the reciprocal of the second fraction (the divisor) to divide two fractions.

51. $\frac{7}{12} + \frac{1}{12} = \frac{7+1}{12}$

$$= \frac{8}{12}$$

$$= \frac{2 \cdot 4}{3 \cdot 4} = \frac{2}{3}$$

53. $\frac{5}{9} + \frac{1}{3}$

Since $9 = 3 \cdot 3$, and 3 is prime, the LCD (least common denominator) is $3 \cdot 3 = 9$.

$$\frac{1}{3} = \frac{1}{3} \cdot \frac{3}{3} = \frac{3}{9}$$

Now add the two fractions with the same denominator.

$$\frac{5}{9} + \frac{1}{3} = \frac{5}{9} + \frac{3}{9} = \frac{8}{9}$$

55. $3\frac{1}{8} + \frac{1}{4}$

$$3\frac{1}{8} = 3 + \frac{1}{8} = \frac{24}{8} + \frac{1}{8} = \frac{25}{8}$$

$$3\frac{1}{8} + \frac{1}{4} = \frac{25}{8} + \frac{1}{4}$$

Since $8 = 2 \cdot 2 \cdot 2$ and $4 = 2 \cdot 2$, the LCD is $2 \cdot 2 \cdot 2$ or 8.

$$3\frac{1}{8} + \frac{1}{4} = \frac{25}{8} + \frac{1 \cdot 2}{4 \cdot 2}$$
$$= \frac{25}{8} + \frac{2}{8}$$
$$= \frac{27}{8} \text{ or } 3\frac{3}{8}$$

57. $\frac{7}{12} - \frac{1}{9}$

Since $12 = 2 \cdot 2 \cdot 3$ and $9 = 3 \cdot 3$, the LCD is $2 \cdot 2 \cdot 3 \cdot 3 = 36$.

$$\frac{7}{12} = \frac{7}{12} \cdot \frac{3}{3} = \frac{21}{36} \text{ and } \frac{1}{9} \cdot \frac{4}{4} = \frac{4}{36}$$

Now subtract fractions with the same denominator.

$$\frac{7}{12} - \frac{1}{9} = \frac{21}{36} - \frac{4}{36} = \frac{17}{36}$$

59. $6\frac{1}{4} - 5\frac{1}{3}$

$$6\frac{1}{4} = 6 + \frac{1}{4} = \frac{24}{4} + \frac{1}{4} = \frac{25}{4}$$
$$5\frac{1}{3} = 5 + \frac{1}{3} = \frac{15}{3} + \frac{1}{3} = \frac{16}{3}$$

Since $4 = 2 \cdot 2$, and 3 is prime, the LCD is $2 \cdot 2 \cdot 3 = 12$.

$$6\frac{1}{4} - 5\frac{1}{3} = \frac{25}{4} - \frac{16}{3}$$
$$= \frac{25 \cdot 3}{4 \cdot 3} - \frac{16 \cdot 4}{3 \cdot 4}$$
$$= \frac{75}{12} - \frac{64}{12}$$
$$= \frac{11}{12}$$

61. $\frac{5}{3} + \frac{1}{6} - \frac{1}{2}$

Since 2 and 3 are prime, and $6 = 2 \cdot 3$, the LCD is $2 \cdot 3 = 6$. Write $\frac{5}{3} \cdot \frac{2}{2} = \frac{10}{6}$ and $\frac{1}{2} = \frac{1}{2} \cdot \frac{3}{3} = \frac{3}{6}$.

Now add and subtract, then write the answer in lowest terms.

$$\frac{5}{3} + \frac{1}{6} - \frac{1}{2} = \frac{10}{6} + \frac{1}{6} - \frac{3}{6}$$
$$= \frac{8}{6} = \frac{4 \cdot 2}{3 \cdot 2} = \frac{4}{3} \text{ or } 1\frac{1}{3}$$

63. Multiply the number of cups of water per serving by the number of servings.

$$\frac{3}{4} \cdot 8 = \frac{3}{4} \cdot \frac{8}{1}$$
$$= \frac{3 \cdot 8}{4 \cdot 1}$$
$$= \frac{3 \cdot 2 \cdot 4}{4 \cdot 1}$$
$$= \frac{3 \cdot 2}{1} = 6 \text{ cups}$$

For 8 microwave servings, 6 cups of water will be needed.

65. The perimeter is the sum of the measures of the 5 sides.

$$196 + 98\frac{3}{4} + 146\frac{1}{2} + 100\frac{7}{8} + 76\frac{5}{8}$$
$$= 196 + 98\frac{6}{8} + 146\frac{4}{8} + 100\frac{7}{8} + 76\frac{5}{8}$$
$$= 196 + 98 + 146 + 100 + 76 + \frac{6+4+7+5}{8}$$
$$= 616 + \frac{22}{8} \quad \left(\frac{22}{8} = 2\frac{6}{8} = 2\frac{3}{4}\right)$$
$$= 618\frac{3}{4} \text{ feet}$$

The perimeter is $618\frac{3}{4}$ feet.

67. The difference between the two measures is found by subtracting, using 16 as the LCD.

$$\frac{3}{4} - \frac{3}{16} = \frac{3 \cdot 4}{4 \cdot 4} - \frac{3}{16}$$
$$= \frac{12}{16} - \frac{3}{16}$$
$$= \frac{12-3}{16} = \frac{9}{16}$$

The difference is $\frac{9}{16}$ inch.

69. Subtract $\frac{3}{8}$ from $\frac{11}{16}$ using 16 as the LCD.

$$\frac{11}{16} - \frac{3}{8} = \frac{11}{16} - \frac{3 \cdot 2}{8 \cdot 2}$$
$$= \frac{11}{16} - \frac{6}{16}$$
$$= \frac{5}{16}$$

Thus, $\frac{3}{8}$ inch is $\frac{5}{16}$ inch smaller than $\frac{11}{16}$ inch.

Chapter R Prealgebra Review

71. Multiply the amount of fabric it takes to make one costume by the number of costumes.

$$2\frac{3}{8} \cdot 7 = \frac{19}{8} \cdot \frac{7}{1}$$
$$= \frac{19 \cdot 7}{8 \cdot 1}$$
$$= \frac{133}{8} \text{ or } 16\frac{5}{8} \text{ yd}$$

For 7 costumes, $16\frac{5}{8}$ yards of fabric would be needed.

73. The sum of the fractions representing immigrants from Latin America, Asia, or Europe is

$$\frac{13}{25} + \frac{3}{10} + \frac{13}{100} = \frac{13 \cdot 4}{25 \cdot 4} + \frac{3 \cdot 10}{10 \cdot 10} + \frac{13}{100}$$
$$= \frac{52 + 30 + 13}{100}$$
$$= \frac{95}{100} = \frac{19}{20}.$$

So the fraction representing immigrants from other regions is

$$1 - \frac{19}{20} = \frac{20}{20} - \frac{19}{20}$$
$$= \frac{1}{20}.$$

75. Multiply the fraction representing immigrants from Europe $\left(\frac{13}{100}\right)$ by the total number of immigrants (more than 8 million).

$$\frac{13}{100} \cdot 8 = \frac{13}{100} \cdot \frac{8}{1} = \frac{104}{100} = \frac{26}{25} = 1\frac{1}{25}$$

There were more than $1\frac{1}{25}$ million immigrants from Europe.

R.2 Decimals and Percents

R.1 Margin Exercises

1. (a) $.8 = \frac{8}{10}$ *One decimal place*   *One zero*

    (b) $.431 = \frac{431}{1000}$ *Three decimal places*   *Three zeros*

    (c) $20.58 = \frac{2058}{100}$ *Two decimal places*   *Two zeros*

2. (a)   68.9   becomes   68.900   *Attach zeros.*
          42.72   42.720
        $+\,8.973$ $+\,8.973$
                                120.593

    (b)   32.5   becomes   32.50   *Attach zero.*
        $-\,21.72$ $-\,21.72$
                                10.78

    (c) $42.83 + 71 + 3.074$ becomes

        42.830   *Attach zeros.*
        71.000
       $+\,3.074$
        116.904

    (d) $351.8 - 2.706$ becomes

        351.800   *Attach zeros.*
      $-\,2.706$
        349.094

3. (a) $2.13 \times .05$ becomes

         2.13 $\leftarrow$ 2 *decimal places*
      $\times\,.05$ $\leftarrow$ 2 *decimal places*
        .1065 $\leftarrow$ 4 *decimal places*

    (b) 9.32×1.4 becomes

         9.32 $\leftarrow$ 2 *decimal places*
      $\times\,1.4$ $\leftarrow$ 1 *decimal place*
        3 728
        9 32
       13.048 $\leftarrow$ 3 *decimal places*

    (c) $300.2 \times .052$ becomes

        300.2 $\leftarrow$ 1 *decimal place*
     $\times\,.052$ $\leftarrow$ 3 *decimal places*
        6004
      15 010
     15.6104 $\leftarrow$ 4 *decimal places*

    (d) $42{,}001 \times .012$ becomes

      42,001 $\leftarrow$ 0 *decimal places*
     $\times\,.012$ $\leftarrow$ 3 *decimal places*
      84 002
      420 01
   504.012 $\leftarrow$ 3 *decimal places*

4. **(a)** $14.9\overline{)451.47}$

To change 14.9 into a whole number, move the decimal point one place to the right. Then, move the decimal point in 451.47 the same number of places to the right, to get 4514.7.

Bring the decimal point straight up and divide as with whole numbers.

```
         3 0. 3
149 ) 4 5 1 4. 7
      4 4 7
      ─────
          4 4 7
          4 4 7
          ─────
              0
```

(b) $.37\overline{)5.476}$

Move the decimal points two places to the right. Bring the decimal point straight up and divide as with whole numbers.

```
        1 4. 8
37 ) 5 4 7. 6
     3 7
     ───
     1 7 7
     1 4 8
     ─────
         2 9 6
         2 9 6
         ─────
             0
```

(c) $375.1 \div 3.001$

Write the problem as follows.

$3.001\overline{)375.1}$

Move the decimal points three places to the right. Divide, bringing the decimal point straight up.

```
              1 2 4. 9 9 1
3001 ) 3 7 5 1 0 0. 0 0 0
       3 0 0 1
       ───────
         7 5 0 0
         6 0 0 2
         ───────
         1 4 9 8 0
         1 2 0 0 4
         ─────────
           2 9 7 6 0
           2 7 0 0 9
           ─────────
             2 7 5 1 0
             2 7 0 0 9
             ─────────
                 5 0 1 0
                 3 0 0 1
                 ───────
                 2 0 0 9
```

Rounded to the nearest hundredth, the quotient is 124.99.

5. **(a)** 294.72×10

Because 10 has 1 zero, move the decimal point 1 place to the right.

$294.72 \times 10 = 2947.2$

(b) 19.5×1000

Because 1000 has 3 zeros, move the decimal point 3 places to the right.

$19.5 \times 1000 = 19{,}500$

(c) $4.793 \div 100$

Because 100 has 2 zeros, move the decimal point 2 places to the left.

$4.793 \div 100 = .04793$

(d) $960.1 \div 10$

Because 10 has 1 zero, move the decimal point 1 place to the left.

$960.1 \div 10 = 96.01$

6. **(a)** Divide the denominator, 9, into the numerator, 2. Attach zeros after the decimal point of the numerator as needed.

```
       . 2 2 2
9 ) 2. 0 0 0 ...
    1 8
    ───
      2 0
      1 8
      ───
        2 0
        1 8
        ───
          2
```

The quotient will keep repeating the digit 2, so $\frac{2}{9} = .\overline{2}$.

To the nearest thousandth, $\frac{2}{9} = .222$.

(b)
```
         . 8 5
20 ) 1 7. 0 0
     1 6 0
     ─────
       1 0 0
       1 0 0
       ─────
           0
```

$\frac{17}{20} = .85$

8 Chapter R Prealgebra Review

(c)
$$11\overline{)1.0000\ldots}.0909$$
$$\phantom{11\overline{)}}\underline{99}$$
$$\phantom{11\overline{)1.}}100$$
$$\phantom{11\overline{)1.}}\underline{99}$$
$$\phantom{11\overline{)1.00}}1$$

The quotient will keep repeating the digits 0 and 9, so $\frac{1}{11} = .\overline{09}$.

To the nearest thousandth, $\frac{1}{11} = .091$.

7. (a) $23\% = 23 \cdot 1\% = 23 \times .01 = .23$

Alternatively, to convert 23% to a decimal, move the decimal point two places to the left and drop the percent sign.

(b) $310\% = 310 \cdot 1\% = 310 \times .01 = 3.10$

(c) $.71 = 71 \times .01 = 71 \times 1\% = 71\%$

Alternatively, to convert .71 to a percent, move the decimal point two places to the right and attach a percent sign.

(d) $1.32 = 132 \times .01 = 132 \cdot 1\% = 132\%$

(e) $.685 = 685 \times .01 = 685 \cdot 1\% = 68.5\%$

R.2 Section Exercises

1. In the decimal 367.9412,
3 is in the hundreds place,
6 is in the tens place,
7 is in the units or ones place,
9 is in the tenths place,
4 is in the hundredths place,
1 is in the thousandths place, and
2 is in the ten thousandths place.

(a) 6   (b) 9   (c) 1   (d) 7   (e) 4

3. (a) Rounding 46.249 to the hundredths place gives 46.25, since 9 is greater than or equal to 5.

(b) Rounding 46.249 to the tenths place gives 46.2, since 4 is less than 5.

(c) Rounding 46.249 to the ones or units place gives 46, since 2 is less than 5.

(d) Rounding 46.249 to the tens place gives 50, since 6 is greater than or equal to 5.

5. Round 35.89 to 36 and 24.1 to 24. Since $36 + 24$ is 60, the best estimate for the sum $35.89 + 24.1$ is 60. The answer is **C**.

7. Since 84.9 is about 85 and 98.3 is about 100, the best estimate for 84.9×98.3 is $85 \times 100 = 8500$, which is choice **B**.

9. $.4 = \frac{4}{10}$ *One decimal place*
               One zero

11. $.64 = \frac{64}{100}$ *Two decimal places*
                  Two zeros

13. $.138 = \frac{138}{1000}$ *Three decimal places*
                      Three zeros

15. $3.805 = \frac{3805}{1000}$ *Three decimal places*
                         Three zeros

17. To get a quick estimate, round the numbers so that only the first digit is not zero.

The estimate for $25.32 + 109.2 + 8.574$ is $30 + 100 + 9 = 139$.

Place the numbers in a column with the decimal points lined up. Attach zeros to make all the numbers the same length.

$$\begin{array}{r} 25.320 \\ 109.200 \\ +8.574 \\ \hline 143.094 \end{array}$$

19. The estimate is $30 - 3 = 27$.

Write the numbers in a column with the decimal points lined up.

$$\begin{array}{r} 28.73 \\ -3.12 \\ \hline 25.61 \end{array}$$

21. The estimate is $40 - 30 = 10$.

Place the numbers in a column with the decimal points lined up. Attach a zero to 43.5 so both numbers will have the same length.

$$\begin{array}{r} 43.50 \\ -28.17 \\ \hline 15.33 \end{array}$$

23. The estimate is $30 + 50 + 2 = 82$.

Attach zeros to make all the numbers the same length.

$$\begin{array}{r} 32.560 \\ 47.356 \\ +1.800 \\ \hline 81.716 \end{array}$$

25. The estimate is $20 - 3 = 17$.

Attach 3 zeros to 18 so both numbers will have the same length.

$$\begin{array}{r} 18.000 \\ -2.789 \\ \hline 15.211 \end{array}$$

R.2 Decimals and Percents

27. The estimate is $.2 \times .03 = .006$.

Multiply as if the numbers were whole numbers.

$$\begin{array}{r} .03 \\ \times\ .2 \\ \hline .006 \end{array} \begin{array}{l} 2\ \textit{decimal places} \\ 1\ \textit{decimal place} \\ 3\ \textit{decimal places} \end{array}$$

29. The estimate for 12.8×9.1 is $10 \times 9 = 90$.

Multiply as if the numbers were whole numbers; then place a decimal point in the product.

$$\begin{array}{r} 12.8 \\ \times\ 9.1 \\ \hline 1\ 28 \\ 115\ 2 \\ \hline 116.48 \end{array} \begin{array}{l} 1\ \textit{decimal place} \\ 1\ \textit{decimal place} \\ \\ \\ 2\ \textit{decimal places} \end{array}$$

31. The estimate is $60 \div 10 = 6$.

Bring the decimal point straight up and divide as with whole numbers.

$$\begin{array}{r} 7.15 \\ 8\overline{)57.20} \\ 56 \\ \hline 1\ 2 \\ 8 \\ \hline 40 \\ 40 \\ \hline 0 \end{array}$$

The quotient is 7.15.

33. The estimate is $20 \div 10 = 2$.

To change 9.74 to a whole number, move the decimal point two places to the right. Move the decimal point in 19.967 the same number of places to the right to get 1996.7. Bring the decimal point straight up and divide as with whole numbers.

$$\begin{array}{r} 2.05 \\ 974\overline{)1996.70} \\ 1948 \\ \hline 48\ 70 \\ 48\ 70 \\ \hline 0 \end{array}$$

The quotient is 2.05.

35. The estimate is $60 \times 100 = 6000$.

To multiply by 100, move the decimal point two places to the right, since 100 has two zeros.

$$57.116 \times 100 = 5711.6$$

37. The estimate is $2 \div 10 = .2$.

Because 10 has 1 zero, move the decimal point one place to the left.

$$1.62 \div 10 = .162$$

39. To add or subtract decimals, line up the decimal points in a column, add or subtract as usual, and move the decimal point straight down in the sum or difference.

41. To change $\frac{1}{8}$ to a decimal, divide 1 by 8. Attach zeros after the decimal point of the numerator as needed.

$$\begin{array}{r} .125 \\ 8\overline{)1.000} \\ 8 \\ \hline 20 \\ 16 \\ \hline 40 \\ 40 \\ \hline 0 \end{array}$$

$\frac{1}{8} = .125$

43. Divide 1 by 4.

$$\begin{array}{r} .25 \\ 4\overline{)1.00} \\ 8 \\ \hline 20 \\ 20 \\ \hline 0 \end{array}$$

$\frac{1}{4} = .25$

45. Divide 5 by 9.

$$\begin{array}{r} .5555 \\ 9\overline{)5.0000} \\ 45 \\ \hline 50 \\ 45 \\ \hline 50 \\ 45 \\ \hline 50 \\ 45 \\ \hline 5 \end{array}$$

Since a 5 is always left after the subtraction, this quotient is a repeating decimal, and $\frac{5}{9} = .\overline{5}$. Rounded to the nearest thousandth, $\frac{5}{9} \approx .556$.

47. Divide 1 by 6.

$$\begin{array}{r} .166 \\ 6\overline{)1.000} \\ \underline{6} \\ 40 \\ \underline{36} \\ 40 \\ \underline{36} \\ 4 \end{array}$$

$\frac{1}{6} = .1\overline{6}$ or, rounding to the nearest thousandth, .167.

49. To convert a decimal to a percent, move the decimal point two places to the right and attach a percent symbol (%).

51. $54\% = 54 \cdot 1\% = 54 \times .01 = .54$

Alternatively, to convert 54% to a decimal, move the decimal point two places to the left and drop the percent sign.

53. $117\% = 117 \cdot 1\% = 117 \times .01 = 1.17$

55. $2.4\% = 2.4 \cdot 1\% = 2.4 \times .01 = .024$

57. $6\frac{1}{4}\% = 6.25\% = 6.25 \cdot 1\% = 6.25 \times .01 = .0625$

59. $.8\% = .8 \cdot 1\% = .8 \times .01 = .008$

61. $.75 = 75 \times .01 = 75 \cdot 1\% = 75\%$

Alternatively, to convert .75 to a percent, move the decimal point two places to the right and attach a percent sign.

63. $.004 = .4 \times .01 = .4 \cdot 1\% = .4\%$

65. $1.28 = 128 \times .01 = 128 \cdot 1\% = 128\%$

67. $.3 = 30 \times .01 = 30 \cdot 1\% = 30\%$

69. Convert $\frac{3}{4}$ to a decimal, and then convert the decimal to a percent.

$$\begin{array}{r} .75 \\ 4\overline{)3.00} \\ \underline{28} \\ 20 \\ \underline{20} \\ 0 \end{array}$$

$\frac{3}{4} = .75 = 75 \times .01 = 75 \cdot 1\% = 75\%$

71. Convert $\frac{5}{6}$ to a decimal, and then convert the decimal to a percent.

$$\begin{array}{r} .833 \\ 6\overline{)5.000} \\ \underline{48} \\ 20 \\ \underline{18} \\ 20 \\ \underline{18} \\ 2 \end{array}$$

$\frac{5}{6} = .8\overline{3} = 83.\overline{3} \times .01 = 83.\overline{3} \cdot 1\% = 83.\overline{3}\%$

73. **(a)** If 80% of the tread is worn off, then 20% remains. Convert 20% to a fraction.

$$\frac{20}{100} = \frac{20 \div 20}{100 \div 20} = \frac{1}{5}$$

Multiply $\frac{10}{32}$ by $\frac{1}{5}$.

$$\frac{10}{32} \cdot \frac{1}{5} = \frac{2 \cdot 5 \cdot 1}{32 \cdot 5} = \frac{2}{32} \cdot \frac{5}{5} = \frac{2}{32}$$

$\frac{2}{32}$ inch of tread indicates that a new tire is needed.

(b) Subtract $\frac{2}{32}$ (from part (a)) from $\frac{4}{32}$.

$$\frac{4}{32} - \frac{2}{32} = \frac{4-2}{32} = \frac{2}{32}$$

There is $\frac{2}{32}$ inch of tread wear remaining.

CHAPTER 1 THE REAL NUMBER SYSTEM

1.1 Exponents, Order of Operations, and Inequality

1.1 Margin Exercises

1. (a) $6^2 = \underline{6 \cdot 6} = 36$
   6 is used as a factor 2 times.

   (b) $3^5 = \underline{3 \cdot 3 \cdot 3 \cdot 3 \cdot 3} = 243$
   3 is used as a factor 5 times.

   (c) $\left(\dfrac{3}{4}\right)^2 = \underline{\dfrac{3}{4} \cdot \dfrac{3}{4}} = \dfrac{9}{16}$
   $\dfrac{3}{4}$ is used as a factor 2 times.

   (d) $\left(\dfrac{1}{2}\right)^4 = \underline{\dfrac{1}{2} \cdot \dfrac{1}{2} \cdot \dfrac{1}{2} \cdot \dfrac{1}{2}} = \dfrac{1}{16}$
   $\dfrac{1}{2}$ is used as a factor 4 times.

   (e) $(.4)^3 = \underline{.4 \cdot .4 \cdot .4} = .064$
   .4 is used as a factor 3 times.

2. (a) $7 + 3 \cdot 8 = 7 + 24$ Multiply.
   $= 31$ Add.

   (b) $2 \cdot 9 + 7 \cdot 3 = 18 + 21$ Multiply.
   $= 39$ Add.

   (c) $7 \cdot 6 - 3(8 + 1)$
   $= 7 \cdot 6 - 3(9)$ Add inside parentheses.
   $= 42 - 27$ Multiply.
   $= 15$ Subtract.

   (d) $2 + 3^2 - 5$
   $= 2 + 9 - 5$ Use the exponent.
   $= 11 - 5$ Add.
   $= 6$ Subtract.

3. (a) $9[(4+8) - 3]$
   $= 9[12 - 3]$ Add inside parentheses.
   $= 9(9)$ Subtract inside parentheses.
   $= 81$ Multiply.

   (b) $\dfrac{2(7+8) + 2}{3 \cdot 5 + 1} = \dfrac{2(15) + 2}{3 \cdot 5 + 1}$ Add inside parentheses.
   $= \dfrac{30 + 2}{15 + 1}$ Multiply.
   $= \dfrac{32}{16}$ Add.
   $= 2$ Divide.

4. (a) $7 < 5$ means "seven is less than five." This statement is *false* since the symbol should point to the lesser number.

   (b) $12 > 6$ means "twelve is greater than six." This statement is *true*.

   (c) $4 \neq 10$ means "four is not equal to ten." This statement is *true*.

   (d) $28 \neq 4 \cdot 7$ means "twenty-eight is not equal to four times seven." Since $4 \cdot 7 = 28$, this statement is *false*.

5. (a) The statement $30 \leq 40$ is *true* since $30 < 40$. Note that if either the $<$ part or the $=$ part is true, then the inequality $\leq$ is true.

   (b) $25 \geq 10$ is a *true* statement since $25 > 10$.

   (c) $40 \leq 10$ is *false* since neither $40 < 10$ nor $40 = 10$ is true.

   (d) $21 \leq 21$ is *true* since $21 = 21$.

   (e) $3 \geq 3$ is *true* since $3 = 3$.

6. (a) "Nine is equal to eleven minus two" is written $9 = 11 - 2$.

   (b) "Seventeen is less than thirty" is written $17 < 30$.

   (c) "Eight is not equal to ten" is written $8 \neq 10$.

   (d) "Fourteen is greater than twelve" is written $14 > 12$.

   (e) "Thirty is less than or equal to fifty" is written $30 \leq 50$.

   (f) "Two is greater than or equal to two" is written $2 \geq 2$.

7. (a) If we write $8 < 10$ with the inequality symbol reversed, $8 < 10$ becomes $10 > 8$.

   (b) $3 > 1$ may be written as $1 < 3$.

   (c) $9 \leq 15$ may be written as $15 \geq 9$.

   (d) $6 \geq 2$ may be written as $2 \leq 6$.

1.1 Section Exercises

1. The statement "Exponents are also called powers" is *true*.

3. False; $4 + 3(8 - 2) = 4 + 3 \cdot 6 = 4 + 18 = 22$. The common error leading to 42 is adding 4 to 3 and then multiplying by 6. One must follow the rules for order of operations.

5. False; the correct interpretation is $4 = 16 - 12$.

7. $7^2 = 7 \cdot 7 = 49$

9. $12^2 = 12 \cdot 12 = 144$

11. $4^3 = 4 \cdot 4 \cdot 4 = 64$

13. $10^3 = 10 \cdot 10 \cdot 10 = 1000$

15. $3^4 = 3 \cdot 3 \cdot 3 \cdot 3 = 81$

17. $4^5 = 4 \cdot 4 \cdot 4 \cdot 4 \cdot 4 = 1024$

19. $\left(\dfrac{2}{3}\right)^4 = \dfrac{2}{3} \cdot \dfrac{2}{3} \cdot \dfrac{2}{3} \cdot \dfrac{2}{3} = \dfrac{16}{81}$

21. $(.04)^3 = (.04)(.04)(.04) = .000064$

23. When evaluating the expression $(4^2 + 3^3)^4$, the last exponent that would be applied is 4, since it is outside the parentheses.

25. $13 + 9 \cdot 5 = 13 + 45$ *Multiply.*
    $ = 58$ *Add.*

27. $20 - 4 \cdot 3 + 5 = 20 - 12 + 5$ *Multiply.*
    $ = 8 + 5$ *Subtract.*
    $ = 13$ *Add.*

29. $9 \cdot 5 - 13 = 45 - 13$ *Multiply.*
    $ = 32$ *Subtract.*

31. $18 - 2 + 3 = 16 + 3$ *Subtract.*
    $ = 19$ *Add.*

33. $\dfrac{1}{4} \cdot \dfrac{2}{3} + \dfrac{2}{5} \cdot \dfrac{11}{3} = \dfrac{1}{6} + \dfrac{22}{15}$ *Multiply.*
    $\phantom{\dfrac{1}{4} \cdot \dfrac{2}{3} + \dfrac{2}{5} \cdot \dfrac{11}{3}} = \dfrac{5}{30} + \dfrac{44}{30}$ *LCD = 30*
    $\phantom{\dfrac{1}{4} \cdot \dfrac{2}{3} + \dfrac{2}{5} \cdot \dfrac{11}{3}} = \dfrac{49}{30}$ or $1\dfrac{19}{30}$ *Add.*

35. $9 \cdot 4 - 8 \cdot 3 = 36 - 24$ *Multiply.*
    $ = 12$ *Subtract.*

37. $2.5(1.9) + 4.3(7.3) = 4.75 + 31.39$ *Multiply.*
    $ = 36.14$ *Add.*

39. $10 + 40 \div 5 \cdot 2 = 10 + 8 \cdot 2$ *Divide.*
    $ = 10 + 16$ *Multiply.*
    $ = 26$ *Add.*

41. $18 - 2(3 + 4) = 18 - 2(7)$ *Add inside parentheses.*
    $ = 18 - 14$ *Multiply.*
    $ = 4$ *Subtract.*

43. $5[3 + 4(2^2)]$
    $= 5[3 + 4(4)]$ *Use the exponent.*
    $= 5[3 + 16]$ *Multiply inside parentheses.*
    $= 5(19)$ *Add.*
    $= 95$ *Multiply.*

45. $\left(\dfrac{3}{2}\right)^2 \left[\left(11 + \dfrac{1}{3}\right) - 6\right]$
    $= \dfrac{9}{4}\left[\left(11 + \dfrac{1}{3}\right) - 6\right]$ *Use the exponent.*
    $= \dfrac{9}{4}\left[\left(\dfrac{33}{3} + \dfrac{1}{3}\right) - 6\right]$ *LCD = 3*
    $= \dfrac{9}{4}\left(\dfrac{34}{3} - 6\right)$ *Add inside parentheses.*
    $= \dfrac{9}{4}\left(\dfrac{34}{3} - \dfrac{18}{3}\right)$ *LCD = 3*
    $= \dfrac{9}{4}\left(\dfrac{16}{3}\right)$ *Subtract inside parentheses.*
    $= \dfrac{144}{12}$ *Multiply.*
    $= 12$

47. $\dfrac{8 + 6(3^2 - 1)}{3 \cdot 2 - 2}$
    $= \dfrac{8 + 6(9 - 1)}{3 \cdot 2 - 2}$ *Use the exponent.*
    $= \dfrac{8 + 6(8)}{3 \cdot 2 - 2}$ *Subtract inside parentheses.*
    $= \dfrac{8 + 48}{6 - 2}$ *Multiply in numerator and denominator.*
    $= \dfrac{56}{4}$ *Add in numerator, subtract in denominator.*
    $= 14$ *Divide.*

49. $\dfrac{4(7 + 2) + 8(8 - 3)}{6(4 - 2) - 2^2}$
    $= \dfrac{4(7 + 2) + 8(8 - 3)}{6(4 - 2) - 4}$ *Use the exponent.*
    $= \dfrac{4(9) + 8(5)}{6(2) - 4}$ *Simplify numerator and denominator separately.*
    $= \dfrac{36 + 40}{12 - 4}$ *Multiply.*
    $= \dfrac{76}{8}$ *Add and subtract.*
    $= \dfrac{19}{2}$ *Reduce.*

51. $8 \geq 17$ means "8 is greater than or equal to 17." Since neither $8 > 17$ nor $8 = 17$ is true, the statement is *false*.

53. $17 \leq 18 - 1$ may be simplified to $17 \leq 17$, which means "17 is less than or equal to 17." Since $17 = 17$, the statement is *true*.

55. $6 \cdot 8 + 6 \cdot 6 \geq 0$ may be simplified to $84 \geq 0$, which means "84 is greater than or equal to 0." Since 84 is greater than 0, the statement is *true*.

57. $6[5 + 3(4 + 2)] \leq 70$
$6[5 + 3(6)] \leq 70$
$6(5 + 18) \leq 70$
$6(23) \leq 70$
$138 \leq 70$

Since $138 > 70$, the statement is *false*.

59. $\dfrac{9(7 - 1) - 8 \cdot 2}{4(6 - 1)} > 3$

$\dfrac{9(6) - 8 \cdot 2}{4(5)} > 3$

$\dfrac{54 - 16}{20} > 3$

$\dfrac{38}{20} > 3$

$1\dfrac{9}{10} > 3$

Since $1\dfrac{9}{10} < 3$, the statement is *false*.

61. $8 \leq 4^2 - 2^2$
$8 \leq 16 - 4$
$8 \leq 12$

Since $8 < 12$, the statement is *true*.

63. "Fifteen is equal to five plus ten" is written
$$15 = 5 + 10.$$

65. "Nine is greater than five minus four" is written
$$9 > 5 - 4.$$

67. "Sixteen is not equal to nineteen" is written
$$16 \neq 19.$$

69. "Two is less than or equal to three" is written
$$2 \leq 3.$$

71. "$7 < 19$" means "seven is less than nineteen." The statement is *true*.

73. $\frac{1}{3} \neq \frac{3}{10}$ means "one-third is not equal to three-tenths." Since $\frac{1}{3}$ can be written as $\frac{10}{30}$ and $\frac{3}{10}$ can be written as $\frac{9}{30}$, the statement is true $\left(\frac{10}{30} \neq \frac{9}{30}\right)$.

75. "$8 \geq 11$" means "eight is greater than or equal to eleven." The statement is *false*.

77. $5 < 30$ becomes $30 > 5$ when the inequality symbol is reversed.

79. $12 \geq 3$ becomes $3 \leq 12$ when the inequality symbol is reversed.

81. The states that had a figure greater than 13.9 are Alaska (16.7), Texas (14.7), California (20.5), and Idaho (17.8).

83. The states that had a figure *not* less than 13.9, which is the same as ≥ 13.9, are Alaska (16.7), Texas (14.7), California (20.5), Idaho (17.8), and Missouri (13.9).

1.2 Variables, Expressions, and Equations

1.2 Margin Exercises

1. Replace p with 3 in each expression.

   (a) $6p = 6(3)$
   $= 18$ Multiply.

   (b) $p + 12 = 3 + 12$
   $= 15$ Add.

   (c) $5p^2 = 5 \cdot 3^2$
   $= 5 \cdot 9$ Square.
   $= 45$ Multiply.

2. Replace x with 6 and y with 9 in each expression.

   (a) $4x + 7y = 4(6) + 7(9)$
   $= 24 + 63$ Multiply.
   $= 87$ Add.

   (b) $\dfrac{4x - 2y}{x + 1} = \dfrac{4(6) - 2(9)}{6 + 1}$
   $= \dfrac{24 - 18}{6 + 1}$ Multiply.
   $= \dfrac{6}{7}$ Subtract and add.

   (c) $2x^2 + y^2 = 2 \cdot 6^2 + 9^2$
   $= 2 \cdot 36 + 81$ Use exponents.
   $= 72 + 81$ Multiply.
   $= 153$ Add.

3. (a) "Sum" is the answer to an addition problem. Using x as the variable to represent the number, "The sum of 5 and a number" translates as $5 + x$ or $x + 5$.

   (b) "Minus" indicates subtraction. Using x as the variable to represent the number, "A number minus 4" translates as $x - 4$.

   (c) Since a number is subtracted *from* 48, write this as $48 - x$ when using x as the variable to represent the number.

   (d) "Product" indicates multiplication. Using x as the variable to represent the number, "The product of 6 and a number" translates as $6 \cdot x$ or $6x$.

   (e) "The sum of a number and 5" suggests a number plus 5. Using x as the variable to represent the number, "9 multiplied by the sum of a number and 5" translates as $9(x + 5)$.

14   Chapter 1   The Real Number System

4. (a) $p - 1 = 3$
    $2 - 1 = 3$ *Replace p with 2.*
    $1 = 3$ *False*

    The number 2 is not a solution of the equation.

    (b) $2k + 3 = 15$
    $2 \cdot 7 + 3 = 15$ *Replace k with 7.*
    $14 + 3 = 15$ *Multiply.*
    $17 = 15$ *False*

    The number 7 is not a solution of the equation.

    (c) $8p - 11 = 5$
    $8 \cdot 2 - 11 = 5$ *Replace p with 2.*
    $16 - 11 = 5$ *Multiply.*
    $5 = 5$ *True*

    The number 2 is a solution of the equation.

5. (a) "The sum of a number and 13" means a number plus 13. With x representing the number, "Three times the sum of a number and 13 is 19" translates as

    $$3(x + 13) = 19.$$

    (b) The word "giving" suggests equals. With x representing the number, "Five times a number is subtracted from 21, giving 15" translates as

    $$21 - 5x = 15.$$

6. (a) $2x + 5y - 7$ has no equals sign, so it is an expression.

    (b) $\dfrac{3x - 1}{5}$ has no equals sign, so it is an expression.

    (c) $2x + 5 = 7$ has an equals sign, so it is an equation.

    (d) $\dfrac{x}{y - 3} = 4x$ has an equals sign, so it is an equation.

1.2   Section Exercises

1. If $x = 3$, then the value of $x + 8$ is $3 + 8$, or 11.

3. The sum of 13 and x is represented by the expression $13 + x$. If $x = 3$, then the value of $13 + x$ is $13 + 3$, or 16.

5. There is no equals sign in $2x + 6$, so it is an expression. There *is* an equals sign in $2x + 6 = 8$, so it is an equation.

7. The statement would read "Nine less than five times a number is 49." "Nine less than" means "nine subtracted from." The equation would be $5x - 9 = 49$.

9. $2x + y = 6$
    $2 \cdot 0 + y = 6$ *Let x = 0.*
    $y = 6$

    $2x + y = 6$
    $2 \cdot 1 + y = 6$ *Let x = 1.*
    $2 + y = 6$
    $y = 4$

    Answers will vary. Two such pairs are $x = 0, y = 6$ and $x = 1, y = 4$. To find a pair, choose one number, substitute it for a variable, then calculate the value for the other variable.

In part (a) of Exercises 11–18, replace x with 4.
In part (b), replace x with 6. Then use the order of operations.

11. (a) $4x^2 = 4 \cdot 4^2$
    $= 4 \cdot 16$
    $= 64$

    (b) $4x^2 = 4 \cdot 6^2$
    $= 4 \cdot 36$
    $= 144$

13. (a) $\dfrac{3x - 5}{2x} = \dfrac{3 \cdot 4 - 5}{2 \cdot 4}$
    $= \dfrac{12 - 5}{8}$
    $= \dfrac{7}{8}$

    (b) $\dfrac{3x - 5}{2x} = \dfrac{3 \cdot 6 - 5}{2 \cdot 6}$
    $= \dfrac{18 - 5}{12}$
    $= \dfrac{13}{12}$

15. (a) $\dfrac{6.459x}{2.7} = \dfrac{6.459 \cdot 4}{2.7}$
    $= \dfrac{25.836}{2.7}$
    $= 9.569$ (to the nearest thousandth)

    (b) $\dfrac{6.459x}{2.7} = \dfrac{6.459 \cdot 6}{2.7}$
    $= \dfrac{38.754}{2.7}$
    $= 14.353$ (to the nearest thousandth)

17. (a) $3x^2 + x = 3 \cdot 4^2 + 4$
    $= 3 \cdot 16 + 4$
    $= 48 + 4 = 52$

(b) $3x^2 + x = 3 \cdot 6^2 + 6$
$= 3 \cdot 36 + 6$
$= 108 + 6 = 114$

19. (a) $3(x + 2y) = 3(2 + 2 \cdot 1)$
$= 3(2 + 2)$
$= 3(4)$
$= 12$

(b) $3(x + 2y) = 3(1 + 2 \cdot 5)$
$= 3(1 + 10)$
$= 3(11)$
$= 33$

21. (a) $x + \dfrac{4}{y} = 2 + \dfrac{4}{1}$
$= 2 + 4$
$= 6$

(b) $x + \dfrac{4}{y} = 1 + \dfrac{4}{5}$
$= \dfrac{5}{5} + \dfrac{4}{5}$
$= \dfrac{9}{5}$

23. (a) $\dfrac{x}{2} + \dfrac{y}{3} = \dfrac{2}{2} + \dfrac{1}{3}$
$= \dfrac{6}{6} + \dfrac{2}{6}$
$= \dfrac{8}{6} = \dfrac{4}{3}$

(b) $\dfrac{x}{2} + \dfrac{y}{3} = \dfrac{1}{2} + \dfrac{5}{3}$
$= \dfrac{3}{6} + \dfrac{10}{6}$
$= \dfrac{13}{6}$

25. (a) $\dfrac{2x + 4y - 6}{5y + 2} = \dfrac{2(2) + 4(1) - 6}{5(1) + 2}$
$= \dfrac{4 + 4 - 6}{5 + 2}$
$= \dfrac{8 - 6}{7}$
$= \dfrac{2}{7}$

(b) $\dfrac{2x + 4y - 6}{5y + 2} = \dfrac{2(1) + 4(5) - 6}{5(5) + 2}$
$= \dfrac{2 + 20 - 6}{25 + 2}$
$= \dfrac{22 - 6}{27}$
$= \dfrac{16}{27}$

27. (a) $2y^2 + 5x = 2 \cdot 1^2 + 5 \cdot 2$
$= 2 \cdot 1 + 5 \cdot 2$
$= 2 + 10$
$= 12$

(b) $2y^2 + 5x = 2 \cdot 5^2 + 5 \cdot 1$
$= 2 \cdot 25 + 5 \cdot 1$
$= 50 + 5$
$= 55$

29. (a) $\dfrac{3x + y^2}{2x + 3y} = \dfrac{3(2) + 1^2}{2(2) + 3(1)}$
$= \dfrac{3(2) + 1}{4 + 3}$
$= \dfrac{6 + 1}{7}$
$= \dfrac{7}{7}$
$= 1$

(b) $\dfrac{3x + y^2}{2x + 3y} = \dfrac{3(1) + 5^2}{2(1) + 3(5)}$
$= \dfrac{3(1) + 25}{2 + 15}$
$= \dfrac{3 + 25}{17}$
$= \dfrac{28}{17}$

31. (a) $.841x^2 + .32y^2$
$= .841 \cdot 2^2 + .32 \cdot 1^2$
$= .841 \cdot 4 + .32 \cdot 1$
$= 3.364 + .32$
$= 3.684$

(b) $.841x^2 + .32y^2$
$= .841 \cdot 1^2 + .32 \cdot 5^2$
$= .841 \cdot 1 + .32 \cdot 25$
$= .841 + 8$
$= 8.841$

16 Chapter 1 The Real Number System

33. "Twelve times a number" translates as $12 \cdot x$ or $12x$.

35. "Two subtracted from a number" translates as $x - 2$.

37. "One-third of a number, subtracted from seven" translates as $7 - \frac{1}{3}x$.

39. "Twice a number" is written as $2x$ and "difference" indicates subtraction, so write this as $2x - 6$.

41. "12 divided by the sum of a number and 3" translates as $\frac{12}{x+3}$.

43. "The product of 6 and four less than a number" translates as $6(x - 4)$.

45. The word *and* does not signify addition here. In the phrase "the product of a number and 6," *and* connects two quantities to be multiplied.

47. $x - 5 = 12$
$7 - 5 = 12$ Let $x = 7$.
$2 = 12$ *False*

The false result shows that 7 is not a solution of the equation.

49. $5x + 2 = 7$
$5(1) + 2 = 7$ Let $x = 1$.
$5 + 2 = 7$
$7 = 7$ *True*

The number 1 is a solution of the equation.

51. $6x + 4x + 9 = 11$
$6\left(\frac{1}{5}\right) + 4\left(\frac{1}{5}\right) + 9 = 11$ Let $x = \frac{1}{5}$.
$\frac{6}{5} + \frac{4}{5} + 9 = 11$
$\frac{10}{5} + 9 = 11$
$2 + 9 = 11$
$11 = 11$ *True*

The true result shows that $\frac{1}{5}$ is a solution of the equation.

53. $2y + 3(y - 2) = 14$
$2 \cdot 3 + 3(3 - 2) = 14$ Let $y = 3$.
$2 \cdot 3 + 3 \cdot 1 = 14$
$6 + 3 = 14$
$9 = 14$ *False*

Because substituting 3 for y results in a false statement, 3 is not a solution of the equation.

55. $\dfrac{z + 4}{2 - z} = \dfrac{13}{5}$

$\dfrac{\frac{1}{3} + 4}{2 - \frac{1}{3}} = \dfrac{13}{5}$ Let $z = \frac{1}{3}$.

$\dfrac{\frac{1}{3} + \frac{12}{3}}{\frac{6}{3} - \frac{1}{3}} = \dfrac{13}{5}$

$\dfrac{\frac{13}{3}}{\frac{5}{3}} = \dfrac{13}{5}$

$\dfrac{13}{3} \cdot \dfrac{3}{5} = \dfrac{13}{5}$

$\dfrac{13}{5} = \dfrac{13}{5}$ *True*

The true result shows that $\frac{1}{3}$ is a solution of the equation.

57. $3r^2 - 2 = 53.47$
$3(4.3)^2 - 2 = 53.47$ Let $r = 4.3$.
$3 \cdot 18.49 - 2 = 53.47$
$55.47 - 2 = 53.47$
$53.47 = 53.47$ *True*

The number 4.3 is a solution of the equation.

59. "The sum of a number and 8 is 18" translates as

$$x + 8 = 18.$$

61. "Five more than twice a number is 5" translates as

$$2x + 5 = 5.$$

63. "Sixteen minus three-fourths of a number is 13" translates as

$$16 - \frac{3}{4}x = 13.$$

65. "Three times a number is equal to 8 more than twice the number" translates as

$$3x = 2x + 8.$$

67. There is no equals sign, so $3x + 2(x - 4)$ is an expression.

69. There is an equals sign, so $7t + 2(t + 1) = 4$ is an equation.

71. Substitute 1997 for x in the equation.

$y = .494x - 974.0$
$y = .494(1997) - 974.0$
$y \approx 986.52 - 974.0$
$y = 12.52$

The approximate hourly earnings in 1997 were $12.52, which is more than $12.49 by $.03.

72. Substitute 2000 for x in the equation.

$y = .494x - 974.0$
$y = .494(2000) - 974.0$
$y = 988 - 974.0$
$y = 14$

The approximate hourly earnings in 2000 were $14.00, which is the same as the actual earnings.

73. Substitute 2001 for x in the equation.

$y = .494x - 974.0$
$y = .494(2001) - 974.0$
$y \approx 988.49 - 974.0$
$y = 14.49$

The approximate hourly earnings in 2001 were $14.49, which is less than $14.53 by $.04.

74. Substitute 2002 for x in the equation.

$y = .494x - 974.0$
$y = .494(2002) - 974.0$
$y \approx 988.99 - 974.0$
$y = 14.99$

The approximate hourly earnings in 2002 were $14.99, which is more than $14.95 by $.04.

1.3 Real Numbers and the Number Line

1.3 Margin Exercises

1. **(a)** Since Erin spends $53 more than she has in her checking account, her balance is -53.

   (b) Since the record high was $134°$ above zero, this temperature is expressed as 134.

   (c) Positive numbers are used to represent gains and negative numbers to represent losses, so a gain of 5 yards is expressed as 5, while a loss of 10 yards is expressed as -10.

2. To compare these numbers and locate them on the number line, rewrite -2.75 and $\frac{17}{8}$ as mixed numbers (or change to decimals).

   $$-2.75 = -2\frac{3}{4} \text{ and } \frac{17}{8} = 2\frac{1}{8}$$

The order of the numbers from smallest to largest is:

$$-3, -2.75, -\frac{3}{4}, 1\frac{1}{2}, \frac{17}{8}$$

3. **(a)** $-2 < 4$ is true because any negative number is less than any positive number.

   (b) $6 > -3$ is true because any positive number is greater than any negative number.

   (c) Since -12 is to the left of -9 on the number line, -12 is less than -9. Therefore, the statement $-9 < -12$ is false.

   (d) Since -4 is to the left of -1 on the number line, -4 is less than -1. Therefore, the statement $-4 \geq -1$ is false.

   (e) $-6 \leq 0$ is true because any negative number is less than 0.

4. **(a)** The opposite of 6 is -6.

   (b) The opposite of 15 is -15.

   (c) The opposite of -9 is $-(-9) = 9$.

   (d) The opposite of -12 is $-(-12) = 12$.

   (e) The opposite of 0 is 0.

5. **(a)** The distance between -6 and 0 on the number line is 6, so $|-6| = 6$.

   (b) $|9| = 9$

   (c) $-|15| = -15$

   (d) $-|-9| = -(9) = -9$

   (e) $|9 - 4| = |5| = 5$

   (f) $-|32 - 2| = -|30| = -30$

1.3 Section Exercises

1. The only integer between 3.6 and 4.6 is 4.

3. There is only one whole number that is not positive and less than 1: the number 0.

5. An irrational number that is between $\sqrt{12}$ and $\sqrt{14}$ is $\sqrt{13}$. There are others.

7. $\left\{-9, -\sqrt{7}, -1\frac{1}{4}, -\frac{3}{5}, 0, \sqrt{5}, 3, 5.9, 7\right\}$

   (a) The natural numbers in the given set are 3 and 7, since they are in the natural number set $\{1, 2, 3, \ldots\}$.

(b) The set of whole numbers includes the natural numbers and 0. The whole numbers in the given set are 0, 3, and 7.

(c) The integers are the set of numbers $\{\ldots, -3, -2, -1, 0, 1, 2, 3, \ldots\}$. The integers in the given set are $-9, 0, 3$, and 7.

(d) Rational numbers are the numbers which can be expressed as the quotient of two integers, with denominators not equal to 0.

We can write numbers from the given set in this form as follows:

$-9 = \frac{-9}{1}, -1\frac{1}{4} = \frac{-5}{4}, -\frac{3}{5} = \frac{-3}{5}, 0 = \frac{0}{1}$

$3 = \frac{3}{1}, 5.9 = \frac{59}{10}$, and $7 = \frac{7}{1}$. Thus, the rational numbers in the given set are $-9, -1\frac{1}{4}, -\frac{3}{5}, 0, 3,$ 5.9, and 7.

(e) Irrational numbers are real numbers that are not rational. $-\sqrt{7}$ and $\sqrt{5}$ can be represented by points on the number line but cannot be written as a quotient of integers. Thus, the irrational numbers in the given set are $-\sqrt{7}$ and $\sqrt{5}$.

(f) Real numbers are all numbers that can be represented on the number line. All the numbers in the given set are real.

9. Positive numbers are used to represent increases, so an increase of 40,000 units is represented as 40,000.

11. Negative numbers are used to represent decreases, so a decrease of 9000 in the population is represented by -9000.

13. Graph $0, 3, -5$, and -6.

Place a dot on the number line at the point that corresponds to each number. The order of the numbers from smallest to largest is $-6, -5, 0, 3$.

15. Graph $-2, -6, -4, 3$, and 4.

17. Graph $\frac{1}{4}, 2\frac{1}{2}, -3\frac{4}{5}, -4$, and $-\frac{13}{8}$.

19. -11 is to the left of -4 on the number line, so -11 is the lesser number.

21. Since any negative number is less than a positive number, -21 is the lesser number.

23. -100 is less than 0 since any negative number is less than 0.

25. In order to compare these two numbers, write them with a common denominator.

$-\frac{2}{3} = -\frac{8}{12}$ and $-\frac{1}{4} = -\frac{3}{12}$

Since

$\frac{8}{12} > \frac{3}{12}$,

$-\frac{2}{3}$ is further to the left of 0 on the number line than $-\frac{1}{4}$, so $-\frac{2}{3}$ is the lesser number.

27. The statement $8 < -16$ is false. Any positive number is greater than any negative number.

29. The statement $-3 < -2$ is true since -3 is to the left of -2 on the number line.

31. **(a)** The opposite of -2 is found by writing the symbol $-$ in front of -2. The opposite of -2 is $-(-2) = 2$.

(b) The absolute value of -2 is the distance between 0 and -2 on the number line. $|-2| = 2$

33. **(a)** The opposite of 6 is found by writing the symbol $-$ in front of 6. The opposite of 6 is -6.

(b) The absolute value of 6 is the distance between 0 and 6 on the number line. $|6| = 6$

35. **(a)** The opposite of $-\frac{3}{4}$ is found by writing the symbol $-$ in front of $-\frac{3}{4}$. The opposite of $-\frac{3}{4}$ is $-(-\frac{3}{4}) = \frac{3}{4}$.

(b) The absolute value of $-\frac{3}{4}$ is the distance between 0 and $-\frac{3}{4}$ on the number line. $|-\frac{3}{4}| = \frac{3}{4}$

37. $|-7| = 7$

39. $-|12| = -(12) = -12$

41. $-|-\frac{2}{3}| = -(\frac{2}{3}) = -\frac{2}{3}$

43. $|13 - 4| = |9| = 9$

45. $|-8| = 8$
$|-8| < 7$ is false since $8 > 7$.

47. $|4| = 4$
$4 \leq |4|$ is true since $4 = 4$.

49. The statement, "The absolute value of a number is always positive" is false because $|0| = 0$ and zero is not a positive number.

51. The greatest decrease shown in the table corresponds to the negative number with the greatest absolute value. This number is -13.3, so the greatest decrease was in fuel/other utilities from 2001 to 2002.

53. $|-.5| = .5$
$|-.8| = .8$
The smallest absolute value is .5, which represents the period from 2000 to 2001.

1.4 Adding Real Numbers

1.4 Margin Exercises

1. **(a)** Start at 0 on a number line. Draw an arrow 1 unit to the right to represent the addition of a positive number. From the right end of this arrow, draw a second arrow 4 units to the right. The number below the end of this second arrow is 5, so $1 + 4 = 5$.

(b) Start at 0 on a number line. Draw an arrow 2 units to the left to represent the addition of a negative number. From the left end of this arrow, draw a second arrow 5 units to the left. The number below the end of this second arrow is -7, so $-2 + (-5) = -7$.

2. **(a)** $-7 + (-3) = -10$

The sum of two negative numbers is negative.

(b) $-12 + (-18) = -30$

The sum of two negative numbers is negative.

(c) $-15 + (-4) = -19$

The sum of two negative numbers is negative.

3. **(a)** Start at 0 on a number line. Draw an arrow 6 units to the right. From the right end of this arrow, draw a second arrow 3 units to the left. The number below the end of this second arrow is 3, so $6 + (-3) = 3$.

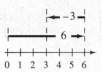

(b) Start at 0 on a number line. Draw an arrow 5 units to the left. From the left end of this arrow, draw a second arrow 1 unit to the right. The number below the end of this second arrow is -4, so $-5 + 1 = -4$.

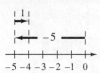

4. **(a)** The number with the greater absolute value is -8, so the sum will be negative. The answer of -6 is correct.

(b) The number with the greater absolute value is -15, so the sum will be negative. The answer of -11 is correct.

(c) The number with the greater absolute value is 17, so the sum will be positive. The answer of 7 is correct.

(d) The number with the greater absolute value is $-1\frac{3}{8} = -\frac{11}{8}$, so the sum will be negative. The answer of $-\frac{5}{8}$ is correct.

(e) The number with the greater absolute value is -9.5, so the sum will be negative. The answer of -5.7 is correct.

5. **(a)** $2 + [7 + (-3)] = 2 + 4$
$= 6$

(b) $6 + [(-2 + 5) + 7] = 6 + (3 + 7)$
$= 6 + 10$
$= 16$

(c) $-9 + [-4 + (-8 + 6)] = -9 + [-4 + (-2)]$
$= -9 + (-6)$
$= -15$

6. **(a)** "4 more than -12" is written $-12 + 4$.
$-12 + 4 = -8$

(b) "The sum of 6 and -7" is written $6 + (-7)$.
$6 + (-7) = -1$

(c) "-12 added to -31" is written $-31 + (-12)$.
$-31 + (-12) = -43$

(d) "7 is increased by the sum of 8 and -3" is written $7 + [8 + (-3)]$.
$7 + [8 + (-3)] = 7 + 5 = 12$

20   Chapter 1   The Real Number System

7. Represent the losses by negative numbers and the gain by positive numbers.
$$-8 + (-5) + 7 = [-8 + (-5)] + 7$$
$$= -13 + 7$$
$$= -6$$
The team lost 6 yards.

1.4 Section Exercises

1. First add -2 and 5 because they are within the innermost grouping symbols.

3. First add -1 and -3 because they are within the innermost grouping symbols.

5. $6 + (-4)$

    Since the numbers have different signs, find the difference between their absolute values:
    $$6 - 4 = 2.$$
    Since 6 has the larger absolute value, the sum is positive, so
    $$6 + (-4) = 2.$$

7. $12 + (-15)$

    Since the numbers have different signs, find the difference between their absolute values:
    $$15 - 12 = 3.$$
    Since -15 has the larger absolute value, the sum is negative, so
    $$12 + (-15) = -3.$$

9. $-7 + (-3)$

    Since the numbers have the same sign, add their absolute values:
    $$7 + 3 = 10.$$
    Since both numbers are negative, the sum is negative, so
    $$-7 + (-3) = -10.$$

11. $-10 + (-3)$

    Since the numbers have the same sign, add their absolute values:
    $$10 + 3 = 13.$$
    Since both numbers are negative, the sum is negative, so
    $$-10 + (-3) = -13.$$

13. $-12.4 + (-3.5) = -15.9$ since the sum of two negative numbers is negative.

15. First work inside the brackets.
$$10 + [-3 + (-2)] = 10 + (-5) = 5$$

17. First work inside the brackets.
$$5 + [14 + (-6)] = 5 + 8 = 13$$

19. First work inside the brackets.
$$-3 + [5 + (-2)] = -3 + 3 = 0$$

21. First work inside the brackets.
$$-8 + [3 + (-1) + (-2)] = -8 + [2 + (-2)]$$
$$= -8 + 0$$
$$= -8$$

23. $\dfrac{9}{10} + \left(-\dfrac{3}{5}\right) = \dfrac{9}{10} + \left(-\dfrac{6}{10}\right) = \dfrac{3}{10}$

25. $-\dfrac{1}{6} + \dfrac{2}{3} = -\dfrac{1}{6} + \dfrac{4}{6} = \dfrac{3}{6} = \dfrac{1}{2}$

27. $2\dfrac{1}{2} + \left(-3\dfrac{1}{4}\right) = \dfrac{5}{2} + \left(-\dfrac{13}{4}\right)$
$$= \dfrac{10}{4} + \left(-\dfrac{13}{4}\right)$$
$$= -\dfrac{3}{4}$$

29. $7.8 + (-9.4) = -1.6$

31. $-7.1 + [3.3 + (-4.9)] = -7.1 + (-1.6)$
$$= -8.7$$

33. $[-8 + (-3)] + [-7 + (-7)] = -11 + (-14)$
$$= -25$$

35. $\left(-\dfrac{1}{2} + .25\right) - \left(-\dfrac{3}{4} + .75\right)$
$$= \left(-\dfrac{1}{2} + \dfrac{1}{4}\right) - \left(-\dfrac{3}{4} + \dfrac{3}{4}\right)$$
$$= \left(-\dfrac{2}{4} + \dfrac{1}{4}\right) - 0$$
$$= -\dfrac{1}{4}$$

37. $-11 + 13 = 13 + (-11)$
$$2 = 2$$
The statement is true.

39. $-10 + 6 + 7 = -3$
Since $-10 + 6 + 7 = 3$, not -3, the statement is false.

41. $\dfrac{7}{3} + \left(-\dfrac{1}{3}\right) + \left(-\dfrac{6}{3}\right) = 0$
$$\dfrac{6}{3} + \left(-\dfrac{6}{3}\right) = 0$$
$$0 = 0$$
The statement is true.

43. $|-8 + 10| = -8 + (-10)$

Since $|-8 + 10| = 2$ and $-8 + (-10) = -18$, the statement is false.

45. $2\frac{1}{5} + \left(-\frac{6}{11}\right) = -\frac{6}{11} + 2\frac{1}{5}$

The statement is true because each side represents the sum of the same two numbers.

47. $-7 + [-5 + (-3)] = [(-7) + (-5)] + 3$

Since $-7 + [-5 + (-3)] = -7 + (-8) = -15$ and $[(-7) + (-5)] + 3 = -12 + 3 = -9$, the statement is false.

49. If the sum of two numbers is negative and you know that one of the numbers is positive, then the other number must be negative and have the larger absolute value.

50. $x + 5 = -7$

If the sum of two numbers is negative and you know that one of the numbers is positive, then the other number must be negative. The sum of a positive number and 5 cannot be -7. Thus, the solution to this equation must be a negative number.

51. If the sum of two numbers is positive and you know that one of the numbers is negative, then the other number must be positive and have the larger absolute value.

52. $x + (-8) = 2$

If the sum of two numbers is positive and you know that one of the numbers is negative, then the other number must be positive. The sum of a negative number and -8 cannot be 2. The solution to this equation must be a positive number.

53. Add the absolute values of the numbers. The sum will be negative.

55. "The sum of -5 and 12 and 6" is written $-5 + 12 + 6$.

$-5 + 12 + 6 = [-5 + 12] + 6 = 7 + 6 = 13$

57. "14 added to the sum of -19 and -4" is written $[-19 + (-4)] + 14$.

$[-19 + (-4)] + 14 = -23 + 14 = -9$

59. "The sum of -4 and -10, increased by 12" is written $[-4 + (-10)] + 12$.

$[-4 + (-10)] + 12 = -14 + 12 = -2$

61. "$\frac{2}{7}$ more than the sum of $\frac{5}{7}$ and $-\frac{9}{7}$" is written

$$\left[\frac{5}{7} + \left(-\frac{9}{7}\right)\right] + \frac{2}{7}.$$

$$\left[\frac{5}{7} + \left(-\frac{9}{7}\right)\right] + \frac{2}{7} = -\frac{4}{7} + \frac{2}{7} = -\frac{2}{7}$$

63. Owing an amount is represented by a negative number, as is borrowing an amount.

$$-10 + (-70) = -80$$

Since Kramer owes $80, his financial status is $-\$80$.

65. $0 + (-130) + (-54) = -130 + (-54)$
$= -184$

Their new altitude is 184 meters below the surface, which can be represented by the signed number -184.

67. Gains are represented by positive numbers and losses by negative numbers.

$$6 + (-12) + 43 = -6 + 43 = 37$$

The total net yardage was 37 yards.

69. The lowest temperature is represented by -29. The highest temperature is represented by $-29 + 149$ or $120°F$.

71. Use negative numbers to represent amounts Dana owes and for purchases. Use a positive number to represent payments.

$$[-153 + (-14)] + 60 = -167 + 60$$
$$= -107$$

Dana's current balance (as a signed number) is $-\$107$.

73. $[-5 + (-4)] + (-3) = -9 + (-3)$
$= -12$

The total number of seats that New York, Pennsylvania, and Ohio are projected to lose is twelve, which can be represented by the signed number -12.

1.5 Subtracting Real Numbers

1.5 Margin Exercises

1. (a) Begin at 0 on a number line and draw an arrow 5 units to the right. From the right end of this arrow, draw a second arrow 1 unit to the left. The number at the end of this second arrow is 4, so $5 - 1 = 4$.

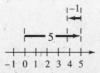

(b) Begin at 0 on a number line and draw an arrow 6 units to the right. From the right end of this arrow, draw a second arrow 2 units to the left. The number at the end of this second arrow is 4, so $6 - 2 = 4$.

2. **(a)** $6 - 10 = 6 + (-10) = -4$

   (b) $-2 - 4 = -2 + (-4) = -6$

   (c) $3 - (-5) = 3 + 5 = 8$

   (d) $-8 - (-12) = -8 + 12 = 4$

   (e) $\dfrac{5}{4} - \left(-\dfrac{3}{7}\right) = \dfrac{5}{4} + \dfrac{3}{7} = \dfrac{35}{28} + \dfrac{12}{28} = \dfrac{47}{28}$

3. **(a)** $2 - [(-3) - (4 + 6)] = 2 - [(-3) - 10]$
   $= 2 - [(-3) + (-10)]$
   $= 2 - (-13)$
   $= 2 + 13$
   $= 15$

   (b) $[(5 - 7) + 3] - 8 = [(-2) + 3] - 8$
   $= 1 - 8$
   $= 1 + (-8)$
   $= -7$

   (c) $6 - [(-1 - 4) - 2] = 6 - [(-5) - 2]$
   $= 6 - (-7)$
   $= 6 + 7$
   $= 13$

4. **(a)** "The difference between -5 and -12" is written $-5 - (-12)$.
   $-5 - (-12) = -5 + 12$
   $= 7$

   (b) "-2 subtracted from the sum of 4 and -4" is written $[4 + (-4)] - (-2)$.
   $[4 + (-4)] - (-2) = 0 - (-2)$
   $= 0 + 2$
   $= 2$

   (c) "7 less than -2" is written $-2 - 7$.
   $-2 - 7 = -2 + (-7)$
   $= -9$

(d) "9, decreased by 10 less than 7" is written $9 - (7 - 10)$.
$9 - (7 - 10) = 9 - [7 + (-10)]$
$= 9 - (-3)$
$= 9 + 3$
$= 12$

5. The difference between the highest and lowest elevations is given by
$6960 - (-40) = 6960 + 40$
$= 7000.$

The difference is 7000 meters.

1.5 Section Exercises

1. By the definition of subtraction, in order to perform the subtraction problem $-6 - (-8)$, we must add the opposite of $\underline{-8}$ to $\underline{-6}$.

3. "The difference between 7 and 12" translates as $\underline{7 - 12}$, while "the difference between 12 and 7" translates as $\underline{12 - 7}$.

5. $-8 - 4 = -8 + \underline{(-4)}$

7. $-7 - 3 = -7 + (-3) = -10$

9. $-10 - 6 = -10 + (-6) = -16$

11. $7 - (-4) = 7 + 4 = 11$

13. $6 - (-13) = 6 + 13 = 19$

15. $-7 - (-3) = -7 + 3 = -4$

17. $3 - (4 - 6) = 3 - [4 + (-6)]$
    $= 3 - (-2)$
    $= 3 + 2$
    $= 5$

19. $-3 - (6 - 9) = -3 - [6 + (-9)]$
    $= -3 - (-3)$
    $= -3 + 3$
    $= 0$

21. $\dfrac{1}{2} - \left(-\dfrac{1}{4}\right) = \dfrac{1}{2} + \dfrac{1}{4}$
    $= \dfrac{2}{4} + \dfrac{1}{4} = \dfrac{3}{4}$

23. $-\dfrac{3}{4} - \dfrac{5}{8} = -\dfrac{3}{4} + \left(-\dfrac{5}{8}\right)$
    $= -\dfrac{6}{8} + \left(-\dfrac{5}{8}\right)$
    $= -\dfrac{11}{8}$ or $-1\dfrac{3}{8}$

25. $\dfrac{5}{8} - \left(-\dfrac{1}{2} - \dfrac{3}{4}\right)$

$= \dfrac{5}{8} - \left[-\dfrac{1}{2} + \left(-\dfrac{3}{4}\right)\right]$

$= \dfrac{5}{8} - \left[-\dfrac{2}{4} + \left(-\dfrac{3}{4}\right)\right]$

$= \dfrac{5}{8} - \left(-\dfrac{5}{4}\right)$

$= \dfrac{5}{8} + \dfrac{5}{4}$

$= \dfrac{5}{8} + \dfrac{10}{8}$

$= \dfrac{15}{8}$ or $1\dfrac{7}{8}$

27. $4.4 - (-9.2) = 4.4 + 9.2 = 13.6$

29. $-7.4 - 4.5 = -7.4 + (-4.5) = -11.9$

31. $-5.2 - (8.4 - 10.8) = -5.2 - [8.4 + (-10.8)]$
$= -5.2 - (-2.4)$
$= -5.2 + 2.4$
$= -2.8$

33. $[(-3.1) - 4.5] - (.8 - 2.1)$
$= [(-3.1) + (-4.5)] - [.8 + (-2.1)]$
$= -7.6 - (-1.3)$
$= -7.6 + 1.3$
$= -6.3$

35. $-12 - [(9 - 2) - (-6 - 3)]$
$= -12 - [7 - (-9)]$
$= -12 - (7 + 9)$
$= -12 - 16$
$= -12 + (-16)$
$= -28$

37. $-8 + [(-3 - 10) - (-4 + 1)]$
$= -8 + [-13 - (-3)]$
$= -8 + (-13 + 3)$
$= -8 + (-10)$
$= -18$

39. $\left(-\dfrac{3}{8} - \dfrac{2}{3}\right) - \left(-\dfrac{9}{8} - 3\right)$

$= \left[-\dfrac{3}{8} + \left(-\dfrac{2}{3}\right)\right] - \left[-\dfrac{9}{8} + (-3)\right]$

$= \left[-\dfrac{9}{24} + \left(-\dfrac{16}{24}\right)\right] - \left[-\dfrac{9}{8} + \left(-\dfrac{24}{8}\right)\right]$

$= -\dfrac{25}{24} - \left(-\dfrac{33}{8}\right)$

$= -\dfrac{25}{24} - \left(-\dfrac{99}{24}\right)$

$= -\dfrac{25}{24} + \dfrac{99}{24}$

$= \dfrac{74}{24}$

$= \dfrac{37}{12}$ or $3\dfrac{1}{12}$

41. $[-12.25 - (8.34 + 3.57)] - 17.88$
$= (-12.25 - 11.91) - 17.88$
$= [-12.25 + (-11.91)] - 17.88$
$= -24.16 - 17.88$
$= -24.16 + (-17.88)$
$= -42.04$

43. For example, let $a = 1, b = 1$ or let $a = 2, b = 2$. In general, choose $a = b$.

45. $8 - (-3) - 9 + 6 = 8 + 3 - 9 + 6$
$= 11 - 9 + 6$
$= 2 + 6$
$= 8$

47. The difference between two negative numbers can be either positive or negative. In the following example, the difference is a negative number.

$-8 - (-2) = -8 + (2) = -6$

49. "The difference between 4 and -8" is written $4 - (-8)$. This expression can be simplified as follows:

$4 - (-8) = 4 + 8 = 12$

51. "8 less than -2" is written $-2 - 8$.

$-2 - 8 = -2 + (-8) = -10$

53. "The sum of 9 and -4, decreased by 7" is written $[9 + (-4)] - 7$.

$[9 + (-4)] - 7 = 5 + (-7) = -2$

55. "12 less than the difference between 8 and -5" is written $[8 - (-5)] - 12$.

$[8 - (-5)] - 12 = (8 + 5) - 12$
$= 13 - 12$
$= 13 + (-12)$
$= 1$

57. 23°F lower than -35°F can be represented as

$-35 - 23 = -35 + (-23)$
$= -58.$

The record low in South Dakota is -58°F.

59. $14,494 - (-282) = 14,494 + 282$
$= 14,776$

The difference between these two elevations is 14,776 feet.

61. Owing money is represented by a negative number, as is borrowing money.
$$-10 + (-70) = -80$$
Ben's financial status is represented by $-\$80$.

63. $86,000 - (-19,000) = 86,000 + (19,000)$
$= 105,000$

The difference is $105,000.

65. Add the scores of the four turns to get the final score.
$-19 + 28 + (-5) + 13 = 9 + (-5) + 13$
$= 4 + 13$
$= 17$

His final score for the four turns was 17.

67. Sum of checks:
$\$35.84 + \$26.14 + \$3.12 = \$61.98 + \$3.12$
$= \$65.10$

Sum of deposits:
$\$85.00 + \$120.76 = \$205.76$

Final balance $=$ Beginning balance $-$ checks $+$ deposits
$= \$904.89 - \$65.10 + \$205.76$
$= \$839.79 + \$205.76$
$= \$1045.55$

Her account balance at the end of August was $1045.55.

69. Represent 426 B.C. as -426.
$-426 - 43 = -426 + (-43)$
$= -469$

His father was born in the year 469 B.C.

71.
```
 -870.00   amount owed
           2 return credits
 +185.90   ($35.90 + $150.00)
 -------
 -684.10
           3 purchases
 -102.50   ($82.50 + $10.00 + $10.00)
 -------
 -786.60
 +500.00   payment
 -------
 -286.60
  -37.23   finance charge
 -------
 -323.83
```

She still owes $323.83.

73. Dean's depth can be represented by -34 ft. Jeff's depth is $-40 + 20 = -20$ ft. The vertical distance between Dean and Jeff is
$$-20 - (-34) = -20 + 34 = 14 \text{ ft.}$$

75. $17,400 - (-23,376) = 17,400 + 23,376$
$= 40,776$

The vertical distance between the top of Mt. Foraker and the bottom of the Java Trench is 40,776 ft.

77. For 2002: -1.5%; for 2001: 0.1%
The percent change in the United States was
$$-1.5\% - 0.1\% = -1.6\%.$$

79. $133,300 - 128,400 = 4900$

The change from 1998 to 1999 was $4900.

81. $147,800 - 139,000 = 8800$

The change from 2000 to 2001 was $8800.

83. The expression $x - y$ would have to be *positive* since subtracting a negative number from a positive number is the same as adding a positive number to a positive number, which is a positive number.

85. $x + |y|$

Since $|y|$ is positive, $x + |y|$ is the sum of two positive numbers, which is *positive*.

1.6 Multiplying and Dividing Real Numbers

1.6 Margin Exercises

1. (a) $3(-3) = -3 + (-3) + (-3) = -9$

(b) $3(-4) = -4 + (-4) + (-4) = -12$

(c) $3(-5) = -5 + (-5) + (-5) = -15$

2. Use the following rule: The product of a positive number and a negative number is a negative number.

(a) $2(-6) = -(2 \cdot 6) = -12$

(b) $7(-8) = -(7 \cdot 8) = -56$

(c) $-9(2) = -(9 \cdot 2) = -18$

(d) $-16\left(\dfrac{5}{32}\right) = -\left(\dfrac{\overset{1}{\cancel{16}}}{1} \cdot \dfrac{5}{\underset{2}{\cancel{32}}}\right)$
$= -\dfrac{5}{2}$

(e) $4.56(-10) = -(4.56 \cdot 10) = -45.6$

1.6 Multiplying and Dividing Real Numbers

3. Use the following rule: The product of two negative numbers is a positive number.
   (a) $-5(-6) = 5 \cdot 6 = 30$
   (b) $-7(-3) = 7 \cdot 3 = 21$
   (c) $-8(-5) = 8 \cdot 5 = 40$
   (d) $-11(-2) = 11 \cdot 2 = 22$
   (e) $-17(3)(-7) = (-51)(-7) = 357$
   (f) $-41(2)(-13) = (-82)(-13) = 1066$

4.
	Number	Reciprocal
(a)	6	$\frac{1}{6}$
(b)	-2	$\frac{1}{-2} = -\frac{1}{2}$
(c)	$\frac{2}{3}$	$\frac{3}{2}$
(d)	$-\frac{1}{4}$	$-\frac{4}{1} = -4$
(e)	0	none

5. (a) $\frac{42}{7} = 42 \cdot \frac{1}{7} = 6$

   (b) $\frac{-36}{(-2)(-3)} = -36 \cdot \frac{1}{6} = -6$

   (c) $\frac{-12.56}{-.4} = 31.4$

   (d) $\frac{10}{7} \div \left(-\frac{24}{5}\right) = \frac{\overset{5}{\cancel{10}}}{7} \cdot \left(-\frac{5}{\underset{12}{\cancel{24}}}\right) = -\frac{25}{84}$

   (e) $\frac{-3}{0}$ is undefined.

   (f) $\frac{0}{-53} = 0$

6. (a) $\frac{-8}{-2} = 4$

   (b) $\frac{-16.4}{2.05} = -8$

   (c) $\frac{1}{4} \div \left(-\frac{2}{3}\right) = \frac{1}{4} \cdot \left(-\frac{3}{2}\right) = -\frac{3}{8}$

   (d) $\frac{12}{-4} = -3$

7. (a) $-3(4) - 2(6) = -12 - 12$
   $= -12 + (-12) = -24$

   (b) $-8[-1 - (-4)(-5)] = -8(-1 - 20)$
   $= -8[-1 + (-20)]$
   $= -8(-21)$
   $= 168$

   (c) $\frac{6(-4) - 2(5)}{3(2-7)} = \frac{-24 - 10}{3(-5)} = \frac{-34}{-15} = \frac{34}{15}$

   (d) $\frac{-6(-8) + 3(9)}{-2[4 - (-3)]} = \frac{48 + 27}{-2(4+3)} = \frac{75}{-2(7)}$
   $= \frac{75}{-14} = -\frac{75}{14}$

8. (a) Replace x with -4 and y with 3.
   $2x - 7(y+1) = 2(-4) - 7(3+1)$
   $= -8 - 7(4)$
   $= -8 - 28$
   $= -36$

   (b) Replace x with -2 and y with -3.
   $2x^2 - 4y^2 = 2(-2)^2 - 4(-3)^2$
   $= 2(4) - 4(9)$
   $= 8 - 36$
   $= -28$

   (c) Replace x with 2 and y with -1.
   $\frac{4x - 2y}{-3x} = \frac{4(2) - 2(-1)}{-3(2)} = \frac{8+2}{-6}$
   $= \frac{10}{-6} = -\frac{5}{3}$

9. (a) "The product of 6 and the sum of -5 and -4" is written $6[(-5) + (-4)]$.
   $6[(-5) + (-4)] = 6(-9) = -54$

   (b) "Twice the difference between 8 and -4" is written $2[8 - (-4)]$.
   $2[8 - (-4)] = 2(8+4) = 2(12) = 24$

   (c) "Three-fifths of the sum of 2 and -7" is written $\frac{3}{5}[2 + (-7)]$.
   $\frac{3}{5}[2 + (-7)] = \frac{3}{5}(-5) = -3$

   (d) "20% of the sum of 9 and -4" is written $.20[9 + (-4)]$.
   $.20[9 + (-4)] = .20(5) = 1$

10. (a) "The quotient of 20 and the sum of 8 and -3" is written $\frac{20}{8 + (-3)}$.
    $\frac{20}{8 + (-3)} = \frac{20}{5} = 4$

    (b) "The product of -9 and 2, divided by the difference between 5 and -1" is written $\frac{-9(2)}{5 - (-1)}$.
    $\frac{-9(2)}{5 - (-1)} = \frac{-18}{6} = -3$

11. (a) "Twice a number is -6" is written
    $2x = -6.$

(b) "The difference between -8 and a number is -11" is written
$$-8 - x = -11.$$

(c) "The sum of 5 and a number is 8" is written
$$5 + x = 8.$$

(d) "The quotient of a number and -2 is 6" is written
$$\frac{x}{-2} = 6.$$

1.6 Section Exercises

1. The product or the quotient of two numbers with the same sign is _greater than 0_, since the product or quotient of two positive numbers is positive and the product or quotient of two negative numbers is positive.

3. If three negative numbers are multiplied together, the product is _less than 0_, since a negative number times a negative number is a positive number, and that positive number times a negative number is a negative number.

5. If a negative number is squared and the result is added to a positive number, the final answer is _greater than 0_, since a negative number squared is a positive number, and a positive number added to another positive number is a positive number.

7. $-7(4) = -(7 \cdot 4) = -28$

9. $-5(-6) = 30$

11. $-8(0) = 0$

13. $-\dfrac{3}{8}\left(-\dfrac{20}{9}\right) = \dfrac{3 \cdot 20}{8 \cdot 9}$
$= \dfrac{3 \cdot 4 \cdot 5}{4 \cdot 2 \cdot 3 \cdot 3} = \dfrac{5}{6}$

15. $-6.8(.35) = -2.38$

17. $-6\left(-\dfrac{1}{4}\right) = \dfrac{6 \cdot 1}{4} = \dfrac{2 \cdot 3}{2 \cdot 2} = \dfrac{3}{2}$

19. $\dfrac{-15}{5} = -3$

    Note that the quotient of two numbers having different signs is negative.

21. $\dfrac{20}{-10} = -2$

23. $\dfrac{-160}{-10} = 16$

    Note that the quotient of two numbers having the same sign is positive.

25. $\dfrac{0}{-3} = 0$

27. $\dfrac{-10.252}{0}$ is undefined.

29. $\left(-\dfrac{3}{4}\right) \div \left(-\dfrac{1}{2}\right) = \left(-\dfrac{3}{4}\right) \cdot \left(-\dfrac{2}{1}\right) = \dfrac{3 \cdot 2}{2 \cdot 2} = \dfrac{3}{2}$

31. Look for the expression that has 0 in the denominator. The expression $\dfrac{5-5}{5-5}$ or $\dfrac{0}{0}$ is undefined. The correct response is **C**.

33. $\dfrac{-5(-6)}{9-(-1)} = \dfrac{30}{10}$
$= \dfrac{3 \cdot 10}{10} = 3$

35. $\dfrac{-21(3)}{-3-6} = \dfrac{-63}{-3+(-6)}$
$= \dfrac{-63}{-9} = 7$

37. $\dfrac{-10(2) + 6(2)}{-3 - (-1)} = \dfrac{-20 + 12}{-3 + 1}$
$= \dfrac{-8}{-2} = 4$

39. $\dfrac{-27(-2) - (-12)(-2)}{-2(3) - 2(2)} = \dfrac{54 - 24}{-6 - 4}$
$= \dfrac{30}{-10}$
$= -\dfrac{3 \cdot 10}{10}$
$= -3$

41. $\dfrac{3^2 - 4^2}{7(-8+9)} = \dfrac{9 - 16}{7(1)} = \dfrac{-7}{7} = -1$

43. If x is negative, $4x$ will be the product of a positive and a negative number, which is negative. If y is negative, $8y$ will likewise be negative. Then $4x + 8y$ will be the sum of two negative numbers, which is negative.

    For $4x - 8y$, we will have a negative number minus a negative number, or a negative number plus a positive number. Without knowing if the negative number is larger or smaller in absolute value than the positive number, it is impossible to determine the sign of $4x - 8y$.

1.6 Multiplying and Dividing Real Numbers

In Exercises 45–55, replace x with 6, y with -4, and a with 3. Then use the order of operations to evaluate the expression.

45. $6x - 5y + 4a = 6(6) - 5(-4) + 4(3)$
$= 36 + 20 + 12$
$= 56 + 12$
$= 68$

47. $(5x - 2y)(-2a) = [5(6) - 2(-4)][-2(3)]$
$= (30 + 8)(-6)$
$= (38)(-6)$
$= -228$

49. $\left(\frac{5}{6}x + \frac{3}{2}y\right)\left(-\frac{1}{3}a\right)$
$= \left[\frac{5}{6}(6) + \frac{3}{2}(-4)\right]\left[-\frac{1}{3}(3)\right]$
$= [5 + (-6)](-1)$
$= (-1)(-1)$
$= 1$

51. $(6 - x)(5 + y)(3 + a)$
$= (6 - 6)[5 + (-4)](3 + 3)$
$= 0(1)(6)$
$= 0$

53. $5x - 4a^2 = 5(6) - 4(3^2)$
$= 30 - 4(9)$
$= 30 - 36$
$= -6$

55. $\dfrac{xy + 9a}{x + y - 2} = \dfrac{6(-4) + 9(3)}{6 + (-4) - 2}$
$= \dfrac{-24 + 27}{2 - 2}$
$= \dfrac{3}{0}$

Division by 0 is undefined.

57. "The product of 4 and -7, added to -12" is written $-12 + 4(-7)$.
$-12 + 4(-7) = -12 + (-28)$
$= -40$

59. "Twice the product of -8 and 2, subtracted from -1" is written $-1 - 2(-8)(2)$.
$-1 - 2[(-8)(2)] = -1 - 2(-16)$
$= -1 - (-32)$
$= -1 + 32$
$= 31$

61. "The product of -3 and the difference between 3 and -7" is written $-3[3 - (-7)]$.
$-3[3 - (-7)] = -3[3 + 7]$
$= -3(10) = -30$

63. "Three-tenths of the sum of -2 and -28" is written $\frac{3}{10}[-2 + (-28)]$.
$\frac{3}{10}[-2 + (-28)] = \frac{3}{10}(-30)$
$= -\frac{90}{10}$
$= -9$

65. "The quotient of -20 and the sum of -8 and -2" is written
$$\frac{-20}{-8 + (-2)},$$
and
$$\frac{-20}{-8 + (-2)} = \frac{-20}{-10} = 2.$$

67. "The sum of -18 and -6, divided by the product of 2 and -4" is written
$$\frac{-18 + (-6)}{2(-4)},$$
and
$$\frac{-18 + (-6)}{2(-4)} = \frac{-24}{-8} = 3.$$

69. "The product of $-\frac{2}{3}$ and $-\frac{1}{5}$, divided by $\frac{1}{7}$" is written $\dfrac{-\frac{2}{3}\left(-\frac{1}{5}\right)}{\frac{1}{7}}$.

$\dfrac{-\frac{2}{3}\left(-\frac{1}{5}\right)}{\frac{1}{7}} = \dfrac{\frac{2}{15}}{\frac{1}{7}}$
$= \frac{2}{15} \cdot \frac{7}{1}$
$= \frac{14}{15}$

71. "Nine times a number is -36" is written
$$9x = -36.$$

73. "The quotient of a number and 4 is -1" is written
$$\frac{x}{4} = -1.$$

75. "$\frac{9}{11}$ less than a number is 5" is written
$$x - \frac{9}{11} = 5.$$

28 Chapter 1 The Real Number System

77. "When 6 is divided by a number, the result is -3" is written
$$\frac{6}{x} = -3.$$

78. $23 + 18 + 13 + (-4) + (-8) = 42$

79. There are 5 numbers in the group.

80. $\frac{42}{5} = 8\frac{2}{5}$

81. $8\frac{2}{5}$ is the average, which is the sum divided the total number of numbers in the group.

82. There are 25 integers between -10 and 14, including both -10 and 14. If we add the integers between -10 and 10, including -10 and 10, we will get zero. Therefore, the average is given by
$$\frac{11 + 12 + 13 + 14}{25} = \frac{50}{25} = 2.$$

83. $\dfrac{-15 + (-14) + (-13) + (-12) + (-11) + (-10)}{6}$
$= \dfrac{-75}{6}$
$= -12\dfrac{3}{6}$
$= -12\dfrac{1}{2}$

Summary Exercises on Operations with Real Numbers

1. $14 - 3 \cdot 10 = 14 - 30$
$ = 14 + (-30)$
$ = -16$

3. $(3 - 8)(-2) - 10 = (-5)(-2) - 10$
$ = 10 - 10$
$ = 0$

5. $7 - (-3)(2 - 10) = 7 - (-3)(-8)$
$ = 7 - (24)$
$ = -17$

7. $(-4)(7) - (-5)(2) = (-28) - (-10)$
$ = -28 + (10)$
$ = -18$

9. $40 - (-2)[8 - 9] = 40 - (-2)[-1]$
$ = 40 - (2)$
$ = 38$

11. $\dfrac{-3 - (-9 + 1)}{-7 - (-6)} = \dfrac{-3 - (-8)}{-7 + 6}$
$\phantom{\dfrac{-3 - (-9 + 1)}{-7 - (-6)}} = \dfrac{-3 + 8}{-1}$
$\phantom{\dfrac{-3 - (-9 + 1)}{-7 - (-6)}} = \dfrac{5}{-1} = -5$

13. $\dfrac{6^2 - 8}{-2(2) + 4(-1)} = \dfrac{36 - 8}{-4 + (-4)}$
$\phantom{\dfrac{6^2 - 8}{-2(2) + 4(-1)}} = \dfrac{28}{-8}$
$\phantom{\dfrac{6^2 - 8}{-2(2) + 4(-1)}} = -\dfrac{4 \cdot 7}{2 \cdot 4} = -\dfrac{7}{2}$ or $-3\dfrac{1}{2}$

15. $\dfrac{9(-6) - 3(8)}{4(-7) + (-2)(-11)} = \dfrac{-54 - 24}{-28 + 22}$
$\phantom{\dfrac{9(-6) - 3(8)}{4(-7) + (-2)(-11)}} = \dfrac{-78}{-6} = 13$

17. $\dfrac{(2 + 4)^2}{(5 - 3)^2} = \dfrac{(6)^2}{(2)^2}$
$\phantom{\dfrac{(2 + 4)^2}{(5 - 3)^2}} = \dfrac{36}{4} = 9$

19. $\dfrac{-9(-6) + (-2)(27)}{3(8 - 9)} = \dfrac{(54) + (-54)}{3(-1)}$
$\phantom{\dfrac{-9(-6) + (-2)(27)}{3(8 - 9)}} = \dfrac{0}{-3} = 0$

21. $\dfrac{6(-10 + 3)}{15(-2) - 3(-9)} = \dfrac{6(-7)}{(-30) - (-27)}$
$\phantom{\dfrac{6(-10 + 3)}{15(-2) - 3(-9)}} = \dfrac{-42}{-30 + 27}$
$\phantom{\dfrac{6(-10 + 3)}{15(-2) - 3(-9)}} = \dfrac{-42}{-3} = 14$

23. $\dfrac{(-10)^2 + 10^2}{-10(5)} = \dfrac{100 + 100}{-50}$
$\phantom{\dfrac{(-10)^2 + 10^2}{-10(5)}} = \dfrac{200}{-50} = -4$

25. $\dfrac{1}{2} \div \left(-\dfrac{1}{2}\right) = \dfrac{1}{2} \cdot \left(-\dfrac{2}{1}\right)$
$\phantom{\dfrac{1}{2} \div \left(-\dfrac{1}{2}\right)} = -\dfrac{2}{2} = -1$

27. $\left[\dfrac{5}{8} - \left(-\dfrac{1}{16}\right)\right] + \dfrac{3}{8} = \left[\dfrac{10}{16} + \dfrac{1}{16}\right] + \dfrac{6}{16}$
$\phantom{\left[\dfrac{5}{8} - \left(-\dfrac{1}{16}\right)\right] + \dfrac{3}{8}} = \left[\dfrac{11}{16}\right] + \dfrac{6}{16}$
$\phantom{\left[\dfrac{5}{8} - \left(-\dfrac{1}{16}\right)\right] + \dfrac{3}{8}} = \dfrac{17}{16}$ or $1\dfrac{1}{16}$

29. $-.9(-3.7) = .9(3.7)$
$ = 3.33$

31. $-3^2 - 2^2 = -(3^2) - (2^2)$
$ = -9 - 4$
$ = -13$

33. $40 - (-2)[-5 - 3] = 40 - (-2)[-8]$
$ = 40 - (16)$
$ = 24$

In Exercises 34–42, replace x with -2, y with 3, and a with 4. Then use the order of operations to evaluate the expression.

35. $(x+6)^3 - y^3 = (-2+6)^3 - 3^3$
$= (4)^3 - 27$
$= 64 - 27$
$= 37$

37. $\left(\dfrac{1}{2}x + \dfrac{2}{3}y\right)\left(-\dfrac{1}{4}a\right) = \left(\dfrac{1}{2}(-2) + \dfrac{2}{3}(3)\right)\left(-\dfrac{1}{4}(4)\right)$
$= (-1+2)(-1)$
$= (1)(-1)$
$= -1$

39. $\dfrac{x^2 - y^2}{x^2 + y^2} = \dfrac{(-2)^2 - 3^2}{(-2)^2 + 3^2}$
$= \dfrac{4-9}{4+9}$
$= \dfrac{-5}{13} = -\dfrac{5}{13}$

41. $\dfrac{-x + 2y}{2x + a} = \dfrac{-(-2) + 2(3)}{2(-2) + 4}$
$= \dfrac{2+6}{-4+4}$
$= \dfrac{8}{0}$, which is undefined.

1.7 Properties of Real Numbers

1.7 Margin Exercises

1. (a) $x + 9 = 9 + \underline{x}$

   (b) $-12(4) = \underline{4}(-12)$

   (c) $5x = x \cdot \underline{5}$

2. (a) $(9+10) + (-3) = 9 + [\underline{10} + (-3)]$

   (b) $-5 + (2+8) = \underline{(-5+2)} + 8$

   (c) $10[-8(-3)] = \underline{[10(-8)]}(-3)$

3. (a) $2(4 \cdot 6) = (2 \cdot 4)6$

   The order of the three numbers is the same on both sides of the equals sign. The only change is in the grouping of the numbers. Therefore, this is an example of an *associative* property.

   (b) $(2 \cdot 4)6 = (4 \cdot 2)6$

   While the same numbers are grouped inside the two pairs of parentheses, the order of the numbers has been changed. This illustrates a *commutative* property.

   (c) $(2+4) + 6 = 4 + (2+6)$

   Both the order and the grouping of the numbers have been changed. This is an example of *both* properties.

4. $5 + 18 + 29 + 31 + 12$
$= (18+12) + (29+31) + 5$
$= 30 + 60 + 5$
$= 95$

5. (a) $9 + 0 = \underline{9}$

   (b) $\underline{0} + (-7) = -7$

   (c) $\underline{5} \cdot 1 = 5$

6. (a) $\dfrac{85}{105} = \dfrac{17 \cdot 5}{21 \cdot 5}$ Factor.
$= \dfrac{17}{21} \cdot \dfrac{5}{5}$ Write as a product.
$= \dfrac{17}{21} \cdot 1$ Property of 1
$= \dfrac{17}{21}$ Identity property

   (b) $\dfrac{9}{10} - \dfrac{53}{50} = \dfrac{9}{10} \cdot 1 - \dfrac{53}{50}$ Identity property
$= \dfrac{9}{10} \cdot \dfrac{5}{5} - \dfrac{53}{50}$ Use $1 = \dfrac{5}{5}$ to get a common denominator.
$= \dfrac{45}{50} - \dfrac{53}{50}$ Multiply.
$= -\dfrac{8}{50}$ Subtract.
$= -\dfrac{4}{25}$ Reduce.

7. (a) $-6 + \underline{6} = 0$ Inverse property

   (b) $\dfrac{4}{3} \cdot \dfrac{3}{\underline{4}} = 1$ Inverse property

   (c) $-\dfrac{1}{9} \cdot \underline{(-9)} = 1$ Inverse property

   (d) $275 + \underline{0} = 275$ Identity property

   (e) $-.75 + \dfrac{3}{4} = \underline{0}$ Inverse property

8. (a) $2(p+5) = 2p + 2 \cdot 5$ Distributive property
$= 2p + 10$ Multiply.

   (b) $-4(y+7) = -4y + (-4)7$
$= -4y + (-28)$
$= -4y - 28$

   (c) $5(m-4) = 5m - 5 \cdot 4$
$= 5m - 20$

   (d) $9 \cdot k + 9 \cdot 5 = 9(k+5)$

(e) $3a - 3b = 3(a - b)$

(f) $7(2y + 7k - 9m) = 7(2y) + 7(7k) - 7(9m)$
$= 14y + 49k - 63m$

9. (a) $-(3k - 5) = -1 \cdot (3k - 5)$
$= -1 \cdot (3k) + (-1)(-5)$
$= -3k + 5$

(b) $-(2 - r) = -1 \cdot (2 - r)$
$= -1 \cdot 2 + (-1)(-r)$
$= -2 + r$

(c) $-(-5y + 8) = -1(-5y + 8)$
$= 5y - 8$

(d) $-(-z + 4) = -1(-z + 4)$
$= z - 4$

1.7 Section Exercises

1. B, since 0 is the identity element for addition.

3. C, since $-a$ is the additive inverse of a.

5. B, since the multiplicative inverse of a number a is $\dfrac{1}{a}$ and the only number that we *cannot* divide by is 0.

7. G, since we can consider $(5 \cdot 4)$ to be one number, $(5 \cdot 4) \cdot 3$ is the same as $3 \cdot (5 \cdot 4)$ by the commutative property.

9. $\dfrac{2}{3}(-4) = -4\left(\dfrac{2}{3}\right)$

The order of the two numbers has been changed, so this is an example of the commutative property of multiplication.

11. $-6 + (12 + 7) = (-6 + 12) + 7$

The numbers are in the same order but grouped differently, so this is an example of the associative property of addition: $(a + b) + c = a + (b + c)$.

13. $-6 + 6 = 0$

The sum of the two numbers is 0, so they are additive inverses (or opposites) of each other. This is an example of the additive inverse property: $a + (-a) = 0$.

15. $\dfrac{2}{3}\left(\dfrac{3}{2}\right) = 1$

The product of the two numbers is 1, so they are multiplicative inverses (or reciprocals) of each other. This is an example of the multiplicative inverse property: $a \cdot \dfrac{1}{a} = 1 \ (a \neq 0)$.

17. $2.34 \cdot 1 = 2.34$

The product of a number and 1 is the original number. This is an example of the identity property of multiplication.

19. $(4 + 17) + 3 = 3 + (4 + 17)$

The order of the numbers has been changed, but not the grouping, so this is an example of the commutative property of addition: $a + b = b + a$.

21. $6(x + y) = 6x + 6y$

The number 6 outside the parentheses is "distributed" over the x and y. This is an example of the distributive property.

23. $-\dfrac{5}{9} = -\dfrac{5}{9} \cdot \dfrac{3}{3} = -\dfrac{15}{27}$

$\dfrac{3}{3}$ is a form of the number 1. We use it to rewrite $-\dfrac{5}{9}$ as $-\dfrac{15}{27}$. This is an example of the identity property of multiplication.

25. $5(2x) + 5(3y) = 5(2x + 3y)$

This is an example of the distributive property. The number 5 is "distributed" over $2x$ and $3y$.

27. (a) 0 is its own additive inverse; $0 + 0 = 0$.

(b) 1 and -1 are their own multiplicative inverses;
$$1 \cdot \dfrac{1}{1} = 1 \quad \text{and} \quad -1 \cdot \dfrac{1}{-1} = 1,$$
where $\dfrac{1}{1} = 1$ and $\dfrac{1}{-1} = -1$.

29. $25 - (6 - 2) = 25 - (4)$
$= 21$
$(25 - 6) - 2 = 19 - 2$
$= 17$

Since $21 \neq 17$, this example shows that subtraction is not associative.

31. $r + 7$; commutative
$$r + 7 = 7 + r$$

33. $s + 0$; identity
$$s + 0 = s$$

35. $-6(x + 7)$; distributive
$-6(x + 7) = -6x + (-6)7$
$= -6x + (-42)$
$= -6x - 42$

37. $(w+5) + (-3)$; associative
$$(w+5) + (-3) = w + [5 + (-3)]$$
$$= w + 2$$

39. We must multiply $\frac{3}{4}$ by 1 in the form $\frac{3}{3}$: $\frac{3}{4} \cdot \frac{3}{3} = \frac{9}{12}$.

41. $-\frac{3}{8} + \frac{2}{5} + \frac{8}{5} + \frac{3}{8} = \left(-\frac{3}{8} + \frac{3}{8}\right) + \left(\frac{2}{5} + \frac{8}{5}\right)$
$$= 0 + \frac{10}{5}$$
$$= 0 + 2$$
$$= 2$$

43. $5 \cdot 3 + 5 \cdot 17 = 5(3 + 17)$
$$= 5(20)$$
$$= 100$$

45. $4(t + 3) = 4t + 4 \cdot 3$
$$= 4t + 12$$

47. $-8(r + 3) = -8(r) + (-8)(3)$
$$= -8r + (-24)$$
$$= -8r - 24$$

49. $-5(y - 4) = -5y + (-5)(-4)$
$$= -5y + 20$$

51. $-\frac{4}{3}(12y + 15z)$
$$= -\frac{4}{3}(12y) + \left(-\frac{4}{3}\right)(15z)$$
$$= \left[\left(-\frac{4}{3}\right) \cdot 12\right]y + \left[\left(-\frac{4}{3}\right) \cdot 15\right]z$$
$$= -16y + (-20)z$$
$$= -16y - 20z$$

53. $8 \cdot z + 8 \cdot w = 8(z + w)$

55. $7(2v) + 7(5r) = 7(2v + 5r)$

57. $8(3r + 4s - 5y)$
$$= 8(3r) + 8(4s) + 8(-5y)$$
Distributive property
$$= (8 \cdot 3)r + (8 \cdot 4)s + [8(-5)]y$$
Associative property
$$= 24r + 32s - 40y \quad \text{Multiply}$$

59. $-3(8x + 3y + 4z)$
$$= -3(8x) + (-3)(3y) + (-3)(4z)$$
Distributive property
$$= (-3 \cdot 8)x + (-3 \cdot 3)y + (-3 \cdot 4)z$$
Associative property
$$= -24x - 9y - 12z \quad \text{Multiply}$$

61. $-(4t + 5m) = -1 \cdot (4t + 5m)$
$$= -1 \cdot (4t) + (-1)(5m)$$
$$= -4t - 5m$$

63. $-(-5c - 4d) = -1 \cdot (-5c - 4d)$
$$= -1(-5c) + (-1)(-4d)$$
$$= 5c + 4d$$

65. $-(-3q + 5r - 8s) = -1(-3q + 5r - 8s)$
$$= 3q - 5r + 8s$$

67. Answers will vary. For example, "putting on your socks" and "putting on your shoes" are everyday operations that are not commutative.

69. The actions "preparing a meal" and "eating a meal" are not commutative because the order in which they are done makes a difference Therefore, the given statement is false.

71. (foreign sales) clerk; foreign (sales clerk)

    The first indicates a clerk who works with foreign sales, while the second indicates a sales clerk who is from another country.

73. $-3[5 + (-5)] = -3(0) = 0$

74. $-3[5 + (-5)] = -3(5) + (-3)(-5)$

75. $-3(5) = -15$

76. The product $-3(-5)$ must equal 15, since it is the additive inverse of -15.

1.8 Simplifying Expressions

1.8 Margin Exercises

1. (a) $9k + 12 - 5 = 9k + (12 - 5) = 9k + 7$

   (b) $7(3p + 2q) = 7(3p) + 7(2q) = 21p + 14q$

   (c) $2 + 5(3z - 1) = 2 + 5(3z) - 5(1)$
   $$= 2 + 15z - 5$$
   $$= 15z - 3$$

   Note: By the order of operations, 5 is multiplied by $(3z - 1)$ first; then 2 is added to the product.

   (d) $-3 - (2 + 5y) = -3 - 1(2 + 5y)$
   $$= -3 - 2 - 5y$$
   $$= -5 - 5y$$

32 Chapter 1 The Real Number System

2. The numerical coefficient is the number in front of the variable or variables.

	Term	Numerical coefficient
(a)	$15q$	15
(b)	$-2m^3$	-2
(c)	$-18m^7q^4$	-18
(d)	$-r = -1 \cdot r$	-1
(e)	$\dfrac{5x}{4} = \dfrac{5}{4}x$	$\dfrac{5}{4}$

3. (a) $9x$ and $4x$ have the same variable, x, and the same exponent, which is understood to be 1, so they are *like* terms.

   (b) $-8y^3$ and $12y^2$ have the same variable but different exponents on y, so they are *unlike* terms.

   (c) $5x^2y^4$ and $5x^4y^2$ have the same variables, but different exponents, so they are *unlike* terms.

   (d) $7x^2y^4$ and $-7x^2y^4$ have the same variables and exponents, so they are *like* terms.

   (e) $13kt$ and $4tk = 4kt$ have the same variables and exponents, so they are *like* terms.

4. (a) $4k + 7k = (4+7)k = 11k$

   (b) $4r - r = 4r - 1r = (4-1)r = 3r$

   (c) $5z + 9z - 4z = (5+9-4)z = 10z$

   (d) $8p + 8p^2$ cannot be simplified. $8p$ and $8p^2$ are unlike terms and cannot be combined.

   (e) $5x - 3y + 2x - 5y - 3$
   $= (5+2)x + (-3-5)y - 3$
   $= 7x - 8y - 3$

5. (a) $10p + 3(5 + 2p) = 10p + 3(5) + 3(2p)$
   $= 10p + 15 + 6p$
   $= 16p + 15$

   (b) $7z - 2 - (1 + z) = 7z - 2 - 1(1 + z)$
   $= 7z - 2 - 1 - z$
   $= 6z - 3$

   (c) $-(3k^2 + 5k) + 7(k^2 - 4k)$
   $= -1(3k^2 + 5k) + 7(k^2 - 4k)$
   $= -3k^2 - 5k + 7k^2 - 28k$
   $= 4k^2 - 33k$

6. (a) "Three times a number, subtracted from the sum of the number and 8" is written $(x+8) - 3x$.
   $(x+8) - 3x = x + 8 - 3x$
   $= -2x + 8$

   (b) "Twice a number added to the sum of 6 and the number" is written $2x + (6+x)$.
   $2x + (6+x) = (2x + x) + 6$
   $= 3x + 6$

1.8 Section Exercises

1. $6x - 2x = (6-2)x = 4x$
   The correct response is **C**.

3. The numerical coefficient of $5x^3y^7$ is 5.
   The correct response is **A**.

5. $3x + 12x = (3 + 12)x$
   $= 15x$

7. $8t - 5t + 2t = (8 - 5 + 2)t$
   $= 5t$

9. $4r + 19 - 8 = 4r + 11$

11. $5 + 2(x - 3y) = 5 + 2(x) + 2(-3y)$
    $= 5 + 2x - 6y$

13. $-2 - (5 - 3p) = -2 - 1(5 - 3p)$
    $= -2 - 1(5) - 1(-3p)$
    $= -2 - 5 + 3p$
    $= -7 + 3p$

15. The numerical coefficient of the term $-12k$ is -12.

17. The numerical coefficient of the term $5m^2$ is 5.

19. Because xw can be written as $1 \cdot xw$, the numerical coefficient of the term xw is 1.

21. Since $-x = -1x$, the numerical coefficient of the term $-x$ is -1.

23. The numerical coefficient of the term 74 is 74.

25. Answers will vary. For example, $-3x$ and $4x$

27. $8r$ and $-13r$ are *like* terms since they have the same variable with the same exponent (which is understood to be 1).

29. $5z^4$ and $9z^3$ are *unlike* terms. Although both have the variable z, the exponents are not the same.

31. All numerical terms (constants) are considered like terms, so 4, 9, and -24 are *like* terms.

33. x and y are *unlike* terms because they do not have the same variable.

35. Apples and oranges are examples of unlike fruits, just like x and y are unlike terms. We cannot add x and y to get an expression any simpler than $x + y$; we cannot add, for example, 2 apples and 3 oranges to obtain 5 fruits that are all alike.

Chapter 1 Review Exercises 33

37. $-5 - 2(x - 3) = -5 - 2x - 2(-3)$
$= -5 - 2x + 6$
$= 1 - 2x$

39. $-\dfrac{4}{3} + 2t + \dfrac{1}{3}t - 8 - \dfrac{8}{3}t$
$= \left(2t + \dfrac{1}{3}t - \dfrac{8}{3}t\right) + \left(-\dfrac{4}{3} - 8\right)$
$= \left(2 + \dfrac{1}{3} - \dfrac{8}{3}\right)t + \left(-\dfrac{4}{3} - 8\right)$
$= \left(\dfrac{6}{3} + \dfrac{1}{3} - \dfrac{8}{3}\right)t + \left(-\dfrac{4}{3} - \dfrac{24}{3}\right)$
$= -\dfrac{1}{3}t - \dfrac{28}{3}$

41. $-5.3r + 4.9 - (2r + .7) + 3.2r$
$= -5.3r + 4.9 - 2r - .7 + 3.2r$
$= (-5.3r - 2r + 3.2r) + (4.9 - .7)$
$= (-5.3 - 2 + 3.2)r + (4.9 - .7)$
$= -4.1r + 4.2$

43. $2y^2 - 7y^3 - 4y^2 + 10y^3$
$= (2y^2 - 4y^2) + (-7y^3 + 10y^3)$
$= (2 - 4)y^2 + (-7 + 10)y^3$
$= -2y^2 + 3y^3$

45. $13p + 4(4 - 8p) = 13p + 4(4) + 4(-8p)$
$= 13p + 16 - 32p$
$= 13p - 32p + 16$
$= -19p + 16$

47. $-\dfrac{4}{3}(y - 12) - \dfrac{1}{6}y$
$= -\dfrac{4}{3}y - \dfrac{4}{3}(-12) - \dfrac{1}{6}y$
$= -\dfrac{4}{3}y + 16 - \dfrac{1}{6}y$
$= -\dfrac{4}{3}y - \dfrac{1}{6}y + 16$
$= \left(-\dfrac{8}{6} - \dfrac{1}{6}\right)y + 16$
$= -\dfrac{3}{2}y + 16 \qquad -\dfrac{9}{6} = -\dfrac{3}{2}$

49. $-5(5y - 9) + 3(3y + 6)$
$= -5(5y) + (-5)(-9) + 3(3y) + 3(6)$
Distributive property
$= -25y + 45 + 9y + 18$
$= (-25y + 9y) + (45 + 18)$
$= (-25 + 9)y + 63$
$= -16y + 63$

51. "Five times a number, added to the sum of the number and three" is written $(x + 3) + 5x$.

$(x + 3) + 5x = x + 3 + 5x$
$= (x + 5x) + 3$
$= 6x + 3$

53. "A number multiplied by -7, subtracted from the sum of 13 and six times the number" is written $(13 + 6x) - (-7x)$.

$(13 + 6x) - (-7x) = 13 + 6x + 7x$
$= 13 + 13x$

55. "Six times a number added to -4, subtracted from twice the sum of three times the number and 4" is written $2(3x + 4) - (-4 + 6x)$.

$2(3x + 4) - (-4 + 6x)$
$= 2(3x + 4) - 1(-4 + 6x)$
$= 6x + 8 + 4 - 6x$
$= 6x + (-6x) + 8 + 4$
$= 0 + 12 = 12$

57. $9x - (x + 2)$

Wording will vary. One example is "the difference between 9 times a number and the sum of the number and 2." Another example is "the sum of a number and 2 subtracted from 9 times a number."

59. For widgets, the fixed cost is $1000 and the variable cost is $5 per widget, so the cost to produce x widgets is

$1000 + 5x$ (dollars).

60. For gadgets, the fixed cost is $750 and the variable cost is $3 per gadget, so the cost to produce y gadgets is

$750 + 3y$ (dollars).

61. The total cost to make x widgets and y gadgets is

$1000 + 5x + 750 + 3y$ (dollars).

62. $1000 + 5x + 750 + 3y$
$= (1000 + 750) + 5x + 3y$
$= 1750 + 5x + 3y$,

so the total cost to make x widgets and y gadgets is

$1750 + 5x + 3y$ (dollars).

Chapter 1 Review Exercises

1. $5^4 = 5 \cdot 5 \cdot 5 \cdot 5 = 625$

2. $(.03)^4 = (.03)(.03)(.03)(.03) = .00000081$

3. $.21^3 = (.21)(.21)(.21) = .009261$

34 Chapter 1 The Real Number System

4. $\left(\dfrac{5}{2}\right)^3 = \dfrac{5}{2} \cdot \dfrac{5}{2} \cdot \dfrac{5}{2} = \dfrac{5 \cdot 5 \cdot 5}{2 \cdot 2 \cdot 2} = \dfrac{125}{8}$

5. $8 \cdot 5 - 13 = 40 - 13 = 27$

6. $5[4^2 + 3(2^3)] = 5[16 + 3(8)]$
    $= 5[16 + 24]$
    $= 5[40]$
    $= 200$

7. $\dfrac{7(3^2 - 5)}{16 - 2 \cdot 6} = \dfrac{7(9 - 5)}{16 - 12} = \dfrac{7(4)}{4} = 7$

8. $\dfrac{3(9 - 4) + 5(8 - 3)}{2^3 - (5 - 3)} = \dfrac{3(5) + 5(5)}{8 - 2}$
    $= \dfrac{15 + 25}{8 - 2}$
    $= \dfrac{40}{6}$
    $= \dfrac{20}{3}$

9. "Thirteen is less than seventeen" is written $13 < 17$.

10. "Five plus two is not equal to 10" is written $5 + 2 \neq 10$.

11. $6 < 15$ is written in words as "six is less than fifteen."

12. One example is $-4 + (-7) \geq \dfrac{12}{-3}$ because $-11 \geq -4$ is a false statement.

In Exercises 13–16, replace x with 6 and y with 3.

13. $2x + 6y = 2(6) + 6(3)$
    $= 12 + 18 = 30$

14. $4(3x - y) = 4[3(6) - 3]$
    $= 4(18 - 3)$
    $= 4(15) = 60$

15. $\dfrac{x}{3} + 4y = \dfrac{6}{3} + 4(3)$
    $= 2 + 12 = 14$

16. $\dfrac{x^2 + 3}{3y - x} = \dfrac{6^2 + 3}{3(3) - 6}$
    $= \dfrac{36 + 3}{9 - 6}$
    $= \dfrac{39}{3} = 13$

17. "Six added to a number" translates as $x + 6$.

18. "A number subtracted from eight" translates as $8 - x$.

19. "Nine subtracted from six times a number" translates as $6x - 9$.

20. "Three-fifths of a number added to 12" translates as $12 + \dfrac{3}{5}x$.

21. $5x + 3(x + 2) = 22;\ 2$
    $5x + 3(x + 2) = 5(2) + 3(2 + 2)$ Let $x = 2$.
    $ = 5(2) + 3(4)$
    $ = 10 + 12 = 22$

Since the left side and the right side are equal, 2 is a solution of the given equation.

22. $\dfrac{x + 5}{3x} = 1;\ 6$
    $\dfrac{x + 5}{3x} = \dfrac{6 + 5}{3(6)}$ Let $x = 6$.
    $\phantom{\dfrac{x + 5}{3x}} = \dfrac{11}{18}$

Since the left side, $\dfrac{11}{18}$, is not equal to the right side, 1, 6 is not a solution of the equation.

23. "Six less than twice a number is 10" is written
    $$2x - 6 = 10.$$

24. "The product of a number and 4 is 8" is written
    $$4x = 8.$$

25. $5r - 8(r + 7) = 2$ is an equation because it has an equals sign.

26. $2y + (5y - 9) + 2$ is an expression because it does not have an equals sign.

27. $-4,\ -\dfrac{1}{2},\ 0,\ 2.5,\ 5$

Graph these numbers on a number line. They are already arranged in order from smallest to largest.

28. $-2,\ -3,\ |-3|,\ |-1|$

Recall that $|-3| = 3$ and $|-1| = 1$. From smallest to largest, the numbers are $-3,\ -2,\ |-1|,\ |-3|$.

29. $-3\frac{1}{4}, \frac{14}{5}, -1\frac{1}{8}, \frac{5}{6}$

From smallest to largest, the numbers are

$$-3\frac{1}{4}, -1\frac{1}{8}, \frac{5}{6}, \frac{14}{5}.$$

30. $|-4|, -|-3|, -|-5|, -6$

Recall that $|-4| = 4$, $-|-3| = -3$, and $-|-5| = -5$.
From smallest to largest, the numbers are

$$-6, -|-5|, -|-3|, |-4|.$$

31. $-10, 5$

Since any negative number is less than any positive number, -10 is the lesser number.

32. $-8, -9$

Since -9 is to the left of -8 on the number line, -9 is the lesser number.

33. $-\frac{2}{3}, -\frac{3}{4}$

To compare these fractions, use a common denominator.

$$-\frac{2}{3} = -\frac{8}{12}, \quad -\frac{3}{4} = -\frac{9}{12}$$

Since $-\frac{9}{12}$ is to the left of $-\frac{8}{12}$ on the number line, $-\frac{3}{4}$ is the lesser number.

34. $0, -|23|$

Since $-|23| = -23$ and $-23 < 0$, $-|23|$ is the lesser number.

35. $12 > -13$

This statement is true since 12 is to the right of -13 on the number line.

36. $0 > -5$

This statement is true since 0 is to the right of -5 on the number line.

37. $-9 < -7$

This statement is true since -9 is to the left of -7 on the number line.

38. $-13 > -13$

This is a false statement since $-13 = -13$.

39. $-|3| = -3$

40. $-|-19| = -[-(-19)] = -19$

41. $-|9 - 2| = -|7| = -7$

42. $|15 - 6| = |9| = 9$

43. $-10 + 4 = -6$

44. $14 + (-18) = -4$

45. $-8 + (-9) = -17$

46. $\frac{4}{9} + \left(-\frac{5}{4}\right) = \frac{4 \cdot 4}{9 \cdot 4} + \left(-\frac{5 \cdot 9}{4 \cdot 9}\right)$ LCD = 36

$$= \frac{16}{36} + \left(-\frac{45}{36}\right)$$

$$= -\frac{29}{36}$$

47. $[-6 + (-8) + 8] + [9 + (-13)]$
$= \{[-6 + (-8)] + 8\} + (-4)$
$= [(-14) + 8] + (-4)$
$= (-6) + (-4) = -10$

48. $(-4 + 7) + (-11 + 3) + (-15 + 1)$
$= (3) + (-8) + (-14)$
$= [3 + (-8)] + (-14)$
$= (-5) + (-14) = -19$

49. "19 added to the sum of -31 and 12" is written

$$(-31 + 12) + 19 = (-19) + 19$$
$$= 0.$$

50. "13 more than the sum of -4 and -8" is written

$$[-4 + (-8)] + 13 = -12 + 13$$
$$= 1.$$

51. $18 + (-26) = -8$

Mohammed's balance is $-\$8$.

52. $93 - 6 = 93 + (-6) = 87$

The new temperature was 87°F.

53. $-7 - 4 = -7 + (-4) = -11$

54. $-12 - (-11) = -12 + (11) = -1$

55. $5 - (-2) = 5 + 2 = 7$

56. $-\frac{3}{7} - \frac{4}{5} = -\frac{3 \cdot 5}{7 \cdot 5} - \frac{4 \cdot 7}{5 \cdot 7}$

$$= -\frac{15}{35} - \frac{28}{35}$$

$$= -\frac{15}{35} + \left(-\frac{28}{35}\right)$$

$$= -\frac{43}{35}$$

36 Chapter 1 The Real Number System

57. $2.56 - (-7.75) = 2.56 + 7.75 = 10.31$

58. $(-10 - 4) - (-2) = [-10 + (-4)] + 2$
$= (-14) + 2$
$= -12$

59. $(-3 + 4) - (-1) = (-3 + 4) + 1$
$= 1 + 1$
$= 2$

60. $|5 - 9| - |-3 + 6| = |-4| - |3|$
$= 4 - 3$
$= 1$

61. "The difference between -4 and -6" is written
$-4 - (-6) = -4 + 6$
$= 2.$

62. "Five less than the sum of 4 and -8" is written
$[4 + (-8)] - 5 = (-4) + (-5)$
$= -9.$

63. Represent a surplus with a positive number and a deficit with a negative number.
$236.4 - (-477.0) = 236.4 + 477.0$
$= 713.4$

The year 2000 budget decreased by $713.4 billion.

64. $35.93 - 7.04 = 28.89$

Her winning time in 1998 was 1 min, 28.89 sec.

65. The first step in performing the subtraction in the problem $-8 - (-6)$ is to change subtracting -6 to adding its opposite, 6. Now the problem becomes $-8 + 6$. The sum is -2.

66. Yes, the difference between two negative numbers can be positive. This will happen whenever the first number is greater than the second number. For example,
$-2 - (-6) = -2 + 6 = 4.$

67. The change from 2000 to 2001 was
$308.0 - 294.5 = 13.5$ billion.

68. The change from 1993 to 1994 was
$279.8 - 291.1 = -11.3$ billion.

69. The change from 1997 to 1998 was
$256.1 - 258.3 = -2.2$ billion.

70. The change from 1996 to 1997 was
$258.3 - 253.3 = 5.0$ billion.

71. $(-12)(-3) = 36$

72. $15(-7) = -(15 \cdot 7)$
$= -105$

73. $\left(-\dfrac{4}{3}\right)\left(-\dfrac{3}{8}\right) = \dfrac{4}{3} \cdot \dfrac{3}{8}$
$= \dfrac{4 \cdot 3}{3 \cdot 8}$
$= \dfrac{4}{8} = \dfrac{1}{2}$

74. $(-4.8)(-2.1) = 10.08$

75. $5(8 - 12) = 5[8 + (-12)]$
$= 5(-4) = -20$

76. $(5 - 7)(8 - 3) = [5 + (-7)][8 + (-3)]$
$= (-2)(5) = -10$

77. $2(-6) - (-4)(-3) = -12 - (12)$
$= -12 + (-12)$
$= -24$

78. $3(-10) - 5 = -30 + (-5) = -35$

79. $\dfrac{-36}{-9} = \dfrac{4 \cdot 9}{9} = 4$

80. $\dfrac{220}{-11} = -\dfrac{20 \cdot 11}{11} = -20$

81. $-\dfrac{1}{2} \div \dfrac{2}{3} = -\dfrac{1}{2} \cdot \dfrac{3}{2} = -\dfrac{3}{4}$

82. $-33.9 \div (-3) = \dfrac{-33.9}{-3} = 11.3$

83. $\dfrac{-5(3) - 1}{8 - 4(-2)} = \dfrac{-15 + (-1)}{8 - (-8)}$
$= \dfrac{-16}{8 + 8}$
$= \dfrac{-16}{16} = -1$

84. $\dfrac{5(-2) - 3(4)}{-2[3 - (-2)] + 10} = \dfrac{-10 - 12}{-2(3 + 2) + 10}$
$= \dfrac{-10 + (-12)}{-2(5) + 10}$
$= \dfrac{-22}{-10 + 10}$
$= \dfrac{-22}{0}$, which is undefined.

85. $\dfrac{10^2 - 5^2}{8^2 + 3^2 - (-2)} = \dfrac{100 - 25}{64 + 9 + 2}$

$= \dfrac{75}{75} = 1$

86. $\dfrac{4^2 - 8 \cdot 2}{(-1.2)^2 - (-.56)} = \dfrac{16 - 16}{1.44 + .56} = \dfrac{0}{2} = 0$

In Exercises 87–90, replace x with -5, y with 4, and z with -3.

87. $6x - 4z = 6(-5) - 4(-3)$
$= -30 - (-12)$
$= -30 + 12 = -18$

88. $5x + y - z = 5(-5) + (4) - (-3)$
$= (-25 + 4) + 3$
$= -21 + 3 = -18$

89. $5x^2 = 5(-5)^2$
$= 5(25)$
$= 125$

90. $z^2(3x - 8y) = (-3)^2[3(-5) - 8(4)]$
$= 9[-15 - 32]$
$= 9[-15 + (-32)]$
$= 9(-47) = -423$

91. "Nine less than the product of -4 and 5" is written

$-4(5) - 9 = -20 + (-9)$
$= -29.$

92. "Five-sixths of the sum of 12 and -6" is written

$\dfrac{5}{6}[12 + (-6)] = \dfrac{5}{6}(6)$
$= 5.$

93. "The quotient of 12 and the sum of 8 and -4" is written

$\dfrac{12}{8 + (-4)} = \dfrac{12}{4} = 3.$

94. "The product of -20 and 12, divided by the difference between 15 and -15" is written

$\dfrac{-20(12)}{15 - (-15)} = \dfrac{-240}{15 + 15}$
$= \dfrac{-240}{30} = -8.$

95. "The quotient of a number and the sum of the number and 5 is -2" is written

$\dfrac{x}{x + 5} = -2.$

96. "3 less than 8 times a number is -7" is written

$8x - 3 = -7.$

97. $6 + 0 = 6$

This is an example of an identity property.

98. $5 \cdot 1 = 5$

This is an example of an identity property.

99. $-\dfrac{2}{3}\left(-\dfrac{3}{2}\right) = 1$

This is an example of an inverse property.

100. $17 + (-17) = 0$

This is an example of an inverse property.

101. $5 + (-9 + 2) = [5 + (-9)] + 2$

This is an example of an associative property.

102. $w(xy) = (wx)y$

This is an example of an associative property.

103. $3x + 3y = 3(x + y)$

This is an example of the distributive property.

104. $(1 + 2) + 3 = 3 + (1 + 2)$

This is an example of a commutative property.

105. $7y + y = 7y + 1y = (7 + 1)y = 8y$

106. $-12(4 - t) = -12(4) - (-12)(t)$
$= -48 + 12t$

107. $3(2s) + 3(4y) = 3(2s + 4y) = 6s + 12y$

108. $-(-4r + 5s) = -1(-4r + 5s)$
$= (-1)(-4r) + (-1)(5s)$
$= 4r - (1)(5s)$
$= 4r - 5s$

109. $16p^2 - 8p^2 + 9p^2 = (16 - 8 + 9)p^2$
$= 17p^2$

110. $4r^2 - 3r + 10r + 12r^2$
$= (4r^2 + 12r^2) + (-3r + 10r)$
$= (4 + 12)r^2 + (-3 + 10)r$
$= 16r^2 + 7r$

111. $-8(5k - 6) + 3(7k + 2)$
$= (-8)(5k) - (-8)(6) + 3(7k) + 3(2)$
$= -40k - (-48) + 21k + 6$
$= (-40 + 21)k + (48 + 6)$
$= -19k + 54$

112. $2s - (-3s + 6) = 2s - 1(-3s + 6)$
$ = 2s + 3s - 6$
$ = 5s - 6$

113. $-7(2t - 4) - 4(3t + 8) - 19(t + 1)$
$= -14t + 28 - 12t - 32 - 19t - 19$
$= (-14t - 12t - 19t) + (28 - 32 - 19)$
$= -45t - 23$

114. $3.6t^2 + 9t - 8.1(6t^2 + 4t)$
$= 3.6t^2 + 9t - (8.1)(6t^2) - (8.1)(4t)$
$= 3.6t^2 + 9t - 48.6t^2 - 32.4t$
$= (3.6t^2 - 48.6t^2) + (9t - 32.4t)$
$= (3.6 - 48.6)t^2 + (9 - 32.4)t$
$= -45t^2 - 23.4t$

115. "Seven times a number, subtracted from the product of -2 and three times the number" is written
$-2(3x) - 7x = -6x - 7x = -13x.$

116. "The quotient of 9 more than a number and 6 less than the number" is written
$\dfrac{x + 9}{x - 6}.$

117. In Exercise 115, the word "and" does not signify addition. The phrase "the product of -2 and ..." means the same thing as "-2 times" and signifies multiplication. Here the word "and" cannot be considered in isolation.

118. Answers may vary. For example, "3 times the difference between 4 times a number and 6"

119. **[1.5]** $[(-2) + 7 - (-5)] + [-4 - (-10)]$
$= \{[(-2) + 7] - (-5)\} + (-4 + 10)$
$= (5 + 5) + 6$
$= 10 + 6 = 16$

120. **[1.6]** $\left(-\dfrac{5}{6}\right)^2 = \left(-\dfrac{5}{6}\right)\left(-\dfrac{5}{6}\right)$
$\phantom{\left(-\dfrac{5}{6}\right)^2} = \dfrac{25}{36}$

121. **[1.6]** $-|(-7)(-4)| - (-2) = -|28| + 2$
$ = -28 + 2 = -26$

122. **[1.6]** $\dfrac{6(-4) + 2(-12)}{5(-3) + (-3)} = \dfrac{-24 + (-24)}{-15 + (-3)}$
$\phantom{\dfrac{6(-4) + 2(-12)}{5(-3) + (-3)}} = \dfrac{-48}{-18} = \dfrac{8 \cdot 6}{3 \cdot 6}$
$\phantom{\dfrac{6(-4) + 2(-12)}{5(-3) + (-3)}} = \dfrac{8}{3} \text{ or } 2\dfrac{2}{3}$

123. **[1.5]** $\dfrac{3}{8} - \dfrac{5}{12} = \dfrac{3 \cdot 3}{8 \cdot 3} - \dfrac{5 \cdot 2}{12 \cdot 2}$
$\phantom{\dfrac{3}{8} - \dfrac{5}{12}} = \dfrac{9}{24} - \dfrac{10}{24}$
$\phantom{\dfrac{3}{8} - \dfrac{5}{12}} = \dfrac{9}{24} + \left(-\dfrac{10}{24}\right)$
$\phantom{\dfrac{3}{8} - \dfrac{5}{12}} = -\dfrac{1}{24}$

124. **[1.6]** $\dfrac{12^2 + 2^2 - 8}{10^2 - (-4)(-15)} = \dfrac{144 + 4 - 8}{100 - (-4)(-15)}$
$\phantom{\dfrac{12^2 + 2^2 - 8}{10^2 - (-4)(-15)}} = \dfrac{148 - 8}{100 - 60}$
$\phantom{\dfrac{12^2 + 2^2 - 8}{10^2 - (-4)(-15)}} = \dfrac{140}{40}$
$\phantom{\dfrac{12^2 + 2^2 - 8}{10^2 - (-4)(-15)}} = \dfrac{7}{2}$

125. **[1.1]** $\dfrac{8^2 + 6^2}{7^2 + 1^2} = \dfrac{64 + 36}{49 + 1}$
$\phantom{\dfrac{8^2 + 6^2}{7^2 + 1^2}} = \dfrac{100}{50} = 2$

126. **[1.6]** $-16(-3.5) - 7.2(-3)$
$= 56 - [(7.2)(-3)]$
$= 56 - (-21.6)$
$= 56 + 21.6$
$= 77.6$

127. **[1.5]** $2\dfrac{5}{6} - 4\dfrac{1}{3} = \dfrac{17}{6} - \dfrac{13}{3}$
$\phantom{2\dfrac{5}{6} - 4\dfrac{1}{3}} = \dfrac{17}{6} - \dfrac{13 \cdot 2}{3 \cdot 2}$
$\phantom{2\dfrac{5}{6} - 4\dfrac{1}{3}} = \dfrac{17}{6} - \dfrac{26}{6}$
$\phantom{2\dfrac{5}{6} - 4\dfrac{1}{3}} = \dfrac{17}{6} + \left(-\dfrac{26}{6}\right)$
$\phantom{2\dfrac{5}{6} - 4\dfrac{1}{3}} = -\dfrac{9}{6} = -\dfrac{3}{2} \text{ or } -1\dfrac{1}{2}$

128. **[1.5]** $-8 + [(-4 + 17) - (-3 - 3)]$
$= -8 + \{(13) - [-3 + (-3)]\}$
$= -8 + [13 - (-6)]$
$= -8 + (13 + 6)$
$= -8 + 19 = 11$

129. [1.6]
$$-\frac{12}{5} \div \frac{9}{7} = -\frac{12}{5} \cdot \frac{7}{9}$$
$$= -\frac{12 \cdot 7}{5 \cdot 9}$$
$$= -\frac{3 \cdot 4 \cdot 7}{5 \cdot 3 \cdot 3}$$
$$= -\frac{28}{15} \text{ or } -1\frac{13}{15}$$

130. [1.6] $(-8-3) - 5(2-9)$
$$= [-8 + (-3)] - 5[2 + (-9)]$$
$$= -11 - 5(-7)$$
$$= -11 - (-35)$$
$$= -11 + 35 = 24$$

131. [1.5] $[-7 + (-2) - (-3)] + [8 + (-13)]$
$$= \{[-7 + (-2)] + 3\} + (-5)$$
$$= (-9 + 3) + (-5)$$
$$= -6 + (-5) = -11$$

132. [1.6] $\frac{15}{2} \cdot \left(-\frac{4}{5}\right) = -\frac{15 \cdot 4}{2 \cdot 5} = -\frac{60}{10} = -6$

133. [1.6] $x + (x - 1400) = 25{,}800$ simplifies to $2x - 1400 = 25{,}800$; where x represents the amount spent in 2002.

134. [1.6] "The quotient of a number and 14 less than three times the number" is written
$$\frac{x}{3x - 14}, \text{ where } x \text{ represents the number.}$$

Chapter 1 Test

1. Evaluate the left side of the statement.
$$4[-20 + 7(-2)] = 4[-20 + (-14)]$$
$$= 4(-34)$$
$$= -136$$

Since $-136 \leq -135$, the statement "$4[-20 + 7(-2)] \leq -135$" is true.

2. Evaluate the left side of the statement.
$$\left(\frac{1}{2}\right)^2 + \left(\frac{2}{3}\right)^2 = \frac{1}{4} + \frac{4}{9}$$
$$= \frac{9}{36} + \frac{16}{36}$$
$$= \frac{25}{36}$$

Evaluate the right side of the statement.

$$\left(\frac{1}{2} + \frac{2}{3}\right)^2 = \left(\frac{3}{6} + \frac{4}{6}\right)^2$$
$$= \left(\frac{7}{6}\right)^2$$
$$= \frac{49}{36}$$

Since $\frac{25}{36} \neq \frac{49}{36}$, the statement "$\left(\frac{1}{2}\right)^2 + \left(\frac{2}{3}\right)^2 = \left(\frac{1}{2} + \frac{2}{3}\right)^2$" is false.

3. $-1, -3, |-4|, |-1|$

Recall that $|-4| = 4$ and $|-1| = 1$. From smallest to largest, the numbers are $-3, -1, |-1|, |-4|$.

$$\xrightarrow{\bullet\bullet\bullet\bullet\bullet\bullet}$$
$$-3\ -2\ -1\ \ 0\ \ 1\ \ 2\ \ 3\ \ 4$$

4. $6, -|-8|$

$-|-8| = -(8) = -8$

Since $-8 < 6$, $-|-8|$ (or -8) is the lesser number.

5. $-.742, -1.277$

Since -1.277 is to the left of $-.742$ on the number line, -1.277 is the lesser number.

6. "The quotient of -6 and the sum of 2 and -8" is written
$$\frac{-6}{2 + (-8)} \text{ and } \frac{-6}{2 + (-8)} = \frac{-6}{-6} = 1.$$

7. $\frac{a + b}{a \cdot b}$

If a and b are both negative, $a + b$ will be negative, and $a \cdot b$ will be positive. Because the quotient of a negative number and a positive number is negative, $\frac{a+b}{a \cdot b}$ would be *negative*.

8. $-2 - (5 - 17) + (-6)$
$$= -2 - [5 + (-17)] + (-6)$$
$$= -2 - (-12) + (-6)$$
$$= (-2 + 12) + (-6)$$
$$= 10 + (-6) = 4$$

9. $-5\frac{1}{2} + 2\frac{2}{3} = -\frac{11}{2} + \frac{8}{3}$
$$= -\frac{11 \cdot 3}{2 \cdot 3} + \frac{8 \cdot 2}{3 \cdot 2}$$
$$= -\frac{33}{6} + \frac{16}{6}$$
$$= -\frac{17}{6} \text{ or } -2\frac{5}{6}$$

10. $-6.2 - [-7.1 + (2.0 - 3.1)]$
    $= -6.2 - [-7.1 + (-1.1)]$
    $= -6.2 - [-8.2]$
    $= -6.2 + 8.2 = 2$

11. $4^2 + (-8) - (2^3 - 6)$
    $= 16 + (-8) - (8 - 6)$
    $= [16 + (-8)] - 2$
    $= 8 - 2 = 6$

12. $(-5)(-12) + 4(-4) + (-8)^2$
    $= (-5)(-12) + 4(-4) + 64$
    $= [60 + (-16)] + 64$
    $= 44 + 64 = 108$

13. $\dfrac{-7 - |-6 + 2|}{-5 - (-4)} = \dfrac{-7 - |-4|}{-5 + 4} = \dfrac{-7 - 4}{-1}$
    $= \dfrac{-11}{-1} = 11$

14. $\dfrac{30(-1 - 2)}{-9[3 - (-2)] - 12(-2)}$
    $= \dfrac{30(-3)}{-9(5) - (-24)}$
    $= \dfrac{-90}{-45 + 24}$
    $= \dfrac{-90}{-21}$
    $= \dfrac{30 \cdot 3}{7 \cdot 3} = \dfrac{30}{7}$ or $4\dfrac{2}{7}$

15. $3x - 4y^2$
    $= 3(-2) - 4(4^2)$ Let $x = -2$, $y = 4$.
    $= 3(-2) - 4(16)$
    $= -6 - 64 = -70$

16. $\dfrac{5x + 7y}{3(x + y)}$
    $= \dfrac{5(-2) + 7(4)}{3(-2 + 4)}$ Let $x = -2$, $y = 4$.
    $= \dfrac{-10 + 28}{3(2)}$
    $= \dfrac{18}{6} = 3$

17. $118 - (-60) = 118 + 60 = 178$

    The difference between the temperatures was $178°F$.

18. $3x + 0 = 3x$

    illustrates an *identity property*. The correct response is **D**.

19. $(5 + 2) + 8 = 8 + (5 + 2)$

    illustrates a *commutative property* because the order of the numbers is changed, but not the grouping. The correct response is **A**.

20. $-3(x + y) = -3x + (-3y)$

    illustrates the *distributive property*. The correct response is **E**.

21. $-5 + (3 + 2) = (-5 + 3) + 2$

    illustrates an *associative property* because the grouping of the numbers is changed, but not the order. The correct response is **B**.

22. $-\dfrac{5}{3}\left(-\dfrac{3}{5}\right) = 1$

    illustrates an *inverse property*. The correct response is **C**.

23. $-2(3x^2 + 4) - 3(x^2 + 2x)$
    $= -2(3x^2) + (-2)(4) - 3(x^2) - 3(2x)$
    $= -6x^2 + (-8) - 3x^2 - 6x$
    $= -9x^2 - 6x - 8$

24. $-(3x + 1) = -1 \cdot (3x + 1)$ $-a = -1 \cdot a$
    $= -1 \cdot 3x + (-1)(1)$ Distributive property
    $= -3x - 1$ Identity property

    So we used the identity and distributive properties.

25. (a) $-6[5 + (-2)] = -6(3) = -18$

    (b) $-6[5 + (-2)] = -6(5) + (-6)(-2)$
    $= -30 + 12 = -18$

    (c) The above two answers must be the same because the distributive property states that $a(b + c) = ab + ac$ is true for all real numbers a, b, and c.

CHAPTER 2 EQUATIONS, INEQUALITIES, AND APPLICATIONS

2.1 The Addition Property of Equality

2.1 Margin Exercises

1. Note: When solving equations we will write "Add 5" as a shorthand notation for "Add 5 to each side" and "Subtract 5" as a notation for "Subtract 5 from each side."

   (a)
   $$\begin{aligned} x - 12 &= 9 & &\text{Given} \\ x - 12 + 12 &= 9 + 12 & &\text{Add 12.} \\ x + 0 &= 21 & &\text{Additive inverse property} \\ x &= 21 & &\text{Additive identity property} \end{aligned}$$

   We check by substituting 21 for x in the *original* equation.

   Check: $\quad \begin{aligned} x - 12 &= 9 & &\text{Original} \\ 21 - 12 &= 9 \;? & &\text{Let } x = 21. \\ 9 &= 9 & &\text{True} \end{aligned}$

   Since a true statement results, 21 is the solution.

   (b)
   $$\begin{aligned} x - 25 &= -18 & &\text{Given} \\ x - 25 + 25 &= -18 + 25 & &\text{Add 25.} \\ x &= 7 \end{aligned}$$

   Check: $\quad \begin{aligned} x - 25 &= -18 & &\text{Original} \\ 7 - 25 &= -18 \;? & &\text{Let } x = 7. \\ -18 &= -18 & &\text{True} \end{aligned}$

   Since a true statement results, 7 is the solution.

2. **(a)**
   $$\begin{aligned} x - 3.7 &= -8.1 \\ x - 3.7 + 3.7 &= -8.1 + 3.7 & &\text{Add 3.7.} \\ x &= -4.4 \end{aligned}$$

   Check $x = -4.4$: $-8.1 = -8.1$

   This is a shorthand notation for showing that if we substitute -4.4 for x, both sides are equal to -8.1, and hence a true statement results. In practice, this is what you will do, especially if you're using a calculator.

   The solution is -4.4.

   (b)
   $$\begin{aligned} a - 4.1 &= 6.3 \\ a - 4.1 + 4.1 &= 6.3 + 4.1 & &\text{Add 4.1.} \\ a &= 10.4 \end{aligned}$$

   Check $a = 10.4$: $6.3 = 6.3$

   The solution is 10.4.

3. **(a)**
   $$\begin{aligned} -3 &= a + 2 \\ -3 - 2 &= a + 2 - 2 & &\text{Subtract 2.} \\ -5 &= a \text{ or } a = -5 \end{aligned}$$

   Check $a = -5$: $-3 = -3$

   The solution is -5.

   (b)
   $$\begin{aligned} 22 &= -16 + r \\ 22 + 16 &= -16 + r + 16 & &\text{Add 16.} \\ 38 &= r \text{ or } r = 38 \end{aligned}$$

   Check $r = 38$: $22 = 22$

   The solution is 38.

4. **(a)**
   $$\begin{aligned} 6m &= 4 + 5m \\ 6m - 5m &= 4 + 5m - 5m & &\text{Subtract 5m.} \\ 1m &= 4 \text{ or } m = 4 \end{aligned}$$

   Check $m = 4$: $24 = 24$

   The solution is 4.

   (b)
   $$\begin{aligned} \frac{7}{2}m + 1 &= \frac{9}{2}m \\ \frac{7}{2}m + 1 - \frac{7}{2}m &= \frac{9}{2}m - \frac{7}{2}m & &\text{Subtract } \tfrac{7}{2}m. \\ 1 &= \frac{2}{2}m & &\text{Combine terms.} \\ 1 &= m \end{aligned}$$

   Check $m = 1$: $\frac{9}{2} = \frac{9}{2}$

   The solution is 1.

   (c) If $-x = 6$, then $x = -6$.
   The solution is -6.

   (d) If $-x = -12$, then $x = 12$.
   The solution is 12.

5. **(a)**
   $$\begin{aligned} 4x + 6 + 2x - 3 &= 9 + 5x - 4 & &\text{Given} \\ 6x + 3 &= 5x + 5 & &\text{Combine terms.} \\ 6x + 3 - 5x &= 5x + 5 - 5x & &\text{Subtract 5x.} \\ x + 3 &= 5 & &\text{Combine terms.} \\ x + 3 - 3 &= 5 - 3 & &\text{Subtract 3.} \\ x &= 2 & &\text{Combine terms.} \end{aligned}$$

   Check $x = 2$: $15 = 15$

   The solution is 2.

(b)
$$9r + 4r + 6 - 2 = 9r + 4 + 3r \quad \text{Given}$$
$$13r + 4 = 12r + 4 \quad \text{Combine terms.}$$
$$13r + 4 - 12r = 12r + 4 - 12r \quad \text{Subtract } 12r.$$
$$r + 4 = 4 \quad \text{Combine terms}$$
$$r + 4 - 4 = 4 - 4 \quad \text{Subtract 4.}$$
$$r = 0 \quad \text{Combine terms.}$$

Check $r = 0$: $4 = 4$

The solution is 0.

6. (a)
$$4(r + 1) - (3r + 5) = 1 \quad \text{Given}$$
$$4(r + 1) - 1(3r + 5) = 1 \quad -a = -1a$$
$$4r + 4 - 3r - 5 = 1 \quad \text{Distributive property}$$
$$r - 1 = 1 \quad \text{Combine terms.}$$
$$r - 1 + 1 = 1 + 1 \quad \text{Add 1.}$$
$$r = 2$$

Check $r = 2$: $1 = 1$

The solution is 2.

(b)
$$-3(m - 4) + 2(5 + 2m) = 29 \quad \text{Given}$$
$$-3m + 12 + 10 + 4m = 29 \quad \text{Distributive property}$$
$$m + 22 = 29 \quad \text{Combine terms.}$$
$$m + 22 - 22 = 29 - 22 \quad \text{Subtract 22.}$$
$$m = 7$$

Check $m = 7$: $29 = 29$

The solution is 7.

2.1 Section Exercises

1. Equations that have exactly the same solutions are **equivalent equations**.

A.
$$x + 2 = 6$$
$$x + 2 - 2 = 6 - 2 \quad \text{Subtract 2.}$$
$$x = 4$$

So $x + 2 = 6$ and $x = 4$ *are* equivalent equations.

B.
$$10 - x = 5$$
$$10 - x - 10 = 5 - 10 \quad \text{Subtract 10.}$$
$$-x = -5$$
$$-1(-x) = -1(-5) \quad \text{Multiply by } -1.$$
$$x = 5$$

So $10 - x = 5$ and $x = -5$ *are not* equivalent equations.

C. Subtract 3 from both sides of $x + 3 = 9$ to get $x = 6$, so $x + 3 = 9$ and $x = 6$ *are* equivalent equations.

D. Subtract 4 from both sides of $4 + x = 8$ to get $x = 4$. The second equation is $x = -4$, so $4 + x = 8$ and $x = -4$ *are not* equivalent equations.

3. A is not linear because of the x^2 term.
B is not linear because of the x^3 term.
C is linear because it is in the form $Ax + B = C$.
D is linear because it can be simplified to be in the form $Ax + B = C$, as follows:

$$7x - 6x = 3 + 9x \quad \text{Given}$$
$$x = 3 + 9x \quad \text{Combine terms.}$$
$$x - 9x = 3 + 9x - 9x \quad \text{Subtract } 9x.$$
$$-8x = 3 \quad \text{Combine terms.}$$
$$-8x - 3 = 3 - 3 \quad \text{Subtract 3.}$$
$$-8x - 3 = 0 \quad \text{Combine terms.}$$

5.
$$x - 4 = 8$$
$$x - 4 + 4 = 8 + 4$$
$$x = 12$$

Check this solution by replacing x with 12 in the original equation.

$$x - 4 = 8$$
$$12 - 4 = 8 \ ? \quad \text{Let } x = 12.$$
$$8 = 8 \quad \text{True}$$

Because the final statement is true, 12 is the solution.

7.
$$x - 5 = -8$$
$$x - 5 + 5 = -8 + 5$$
$$x = -3$$

Checking yields a true statement, so -3 is the solution.

9.
$$r + 9 = 13$$
$$r + 9 - 9 = 13 - 9$$
$$r = 4$$

Checking yields a true statement, so 4 is the solution.

2.1 The Addition Property of Equality

11.
$$x + 26 = 17$$
$$x + 26 - 26 = 17 - 26$$
$$x = -9$$

Checking yields a true statement, so -9 is the solution.

13.
$$x - 8.4 = -2.1$$
$$x - 8.4 + 8.4 = -2.1 + 8.4$$
$$x = 6.3$$

Check $x = 6.3$: $-2.1 = -2.1$

The solution is 6.3.

15.
$$t + 12.3 = -4.6$$
$$t + 12.3 - 12.3 = -4.6 - 12.3$$
$$t = -16.9$$

Check $t = -16.9$: $-4.6 = -4.6$

The solution is -16.9.

17.
$$7 + r = -3$$
$$r + 7 = -3$$
$$r + 7 - 7 = -3 - 7$$
$$r = -10$$

Check $r = -10$: $-3 = -3$

The solution is -10.

19.
$$2 = p + 15$$
$$2 - 15 = p + 15 - 15$$
$$-13 = p$$

Check $p = -13$: $2 = 2$

The solution is -13.

21.
$$-2 = x - 12$$
$$-2 + 12 = x - 12 + 12$$
$$10 = x$$

Check $x = 10$: $-2 = -2$

The solution is 10.

23.
$$3x = 2x + 7$$
$$3x - 2x = 2x + 7 - 2x \quad \text{Subtract } 2x.$$
$$1x = 7 \quad \text{or} \quad x = 7$$

Check $x = 7$: $21 = 21$

The solution is 7.

25.
$$10x + 4 = 9x$$
$$10x + 4 - 9x = 9x - 9x \quad \text{Subtract } 9x.$$
$$1x + 4 = 0$$
$$x + 4 - 4 = 0 - 4 \quad \text{Subtract } 4.$$
$$x = -4$$

Check $x = -4$: $-36 = -36$

The solution is -4.

27.
$$\frac{9}{7}r - 3 = \frac{2}{7}r$$
$$\frac{9}{7}r - 3 - \frac{2}{7}r = \frac{2}{7}r - \frac{2}{7}r \quad \text{Subtract } \frac{2}{7}r.$$
$$\frac{7}{7}r - 3 = 0$$
$$r - 3 + 3 = 0 + 3 \quad \text{Add } 3.$$
$$r = 3$$

Check $r = 3$: $\frac{6}{7} = \frac{6}{7}$

The solution is 3.

29.
$$5.6x + 2 = 4.6x$$
$$5.6x + 2 - 4.6x = 4.6x - 4.6x$$
$$1.0x + 2 = 0$$
$$x + 2 - 2 = 0 - 2$$
$$x = -2$$

Check $x = -2$: $-9.2 = -9.2$

The solution is -2.

31.
$$3p + 6 = 10 + 2p$$
$$3p + 6 - 2p = 10 + 2p - 2p$$
$$p + 6 = 10$$
$$p + 6 - 6 = 10 - 6$$
$$p = 4$$

Check $p = 4$: $18 = 18$

The solution is 4.

33.
$$3x + 6 - 10 = 2x - 2$$
$$3x - 4 = 2x - 2$$
$$3x - 4 - 2x = 2x - 2 - 2x$$
$$x - 4 = -2$$
$$x - 4 + 4 = -2 + 4$$
$$x = 2$$

Check $x = 2$: $2 = 2$

The solution is 2.

35.
$$6x + 5 + 7x + 3 = 12x + 4$$
$$13x + 8 = 12x + 4$$
$$13x + 8 - 12x = 12x + 4 - 12x$$
$$x + 8 = 4$$
$$x + 8 - 8 = 4 - 8$$
$$x = -4$$

Check $x = -4$: $-44 = -44$

The solution is -4.

44 Chapter 2 Equations, Inequalities, and Applications

37. $10x + 5x + 7 - 8 = 12x + 3 + 2x$
$15x - 1 = 14x + 3$
$15x - 1 - 14x = 14x + 3 - 14x$
$x - 1 = 3$
$x - 1 + 1 = 3 + 1$
$x = 4$

   Check $x = 4$: $59 = 59$

   The solution is 4.

39. $5.2q - 4.6 - 7.1q = -.9q - 4.6$
$-1.9q - 4.6 = -.9q - 4.6$
$-1.9q - 4.6 + .9q = -.9q - 4.6 + .9q$
$-1.0q - 4.6 = -4.6$
$-1.0q - 4.6 + 4.6 = -4.6 + 4.6$
$-q = 0$
$q = 0$

   Check $q = 0$: $-4.6 = -4.6$

   The solution is 0.

41. $\frac{5}{7}x + \frac{1}{3} = \frac{2}{5} - \frac{2}{7}x + \frac{2}{5}$

   $\frac{5}{7}x + \frac{1}{3} = \frac{4}{5} - \frac{2}{7}x$

   $\frac{5}{7}x + \frac{2}{7}x + \frac{1}{3} = \frac{4}{5} - \frac{2}{7}x + \frac{2}{7}x$ *Add* $\frac{2}{7}x$.

   $\frac{7}{7}x + \frac{1}{3} = \frac{4}{5}$ *Combine like terms.*

   $1x + \frac{1}{3} - \frac{1}{3} = \frac{4}{5} - \frac{1}{3}$ *Subtract* $\frac{1}{3}$.

   $x = \frac{12}{15} - \frac{5}{15}$ *LCD = 15*

   $x = \frac{7}{15}$

   Check: $x = \frac{7}{15}$: $\frac{2}{3} = \frac{2}{3}$

   The solution is $\frac{7}{15}$.

43. $(5x + 6) - (3 + 4x) = 10$
$5x + 6 - 3 - 4x = 10$ *Distributive property*
$x + 3 = 10$ *Combine terms.*
$x + 3 - 3 = 10 - 3$ *Subtract 3.*
$x = 7$

   Check $x = 7$: $10 = 10$

   The solution is 7.

45. $2(p + 5) - (9 + p) = -3$
$2p + 10 - 9 - p = -3$
$p + 1 = -3$
$p + 1 - 1 = -3 - 1$
$p = -4$

   Check $p = -4$: $-3 = -3$

   The solution is -4.

47. $-6(2x + 1) + (13x - 7) = 0$
$-12x - 6 + 13x - 7 = 0$
$x - 13 = 0$
$x - 13 + 13 = 0 + 13$
$x = 13$

   Check $x = 13$: $0 = 0$

   The solution is 13.

49. $10(-2x + 1) = -19(x + 1)$
$-20x + 10 = -19x - 19$
$-20x + 10 + 19x = -19x - 19 + 19x$
$-x + 10 = -19$
$-x + 10 - 10 = -19 - 10$
$-x = -29$
$x = 29$

   Check $x = 29$: $-570 = -570$

   The solution is 29.

51. $-2(8p + 2) - 3(2 - 7p) = 2(4 + 2p)$
$-16p - 4 - 6 + 21p = 8 + 4p$
$5p - 10 = 8 + 4p$
$5p - 10 - 4p = 8 + 4p - 4p$
$p - 10 = 8$
$p - 10 + 10 = 8 + 10$
$p = 18$

   Check $p = 18$: $80 = 80$

   The solution is 18.

53. Answers will vary. One example is $x - 6 = -8$.

2.2 The Multiplication Property of Equality

2.2 Margin Exercises

1. To check that 5 is the solution of $3x = 15$, replace x with 5 in the given equation.

   $3x = 15$ *Given equation*
   $3(5) = 15$? *Let x = 5.*
   $15 = 15$ *True*

   The solution 5 is correct.

2. (a) $-6p = -14$

   $\frac{-6p}{-6} = \frac{-14}{-6}$ *Divide by -6.*

   $p = \frac{7}{3}$ *Reduce.*

   Check $p = \frac{7}{3}$: $-14 = -14$

   The solution is $\frac{7}{3}$.

2.2 The Multiplication Property of Equality

(b) $3r = -12$
$\dfrac{3r}{3} = \dfrac{-12}{3}$ *Divide by 3.*
$r = -4$

Check $r = -4$: $-12 = -12$

The solution is -4.

(c) $-2m = 16$
$\dfrac{-2m}{-2} = \dfrac{16}{-2}$ *Divide by -2.*
$m = -8$

Check $m = -8$: $16 = 16$

The solution is -8.

3. (a) $-.7m = -5.04$
$\dfrac{-.7m}{-.7} = \dfrac{-5.04}{-.7}$ *Divide by $-.7$.*
$m = 7.2$

Check $m = 7.2$: $-5.04 = -5.04$

The solution is 7.2.

(b) $12.5k = -63.75$
$\dfrac{12.5k}{12.5} = \dfrac{-63.75}{12.5}$ *Divide by 12.5.*
$k = -5.1$

Check $k = -5.1$: $-63.75 = -63.75$

The solution is -5.1.

4. (a) $\dfrac{y}{5} = 5$
$\dfrac{1}{5}y = 5$
$5 \cdot \dfrac{1}{5}y = 5 \cdot 5$ *Multiply by 5, the reciprocal of $\frac{1}{5}$.*
$y = 25$

Check $y = 25$: $5 = 5$

The solution is 25.

(b) $\dfrac{p}{4} = -6$
$\dfrac{1}{4}p = -6$
$4 \cdot \dfrac{1}{4}p = 4(-6)$ *Multiply by 4, the reciprocal of $\frac{1}{4}$.*
$p = -24$

Check $p = -24$: $-6 = -6$

The solution is -24.

5. (a) $-\dfrac{5}{6}t = -15$
$-\dfrac{6}{5}\left(-\dfrac{5}{6}t\right) = -\dfrac{6}{5}(-15)$ *Multiply by $-\frac{6}{5}$.*
$t = 18$

Check $t = 18$: $-15 = -15$

The solution is 18.

(b) $\dfrac{3}{4}k = -21$
$\dfrac{4}{3} \cdot \dfrac{3}{4}k = \dfrac{4}{3}(-21)$ *Multiply by $\frac{4}{3}$.*
$k = -28$

Check $k = -28$: $-21 = -21$

The solution is -28.

6. (a) $-m = 2$
$-1 \cdot m = 2$ $-m = -1 \cdot m$
$(-1)(-1 \cdot m) = -1 \cdot 2$ *Multiply by -1.*
$1 \cdot m = -2$
$m = -2$

Check $m = -2$: $2 = 2$

The solution is -2.

(b) $-p = -7$
$-1 \cdot p = -7$ $-p = -1 \cdot p$
$(-1)(-1) \cdot p = (-1)(-7)$ *Multiply by -1.*
$1 \cdot p = 7$
$p = 7$

Check $p = 7$: $-7 = -7$

The solution is 7.

7. (a) $4r - 9r = 20$
$-5r = 20$ *Combine terms.*
$\dfrac{-5r}{-5} = \dfrac{20}{-5}$ *Divide by -5.*
$r = -4$

Check $r = -4$: $20 = 20$

The solution is -4.

(b) $7m - 5m = -12$
$2m = -12$ *Combine terms.*
$\dfrac{2m}{2} = \dfrac{-12}{2}$ *Divide by 2.*
$m = -6$

Check $m = -6$: $-12 = -12$

The solution is -6.

2.2 Section Exercises

1. $\frac{2}{3}x = 8$

   To get just x on the left side, multiply both sides of the equation by the reciprocal of $\frac{2}{3}$, which is $\frac{3}{2}$.

3. $.1x = 3$

   This equation is equivalent to $\frac{1}{10}x = 3$. To get just x on the left side, multiply both sides of the equation by the reciprocal of $\frac{1}{10}$, which is 10.

5. $-\frac{9}{2}x = -4$

   To get just x on the left side, multiply both sides of the equation by the reciprocal of $-\frac{9}{2}$, which is $-\frac{2}{9}$.

7. $-x = .36$

   This equation is equivalent to $-1x = .36$. To get just x on the left side, multiply both sides of the equation by the reciprocal of -1, which is -1.

9. $6x = 5$

   To get just x on the left side, divide both sides of the equation by the coefficient of x, which is 6.

11. $-4x = 13$

    To get just x on the left side, divide both sides of the equation by the coefficient of x, which is -4.

13. $.12x = 48$

    To get just x on the left side, divide both sides of the equation by the coefficient of x, which is .12.

15. $-x = 23$

    This equation is equivalent to $-1x = 23$. To get just x on the left side, divide both sides of the equation by the coefficient of x, which is -1.

17. To get x alone on the left side, divide each side by 4, the coefficient of x.

19. $5x = 30$
    $$\frac{5x}{5} = \frac{30}{5} \quad \text{Divide by 5.}$$
    $1x = 6$
    $x = 6$

    Check $x = 6$: $30 = 30$

    The solution is 6.

21. $2m = 15$
    $$\frac{2m}{2} = \frac{15}{2} \quad \text{Divide by 2.}$$
    $m = \frac{15}{2}$

    Check $m = \frac{15}{2}$: $15 = 15$

    The solution is $\frac{15}{2}$.

23. $3a = -15$
    $$\frac{3a}{3} = \frac{-15}{3} \quad \text{Divide by 3.}$$
    $a = -5$

    Check $a = -5$: $-15 = -15$

    The solution is -5.

25. $10t = -36$
    $$\frac{10t}{10} = \frac{-36}{10} \quad \text{Divide by 10.}$$
    $$t = -\frac{36}{10} = -\frac{18}{5} \quad \text{Reduce.}$$

    Check $t = -\frac{18}{5}$: $-36 = -36$

    The solution is $-\frac{18}{5}$.

27. $-6x = -72$
    $$\frac{-6x}{-6} = \frac{-72}{-6} \quad \text{Divide by } -6.$$
    $x = 12$

    Check $x = 12$: $-72 = -72$

    The solution is 12.

29. $2r = 0$
    $$\frac{2r}{2} = \frac{0}{2} \quad \text{Divide by 2.}$$
    $r = 0$

    Check $r = 0$: $0 = 0$

    The solution is 0.

31. $-y = 12$
    $$\frac{-y}{-1} = \frac{12}{-1} \quad \text{Divide by } -1.$$
    $y = -12$

    Check $y = -12$: $12 = 12$

    The solution is -12.

33. $.2t = 8$
    $$\frac{.2t}{.2} = \frac{8}{.2}$$
    $t = 40$

    Check $t = 40$: $8 = 8$

    The solution is 40.

35. $-2.1m = 25.62$
$$\frac{-2.1m}{-2.1} = \frac{25.62}{-2.1}$$
$$m = -12.2$$
Check $m = -12.2$: $25.62 = 25.62$
The solution is -12.2.

37. $\frac{1}{4}y = -12$
$$4 \cdot \frac{1}{4}y = 4(-12) \quad \text{Multiply by 4.}$$
$$1y = -48$$
$$y = -48$$
Check $y = -48$: $-12 = -12$
The solution is -48.

39. $\frac{x}{7} = -5$
$$\frac{1}{7}x = -5$$
$$7\left(\frac{1}{7}x\right) = 7(-5)$$
$$x = -35$$
Check $x = -35$: $-5 = -5$
The solution is -35.

41. $\frac{z}{6} = 12$
$$\frac{1}{6}z = 12$$
$$6 \cdot \frac{1}{6}z = 6 \cdot 12$$
$$z = 72$$
Check $z = 72$: $12 = 12$
The solution is 72.

43. $\frac{2}{7}p = 4$
$$\frac{7}{2}\left(\frac{2}{7}p\right) = \frac{7}{2}(4) \quad \begin{array}{l}\text{Multiply by}\\ \text{the reciprocal}\\ \text{of } \frac{2}{7}.\end{array}$$
$$p = 14$$
Check $p = 14$: $4 = 4$
The solution is 14.

45. $-\frac{7}{9}c = \frac{3}{5}$
$$-\frac{9}{7}\left(-\frac{7}{9}c\right) = -\frac{9}{7} \cdot \frac{3}{5} \quad \begin{array}{l}\text{Multiply by}\\ \text{the reciprocal}\\ \text{of } -\frac{7}{9}.\end{array}$$
$$c = -\frac{27}{35}$$
Check $c = -\frac{27}{35}$: $\frac{3}{5} = \frac{3}{5}$
The solution is $-\frac{27}{35}$.

47. $4x + 3x = 21$
$$7x = 21$$
$$\frac{7x}{7} = \frac{21}{7}$$
$$x = 3$$
Check $x = 3$: $21 = 21$
The solution is 3.

49. $3r - 5r = 10$
$$-2r = 10$$
$$\frac{-2r}{-2} = \frac{10}{-2}$$
$$r = -5$$
Check $r = -5$: $10 = 10$
The solution is -5.

51. $5m + 6m - 2m = 63$
$$9m = 63$$
$$\frac{9m}{9} = \frac{63}{9}$$
$$m = 7$$
Check $m = 7$: $63 = 63$
The solution is 7.

53. $-6x + 4x - 7x = 0$
$$-9x = 0$$
$$\frac{-9x}{-9} = \frac{0}{-9}$$
$$x = 0$$
Check $x = 0$: $0 = 0$
The solution is 0.

55. $.9w - .5w + .1w = -3$
$$.5w = -3 \quad \text{Combine terms.}$$
$$\frac{.5w}{.5} = \frac{-3}{.5} \quad \text{Divide by .5.}$$
$$w = -6$$
Check $w = -6$: $-3 = -3$
The solution is -6.

48 Chapter 2 Equations, Inequalities, and Applications

57. Answers will vary. One example is
$$\frac{3}{2}x = -6.$$

59. "When a number is multiplied by -4, the result is 10."
$$-4x = 10$$
$$\frac{-4x}{-4} = \frac{10}{-4}$$
$$x = -\frac{10}{4} = -\frac{5}{2}$$

The number is $-\frac{5}{2}$.

2.3 More on Solving Linear Equations

2.3 Margin Exercises

1. (a) *Step 1*
$$5y - 7y + 6y - 9 = 3 + 2y$$
$$4y - 9 = 3 + 2y \quad \text{Combine terms.}$$

Step 2
$$4y - 9 + 9 = 3 + 2y + 9 \quad \text{Add 9.}$$
$$4y = 12 + 2y \quad \text{Combine terms.}$$
$$4y - 2y = 12 + 2y - 2y \quad \text{Subtract } 2y.$$
$$2y = 12$$

Step 3
$$\frac{2y}{2} = \frac{12}{2} \quad \text{Divide by 2.}$$
$$y = 6$$

Step 4
Check $y = 6$: $15 = 15$

The solution is 6.

(b) *Step 1*
$$-3k - 5k - 6 + 11 = 2k - 5$$
$$-8k + 5 = 2k - 5 \quad \text{Combine terms.}$$

Step 2
$$-8k + 5 + 5 = 2k - 5 + 5 \quad \text{Add 5.}$$
$$-8k + 10 = 2k$$
$$-8k + 10 + 8k = 2k + 8k \quad \text{Add } 8k.$$
$$10 = 10k$$

Step 3
$$\frac{10}{10} = \frac{10k}{10} \quad \text{Divide by 10.}$$
$$1 = k$$

Step 4
Check $k = 1$: $-3 = -3$

The solution is 1.

2. (a) *Step 1*
$$7(p - 2) + p = 2p + 4$$
$$7p - 14 + p = 2p + 4 \quad \text{Distributive property}$$
$$8p - 14 = 2p + 4 \quad \text{Combine terms.}$$

Step 2
$$8p - 14 + 14 = 2p + 4 + 14 \quad \text{Add 14.}$$
$$8p = 2p + 18$$
$$8p - 2p = 2p + 18 - 2p \quad \text{Subtract } 2p.$$
$$6p = 18$$

Step 3
$$\frac{6p}{6} = \frac{18}{6} \quad \text{Divide by 6.}$$
$$p = 3$$

Step 4
Check $p = 3$: $10 = 10$

The solution is 3.

(b) *Step 1*
$$11 + 3(x + 1) = 5x + 16$$
$$11 + 3x + 3 = 5x + 16 \quad \text{Distributive property}$$
$$3x + 14 = 5x + 16 \quad \text{Combine terms.}$$

Step 2
$$3x + 14 - 14 = 5x + 16 - 14 \quad \text{Subtract 14.}$$
$$3x = 5x + 2$$
$$3x - 5x = 5x + 2 - 5x \quad \text{Subtract } 5x.$$
$$-2x = 2$$

Step 3
$$\frac{-2x}{-2} = \frac{2}{-2} \quad \text{Divide by } -2.$$
$$x = -1$$

Step 4
Check $x = -1$: $11 = 11$

The solution is -1.

3. (a) *Step 1*
$$7m - (2m - 9) = 39$$
$$7m - 2m + 9 = 39 \quad \text{Distributive property}$$
$$5m + 9 = 39$$

Step 2
$$5m + 9 - 9 = 39 - 9 \quad \text{Subtract 9.}$$
$$5m = 30$$

Step 3
$$\frac{5m}{5} = \frac{30}{5} \quad \text{Divide by 5.}$$
$$m = 6$$

Step 4
Check $m = 6$: $39 = 39$

The solution is 6.

(b) *Step 1*
$$4x - (x + 7) = 9$$
$$4x - x - 7 = 9 \quad \text{Distributive property}$$
$$3x - 7 = 9$$

Step 2
$$3x - 7 + 7 = 9 + 7 \quad \text{Add 7.}$$
$$3x = 16$$

Step 3
$$\frac{3x}{3} = \frac{16}{3} \quad \text{Divide by 3.}$$
$$x = \frac{16}{3}$$

Step 4
Check $x = \frac{16}{3}$: $9 = 9$

The solution is $\frac{16}{3}$.

4. (a) *Step 1*
$$2(4 + 3r) = 3(r + 1) + 11$$
$$8 + 6r = 3r + 3 + 11 \quad \text{Dist. prop.}$$
$$8 + 6r = 3r + 14$$

Step 2
$$8 + 6r - 8 = 3r + 14 - 8 \quad \text{Subtract 8.}$$
$$6r = 3r + 6$$
$$6r - 3r = 3r + 6 - 3r \quad \text{Subtract 3r.}$$
$$3r = 6$$

Step 3
$$\frac{3r}{3} = \frac{6}{3} \quad \text{Divide by 3.}$$
$$r = 2$$

Step 4
Check $r = 2$: $20 = 20$

The solution is 2.

(b) *Step 1*
$$2 - 3(2 + 6z) = 4(z + 1) + 18$$
$$2 - 6 - 18z = 4z + 4 + 18 \quad \text{Dist. prop.}$$
$$-4 - 18z = 4z + 22$$

Step 2
$$-4 - 18z + 4 = 4z + 22 + 4 \quad \text{Add 4.}$$
$$-18z = 4z + 26$$
$$-18z - 4z = 4z + 26 - 4z \quad \text{Subtract 4z.}$$
$$-22z = 26$$

Step 3
$$\frac{-22z}{-22} = \frac{26}{-22} \quad \text{Divide by } -22.$$
$$z = -\frac{13}{11}$$

Step 4
Check $z = -\frac{13}{11}$: $\frac{190}{11} = \frac{190}{11}$

The solution is $-\frac{13}{11}$.

5.
$$\frac{1}{4}x - 4 = \frac{3}{2}x + \frac{3}{4}x$$

The LCD of all the fractions in the equation is 4, so multiply each side by 4 to clear the fractions.

$$4\left(\frac{1}{4}x - 4\right) = 4\left(\frac{3}{2}x + \frac{3}{4}x\right)$$

$$4\left(\frac{1}{4}x\right) - 4(4) = 4\left(\frac{3}{2}x\right) + 4\left(\frac{3}{4}x\right)$$

Distributive property
$$x - 16 = 6x + 3x$$

Step 1
$$x - 16 = 9x \quad \text{Combine terms.}$$

Step 2
$$x - 16 - x = 9x - x \quad \text{Subtract x.}$$
$$-16 = 8x$$

Step 3
$$\frac{-16}{8} = \frac{8x}{8} \quad \text{Divide by 8.}$$
$$-2 = x$$

Step 4
Check $x = -2$: $-\frac{9}{2} = -\frac{9}{2}$

The solution is -2.

6.
$$.06(100 - x) + .04x = .05(92)$$

To clear decimals, multiply both sides by 100.
$$100[.06(100 - x) + .04x] = 100[.05(92)]$$
$$6(100 - x) + 4x = 5(92)$$

Step 1
$$600 - 6x + 4x = 460$$
$$600 - 2x = 460$$

Step 2
$$600 - 2x - 600 = 460 - 600$$
$$-2x = -140$$

Step 3
$$\frac{-2x}{-2} = \frac{-140}{-2}$$
$$x = 70$$

Step 4
Check $x = 70$: $4.6 = 4.6$

The solution is 70.

Chapter 2 Equations, Inequalities, and Applications

7. (a)
$$2(x-6) = 2x - 12$$
$$2x - 12 = 2x - 12 \quad \text{Distributive property}$$
$$2x - 12 + 12 = 2x - 12 + 12 \quad \text{Add 12.}$$
$$2x = 2x$$
$$2x - 2x = 2x - 2x \quad \text{Subtract } 2x.$$
$$0 = 0 \quad \text{True}$$

The variable x has "disappeared," and a *true* statement has resulted. The original equation is an identity. This means that for every real number value of x, the equation is true. Thus, **all real numbers** are solutions of the equation.

(b) $3x + 6(x+1) = 9x - 4$
$$3x + 6x + 6 = 9x - 4 \quad \text{Distributive property}$$
$$9x + 6 = 9x - 4 \quad \text{Combine terms.}$$
$$9x + 6 - 9x = 9x - 4 - 9x \quad \text{Subtract } 9x.$$
$$6 = -4 \quad \text{False}$$

The variable x has "disappeared," and a *false* statement has resulted. This means that for every real number value of x, the equation is false. Thus, the equation has **no solution.**

8. If one number is represented by x, then twice that number is represented by $2x$, and 5 more than twice that number is represented by $2x + 5$.

2.3 Section Exercises

Use the four-step method for solving linear equations as given in the text. The details of these steps will only be shown for a few of the exercises.

1. $5m + 8 = 7 + 4m$

For this equation, step 1 is not needed.

Step 2
$$5m + 8 - 8 = 7 + 4m - 8 \quad \text{Subtract 8.}$$
$$5m = 4m - 1$$
$$5m - 4m = 4m - 1 - 4m \quad \text{Subtract } 4m.$$
$$m = -1$$

For this equation, step 3 is not needed since the coefficient of m is already 1.

Step 4
Substitute -1 for m in the original equation.
$$5m + 8 = 7 + 4m$$
$$5(-1) + 8 = 7 + 4(-1) \quad ? \quad \text{Let } m = -1.$$
$$-5 + 8 = 7 + (-4) \quad ?$$
$$3 = 3 \quad \text{True}$$

The solution is -1.

3. $10p + 6 = 12p - 4$

For this equation, step 1 is not needed.

Step 2
$$10p + 6 - 12p = 12p - 4 - 12p \quad \text{Subtract } 12p.$$
$$-2p + 6 = -4$$
$$-2p + 6 - 6 = -4 - 6 \quad \text{Subtract 6.}$$
$$-2p = -10$$

Step 3
$$\frac{-2p}{-2} = \frac{-10}{-2} \quad \text{Divide by } -2.$$
$$p = 5$$

Step 4
Check $p = 5$: $56 = 56$

The solution is 5.

5. Step 1
$$7r - 5r + 2 = 5r - r$$
$$2r + 2 = 4r \quad \text{Combine terms.}$$

Step 2
$$2r + 2 - 2 = 4r - 2 \quad \text{Subtract 2.}$$
$$2r = 4r - 2$$
$$2r - 4r = 4r - 2 - 4r \quad \text{Subtract } 4r.$$
$$-2r = -2$$

Step 3
$$\frac{-2r}{-2} = \frac{-2}{-2} \quad \text{Divide by } -2.$$
$$r = 1$$

Step 4
Check $r = 1$: $4 = 4$

The solution is 1.

7. Step 1
$$x + 3 = -(2x + 2)$$
$$x + 3 = -1(2x + 2)$$
$$x + 3 = -2x - 2 \quad \text{Distributive property}$$

Step 2
$$x + 3 + 2x = -2x - 2 + 2x \quad \text{Add } 2x.$$
$$3x + 3 = -2$$
$$3x + 3 - 3 = -2 - 3 \quad \text{Subtract 3.}$$
$$3x = -5$$

Step 3
$$\frac{3x}{3} = \frac{-5}{3} \quad \text{Divide by 3.}$$
$$x = -\frac{5}{3}$$

Step 4
Check $x = -\frac{5}{3}$: $\frac{4}{3} = \frac{4}{3}$

The solution is $-\frac{5}{3}$.

9. $4(2x - 1) = -6(x + 3)$
$8x - 4 = -6x - 18$
$14x - 4 = -18$ \quad Add $6x$.
$14x = -14$ \quad Add 4.
$x = -1$ \quad Divide by 14.

Check $x = -1$: $-12 = -12$

The solution is -1.

11. $6(4x - 1) = 12(2x + 3)$
$24x - 6 = 24x + 36$
$-6 = 36$ \quad Subtract $24x$.

The variable has "disappeared," and the resulting equation is *false*. Therefore, the equation has no solution.

13. $3(2x - 4) = 6(x - 2)$
$6x - 12 = 6x - 12$
$-12 = -12$ \quad Subtract $6x$.

The variable has "disappeared." Since the resulting statement is a *true* one, *any* real number is a solution. We indicate this by writing *all real numbers*.

15. Equations **A**, **B**, and **C** each have all real numbers for their solution. However, equation **D** gives

$$3x = 2x$$
$$3x - 2x = 2x - 2x$$
$$x = 0.$$

The only solution of this equation is 0, so the correct choice is **D**.

17. Simplify each side separately. Use the addition property to get all variable terms on one side of the equation and all numbers on the other, and then combine like terms. Use the multiplication property to get the equation in the form $x = $ a number. Check the solution.

19. $-\frac{2}{7}r + 2r = \frac{1}{2}r + \frac{17}{2}$

The least common denominator of all the fractions in the equation is 14, so multiply both sides by 14 and solve for r.

$$14\left(-\frac{2}{7}r + 2r\right) = 14\left(\frac{1}{2}r + \frac{17}{2}\right)$$
$$-4r + 28r = 7r + 119$$
$$24r = 7r + 119$$
$$17r = 119$$
$$r = \frac{119}{17} = 7$$

Check $r = 7$: $12 = 12$

The solution is 7.

21. $\frac{1}{9}(x + 18) + \frac{1}{3}(2x + 3) = x + 3$

The least common denominator of all the fractions in the equation is 9, so multiply both sides by 9 and solve for x.

$$9\left[\frac{1}{9}(x + 18) + \frac{1}{3}(2x + 3)\right] = 9(x + 3)$$
$$x + 18 + 3(2x + 3) = 9x + 27$$
$$x + 18 + 6x + 9 = 9x + 27$$
$$7x + 27 = 9x + 27$$
$$-2x + 27 = 27$$
$$-2x = 0$$
$$\frac{-2x}{-2} = \frac{0}{-2}$$
$$x = 0$$

Check $x = 0$: $3 = 3$

The solution is 0.

23. $-\frac{5}{6}q - \left(q - \frac{1}{2}\right) = \frac{1}{4}(q + 1)$

The least common denominator is 12.

$$12\left[-\frac{5}{6}q - \left(q - \frac{1}{2}\right)\right] = 12\left[\frac{1}{4}(q + 1)\right]$$
\quad *Multiply by 12.*
$$12\left(-\frac{5}{6}q\right) - 12\left(q - \frac{1}{2}\right) = 12\left[\frac{1}{4}(q + 1)\right]$$
$$-10q - 12q + 6 = 3(q + 1)$$
$$-22q + 6 = 3q + 3$$
$$-25q + 6 = 3$$
$$-25q = -3$$
$$q = \frac{-3}{-25} = \frac{3}{25}$$

Check $q = \frac{3}{25}$: $\frac{7}{25} = \frac{7}{25}$

The solution is $\frac{3}{25}$.

25. $.30(30) + .15x = .20(30 + x)$
$100[.30(30) + .15x] = 100[.20(30 + x)]$
\quad *Multiply both sides by 100.*
$30(30) + 15x = 20(30 + x)$
$900 + 15x = 600 + 20x$
$900 - 5x = 600$
$-5x = -300$
$x = 60$

Check $x = 60$: $18 = 18$

The solution is 60.

27.
$$.92x + .98(12 - x) = .96(12)$$
$$100[.92x + .98(12 - x)] = 100[.96(12)]$$
$$\text{Multiply both sides by 100.}$$
$$92x + 98(12 - x) = 96(12)$$
$$92x + 1176 - 98x = 1152$$
$$-6x + 1176 = 1152$$
$$-6x = -24$$
$$x = \frac{-24}{-6} = 4$$

Check $x = 4$: $11.52 = 11.52$

The solution is 4.

29.
$$.02(5000) + .03x = .025(5000 + x)$$
$$1000[.02(5000) + .03x] =$$
$$1000[.025(5000 + x)]$$
$$\text{Multiply both sides by 1000, not 100.}$$
$$20(5000) + 30x = 25(5000 + x)$$
$$100{,}000 + 30x = 125{,}000 + 25x$$
$$5x + 100{,}000 = 125{,}000$$
$$5x = 25{,}000$$
$$x = \frac{25{,}000}{5} = 5000$$

Check $x = 5000$: $250 = 250$

The solution is 5000.

31. If $a = 2$ and $b = 4$,
$$100ab = 100(2 \cdot 4)$$
$$= 800.$$

32. If $a = 2$ and $b = 4$,
$$(100a)b = (100 \cdot 2) \cdot 4$$
$$= 800.$$

Yes, the answer is the same as in Exercise 31. This is a result of the associative property of multiplication.

33. $(100a)(100b) = 100 \cdot 100 \cdot a \cdot b$
$$= 10{,}000ab$$
$$\neq 100ab$$

No, $(100a)(100b)$ is not equivalent to $100ab$.

34. In Exercise 33, the only operation is multiplication. The distributive property involves the operation of addition as well as multiplication.

35. $100[.05(x + 2)] = 100(.05)(x + 2)$
$$= [100(.05)](x + 2)$$

Yes, the expressions are equivalent. The associative property of multiplication is used.

36. No, it is not correct to "distribute" the 100 to both .05 and $(x + 2)$ because .05 and $(x + 2)$ are multiplied, not added, in the expression $.05(x + 2)$.

37.
$$-3(5z + 24) + 2 = 2(3 - 2z) - 4$$
$$-15z - 72 + 2 = 6 - 4z - 4$$
$$-15z - 70 = 2 - 4z$$
$$-70 = 2 + 11z \quad \text{Add } 15z.$$
$$-72 = 11z \quad \text{Subtract 2.}$$
$$-\frac{72}{11} = \frac{11z}{11} \quad \text{Divide by 11.}$$
$$-\frac{72}{11} = z$$

Check $z = -\frac{72}{11}$: $\frac{310}{11} = \frac{310}{11}$

The solution is $-\frac{72}{11}$.

39.
$$-(6k - 5) - (-5k + 8) = -3$$
$$-1(6k - 5) - 1(-5k + 8) = -3$$
$$-6k + 5 + 5k - 8 = -3$$
$$-k - 3 = -3$$
$$-k = 0$$
$$k = 0$$

Check $k = 0$: $-3 = -3$

The solution is 0.

41.
$$\frac{1}{3}(x + 3) + \frac{1}{6}(x - 6) = x + 3$$

To clear fractions, multiply both sides by the LCD, which is 6.

$$6\left[\frac{1}{3}(x + 3) + \frac{1}{6}(x - 6)\right] = 6(x + 3)$$
$$6\left(\frac{1}{3}\right)(x + 3) + 6\left(\frac{1}{6}\right)(x - 6) = 6(x + 3)$$
$$2(x + 3) + 1(x - 6) = 6(x + 3)$$
$$2x + 6 + x - 6 = 6x + 18$$
$$3x = 6x + 18$$
$$-3x = 18$$
$$x = \frac{18}{-3} = -6$$

Check $x = -6$: $-3 = -3$

The solution is -6.

43.
$$.30(x + 15) + .40(x + 25) = 25$$

To clear the decimals, multiply both sides by 10.
$$3(x + 15) + 4(x + 25) = 250$$
$$3x + 45 + 4x + 100 = 250$$
$$7x + 145 = 250$$
$$7x = 105$$
$$x = 15$$

Check $x = 15$: $25 = 25$

The solution is 15.

45. $4(x+3) = 2(2x+8) - 4$
$4x + 12 = 4x + 16 - 4$
$4x + 12 = 4x + 12$
$12 = 12$

Since $12 = 12$ is a *true* statement, all real numbers are solutions of the equation.

47. $8(t-3) + 4t = 6(2t+1) - 10$
$8t - 24 + 4t = 12t + 6 - 10$
$12t - 24 = 12t - 4$
$-24 = -4$

Since $-24 = -4$ is a *false* statement, the equation has no solution.

49. The sum of q and the other number is 12. To find the other number, you would subtract q from 12, so the other number is $12 - q$.

51. If Mary is a years old now, in 12 years she will be $a + 12$ years old. Five years ago she was $a - 5$ years old.

53. Since each bill is worth 10 dollars, the number of bills is $\dfrac{t}{10}$.

Summary Exercises on Solving Linear Equations

1. $a + 2 = -3$
$a = -5$ *Subtract 2.*

Check $a = -5$: $-3 = -3$

The solution is -5.

3. $16.5k = -84.15$
$k = \dfrac{-84.15}{16.5}$ *Divide by 16.5.*
$= -5.1$

Check $k = -5.1$: $-84.15 = -84.15$

The solution is -5.1.

5. $\dfrac{4}{5}x = -20$
$x = \left(\dfrac{5}{4}\right)(-20)$ *Multiply by $\dfrac{5}{4}$.*
$= -25$

Check $x = -25$: $-20 = -20$

The solution is -25.

7. $5x - 9 = 4(x - 3)$
$5x - 9 = 4x - 12$ *Distributive property*
$x - 9 = -12$ *Subtract $4x$.*
$x = -3$ *Add 9.*

Check $x = -3$: $-24 = -24$

The solution is -3.

9. $-3(t - 5) + 2(7 + 2t) = 36$
$-3t + 15 + 14 + 4t = 36$
$t + 29 = 36$
$t = 7$

Check $t = 7$: $36 = 36$

The solution is 7.

11. $.08x + .06(x + 9) = 1.24$

To eliminate the decimals, multiply both sides by 100.

$100[.08x + .06(x + 9)] = 100(1.24)$
$8x + 6(x + 9) = 124$
$8x + 6x + 54 = 124$
$14x + 54 = 124$
$14x = 70$
$x = 5$

Check $x = 5$: $.4 + .84 = 1.24$

The solution is 5.

13. $4x + 2(3 - 2x) = 6$
$4x + 6 - 4x = 6$
$6 = 6$

Since $6 = 6$ is a *true* statement, the solution is all real numbers.

15. $-x = 16$
$x = -16$ *Multiply by -1.*

Check $x = -16$: $16 = 16$

The solution is -16.

17. $10m - (5m - 9) = 39$
$10m - 5m + 9 = 39$
$5m + 9 = 39$
$5m = 30$
$m = 6$

Check $m = 6$: $39 = 39$

The solution is 6.

54 Chapter 2 Equations, Inequalities, and Applications

19. $-2t + 5t - 9 = 3(t - 4) - 5$
$-2t + 5t - 9 = 3t - 12 - 5$
$3t - 9 = 3t - 17$
$-9 = -17$

Because $-9 = -17$ is a *false* statement, the equation has no solution.

21. $.02(50) + .08r = .04(50 + r)$

To eliminate the decimals, multiply both sides by 100.
$100[.02(50) + .08r] = 100[.04(50 + r)]$
$2(50) + 8r = 4(50 + r)$
$100 + 8r = 200 + 4r$
$100 + 4r = 200$
$4r = 100$
$r = 25$

Check $r = 25$: $1 + 2 = 3$ ✓

The solution is 25.

23. $2(3 + 7x) - (1 + 15x) = 2$
$6 + 14x - 1 - 15x = 2$
$-x + 5 = 2$
$-x = -3$
$x = 3$

Check $x = 3$: $48 - 46 = 2$ ✓

The solution is 3.

25. $2(5 + 3x) = 3(x + 1) + 13$
$10 + 6x = 3x + 3 + 13$
$6x + 10 = 3x + 16$
$6x = 3x + 6$
$3x = 6$
$x = 2$

Check $x = 2$: $22 = 22$ ✓

The solution is 2.

27. $\frac{5}{6}x + \frac{1}{3} = 2x + \frac{2}{3}$

To clear fractions, multiply both sides by the LCD, which is 6.
$6\left(\frac{5}{6}x + \frac{1}{3}\right) = 6\left(2x + \frac{2}{3}\right)$
$5x + 2 = 12x + 4$
$-7x + 2 = 4$
$-7x = 2$
$x = -\frac{2}{7}$

Check $x = -\frac{2}{7}$: $\frac{2}{21} = \frac{2}{21}$ ✓

The solution is $-\frac{2}{7}$.

29. $\frac{3}{4}(a - 2) - \frac{1}{3}(5 - 2a) = -2$

To clear fractions, multiply both sides by the LCD, which is 12.
$12\left[\frac{3}{4}(a - 2) - \frac{1}{3}(5 - 2a)\right] = 12(-2)$
$9(a - 2) - 4(5 - 2a) = -24$
$9a - 18 - 20 + 8a = -24$
$17a - 38 = -24$
$17a = 14$
$a = \frac{14}{17}$

Check $a = \frac{14}{17}$: $-\frac{15}{17} - \frac{19}{17} = -2$ ✓

The solution is $\frac{14}{17}$.

31. $5.2x - 4.6 - 7.1x = -2.1 - 1.9x - 2.5$

Multiply both sides by 10.
$52x - 46 - 71x = -21 - 19x - 25$
$-19x - 46 = -19x - 46$
$-46 = -46$

Since $-46 = -46$ is a *true* statement, the solution is all real numbers.

2.4 An Introduction to Applications of Linear Equations

2.4 Margin Exercises

1. *Step 2*
Let $x = $ the number.

Step 3

If 5 added is	the product of 9 and a number,	the result is	19 less than the number.
↓	↓	↓	↓
5 +	9x	=	x − 19

Step 4
Solve the equation.
$5 + 9x = x - 19$
$5 + 9x - 5 = x - 19 - 5$ *Subtract 5.*
$9x = x - 24$
$9x - x = x - 24 - x$ *Subtract x.*
$8x = -24$
$\frac{8x}{8} = \frac{-24}{8}$ *Divide by 8.*
$x = -3$

Step 5
The number is -3.

Step 6
9 times -3 is -27. 5 added to -27 is -22, which is 19 less than -3, so -3 is the number.

2.4 An Introduction to Applications of Linear Equations

2. *Step 2*
Let $x = $ the number of miles Jim drove.
Then $3x = $ the number of miles Annie drove.

Step 3

Number of miles Jim drove	plus	number of miles Annie drove	is	84 miles altogether
↓	↓	↓	↓	↓
x	$+$	$3x$	$=$	84

Step 4
Solve this equation.
$$x + 3x = 84$$
$$4x = 84$$
$$\frac{4x}{4} = \frac{84}{4} \quad \text{Divide by 4.}$$
$$x = 21$$

Step 5
Jim drove 21 miles and Annie drove $3 \cdot 21 = 63$ miles.

Step 6
63 is 3 times 21 and the sum of 21 and 63 is 84.

3. *Step 2*
Let $x = $ the number of members.
Then $2x = $ the number of nonmembers.
(If each member brought two nonmembers, there would be twice as many nonmembers as members.)

Step 3

Number of members	plus	Number of nonmembers	is	the total in attendance
↓	↓	↓	↓	↓
x	$+$	$2x$	$=$	27

Step 4
Solve this equation.
$$x + 2x = 27$$
$$3x = 27$$
$$\frac{3x}{3} = \frac{27}{3}$$
$$x = 9$$

Step 5
There were 9 members and $2 \cdot 9 = 18$ nonmembers.

Step 6
18 is twice as much as 9 and the sum of 9 and 18 is 27.

4. *Step 2*
Let $x = $ the length of the middle-sized piece.
Then $x + 10 = $ the length of the longest piece and $x - 5 = $ the length of the shortest piece.

Step 3

Length of shortest piece	plus	length of middle-sized piece
↓	↓	↓
$x - 5$	$+$	x

plus	length of longest piece	is	total length of pipe.
↓	↓	↓	↓
$+$	$x + 10$	$=$	50

Step 4
Solve this equation.
$$(x - 5) + x + (x + 10) = 50$$
$$3x + 5 = 50$$
$$3x = 45$$
$$\frac{3x}{3} = \frac{45}{3}$$
$$x = 15$$

Step 5
The middle-sized piece is 15 inches long, the longest piece is $15 + 10 = 25$ inches long, and the shortest piece is $15 - 5 = 10$ inches long.

Step 6
Since 25 inches is 10 inches longer than 15 inches, 15 inches is 5 inches longer than 10 inches, and $15 + 25 + 10 = 50$ inches (the length of the pipe), the answers are correct.

5. **(a)** Let $x = $ the degree measure of the angle. Because the sum of the measures of an angle and its supplement is $180°$, we have
$$92 + x = \underline{180}$$
$$x = 88°.$$

The supplement of a $92°$ angle is an $88°$ angle.

(b) Let $x = $ the degree measure of the angle. Then $2x = $ the degree measure of its complement.

From the given information, we have
$$2x + x = \underline{90}.$$

Solve this equation.
$$3x = 90$$
$$\frac{3x}{3} = \frac{90}{3}$$
$$x = 30$$

The measure of the angle is $30°$.

56 Chapter 2 Equations, Inequalities, and Applications

(c) Let $x=$ the degree measure of the angle. Then $90-x=$ the degree measure of its complement, and $180-x=$ the degree measure of its supplement.

Twice the complement	is	the supplement	less	30°
↓	↓	↓	↓	↓
$2(90-x)$	$=$	$(180-x)$	$-$	30

Solve this equation.
$$180 - 2x = 180 - x - 30$$
$$180 - 2x = 150 - x$$
$$180 = 150 + x$$
$$30 = x$$

The measure of the angle is 30°.

6. Let $x=$ the smallest integer. Then $x+1=$ the next consecutive integer.

Because the sum of the two integers is -45, we have
$$x + (x+1) = -45$$
$$2x + 1 = -45$$
$$2x = -46$$
$$x = -23$$

The first integer is -23 and the next consecutive integer is $-23+1=-22$. Their sum is -45.

7. Let $x=$ the smallest even integer. Then $x+2=$ the next even integer.

From the given information, we have
$$6 \cdot x + (x+2) = 86.$$

Solve this equation.
$$7x + 2 = 86$$
$$7x = 84$$
$$x = 12$$

The smallest even integer is 12 and the next consecutive even integer is $12+2=14$. Six times 12 is 72 and 72 plus 14 is 86.

2.4 Section Exercises

1. It is important to read the problem carefully before you write anything down. You have to pick a variable to stand for the unknown number and then express any other unknown quantities in terms of the same variable. Be sure to write these things down. Sometimes a figure or diagram will help you to write an equation for the problem. Once you've written the equation, solve it by the methods covered earlier in this chapter. Use your solution to answer the question that the problem asked. Finally, check your solution in the words of the original problem and make sure your answer makes sense.

3. Choice **D**, $6\frac{1}{2}$, is *not* a reasonable answer in an applied problem that requires finding the number of cars on a dealer's lot, since you cannot have $\frac{1}{2}$ of a car. The number of cars must be a whole number.

5. Choice **A**, -10, is *not* a reasonable answer since distance cannot be negative.

The applied problems in this section should be solved by using the six-step method shown in the text. These steps will only be listed in a few of the solutions, but all of the solutions are based on this method.

7. *Step 2*
Let $x=$ the number.

Step 3

The product of 8,		and a number increased by 6,	is	104.
↓		↓	↓	↓
8	$\cdot$	$(x+6)$	$=$	104

Step 4
Solve this equation.
$$8(x+6) = 104$$
$$8x + 48 = 104$$
$$8x = 56$$
$$x = 7$$

Step 5
The number is 7.

Step 6
7 increased by 6 is 13. The product of 8 and 13 is 104, so 7 is the number.

9. *Step 2*
Let $x=$ the number.

Step 3

Two less than three times a number	is equal to	14 more than five times the number.
↓	↓	↓
$3x - 2$	$=$	$5x + 14$

Step 4
Solve this equation.
$$3x = 5x + 16$$
$$-2x = 16$$
$$x = -8$$

Step 5
The number is -8.

Step 6
If the number is -8, then three times the number is -24, and two less than three times the number is -26. Five times the number is $5(-8)=-40$ and 14 more than five times the number is -26, so -8 is the number.

2.4 An Introduction to Applications of Linear Equations

11. *Step 2*

Let $x =$ the unknown number. Then $x - 2$ is two subtracted from the number, $3(x - 2)$ is triple the difference, and $x + 6$ is six more than the number.

Step 3 $\quad 3(x - 2) = x + 6$

Step 4
$$3x - 6 = x + 6$$
$$2x - 6 = 6$$
$$2x = 12$$
$$x = 6$$

Step 5
The number is 6.

Step 6
Check that 6 is the correct answer by substituting this result into the words of the original problem. Two subtracted from the number is $6 - 2 = 4$. Triple this difference is $3(4) = 12$, which is equal to 6 more than the number, since $6 + 6 = 12$.

13. *Step 2*

Let $x =$ the unknown number. Then $3x$ is three times the number, $x + 7$ is 7 more than the number, and the sum is $3x + (x + 7)$; $2x$ is twice the number, and $-11 - 2x$ is the difference between -11 and twice the number.

Step 3 $\quad 3x + (x + 7) = -11 - 2x$

Step 4
$$4x + 7 = -11 - 2x$$
$$6x + 7 = -11$$
$$6x = -18$$
$$x = -3$$

Step 5
The number is -3.

Step 6
Check that -3 is the correct answer by substituting this result into the words of the original problem. The sum of three times a number and 7 more than the number is $3(-3) + (-3 + 7) = -5$ and the difference between -11 and twice the number is $-11 - 2(-3) = -5$. The values are equal, so the number -3 is the correct answer.

15. Let $x =$ the number of drive-in movie screens in New York.

Then $x + 11 =$ the number of drive-in movie screens in California.

Since the total number of screens was 107, we can write the equation
$$x + (x + 11) = 107.$$

Solve this equation.
$$2x + 11 = 107$$
$$2x = 96$$
$$x = 48$$

Since $x = 48$, $x + 11 = 48 + 11 = 59$.

There were 48 drive-in movie screens in New York and 59 in California. Since 59 is 11 more than 48 and $48 + 59 = 107$, this answer checks.

17. Let $x =$ the number of Democrats.
Then $x + 3 =$ the number of Republicans.

The number of Democrats	plus	the number of Republicans
↓	↓	↓
x	$+$	$(x + 3)$

equals	the number of members of the Senate.
↓	↓
$=$	99

Solve the equation.
$$x + (x + 3) = 99$$
$$2x + 3 = 99$$
$$2x = 96$$
$$x = 48$$

There were 48 Democrats and $48 + 3 = 51$ Republicans.

19. Let $x =$ revenue from ticket sales for Bruce Springsteen and the E Street Band.

Then $x - 35.4 =$ revenue from ticket sales for Céline Dion.

Since the total revenue from ticket sales was $196.4 (all numbers in millions), we can write the equation
$$x + (x - 35.4) = 196.4.$$
Solve this equation.
$$2x - 35.4 = 196.4$$
$$2x = 231.8$$
$$x = 115.9$$

Since $x = 115.9$, $x - 35.4 = 80.5$.

Bruce Springsteen and the E Street Band took in $115.9 million and Céline Dion took in $80.5 million. Since 80.5 is 35.4 less than 115.9 and $115.9 + 80.5 = 196.4$, this answer checks.

21. Let $x =$ the number of games the Kings lost.
Then $2x + 13 =$ the number of games the Kings won.

Since the total number of games played was 82, we can write the equation
$$x + (2x + 13) = 82.$$
Solve this equation
$$3x + 13 = 82$$
$$3x = 69$$
$$x = 23$$

Since $x = 23$, $2x + 13 = 59$.

The Kings won 59 games and lost 23 games. Since 59 is 13 more than two times 23 and $59 + 23 = 82$, this answer checks.

23. Let $x =$ the value of the 1945 nickel.
Then $\frac{8}{7}x =$ the value of the 1950 nickel.

The total value of the two coins is $15.00, so
$$x + \frac{8}{7}x = 15.$$
Solve this equation. First multiply both sides by 7 to clear fractions.
$$7\left(x + \frac{8}{7}x\right) = 7(15)$$
$$7x + 8x = 105$$
$$15x = 105$$
$$x = 7 \quad \text{Divide by 15.}$$

Since $x = 7$, $\frac{8}{7}x = \frac{8}{7}(7) = 8$.

The value of the 1945 Philadelphia nickel is $7.00 and the value of the 1950 Denver nickel is $8.00.

25. Let $x =$ the number of pounds of topping.
Then $83.2x =$ the number of pounds of ice cream.

The total weight of the toppings and ice cream was 45,225 pounds, so
$$x + 83.2x = 45,225.$$
Solve this equation.
$$1x + 83.2x = 45,225$$
$$84.2x = 45,225$$
$$x = \frac{45,225}{84.2} \approx 537.1$$

Now $83.2x = 83.2\left(\frac{45,225}{84.2}\right) \approx 44,687.9$.

To the nearest tenth of a pound, there were 44,687.9 pounds of ice cream and 537.1 pounds of topping.

27. Let $x =$ the amount of gravel. Then $3x =$ the amount of cement mix.

The mixture weighs 140 pounds, so
$$x + 3x = 140$$
$$4x = 140$$
$$x = 35.$$

There are 35 pounds of gravel and $3(35) = 105$ pounds of cement mix. The total number of pounds is $35 + 105 = 140$.

29. Let $x =$ the number of packages delivered by Airborne Express.
Then $3x =$ the number of packages delivered by Federal Express, and
$x - 2 =$ the number of packages delivered by United Parcel Service.

Since the total number of packages received was 13,
$$x + 3x + (x - 2) = 13.$$
Solve this equation.
$$5x - 2 = 13$$
$$5x = 15$$
$$x = 3$$

Since $x = 3$, $3x = 3(3) = 9$ and $x - 2 = 3 - 2 = 1$.

One package was delivered by United Parcel Service, 3 were delivered by Airborne Express, and 9 were delivered by Federal Express.

31. Let $x =$ the number of gold medals.
Then $x + 4 =$ the number of silver medals, and $x - 6 =$ the number of bronze medals.

The total number of medals earned by the United States was 103, so
$$x + (x + 4) + (x - 6) = 103.$$
Solve this equation.
$$3x - 2 = 103$$
$$3x = 105$$
$$x = 35$$

Since $x = 35$, $x + 4 = 39$, and $x - 6 = 29$.

The United States earned 35 gold medals, 39 silver medals, and 29 bronze medals. The answer checks since
$$35 + 39 + 29 = 103.$$

33. Let $x =$ the distance of Mercury from the sun.
 Then $x + 31.2 =$ the distance of Venus from the sun, and
 $x + 57 =$ the distance of Earth from the sun.

 Since the total of the distances from these three planets is 196.2 (all distances in millions of miles), we can write the equation

 $$x + (x + 31.2) + (x + 57) = 196.2.$$

 Solve this equation.

 $$3x + 88.2 = 196.2$$
 $$3x = 108$$
 $$x = 36$$

 Mercury is 36 million miles from the sun, Venus is $36 + 31.2 = 67.2$ million miles from the sun, and Earth is $36 + 57 = 93$ million miles from the sun. The answer checks since

 $$36 + 67.2 + 93 = 196.2.$$

35. Let $x =$ the measure of angles A and B.
 Then $x + 60 =$ the measure of angle C.

 The sum of the measures of the angles of any triangle is $180°$, so

 $$x + x + (x + 60) = 180.$$

 Solve this equation.

 $$3x + 60 = 180$$
 $$3x = 120$$
 $$x = 40$$

 Angles A and B have measures of 40 degrees, and angle C has a measure of $40 + 60 = 100$ degrees. The answer checks since

 $$40 + 40 + 100 = 180.$$

37. An angle cannot have its supplement equal to its complement. The sum of an angle and its supplement equals $180°$, while the sum of an angle and its complement equals $90°$. If we try to solve the equation

 $$90 - x = 180 - x,$$

 we will get

 $$90 - x + x = 180 - x + x$$
 $$90 = 180 \quad \textit{False}$$

 so this equation has no solution.

39. The next smaller consecutive integer is one less than a number. Thus, if x represents an integer, the next smaller consecutive integer is $x - 1$.

41. Let $x =$ the measure of the angle.
 Then $90 - x =$ the measure of its complement.

 The "complement is four times its measure" can be written as

 $$90 - x = 4x.$$

 Solve this equation.

 $$90 = 5x$$
 $$x = \frac{90}{5} = 18$$

 The measure of the angle is $18°$. The complement is $90° - 18° = 72°$, which is four times $18°$.

43. Let $x =$ the measure of the angle. Then
 $90 - x =$ the measure of its complement, and
 $180 - x =$ the measure of its supplement.

Its supplement	measures	$39°$
↓	↓	↓
$180 - x$	$=$	39
more than	twice its complement.	
↓	↓	
$+$	$2(90 - x)$	

 Solve the equation.

 $$180 - x = 39 + 2(90 - x)$$
 $$180 - x = 39 + 180 - 2x$$
 $$180 - x = 219 - 2x$$
 $$x + 180 = 219$$
 $$x = 39$$

 The measure of the angle is $39°$. The complement is $90° - 39° = 51°$. Now $39°$ more than twice its complement is $39° + 2(51°) = 141°$, which is the supplement of $39°$ since $180° - 39° = 141°$.

45. Let $x =$ the measure of the angle. Then
 $180 - x =$ the measure of its supplement, and
 $90 - x =$ the measure of its complement.

The difference between the measure of its supplement and		three times the measure of its complement
↓	↓	↓
$(180 - x)$	$-$	$3(90 - x)$
is	$10°$.	
↓	↓	
$=$	10	

continued

Solve the equation.

$$(180 - x) - 3(90 - x) = 10$$
$$180 - x - 270 + 3x = 10$$
$$2x - 90 = 10$$
$$2x = 100$$
$$x = 50$$

The measure of the angle is 50°. The supplement is $180° - 50° = 130°$ and the complement is $90° - 50° = 40°$. The answer checks since $130° - 3(40°) = 10°$.

47. Let $x =$ the number on the first locker.
Then $x + 1 =$ the number on the next locker.

Since the numbers have a sum of 137, we can write the equation

$$x + (x + 1) = 137.$$

Solve the equation.

$$2x + 1 = 137$$
$$2x = 136$$
$$x = \frac{136}{2} = 68$$

Since $x = 68$, $x + 1 = 69$.

The lockers have numbers 68 and 69. Since $68 + 69 = 137$, this answer checks.

49. Because the two pages are back-to-back, they must have page numbers that are consecutive integers.

Let $x =$ the smaller page number.
Then $x + 1 =$ the larger page number.

$$x + (x + 1) = 293$$
$$2x + 1 = 293$$
$$2x = 292$$
$$x = 146$$

Since $x = 146$, $x + 1 = 147$.

The page numbers are 146 and 147. This answer checks since the sum is 293.

51. Let $x =$ the smaller even integer.
Then $x + 2 =$ the larger even integer.

The smaller added to three times the larger gives a sum of 46 can be written as

$$x + 3(x + 2) = 46.$$
$$x + 3x + 6 = 46$$
$$4x + 6 = 46$$
$$4x = 40$$
$$x = 10$$

Since $x = 10$, $x + 2 = 12$.

The integers are 10 and 12. This answer checks since $10 + 3(12) = 46$.

53. Let $x =$ the smaller integer.
Then $x + 1 =$ the larger integer.

$$x + 3(x + 1) = 43$$
$$x + 3x + 3 = 43$$
$$4x + 3 = 43$$
$$4x = 40$$
$$x = 10$$

Since $x = 10$, $x + 1 = 11$.

The integers are 10 and 11. This answer checks since $10 + 3(11) = 43$.

55. Let $x =$ the amount of Head Start funding in the first year (in billions of dollars).
Then $x + .61 =$ the amount of funding in the second year, and
$(x + .61) + .93 = x + 1.54 =$ the amount of funding in the third year.

The total funding was 16.13 billion dollars, so

$$x + (x + .61) + (x + 1.54) = 16.13.$$

Solve this equation.

$$3x + 2.15 = 16.13$$
$$3x = 13.98$$
$$x = 4.66$$

Since $x = 4.66$, $x + .61 = 5.27$, and $x + 1.54 = 6.20$.

The Head Start funding was 4.66 billion dollars in the first year, 5.27 billion dollars in the second year, and 6.20 billion dollars in the third year. The answer checks since the sum is 16.13.

2.5 Formulas and Applications from Geometry

2.5 Margin Exercises

1. (a) Given the formula $I = prt$, the problem is to find p when $I = \$246$, $r = .06$, and $t = 2$. Substitute these values into the formula, and solve for p.

$$I = prt$$
$$\$246 = p(.06)(2)$$
$$\$246 = .12p \quad \text{Multiply.}$$
$$\frac{\$246}{.12} = \frac{.12p}{.12} \quad \text{Divide by .12.}$$
$$\$2050 = p$$

2.5 Formulas and Applications from Geometry

(b) Using $P = 2L + 2W$ as the formula, find L when $P = 126$ and $W = 25$. Substitute these values into the formula, and solve for L.

$$P = 2L + 2W$$
$$126 = 2L + 2(25)$$
$$126 = 2L + 50$$
$$126 - 50 = 2L + 50 - 50 \quad \textit{Subtract 50.}$$
$$76 = 2L$$
$$\frac{76}{2} = \frac{2L}{2} \quad \textit{Divide by 2.}$$
$$38 = L$$

2. The fence will enclose the perimeter of the rectangular field, so use the formula for the perimeter of a rectangle. Find the length of the field by substituting $P = 800$ and $W = 175$ into the formula and solving for L.

$$P = 2L + 2W$$
$$800 = 2L + 2(175)$$
$$800 = 2L + 350$$
$$450 = 2L$$
$$225 = L$$

The length of the field is 225 meters.

3. Use the formula for the area of a triangle.

$$A = \frac{1}{2}bh$$
$$120 = \frac{1}{2}b(24) \quad \textit{Let A = 120, h = 24.}$$
$$120 = 12b$$
$$10 = b$$

The length of the base is 10 meters.

4. (a) Because the angles are vertical angles, they have the same measure.

$$2x + 24 = 4x - 40$$
$$24 = 2x - 40$$
$$64 = 2x$$
$$32 = x$$

If $x = 32$, $2x + 24 = 2(32) + 24 = 88$ and $4x - 40 = 4(32) - 40 = 88$.

Both angles measure 88°.

(b) The sum of the measures of the two angles is 180° because together they form a straight angle.

$$(5x + 12) + 3x = 180$$
$$8x + 12 = 180$$
$$8x = 168$$
$$x = 21$$

If $x = 21$, $5x + 12 = 5(21) + 12 = 117$ and $3x = 3(21) = 63$.

The measures of the angles are 117° and 63°.

5. Solve $I = prt$ for t.

$$I = prt$$
$$\frac{I}{pr} = \frac{prt}{pr} \quad \textit{Divide by pr.}$$
$$\frac{I}{pr} = t \quad \text{or} \quad t = \frac{I}{pr}$$

6. Solve $P = a + b + c$ for a.

$$P = a + b + c$$
$$P - (b + c) = a + b + c - (b + c) \quad \textit{Subtract (b + c).}$$
$$P - b - c = a \quad \text{or} \quad a = P - b - c$$

7. (a) Solve $A = p + prt$ for t.

$$A = p + prt$$
$$A - p = p + prt - p \quad \textit{Subtract p.}$$
$$A - p = prt$$
$$\frac{A - p}{pr} = \frac{prt}{pr} \quad \textit{Divide by pr.}$$
$$\frac{A - p}{pr} = t \quad \text{or} \quad t = \frac{A - p}{pr}$$

(b) Solve $Ax + By = C$ for y.

$$Ax + By = C$$
$$Ax + By - Ax = C - Ax \quad \textit{Subtract Ax.}$$
$$By = C - Ax$$
$$\frac{By}{B} = \frac{C - Ax}{B} \quad \textit{Divide by B.}$$
$$y = \frac{C - Ax}{B}$$

2.5 Section Exercises

1. (a) The perimeter of a plane geometric figure is the distance around the figure. It can be found by adding up the lengths of all the sides. Perimeter is a one-dimensional (linear) measurement, so it is given in linear units (inches, centimeters, feet, etc.).

(b) The area of a plane geometric figure is the measure of the surface covered or enclosed by the figure. Area is a two-dimensional measurement, so it is given in square units (square centimeters, square feet, etc.).

3. You would need to be given 4 values in a formula with 5 variables to find the value of any one variable.

5. Sod for a lawn covers the surface of the lawn, so *area* would be used.

7. The baseboards of a living room go around the edges of the room. The amount of baseboard needed will be the sum of the lengths of the sides of the room, so *perimeter* would be used.

9. Fertilizer for a garden covers the surface of the garden, so *area* would be used.

11. Determining the cost for planting rye grass in a lawn for the winter requires finding the amount of surface to be covered, so *area* would be used.

In Exercises 13–26, substitute the given values into the formula and then solve for the remaining variable.

13. $P = 2L + 2W$; $L = 8$, $W = 5$

$$P = 2L + 2W$$
$$= 2(8) + 2(5)$$
$$= 16 + 10$$
$$P = 26$$

15. $A = \frac{1}{2}bh$; $b = 8$, $h = 16$

$$A = \frac{1}{2}bh$$
$$= \frac{1}{2}(8)(16)$$
$$A = 64$$

17. $P = a + b + c$; $P = 12$, $a = 3$, $c = 5$

$$P = a + b + c$$
$$12 = 3 + b + 5$$
$$12 = b + 8$$
$$4 = b$$

19. $d = rt$; $d = 252$, $r = 45$

$$d = rt$$
$$252 = 45t$$
$$\frac{252}{45} = \frac{45t}{45}$$
$$5.6 = t$$

21. $I = prt$; $p = 7500$, $r = .035$, $t = 6$

$$I = prt$$
$$= (7500)(.035)(6)$$
$$I = 1575$$

23. $C = 2\pi r$; $C = 16.328$, $\pi \approx 3.14$

$$C = 2\pi r$$
$$16.328 \approx 2(3.14)r$$
$$16.328 = 6.28r$$
$$2.6 = r$$

25. $A = \pi r^2$; $r = 4$, $\pi \approx 3.14$

$$A = \pi r^2$$
$$\approx 3.14(4)^2$$
$$= 3.14(16)$$
$$A = 50.24$$

In Exercises 27–32, substitute the given values into the formula and then solve for V.

27. $V = LWH$; $L = 10$, $W = 5$, $H = 3$

$$V = LWH$$
$$= (10)(5)(3)$$
$$V = 150$$

29. $V = \frac{1}{3}Bh$; $B = 12$, $h = 13$

$$V = \frac{1}{3}Bh$$
$$= \frac{1}{3}(12)(13)$$
$$V = 52$$

31. $V = \frac{4}{3}\pi r^3$; $r = 12$, $\pi \approx 3.14$

$$V = \frac{4}{3}\pi r^3$$
$$\approx \frac{4}{3}(3.14)(12)^3$$
$$= \frac{4}{3}(3.14)(1728)$$
$$V = 7234.56$$

33. The diameter of the circle is 443 feet, so its radius is $\frac{443}{2} = 221.5$ ft. Use the area of a circle formula to find the enclosed area.

$$A = \pi r^2$$
$$= \pi(221.5)^2$$
$$\approx 154{,}133.6 \text{ ft}^2,$$

or about 154,000 ft². (If 3.14 is used for π, the value is 154,055.465.)

35. The page is a rectangle with length 3.5 inches and width 3 inches, so use the formulas for the perimeter and area of a rectangle.

$$P = 2L + 2W$$
$$= 2(3.5) + 2(3)$$
$$= 7 + 6$$
$$P = 13 \text{ inches}$$

$$A = LW$$
$$= (3.5)(3)$$
$$A = 10.5 \text{ square inches}$$

2.5 Formulas and Applications from Geometry

37. To find the area of the drum face, use the formula for the area of a circle, $A = \pi r^2$. Since the diameter of the circle is 15.74 feet, the radius is $\left(\frac{1}{2}\right)(15.74) = 7.87$ feet.

$$\begin{aligned} A &= \pi r^2 \\ &\approx (3.14)(7.87)^2 \\ &= (3.14)(61.9369) \\ A &\approx 194.48 \end{aligned}$$

The area of the drum face is about 194.48 square feet.

39. Use the formula for the area of a trapezoid with $B = 115.80$, $b = 171.00$, and $h = 165.97$.

$$\begin{aligned} A &= \frac{1}{2}(B+b)h \\ &= \frac{1}{2}(115.80 + 171.00)(165.97) \\ &= \frac{1}{2}(286.80)(165.97) \\ &= 23{,}800.098 \end{aligned}$$

To the nearest hundredth of a square foot, the combined area of the two lots is 23,800.10 square feet.

41. The girth is $4 \cdot 18 = 72$ inches. Since the length plus the girth is 108, we have

$$\begin{aligned} L + G &= 108 \\ L + 72 &= 108 \\ L &= 36 \text{ in.} \end{aligned}$$

The volume of the box is

$$\begin{aligned} V &= LWH \\ &= (36)(18)(18) \\ &= 11{,}664 \text{ in.}^3 \end{aligned}$$

43. The two angles are supplementary, so the sum of their measures is 180°.

$$\begin{aligned} (x+1) + (4x - 56) &= 180 \\ 5x - 55 &= 180 \\ 5x &= 235 \\ x &= 47 \end{aligned}$$

Since $x = 47$, $x + 1 = 47 + 1 = 48$, and $4x - 56 = 4(47) - 56 = 132$.

The measures of the angles are 48° and 132°.

45. The two angles are vertical angles, which have equal measures. Set their measures equal to each other and solve for x.

$$\begin{aligned} 5x - 129 &= 2x - 21 \\ 3x - 129 &= -21 \\ 3x &= 108 \\ x &= 36 \end{aligned}$$

Since $x = 36$, $5x - 129 = 5(36) - 129 = 51$, and $2x - 21 = 2(36) - 21 = 51$.

The measure of each angle is 51°.

47. The angles are vertical angles, so their measures are equal.

$$\begin{aligned} 12x - 3 &= 10x + 15 \\ 2x - 3 &= 15 \\ 2x &= 18 \\ x &= 9 \end{aligned}$$

Since $x = 9$, $12x - 3 = 12(9) - 3 = 105$, and $10x + 15 = 10(9) + 15 = 105$.

The measure of each angle is 105°.

49. $d = rt$ for t

$$\begin{aligned} d &= rt \\ \frac{d}{r} &= \frac{rt}{r} \quad \text{Divide by } r. \\ \frac{d}{r} &= t \quad \text{or} \quad t = \frac{d}{r} \end{aligned}$$

51. $V = LWH$ for H

$$\begin{aligned} V &= LWH \\ \frac{V}{LW} &= \frac{LWH}{LW} \quad \text{Divide by } LW. \\ \frac{V}{LW} &= H \quad \text{or} \quad H = \frac{V}{LW} \end{aligned}$$

53. $P = a + b + c$ for b

$$\begin{aligned} P - a - c &= a + b + c - a - c \\ &\qquad \text{Subtract } a \text{ and } c. \\ P - a - c &= b \quad \text{or} \quad b = P - a - c \end{aligned}$$

55. $I = prt$ for r

$$\begin{aligned} I &= prt \\ \frac{I}{pt} &= \frac{prt}{pt} \quad \text{Divide by } pt. \\ \frac{I}{pt} &= r \quad \text{or} \quad r = \frac{I}{pt} \end{aligned}$$

64 Chapter 2 Equations, Inequalities, and Applications

57. $A = \frac{1}{2}bh$ for h

$$2A = 2\left(\frac{1}{2}bh\right) \quad \text{Multiply by 2.}$$
$$2A = bh$$
$$\frac{2A}{b} = \frac{bh}{b} \quad \text{Divide by } b.$$
$$\frac{2A}{b} = h \quad \text{or} \quad h = \frac{2A}{b}$$

59. $P = 2L + 2W$ for W

$$P - 2L = 2L + 2W - 2L \quad \text{Subtract } 2L.$$
$$P - 2L = 2W$$
$$\frac{P - 2L}{2} = \frac{2W}{2} \quad \text{Divide by 2.}$$
$$\frac{P - 2L}{2} = W \quad \text{or} \quad W = \frac{P}{2} - L$$

61. $V = \frac{1}{3}\pi r^2 h$ for h

$$3V = 3\left(\frac{1}{3}\right)\pi r^2 h \quad \text{Multiply by 3.}$$
$$3V = \pi r^2 h$$
$$\frac{3V}{\pi r^2} = \frac{\pi r^2 h}{\pi r^2} \quad \text{Divide by } \pi r^2.$$
$$\frac{3V}{\pi r^2} = h \quad \text{or} \quad h = \frac{3V}{\pi r^2}$$

63. $y = mx + b$ for m

$$y - b = mx + b - b \quad \text{Subtract } b.$$
$$y - b = mx$$
$$\frac{y - b}{x} = \frac{mx}{x} \quad \text{Divide by } x.$$
$$\frac{y - b}{x} = m \quad \text{or} \quad m = \frac{y - b}{x}$$

65. $M = C(1 + r)$ for r

$$M = C + Cr \quad \text{Distributive Property}$$
$$M - C = Cr \quad \text{Subtract } C.$$
$$\frac{M - C}{C} = \frac{Cr}{C} \quad \text{Divide by } C.$$
$$\frac{M - C}{C} = r \quad \text{or} \quad r = \frac{M}{C} - 1$$

2.6 Ratio, Proportion, and Percent

2.6 Margin Exercises

1. (a) The ratio of 9 women to 5 women is

$$\frac{9 \text{ women}}{5 \text{ women}} = \frac{9}{5}.$$

(b) To find the ratio of 4 inches to 1 foot, first convert 1 foot to 12 inches. The ratio of 4 inches to 1 foot is then

$$\frac{4 \text{ inches}}{1 \text{ foot}} = \frac{4 \text{ inches}}{12 \text{ inches}} = \frac{4}{12} = \frac{1}{3}.$$

2. The results in the following table are rounded to the nearest thousandth.

Size	Unit Cost (dollars per oz)
36-oz	$\frac{\$3.89}{36} = \$.108$ (∗)
24-oz	$\frac{\$2.79}{24} = \$.116$
12-oz	$\frac{\$1.89}{12} = \$.158$

Because the 36-oz size produces the lowest unit cost, it is the best buy. The unit cost, to the nearest thousandth, is $.108 per oz.

3. (a) $\dfrac{y}{6} = \dfrac{35}{42}$

$$42y = 6 \cdot 35 \quad \text{Cross products}$$
$$y = \frac{6 \cdot 35}{42} \quad \text{Divide by 42.}$$
$$= \frac{6 \cdot 5 \cdot 7}{6 \cdot 7} \quad \text{Factor.}$$
$$= 5 \quad \text{Cancel.}$$

Note: We could have multiplied $6 \cdot 35$ to get 210 and then divided 210 by 42 to get 5. This may be the best approach if you are doing these calculations on a calculator. The factor and cancel method is preferable if you're not using a calculator.

(b) $\dfrac{a}{24} = \dfrac{15}{16}$

$$16a = 24 \cdot 15 \quad \text{Cross products}$$
$$a = \frac{24 \cdot 15}{16} \quad \text{Divide by 16.}$$
$$= \frac{3 \cdot 8 \cdot 15}{2 \cdot 8} \quad \text{Factor.}$$
$$= \frac{45}{2} \quad \text{Cancel.}$$

4. (a) $\dfrac{z}{2} = \dfrac{z + 1}{3}$

$$3z = 2(z + 1) \quad \text{Find the cross products; use parentheses.}$$
$$3z = 2z + 2 \quad \text{Distributive property}$$
$$z = 2 \quad \text{Subtract } 2z.$$

(b) $\dfrac{p + 3}{3} = \dfrac{p - 5}{4}$

$$4(p + 3) = 3(p - 5) \quad \text{Cross products}$$
$$4p + 12 = 3p - 15$$
$$4p = 3p - 27$$
$$p = -27$$

5. Let $x =$ the cost for 16.5 gallons.

$$\frac{\$20.88}{12} = \frac{x}{16.5}$$

$$12x = 16.5(20.88) \quad \text{Cross products}$$
$$12x = 344.52 \quad \text{Multiply.}$$
$$x = 28.71 \quad \text{Divide by 12.}$$

It would cost $28.71.

6. (a) Find 20% of 70.

Here, the base b is 70, the percent p is 20, and we must find the amount a (or percentage). Substitute $b = 70$ and $p = 20$ into the percent proportion; then find a.

$$\frac{\text{amount } a}{\text{base } b} = \frac{\text{percent } p}{100}$$

$$\frac{a}{70} = \frac{20}{100}$$

$$100a = 70(20) \quad \text{Cross products}$$

$$a = \frac{70(20)}{100} \quad \text{Divide by 100.}$$

$$a = 14$$

Thus, 20% of 70 is 14.

(b) First, find the amount of discount, which is 25% of $270. Substitute $b = 270$ and $p = 25$ into the percent proportion; then find a.

$$\frac{\text{amount } a}{\text{base } b} = \frac{\text{percent } p}{100}$$

$$\frac{a}{270} = \frac{25}{100}$$

$$100a = 270(25) \quad \text{Cross products}$$

$$a = \frac{270(25)}{100} \quad \text{Divide by 100.}$$

$$a = 67.5$$

The amount of discount is $67.50. The sale price is $270.00 − $67.50 = $202.50.

7. (a) 90 is what percent of 360?

Here, the amount (or percentage) is 90, the base is 360, and we must find the percent. Substitute $a = 90$ and $b = 360$ into the percent proportion; then find p.

$$\frac{\text{amount } a}{\text{base } b} = \frac{\text{percent } p}{100}$$

$$\frac{90}{360} = \frac{p}{100}$$

$$360p = 100(90) \quad \text{Cross products}$$

$$p = \frac{100(90)}{360} \quad \text{Divide by 360.}$$

$$p = 25$$

Thus, 90 is 25% of 360.

(b) The problem may be stated as follows: "What percent of $11,000 is $682?" Substitute $a = 682$ and $b = 11{,}000$ into the percent proportion; then find p.

$$\frac{\text{amount } a}{\text{base } b} = \frac{\text{percent } p}{100}$$

$$\frac{682}{11{,}000} = \frac{p}{100}$$

$$11{,}000p = 100(682)$$

$$p = \frac{100(682)}{11{,}000}$$

$$p = 6.2$$

The interest rate was 6.2%.

2.6 Section Exercises

1. (a) 75 to 100 is $\frac{75}{100} = \frac{3}{4}$ or 3 to 4.

The answer is **C**.

(b) 5 to 4 or $\frac{5}{4} = \frac{5 \cdot 3}{4 \cdot 3} = \frac{15}{12}$ or 15 to 12.

The answer is **D**.

(c) $\frac{1}{2} = \frac{1 \cdot 50}{2 \cdot 50} = \frac{50}{100}$ or 50 to 100

The answer is **B**.

(d) 4 to 5 or $\frac{4}{5} = \frac{4 \cdot 20}{5 \cdot 20} = \frac{80}{100}$ or 80 to 100.

The answer is **A**.

3. The ratio of 60 feet to 70 feet is

$$\frac{60 \text{ feet}}{70 \text{ feet}} = \frac{60}{70} = \frac{6}{7}.$$

5. The ratio of 72 dollars to 220 dollars is

$$\frac{72 \text{ dollars}}{220 \text{ dollars}} = \frac{72}{220} = \frac{18 \cdot 4}{55 \cdot 4} = \frac{18}{55}.$$

7. First convert 8 feet to inches.

$$8 \text{ feet} = 8 \cdot 12 = 96 \text{ inches}$$

The ratio of 30 inches to 8 feet is then

$$\frac{30 \text{ inches}}{96 \text{ inches}} = \frac{30}{96} = \frac{5 \cdot 6}{16 \cdot 6} = \frac{5}{16}.$$

9. To find the ratio of 16 minutes to 1 hour, first convert 1 hour to minutes.

$$1 \text{ hour} = 60 \text{ minutes}$$

The ratio of 16 minutes to 1 hour is then

$$\frac{16 \text{ minutes}}{60 \text{ minutes}} = \frac{16}{60} = \frac{4 \cdot 4}{15 \cdot 4} = \frac{4}{15}.$$

Chapter 2 Equations, Inequalities, and Applications

In Exercises 11–18, to find the best buy, divide the price by the number of units to get the unit cost. Each result was found by using a calculator and rounding the answer to three decimal places. The *best buy* (based on price per unit) is the smallest unit cost.

11.

Size	Unit Cost (dollars per lb)
4-lb	$\frac{\$1.78}{4} = \$.445$
10-lb	$\frac{\$4.39}{10} = \$.439$ (*)

The 10-lb size is the best buy.

13.

Size	Unit Cost (dollars per oz)
16-oz	$\frac{\$2.44}{16} = \$.153$
32-oz	$\frac{\$2.98}{32} = \$.093$ (*)
48-oz	$\frac{\$4.95}{48} = \$.103$

The 32-oz size is the best buy.

15.

Size	Unit Cost (dollars per oz)
16-oz	$\frac{\$1.54}{16} = \$.096$
24-oz	$\frac{\$2.08}{24} = \$.087$
64-oz	$\frac{\$3.63}{64} = \$.057$
128-oz	$\frac{\$5.65}{128} = \$.044$ (*)

The 128-oz size is the best buy.

17.

Size	Unit Cost (dollars per oz)
14-oz	$\frac{\$1.39}{14} = \$.099$
24-oz	$\frac{\$1.55}{24} = \$.065$
36-oz	$\frac{\$1.78}{36} = \$.049$ (*)
64-oz	$\frac{\$3.99}{64} = \$.062$

The 36-oz size is the best buy.

19. A percent is a ratio where the basis of comparison is 100. For example, 27% represents the ratio of 27 to 100.

21. $\dfrac{k}{4} = \dfrac{175}{20}$

$20k = 4(175)$ Cross products

$20k = 700$

$\dfrac{20k}{20} = \dfrac{700}{20}$ Divide by 20.

$k = 35$

The solution is 35.

23. $\dfrac{49}{56} = \dfrac{z}{8}$

$56z = 49(8)$ Cross products

$56z = 392$

$\dfrac{56z}{56} = \dfrac{392}{56}$ Divide by 56.

$z = 7$

The solution is 7.

25. $\dfrac{z}{4} = \dfrac{z+1}{6}$

$6z = 4(z+1)$ Cross products

$6z = 4z + 4$ Distributive property

$2z = 4$ Subtract $4z$.

$z = 2$

The solution is 2.

27. $\dfrac{3y-2}{5} = \dfrac{6y-5}{11}$

$11(3y-2) = 5(6y-5)$ Cross products

$33y - 22 = 30y - 25$ Distributive property

$3y - 22 = -25$ Subtract $30y$.

$3y = -3$ Add 22.

$y = -1$ Divide by 3.

The solution is -1.

29. $\dfrac{5k+1}{6} = \dfrac{3k-2}{3}$

$3(5k+1) = 6(3k-2)$ Cross products

$15k + 3 = 18k - 12$ Distributive property

$-3k + 3 = -12$ Subtract $18k$.

$-3k = -15$ Subtract 3.

$k = 5$ Divide by -3.

The solution is 5.

31. $\dfrac{2p+7}{3} = \dfrac{p-1}{4}$

$4(2p+7) = 3(p-1)$ Cross products

$8p + 28 = 3p - 3$ Distributive property

$5p + 28 = -3$ Subtract $3p$.

$5p = -31$ Subtract 28.

$p = -\dfrac{31}{5}$ Divide by 5.

The solution is $-\dfrac{31}{5}$.

2.6 Ratio, Proportion, and Percent

33. Let $x =$ the cost for filling a 15-gallon tank.
Set up a proportion.

$$\frac{x \text{ dollars}}{\$11.34} = \frac{15 \text{ gallons}}{6 \text{ gallons}}$$
$$\frac{x}{11.34} = \frac{15}{6}$$
$$6x = 15(11.34)$$
$$6x = 170.1$$
$$x = 28.35$$

It would cost $28.35 to completely fill a 15-gallon tank.

35. Let $x =$ the distance between Memphis and Philadelphia on the map (in feet).

Set up a proportion with one ratio involving map distances and the other involving actual distances.

$$\frac{x \text{ feet}}{2.4 \text{ feet}} = \frac{1000 \text{ miles}}{600 \text{ miles}}$$
$$\frac{x}{2.4} = \frac{1000}{600}$$
$$600x = (2.4)(1000)$$
$$600x = 2400$$
$$x = 4$$

The distance on the map between Memphis and Philadelphia would be 4 feet.

37. Let $x =$ the number of vehicles for the 107 million U.S. households.

$$\frac{19 \text{ vehicles}}{10 \text{ households}} = \frac{x \text{ vehicles}}{107,000,000 \text{ households}}$$
$$\frac{19}{10} = \frac{x}{107,000,000}$$
$$10x = 19(107,000,000)$$
$$10x = 2,033,000,000$$
$$x = 203,300,000$$

There were 203.3 million vehicles.

39. Let $x =$ the number of fluid ounces of oil required to fill the tank.

Set up a proportion with one ratio involving the number of ounces of oil and the other involving the number of gallons of gasoline.

$$\frac{2.5 \text{ ounces}}{x \text{ ounces}} = \frac{1 \text{ gallon}}{2.75 \text{ gallons}}$$
$$\frac{2.5}{x} = \frac{1}{2.75}$$
$$x \cdot 1 = 2.5(2.75)$$
$$x = 6.875$$

To fill the tank of the chain saw, 6.875 fluid ounces of oil are required.

41. Let $x =$ the number of U.S. dollars Margaret exchanged.

Set up a proportion.

$$\frac{\$1.8329}{x \text{ dollars}} = \frac{1 \text{ pound}}{400 \text{ pounds}}$$
$$\frac{1.8329}{x} = \frac{1}{400}$$
$$x \cdot 1 = 1.8329(400)$$
$$x = 733.16$$

Margaret exchanged $733.16.

43. What is 48.6% of 19?

Here, the base is 19, the percent is 48.6 and we must find the amount (or percentage). Substitute $b = 19$ and $p = 48.6$ into the percent proportion; then find a.

$$\frac{\text{amount } a}{\text{base } b} = \frac{\text{percent } p}{100}$$
$$\frac{a}{19} = \frac{48.6}{100}$$
$$100a = 19(48.6)$$
$$100a = 923.4$$
$$a = 9.234$$

Thus, 48.6% of 19 is 9.234.

45. What percent of 48 is 96?

Substitute $a = 96$ and $b = 48$ into the percent proportion; then find p.

$$\frac{\text{amount } a}{\text{base } b} = \frac{\text{percent } p}{100}$$
$$\frac{96}{48} = \frac{p}{100}$$
$$48p = 96(100)$$
$$48p = 9600$$
$$p = 200$$

Thus, 96 is 200% of 48.

47. 12% of what number is 3600?

Here, the amount is 3600, the percent is 12, and we must find the base. Substitute $a = 3600$ and $p = 12$ into the percent proportion; then find b.

$$\frac{\text{amount } a}{\text{base } b} = \frac{\text{percent } p}{100}$$
$$\frac{3600}{b} = \frac{12}{100}$$
$$12b = 100(3600)$$
$$12b = 360,000$$
$$b = 30,000$$

Thus, 12% of 30,000 is 3600.

68 Chapter 2 Equations, Inequalities, and Applications

49. 78.84 is what percent of 292?

Substitute $a = 78.84$ and $b = 292$ into the percent proportion; then find p.

$$\frac{\text{amount } a}{\text{base } b} = \frac{\text{percent } p}{100}$$
$$\frac{78.84}{292} = \frac{p}{100}$$
$$292p = 100(78.84)$$
$$292p = 7884$$
$$p = 27$$

Thus, 78.84 is 27% of 292.

51. Mentally, think:

26% is about 25% or $\frac{1}{4}$.
4,447,000 is about 4,000,000.
$\frac{1}{4}$ of 4,000,000 is 1,000,000, or **C**.

53. The problem can be stated as follows:

"8,378,000 is what percent of 144,863,000?"

Substitute $a = 8,378,000$ and $b = 144,863,000$ into the percent proportion; then find p.

$$\frac{a}{b} = \frac{p}{100}$$
$$\frac{8,378,000}{144,863,000} = \frac{p}{100}$$
$$144,863,000p = 8,378,000(100)$$
$$p = \frac{8,378,000(100)}{144,863,000}$$
$$p \approx 5.8$$

The percent of unemployment was about 5.8%.

55. Substitute $a = 177,000$ and $b = 844,000$ into the percent proportion; then find p.

$$\frac{177,000}{844,000} = \frac{p}{100}$$
$$844,000p = 177,000(100)$$
$$p = \frac{177,000(100)}{844,000}$$
$$p \approx 21.0$$

If about 21.0% of the eateries served fast food, then about

$$100\% - 21.0\% = 79.0\%$$

were *not* fast food establishments.

57. The problem may be stated as follows:

"What is 8% of $3800?"

Substitute $b = 3800$ and $p = 8$ into the percent proportion; then find a.

$$\frac{a}{b} = \frac{p}{100}$$
$$\frac{a}{3800} = \frac{8}{100}$$
$$100a = 3800(8)$$
$$a = \frac{3800(8)}{100}$$
$$a = 304$$

The family plans to spend $304 on entertainment.

59. The increase in value was

$$\$2400 - \$625 = \$1775.$$

We can state the problem as follows:

"What percent of $625 is $1775?"

Substitute $a = 1775$ and $b = 625$ into the percent proportion; then find p.

$$\frac{a}{b} = \frac{p}{100}$$
$$\frac{1775}{625} = \frac{p}{100}$$
$$625p = 100(1775)$$
$$p = \frac{100(1775)}{625}$$
$$p = 284$$

The percent increase in the value of the coin was 284%.

61. Let $x =$ the 1995 price of electricity.

$$\frac{1991 \text{ price}}{1991 \text{ index}} = \frac{1995 \text{ price}}{1995 \text{ index}}$$
$$\frac{225}{136.2} = \frac{x}{152.4}$$
$$136.2x = 225(152.4)$$
$$x = \frac{225(152.4)}{136.2} \approx 251.76$$

The 1995 price would be about $252.

63. Let $x =$ the 2001 price of electricity.

$$\frac{1991 \text{ price}}{1991 \text{ index}} = \frac{2001 \text{ price}}{2001 \text{ index}}$$
$$\frac{225}{136.2} = \frac{x}{177.1}$$
$$136.2x = 225(177.1)$$
$$x = \frac{225(177.1)}{136.2} \approx 292.57$$

The 2001 price would be about $293.

65. $\dfrac{x}{6} = \dfrac{2}{5}$

The least common denominator of the two fractions is $6 \cdot 5 = 30$.

66. (a) $\dfrac{x}{6} = \dfrac{2}{5}$

$$30\left(\dfrac{x}{6}\right) = 30\left(\dfrac{2}{5}\right)$$
$$5x = 12$$

(b) $\dfrac{5x}{5} = \dfrac{12}{5}$

$$x = \dfrac{12}{5}$$

The solution is $\dfrac{12}{5}$.

67. $\dfrac{x}{6} = \dfrac{2}{5}$
$$5x = 2 \cdot 6$$
$$5x = 12$$
$$x = \dfrac{12}{5}, \text{ so the solution is } \dfrac{12}{5}.$$

68. The results are the same. Solving by cross products yields the same solution as multiplying by the LCD.

Summary Exercises on Solving Applied Problems

The following problems should be solved using the six-step method shown in the text. These steps will only be listed for the first solution, but all of the solutions are based on this method.

1. *Step 1*
   Find how many votes Smith received.

   Step 2
   Let $x = $ number of votes Smith received.
   Then $x + 30 = $ number of votes Nevaraz received.

   Step 3
   Since the total number of votes is 516, we have the following equation:
   $$x + (x + 30) = 516$$

   Step 4
   Solve the equation.
   $$2x + 30 = 516$$
   $$2x = 486$$
   $$x = 243$$

   Step 5
   Smith received 243 votes.

   Step 6
   If Smith received 243 votes, then Nevaraz received $243 + 30 = 273$ votes. Since $243 + 273 = 516$, the answer checks.

3. Let $x = $ the number of tanks of gas to cut 30 acres.
$$\dfrac{3 \text{ tanks}}{10 \text{ acres}} = \dfrac{x \text{ tanks}}{30 \text{ acres}}$$
$$10x = 3(30)$$
$$10x = 90$$
$$x = 9$$

9 tanks of gas would be needed to cut 30 acres of lawn.

5. Let $x = $ the length of a side. Note that $P = 4x$. From the given information, we have
$$4x = 7x - 12.$$
Solve the equation.
$$4x - 7x = 7x - 12 - 7x$$
$$-3x = -12$$
$$x = 4$$

The length of a side is 4.

7. Let $x = $ the measure of the angle. Then $90 - x = $ the measure of the complement of the angle.

From the given information, we have
$$x = (90 - x) + 70.$$
Solve the equation.
$$x = 160 - x$$
$$2x = 160$$
$$x = 80$$

The angle's measure is 80°.

9. Let $x = $ the number. Then $5x + 2$ is two added to five times the number, and $4x + 5$ is five more than four times the number.

From the given information, we have
$$5x + 2 = 4x + 5.$$
Solve the equation.
$$x + 2 = 5 \quad \text{Subtract } 4x.$$
$$x = 3 \quad \text{Subtract 2.}$$

The number is 3.

11. Let x = the smallest integer.
Then $x + 1$ = the next consecutive integer, and $x + 2$ = the largest integer.

From the given information, we have
$$x + 2(x + 2) = 4(x + 1) - 15.$$

Solve the equation.
$$x + 2x + 4 = 4x + 4 - 15$$
$$3x + 4 = 4x - 11$$
$$3x + 15 = 4x$$
$$15 = x$$

The smallest integer is 15.

13. Let x = the number of pint cartons.
Then $6x$ = the number of quart cartons.
Since 2 pints = 1 quart, 1 pint = $\frac{1}{2}$ quart.

There is a total of 39 quarts, so we have
$$\frac{1}{2}x + 6x = 39.$$

Solve the equation.
$$x + 12x = 78 \quad \textit{Multiply by 2.}$$
$$13x = 78 \quad \textit{Combine terms.}$$
$$x = 6 \quad \textit{Divide by 13.}$$

There are $6x = 36$ quart cartons.

15. The measures of the angles are equal since the angles are vertical.
$$9x - 4 = 6x + 32$$
$$9x = 6x + 36 \quad \textit{Add 4.}$$
$$3x = 36 \quad \textit{Subtract 6x.}$$
$$x = 12 \quad \textit{Divide by 3.}$$

$9x - 4 = 9(12) - 4 = 104$
$6x + 32 = 6(12) + 32 = 104$

Each angle measures 104°.

17. Let L = the length of the painting.
Use $P = 2L + 2W$ with $P = 108.44$ and $W = 29.88$.
$$P = 2L + 2W$$
$$108.44 = 2L + 2(29.88)$$
$$108.44 = 2L + 59.76$$
$$48.68 = 2L$$
$$24.34 = L$$

For these values, the length is 24.34 inches.

19. Let r = the radius.
Then $3r$ = the radius tripled.

From the given information, we have
$$3r + 8.2 = 2\pi r.$$

Use 3.14 for π and solve the equation.
$$3r + 8.2 \approx 2(3.14)r$$
$$3r + 8.2 = 6.28r$$
$$8.2 = 3.28r$$
$$\frac{8.2}{3.28} = \frac{3.28r}{3.28}$$
$$2.5 = r$$

The radius is about 2.5 cm.

21. Let x = Mrs. Silvester's age now.
Then $x + 5$ = Mr. Silvester's age now.

From the given information, we have
$$(x + 5) - 5 = \frac{4}{3}(x - 5).$$

Solve the equation.
$$x = \frac{4}{3}(x - 5)$$
$$3x = 4(x - 5) \quad \textit{Multiply by 3.}$$
$$3x = 4x - 20$$
$$20 = x$$

Mr. Silvester is $x + 5 = 25$ years old and Mrs. Silvester is 20 years old.

23. $180 - $150 = 30 discount

The problem may be stated as follows: "$30 is what percent of $180?"

Substitute $a = 30$ and $b = 180$ into the percent proportion; then find p.
$$\frac{a}{b} = \frac{p}{100}$$
$$\frac{30}{180} = \frac{p}{100}$$
$$180p = 30(100)$$
$$180p = 3000$$
$$p = 16\frac{2}{3}$$

The percent discount is $16\frac{2}{3}\%$.

25. Let x = the number of calories a 175-pound athlete can consume.
Set up a proportion with one ratio involving calories and the other involving pounds.
$$\frac{x \text{ calories}}{50 \text{ calories}} = \frac{175 \text{ pounds}}{2.2 \text{ pounds}}$$
$$2.2x = 50(175)$$
$$x = \frac{8750}{2.2} \approx 3977.3$$

To the nearest hundred calories, a 175-pound athlete in a vigorous training program can consume 4000 calories per day.

27. Let $x =$ the weight of the cherry.
Then $5x - 200 =$ the weight of the spoon. Since the total weight is 7000 lb,
$$x + (5x - 200) = 7000.$$
Solve the equation.
$$6x - 200 = 7000$$
$$6x = 7200$$
$$x = 1200$$
The cherry weighs 1200 lb and the spoon weighs $5(1200) - 200 = 5800$ lb.

29. Let $x =$ the number of calories in 12 slices of bacon.
$$\frac{2 \text{ slices}}{85 \text{ calories}} = \frac{12 \text{ slices}}{x \text{ calories}}$$
$$2x = 12(85)$$
$$2x = 1020$$
$$x = 510$$
There are 510 calories in 12 slices of bacon.

31. First find 5% of $34.56.
5% of $34.56 = .05(34.56) = 1.728 \approx 1.73$
Then add this amount to $34.56.
$$\$34.56 + \$1.73 = \$36.29$$
The price of the dinner was $36.29.
Alternatively, we could use the percent proportion
$$\frac{x}{\$34.56} = \frac{105}{100}$$
and get $x = \$36.29$.

33. Let $x =$ the number of miles traveled in 120 days.
$$\frac{2226 \text{ mi}}{103 \text{ days}} = \frac{x \text{ mi}}{120 \text{ days}}$$
$$103x = 2226(120)$$
$$x = \frac{2226(120)}{103}$$
$$x \approx 2593$$
They would have traveled about 2593 miles in 120 days.

35. The results in the following table are rounded to the nearest thousandth.

Size	Unit Cost (dollars per oz)
$15\frac{1}{2}$-oz	$\frac{\$1.19}{15.5} = \$.077$
32-oz	$\frac{\$1.69}{32} = \$.053$ (*)
48-oz	$\frac{\$2.69}{48} = \$.056$

The 32-oz size is the best buy.

2.7 Solving Linear Inequalities

2.7 Margin Exercises

1. **(a)** $x \leq 3$

   The statement $x \leq 3$ says that x can be any real number less than or equal to 3. To graph this inequality, place a solid dot at 3 on a number line and draw an arrow extending from the dot to the left.

   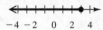

   (b) $x > -4$

   x can be any real number greater than -4, but not equal to -4. To graph this inequality, place an open circle at -4 on a number line and draw an arrow to the right.

   (c) $-4 \geq x$

   $-4 \geq x$ is the same as $x \leq -4$. Graph this inequality by placing a solid dot at -4 on a number line and drawing an arrow to the left.

   (d) $0 < x$

   $0 < x$ is the same as $x > 0$. Graph this inequality by placing an open circle at 0 on a number line and drawing an arrow to the right.

2. **(a)** $-7 < x < -2$

   To graph this inequality, place open circles at -7 and -2 on a number line. Then draw a line segment between the two circles.

   (b) $-6 < x \leq -4$

   To graph this inequality, draw an open circle at -6 and a solid dot at -4 on a number line. Then draw a line segment between the circle and the dot.

   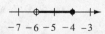

72 Chapter 2 Equations, Inequalities, and Applications

3. **(a)** $-1 + 8r < 7r + 2$
$-1 + 8r + 1 < 7r + 2 + 1$ Add 1.
$8r < 7r + 3$
$8r - 7r < 7r + 3 - 7r$ Subtract $7r$.
$r < 3$

To graph this inequality, place an open circle at 3 on a number line and draw an arrow to the left.

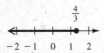
$-4\ -2\ \ 0\ \ 2\ 3\ 4$

(b) $5m - \dfrac{4}{3} \le 4m$

$m - \dfrac{4}{3} \le 0$ Subtract $4m$.

$m \le \dfrac{4}{3}$ Add $\dfrac{4}{3}$.

To graph this inequality, place a solid dot at $\dfrac{4}{3}$ and draw an arrow to the left.

$\dfrac{4}{3}$
$-2\ -1\ \ 0\ \ 1\ \ 2$

4. **(a) (i)** $-2 < 8$
$6(-2) < 6(8)$ Multiply by 6.
$-12 < 48$ True

(ii) $-2 < 8$
$-5(-2) > -5(8)$ Multiply by -5; reverse the symbol.
$10 > -40$ True

(b) (i) $-4 > -9$
$2(-4) > 2(-9)$ Multiply by 2.
$-8 > -18$ True

(ii) $-4 > -9$
$-8(-4) < -8(-9)$ Multiply by -8; reverse the symbol.
$32 < 72$ True

5. **(a)** $9x < -18$
$\dfrac{9x}{9} < \dfrac{-18}{9}$ Divide by 9.
$x < -2$

$-5\ -4\ -3\ -2\ -1\ \ 0\ \ 1$

(b) $-2r > -12$
$\dfrac{-2r}{-2} < \dfrac{-12}{-2}$ Divide by -2; reverse the symbol.
$r < 6$

$3\ 4\ 5\ 6\ 7\ 8\ 9$

(c) $-5p \le 0$
$\dfrac{-5p}{-5} \ge \dfrac{0}{-5}$ Divide by -5; reverse the symbol.
$p \ge 0$

$-3\ -2\ -1\ \ 0\ \ 1\ \ 2\ \ 3$

6. **(a)** $5r - r + 2 < 7r - 5$
$4r + 2 < 7r - 5$
$4r + 2 - 7r < 7r - 5 - 7r$ Subtract $7r$.
$-3r + 2 < -5$
$-3r + 2 - 2 < -5 - 2$ Subtract 2.
$-3r < -7$
$\dfrac{-3r}{-3} > \dfrac{-7}{-3}$ Divide by -3; reverse the symbol.
$r > \dfrac{7}{3}$

$\dfrac{7}{3}$
$-1\ \ 0\ \ 1\ \ 2\ \ 3\ \ 4\ \ 5$

(b) $4(x - 1) - 3x > -15 - (2x + 1)$
$4x - 4 - 3x > -15 - 2x - 1$ Distributive property
$x - 4 > -16 - 2x$
$x - 4 + 2x > -16 - 2x + 2x$ Add $2x$.
$3x - 4 > -16$
$3x - 4 + 4 > -16 + 4$ Add 4.
$3x > -12$
$\dfrac{3x}{3} > \dfrac{-12}{3}$ Divide by 3.
$x > -4$

$-6\ -5\ -4\ -3\ -2\ -1\ \ 0\ \ 1$

7. Let $x =$ Maggie's score on the fourth test.

The average   is at least   90.
↓   ↓   ↓
$\dfrac{98 + 86 + 88 + x}{4} \ge 90$

Solve the inequality.

$4\left(\dfrac{272 + x}{4}\right) \ge 4(90)$ Add in the numerator; multiply by 4.

$272 + x \ge 360$
$272 + x - 272 \ge 360 - 272$ Subtract 272.
$x \ge 88$ Combine terms.

She must score 88 or more on the fourth test to have an average of *at least* 90.

2.7 Section Exercises

1. x can take on any value greater than -4.
$$x > -4$$

3. x can take on any value less than or equal to 4.
$$x \leq 4$$

5. x can take on any value greater than -1 and less than 2.
$$-1 < x < 2$$

7. x can take on any value greater than -1 and less than or equal to 2.
$$-1 < x \leq 2$$

9. Use an open circle if the symbol is $>$ or $<$. Use a closed circle if the symbol is $\geq$ or $\leq$.

11. $k \leq 4$

On a number line, place a solid circle at 4 and shade everything to the left.

13. $x > -3$

This statement $x > -3$ says that x can take any value greater than -3. Place an open circle at -3 (because -3 is not part of the graph) and draw an arrow extending to the right.

15. $8 \leq x \leq 10$

On a number line, place a solid circle at 8 and a solid circle at 10 and shade everything between them.

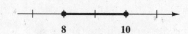

17. $0 < x \leq 10$

Place an open circle at 0 (because 0 is not part of the graph) and a solid dot at 10 (because 10 is part of the graph); then draw a line segment between the two circles.

19. $3 < x < -2$ implies that 3 is less than -2, which is false. This inequality should be written as
$$-2 < x < 3.$$

21. $z - 8 \geq -7$
$z - 8 + 8 \geq -7 + 8$ Add 8.
$z \geq 1$

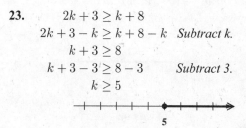

23. $2k + 3 \geq k + 8$
$2k + 3 - k \geq k + 8 - k$ Subtract k.
$k + 3 \geq 8$
$k + 3 - 3 \geq 8 - 3$ Subtract 3.
$k \geq 5$

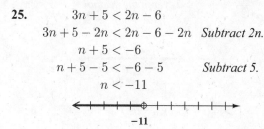

25. $3n + 5 < 2n - 6$
$3n + 5 - 2n < 2n - 6 - 2n$ Subtract 2n.
$n + 5 < -6$
$n + 5 - 5 < -6 - 5$ Subtract 5.
$n < -11$

27. When solving an inequality, the inequality symbol must be reversed when multiplying or dividing by a negative number.

29. $3x < 18$
$\dfrac{3x}{3} < \dfrac{18}{3}$ Divide by 3.
$x < 6$

31. $2x \geq -20$
$\dfrac{2x}{2} \geq \dfrac{-20}{2}$ Divide by 2.
$x \geq -10$

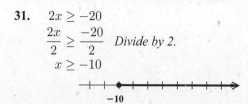

33. $-8t > 24$
$\dfrac{-8t}{-8} < \dfrac{24}{-8}$ Divide by -8; reverse the symbol from $>$ to $<$.
$t < -3$

35. $-x \geq 0$
$-1x \geq 0$
$\dfrac{-1x}{-1} \leq \dfrac{0}{-1}$ Divide by -1; reverse the symbol from $\geq$ to $\leq$.
$x \leq 0$

74 Chapter 2 Equations, Inequalities, and Applications

37. $-\dfrac{3}{4}r < -15$

$\left(-\dfrac{4}{3}\right)\left(-\dfrac{3}{4}r\right) > \left(-\dfrac{4}{3}\right)(-15)$

Multiply by $-\dfrac{4}{3}$ (the reciprocal of $-\dfrac{3}{4}$); reverse the symbol from $<$ to $>$

$r > 20.$

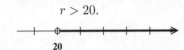

39. $-.02x \leq .06$

$\dfrac{-.02x}{-.02} \geq \dfrac{.06}{-.02}$ Divide by $-.02$; reverse the symbol from $\leq$ to $\geq$.

$x \geq -3$

41. $5r + 1 \geq 3r - 9$

$2r + 1 \geq -9$ Subtract $3r$.

$2r \geq -10$ Subtract 1.

$r \geq -5$ Divide by 2.

43. $6x + 3 + x < 2 + 4x + 4$

$7x + 3 < 4x + 6$ Combine like terms.

$3x + 3 < 6$ Subtract $4x$.

$3x < 3$ Subtract 3.

$x < 1$ Divide by 3.

45. $-x + 4 + 7x \leq -2 + 3x + 6$

$6x + 4 \leq 4 + 3x$

$3x + 4 \leq 4$

$3x \leq 0$

$x \leq 0$

47. $5(x + 3) - 6x \leq 3(2x + 1) - 4x$

$5x + 15 - 6x \leq 6x + 3 - 4x$

$-x + 15 \leq 2x + 3$

$-3x + 15 \leq 3$

$-3x \leq -12$

$\dfrac{-3x}{-3} \geq \dfrac{-12}{-3}$ Divide by -3; reverse the symbol.

$x \geq 4$

49. $\dfrac{2}{3}(p + 3) > \dfrac{5}{6}(p - 4)$

$6\left(\dfrac{2}{3}\right)(p + 3) > 6\left(\dfrac{5}{6}\right)(p - 4)$

Multiply by 6, the LCD.

$4(p + 3) > 5(p - 4)$

$4p + 12 > 5p - 20$

$-p + 12 > -20$

$-p > -32$

$\dfrac{-p}{-1} < \dfrac{-32}{-1}$ Divide by -1; reverse the symbol.

$p < 32$

51. $4x - (6x + 1) \leq 8x + 2(x - 3)$

$4x - 6x - 1 \leq 8x + 2x - 6$

$-2x - 1 \leq 10x - 6$

$-12x - 1 \leq -6$

$-12x \leq -5$

$\dfrac{-12x}{-12} \geq \dfrac{-5}{-12}$ Divide by -12; reverse the symbol.

$x \geq \dfrac{5}{12}$

53. $5(2k + 3) - 2(k - 8) > 3(2k + 4) + k - 2$

$10k + 15 - 2k + 16 > 6k + 12 + k - 2$

$8k + 31 > 7k + 10$

$k + 31 > 10$

$k > -21$

55. Let $x =$ the score on the third test.

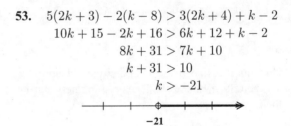

$\dfrac{84 + 98 + x}{3} \geq 90$

$3\left(\dfrac{84 + 98 + x}{3}\right) \geq 3(90)$

$84 + 98 + x \geq 270$

$182 + x \geq 270$

$x \geq 88$

John must score 88 or more on his third test so that his average is at least 90.

57. Let n = the number.
"When 2 is added to the difference between six times a number and 5, the result is greater than 13 added to five times the number" translates to

$$(6n - 5) + 2 > 5n + 13.$$

Solve the inequality.

$$6n - 5 + 2 > 5n + 13$$
$$6n - 3 > 5n + 13$$
$$n - 3 > 13 \quad \text{Subtract } 5n.$$
$$n > 16 \quad \text{Add 3.}$$

All numbers greater than 16 satisfy the given condition.

59. The Celsius temperature must give a Fahrenheit temperature that is less than or equal to 104 degrees.

$$F = \frac{9}{5}C + 32 \leq 104$$
$$\frac{9}{5}C \leq 72$$
$$\frac{5}{9}\left(\frac{9}{5}C\right) \leq \frac{5}{9}(72)$$
$$C \leq 40$$

The temperature of Providence, Rhode Island has never exceeded 40° Celsius.

61. $P = 2L + 2W; P \geq 400$
From the figure, we have $L = 4x + 3$ and $W = x + 37$. Thus, we have the inequality

$$2(4x + 3) + 2(x + 37) \geq 400.$$

Solve this inequality.

$$8x + 6 + 2x + 74 \geq 400$$
$$10x + 80 \geq 400$$
$$10x \geq 320$$
$$x \geq 32$$

The rectangle will have a perimeter of at least 400 if the value of x is 32 or greater.

63. He can spend at most $5.60.

$$2 + .30x \leq 5.60$$
$$2 + .30x - 2 \leq 5.60 - 2$$
$$.30x \leq 3.60$$
$$\frac{.30x}{.30} \leq \frac{3.60}{.30}$$
$$x \leq 12$$

Since x represents the length of the call after the first three minutes, he can use the phone a maximum of $12 + 3 = 15$ min.

65. $3x + 2 = 14$
$3x = 12$
$x = 4$

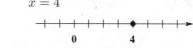

66. $3x + 2 < 14$
$3x < 12$
$x < 4$

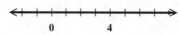

67. $3x + 2 > 14$
$3x > 12$
$x > 4$

68. The graph would include all points on the number line.

It is the set of all real numbers.

69. If you were to graph the solutions of

$$-4x + 3 = -1,$$
$$-4x + 3 > -1,$$
$$\text{and } -4x + 3 < -1,$$

the graph would also be the complete number line, that is, all real numbers.

Chapter 2 Review Exercises

In Exercises 1–10, all solutions should be checked by substituting into the original equation. Checks will be shown here for only a few of the exercises.

1. $x - 7 = 2$
$x - 7 + 7 = 2 + 7 \quad \text{Add 7.}$
$x = 9$

Check $x = 9$: $\cdot 2 = 2$

The solution is 9.

2. $4r - 6 = 10$
$4r = 16 \quad \text{Add 6.}$
$\frac{4r}{4} = \frac{16}{4} \quad \text{Divide by 4.}$
$r = 4$

Check $r = 4$: $10 = 10$

The solution is 4.

3. $5x + 8 = 4x + 2$
    $x + 8 = 2$ Subtract $4x$.
    $x = -6$ Subtract 8.

    Check $x = -6$: $-22 = -22$

    The solution is -6.

4. $8t = 7t + \dfrac{3}{2}$

    $t = \dfrac{3}{2}$ Subtract $7t$.

    Check $t = \frac{3}{2}$: $12 = 12$

    The solution is $\frac{3}{2}$.

5. $(4r - 8) - (3r + 12) = 0$
    $1(4r - 8) - 1(3r + 12) = 0$
    $4r - 8 - 3r - 12 = 0$
    $r - 20 = 0$
    $r = 20$

    Check $r = 20$: $0 = 0$

    The solution is 20.

6. $7(2x + 1) = 6(2x - 9)$
    $14x + 7 = 12x - 54$
    $2x + 7 = -54$ Subtract $12x$.
    $2x = -61$ Subtract 7.
    $x = -\dfrac{61}{2}$ Divide by 2.

    Check $x = -\frac{61}{2}$: $-420 = -420$

    The solution is $-\frac{61}{2}$.

7. $-\dfrac{6}{5}y = -18$

    $\left(-\dfrac{5}{6}\right)\left(-\dfrac{6}{5}y\right) = \left(-\dfrac{5}{6}\right)(-18)$

    $y = 15$

    Check $y = 15$: $-18 = -18$

    The solution is 15.

8. $\dfrac{1}{2}r - \dfrac{1}{6}r + 3 = 2 + \dfrac{1}{6}r + 1$

    The least common denominator is 6.

    $6\left(\dfrac{1}{2}r - \dfrac{1}{6}r + 3\right) = 6\left(2 + \dfrac{1}{6}r + 1\right)$

    $3r - r + 18 = 12 + r + 6$
    $2r + 18 = 18 + r$
    $r + 18 = 18$
    $r = 0$

    Check $r = 0$: $3 = 3$

    The solution is 0.

9. $3x - (-2x + 6) = 4(x - 4) + x$
    $3x + 2x - 6 = 4x - 16 + x$
    $5x - 6 = 5x - 16$
    $-6 = -16$

    This statement is *false*, so there is no solution.

10. $.10(x + 80) + .20x = 8 + .30x$

    To clear decimals, multiply both sides by 10.

    $10[.10(x + 80) + .20x] = 10(8 + .30x)$
    $1(x + 80) + 2x = 80 + 3x$
    $x + 80 + 2x = 80 + 3x$
    $3x + 80 = 80 + 3x$
    $80 = 80$

    Since this is a *true* statement, all real numbers satisfy this equation.

11. Let x represent the number.

    $5x + 7 = 3x$
    $7 = -2x$ Subtract $5x$.
    $-\dfrac{7}{2} = x$ Divide by -2.

    The number is $-\frac{7}{2}$.

12. Let x represent the number.

    $2x - 4 = 36$
    $2x = 40$
    $x = 20$

    The number is 20.

13. *Step 2*
    Let $x =$ the land area of Rhode Island.
    Then $x + 5213 =$ land area of Hawaii.

    Step 3
    The areas total 7637 square miles, so
    $$x + (x + 5213) = 7637.$$

    Step 4
    $$2x + 5213 = 7637$$
    $$2x = 2424$$
    $$x = 1212$$

    Step 5
    Since $x = 1212$, $x + 5213 = 6425$. The land area of Rhode Island is 1212 square miles and that of Hawaii is 6425 square miles.

    Step 6
    The land area of Hawaii is 5213 square miles greater than the land area of Rhode Island and the total is 7637 square miles.

14. *Step 2*
Let $x =$ the height of Twin Falls.
Then $\frac{5}{2}x =$ the height of Seven Falls.

Step 3
The sum of the heights is 420 feet, so
$$x + \frac{5}{2}x = 420.$$

Step 4
$$2\left(x + \frac{5}{2}x\right) = 2(420)$$
$$2x + 5x = 840$$
$$7x = 840$$
$$x = 120$$

Step 5
Since $x = 120$, $\frac{5}{2}x = \frac{5}{2}(120) = 300$. The height of Twin Falls is 120 feet and that of Seven Falls is 300 feet.

Step 6
The height of Seven Falls is $\frac{5}{2}$ the height of Twin Falls and the sum is 420.

15. *Step 2*
Let $x =$ the measure of the angle.
Then $90 - x =$ the measure of its complement and $180 - x =$ the measure of its supplement.

Step 3 $\quad 180 - x = 10(90 - x)$

Step 4
$$180 - x = 900 - 10x$$
$$9x + 180 = 900$$
$$9x = 720$$
$$x = 80$$

Step 5
The measure of the angle is 80°.
Its complement measures $90° - 80° = 10°$, and its supplement measures $180° - 80° = 100°$.

Step 6
The measure of the supplement is 10 times the measure of the complement.

16. *Step 2*
Let $x =$ smaller odd integer.
Then $x + 2 =$ larger odd integer.

Step 3 $\quad x + 2(x + 2) = (x + 2) + 24$

Step 4
$$x + 2x + 4 = x + 26$$
$$3x + 4 = x + 26$$
$$2x + 4 = 26$$
$$2x = 22$$
$$x = 11$$

Step 5
Since $x = 11$, $x + 2 = 13$. The consecutive odd integers are 11 and 13.

Step 6
The smaller plus twice the larger is $11 + 2(13) = 37$, which is 24 more than the larger.

In Exercises 17–20, substitute the given values into the given formula and then solve for the remaining variable.

17. $A = \frac{1}{2}bh$; $A = 44$, $b = 8$
$$A = \frac{1}{2}bh$$
$$44 = \frac{1}{2}(8)h$$
$$44 = 4h$$
$$11 = h$$

18. $A = \frac{1}{2}h(b + B)$; $b = 3$, $B = 4$, $h = 8$
$$A = \frac{1}{2}h(b + B)$$
$$A = \frac{1}{2}(8)(3 + 4)$$
$$= \frac{1}{2}(8)(7)$$
$$= (4)(7)$$
$$A = 28$$

19. $C = 2\pi r$; $C = 29.83$, $\pi \approx 3.14$
$$C = 2\pi r$$
$$29.83 \approx 2(3.14)r$$
$$29.83 = 6.28r$$
$$\frac{29.83}{6.28} = \frac{6.28r}{6.28}$$
$$4.75 = r$$

20. $V = \frac{4}{3}\pi r^3$; $r = 9$, $\pi \approx 3.14$

$$V = \frac{4}{3}\pi r^3$$
$$\approx \frac{4}{3}(3.14)(9)^3$$
$$= \frac{4}{3}(3.14)(729)$$
$$= \frac{4}{3}(2289.06)$$
$$V = 3052.08$$

21. $A = LW$ for W

$$\frac{A}{L} = \frac{LW}{L} \quad \text{Divide by } L.$$
$$\frac{A}{L} = W \quad \text{or} \quad W = \frac{A}{L}$$

22. $A = \frac{1}{2}h(b+B)$ for h

$$2A = 2\left[\frac{1}{2}h(b+B)\right] \quad \text{Multiply by 2.}$$
$$2A = h(b+B)$$
$$\frac{2A}{(b+B)} = \frac{h(b+B)}{(b+B)} \quad \text{Divide by } b+B.$$
$$\frac{2A}{b+B} = h \quad \text{or} \quad h = \frac{2A}{b+B}$$

23. Because the two angles are supplementary,
$$(8x - 1) + (3x - 6) = 180.$$
$$11x - 7 = 180$$
$$11x = 187$$
$$x = 17$$

Since $x = 17$, $8x - 1 = 135$, and $3x - 6 = 45$.

The measures of the two angles are 135° and 45°.

24. The angles are vertical angles, so their measures are equal.
$$3x + 10 = 4x - 20$$
$$10 = x - 20$$
$$30 = x$$

Since $x = 30$, $3x + 10 = 100$, and $4x - 20 = 100$.

Each angle has a measure of 100°.

25. The screen is a rectangle with length 92.75 feet and width 70.5 feet, so substitute these values in the formulas for the perimeter and area of a rectangle.

$$P = 2L + 2W$$
$$= 2(92.75) + 2(70.5)$$
$$= 185.5 + 141$$
$$P = 326.5$$

The perimeter is 326.5 feet.

$$A = LW$$
$$= (92.75)(70.5)$$
$$A = 6538.875$$

The area is 6538.875 square feet.

26. Use $C = \pi d$.
$$C = \pi d$$
$$146.9 \approx 3.14d$$
$$\frac{146.9}{3.14} = d$$
$$d \approx 46.78$$

The diameter of the tree is about 46.78 ft and the radius is $\frac{1}{2}(46.78) = 23.39$ ft.

27. The ratio of 60 centimeters to 40 centimeters is
$$\frac{60 \text{ cm}}{40 \text{ cm}} = \frac{3 \cdot 20}{2 \cdot 20} = \frac{3}{2}.$$

28. To find the ratio of 5 days to 2 weeks, first convert 2 weeks to days.
$$2 \text{ weeks} = 2 \cdot 7 = 14 \text{ days}$$

Thus, the ratio of 5 days to 2 weeks is $\frac{5}{14}$.

29. To find the ratio of 90 inches to 10 feet, first convert 10 feet to inches.
$$10 \text{ feet} = 10 \cdot 12 = 120 \text{ inches}$$

Thus, the ratio of 90 inches to 10 feet is
$$\frac{90}{120} = \frac{3 \cdot 30}{4 \cdot 30} = \frac{3}{4}.$$

30. To find the ratio of 3 months to 3 years, first convert 3 years to months.
$$3 \text{ years} = 3 \cdot 12 = 36 \text{ months}$$

Thus, the ratio of 3 months to 3 years is
$$\frac{3}{36} = \frac{1 \cdot 3}{12 \cdot 3} = \frac{1}{12}.$$

31. $\dfrac{p}{21} = \dfrac{5}{30}$

$30p = 105$ *Cross products are equal.*

$\dfrac{30p}{30} = \dfrac{105}{30}$ *Divide by 30.*

$p = \dfrac{105}{30} = \dfrac{7 \cdot 15}{2 \cdot 15} = \dfrac{7}{2}$

The solution is $\dfrac{7}{2}$.

32. $\dfrac{5+x}{3} = \dfrac{2-x}{6}$

$6(5+x) = 3(2-x)$ *Cross products are equal.*

$30 + 6x = 6 - 3x$ *Distributive property*

$30 + 9x = 6$ *Add 3x.*

$9x = -24$ *Subtract 30.*

$x = \dfrac{-24}{9} = -\dfrac{8}{3}$

The solution is $-\dfrac{8}{3}$.

33. $\dfrac{y}{5} = \dfrac{6y-5}{11}$

$11y = 5(6y - 5)$

$11y = 30y - 25$

$-19y = -25$

$y = \dfrac{-25}{-19} = \dfrac{25}{19}$

The solution is $\dfrac{25}{19}$.

34. 40% means $\dfrac{40}{100}$ or $\dfrac{2}{5}$. It is the same as the ratio of 2 to 5.

35. Let x = the number of pounds of fertilizer needed to cover 500 square feet.

$\dfrac{x \text{ pounds}}{2 \text{ pounds}} = \dfrac{500 \text{ square feet}}{150 \text{ square feet}}$

$\dfrac{x}{2} = \dfrac{500}{150}$

$150x = 2(500)$

$x = \dfrac{1000}{150} = \dfrac{20 \cdot 50}{3 \cdot 50}$

$= \dfrac{20}{3}$ or $6\dfrac{2}{3}$

$6\dfrac{2}{3}$ pounds of fertilizer will cover 500 square feet.

36. Let x = the number of ounces of medicine to mix with 90 ounces of water.

$\dfrac{x \text{ ounces of medicine}}{8 \text{ ounces of medicine}} = \dfrac{90 \text{ ounces of water}}{20 \text{ ounces of water}}$

$\dfrac{x}{8} = \dfrac{90}{20}$

$20x = 720$

$x = 36$

36 ounces of medicine should be mixed with 90 ounces of water.

37. Let x = the number of fish in Willow Lake.

Set up a proportion with one ratio involving the sample and the other involving the total number of fish.

$\dfrac{7 \text{ tagged fish}}{700 \text{ fish}} = \dfrac{500 \text{ tagged fish}}{x \text{ fish}}$

$\dfrac{7}{700} = \dfrac{500}{x}$

$7x = 700 \cdot 500$

$7x = 350,000$

$x = 50,000$

We estimate that there are 50,000 fish in Willow Lake.

38. Let x = the actual distance between the second pair of cities (in kilometers).

Set up a proportion with one ratio involving map distances and the other involving actual distances.

$\dfrac{x \text{ kilometers}}{150 \text{ kilometers}} = \dfrac{80 \text{ centimeters}}{32 \text{ centimeters}}$

$\dfrac{x}{150} = \dfrac{80}{32}$

$32x = (150)(80) = 12,000$

$x = \dfrac{12,000}{32} = 375$

The cities are 375 kilometers apart.

39. What is 23% of 76?

Substitute $b = 76$ and $p = 23$ into the percent proportion; then find a.

$\dfrac{a}{b} = \dfrac{p}{100}$

$\dfrac{a}{76} = \dfrac{23}{100}$

$100a = 76(23)$ *Cross multiply.*

$a = \dfrac{76(23)}{100}$

$a = 17.48$

Thus, 17.48 is 23% of 76.

40. What percent of 12 is 21?

Substitute $a = 21$ and $b = 12$ into the percent proportion; then find p.

$\dfrac{21}{12} = \dfrac{p}{100}$

$12p = 100(21)$

$p = \dfrac{100(21)}{12}$

$p = 175$

Thus, 21 is 175% of 12.

80   Chapter 2 Equations, Inequalities, and Applications

41. 6 is what percent of 18?

Substitute $a = 6$ and $b = 18$ into the percent proportion; then find p.

$$\frac{a}{b} = \frac{p}{100}$$
$$\frac{6}{18} = \frac{p}{100}$$
$$18p = 100(6)$$
$$p = \frac{100(6)}{18}$$
$$p = 33\frac{1}{3}$$

Thus, 6 is $33\frac{1}{3}\%$ of 18.

42. 36% of what number is 900?

Substitute $a = 900$ and $p = 36$ into the percent proportion; then find b.

$$\frac{a}{b} = \frac{p}{100}$$
$$\frac{900}{b} = \frac{36}{100}$$
$$36b = 100(900)$$
$$b = \frac{100(900)}{36}$$
$$b = 2500$$

Thus, 900 is 36% of 2500.

43. The $18,690 represents 105% of the actual price of the car. Substitute $a = 18{,}690$ and $p = 105$ into the percent proportion; then find b.

$$\frac{18{,}690}{b} = \frac{105}{100}$$
$$105b = 18{,}690(100)$$
$$b = \frac{18{,}690(100)}{105}$$
$$b = 17{,}800$$

The actual price of the car was $17,800.

44. The total price represents 115% of the bill. Substitute $b = 304.75$ and $p = 115$ into the percent proportion; then find a.

$$\frac{a}{304.75} = \frac{115}{100}$$
$$100a = 304.75(115)$$
$$a = \frac{304.75(115)}{100}$$
$$a = 350.4625$$

The total price (including a 15% tip) was $350.46.

45. $p \geq -4$

Place a solid dot at -4 (to show that -4 is part of the graph) and draw an arrow extending to the right.

46. $x < 7$

Place an open circle at 7 (to show that 7 is not part of the graph) and draw an arrow extending to the left.

47. $-5 \leq k < 6$

Place a solid dot at -5 and an open circle at 6; then draw a line segment between them.

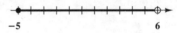

48. $r \geq \frac{1}{2}$

Place a solid dot at $\frac{1}{2}$ and draw an arrow extending to the right.

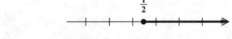

49.
$$x + 6 \geq 3$$
$$x + 6 - 6 \geq 3 - 6 \quad \text{Subtract 6.}$$
$$x \geq -3$$

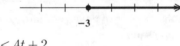

50.
$$5t < 4t + 2$$
$$t < 2 \quad \text{Subtract } 4t.$$

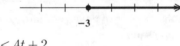

51.
$$-6x \leq -18$$
$$\frac{-6x}{-6} \geq \frac{-18}{-6} \quad \text{Divide by } -6;\ \text{reverse the symbol.}$$
$$x \geq 3$$

52.
$$8(k - 5) - (2 + 7k) \geq 4$$
$$8k - 40 - 2 - 7k \geq 4$$
$$k - 42 \geq 4$$
$$k \geq 46$$

53. $4x - 3x > 10 - 4x + 7x$
    $x > 10 + 3x$
    $-2x > 10$
    $\dfrac{-2x}{-2} < \dfrac{10}{-2}$ *Divide by –2; reverse the symbol.*
    $x < -5$

    <--+--+--+--o--+--+-->
              -5

54. $3(2w + 5) + 4(8 + 3w) < 5(3w + 2) + 2w$
    $6w + 15 + 32 + 12w < 15w + 10 + 2w$
    $18w + 47 < 17w + 10$
    $w + 47 < 10$
    $w < -37$

    <--+--+--o--+--+--+-->
          -37

55. Let $x =$ the score on the third test.

The average of the three tests	is at least	90.
↓	↓	↓
$\dfrac{94 + 88 + x}{3}$	$\geq$	90

    $\dfrac{182 + x}{3} \geq 90$

    $3\left(\dfrac{182 + x}{3}\right) \geq 3(90)$

    $182 + x \geq 270$
    $x \geq 88$

    In order to average at least 90, Justin's score on his third test must be 88 or more.

56. Let $n =$ the number.
    "If nine times a number is added to 6, the result is at most 3" can be written as

    $9n + 6 \leq 3.$

    Solve the inequality.

    $9n \leq -3$ *Subtract 6.*
    $n \leq \dfrac{-3}{9}$ *Divide by 9.*

    All numbers less or equal to $-\frac{1}{3}$ satisfy the given condition.

57. [2.6] $\dfrac{y}{7} = \dfrac{y - 5}{2}$
    $2y = 7(y - 5)$ *Cross products are equal.*
    $2y = 7y - 35$
    $-5y = -35$
    $y = 7$

    The solution is 7.

58. [2.5] $C = \pi d$ for d
    $\dfrac{C}{\pi} = \dfrac{\pi d}{\pi}$
    $\dfrac{C}{\pi} = d$ or $d = \dfrac{C}{\pi}$

59. [2.7] $-2x > -4$
    $\dfrac{-2x}{-2} < \dfrac{-4}{-2}$ *Divide by –2; reverse the symbol.*
    $x < 2$

60. [2.2] $2k - 5 = 4k + 13$
    $-2k - 5 = 13$ *Subtract 4k.*
    $-2k = 18$ *Add 5.*
    $k = -9$ *Divide by –2.*

    The solution is -9.

61. [2.3] $.05x + .02x = 4.9$
    To clear decimals, multiply both sides by 100.

    $100(.05x + .02x) = 100(4.9)$
    $5x + 2x = 490$
    $7x = 490$
    $x = 70$

    The solution is 70.

62. [2.3] $2 - 3(t - 5) = 4 + t$
    $2 - 3t + 15 = 4 + t$
    $17 - 3t = 4 + t$
    $17 - 4t = 4$
    $-4t = -13$
    $t = \dfrac{-13}{-4} = \dfrac{13}{4}$

    The solution is $\frac{13}{4}$.

63. [2.3] $9x - (7x + 2) = 3x + (2 - x)$
    $9x - 7x - 2 = 3x + 2 - x$
    $2x - 2 = 2x + 2$
    $-2 = 2$

    Because $-2 = 2$ is a *false* statement, the given equation has no solution.

64. [2.3] $\dfrac{1}{3}s + \dfrac{1}{2}s + 7 = \dfrac{5}{6}s + 5 + 2$
    $\dfrac{1}{3}s + \dfrac{1}{2}s = \dfrac{5}{6}s$ *Subtract 7.*

    The least common denominator is 6.

    $6\left(\dfrac{1}{3}s + \dfrac{1}{2}s\right) = 6\left(\dfrac{5}{6}s\right)$
    $2s + 3s = 5s$
    $5s = 5s$
    $0 = 0$

    Because $0 = 0$ is a *true* statement, the solution is all real numbers.

65. [2.6] Let $x =$ the number of fish in Argyle Lake.

Set up a proportion with one ratio involving the sample and the other involving the total number of fish.

$$\frac{18 \text{ tagged fish}}{1000 \text{ fish}} = \frac{840 \text{ tagged fish}}{x \text{ fish}}$$

$$\frac{18}{1000} = \frac{840}{x}$$

$$18x = (1000)(840)$$

$$18x = 840{,}000$$

$$x = 46{,}666.\overline{6}$$

The approximate fish population of Argyle Lake, to the nearest hundred, is 46,700 fish.

66. [2.4] Let $x =$ the number.

Two-thirds of a number	added to	the number	is	10.
↓	↓	↓	↓	↓
$\frac{2}{3}x$	$+$	x	$=$	10

Solve the equation.

$$\frac{2}{3}x + \frac{3}{3}x = 10$$

$$\frac{5}{3}x = 10$$

$$\frac{3}{5}\left(\frac{5}{3}x\right) = \frac{3}{5}(10)$$

$$x = 6$$

The number is 6.

67. [2.4] Let $x =$ the number of medals earned by Canada. Then $2x =$ the number of medals earned by the United States.

The total number of medals was 51.

$$x + 2x = 51$$

Solve the equation.

$$3x = 51$$

$$x = 17$$

Canada earned 17 medals and the United States earned $2(17) = 34$ medals.

68. [2.4] Let $x =$ the number of gold medals. Then $x + 9 =$ the number of silver medals, and $(x+9) - 8 = x + 1 =$ the number of bronze medals.

Since there were 58 medals total, we have the equation

$$x + (x+9) + (x+1) = 58.$$

Solve the equation.

$$x + x + 9 + x + 1 = 58$$

$$3x + 10 = 58$$

$$3x = 48$$

$$x = 16$$

There were 16 gold medals, $16 + 9 = 25$ silver medals, and $16 + 1 = 17$ bronze medals.

69. [2.5] Let $x =$ the length of the first side. Then $2x =$ the length of the second side.

Use the formula for the perimeter of a triangle, $P = a + b + c$, with perimeter 96 and third side 30.

$$x + 2x + 30 = 96$$

$$3x + 30 = 96$$

$$3x = 66$$

$$x = 22$$

The sides have lengths 22 meters, 44 meters, and 30 meters. The length of the longest side is 44 meters.

70. [2.5] Let W represent the width of the rectangle. Then the length L is $W + 4$. The formula for the perimeter of a rectangle is $P = 2L + 2W$.

The perimeter	is	288.
↓	↓	↓
$2(W+4) + 2W$	$=$	288

$$2W + 8 + 2W = 288$$

$$4W + 8 = 288$$

$$4W = 280$$

$$W = 70$$

The width is 70 feet.

71. [2.6]

Size	Unit Cost (dollars per oz)
87-oz	$\frac{\$7.88}{87} = \$.091$
131-oz	$\frac{\$10.98}{131} = \$.084$
263-oz	$\frac{\$19.96}{263} = \$.076$ (∗)

The 263-oz size is the best buy.

72. [2.6] Set up a proportion.

$$\frac{\$121.50}{9 \text{ pairs}} = \frac{\$x}{5 \text{ pairs}}$$

$$\frac{121.50}{9} = \frac{x}{5}$$

$$9x = 5(121.50)$$

$$x = \frac{5(121.50)}{9}$$

$$x = 67.5$$

Five pairs of jeans cost $67.50.

73. **[2.7]** Let $x =$ Latarsha's grade on her third test.

$$\frac{82 + 96 + x}{3} \geq 90$$

$$\frac{178 + x}{3} \geq 90$$

$$3\left(\frac{178 + x}{3}\right) \geq 3(90)$$

$$178 + x \geq 270$$

$$x \geq 92$$

Latarsha must make 92 or more on the third test so that her average will be at least 90.

74. **[2.5]** The angles are vertical angles, so their measures are equal.

$$3(4x + 1) = 10x + 11$$
$$12x + 3 = 10x + 11$$
$$2x + 3 = 11$$
$$2x = 8$$
$$x = 4$$

If $x = 4$, $3(4x + 1) = 3(17) = 51$, and $10x + 11 = 40 + 11 = 51$.

The measure of each angle is 51°.

Chapter 2 Test

1. $3x - 7 = 11$
$\quad\ \ 3x = 18 \quad$ *Add 7.*
$\quad\ \ \ \ x = 6 \quad$ *Divide by 3.*

Check $x = 6$: $11 = 11$

The solution is 6.

2. $\quad 5x + 9 = 7x + 21$
$\quad -2x + 9 = 21 \qquad$ *Subtract 7x.*
$\quad\ -2x = 12 \qquad$ *Subtract 9.*
$\qquad\ \ x = -6 \qquad$ *Divide by –2.*

Check $x = -6$: $-21 = -21$

The solution is -6.

3. $2 - 3(x - 5) = 3 + (x + 1)$
$\quad 2 - 3x + 15 = 3 + x + 1$
$\quad -3x + 17 = x + 4$
$\quad -4x + 17 = 4 \qquad$ *Subtract x.*
$\quad -4x = -13 \qquad$ *Subtract 17.*
$\qquad x = \frac{13}{4} \qquad$ *Divide by –4.*

Check $x = \frac{13}{4}$: $\frac{29}{4} = \frac{29}{4}$

The solution is $\frac{13}{4}$.

4. $2.3x + 13.7 = 1.3x + 2.9$
$\quad x + 13.7 = 2.9 \qquad$ *Subtract 1.3x.*
$\quad\ \ \ \ x = -10.8 \qquad$ *Subtract 13.7.*

Check $x = -10.8$: $-11.14 = -11.14$

The solution is -10.8.

5. $7 - (m - 4) = -3m + 2(m + 1)$
$\quad 7 - m + 4 = -3m + 2m + 2$
$\quad -m + 11 = -m + 2$
$\qquad\ \ 11 = 2$

Because the last statement is *false*, the equation has no solution.

6. $\qquad -\frac{4}{7}x = -12$

$\quad \left(-\frac{7}{4}\right)\left(-\frac{4}{7}x\right) = \left(-\frac{7}{4}\right)(-12)$

$\qquad\qquad x = 21$

Check $x = 21$: $-12 = -12$

The solution is 21.

7. $.06(x + 20) + .08(x - 10) = 4.6$

To clear decimals, multiply both sides by 100.

$100[.06(x + 20) + .08(x - 10)] = 100(4.6)$
$\quad 6(x + 20) + 8(x - 10) = 460$
$\quad 6x + 120 + 8x - 80 = 460$
$\quad 14x + 40 = 460$
$\quad 14x = 420$
$\quad\ \ x = 30$

Check $x = 30$: $4.6 = 4.6$

The solution is 30.

8. $-8(2x + 4) = -4(4x + 8)$
$\quad -16x - 32 = -16x - 32$
$\qquad -32 = -32$

Because the last statement is *true*, the solution is all real numbers.

9. **(a)** Let $x =$ the number of points for the Nets. Then $x + 11 =$ the number of points for the Spurs.

Since the total score was 165, we have

$$x + (x + 11) = 165.$$

Solve the equation.

$$2x + 11 = 165$$
$$2x = 154$$
$$x = 77$$

The Nets had 77 points and the Spurs had $77 + 11 = 88$ points.

(b) Let $x =$ the number of field goals for Kidd. Then $x + 10 =$ the number of field goals for Duncan.

Since the total number of field goals was 98, we have
$$x + (x + 10) = 98.$$

Solve the equation.
$$2x + 10 = 98$$
$$2x = 88$$
$$x = 44$$

Kidd made 44 field goals.

10. Let $x =$ the area of Kauai (in square miles). Then $x + 177 =$ the area of Maui (in square miles), and $(x + 177) + 3293 = x + 3470 =$ the area of Hawaii.

$$x + (x + 177) + (x + 3470) = 5300$$
$$3x + 3647 = 5300$$
$$3x = 1653$$
$$x = 551$$

Since $x = 551$, $x + 177 = 728$, and $x + 3470 = 4021$.

The area of Hawaii is 4021 square miles, the area of Maui is 728 square miles, and the area of Kauai is 551 square miles.

11. Let $x =$ the measure of the angle. Then $90 - x =$ the measure of its complement, and $180 - x =$ the measure of its supplement.

$$180 - x = 3(90 - x) + 10$$
$$180 - x = 270 - 3x + 10$$
$$180 - x = 280 - 3x$$
$$180 + 2x = 280$$
$$2x = 100$$
$$x = 50$$

The measure of the angle is 50°. The measure of its supplement, 130°, is 10° more than three times its complement, 40°.

12. **(a)** Solve $P = 2L + 2W$ for W.
$$P - 2L = 2W$$
$$\frac{P - 2L}{2} = W$$
$$W = \frac{P - 2L}{2} \quad \text{or} \quad W = \frac{P}{2} - L$$

(b) Substitute 116 for P and 40 for L in either form of the formula obtained in (a).

$$W = \frac{P - 2L}{2}$$
$$= \frac{116 - 2(40)}{2}$$
$$= \frac{116 - 80}{2}$$
$$= \frac{36}{2} = 18$$

13. The angles are supplementary, so
$$(3x + 55) + (7x - 25) = 180.$$

Solve this equation.
$$10x + 30 = 180$$
$$10x = 150$$
$$x = 15$$

If $x = 15$, $3x + 55 = 100$, and $7x - 25 = 80$.

The measures of the angles are 100° and 80°.

14. The angles are vertical angles, so their measures are equal.
$$3x + 15 = 4x - 5$$
$$15 = x - 5$$
$$20 = x$$

Since $x = 20$, $3x + 15 = 75$ and $4x - 5 = 75$.

Both angles have measures of 75°.

15.
$$\frac{z}{8} = \frac{12}{16}$$
$$16z = 8(12) \quad \text{Cross products are equal.}$$
$$16z = 96$$
$$\frac{16z}{16} = \frac{96}{16} \quad \text{Divide by 16.}$$
$$z = 6$$

The solution is 6.

16.
$$\frac{x + 5}{3} = \frac{x - 3}{4}$$
$$4(x + 5) = 3(x - 3)$$
$$4x + 20 = 3x - 9$$
$$x + 20 = -9$$
$$x = -29$$

The solution is -29.

17. The results in the following table are rounded to the nearest thousandth.

Size	Unit Cost (dollars per oz)
12-oz	$\frac{\$1.99}{12} = \$.166$
16-oz	$\frac{\$2.49}{16} = \$.156$
24-oz	$\frac{\$3.49}{24} = \$.145$ (∗)

    The best buy is the 24-oz size.

18. Let $x =$ the actual distance between Seattle and Cincinnati.
    $$\frac{x \text{ miles}}{1050 \text{ miles}} = \frac{92 \text{ inches}}{42 \text{ inches}}$$
    $$42x = 92(1050)$$
    $$x = \frac{96,600}{42} = 2300$$
    The actual distance between Seattle and Cincinnati is 2300 miles.

19. The problem can be stated as follows: "What percent of 69 is 95?"

    Substitute $b = 69$ and $a = 95$ into the percent proportion; then find p.
    $$\frac{a}{b} = \frac{p}{100}$$
    $$\frac{95}{69} = \frac{p}{100}$$
    $$69p = 95(100)$$
    $$p = \frac{9500}{69}$$
    $$p \approx 137.7$$

    Thus, 95 points is about 137.7% of 69 points, so this is about a 37.7% increase in points.

    Alternatively, we could use $a = 95 - 69 = 26$, which leads us directly to the 37.7% figure.

20. (a) x can take on any value less than 0.
    $$x < 0$$
    (b) x can take on any value greater than -2 and less than or equal to 3.
    $$-2 < x \leq 3$$

21. $$-3x > -33$$
    $$\frac{-3x}{-3} < \frac{-33}{-3} \quad \text{Divide by } -3;$$
    $$\text{reverse the symbol.}$$
    $$x < 11$$

    <--+--+--+--○--+--+-->
    $\qquad\qquad 11$

22. $$-.04x \leq .12$$
    $$\frac{-.04x}{-.04} \geq \frac{.12}{-.04} \quad \text{Divide by } -.04;$$
    $$\text{reverse the symbol.}$$
    $$x \geq -3$$

    <--+--+--●--+--+--+-->
    $\quad -3$

23. $$-4x + 2(x - 3) \geq 4x - (3 + 5x) - 7$$
    $$-4x + 2x - 6 \geq 4x - 3 - 5x - 7$$
    $$-2x - 6 \geq -x - 10$$
    $$-x - 6 \geq -10$$
    $$-x \geq -4$$
    $$\frac{-1x}{-1} \leq \frac{-4}{-1}$$
    Divide by -1; reverse the symbol.
    $$x \leq 4$$

    <--+--+--+--●--+--+-->
    $\qquad\qquad 4$

24. Let $x =$ the score on the third test.

    The average of the three tests | is at least | 80.
    $\downarrow$ | $\downarrow$ | $\downarrow$
    $\frac{76 + 81 + x}{3}$ | $\geq$ | 80

    $$\frac{157 + x}{3} \geq 80$$
    $$3\left(\frac{157 + x}{3}\right) \geq 3(80)$$
    $$157 + x \geq 240$$
    $$x \geq 83$$

    In order to average at least 80, Shania's score on her third test must be 83 or more.

25. When both sides of an inequality are multiplied or divided by a negative number, the direction of the inequality symbol must be reversed.

Cumulative Review Exercises (Chapters R–2)

1. $\frac{15}{40} = \frac{3 \cdot 5}{8 \cdot 5} = \frac{3}{8}$

2. $\frac{108}{144} = \frac{3 \cdot 36}{4 \cdot 36} = \frac{3}{4}$

3. $\frac{5}{6} + \frac{1}{4} + \frac{7}{15} = \frac{50}{60} + \frac{15}{60} + \frac{28}{60}$
   $= \frac{50 + 15 + 28}{60}$
   $= \frac{93}{60} = \frac{31 \cdot 3}{20 \cdot 3} = \frac{31}{20}$

4. $16\dfrac{7}{8} - 3\dfrac{1}{10} = \dfrac{135}{8} - \dfrac{31}{10}$

$= \dfrac{675}{40} - \dfrac{124}{40}$

$= \dfrac{551}{40}$ or $13\dfrac{31}{40}$

5. $\dfrac{9}{8} \cdot \dfrac{16}{3} - \dfrac{3 \cdot 3 \cdot 2 \cdot 8}{8 \cdot 3} - \dfrac{3 \cdot 2}{1} = 6$

6. $\dfrac{3}{4} \div \dfrac{5}{8} = \dfrac{3}{4} \cdot \dfrac{8}{5} = \dfrac{3 \cdot 2 \cdot 4}{4 \cdot 5} = \dfrac{3 \cdot 2}{5} = \dfrac{6}{5}$

7. $4.8 + 12.5 + 16.73$

$\begin{array}{r}\overset{1\;2}{4.80}\\12.50\\+16.73\\\hline 34.03\end{array}$

8. $56.3 - 28.99$

$\begin{array}{r}\overset{15\;12}{4\;\cancel{5}\;\cancel{2}\;10}\\\cancel{5}\;\cancel{6}.\cancel{3}\;\cancel{0}\\-2\;8.9\;9\\\hline 2\;7.3\;1\end{array}$

9. $67.8(.45)$

$\begin{array}{r}67.8\quad\leftarrow\;1\;decimal\;place\\\times\;.45\quad\leftarrow\;2\;decimal\;places\\\hline 3\;390\\27\;12\quad\\\hline 30.510\quad\leftarrow\;3\;decimal\;places\end{array}$

$67.8(.45) = 30.51$

10. $236.46 \div 4.2$

Move the decimal point in both dividend and divisor one place to the right; then move the decimal point straight up into the divisor.

$\begin{array}{r}56.3\\42.\overline{)2364.6}\\\underline{210}\\264\\\underline{252}\\12\;6\\\underline{12\;6}\\0\end{array}$

$236.46 \div 4.2 = 56.3$

11. $56\left(\dfrac{5}{8}\right) = \dfrac{56}{1} \cdot \dfrac{5}{8} = \dfrac{7 \cdot 8 \cdot 5}{1 \cdot 8} = 35$

To make 56 dresses, 35 yards of trim would be used.

12. Let $x =$ the number of cups of water needed to serve 10 people.

$\dfrac{3 \text{ cups}}{4 \text{ people}} = \dfrac{x \text{ cups}}{10 \text{ people}}$

$\dfrac{3}{4} = \dfrac{x}{10}$

$4x = 3(10)$

$4x = 30$

$x = \dfrac{30}{4}$ or $7\dfrac{1}{2}$

$7\dfrac{1}{2}$ cups of water are needed to serve 10 people.

An alternative solution follows:

If the recipe for 4 people requires 3 cups of water, the amount needed for 1 person is

$3 \div 4 = \dfrac{3}{1} \div \dfrac{4}{1} = \dfrac{3}{1} \cdot \dfrac{1}{4} = \dfrac{3}{4}$ cup.

For 10 people, the amount of water needed is

$10\left(\dfrac{3}{4}\right) = \dfrac{5 \cdot 2 \cdot 3}{2 \cdot 2} = \dfrac{15}{2} = 7\dfrac{1}{2}$ cups.

13. First dog's weight plus second dog's weight

$\downarrow \qquad \downarrow \qquad \downarrow$

$27\dfrac{1}{4} \quad + \quad 39\dfrac{7}{8}$

is the total weight of both dogs.

$27\dfrac{1}{4} + 39\dfrac{7}{8} = \dfrac{109}{4} + \dfrac{319}{8}$

$= \dfrac{218}{8} + \dfrac{319}{8}$

$= \dfrac{537}{8}$ or $67\dfrac{1}{8}$

The total weight is $67\dfrac{1}{8}$ pounds.

14. The three computer workstations cost

$\$329.99 + \$379.99 + \$439.99 = \$1149.97$.

The three ergonomic office chairs cost

$3(\$249.99) = \$749.97$.

The total cost is

$\$1149.97 + \$749.97 = \$1899.94$.

15. $\dfrac{8(7) - 5(6+2)}{3 \cdot 5 + 1} \geq 1$

$\dfrac{8(7) - 5(8)}{3 \cdot 5 + 1} \geq 1$

$\dfrac{56 - 40}{15 + 1} \geq 1$

$\dfrac{16}{16} \geq 1$

$1 \geq 1$

The statement is true.

16.
$$\frac{4(9+3) - 8(4)}{2 + 3 - 3} \geq 2$$
$$\frac{4(12) - 8(4)}{2 + [3 + (-3)]} \geq 2$$
$$\frac{48 - 32}{2 + 0} \geq 2$$
$$\frac{16}{2} \geq 2$$
$$8 \geq 2$$

The statement is true.

17. $-11 + 20 + (-2) = 9 + (-2) = 7$

18. $13 + (-19) + 7 = -6 + 7 = 1$

19. $9 - (-4) = 9 + 4 = 13$

20. $-2(-5)(-4) = 10(-4) = -40$

21. $\dfrac{4 \cdot 9}{-3} = \dfrac{4 \cdot 3 \cdot 3}{-3} = -12$

22. $\dfrac{8}{7-7} = \dfrac{8}{0}$, which is undefined.

23. $(-5 + 8) + (-2 - 7) = 3 + (-9) = -6$

24. $(-7 - 1)(-4) + (-4) = (-8)(-4) + (-4)$
$$= 32 + (-4)$$
$$= 28$$

25. $\dfrac{-3 - (-5)}{1 - (-1)} = \dfrac{-3 + 5}{1 + 1} = \dfrac{2}{2} = 1$

26. $\dfrac{6(-4) - (-2)(12)}{3^2 + 7^2} = \dfrac{(-24) - (-24)}{9 + 49}$
$$= \dfrac{(-24) + 24}{58}$$
$$= \dfrac{0}{58}$$
$$= 0$$

27. $\dfrac{(-3)^2 - (-4)(2^4)}{5 \cdot 2 - (-2)^3} = \dfrac{9 - (-4)(16)}{5 \cdot 2 - (-8)}$
Exponents first
$$= \dfrac{9 - (-64)}{10 - (-8)}$$
$$= \dfrac{9 + 64}{10 + 8}$$
$$= \dfrac{73}{18} \text{ or } 4\dfrac{1}{18}$$

28. $\dfrac{-2(5^3) - 6}{4^2 + 2(-5) + (-2)} = \dfrac{-2(125) - 6}{16 + (-10) + (-2)}$
$$= \dfrac{-250 - 6}{6 + (-2)}$$
$$= \dfrac{-256}{4}$$
$$= -64$$

In Exercises 29 and 30, replace x with -2, y with -4, and z with 3.

29. $xz^3 - 5y^2 = -2(3)^3 - 5(-4)^2$
$$= -2(27) + (-5)(16)$$
$$= -54 + (-80)$$
$$= -134$$

30. $\dfrac{xz - y^3}{-4z} = \dfrac{-2(3) - (-4)^3}{-4(3)}$
$$= \dfrac{-6 - (-64)}{-12}$$
$$= \dfrac{-6 + 64}{-12}$$
$$= \dfrac{58}{-12}$$
$$= -\dfrac{29}{6} \text{ or } -4\dfrac{5}{6}$$

31. $7(k + m) = 7k + 7m$

The multiplication of 7 is distributed over the sum, which illustrates the distributive property.

32. $3 + (5 + 2) = 3 + (2 + 5)$

The order of the numbers added in the parentheses is changed, which illustrates the commutative property.

33. $7 + (-7) = 0$

Inverses are added to give 0, which illustrates the inverse property.

34. $3.5(1) = 3.5$

The value of the number 3.5 is unchanged by multiplying by one. This illustrates the identity property.

35. $4p - 6 + 3p - 8 = 4p + 3p - 6 - 8$
$$= 7p - 14$$

36. $-4(k + 2) + 3(2k - 1)$
$$= (-4)(k) + (-4)(2) + (3)(2k) + (3)(-1)$$
$$= -4k - 8 + 6k - 3$$
$$= -4k + 6k - 8 - 3$$
$$= 2k - 11$$

37. $2r - 6 = 8$
$\quad\quad 2r = 14 \quad$ *Add 6.*
$\quad\quad\; r = 7 \quad$ *Divide by 2.*

Check $r = 7$: $\;8 = 8$

The solution is 7.

88 Chapter 2 Equations, Inequalities, and Applications

38. $2(p - 1) = 3p + 2$
    $2p - 2 = 3p + 2$
    $-2 = p + 2$ Subtract $2p$.
    $-4 = p$ Subtract 2.

    Check $p = -4$: $-10 = -10$

    The solution is -4.

39. $4 - 5(a + 2) = 3(a + 1) - 1$
    $4 - 5a - 10 = 3a + 3 - 1$
    $-5a - 6 = 3a + 2$
    $-8a - 6 = 2$
    $-8a = 8$
    $a = -1$

    Check $a = -1$: $-1 = -1$

    The solution is -1.

40. $2 - 6(z + 1) = 4(z - 2) + 10$
    $2 - 6z - 6 = 4z - 8 + 10$
    $-6z - 4 = 4z + 2$ Combine terms.
    $-10z - 4 = 2$ Subtract $4z$.
    $-10z = 6$ Add 4.
    $z = -\frac{6}{10}$ Divide by -10.
    $z = -\frac{3}{5}$ Reduce.

    Check $z = -\frac{3}{5}$: $-\frac{2}{5} = -\frac{2}{5}$

    The solution is $-\frac{3}{5}$.

41. $-(m - 1) = 3 - 2m$
    $-1(m - 1) = 3 - 2m$ $-a = -1a$
    $-m + 1 = 3 - 2m$
    $m + 1 = 3$ Add $2m$.
    $m = 2$ Subtract 1.

    Check $m = 2$: $-1 = -1$

    The solution is 2.

42. $\dfrac{x - 2}{3} = \dfrac{2x + 1}{5}$
    $(x - 2)(5) = (3)(2x + 1)$ Cross products
    $5x - 10 = 6x + 3$
    $-10 = x + 3$ Subtract $5x$.
    $-13 = x$ Subtract 3.

    Check $x = -13$: $-5 = -5$

    The solution is -13.

43. $\dfrac{2x + 3}{5} = \dfrac{x - 4}{2}$
    $(2x + 3)(2) = (5)(x - 4)$
    $4x + 6 = 5x - 20$
    $6 = x - 20$
    $26 = x$

    Check $x = 26$: $11 = 11$

    The solution is 26.

44. $\dfrac{2}{3}x + \dfrac{3}{4}x = -17$
    $12\left(\dfrac{2}{3}x + \dfrac{3}{4}x\right) = 12(-17)$ LCD = 12
    $8x + 9x = -204$
    $17x = -204$
    $x = -12$

    Check $y = -12$: $-17 = -17$

    The solution is -12.

45. $P = a + b + c$ for c
    $P - a - b = a + b + c - a - b$
    Subtract a and b.
    $P - a - b = c$ or $c = P - a - b$

46. $P = 4s$ for s
    $\dfrac{P}{4} = \dfrac{4s}{4}$ Divide by 4.
    $\dfrac{P}{4} = s$ or $s = \dfrac{P}{4}$

47. $-5z \geq 4z - 18$
    $-9z \geq -18$ Subtract $4z$.
    $\dfrac{-9z}{-9} \leq \dfrac{-18}{-9}$ Divide by -9; reverse the symbol.
    $z \leq 2$

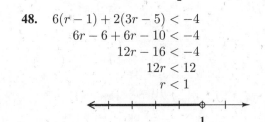

48. $6(r - 1) + 2(3r - 5) < -4$
    $6r - 6 + 6r - 10 < -4$
    $12r - 16 < -4$
    $12r < 12$
    $r < 1$

49. In Exercise 14, we found that the total bill without tax was $1899.94. To find the final bill, including tax, first find the amount of sales tax.

To find $6\frac{1}{4}\%$ of $1899.94, substitute $b = 1899.94$ and $p = 6\frac{1}{4} = 6.25$ into the percent proportion; then solve for a.

$$\frac{a}{b} = \frac{p}{100}$$

$$\frac{a}{1899.94} = \frac{6.25}{100}$$

$$a = \frac{1899.94(6.25)}{100}$$

$$\approx 118.75 \text{ (to the nearest hundredth)}$$

The sales tax on the purchase was $118.75, so the final bill, including tax, was

$$\$1899.94 + \$118.75 = \$2018.69.$$

50. First, find the discount with the trade-in. To find 25% of 5000, substitute $b = 5000$ and $p = 25$ into the percent proportion; then solve for a.

$$\frac{a}{b} = \frac{p}{100}$$

$$\frac{a}{5000} = \frac{25}{100}$$

$$100a = 5000(25)$$

$$a = \frac{5000(25)}{100}$$

$$a = 1250$$

The price of the car with the trade-in is

$$\$5000 - \$1250 = \$3750.$$

51. Let x be the price of the books. Write 6% as .06. Then

Price of books and sales tax is total cost.
↓   ↓   ↓   ↓   ↓
x + $.06x$ =   244.33

$$x + .06x = 244.33$$
$$1.06x = 244.33$$
$$\frac{1.06x}{1.06} = \frac{244.33}{1.06}$$
$$x = 230.50$$

The cost of the textbooks is $230.50.

Alternatively, use $a = 244.33$ and $p = 106$ in the percent proportion; then find b.

52. Let p be the amount of his purchases.

Use the formula $I = prt$ with $r = 1\frac{1}{2}\%$ and $t = 1$ to find the interest. Write $1\frac{1}{2}\%$ as .015.

Total bill is
↓   ↓
104.93 =

purchase plus interest plus late charge.
↓   ↓   ↓   ↓   ↓
p + $.015p$ +   5

$$104.93 = 1p + .015p + 5$$
$$104.93 = p(1.015) + 5$$
$$99.93 = 1.015p \qquad \textit{Subtract 5.}$$
$$98.45 \approx p \qquad \textit{Divide by 1.015.}$$

His purchases amounted to $98.45.

53. To find the length, substitute $P = 98$ and $W = 19$ in the formula for the perimeter of a rectangle.

$$P = 2L + 2W$$
$$98 = 2L + 2(19)$$
$$98 = 2L + 38$$
$$60 = 2L \qquad \textit{Subtract 38.}$$
$$30 = L \qquad \textit{Divide by 2.}$$

The length is 30 centimeters.

54. To find the height, substitute $A = 104$ and $b = 13$ in the formula for the area of a triangle.

$$A = \frac{1}{2}bh$$
$$104 = \frac{1}{2}(13)h$$
$$104 = \left(\frac{13}{2}\right)h$$
$$\left(\frac{2}{13}\right)(104) = \left(\frac{2}{13}\right)\left(\frac{13}{2}\right)h$$
$$16 = h$$

The height is 16 inches.

CHAPTER 3 GRAPHS OF LINEAR EQUATIONS AND INEQUALITIES IN TWO VARIABLES

3.1 Reading Graphs; Linear Equations in Two Variables

3.1 Margin Exercises

1. **(a)** In the circle graph, the sector for C-Band is the smallest, so C-Band had the smallest market share in December 2003.

   (b) A market share of 55% can be rounded to 60%, or .60. We multiply .60 by the total number of subscribers, 22 million. A good estimate of the number of DirecTV subscribers would be

   $$.60(22) = 13.2 \text{ million.}$$

   (c) To find the answer, we multiply the actual percent from the graph for DirecTV, 55% or .55, by the number of subscribers, 22 million:

   $$.55(22) = 12.1$$

   Thus, 12.1 million homes subscribed to DirecTV.

2. **(a)** Locate 200 on the vertical scale and follow the line across to the right. Three years, 1999, 2000, and 2001, have bars that extend below the line for 200, so they had savings less than $200 billion.

   (b) Locate the top of the bar for 2000 and move horizontally across to the vertical scale to see that the savings in 2000 was about $60 billion.

   Follow the top of the bar for 2001 across to the vertical scale to see that this bar is about 120. So the savings in 2001 was about $120 billion.

3. **(a)** Locate the highest point on the graph and move down to read the year on the horizontal scale. 2005 had the greatest amount of Medicare funds.

   (b) Move up from 2008 on the horizontal scale to the point plotted for 2008. Then move across to the vertical scale. There appears to be a surplus of $5 billion in 2008.

   (c) The values for 2006 and 2011 are about 9.5 and $-.5$. The difference is

   $$-.5 - 9.5 = -10.$$

   Thus, the funds are projected to decrease about $10 billion from 2006 to 2011.

4. **(a)** $x = 5$ and $y = 7$ is written as the ordered pair $(5, 7)$.

   (b) $y = 6$ and $x = -1$ is written as the ordered pair $(-1, 6)$.

   (c) $y = 4$ and $x = -3$ is written as the ordered pair $(-3, 4)$.

   (d) $x = 3$ and $y = -12$ is written as the ordered pair $(3, -12)$.

5. **(a)** $(0, 10)$

   $$5x + 2y = 20$$
   $$5(\underline{0}) + 2(\underline{10}) = 20 \quad ? \quad \textit{Let } x = 0, y = 10.$$
   $$\underline{0} + 20 = 20 \quad ?$$
   $$\underline{20} = 20 \quad \textit{True}$$

   Yes, $(0, 10)$ is a solution.

   (b) $(2, -5)$

   $$5x + 2y = 20$$
   $$5(2) + 2(-5) = 20 \quad ? \quad \textit{Let } x = 2, y = -5.$$
   $$10 + (-10) = 20 \quad ?$$
   $$0 = 20 \quad \textit{False}$$

   No, $(2, -5)$ is not a solution.

   (c) $(3, 2)$

   $$5x + 2y = 20$$
   $$5(3) + 2(2) = 20 \quad ? \quad \textit{Let } x = 3, y = 2.$$
   $$15 + 4 = 20 \quad ?$$
   $$19 = 20 \quad \textit{False}$$

   No, $(3, 2)$ is not a solution.

   (d) $(-4, 20)$

   $$5x + 2y = 20$$
   $$5(-4) + 2(20) = 20 \quad ? \quad \textit{Let } x = -4, y = 20.$$
   $$-20 + 40 = 20 \quad ?$$
   $$20 = 20 \quad \textit{True}$$

   Yes, $(-4, 20)$ is a solution.

6. **(a)** In the ordered pair $(5, \ )$, $x = 5$. Find the corresponding value of y by replacing x with 5 in the given equation.

   $$y = 2x - 9$$
   $$y = 2(\underline{5}) - 9 \quad \textit{Let } x = 5.$$
   $$y = \underline{10} - 9$$
   $$y = \underline{1}$$

   The ordered pair is $\underline{(5, 1)}$.

   (b) In the ordered pair $(2, \ )$, $x = 2$.

   $$y = 2x - 9$$
   $$y = 2(2) - 9 \quad \textit{Let } x = 2.$$
   $$y = 4 - 9$$
   $$y = -5$$

   The ordered pair is $(2, -5)$.

92 Chapter 3 Graphs of Linear Equations and Inequalities in Two Variables

(c) In the ordered pair (, 7), $y = 7$. Find the corresponding value of x by replacing y with 7 in the given equation.

$$y = 2x - 9$$
$$7 = 2x - 9 \quad \text{Let } y = 7.$$
$$16 = 2x \quad \text{Add 9.}$$
$$8 = x \quad \text{Divide by 2.}$$

The ordered pair is $(8, 7)$.

(d) In the ordered pair (, -13), $y = -13$.

$$y = 2x - 9$$
$$-13 = 2x - 9 \quad \text{Let } y = -13.$$
$$-4 = 2x \quad \text{Add 9.}$$
$$-2 = x \quad \text{Divide by 2.}$$

The ordered pair is $(-2, -13)$.

7. **(a)** To complete the first ordered pair, let $x = 0$.

$$2x - 3y = 12$$
$$2(0) - 3y = 12 \quad \text{Let } x = 0.$$
$$-3y = 12$$
$$y = -4$$

The first ordered pair is $(0, -4)$.

To complete the second ordered pair, let $y = 0$.

$$2x - 3y = 12$$
$$2x - 3(0) = 12 \quad \text{Let } y = 0.$$
$$2x = 12$$
$$x = 6$$

The second ordered pair is $(6, 0)$.

To complete the third ordered pair, let $x = 3$.

$$2x - 3y = 12$$
$$2(3) - 3y = 12 \quad \text{Let } x = 3.$$
$$6 - 3y = 12$$
$$-3y = 6$$
$$y = -2$$

The third ordered pair is $(3, -2)$.

To complete the last ordered pair, let $y = -3$.

$$2x - 3y = 12$$
$$2x - 3(-3) = 12 \quad \text{Let } y = -3.$$
$$2x + 9 = 12$$
$$2x = 3$$
$$x = \frac{3}{2}$$

The last ordered pair is $\left(\frac{3}{2}, -3\right)$.

The completed table of ordered pairs follows.

x	y
0	-4
6	0
3	-2
$\frac{3}{2}$	-3

(b) The given equation is $y = 4$. No matter which value of x might be chosen, the value of y is always the same, 4. Each ordered pair can be completed by placing 4 in the second position.

x	y
-3	4
2	4
5	4

8. A is in quadrant II, B is in quadrant IV, C is in quadrant I, and D is in quadrant II. E is on the y-axis, so it is not in any quadrant.

9. To plot the ordered pairs, start at the origin in each case.

(a) $(3, 5)$

Go 3 units to the right along the x-axis; then go up 5 units, parallel to the y-axis.

(b) $(-2, 6)$

Go 2 units to the left along the x-axis; then go up 6 units.

(c) $(-4, 0)$

Go 4 units to the left. This point is on the x-axis since the y-coordinate is 0.

(d) $(-5, -2)$

Go 5 units to the left; then go down 2 units.

(e) $(5, -2)$

Go 5 units to the right; then go down 2 units.

(f) $(0, -6)$

Go down 6 units. This point is on the y-axis since the x-coordinate is 0.

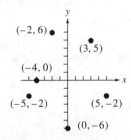

3.1 Reading Graphs; Linear Equations in Two Variables

10. (a) To find y when $x = 1996$, substitute 1996 for x into the equation and use a calculator.

$$y = 34.3(1996) - 67{,}693$$
$$y = 770$$

This means that in 1996, Americans each spent about $770 on doctors' visits.

(b) To find y when $x = 2000$, substitute 2000 for x into the equation and use a calculator.

$$y = 34.3(2000) - 67{,}693$$
$$y = 907$$

This means that in 2000, Americans each spent about $907 on doctors' visits.

3.1 Section Exercises

1. In the circle graph, the sector for Snoopy is the largest, 31%, so Snoopy was the most popular at 31%.

3. The sector for Charlie Brown is 26%. The sector for Linus is 13%. Since 26% is twice as much as 13%, we can expect twice as many adults to favor Charlie Brown.

5. Locate the two tallest bars. Follow the bar down to see which state it is and move across from the top of the bar to the vertical scale to estimate their production. Ohio (OH) and Iowa (IA) are the top two egg-producing states. Ohio produced about 8000 million eggs and Iowa produced about 10,000 million eggs.

7. Locate the shortest bar. It's above North Carolina (NC). Follow the top of the bar across to the vertical scale to estimate North Carolina's production, which is about 2500 million eggs.

9. The line between 1975 and 1980 shows the steepest rise, so from 1975 to 1980 the greatest increase in the price of a gallon of gas occurred. Subtract the price per gallon for 1975 from the price per gallon for 1980.

$$2.50 - 1.75 = .75$$

The increase was about $.75.

11. The line between 1980 and 1995 falls, so the price of a gallon of gas was decreasing from 1980 to 1995.

13. The symbol (x, y) *does* represent an ordered pair, while the symbols $[x, y]$ and $\{x, y\}$ *do not* represent ordered pairs. (Note that only parentheses are used to write ordered pairs.)

15. All points having x-coordinate 0 lie on the y-axis, so the point whose graph has coordinates $(0, 5)$ lies on the _y_-axis.

17. All ordered pairs that are solutions of the equation $x = 6$ have x-coordinates equal to 6, so the ordered pair (_6_ , -2) is a solution of the equation $x = 6$.

19. $x + y = 9$; $(0, 9)$

To determine whether $(0, 9)$ is a solution of the given equation, substitute 0 for x and 9 for y.

$$x + y = 9$$
$$0 + 9 = 9 \quad ? \quad \text{Let } x = 0, y = 9.$$
$$9 = 9 \quad \text{True}$$

The result is true, so $(0, 9)$ is a solution of the given equation $x + y = 9$.

21. $2x - y = 6$; $(4, 2)$

Substitute 4 for x and 2 for y.

$$2x - y = 6$$
$$2(4) - 2 = 6 \quad ? \quad \text{Let } x = 4, y = 2.$$
$$8 - 2 = 6 \quad ?$$
$$6 = 6 \quad \text{True}$$

The result is true, so $(4, 2)$ is a solution of $2x - y = 6$.

23. $4x - 3y = 6$; $(2, 1)$

Substitute 2 for x and 1 for y.

$$4x - 3y = 6$$
$$4(2) - 3(1) = 6 \quad ? \quad \text{Let } x = 2, y = 1.$$
$$8 - 3 = 6 \quad ?$$
$$5 = 6 \quad \text{False}$$

The result is false, so $(2, 1)$ is not a solution of $4x - 3y = 6$.

25. $y = \frac{2}{3}x$; $(-6, -4)$

Substitute -6 for x and -4 for y.

$$y = \frac{2}{3}x$$
$$-4 = \frac{2}{3}(-6) \quad ? \quad \text{Let } x = -6, y = -4.$$
$$-4 = -4 \quad \text{True}$$

The result is true, so $(-6, -4)$ is a solution of $y = \frac{2}{3}x$.

27. $x = -6$; $(5, -6)$

Since y does not appear in the equation, we just substitute 5 for x.

$$x = -6$$
$$5 = -6 \quad \text{Let } x = 5.\ \text{False}$$

The result is false, so $(5, -6)$ is not a solution of $x = -6$.

29. The ordered pairs $(4, -1)$ and $(-1, 4)$ are different ordered pairs because the order of the x- and y- coordinates has been reversed. Two *ordered* pairs are equal only if they have the same numbers in the same *order*.

31. $y = 2x + 7; (2, \ )$

In this ordered pair, $x = 2$. Find the corresponding value of y by replacing x with 2 in the given equation.

$$y = 2x + 7$$
$$y = 2(2) + 7 \quad \text{Let } x = 2.$$
$$y = 4 + 7$$
$$y = 11$$

The ordered pair is $(2, 11)$.

33. $y = 2x + 7; (\ , 0)$

In this ordered pair, $y = 0$. Find the corresponding value of x by replacing y with 0 in the given equation.

$$y = 2x + 7$$
$$0 = 2x + 7 \quad \text{Let } y = 0.$$
$$-7 = 2x$$
$$\frac{-7}{2} = x$$

The ordered pair is $\left(-\frac{7}{2}, 0\right)$.

35. $y = -4x - 4; (0, \ )$

$$y = -4x - 4$$
$$y = -4(0) - 4 \quad \text{Let } x = 0.$$
$$y = 0 - 4$$
$$y = -4$$

The ordered pair is $(0, -4)$.

37. $y = -4x - 4; (\ , 16)$

$$y = -4x - 4$$
$$16 = -4x - 4 \quad \text{Let } y = 16.$$
$$20 = -4x$$
$$-5 = x$$

The ordered pair is $(-5, 16)$.

39. $2x + 3y = 12$

If $x = 0$,

$$2(0) + 3y = 12$$
$$0 + 3y = 12$$
$$3y = 12$$
$$y = 4. \quad (0, 4)$$

If $y = 0$,

$$2x + 3(0) = 12$$
$$2x + 0 = 12$$
$$2x = 12$$
$$x = 6. \quad (6, 0)$$

If $y = 8$,

$$2x + 3(8) = 12$$
$$2x + 24 = 12$$
$$2x = -12$$
$$x = -6. \quad (-6, 8)$$

The completed table of values is shown below.

x	y
0	4
6	0
-6	8

41. $3x - 5y = -15$

If $x = 0$,

$$3(0) - 5y = -15$$
$$0 - 5y = -15$$
$$-5y = -15$$
$$y = 3. \quad (0, 3)$$

If $y = 0$,

$$3x - 5(0) = -15$$
$$3x - 0 = -15$$
$$3x = -15$$
$$x = -5. \quad (-5, 0)$$

If $y = -6$,

$$3x - 5(-6) = -15$$
$$3x + 30 = -15$$
$$3x = -45$$
$$x = -15. \quad (-15, -6)$$

The completed table of values is shown below.

x	y
0	3
-5	0
-15	-6

43. $x = -9$

No matter which value of y is chosen, the value of x will always be -9. Each ordered pair can be completed by placing -9 in the first position.

x	y
-9	6
-9	2
-9	-3

3.1 Reading Graphs; Linear Equations in Two Variables

45. $y = -6$

No matter which value of x is chosen, the value of y will always be -6. Each ordered pair can be completed by placing -6 in the second position.

x	y
8	-6
4	-6
-2	-6

47. $x - 8 = 0$; (,8) (,3) (,0)

The given equation may be written $x = 8$. For any value of y, the value of x will always be 8. The ordered pairs are $(8, 8)$, $(8, 3)$, and $(8, 0)$.

49. Point A was plotted by starting at the origin, going 2 units to the right along the x-axis, and then going up 4 units on a line parallel to the y-axis. The ordered pair for this point is $(2, 4)$.

51. Point C was plotted by starting at the origin, going 5 units to the left along the x-axis, and then going up 4 units on a line parallel to the y-axis. The ordered pair for this point is $(-5, 4)$.

53. Point E was plotted by starting at the origin, going 3 units to the right along the x-axis. The ordered pair for this point is $(3, 0)$.

55. The point with coordinates (x, y) is in quadrant III if x is *negative* and y is *negative*.

57. The point with coordinates (x, y) is in quadrant IV if x is *positive* and y is *negative*.

59. If $xy < 0$, then either $x < 0$ and $y > 0$ or $x > 0$ and $y < 0$. If $x < 0$ and $y > 0$, then the point lies in quadrant II. If $x > 0$ and $y < 0$, then the point lies in quadrant IV.

For Exercises 61–70, the ordered pairs are plotted on the graph following the solution for Exercise 69.

61. To plot $(6, 2)$, start at the origin, go 6 units to the right, and then go up 2 units.

63. To plot $(-4, 2)$, start at the origin, go 4 units to the left, and then go up 2 units.

65. To plot $\left(-\frac{4}{5}, -1\right)$, start at the origin, go $\frac{4}{5}$ unit to the left, and then go down 1 unit.

67. To plot $(0, 4)$, start at the origin and go up 4 units. The point lies on the y-axis.

69. To plot $(4, 0)$, start at the origin and go 4 units to the right along the x-axis. The point lies on the x-axis.

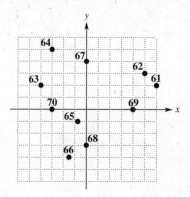

71. $x - 2y = 6$

x	y
0	
	0
2	
	-1

Substitute the given values to complete the ordered pairs.

$x - 2y = 6$
$0 - 2y = 6$ *Let $x = 0$.*
$-2y = 6$
$y = -3$

$x - 2y = 6$
$x - 2(0) = 6$ *Let $y = 0$.*
$x - 0 = 6$
$x = 6$

$x - 2y = 6$
$2 - 2y = 6$ *Let $x = 2$.*
$-2y = 4$
$y = -2$

$x - 2y = 6$
$x - 2(-1) = 6$ *Let $y = -1$.*
$x + 2 = 6$
$x = 4$

The completed table of values follows.

x	y
0	-3
6	0
2	-2
4	-1

Plot the points $(0, -3)$, $(6, 0)$, $(2, -2)$, and $(4, -1)$ on a coordinate system.

continued

96 Chapter 3 Graphs of Linear Equations and Inequalities in Two Variables

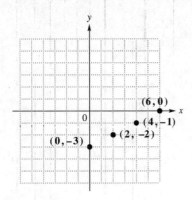

73. $3x - 4y = 12$

x	y
0	
	0
-4	
	-4

Substitute the given values to complete the ordered pairs.

$3x - 4y = 12$
$3(0) - 4y = 12$ *Let x = 0.*
$0 - 4y = 12$
$-4y = 12$
$y = -3$

$3x - 4y = 12$
$3x - 4(0) = 12$ *Let y = 0.*
$3x - 0 = 12$
$3x = 12$
$x = 4$

$3x - 4y = 12$
$3(-4) - 4y = 12$ *Let x = −4.*
$-12 - 4y = 12$
$-4y = 24$
$y = -6$

$3x - 4y = 12$
$3x - 4(-4) = 12$ *Let y = −4.*
$3x + 16 = 12$
$3x = -4$
$x = -\dfrac{4}{3}$

The completed table is as follows.

x	y
0	-3
4	0
-4	-6
$-\dfrac{4}{3}$	-4

Plot the points $(0, -3)$, $(4, 0)$, $(-4, -6)$, and $\left(-\dfrac{4}{3}, -4\right)$ on a coordinate system.

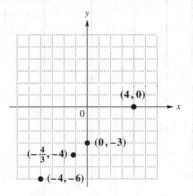

75. The given equation, $y + 4 = 0$, can be written as $y = -4$. So regardless of the value of x, the value of y is -4.

x	y
0	-4
5	-4
-2	-4
-3	-4

Plot the points $(0, -4)$, $(5, -4)$, $(-2, -4)$, and $(-3, -4)$ on a coordinate system.

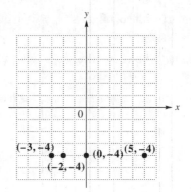

77. The points in each graph appear to lie on a straight line.

79. **(a)** We can write the results from the table as ordered pairs (x, y).

$(1996, 53.3)$, $(1997, 52.8)$, $(1998, 52.1)$, $(1999, 51.6)$, $(2000, 51.2)$, $(2001, 50.9)$

(b) $(2002, 51.0)$ indicates that 51.0 percent of college students in 2002 graduated within 5 years.

3.2 Graphing Linear Equations in Two Variables

(c)

4-YEAR COLLEGE STUDENTS GRADUATING WITHIN 5 YEARS

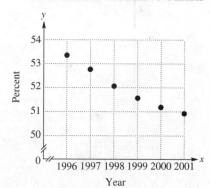

(d) The points appear to be approximated by a straight line. Graduation rates for 4-year college students that graduate within 5 years are decreasing.

81. (a) Substitute $x = 20, 40, 60, 80$ in the equation $y = -.8x + 173$.

    $y = -.8(20) + 173 = -16 + 173 = 157$
    $y = -.8(40) + 173 = -32 + 173 = 141$
    $y = -.8(60) + 173 = -48 + 173 = 125$
    $y = -.8(80) + 173 = -64 + 173 = 109$

    The completed table follows.

Age	Heartbeats (per minute)
20	157
40	141
60	125
80	109

    (b) x represents age and y represents heartbeats in the ordered pairs (x, y).
    $(20, 157), (40, 141), (60, 125), (80, 109)$

    (c) **TARGET HEART RATE ZONE (Upper Limit)**

    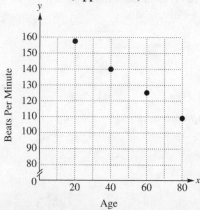

    The points lie in a linear pattern.

3.2 Graphing Linear Equations in Two Variables

3.2 Margin Exercises

1. (a) $(-3,)$

    $x + 2y = 7$
    $-3 + 2y = 7$ Let $x = -3$.
    $2y = 10$
    $y = 5$

    The ordered pair is $(-3, 5)$.

    (b) $(3,)$

    $x + 2y = 7$
    $3 + 2y = 7$ Let $x = 3$.
    $2y = 4$
    $y = 2$

    The ordered pair is $(3, 2)$.

    (c) $(-1,)$

    $x + 2y = 7$
    $-1 + 2y = 7$ Let $x = -1$.
    $2y = 8$
    $y = 4$

    The ordered pair is $(-1, 4)$.

    (d) $(7,)$

    $x + 2y = 7$
    $7 + 2y = 7$ Let $x = 7$.
    $2y = 0$
    $y = 0$

    The ordered pair is $(7, 0)$.

2. $x + y = 6$

x	y
0	
	0
2	

    Find the first missing value in the table by substituting 0 for x in the equation and solving for y.

    $x + y = 6$
    $0 + y = 6$ Let $x = 0$.
    $y = 6$

    The first ordered pair is $(0, 6)$.

    continued

98 Chapter 3 Graphs of Linear Equations and Inequalities in Two Variables

Find the second missing value.

$$x + y = 6$$
$$x + 0 = 6 \quad \text{Let } y = 0.$$
$$x = 6$$

The second ordered pair is $(6, 0)$.

Find the third missing value.

$$x + y = 6$$
$$2 + y = 6 \quad \text{Let } x = 2.$$
$$y = 4$$

The last ordered pair is $(2, 4)$.

The completed table of ordered pairs follows.

x	y
0	6
6	0
2	4

To graph the line, plot the corresponding points and draw a line through them.

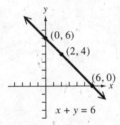

3. Ordered pairs may vary; examples are given here.

$$2x = 3y + 6$$
$$2(0) = 3y + 6 \quad \text{Let } x = 0.$$
$$0 = 3y + 6$$
$$-3y = 6$$
$$y = -2$$

This gives the ordered pair $(0, -2)$.

$$2x = 3y + 6$$
$$2x = 3(0) + 6 \quad \text{Let } y = 0.$$
$$2x = 0 + 6$$
$$2x = 6$$
$$x = 3$$

This gives the ordered pair $(3, 0)$.

$$2x = 3y + 6$$
$$2(6) = 3y + 6 \quad \text{Let } x = 6.$$
$$12 = 3y + 6$$
$$-3y = -6$$
$$y = 2$$

This gives the ordered pair $(6, 2)$.

Graph the equation by plotting these three points and drawing a line through them.

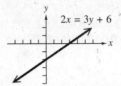

4. $5x + 2y = 10$

Find the x-intercept by letting $y = 0$.

$$5x + 2y = 10$$
$$5x + 2(0) = 10 \quad \text{Let } y = 0.$$
$$5x + 0 = 10$$
$$5x = 10$$
$$x = 2$$

The x-intercept is $(2, 0)$.

Find the y-intercept by letting $x = 0$.

$$5x + 2y = 10$$
$$5(0) + 2y = 10 \quad \text{Let } x = 0.$$
$$0 + 2y = 10$$
$$2y = 10$$
$$y = 5$$

The y-intercept is $(0, 5)$.

Get a third point as a check. For example, choosing $x = 4$ gives $y = -5$. Plot $(2, 0)$, $(0, 5)$, and $(4, -5)$ and draw a line through them.

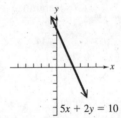

5. Ordered pairs may vary. Examples are given here.

(a) $2x - y = 0$

Three ordered pairs that can be used are shown in the following table.

x	y
0	0
1	2
-1	-2

Graph the equation by plotting these points and drawing a line through them.

3.2 Graphing Linear Equations in Two Variables

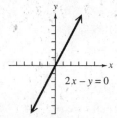

(b) $x = -4y$

Three ordered pairs that can be used are shown in the following table.

x	y
0	0
-4	1
4	-1

Graph the equation by plotting these points and drawing a line through them.

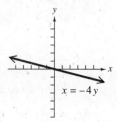

6. (a) $y = -5$

No matter what value we choose for x, y is always -5, as shown in the table of ordered pairs.

x	y
0	-5
-5	-5
3	-5

The graph is a horizontal line through $(0, -5)$.

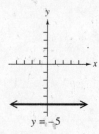

(b) $x + 4 = 6$ or $x = 2$

x is always 2 regardless of the value of y, as shown in the table of ordered pairs.

x	y
2	0
2	-5
2	5

The graph is a vertical line through $(2, 0)$.

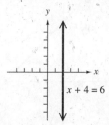

7. (a) If the line is horizontal, it has the form $y = k$. Choice **C**, $y - 2 = 3$, or $y = 5$, is the correct answer.

(b) Substitute $(0, 0)$ in $x + 4y = 0$.
$$0 + 4(0) = 0$$
$$0 + 0 = 0$$
$$0 = 0$$

Choice **D**, $x + 4y = 0$, is the correct answer.

(c) If the line is vertical, it has the form $x = k$. Choice **A**, $x = 5$, is the correct answer.

(d) Substitute $(9, 2)$ in $2x - 5y = 8$.
$$2(9) - 5(2) = 8$$
$$18 - 10 = 8$$
$$8 = 8$$

Choice **B**, $2x - 5y = 8$, is the correct answer.

8. Substitute $x = 1997 - 1992 = 5$ into the equation
$$y = 47.3x + 281.$$
$$y = 47.3(5) + 281 \quad \text{Let } x = 5.$$
$$y = 517.5$$

From the graph, the approximate credit card debt in 1997 is about 510 billion dollars. From the equation, it is 517.5 billion dollars.

3.2 Section Exercises

1. $y = -x + 5$

$(0, \ ), (\ , 0), (2, \ )$

If $x = 0$,   If $y = 0$,
$y = -0 + 5$ $0 = -x + 5$
$y = 5.$ $x = 5.$

If $x = 2$,
$y = -2 + 5$
$y = 3.$

The ordered pairs are $(0, 5)$, $(5, 0)$, and $(2, 3)$. Plot the corresponding points and draw a line through them.

continued

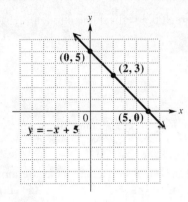

3. $y = \dfrac{2}{3}x + 1$

   $(0, \ ), (3, \ ), (-3, \ )$

   If $x = 0$,   If $x = 3$,

   $y = \dfrac{2}{3}(0) + 1$ $y = \dfrac{2}{3}(3) + 1$

   $y = 0 + 1$ $y = 2 + 1$

   $y = 1.$ $y = 3.$

   If $x = -3$,

   $y = \dfrac{2}{3}(-3) + 1$

   $y = -2 + 1$

   $y = -1.$

   The ordered pairs are $(0, 1)$, $(3, 3)$, and $(-3, -1)$. Plot the corresponding points and draw a line through them.

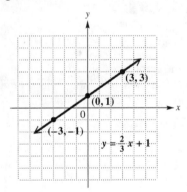

5. $3x = -y - 6$

   $(0, \ ), (\ , 0), \left(-\dfrac{1}{3}, \ \right)$

   If $x = 0$,   If $y = 0$,

   $3(0) = -y - 6$ $3x = -0 - 6$

   $0 = -y - 6$ $3x = -6$

   $y = -6.$ $x = -2.$

   If $x = -\dfrac{1}{3}$,

   $3\left(-\dfrac{1}{3}\right) = -y - 6$

   $-1 = -y - 6$

   $y - 1 = -6$

   $y = -5.$

   The ordered pairs are $(0, -6)$, $(-2, 0)$, and $\left(-\dfrac{1}{3}, -5\right)$. Plot the corresponding points and draw a line through them.

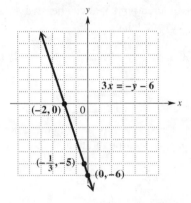

7. To determine which equation has y-intercept $(0, -4)$, set x equal to 0 and see which equation is equivalent to $y = -4$. Choice **A** is correct since

   $3x + y = -4$

   $3(0) + y = -4$

   $y = -4$

9. Choice **D** is correct since the graph of $y = 4$ is a horizontal line.

11. To find the x-intercept, let $y = 0$.

    $2x - 3y = 24$

    $2x - 3(0) = 24$

    $2x - 0 = 24$

    $2x = 24$

    $x = 12$

    The x-intercept is $(12, 0)$.

    To find the y-intercept, let $x = 0$.

    $2x - 3y = 24$

    $2(0) - 3y = 24$

    $0 - 3y = 24$

    $-3y = 24$

    $y = -8$

    The y-intercept is $(0, -8)$.

13. To find the x-intercept, let $y = 0$.

$$x + 6y = 0$$
$$x + 6(0) = 0$$
$$x + 0 = 0$$
$$x = 0$$

The x-intercept is $(0, 0)$. Since we have found the point with x equal to 0, this is also the y-intercept.

15. Her next step would be to let x be some number other than 0 and find the corresponding value of y. For example, if $x = -5$, $y = 4$. This will give a second point, $(-5, 4)$.

17. Begin by finding the intercepts.

$$x = y + 2$$
$$x = 0 + 2 \quad \textit{Let y = 0.}$$
$$x = 2$$

$$x = y + 2$$
$$0 = y + 2 \quad \textit{Let x = 0.}$$
$$-2 = y$$

The x-intercept is $(2, 0)$ and the y-intercept is $(0, -2)$. To find a third point, choose $y = 1$.

$$x = y + 2$$
$$x = 1 + 2 \quad \textit{Let y = 1.}$$
$$x = 3$$

This gives the ordered pair $(3, 1)$. Plot $(2, 0)$, $(0, -2)$, and $(3, 1)$ and draw a line through them.

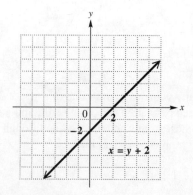

19. Find the intercepts.

$$x - y = 4$$
$$x - 0 = 4 \quad \textit{Let y = 0.}$$
$$x = 4$$

$$x - y = 4$$
$$0 - y = 4 \quad \textit{Let x = 0.}$$
$$y = -4$$

The x-intercept is $(4, 0)$ and the y-intercept is $(0, -4)$. To find a third point, choose $y = 1$.

$$x - y = 4$$
$$x - 1 = 4 \quad \textit{Let y = 1.}$$
$$x = 5$$

This gives the ordered pair $(5, 1)$. Plot $(4, 0)$, $(0, -4)$, and $(5, 1)$ and draw a line through them.

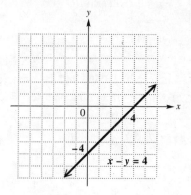

21. Find the intercepts.

$$2x + y = 6$$
$$2x + 0 = 6 \quad \textit{Let y = 0.}$$
$$2x = 6$$
$$x = 3$$

$$2x + y = 6$$
$$2(0) + y = 6 \quad \textit{Let x = 0.}$$
$$0 + y = 6$$
$$y = 6$$

The x-intercept is $(3, 0)$ and the y-intercept is $(0, 6)$. To find a third point, choose $x = 1$.

$$2x + y = 6$$
$$2(1) + y = 6 \quad \textit{Let x = 1.}$$
$$2 + y = 6$$
$$y = 4$$

This gives the ordered pair $(1, 4)$. Plot $(3, 0)$, $(0, 6)$, and $(1, 4)$ and draw a line through them.

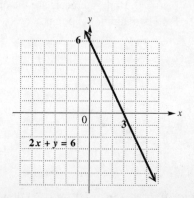

23. Find the intercepts.

$$3x + 7y = 14$$
$$3x + 7(0) = 14 \quad \text{Let } y = 0.$$
$$3x + 0 = 14$$
$$3x = 14$$
$$x = \tfrac{14}{3}$$

$$3x + 7y = 14$$
$$3(0) + 7y = 14 \quad \text{Let } x = 0.$$
$$0 + 7y = 14$$
$$7y = 14$$
$$y = 2$$

The x-intercept is $\left(\tfrac{14}{3}, 0\right)$ and the y-intercept is $(0, 2)$. To find a third point, choose $x = 2$.

$$3x + 7y = 14$$
$$3(2) + 7y = 14 \quad \text{Let } x = 2.$$
$$6 + 7y = 14$$
$$7y = 8$$
$$y = \tfrac{8}{7}$$

This gives the ordered pair $\left(2, \tfrac{8}{7}\right)$. Plot $\left(\tfrac{14}{3}, 0\right)$, $(0, 2)$, and $\left(2, \tfrac{8}{7}\right)$. Writing $\tfrac{14}{3}$ as the mixed number $4\tfrac{2}{3}$ and $\tfrac{8}{7}$ as $1\tfrac{1}{7}$ will be helpful for plotting. Draw a line through these three points.

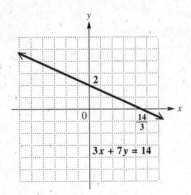

25. $y - 2x = 0$

If $y = 0$, $x = 0$. Both intercepts are the origin, $(0, 0)$. Find two additional points.

$$y - 2x = 0$$
$$y - 2(1) = 0 \quad \text{Let } x = 1.$$
$$y - 2 = 0$$
$$y = 2$$

$$y - 2x = 0$$
$$y - 2(-3) = 0 \quad \text{Let } x = -3.$$
$$y + 6 = 0$$
$$y = -6$$

Plot $(0, 0)$, $(1, 2)$, and $(-3, -6)$ and draw a line through them.

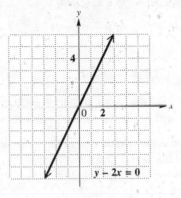

27. $y = -6x$

Find three points on the line.

If $x = 0$, $y = -6(0) = 0$.

If $x = 1$, $y = -6(1) = -6$.

If $x = -1$, $y = -6(-1) = 6$.

Plot $(0, 0)$, $(1, -6)$, and $(-1, 6)$ and draw a line through these points.

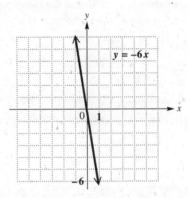

29. $x = -2$

For any value of y, the value of x is -2. Three ordered pairs are $(-2, 3)$, $(-2, 0)$, and $(-2, -4)$. Plot these points and draw a line through them. The graph is a vertical line.

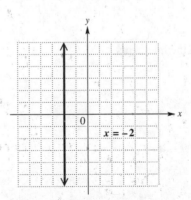

31. $y - 3 = 0$

$y = 3$

For any value of x, the value of y is 3. Three ordered pairs are $(-2, 3)$, $(0, 3)$, and $(4, 3)$. Plot these points and draw a line through them. The graph is a horizontal line.

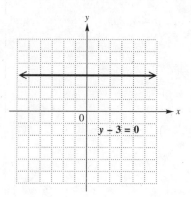

33. (a) $y = 3.9x + 73.5$

Let $x = 20$.
$y = 3.9(20) + 73.5 = 151.5$

Let $x = 26$.
$y = 3.9(26) + 73.5 = 174.9$

Let $x = 22$.
$y = 3.9(22) + 73.5 = 159.3$

The approximate heights of women with radius bones of lengths 20 cm, 26 cm, and 22 cm are 151.5 cm, 174.9 cm, and 159.3 cm, respectively.

(b) Plot the points $(20, 151.5)$, $(26, 174.9)$, and $(22, 159.3)$ and connect them with a smooth line.

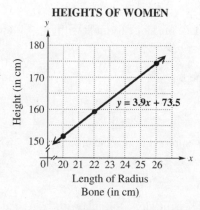

(c) Locate 167 on the vertical scale, then move across to the line, then down to the horizontal scale. From the graph, the radius bone in a woman who is 167 cm tall is about 24 cm.

Now substitute 167 for y in the equation.

$y = 3.9x + 73.5$
$167 = 3.9x + 73.5$
$93.5 = 3.9x$
$x = \dfrac{93.5}{3.9} \approx 23.97$

From the equation, the length of the radius bone is 24 cm to the nearest centimeter.

35.

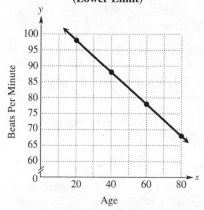

(a) From the graph, for age 30, the lower limit of the target heart rate zone is about 93.

(b) Let $x = 30$.

$y = -.5x + 108$
$y = -.5(30) + 108$
$y = 93$

From the equation, for age 30, the lower limit of the target heart rate zone is 93.

(c) They are the same.

37. From Exercises 35(b) and 36(b), the target heart rate zone for age 30 is between 93 and 149.

39. (a) $y = 1.13x + 20.67$

For 1995, let $x = 0$.

$y = 1.13(0) + 20.67 \approx 20.7$

For 1996, let $x = 1$.

$y = 1.13(1) + 20.67 = 21.8$

For 1998, let $x = 3$.

$y = 1.13(3) + 20.67 = 24.06 \approx 24.1$

The approximate consumptions for 1995, 1996, and 1998 are 20.7 gal, 21.8 gal, and 24.1 gal, respectively.

(b) Locate 0, 1, and 3 on the horizontal scale, then find the corresponding value on the vertical scale.

The approximate consumptions for 1995, 1996, and 1998 are 20.3 gal, 22.1 gal, and 23.9 gal, respectively.

(c) They differ by .4, .3, and .2 gallon, respectively.

41. (a) At $x = 0$, $y = 30{,}000$.
The initial value of the SUV is \$30,000.

(b) At $x = 3$, $y = 15{,}000$.
$30{,}000 - 15{,}000 = 15{,}000$
The depreciation after the first 3 years is \$15,000.

(c) The line connecting consecutive years from 0 to 1, 1 to 2, 2 to 3, and so on, drops 5000 for each segment. Therefore, the annual depreciation in each of the first 5 years is \$5000.

(d) $(5, 5000)$ means after 5 years the SUV has a value of \$5000.

43. (a) The equation is a fairly good model.

(b) The actual debt for 1996 is about 500 billion dollars; this is about 30 billion dollars more than the amount given by the equation.

(c) No. Data for future years might not follow the same pattern, so the linear equation would not be a reliable model.

3.3 Slope of a Line

3.3 Margin Exercises

1. (a) The indicated points have coordinates $(-1, -4)$ and $(1, -1)$.

$$\frac{\text{rise}}{\text{run}} = \frac{\text{vertical change}}{\text{horizontal change}}$$

$$= \frac{-1 - (-4)}{1 - (-1)}$$

$$= \frac{-1 + 4}{1 + 1}$$

$$= \frac{3}{2}$$

(b) The indicated points have coordinates $(2, 2)$ and $(4, -2)$.

$$\frac{\text{rise}}{\text{run}} = \frac{\text{vertical change}}{\text{horizontal change}}$$

$$= \frac{-2 - 2}{4 - 2}$$

$$= \frac{-4}{2} = -2$$

2. (a) For $y_2 = 4$, $y_1 = -1$, $x_2 = 3$, and $x_1 = 4$,

$$\frac{y_2 - y_1}{x_2 - x_1} = \frac{4 - (-1)}{3 - 4}$$

$$= \frac{4 + 1}{3 - 4}$$

$$= \frac{5}{-1} = -5.$$

(b) For $x_1 = 3$, $x_2 = -5$, $y_1 = 7$, and $y_2 = -9$,

$$\frac{y_2 - y_1}{x_2 - x_1} = \frac{-9 - 7}{-5 - 3}$$

$$= \frac{-16}{-8}$$

$$= 2.$$

(c) For $x_1 = 2$, $x_2 = 7$, $y_1 = 4$, and $y_2 = 9$,

$$\frac{y_2 - y_1}{x_2 - x_1} = \frac{9 - 4}{7 - 2}$$

$$= \frac{5}{5}$$

$$= 1.$$

3. (a) Use the slope formula with
$(6, -2) = (x_1, y_1)$ and $(5, 4) = (x_2, y_2)$.

$$\text{slope } m = \frac{\text{change in } y}{\text{change in } x} = \frac{y_2 - y_1}{x_2 - x_1}$$

$$= \frac{4 - (-2)}{5 - 6} = \frac{6}{-1} = -6$$

(b) Use the slope formula with $(-3, 5)$ and $(-4, -7)$.

$$\text{slope } m = \frac{\text{change in } y}{\text{change in } x} = \frac{y_2 - y_1}{x_2 - x_1}$$

$$= \frac{-7 - 5}{-4 - (-3)} = \frac{-12}{-1} = 12$$

(c) Use the slope formula with $(6, -8)$ and $(-2, 4)$.

$$\text{slope } m = \frac{\text{change in } y}{\text{change in } x} = \frac{y_2 - y_1}{x_2 - x_1}$$

$$= \frac{4 - (-8)}{-2 - 6} = \frac{12}{-8} = -\frac{3}{2}$$

Now subtract in reverse order:

$$m = \frac{-8 - 4}{6 - (-2)} = \frac{-12}{8} = -\frac{3}{2}$$

4. (a) Use the slope formula with $(2, 5)$ and $(-1, 5)$.

$$\text{slope } m = \frac{\text{change in } y}{\text{change in } x} = \frac{5 - 5}{-1 - 2} = \frac{0}{-3} = 0$$

(b) Use the slope formula with $(3, 1)$ and $(3, -4)$.

$$\text{slope } m = \frac{\text{change in } y}{\text{change in } x} = \frac{-4 - 1}{3 - 3} = \frac{-5}{0},$$

which is undefined.

(c) With equation $y = -1$

All lines with an equation of the form $y = k$ are horizontal and have a slope of 0.

(d) With equation $x - 4 = 0$

The equation $x - 4 = 0$ is equivalent to the equation $x = 4$. All lines with equation $x = k$ are vertical and have undefined slope.

5. (a) $y = -\frac{7}{2}x + 1$

The slope is the coefficient of x, $-\frac{7}{2}$.

(b) $3x + 2y = 9$

Solve for y.

$$\begin{aligned} 3x + 2y &= 9 \\ 2y &= -3x + 9 \quad \text{Subtract } 3x. \\ y &= -\frac{3}{2}x + \frac{9}{2} \quad \text{Divide by 2.} \end{aligned}$$

The slope is the coefficient of x, $-\frac{3}{2}$.

(c) $y + 4 = 0$
$\quad\quad y = -4 \quad$ Solve for y.

All lines with an equation of the form $y = k$ are horizontal and have a slope of 0.

(d) $x + 3 = 7$
$\quad\quad x = 4 \quad$ Solve for x.

All lines with an equation of the form $x = k$ are vertical and have undefined slope.

6. Find the slope of each line by first solving each equation for y.

(a) $\quad x + y = 6 \quad\quad\quad x + y = 1$
$\quad\quad\quad y = -x + 6 \quad\quad y = -x + 1$
$\quad\quad\quad \text{Slope is } -1. \quad\quad \text{Slope is } -1.$

The lines have the same slope, so they are parallel.

(b) $\quad 3x - y = 4 \quad\quad\quad x + 3y = 9$
$\quad\quad\quad -y = -3x + 4 \quad\quad 3y = -x + 9$
$\quad\quad\quad y = 3x - 4 \quad\quad\quad y = -\frac{1}{3}x + 3$
$\quad\quad\quad \text{Slope is 3.} \quad\quad\quad \text{Slope is } -\frac{1}{3}.$

The product of the slopes is $3\left(-\frac{1}{3}\right) = -1$, so the lines are perpendicular.

(c) $\quad 2x - y = 5 \quad\quad\quad 2x + y = 3$
$\quad\quad\quad -y = -2x + 5 \quad\quad y = -2x + 3$
$\quad\quad\quad y = 2x - 5 \quad\quad\quad \text{Slope is } -2.$
$\quad\quad\quad \text{Slope is 2.}$

The slopes are not equal, and the product of the slopes is $2(-2) = -4$, not -1. The lines are neither parallel nor perpendicular.

(d) $\quad 3x - 7y = 35 \quad\quad 7x - 3y = -6$
$\quad\quad\quad -7y = -3x + 35 \quad -3y = -7x - 6$
$\quad\quad\quad y = \frac{3}{7}x - 5 \quad\quad\quad y = \frac{7}{3}x + 2$
$\quad\quad\quad \text{Slope is } \frac{3}{7}. \quad\quad\quad \text{Slope is } \frac{7}{3}.$

The slopes are not equal, and the product of the slopes is $\left(\frac{3}{7}\right)\left(\frac{7}{3}\right) = 1$, not -1. The lines are neither parallel nor perpendicular.

3.3 Section Exercises

1. The indicated points have coordinates $(-1, -4)$ and $(3, 2)$.

$$\frac{\text{rise}}{\text{run}} = \frac{\text{vertical change}}{\text{horizontal change}}$$

$$= \frac{2 - (-4)}{3 - (-1)}$$

$$= \frac{6}{4} = \frac{3}{2}$$

3. The indicated points have coordinates $(-3, 5)$ and $(1, -2)$.

$$\frac{\text{rise}}{\text{run}} = \frac{\text{vertical change}}{\text{horizontal change}}$$

$$= \frac{-2 - 5}{1 - (-3)}$$

$$= \frac{-7}{4} = -\frac{7}{4}$$

5. The indicated points have coordinates $(1, 3)$ and $(4, 3)$.

$$\frac{\text{rise}}{\text{run}} = \frac{\text{vertical change}}{\text{horizontal change}}$$

$$= \frac{3 - 3}{1 - 4}$$

$$= \frac{0}{-3} = 0$$

7. Rise is the vertical change between two different points on a line.

Run is the horizontal change between two different points on a line.

9. Yes, the slope will be the same. If we *start* at $(-1, -4)$ and *end* at $(3, 2)$, the vertical change will be 6 (6 units up) and the horizontal change will be 4 (4 units to the right), giving a slope of
$$m = \frac{6}{4} = \frac{3}{2}.$$
If we *start* at $(3, 2)$ and *end* at $(-1, -4)$, the vertical change will be -6 (6 units down) and the horizontal change will be -4 (4 units to the left), giving a slope of
$$m = \frac{-6}{-4} = \frac{3}{2}.$$

11. Positive slope

    Sketches will vary. The line must rise from left to right. One such line is shown in the following graph.

    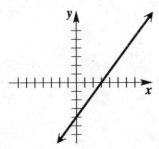

13. Zero slope

    Sketches will vary. The line must be horizontal. One such line is shown in the following graph.

    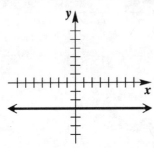

15. His answer is incorrect. Because he found the difference $3 - 5 = -2$ in the numerator, he should have subtracted in the same order in the denominator to get $-1 - 2 = -3$. The correct slope is $\frac{-2}{-3} = \frac{2}{3}$. Note that these slopes are opposites of one another.

17. Use the slope formula with $(1, -2) = (x_1, y_1)$ and $(-3, -7) = (x_2, y_2)$.
$$\text{slope } m = \frac{\text{change in } y}{\text{change in } x}$$
$$= \frac{y_2 - y_1}{x_2 - x_1}$$
$$= \frac{-7 - (-2)}{-3 - 1}$$
$$= \frac{-5}{-4} = \frac{5}{4}$$

19. Use the slope formula with $(0, 3) = (x_1, y_1)$ and $(-2, 0) = (x_2, y_2)$.
$$\text{slope } m = \frac{\text{change in } y}{\text{change in } x}$$
$$= \frac{y_2 - y_1}{x_2 - x_1}$$
$$= \frac{0 - 3}{-2 - 0}$$
$$= \frac{-3}{-2} = \frac{3}{2}$$

21. Use the slope formula with $(-2, 4) = (x_1, y_1)$ and $(-3, 7) = (x_2, y_2)$.
$$\text{slope } m = \frac{\text{change in } y}{\text{change in } x}$$
$$= \frac{y_2 - y_1}{x_2 - x_1}$$
$$= \frac{7 - 4}{-3 - (-2)}$$
$$= \frac{3}{-1} = -3$$

23. Use the slope formula with $(4, 3) = (x_1, y_1)$ and $(-6, 3) = (x_2, y_2)$.
$$\text{slope } m = \frac{\text{change in } y}{\text{change in } x}$$
$$= \frac{y_2 - y_1}{x_2 - x_1}$$
$$= \frac{3 - 3}{-6 - 4}$$
$$= \frac{0}{-10} = 0$$

25. Use the slope formula with $(-12, 3) = (x_1, y_1)$ and $(-12, -7) = (x_2, y_2)$.

$$\text{slope } m = \frac{\text{change in } y}{\text{change in } x}$$
$$= \frac{y_2 - y_1}{x_2 - x_1}$$
$$= \frac{-7 - 3}{-12 - (-12)}$$
$$= \frac{-10}{0},$$

which is undefined.

27. Use the slope formula with $\left(-\frac{7}{5}, \frac{3}{10}\right) = (x_1, y_1)$ and $\left(\frac{1}{5}, -\frac{1}{2}\right) = (x_2, y_2)$.

$$\text{slope } m = \frac{\text{change in } y}{\text{change in } x}$$
$$= \frac{y_2 - y_1}{x_2 - x_1}$$
$$= \frac{-\frac{1}{2} - \frac{3}{10}}{\frac{1}{5} - \left(-\frac{7}{5}\right)}$$
$$= \frac{-\frac{5}{10} - \frac{3}{10}}{\frac{1}{5} + \frac{7}{5}}$$
$$= \frac{-\frac{8}{10}}{\frac{8}{5}}$$
$$= \left(-\frac{8}{10}\right)\left(\frac{5}{8}\right)$$
$$= -\frac{5}{10} = -\frac{1}{2}$$

29. $y = 5x + 12$

Since the equation is already solved for y, the slope is given by the coefficient of x, which is 5. Thus, the slope of the line is 5.

31. Solve the equation for y.

$$4y = x + 1$$
$$y = \frac{1}{4}x + \frac{1}{4} \quad \text{Divide by 4.}$$

The slope of the line is given by the coefficient of x, so the slope is $\frac{1}{4}$.

33. Solve the equation for y.

$$3x - 2y = 3$$
$$-2y = -3x + 3 \quad \text{Subtract } 3x.$$
$$y = \frac{3}{2}x - \frac{3}{2} \quad \text{Divide by } -2.$$

The slope of the line is given by the coefficient of x, so the slope is $\frac{3}{2}$.

35. $y = 6$

This is an equation of a horizontal line. Its slope is 0. (This equation may be rewritten in the form $y = 0x + 6$, where the coefficient of x gives the slope.)

37. $x = -2$

This is an equation of a vertical line. Its slope is *undefined*.

39. Solve the equation for y.

$$x - y = 0$$
$$x = y \quad \text{Add } y.$$

The slope of the line is given by the coefficient of x, so the slope is 1.

41. (a) Because the line *falls* from left to right, its slope is *negative*.

(b) Because the line intersects the y-axis *at* the origin, the y-value of its y-intercept is *zero*.

43. (a) Because the line *rises* from left to right, its slope is *positive*.

(b) Because the line intersects the y-axis *below* the origin, the y-value of its y-intercept is *negative*.

45. (a) The line is *horizontal*, so its slope is *zero*.

(b) The line intersects the y-axis *below* the origin, so the y-value of its y-intercept is *negative*.

47. Find the slope of each line by solving the equations for y.

$$-4x + 3y = 4$$
$$3y = 4x + 4 \quad \text{Add } 4x.$$
$$y = \frac{4}{3}x + \frac{4}{3} \quad \text{Divide by 3.}$$

The slope of the first line is $\frac{4}{3}$.

$$-8x + 6y = 0$$
$$6y = 8x \quad \text{Add } 8x.$$
$$y = \frac{8}{6}x \quad \text{Divide by 6.}$$
$$y = \frac{4}{3}x \quad \text{Reduce.}$$

The slope of the second line is $\frac{4}{3}$.

The slopes are equal, so the lines are *parallel*.

49. Find the slope of each line by solving the equations for y.

$$5x - 3y = -2$$
$$-3y = -5x - 2 \quad \text{Subtract } 5x.$$
$$y = \frac{5}{3}x + \frac{2}{3} \quad \text{Divide by } -3.$$

The slope of the first line is $\frac{5}{3}$.

$$3x - 5y = -8$$
$$-5y = -3x - 8 \quad \text{Subtract } 3x.$$
$$y = \frac{3}{5}x + \frac{8}{5} \quad \text{Divide by } -5.$$

The slope of the second line is $\frac{3}{5}$.

The slopes are not equal, so the lines are not parallel. The product of the slopes of the (nonvertical) lines is not -1, so the lines are not perpendicular. Thus, the lines are *neither* parallel nor perpendicular.

51. Find the slope of each line by solving the equations for y.

$$3x - 5y = -1$$
$$-5y = -3x - 1 \quad \text{Subtract } 3x.$$
$$y = \frac{3}{5}x + \frac{1}{5} \quad \text{Divide by } -5.$$

The slope of the first line is $\frac{3}{5}$.

$$5x + 3y = 2$$
$$3y = -5x + 2 \quad \text{Subtract } 5x.$$
$$y = -\frac{5}{3}x + \frac{2}{3} \quad \text{Divide by } 3.$$

The slope of the second line is $-\frac{5}{3}$.

The product of the slopes is

$$\frac{3}{5}\left(-\frac{5}{3}\right) = -1,$$

so the lines are *perpendicular*.

53. The slope (or grade) of the hill is the ratio of the rise to the run, or the ratio of the vertical change to the horizontal change. Since the rise is 32 and the run is 108, the slope is

$$\frac{32}{108} = \frac{8 \cdot 4}{27 \cdot 4} = \frac{8}{27}.$$

54. We use the points with coordinates (1990, 11,338) and (2005, 14,818).

$$m = \frac{14{,}818 \text{ thousand} - 11{,}338 \text{ thousand}}{2005 - 1990}$$
$$= \frac{3480 \text{ thousand}}{15}$$
$$= 232 \text{ thousand}$$

or 232,000.

55. The slope of the line in Figure A is *positive*. This means that during the period represented, enrollment *increased* in grades 9–12.

56. The increase is approximately 232,000 students per year.

57. We use the points with coordinates (1990, 22) and (2000, 5.4).

$$m = \frac{5.4 - 22}{2000 - 1990}$$
$$= \frac{-16.6}{10}$$
$$= -1.66 \text{ students per computer}$$

58. The slope of the line in Figure B is *negative*. This means that during the period represented, the number of students per computer *decreased*.

59. The decrease is 1.66 students per computer per year.

3.4 Equations of Lines

3.4 Margin Exercises

1. (a) slope $\frac{1}{2}$; y-intercept $(0, -4)$

Use the slope-intercept form of a line with $m = \frac{1}{2}$ and $b = -4$.

$$y = mx + b$$
$$y = \frac{1}{2}x - 4$$

(b) $m = -1; b = 8$

$$y = -1x + 8 \quad \text{or} \quad y = -x + 8$$

(c) $m = 3; b = 0$

$$y = 3x + 0 \quad \text{or} \quad y = 3x$$

(d) $m = 0; b = 2$

$$y = (0)x + 2 \quad \text{or} \quad y = 2$$

3.4 Equations of Lines 109

2. *Step 1*
Begin by solving for y.

$$3x - 4y = 8 \quad \text{Given equation}$$
$$-4y = -3x + 8 \quad \text{Subtract } 3x.$$
$$y = \frac{3}{4}x - 2 \quad \text{Divide by } -4.$$

Step 2
The y-intercept is $(0, -2)$. Graph this point.

Step 3
The slope is $\frac{3}{4} = \frac{\text{change in } y}{\text{change in } x}$.

Starting at the y-intercept, we count 3 units up and 4 units right to obtain another point on the graph, $(4, 1)$.

Step 4
Draw the line through the points $(0, -2)$ and $(4, 1)$ to obtain the graph of $3x - 4y = 8$.

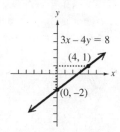

3. Through $(2, -3)$, with slope $-\frac{1}{3}$

To graph the line, write the slope as

$$m = \frac{\text{difference in } y\text{-values}}{\text{difference in } x\text{-values}} = \frac{-1}{3}.$$

(This could also be $\frac{1}{-3}$.)

Locate $(2, -3)$. Count 1 unit down and 3 units to the right. Draw a line through this point, $(5, -4)$, and $(2, -3)$.

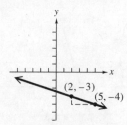

4. **(a)** Through $(-1, 3)$, with slope -2

Use the point-slope form of the equation of a line with $x_1 = -1$, $y_1 = 3$, and $m = -2$.

$$y - y_1 = m(x - x_1)$$
$$y - \underline{3} = \underline{-2}\,[x - (\underline{-1})]$$
$$y - 3 = -2(x + \underline{1})$$
$$y - 3 = -2x - \underline{2}$$
$$y = \underline{-2x + 1}$$

(b) Through $(5, 2)$, with slope $-\frac{1}{3}$

Use the point-slope form of the equation of a line with $x_1 = 5$, $y_1 = 2$, and $m = -\frac{1}{3}$.

$$y - y_1 = m(x - x_1)$$
$$y - 2 = -\frac{1}{3}(x - 5)$$
$$y - 2 = -\frac{1}{3}x + \frac{5}{3}$$
$$y = -\frac{1}{3}x + \frac{5}{3} + \frac{6}{3}$$
$$y = -\frac{1}{3}x + \frac{11}{3}$$

5. **(a)** $(-3, 1)$ and $(2, 4)$

First, find the slope of the line.

$$m = \frac{4 - 1}{2 - (-3)} = \frac{3}{5}$$

Now use the point $(-3, 1)$ for (x_1, y_1) and $m = \frac{3}{5}$ in the point-slope form.

$$y - y_1 = m(x - x_1)$$
$$y - 1 = \frac{3}{5}[x - (-3)]$$
$$y - 1 = \frac{3}{5}(x + 3)$$
$$y - 1 = \frac{3}{5}x + \frac{9}{5}$$
$$y = \frac{3}{5}x + \frac{9}{5} + \frac{5}{5}$$
$$y = \frac{3}{5}x + \frac{14}{5}$$

(b) $(2, 5)$ and $(-1, 6)$

First, find the slope of the line.

$$m = \frac{6 - 5}{-1 - 2} = \frac{1}{-3} = -\frac{1}{3}$$

Now use the point $(2, 5)$ for (x_1, y_1) and $m = -\frac{1}{3}$ in the point-slope form.

$$y - y_1 = m(x - x_1)$$
$$y - 5 = -\frac{1}{3}(x - 2)$$
$$y - 5 = -\frac{1}{3}x + \frac{2}{3}$$
$$y = -\frac{1}{3}x + \frac{2}{3} + \frac{15}{3}$$
$$y = -\frac{1}{3}x + \frac{17}{3}$$

6. Let $(x_1, y_1) = (1, 2537)$ and $(x_2, y_2) = (9, 3746)$.

$$m = \frac{y_2 - y_1}{x_2 - x_1}$$
$$= \frac{3746 - 2537}{9 - 1}$$
$$= \frac{1209}{8}$$
$$= 151.125$$

Now use this slope and the point $(1, 2537)$ in the point-slope form to find an equation of the line.

$$y - y_1 = m(x - x_1)$$
$$y - 2537 = 151.125(x - 1)$$
$$y - 2537 = 151.125x - 151.125$$
$$y = 151.125x + 2385.875$$

For 1998, let $x = 5$.

$$y = 151.125(5) + 2385.875$$
$$y = 3141.5 \approx 3142$$

The equation gives $y \approx 3142$ when $x = 5$, which is a very good approximation of the cost given in the table, 3110.

3.4 Section Exercises

1. The point-slope form of the equation of a line with slope -2 going through the point $(4, 1)$ is

$$y - 1 = -2(x - 4).$$

Choice **D** is correct.

3. A line that passes through the points $(0, 0)$ [its y-intercept] and $(4, 1)$ has slope

$$m = \frac{\text{rise}}{\text{run}} = \frac{1 - 0}{4 - 0} = \frac{1}{4}.$$

Its slope-intercept form is

$$y = \frac{1}{4}x + 0 \quad \text{or} \quad y = \frac{1}{4}x.$$

Choice **B** is correct.

5. The rise is 3 and the run is 1, so the slope is given by

$$m = \frac{\text{rise}}{\text{run}} = \frac{3}{1} = 3.$$

The y-intercept is $(0, -3)$, so $b = -3$. The equation of the line, written in slope-intercept form, is

$$y = 3x - 3.$$

7. Since the line falls from left to right, the "rise" is negative. For this line, the rise is -3 and the run is 3, so the slope is

$$m = \frac{\text{rise}}{\text{run}} = \frac{-3}{3} = -1.$$

The y-intercept is $(0, 3)$, so $b = 3$. The slope-intercept form of the equation of the line is

$$y = -1x + 3$$
$$y = -x + 3.$$

9. slope 4; y-intercept $(0, -3)$

Since the y-intercept is $(0, -3)$, we have $b = -3$. Use the slope-intercept form.

$$y = mx + b$$
$$y = 4x + (-3)$$
$$y = 4x - 3$$

11. slope 0; y-intercept $(0, 3)$

Since the y-intercept is $(0, 3)$, we have $b = 3$. Use the slope-intercept form.

$$y = mx + b$$
$$y = 0x + 3$$
$$y = 3$$

13. A vertical line has undefined slope, so there is no value for m. Also, there is no y-intercept, so there can be no value for b.

15. $y = 3x + 2$

The slope is $3 = \frac{3}{1} = \frac{\text{change in } y}{\text{change in } x}$, and the y-intercept is $(0, 2)$. Graph that point and count up 3 units and right 1 unit to get to the point $(1, 5)$. Draw a line through the points.

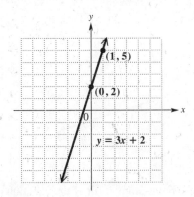

17. $2x + y = -5$, so $y = -2x - 5$

The slope is $-2 = \dfrac{-2}{1} = \dfrac{\text{change in } y}{\text{change in } x}$, and the y-intercept is $(0, -5)$. Graph that point and count down 2 units and right 1 unit to get to the point $(1, -7)$. Draw a line through the points.

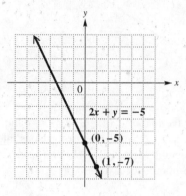

19. $x + 2y = 4$

Solve the given equation for y.

$2y = -x + 4$ Subtract x.

$y = -\dfrac{1}{2}x + 2$ Divide by 2.

The slope is $-\dfrac{1}{2} = \dfrac{-1}{2} = \dfrac{\text{change in } y}{\text{change in } x}$, and the y-intercept is $(0, 2)$. Graph that point and count down 1 unit and right 2 units to get to the point $(2, 1)$. Draw a line through the points.

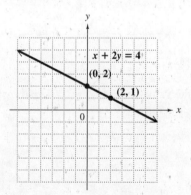

21. $(-2, 3)$, $m = \dfrac{1}{2}$

First, locate the point $(-2, 3)$. Write the slope as

$$m = \dfrac{\text{rise}}{\text{run}} = \dfrac{1}{2}.$$

Locate another point by counting 1 unit up and then 2 units to the right. Draw a line through this new point, $(0, 4)$, and the given point $(-2, 3)$. The slope-intercept form of the equation of this line is $y = \dfrac{1}{2}x + 4$.

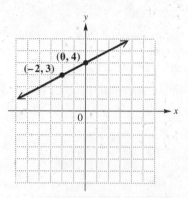

23. $(1, -5)$, $m = -\dfrac{2}{5}$

First, locate the point $(1, -5)$. Write the slope as

$$m = \dfrac{\text{rise}}{\text{run}} = \dfrac{-2}{5}.$$

Locate another point by counting 2 units down (because of the negative sign) and then 5 units to the right. Draw a line through this new point, $(6, -7)$, and the given point, $(1, -5)$.

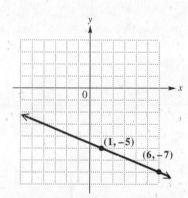

To find the slope-intercept form of the equation of this line, we will use the point-slope form and solve for y.

$y - y_1 = m(x - x_1)$

$y - (-5) = -\dfrac{2}{5}(x - 1)$ $(x_1, y_1) = (1, -5)$, $m = -\dfrac{2}{5}$

$y + 5 = -\dfrac{2}{5}x + \dfrac{2}{5}$

$y = -\dfrac{2}{5}x + \dfrac{2}{5} - \dfrac{25}{5}$ Subtract $5 = \dfrac{25}{5}$.

$y = -\dfrac{2}{5}x - \dfrac{23}{5}$

25. $(3, 2)$, $m = 0$

First, locate the point $(3, 2)$. Since the slope is 0, the line will be horizontal. Draw a horizontal line through the point $(3, 2)$.

The slope-intercept form of the equation of this line is $y = 0x + 2$, or $y = 2$.

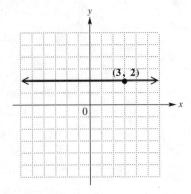

27. $(3, -2)$, undefined slope

First, locate the point $(3, -2)$. Since the slope is undefined, the line will be vertical. Draw a vertical line through the point $(3, -2)$.

An equation of this line is $x = 3$. There is no slope-intercept form.

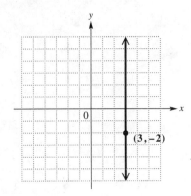

29. $(0, 0)$, $m = \dfrac{2}{3}$

Start at the origin and locate another point by counting 2 units up and then 3 units to the right. Draw the line through this new point, $(3, 2)$, and the given point, $(0, 0)$. The slope-intercept form of the equation of this line is $y = \dfrac{2}{3}x$.

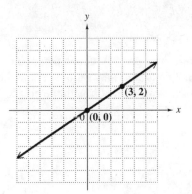

31. $(4, 1)$, $m = 2$

The given point is $(4, 1)$, so $x_1 = 4$ and $y_1 = 1$. Also, $m = 2$. Substitute these values into the point-slope form. Then solve for y to obtain the slope-intercept form.

$$y - y_1 = m(x - x_1)$$
$$y - 1 = 2(x - 4)$$
$$y - 1 = 2x - 8 \quad \textit{Distributive property}$$
$$y = 2x - 7 \quad \textit{Add 1.}$$

33. $(3, -10)$, $m = -2$

Use the values $x_1 = 3$, $y_1 = -10$, and $m = -2$ in the point-slope form.

$$y - y_1 = m(x - x_1)$$
$$y - (-10) = -2(x - 3)$$
$$y + 10 = -2x + 6 \quad \textit{Distributive property}$$
$$y = -2x - 4 \quad \textit{Subtract 10.}$$

35. $(-2, 5)$, $m = \dfrac{2}{3}$

Use the values $x_1 = -2$, $y_1 = 5$, and $m = \dfrac{2}{3}$ in the point-slope form.

$$y - y_1 = m(x - x_1)$$
$$y - 5 = \dfrac{2}{3}[x - (-2)]$$
$$y - 5 = \dfrac{2}{3}(x + 2)$$
$$y - 5 = \dfrac{2}{3}x + \dfrac{4}{3}$$
$$y = \dfrac{2}{3}x + \dfrac{19}{3} \quad \textit{Add } 5 = \dfrac{15}{3}.$$

37. $(8, 5)$ and $(9, 6)$

First, find the slope of the line.
$$m = \frac{6-5}{9-8} = \frac{1}{1} = 1$$

Now use the point $(8, 5)$ for (x_1, y_1) and $m = 1$ in the point-slope form.
$$y - y_1 = m(x - x_1)$$
$$y - 5 = 1(x - 8)$$
$$y - 5 = x - 8$$
$$y = x - 3$$

The same result would be found by using $(9, 6)$ for (x_1, y_1).

39. $(-1, -7)$ and $(-8, -2)$

First, find the slope of the line.
$$m = \frac{-2 - (-7)}{-8 - (-1)} = \frac{5}{-7} = -\frac{5}{7}$$

Now use the point $(-1, -7)$ for (x_1, y_1) and $m = -\frac{5}{7}$ in the point-slope form.
$$y - y_1 = m(x - x_1)$$
$$y - (-7) = -\frac{5}{7}[x - (-1)]$$
$$y + 7 = -\frac{5}{7}(x + 1)$$
$$y + 7 = -\frac{5}{7}x - \frac{5}{7}$$
$$y = -\frac{5}{7}x - \frac{54}{7} \quad \text{Subtract } 7 = \frac{49}{7}.$$

41. $(0, -2)$ and $(-3, 0)$

First, find the slope of the line.
$$m = \frac{0 - (-2)}{-3 - 0} = \frac{2}{-3} = -\frac{2}{3}$$

Since one of the given points is the y-intercept, use the slope-intercept form of a line.
$$y = mx + b$$
$$y = -\frac{2}{3}x - 2$$

43. $(3, 5)$ and $(3, -2)$
$$m = \frac{-2 - 5}{3 - 3} = \frac{-7}{0}$$

The slope is undefined, so the equation has the form $x = k$ and since $x = 3$ in both of the given points, the equation is $x = 3$ (no slope-intercept form).

45. $\left(\frac{1}{2}, \frac{3}{2}\right), \left(-\frac{1}{4}, \frac{5}{4}\right)$

First, find the slope of the line.

$$m = \frac{\frac{5}{4} - \frac{3}{2}}{-\frac{1}{4} - \frac{1}{2}} = \frac{\frac{5}{4} - \frac{6}{4}}{-\frac{1}{4} - \frac{2}{4}}$$

$$= \frac{-\frac{1}{4}}{-\frac{3}{4}} = \left(-\frac{1}{4}\right)\left(-\frac{4}{3}\right) = \frac{1}{3}$$

Now use the point $\left(\frac{1}{2}, \frac{3}{2}\right)$ for (x_1, y_1) and $m = \frac{1}{3}$ in the point-slope form.
$$y - y_1 = m(x - x_1)$$
$$y - \frac{3}{2} = \frac{1}{3}\left(x - \frac{1}{2}\right)$$
$$y - \frac{3}{2} = \frac{1}{3}x - \frac{1}{6}$$
$$y = \frac{1}{3}x - \frac{1}{6} + \frac{9}{6} \quad \left(\frac{3}{2} = \frac{9}{6}\right)$$
$$y = \frac{1}{3}x + \frac{4}{3} \quad \left(\frac{8}{6} = \frac{4}{3}\right)$$

47. Solve the equation for y.
$$3x = 4y + 5$$
$$-4y = -3x + 5$$
$$y = \frac{3}{4}x - \frac{5}{4}$$

The slope is $\frac{3}{4}$. A line parallel to this line has the same slope. Now use the point-slope form with $m = \frac{3}{4}$ and $(x_1, y_1) = (2, -3)$.
$$y - y_1 = m(x - x_1)$$
$$y - (-3) = \frac{3}{4}(x - 2)$$
$$y + 3 = \frac{3}{4}x - \frac{3}{2}$$
$$y = \frac{3}{4}x - \frac{3}{2} - \frac{6}{2}$$
$$y = \frac{3}{4}x - \frac{9}{2}$$

49. Solve the equation for y.
$$x - 2y = 7$$
$$-2y = -x + 7$$
$$y = \frac{1}{2}x - \frac{7}{2}$$

The slope is $\frac{1}{2}$. A line perpendicular to this line has slope -2 (since $\frac{1}{2}(-2) = -1$).
Now use the slope-intercept form with $m = -2$ and y-intercept $(0, -3)$.
$$y = mx + b$$
$$y = -2x - 3$$

51. When $C = 0$, $F = 32$. This gives the ordered pair $(0, 32)$. When $C = 100$, $F = 212$. This gives the ordered pair $(100, 212)$.

52. Use the two points $(0, 32)$ and $(100, 212)$.
$$m = \frac{212 - 32}{100 - 0} = \frac{180}{100} = \frac{9}{5}$$

53. Write the point-slope form as
$$F - F_1 = m(C - C_1).$$
We may use either the point $(0, 32)$ or the point $(100, 212)$ with the slope $\frac{9}{5}$, which we found in Exercise 52. Using $F_1 = 32$, $C_1 = 0$, and $m = \frac{9}{5}$, we obtain
$$F - 32 = \frac{9}{5}(C - 0).$$
Using $F_1 = 212$, $C_1 = 100$, and $m = \frac{9}{5}$, we obtain
$$F - 212 = \frac{9}{5}(C - 100).$$

54. We want to write an equation in the form
$$F = mC + b.$$
From Exercise 52, we have $m = \frac{9}{5}$. The point $(0, 32)$ is the y-intercept of the graph of the desired equation, so we have $b = 32$. Thus, an equation for F in terms of C is
$$F = \frac{9}{5}C + 32.$$

55. The expression for F in terms of C obtained in Exercise 54 is
$$F = \frac{9}{5}C + 32.$$
To obtain an expression for C in terms of F, solve this equation for C.
$$F - 32 = \frac{9}{5}C \quad \text{Subtract 32.}$$
$$\frac{5}{9}(F - 32) = \frac{5}{9} \cdot \frac{9}{5}C \quad \text{Multiply by } \frac{5}{9}, \text{ the reciprocal of } \frac{9}{5}.$$
$$\frac{5}{9}(F - 32) = C \quad \text{or} \quad C = \frac{5}{9}F - \frac{160}{9}$$

56. $F = \frac{9}{5}C + 32$

If $C = 30$,
$$F = \frac{9}{5}(30) + 32$$
$$= 54 + 32 = 86.$$
Thus, when $C = 30°$, $F = 86°$.

57. $C = \frac{5}{9}(F - 32)$

When $F = 50$,
$$C = \frac{5}{9}(50 - 32)$$
$$= \frac{5}{9}(18) = 10.$$
Thus, when $F = 50°$, $C = 10°$.

58. Let $F = C$ in the equation obtained in Exercise 54.
$$F = \frac{9}{5}C + 32$$
$$C = \frac{9}{5}C + 32 \quad \text{Let } F = C.$$
$$5C = 5\left(\frac{9}{5}C + 32\right) \quad \text{Multiply by 5.}$$
$$5C = 9C + 160$$
$$-4C = 160 \quad \text{Subtract } 9C.$$
$$C = -40 \quad \text{Divide by } -4.$$
(The same result may be found by using either form of the equation obtained in Exercise 55.) The Celsius and Fahrenheit temperatures are equal ($F = C$) at -40 degrees.

59. (a) The fixed cost is $400.

(b) The variable cost is $.25.

(c) Substitute $m = .25$ and $b = 400$ into $y = mx + b$ to get the cost equation
$$y = .25x + 400.$$

(d) Let $x = 100$ in the cost equation.
$$y = .25(100) + 400$$
$$y = 25 + 400$$
$$y = 425$$
The cost to produce 100 snow cones will be $425.

(e) Let $y = 775$ in the cost equation.
$$775 = .25x + 400$$
$$375 = .25x \quad \text{Subtract 400.}$$
$$x = \frac{375}{.25} = 1500 \quad \text{Divide by .25.}$$
If the total cost is $775, 1500 snow cones will be produced.

61. (a) x represents the year and y represents the cost in the ordered pairs (x, y). The ordered pairs are $(1, 1125)$, $(3, 1239)$, $(5, 1314)$, $(7, 1338)$, and $(9, 1379)$.

(b)

AVERAGE ANNUAL COSTS AT 2-YEAR COLLEGES

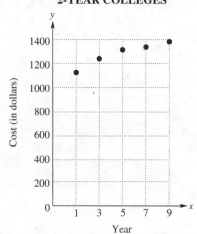

Yes, the points lie approximately in a straight line.

(c) Find the slope using $(x_1, y_1) = (3, 1239)$ and $(x_2, y_2) = (9, 1379)$.

$$m = \frac{y_2 - y_1}{x_2 - x_1} = \frac{1379 - 1239}{9 - 3} = \frac{140}{6} = \frac{70}{3}$$

Now use the point-slope form with $m = \frac{70}{3}$ and $(x_1, y_1) = (3, 1239)$.

$$y - y_1 = m(x - x_1)$$
$$y - 1239 = \frac{70}{3}(x - 3)$$
$$y - 1239 = \frac{70}{3}x - 70$$
$$y = \frac{70}{3}x + 1169$$

[or $y = 23.33x + 1169$ (rounded)]

(d) Since year 1 represents 1994, year 0 represents 1993.

For 2004, $x = 2004 - 1993 = 11$.

$$y = 23.33x + 1169$$
$$y = 23.33(11) + 1169$$
$$y = 1425.63$$

In 2004, the estimate of the average annual cost at 2-year colleges is $1426.

Summary Exercises on Graphing Linear Equations

1. Solve the equation for y.

$3x + y = -6$ *Given equation*
$y = -3x - 6$ *Subtract 3x.*

The slope of the line is given by the coefficient of x, so the slope is -3. The y-intercept is $(0, -6)$.

3. Solve the equation for y.

$-4x - y = 3$ *Given equation*
$-y = 4x + 3$ *Add 4x.*
$y = -4x - 3$ *Multiply by -1.*

The slope of the line is given by the coefficient of x, so the slope is -4. The y-intercept is $(0, -3)$.

5. Solve the equation for y.

$-3x + 2y = 12$ *Given equation*
$2y = 3x + 12$ *Add 3x.*
$y = \frac{3}{2}x + 6$ *Divide by 2.*

The slope of the line is given by the coefficient of x, so the slope is $\frac{3}{2}$. The y-intercept is $(0, 6)$.

7. $m = 1$, $b = -2$

The slope is $1 = \frac{1}{1} = \frac{\text{change in } y}{\text{change in } x}$, and the y-intercept is $(0, -2)$. Graph that point and count up 1 unit and right 1 unit to get to the point $(1, -1)$. Draw a line through the points.

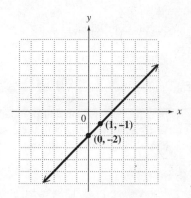

9. $m = -2$, $b = 6$

The slope is $-2 = \frac{-2}{1} = \frac{\text{change in } y}{\text{change in } x}$, and the y-intercept is $(0, 6)$. Graph that point and count down 2 units and right 1 unit to get to the point $(1, 4)$. Draw a line through the points.

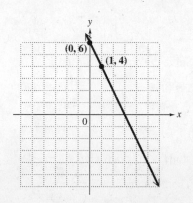

11. $m = -\frac{2}{3}, b = -2$

The slope is $-\frac{2}{3} = \frac{-2}{3} = \frac{\text{change in } y}{\text{change in } x}$, and the y-intercept is $(0, -2)$. Graph that point and count down 2 units and right 3 units to get to the point $(3, -4)$. Draw a line through the points.

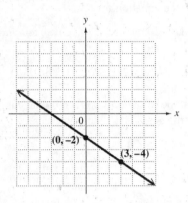

13. (a) Solve the equation for y.

$-x + y = -3$ *Given equation*
$y = x - 3$ *Add x.*

The slope is the coefficient of x, 1.

(b) The y-intercept is the point $(0, -3)$.

(c) The slope is $1 = \frac{1}{1} = \frac{\text{change in } y}{\text{change in } x}$, and the y-intercept is $(0, -3)$. Graph that point and count up 1 unit and right 1 unit to get to the point $(1, -2)$. Draw a line through the points.

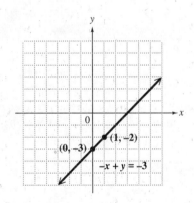

15. (a) Solve the equation for y.

$x + 2y = 4$ *Given equation*
$2y = -x + 4$ *Subtract x.*
$y = -\frac{1}{2}x + 2$ *Divide by 2.*

The slope is the coefficient of x, $-\frac{1}{2}$.

(b) The y-intercept is the point $(0, 2)$.

(c) The slope is $-\frac{1}{2} = \frac{-1}{2} = \frac{\text{change in } y}{\text{change in } x}$, and the y-intercept is $(0, 2)$. Graph that point and count down 1 unit and right 2 units to get to the point $(2, 1)$. Draw a line through the points.

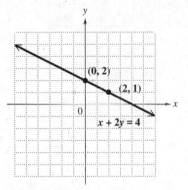

17. (a) Solve the equation for y.

$4x - 5y = 20$ *Given equation*
$-5y = -4x + 20$ *Subtract 4x.*
$y = \frac{4}{5}x - 4$ *Divide by -5.*

The slope is the coefficient of x, $\frac{4}{5}$.

(b) The y-intercept is the point $(0, -4)$.

(c) The slope is $\frac{4}{5} = \frac{\text{change in } y}{\text{change in } x}$, and the y-intercept is $(0, -4)$. Graph that point and count up 4 units and right 5 units to get to the point $(5, 0)$. Draw a line through the points.

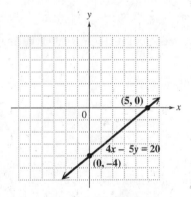

19. (a) Solve the equation for y.

$2x + 3y = 12$ *Given equation*
$3y = -2x + 12$ *Subtract 2x.*
$y = -\frac{2}{3}x + 4$ *Divide by 3.*

The slope is the coefficient of x, $-\frac{2}{3}$.

(b) The y-intercept is the point $(0, 4)$.

(c) The slope is $-\dfrac{2}{3} = \dfrac{-2}{3} = \dfrac{\text{change in } y}{\text{change in } x}$, and the y-intercept is $(0, 4)$. Graph that point and count down 2 units and right 3 units to get to the point $(3, 2)$. Draw a line through the points.

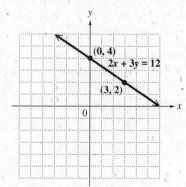

21. (a) Solve the equation for y.

$$x - 3y = 6 \quad \text{Given equation}$$
$$-3y = -x + 6 \quad \text{Subtract } x.$$
$$y = \dfrac{1}{3}x - 2 \quad \text{Divide by } -3.$$

The slope is the coefficient of x, $\dfrac{1}{3}$.

(b) The y-intercept is the point $(0, -2)$.

(c) The slope is $\dfrac{1}{3} = \dfrac{\text{change in } y}{\text{change in } x}$, and the y-intercept is $(0, -2)$. Graph that point and count up 1 unit and right 3 units to get to the point $(3, -1)$. Draw a line through the points.

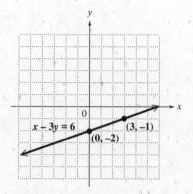

23. (a) Solve the equation for y.

$$x - 4y = 0 \quad \text{Given equation}$$
$$-4y = -x \quad \text{Subtract } x.$$
$$y = \dfrac{1}{4}x \quad \text{Divide by } -4.$$

The slope is the coefficient of x, $\dfrac{1}{4}$.

(b) The y-intercept is the point $(0, 0)$.

(c) The slope is $\dfrac{1}{4} = \dfrac{\text{change in } y}{\text{change in } x}$, and the y-intercept is $(0, 0)$. Graph that point and count up 1 unit and right 4 units to get to the point $(4, 1)$. Draw a line through the points.

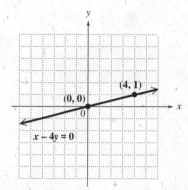

3.5 Graphing Linear Inequalities in Two Variables

3.5 Margin Exercises

1. (a) $x + 2y \geq 6$

$$2y \geq -x + 6 \quad \text{Subtract } x.$$
$$y \geq -\dfrac{1}{2}x + 3 \quad \text{Divide by 2.}$$

The ordered pairs for which y is equal to $-\dfrac{1}{2}x + 3$ are on the line with equation

$$y = -\dfrac{1}{2}x + 3.$$

Graph this line with y-intercept $(0, 3)$ and slope $-\dfrac{1}{2}$. The ordered pairs for which y is *greater than* $-\dfrac{1}{2}x + 3$ are *above* this line. Indicate the solution by shading the region above the line.

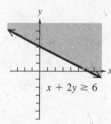

(b) $3x + 4y \leq 12$

$$4y \leq -3x + 12 \quad \text{Subtract } 3x.$$
$$y \leq -\dfrac{3}{4}x + 3 \quad \text{Divide by 4.}$$

The ordered pairs for which y is equal to $-\dfrac{3}{4}x + 3$ are on the line with equation

$$y = -\dfrac{3}{4}x + 3.$$

continued

118 Chapter 3 Graphs of Linear Equations and Inequalities in Two Variables

Graph this line with y-intercept $(0, 3)$ and slope $-\frac{3}{4}$. The ordered pairs for which y is *less than* $-\frac{3}{4}x + 3$ are *below* this line. Indicate the solution by shading the region below the line.

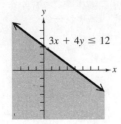

2. $4x - 5y \leq 20$
$4(0) - 5(0) \leq 20$? Let $x = 0, y = 0$.
$0 - 0 \leq 20$?
$0 \leq 20$ True

Since the last statement is true, shade the region that includes the test point $(0, 0)$, that is, the region above the line.

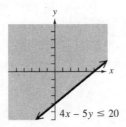

3. $3x + 5y > 15$
$3(1) + 5(1) > 15$? Let $x = 1, y = 1$.
$3 + 5 > 15$?
$8 > 15$ False

Since the last statement is false, shade the region that does *not* include the test point, $(1, 1)$, that is, the region above the line.

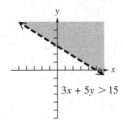

4. Graph $2x - y \geq -4$.

Start by graphing the equation
$$2x - y = -4.$$

The intercepts are $(-2, 0)$ and $(0, 4)$. Draw a solid line through these points to show that the points on the line are solutions to the inequality
$$2x - y \geq -4.$$

Choose a test point not on the line, such as $(0, 0)$.

$2x - y \geq -4$
$2(0) - 0 \geq -4$? Let $x = 0, y = 0$.
$0 \geq -4$ True

Since this statement is true, shade the region that includes the test point $(0, 0)$, that is, the region below the line.

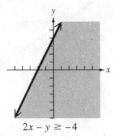

5. $y < 4$

First, graph $y = 4$, a horizontal line through the point $(0, 4)$. Use a dashed line because of the $<$ symbol. Choose $(0, 0)$ as a test point.

$y < 4$
$0 < 4$ Let $y = 0$. True

Since the last statement is true, shade the region that includes the test point $(0, 0)$, that is, the region below the line.

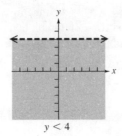

6. $x \geq -3y$

Graph $x = -3y$ as a solid line through $(0, 0)$ and $(-3, 1)$. We cannot use $(0, 0)$ as a test point because $(0, 0)$ is on the line $x = -3y$. Instead, we choose a test point off the line, $(0, 1)$.

$x \geq -3y$
$0 \geq -3(1)$? Let $x = 0, y = 1$.
$0 \geq -3$ True

Since the last statement is true, shade the region that includes the test point $(0, 1)$, that is, the region above the line.

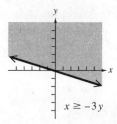

Because the last statement is *false*, we shade the region that *does not* include the test point $(0, 0)$. This is the region above the line.

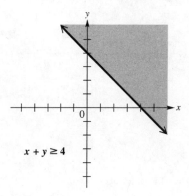

3.5 Section Exercises

1. For the point $(4, 0)$, substitute 4 for x and 0 for y in the inequality.

    $$3x - 4y < 12$$
    $$3(4) - 4(0) < 12 \text{ ?}$$
    $$12 - 0 < 12 \text{ ?}$$
    $$12 < 12 \quad False$$

    The *false* result shows that $(4, 0)$ *is not* a solution of the given inequality.

3. $3x - 2y \geq 0$

    (i) For the point $(4, 1)$, substitute 4 for x and 1 for y in the given inequality.

    $$3x - 2y \geq 0$$
    $$3(4) - 2(1) \geq 0 \text{ ?}$$
    $$12 - 2 \geq 0 \text{ ?}$$
    $$10 \geq 0 \quad True$$

    The *true* result shows that $(4, 1)$ *is* a solution of the given inequality.

    (ii) For the point $(0, 0)$, substitute 0 for x and 0 for y in the given inequality.

    $$3x - 2y \geq 0$$
    $$3(0) - 2(0) \geq 0 \text{ ?}$$
    $$0 \geq 0 \quad True$$

    The *true* result shows that $(0, 0)$ *is* a solution of the given inequality.

    Since (i) and (ii) are true, the given statement is *true*.

5. The key phrase is "more than," so use the symbol $>$.

7. The key phrase is "at most," so use the symbol $\leq$.

9. $x + y \geq 4$

    Use $(0, 0)$ as a test point.

    $$x + y \geq 4$$
    $$0 + 0 \geq 4 \text{ ?} \quad Let\ x = 0,\ y = 0.$$
    $$0 \geq 4 \quad False$$

11. $x + 2y \geq 7$

    Use $(0, 0)$ as a test point.

    $$x + 2y \geq 7$$
    $$0 + 2(0) \geq 7 \text{ ?} \quad Let\ x = 0,\ y = 0.$$
    $$0 \geq 7 \quad False$$

    Because the last statement is *false*, we shade the region that *does not* include the test point $(0, 0)$. This is the region above the line.

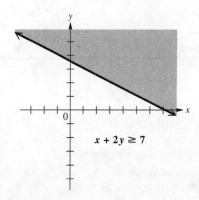

13. $-3x + 4y > 12$

    Use $(0, 0)$ as a test point.

    $$-3x + 4y > 12$$
    $$-3(0) + 4(0) > 12 \text{ ?} \quad Let\ x = 0,\ y = 0.$$
    $$0 > 12 \quad False$$

    Because the last statement is *false*, we shade the region that *does not* include the test point $(0, 0)$. This is the region above the line.

    continued

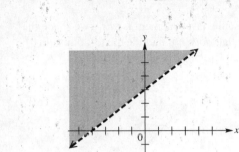

$-3x + 4y > 12$

15. $x > 4$

Use $(0, 0)$ as a test point.

$x > 4$
$0 > 4$? *Let x = 0. False*

Because $0 > 4$ is *false*, shade the region *not* containing $(0, 0)$. This is the region to the right of the line.

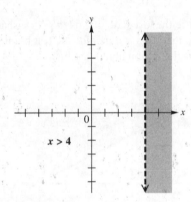

$x > 4$

17. Use a dashed line if the symbol is $<$ or $>$. Use a solid line if the symbol is $\leq$ or $\geq$.

19. $x + y \leq 5$

Step 1
Graph the boundary of the region, the line with equation $x + y = 5$.

If $y = 0$, $x = 5$, so the x-intercept is $(5, 0)$.
If $x = 0$, $y = 5$, so the y-intercept is $(0, 5)$.

Draw the line through these intercepts. Make the line solid because of the $\leq$ sign.

Step 2
Choose the point $(0, 0)$ as a test point.

$x + y \leq 5$
$0 + 0 \leq 5$? *Let x = 0, y = 0.*
$0 \leq 5$ *True*

Because $0 \leq 5$ is true, shade the region containing the origin. The shaded region, along with the boundary, is the desired graph.

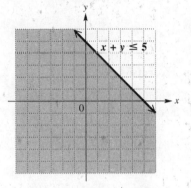

$x + y \leq 5$

21. $x + 2y < 4$

Start by graphing the boundary, the line with equation $x + 2y = 4$, through its intercepts $(0, 2)$ and $(4, 0)$. Make the line dashed because of the $<$ sign. Choose $(0, 0)$ as a test point.

$x + 2y < 4$
$0 + 2(0) < 4$? *Let x = 0, y = 0.*
$0 < 4$ *True*

Because the last statement is *true*, shade the region containing the origin. The dashed line indicates that the boundary is not part of the graph.

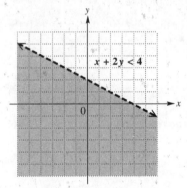

$x + 2y < 4$

23. $2x + 6 > -3y$

Start by graphing the boundary, the line with equation $2x + 6 = -3y$, through its intercepts $(-3, 0)$ and $(0, -2)$. Make the line dashed because of the $>$ sign.

Choose $(0, 0)$ as a test point.

$2x + 6 > -3y$
$2(0) + 6 > -3(0)$? *Let x = 0, y = 0.*
$6 > 0$ *True*

Because the last statement is *true*, shade the region containing the origin. The dashed line indicates that the boundary is not part of the graph.

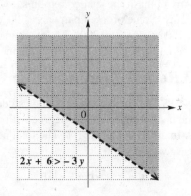

25. $y \geq 2x + 1$

The boundary is the line with equation $y = 2x + 1$. This line has slope 2 and y-intercept $(0, 1)$. It may be graphed by starting at $(0, 1)$ and going 2 units up and then 1 unit to the right to reach the point $(1, 3)$. Draw a solid line through $(0, 1)$ and $(1, 3)$.

The ordered pairs for which y is *greater than* $2x + 1$ are *above* this line. Indicate the solution by shading the region above the line. The solid line shows that the boundary is part of the graph.

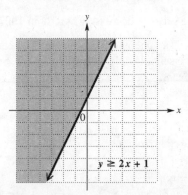

27. $x \leq -2$

The boundary is the line with equation $x = -2$. This is a vertical line through $(-2, 0)$. Make this line solid because of the $\leq$ sign.

Using $(0, 0)$ as a test point will result in the inequality $0 \leq -2$, which is false. Shade the region not containing the origin. This is the region to the left of the boundary. The solid line shows that the boundary is part of the graph.

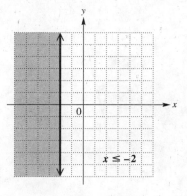

29. $y < 5$

The boundary is the line with equation $y = 5$. This is the horizontal line through $(0, 5)$. Make this line dashed because of the $<$ sign.

The ordered pairs for which y is *less than* 5 are *below* this line. Indicate the solution by shading the region below the line. The dashed line shows that the boundary is not part of the graph.

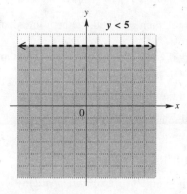

31. $y \geq 4x$

The boundary has the equation $y = 4x$. This line goes through the points $(0, 0)$ and $(1, 4)$. Make the line solid because of the $\geq$ sign. Because the boundary passes through the origin, we cannot use $(0, 0)$ as a test point.

The ordered pairs for which y is *greater than* $4x$ are *above* this line. Indicate the solution by shading the region above the line. The solid line shows that the boundary is part of the graph.

continued

122 Chapter 3 Graphs of Linear Equations and Inequalities in Two Variables

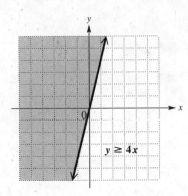

33. Every point in quadrant IV has a positive x-value and a negative y-value. Substituting into $y > x$ would imply that a negative number is greater than a positive number, which is always false. Thus, the graph of $y > x$ cannot lie in quadrant IV.

35. $x + y \geq 500$

    (a) Graph the inequality.

    Step 1
    Graph the line $x + y = 500$.

    If $x = 0$, then $y = 500$, so the y-intercept is $(0, 500)$.
    If $y = 0$, then $x = 500$, so the x-intercept is $(500, 0)$.

    Graph the line with these intercepts.

    The line is solid because of the $\geq$ sign.

    Step 2
    Use $(0, 0)$ as a test point.

    $x + y \geq 500$ *Original inequality*
    $0 + 0 \geq 500$?   *Let x = 0, y = 0.*
    $0 \geq 500$ *False*

    Since $0 \geq 500$ is false, shade the side of the boundary not containing $(0, 0)$. Because of the restrictions $x \geq 0$ and $y \geq 0$ in this applied problem, only the portion of the graph that lies in quadrant I is included.

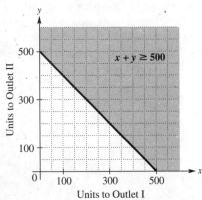

(b) Any point in the shaded region satisfies the inequality. Some ordered pairs are $(500, 0)$, $(200, 400)$, and $(400, 200)$. There are many other ordered pairs that will also satisfy the inequality.

Chapter 3 Review Exercises

1. In 1996, 57.1% of four-year college students in private institutions earned a degree within five years. In 2001, it was 55.1%. The total percent decrease was

    $$57.1\% - 55.1\% = 2.0\%.$$

2. In this case, x represents the year and y represents the percent. Ordered pairs that reflect the data shown in the graph are $(1996, 57.1)$, $(1997, 56.6)$, $(1998, 56.2)$, $(1999, 55.8)$, $(2000, 55.5)$, and $(2001, 55.1)$.

3. Create a table to determine the differences.

From	to	the difference was
1996	1997	$56.6\% - 57.1\% = -.5\%$
1997	1998	$56.2\% - 56.6\% = -.4\%$
1998	1999	$55.8\% - 56.2\% = -.4\%$
1999	2000	$55.5\% - 55.8\% = -.3\%$
2000	2001	$55.1\% - 55.5\% = -.4\%$

    So the greatest decrease was .5% in 1997.

4. There is a general trend of decreasing percents of private school students earning a degree within five years.

5. $y = 3x + 2$; $(-1, \ )\ (0, \ )\ (\ , 5)$

    $y = 3x + 2$
    $y = 3(-1) + 2$ *Let x = -1.*
    $y = -3 + 2$
    $y = -1$

    $y = 3x + 2$
    $y = 3(0) + 2$ *Let x = 0.*
    $y = 0 + 2$
    $y = 2$

    $y = 3x + 2$
    $5 = 3x + 2$ *Let y = 5.*
    $3 = 3x$
    $1 = x$

    The ordered pairs are $(-1, -1)$, $(0, 2)$, and $(1, 5)$.

6. $4x + 3y = 6$; $(0, \ )\ (\ , 0)\ (-2, \ )$

    $4x + 3y = 6$
    $4(0) + 3y = 6$ *Let x = 0.*
    $3y = 6$
    $y = 2$

$$4x + 3y = 6$$
$$4x + 3(0) = 6 \quad \text{Let } y = 0.$$
$$4x = 6$$
$$x = \frac{6}{4} = \frac{3}{2}$$

$$4x + 3y = 6$$
$$4(-2) + 3y = 6 \quad \text{Let } x = -2.$$
$$-8 + 3y = 6$$
$$3y = 14$$
$$y = \frac{14}{3}$$

The ordered pairs are $(0, 2)$, $\left(\frac{3}{2}, 0\right)$, and $\left(-2, \frac{14}{3}\right)$.

7. $x = 3y$; $(0, \ )$ $(8, \ )$ $(\ , -3)$

$$x = 3y$$
$$0 = 3y \quad \text{Let } x = 0.$$
$$0 = y$$

$$x = 3y$$
$$8 = 3y \quad \text{Let } x = 8.$$
$$\frac{8}{3} = y$$

$$x = 3y$$
$$x = 3(-3) \quad \text{Let } y = -3.$$
$$x = -9$$

The ordered pairs are $(0, 0)$, $\left(8, \frac{8}{3}\right)$, and $(-9, -3)$.

8. $x - 7 = 0$; $(\ , -3)$ $(\ , 0)$ $(\ , 5)$

The given equation may be written $x = 7$. For any value of y, the value of x will always be 7. The ordered pairs are $(7, -3)$, $(7, 0)$, and $(7, 5)$.

9. $x + y = 7$; $(2, 5)$

Substitute 2 for x and 5 for y in the given equation.

$$x + y = 7$$
$$2 + 5 = 7 \quad ?$$
$$7 = 7 \quad \text{True}$$

Yes, $(2, 5)$ is a solution of the given equation.

10. $2x + y = 5$; $(-1, 3)$

Substitute -1 for x and 3 for y in the given equation.

$$2x + y = 5$$
$$2(-1) + 3 = 5 \quad ?$$
$$-2 + 3 = 5 \quad ?$$
$$1 = 5 \quad \text{False}$$

No, $(-1, 3)$ is not a solution of the given equation.

11. $3x - y = 4$; $\left(\frac{1}{3}, -3\right)$

Substitute $\frac{1}{3}$ for x and -3 for y in the given equation.

$$3x - y = 4$$
$$3\left(\frac{1}{3}\right) - (-3) = 4 \quad ?$$
$$1 + 3 = 4 \quad ?$$
$$4 = 4 \quad \text{True}$$

Yes, $\left(\frac{1}{3}, -3\right)$ is a solution of the given equation.

For Exercises 12–15, the ordered pairs are plotted on the graph following the solution for Exercise 15.

12. To plot $(2, 3)$, start at the origin, go 2 units to the right, and then go up 3 units. The point lies in quadrant I.

13. To plot $(-4, 2)$, start at the origin, go 4 units to the left, and then go up 2 units. The point lies in quadrant II.

14. To plot $(3, 0)$, start at the origin, go 3 units to the right. The point lies on the x-axis (not in any quadrant).

15. To plot $(0, -6)$, start at the origin, go down 6 units. The point lies on the y-axis (not in any quadrant).

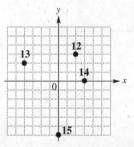

16. x is positive in quadrants I and IV; y is negative in quadrants III and IV. Thus, if x is positive and y is negative, (x, y) must lie in quadrant IV.

17. In the ordered pair $(k, 0)$, the y-value is 0, so the point lies on the x-axis. In the ordered pair $(0, k)$, the x-value is 0, so the point lies on the y-axis.

18. The point $(-2, 3)$ has a negative x-coordinate and a positive y-coordinate, so it lies in quadrant II.

19. The point $(-1, -4)$ has a negative x-coordinate and a negative y-coordinate, so it lies in quadrant III.

20. The point $\left(0, -5\frac{1}{2}\right)$ has an x-coordinate of 0, so it lies on the y-axis. It does not lie in any quadrant.

21. To find the x-intercept, let $y = 0$.

$$y = 2x + 5$$
$$0 = 2x + 5 \quad \text{Let } y = 0.$$
$$-2x = 5$$
$$x = -\frac{5}{2}$$

The x-intercept is $\left(-\frac{5}{2}, 0\right)$.

To find the y-intercept, let $x = 0$.

$$y = 2x + 5$$
$$y = 2(0) + 5 \quad \text{Let } x = 0.$$
$$y = 5$$

The y-intercept is $(0, 5)$.

22. To find the x-intercept, let $y = 0$.

$$2x + y = -7$$
$$2x + 0 = -7 \quad \text{Let } y = 0.$$
$$2x = -7$$
$$x = -\frac{7}{2}$$

The x-intercept is $\left(-\frac{7}{2}, 0\right)$.

To find the y-intercept, let $x = 0$.

$$2x + y = -7$$
$$2(0) + y = -7 \quad \text{Let } x = 0.$$
$$y = -7$$

The y-intercept is $(0, -7)$.

23. To find the x-intercept, let $y = 0$.

$$3x + 2y = 8$$
$$3x + 2(0) = 8 \quad \text{Let } y = 0.$$
$$3x = 8$$
$$x = \frac{8}{3}$$

The x-intercept is $\left(\frac{8}{3}, 0\right)$.

To find the y-intercept, let $x = 0$.

$$3x + 2y = 8$$
$$3(0) + 2y = 8 \quad \text{Let } x = 0.$$
$$2y = 8$$
$$y = 4$$

The y-intercept is $(0, 4)$.

24. Begin by finding the intercepts.

$$2x - y = 3$$
$$2x - 0 = 3 \quad \text{Let } y = 0.$$
$$2x = 3$$
$$x = \frac{3}{2}$$

$$2x - y = 3$$
$$2(0) - y = 3 \quad \text{Let } x = 0.$$
$$0 - y = 3$$
$$y = -3$$

The x-intercept is $\left(\frac{3}{2}, 0\right)$ and the y-intercept is $(0, -3)$. To find a third point, choose $y = 3$.

$$2x - y = 3$$
$$2x - 3 = 3 \quad \text{Let } y = 3.$$
$$2x = 6$$
$$x = 3$$

This gives the ordered pair $(3, 3)$. Plot $\left(\frac{3}{2}, 0\right)$, $(0, -3)$, and $(3, 3)$ and draw a line through them.

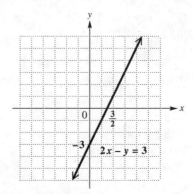

25. $x + 2y = -4$

Find the intercepts.

If $y = 0$, $x = -4$, so the x-intercept is $(-4, 0)$.
If $x = 0$, $y = -2$, so the y-intercept is $(0, -2)$.

To find a third point, choose $x = 2$.

$$x + 2y = -4$$
$$2 + 2y = -4 \quad \text{Let } x = 2.$$
$$2y = -6$$
$$y = -3$$

This gives the ordered pair $(2, -3)$. Plot $(-4, 0)$, $(0, -2)$, and $(2, -3)$ and draw a line through them.

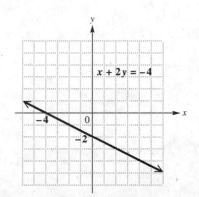

26. $x + y = 0$

   This line goes through the origin, so both intercepts are $(0,0)$. Two other points on the line are $(2,-2)$ and $(-2,2)$.

   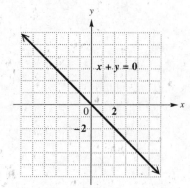

27. Let $(2,3) = (x_1, y_1)$ and $(-4, 6) = (x_2, y_2)$.

   $$\text{slope } m = \frac{y_2 - y_1}{x_2 - x_1} = \frac{6 - 3}{-4 - 2}$$
   $$= \frac{3}{-6} = -\frac{1}{2}$$

28. Let $(0,0) = (x_1, y_1)$ and $(-3, 2) = (x_2, y_2)$.

   $$\text{slope } m = \frac{2 - 0}{-3 - 0} = \frac{2}{-3} = -\frac{2}{3}$$

29. Let $(0,6) = (x_1, y_1)$ and $(1, 6) = (x_2, y_2)$.

   $$\text{slope } m = \frac{6 - 6}{1 - 0} = \frac{0}{1} = 0$$

30. Let $(2,5) = (x_1, y_1)$ and $(2, 8) = (x_2, y_2)$.

   $$\text{slope } m = \frac{8 - 5}{2 - 2}$$
   $$= \frac{3}{0}, \text{ which is } undefined.$$

31. $y = 3x - 4$

   The equation is already solved for y, so the slope of the line is given by the coefficient of x. Thus, the slope is 3.

32. $y = \frac{2}{3}x + 1$

   The equation is already solved for y, so the slope is given by the coefficient of x. Thus, the slope is $\frac{2}{3}$.

33. The indicated points have coordinates $(0, -2)$ and $(2, 1)$. Use the definition of slope with $(0, -2) = (x_1, y_1)$ and $(2, 1) = (x_2, y_2)$.

   $$m = \frac{y_2 - y_1}{x_2 - x_1}$$
   $$= \frac{1 - (-2)}{2 - 0}$$
   $$= \frac{3}{2}$$

34. The indicated points have coordinates $(0, 1)$ and $(3, 0)$. Use the definition of slope with $(0, 1) = (x_1, y_1)$ and $(3, 0) = (x_2, y_2)$.

   $$m = \frac{y_2 - y_1}{x_2 - x_1}$$
   $$= \frac{0 - 1}{3 - 0}$$
   $$= \frac{-1}{3} = -\frac{1}{3}$$

35. $x = 0$ is the equation of the y-axis, which is a vertical line. Its slope is undefined.

36. $y = 4$ is the equation of a horizontal line. Its slope is 0.

37. From the table, we choose two points $(0, 1)$ and $(2, 4)$. Therefore,

   $$\text{slope } m = \frac{4 - 1}{2 - 0} = \frac{3}{2}.$$

   Any points from the table will give the same result.

38. (a) Because parallel lines have equal slopes and the slope of the graph of $y = 2x + 3$ is 2, the slope of a line parallel to it will also be 2.

   (b) Because perpendicular lines have slopes whose product is -1 and the slope of the graph of $y = -3x + 3$ is -3, the slope of a line perpendicular to it will be $\frac{1}{3}$ (since $-3\left(\frac{1}{3}\right) = -1$).

39. Find the slope of each line by solving the equations for y.

   $$3x + 2y = 6$$
   $$2y = -3x + 6 \quad \textit{Subtract } 3x.$$
   $$y = -\frac{3}{2}x + 3 \quad \textit{Divide by 2.}$$

   The slope of the first line is $-\frac{3}{2}$.

   $$6x + 4y = 8$$
   $$4y = -6x + 8 \quad \textit{Subtract } 6x.$$
   $$y = -\frac{6}{4}x + 2 \quad \textit{Divide by 4.}$$
   $$y = -\frac{3}{2}x + 2 \quad \textit{Reduce.}$$

   The slope of the second line is $-\frac{3}{2}$. The slopes are equal so the lines are *parallel*.

40. Find the slope of each line by solving the equations for y.

$$x - 3y = 1$$
$$-3y = -x + 1 \quad \text{Subtract } x.$$
$$y = \frac{1}{3}x - \frac{1}{3} \quad \text{Divide by } -3.$$

The slope of the first line is $\frac{1}{3}$.

$$3x + y = 4$$
$$y = -3x + 4 \quad \text{Subtract } 3x.$$

The slope of the second line is -3.

The product of the slopes is

$$\frac{1}{3}(-3) = -1,$$

so the lines are *perpendicular*.

41. Find the slope of each line by solving the equations for y.

$$x - 2y = 8$$
$$-2y = -x + 8 \quad \text{Subtract } x.$$
$$y = \frac{1}{2}x - 4 \quad \text{Divide by } -2.$$

The slope of the first line is $\frac{1}{2}$.

$$x + 2y = 8$$
$$2y = -x + 8 \quad \text{Subtract } x.$$
$$y = -\frac{1}{2}x + 4 \quad \text{Divide by } 2.$$

The slope of the second line is $-\frac{1}{2}$.

The slopes are not equal and their product is

$$\left(\frac{1}{2}\right)\left(-\frac{1}{2}\right) = -\frac{1}{4} \neq -1,$$

so the lines are *neither* parallel nor perpendicular.

42. A line with undefined slope is vertical. Any line perpendicular to a vertical line is a horizontal line, which has a slope of 0.

43. $m = -1$, $b = \frac{2}{3}$

Use the slope-intercept form.

$$y = mx + b$$
$$y = -1 \cdot x + \frac{2}{3}$$
$$y = -x + \frac{2}{3}$$

44. The line in Exercise 34 has slope $-\frac{1}{3}$ and y-intercept $(0, 1)$, so we may write its equation in slope-intercept form as

$$y = -\frac{1}{3}x + 1.$$

45. Through $(4, -3)$, $m = 1$

Use the point-slope form.

$$y - y_1 = m(x - x_1)$$
$$y - (-3) = 1(x - 4)$$
$$y + 3 = x - 4$$
$$y = x - 7$$

46. Through $(-1, 4)$, $m = \frac{2}{3}$

Use the point-slope form.

$$y - y_1 = m(x - x_1)$$
$$y - 4 = \frac{2}{3}[x - (-1)]$$
$$y - 4 = \frac{2}{3}(x + 1)$$
$$y - 4 = \frac{2}{3}x + \frac{2}{3}$$
$$y = \frac{2}{3}x + \frac{14}{3}$$

47. Through $(1, -1)$, $m = -\frac{3}{4}$

Use the point-slope form.

$$y - (-1) = -\frac{3}{4}(x - 1)$$
$$y + 1 = -\frac{3}{4}x + \frac{3}{4}$$
$$y = -\frac{3}{4}x - \frac{1}{4}$$

48. Through $(2, 1)$ and $(-2, 2)$

First, find the slope of the line.

$$m = \frac{2 - 1}{-2 - 2} = \frac{1}{-4} = -\frac{1}{4}$$

Using $(2, 1)$ and $m = -\frac{1}{4}$, substitute into the point-slope form.

$$y - y_1 = m(x - x_1)$$
$$y - 1 = -\frac{1}{4}(x - 2)$$
$$y - 1 = -\frac{1}{4}x + \frac{1}{2}$$
$$y = -\frac{1}{4}x + \frac{1}{2} + \frac{2}{2}$$
$$y = -\frac{1}{4}x + \frac{3}{2}$$

49. Through $(-4, 1)$ with slope 0

Horizontal lines have 0 slope and equations of the form $y = k$. In this case, k must equal 1 since the line goes through $(-4, 1)$, so the equation is

$$y = 1.$$

50. Through $\left(\frac{1}{3}, -\frac{3}{4}\right)$ with undefined slope

Vertical lines have undefined slope and equations of the form $x = k$. In this case, k must equal $\frac{1}{3}$ since the line goes through $\left(\frac{1}{3}, -\frac{3}{4}\right)$, so the equation is

$$x = \tfrac{1}{3}.$$

It is not possible to express this equation as $y = mx + b$.

51. (a) Solve the equation for y.

$$\begin{aligned} x + 3y &= 15 & \text{Given equation} \\ 3y &= -x + 15 & \text{Subtract } x. \\ y &= -\tfrac{1}{3}x + 5 & \text{Divide by 3.} \end{aligned}$$

(b) The slope is $-\frac{1}{3}$ and the y-intercept is $(0, 5)$.

(c) The slope is $-\dfrac{1}{3} = \dfrac{-1}{3} = \dfrac{\text{change in } y}{\text{change in } x}$, and the y-intercept is $(0, 5)$. Graph that point and count down 1 unit and right 3 units to get to the point $(3, 4)$. Draw a line through the points.

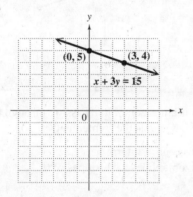

52. $3x + 5y > 9$

To graph the boundary, which is the line $3x + 5y = 9$, find its intercepts.

$$\begin{array}{ll} 3x + 5y = 9 & 3x + 5y = 9 \\ 3x + 5(0) = 9 & 3(0) + 5y = 9 \\ \quad \text{Let } y = 0. & \quad \text{Let } x = 0. \\ 3x = 9 & 5y = 9 \\ x = 3 & y = \tfrac{9}{5} \end{array}$$

The x-intercept is $(3, 0)$ and the y-intercept is $\left(0, \frac{9}{5}\right)$. (A third point may be used as a check.) Draw a dashed line through these points. In order to determine which side of the line should be shaded, use $(0, 0)$ as a test point. Substituting 0 for x and 0 for y will result in the inequality $0 > 9$, which is false. Shade the region *not* containing the origin. This is the region above the line. The dashed line shows that the boundary is not part of the graph.

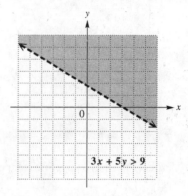

53. $2x - 3y > -6$

Use intercepts to graph the boundary, $2x - 3y = -6$.

If $y = 0$, $x = -3$, so the x-intercept is $(-3, 0)$.

If $x = 0$, $y = 2$, so the y-intercept is $(0, 2)$.

Draw a dashed line through $(-3, 0)$ and $(0, 2)$.

Using $(0, 0)$ as a test point will result in the inequality $0 > -6$, which is true. Shade the region containing the origin. This is the region below the line. The dashed line shows that the boundary is not part of the graph.

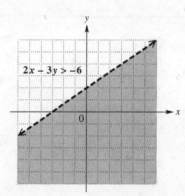

54. $x \geq -4$

The boundary line has the equation $x = -4$. This is the vertical line through $(-4, 0)$. Make this line solid because of the $\geq$ sign. Using $(0, 0)$ as a test point will result in the inequality $0 \geq -4$, which is true. Shade the region containing the origin. The solid line shows that the boundary is part of the graph.

continued

128 Chapter 3 Graphs of Linear Equations and Inequalities in Two Variables

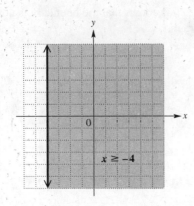

55. [3.3] Vertical lines have undefined slopes. The answer is **A**.

56. [3.2] Two graphs pass through $(0, -3)$. **C** and **D** are the answers.

57. [3.2] Three graphs pass through the point $(-3, 0)$. **A**, **B**, and **D** are the answers.

58. [3.3] Lines that fall from left to right have negative slope. The answer is **D**.

59. [3.2] $y = -3$ is a horizontal line passing through $(0, -3)$. **C** is the answer.

60. [3.3] **B** is the only graph that has a positive slope, so it is the only one we need to investigate. **B** passes through the points $(0, 3)$ and $(-3, 0)$. Find the slope.

$$m = \frac{0-3}{-3-0} = \frac{-3}{-3} = 1$$

B is the answer.

61. [3.3] $y = -2x - 5$

The equation is in the slope-intercept form, so the slope is -2 and the y-intercept is $(0, -5)$. To find the x-intercept, let $y = 0$.

$$0 = -2x - 5 \quad \text{Let } y = 0.$$
$$2x = -5$$
$$x = -\frac{5}{2}$$

The x-intercept is $\left(-\frac{5}{2}, 0\right)$.

Graph the line using the intercepts.

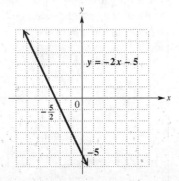

62. [3.3] $x + 3y = 0$

Solve the equation for y.

$$x + 3y = 0$$
$$3y = -x \quad \text{Subtract } x.$$
$$y = -\frac{1}{3}x \quad \text{Divide by 3.}$$

From this slope-intercept form, we see that the slope is $-\frac{1}{3}$ and the y-intercept is $(0, 0)$, which is also the x-intercept. To find another point, let $y = 1$.

$$x + 3(1) = 0 \quad \text{Let } y = 1.$$
$$x = -3$$

So the point $(-3, 1)$ is on the graph. Graph the line through $(0, 0)$ and $(-3, 1)$.

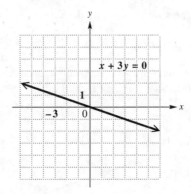

63. [3.3] $y - 5 = 0$ or $y = 5$

This is a horizontal line passing through the point $(0, 5)$, which is the y-intercept.

There is no x-intercept.
Horizontal lines have slopes of 0.

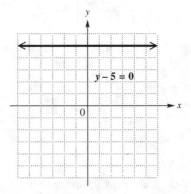

64. [3.4] $m = -\frac{1}{4}, b = -\frac{5}{4}$

Substitute the values $m = -\frac{1}{4}$ and $b = -\frac{5}{4}$ into the slope-intercept form.

$$y = mx + b$$
$$y = -\frac{1}{4}x - \frac{5}{4}$$

65. [3.4] Through $(8,6)$; $m = -3$

Use the point-slope form with $(x_1, y_1) = (8, 6)$ and $m = -3$.

$$y - y_1 = m(x - x_1)$$
$$y - 6 = -3(x - 8)$$
$$y - 6 = -3x + 24$$
$$y = -3x + 30$$

66. [3.4] Through $(3, -5)$ and $(-4, -1)$

First, find the slope of the line.

$$m = \frac{-1 - (-5)}{-4 - 3} = \frac{4}{-7} = -\frac{4}{7}$$

Now use either point and the slope in the point-slope form. If we use $(3, -5)$, we get the following.

$$y - y_1 = m(x - x_1)$$
$$y - (-5) = -\frac{4}{7}(x - 3)$$
$$y + 5 = -\frac{4}{7}x + \frac{12}{7}$$
$$y = -\frac{4}{7}x - \frac{23}{7} \quad \text{Subtract } 5 = \frac{35}{7}.$$

67. [3.5] $y < -4x$

This is a linear inequality, so its graph will be a shaded region. Graph the boundary, $y = -4x$, as a dashed line through $(0, 0)$, $(1, -4)$, and $(-1, 4)$.

The ordered pairs for which y is *less than* $-4x$ are *below* this line. Indicate the solution by shading the region below the line. The dashed line shows that the boundary is not part of the graph.

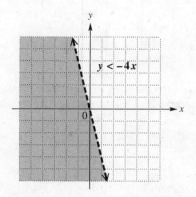

68. [3.5] $x - 2y \leq 6$

This is a linear inequality, so its graph will be a shaded region. Graph the boundary, $x - 2y = 6$, as a solid line through the intercepts $(0, -3)$ and $(6, 0)$.

Using $(0, 0)$ as a test point results in the true statement $0 \leq 6$, so shade the region containing the origin. This is the region above the line. The solid line shows that the boundary is part of the graph.

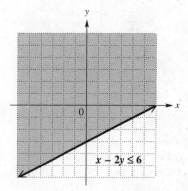

69. In 1997, 44.2% of four-year college students in public institutions earned a degree within five years. In 2002, it was 41.2%. The total decrease in percent was

$$44.2\% - 41.2\% = 3.0\%.$$

70. Since the graph falls from left to right, the slope is negative.

71. Let x represent the year and y represent the percent in the ordered pairs (x, y). The ordered pairs are $(1997, 44.2)$ and $(2002, 41.2)$.

72. Find the slope using $(x_1, y_1) = (1997, 44.2)$ and $(x_2, y_2) = (2002, 41.2)$.

$$m = \frac{y_2 - y_1}{x_2 - x_1}$$
$$= \frac{41.2 - 44.2}{2002 - 1997}$$
$$= \frac{-3}{5}$$
$$= -.6$$

Now use the point-slope form with $m = -.6$ and $(x_1, y_1) = (1997, 44.2)$.

$$y - y_1 = m(x - x_1)$$
$$y - 44.2 = -.6(x - 1997)$$
$$y - 44.2 = -.6x + 1198.2$$
$$y = -.6x + 1242.4$$

73. From the slope-intercept form, $y = mx + b$, the slope of the line is $-.6$. Yes, the slope agrees with the answer in Exercise 70 because the slope is negative.

130 Chapter 3 Graphs of Linear Equations and Inequalities in Two Variables

74. Substitute 1998, 1999, 2000, and 2001 for x in $y = -.6x + 1242.4$ to complete the table.

Year (x)	Percent (y)
1998	43.6
1999	43.0
2000	42.4
2001	41.8

75. For 2003, let $x = 2003$.

$y = -.6(2003) + 1242.4$
$= -1201.8 + 1242.4$
$= 40.6$

The prediction is 40.6% for 2003. We cannot be sure that this prediction is accurate since the equation is based on data only from 1997 through 2002.

Chapter 3 Test

1. The dot above 1998 appears to be slightly closer to 1.10 than to 1.00, so we estimate the cost of a gallon of gasoline in 1998 to be $1.06.

2. We estimate the cost in 2000 to be $1.51, so the increase from 1998 to 2000 is

$$\$1.51 - \$1.06 = \$.45.$$

3. The graph falls from 2000 to 2001 and from 2001 to 2002, so the price decreases from 2000 to 2002.

4. $3x + y = 6$

If $y = 0$, $x = 2$, so the x-intercept is $(2, 0)$.
If $x = 0$, $y = 6$, so the y-intercept is $(0, 6)$.

A third point, such as $(1, 3)$, can be used as a check. Draw a line through $(0, 6)$, $(1, 3)$, and $(2, 0)$.

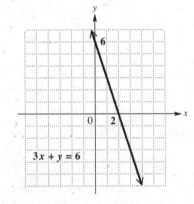

5. $y - 2x = 0$

Solving for y gives us the slope-intercept form of the line, $y = 2x$. We see that the y-intercept is $(0, 0)$ and so the x-intercept is also $(0, 0)$. The slope is 2 and can be written as

$$m = \frac{\text{rise}}{\text{run}} = \frac{2}{1}.$$

Starting at the origin and moving to the right 1 unit and then up 2 units gives us the point $(1, 2)$. Draw a line through $(0, 0)$ and $(1, 2)$.

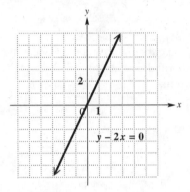

6. $x + 3 = 0$ can also be written as $x = -3$. Its graph is a vertical line with x-intercept $(-3, 0)$. There is no y-intercept.

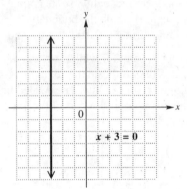

7. $y = 1$ is the graph of a horizontal line with y-intercept $(0, 1)$. There is no x-intercept.

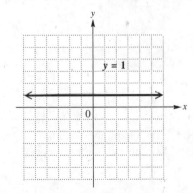

8. $x - y = 4$

   If $y = 0$, $x = 4$, so the x-intercept is $(4, 0)$.
   If $x = 0$, $y = -4$, so the y-intercept is $(0, -4)$.

   A third point, such as $(2, -2)$, can be used as a check. Draw a line through $(0, -4)$, $(2, -2)$, and $(4, 0)$.

   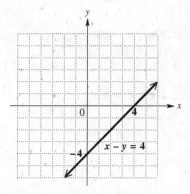

9. Through $(-4, 6)$ and $(-1, -2)$

   Use the definition of slope with $(x_1, y_1) = (-4, 6)$ and $(x_2, y_2) = (-1, -2)$.

   $$\text{slope } m = \frac{y_2 - y_1}{x_2 - x_1}$$
   $$= \frac{-2 - 6}{-1 - (-4)}$$
   $$= \frac{-8}{3} = -\frac{8}{3}$$

10. $2x + y = 10$

    To find the slope, solve the given equation for y.

    $$2x + y = 10$$
    $$y = -2x + 10$$

    The equation is now written in $y = mx + b$ form, so the slope is given by the coefficient of x, which is -2.

11. $x + 12 = 0$ can also be written as $x = -12$. Its graph is a vertical line with x-intercept $(-12, 0)$. The slope is undefined.

12. The indicated points are $(0, -4)$ and $(2, 1)$. Use the definition of slope with $(x_1, y_1) = (0, -4)$ and $(x_2, y_2) = (2, 1)$.

    $$\text{slope } m = \frac{y_2 - y_1}{x_2 - x_1}$$
    $$= \frac{1 - (-4)}{2 - 0}$$
    $$= \frac{5}{2}$$

13. $y - 4 = 6$ can also be written as $y = 10$. Its graph is a horizontal line with y-intercept $(0, 10)$. Its slope is 0, as is the slope of any line parallel to it.

14. Through $(-1, 4)$; $m = 2$

    Let $x_1 = -1$, $y_1 = 4$, and $m = 2$ in the point-slope form.

    $$y - y_1 = m(x - x_1)$$
    $$y - 4 = 2[x - (-1)]$$
    $$y - 4 = 2(x + 1)$$
    $$y - 4 = 2x + 2$$
    $$y = 2x + 6$$

15. The indicated points are $(0, -4)$ and $(2, 1)$. The slope of the line through those points is

    $$m = \frac{1 - (-4)}{2 - 0} = \frac{5}{2}.$$

    The y-intercept is $(0, -4)$, so the slope-intercept form is

    $$y = \frac{5}{2}x - 4.$$

16. Through $(2, -6)$ and $(1, 3)$

    The slope of the line through these points is

    $$m = \frac{3 - (-6)}{1 - 2} = \frac{9}{-1} = -9.$$

    Use the point-slope form of a line with $(1, 3) = (x_1, y_1)$ and $m = -9$.

    $$y - y_1 = m(x - x_1)$$
    $$y - 3 = -9(x - 1)$$
    $$y - 3 = -9x + 9$$
    $$y = -9x + 12$$

17. x-intercept: $(3, 0)$; y-intercept: $\left(0, \frac{9}{2}\right)$

    First, find the slope of the line.

    $$m = \frac{\frac{9}{2} - 0}{0 - 3} = \frac{\frac{9}{2}}{-3} = \frac{9}{2} \cdot \left(-\frac{1}{3}\right) = -\frac{3}{2}$$

    Since one of the given points is the y-intercept, use the slope-intercept form of a line.

    $$y = mx + b$$
    $$y = -\frac{3}{2}x + \frac{9}{2}$$

18. $x + y \leq 3$

    Graph the boundary, $x + y = 3$, as a solid line through the intercepts $(3, 0)$ and $(0, 3)$.

    Using $(0, 0)$ as a test point results in the true statement $0 \leq 3$, so shade the region containing the origin. This is the region below the line. The solid line shows that the boundary is part of the graph.

    continued

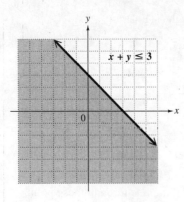

19. $3x - y > 0$

The boundary, $3x - y = 0$, goes through the origin, so both intercepts are $(0, 0)$. Two other points on this line are $(1, 3)$ and $(-1, -3)$. Draw the boundary as a dashed line.

Choose a test point which is not on the boundary. Using $(3, 0)$ results in the true statement $9 > 0$, so shade the region containing $(3, 0)$. This is the region below the line. The dashed line shows that the boundary is not part of the graph.

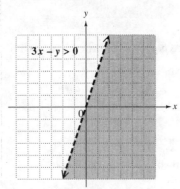

20. The slope is positive since food and drink sales are increasing, indicated by the line which rises from left to right.

21. Two ordered pairs are $(0, 43)$ and $(30, 376)$. Use these points to find the slope.

$$m = \frac{y_2 - y_1}{x_2 - x_1}$$
$$= \frac{376 - 43}{30 - 0}$$
$$= \frac{333}{30}$$
$$= 11.1$$

The slope is 11.1.

22. (a) For 1990, $x = 1990 - 1970 = 20$.

$$y = 11.1x + 43$$
$$y = 11.1(20) + 43$$
$$y = 265$$

For 1995, $x = 1995 - 1970 = 25$.

$$y = 11.1x + 43$$
$$y = 11.1(25) + 43$$
$$y = 320.5$$

The approximate food and drink sales for 1990 and 1995 were $265 billion and $320.5 billion, respectively.

(b) $(30, 376)$; $x = 30$ represents $1970 + 30 = 2000$.
In 2000, food and drink sales were $376 billion.

Cumulative Review Exercises (Chapters R–3)

1. $10\frac{5}{8} - 3\frac{1}{10} = \frac{85}{8} - \frac{31}{10}$

$$= \frac{425}{40} - \frac{124}{40}$$

$$= \frac{301}{40} \quad \text{or} \quad 7\frac{21}{40}$$

2. $\frac{3}{4} \div \frac{1}{8} = \frac{3}{4} \cdot \frac{8}{1} = \frac{3 \cdot 2 \cdot 4}{4 \cdot 1} = 3 \cdot 2 = 6$

3. $5 - (-4) + (-2) = 9 + (-2) = 7$

4. $\dfrac{(-3)^2 - (-4)(2^4)}{5(2) - (-2)^3}$

$$= \frac{9 - (-4)(16)}{10 - (-8)} \quad \begin{array}{l}\text{Do exponents}\\ \text{first.}\end{array}$$

$$= \frac{9 - (-64)}{10 - (-8)} \quad \text{Multiply.}$$

$$= \frac{9 + 64}{10 + 8} = \frac{73}{18} \quad \text{or} \quad 4\frac{1}{18}$$

5. $\dfrac{4(3 - 9)}{2 - 6} \geq 6$?

$\dfrac{4(-6)}{-4} \geq 6$?

$\dfrac{-24}{-4} \geq 6$?

$6 \geq 6$

The statement is *true* since $6 = 6$.

6. $xz^3 - 5y^2 = (-2)(-1)^3 - 5(-3)^2$
Let $x = -2, y = -3, z = -1$.
$= (-2)(-1) + (-5)(9)$
$= 2 + (-45)$
$= -43$

7. $3(-2 + x) = 3 \cdot (-2) + 3(x)$
$= -6 + 3x$

illustrates the *distributive property*.

8. $-4p - 6 + 3p + 8 = (-4p + 3p) + (-6 + 8)$
$= -p + 2$

9. $V = \dfrac{1}{3}\pi r^2 h$
$3V = \pi r^2 h$ Multiply by 3.
$\dfrac{3V}{\pi r^2} = h$ Divide by πr^2.

10. $6 - 3(1 + a) = 2(a + 5) - 2$
$6 - 3 - 3a = 2a + 10 - 2$
$3 - 3a = 2a + 8$
$-5a = 5$
$a = -1$

The solution is -1.

11. $-(m - 3) = 5 - 2m$
$-m + 3 = 5 - 2m$ *Distributive property*
$m + 3 = 5$ *Add 2m.*
$m = 2$ *Subtract 3.*

The solution is 2.

12. $\dfrac{y - 2}{3} = \dfrac{2y + 1}{5}$
$(y - 2)(5) = (3)(2y + 1)$ *Cross products*
$5y - 10 = 6y + 3$
$-10 = y + 3$
$-13 = y$

The solution is -13.

13. $-2.5x < 6.5$
$\dfrac{-2.5x}{-2.5} > \dfrac{6.5}{-2.5}$
Divide by -2.5; reverse the symbol.
$x > -2.6$

<------o------>
       -2.6

14. $4(x + 3) - 5x < 12$
$4x + 12 - 5x < 12$ *Distributive property*
$-x + 12 < 12$ *Combine like terms.*
$-x < 0$ *Subtract 12.*
$x > 0$ *Divide by -1; reverse the symbol.*

<------o------>
       0

15. $\dfrac{2}{3}t - \dfrac{1}{6}t \leq -2$

$6\left(\dfrac{2}{3}t - \dfrac{1}{6}t\right) \leq 6(-2)$ *Multiply by 6 to clear fractions.*

$6\left(\dfrac{2}{3}t\right) - 6\left(\dfrac{1}{6}t\right) \leq 6(-2)$ *Distributive property*

$4t - t \leq -12$
$3t \leq -12$
$t \leq -4$

<------●------>
   -4

16. Let x = average annual earnings for a person with a high school diploma.
Then $x + 29{,}200$ = average annual earnings for a person with a bachelor's degree.

$x + (x + 29{,}200) = 102{,}644$
$2x + 29{,}200 = 102{,}644$
$2x = 73{,}444$
$x = 36{,}722$
$x + 29{,}200 = 65{,}922$

A person with a high school diploma can expect to earn \$36,722/year while a person with a bachelor's degree can expect to earn \$65,922/year.

17. $C = 2\pi r$ *Circumference formula*
$80 = 2\pi r$ *Let $C = 80$.*
$\dfrac{80}{2\pi} = r$ *Divide by 2π.*
$r \approx 13$

The radius is about 13 miles.

18. **(a)** $y = -.5075x + 95.4179$

Let $x = 12$.
$y = -.5075(12) + 95.4179 = 89.33$

Let $x = 28$.
$y = -.5075(28) + 95.4179 = 81.21$

Let $x = 34$.
$y = -.5075(34) + 95.4179 = 78.16$

The completed table follows.

x	y
12	89.33
28	81.21
34	78.16

(b) $(20, 85.27)$; $x = 20$ represents $1960 + 20 = 1980$. In 1980, the winning time was approximately 85.27 sec.

19. **(a)** Multiply 14% (or .14) by the total of $50,000.

$$.14(50,000) = 7000$$

$7000 is expected to go toward home purchase.

(b) Multiply 20% (or .20) by the total of $50,000.

$$.20(50,000) = 10,000$$

$10,000 is expected to go toward retirement.

(c) Since the sector for paying off debt or funding children's education is about three times larger than the sector for retirement, 3($10,000) or about $30,000 is expected to go toward paying off debt or funding children's education.

20. To find the x-intercept, let $y = 0$.

$$-3x + 4y = 12$$
$$-3x + 4(0) = 12$$
$$-3x = 12$$
$$x = -4$$

The x-intercept is $(-4, 0)$.

To find the y-intercept, let $x = 0$.

$$-3x + 4y = 12$$
$$-3(0) + 4y = 12$$
$$4y = 12$$
$$y = 3$$

The y-intercept is $(0, 3)$.

21. To find the slope of the line, solve the equation for y.

$$-3x + 4y = 12$$
$$4y = 3x + 12$$
$$y = \frac{3}{4}x + 3$$

The slope is the coefficient of x, $\frac{3}{4}$.

22. To find a third point, let $x = 4$.

$$-3x + 4y = 12$$
$$-3(4) + 4y = 12$$
$$4y = 12 + 12 = 24$$
$$y = 6$$

Plot the points $(-4, 0)$, $(0, 3)$, and $(4, 6)$ and draw a line through them.

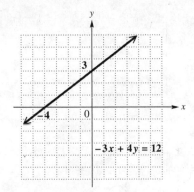

23.
$$-3x + 4y = 12$$
$$-3(4) + 4y = 12 \quad \text{Let } x = 4.$$
$$-12 + 4y = 12$$
$$4y = 24$$
$$y = 6$$

The y-value of the point having x-value 4 is 6.

24.
$$x + 5y = -6$$
$$5y = -x - 6$$
$$y = -\frac{1}{5}x - \frac{6}{5}$$

The slope of the first line is $-\frac{1}{5}$.
The slope of the second line, $y = 5x - 8$, is 5.
Since $-\frac{1}{5}(5) = -1$, the lines are *perpendicular*.

25. Through $(2, -5)$ with slope 3

Use the point-slope form of a line.

$$y - y_1 = m(x - x_1)$$
$$y - (-5) = 3(x - 2)$$
$$y + 5 = 3x - 6$$
$$y = 3x - 11$$

26. Through $(0, 4)$ and $(2, 4)$

$$\text{slope } m = \frac{4 - 4}{2 - 0} = \frac{0}{2} = 0$$

Since the slope is 0, the line is horizontal. Horizontal lines have equations of the form $y = k$. An equation of the line is $y = 4$.

CHAPTER 4 SYSTEMS OF LINEAR EQUATIONS AND INEQUALITIES

4.1 Solving Systems of Linear Equations by Graphing

4.1 Margin Exercises

1. **(a)** $(2, 5)$

    $3x - 2y = -4$
    $5x + y = 15$

    To decide whether $(2, 5)$ is a solution, substitute 2 for x and 5 for y in each equation.

    $3x - 2y = -4$
    $3(\underline{2}) - 2(\underline{5}) = -4$?
    $6 - 10 = -4$?
    $-4 = -4$ *True*

    $5x + y = 15$
    $5(2) + 5 = 15$?
    $10 + \underline{5} = \underline{15}$?
    $15 = 15$ *True*

    Since $(2, 5)$ satisfies both equations, it <u>is</u> a solution of the system.

    (b) $(1, -2)$

    $x - 3y = 7$
    $4x + y = 5$

    To decide whether $(1, -2)$ is a solution, substitute 1 for x and -2 for y in each equation.

    $x - 3y = -7$
    $1 - 3(-2) = 7$?
    $1 + 6 = 7$?
    $7 = 7$ *True*

    $4x + y = 5$
    $4(1) + (-2) = 5$?
    $4 - 2 = 5$?
    $2 = 5$ *False*

    Since $(1, -2)$ does not satisfy the second equation, it <u>is not</u> a solution of the system.

2. **(a)** $5x - 3y = 9$
    $x + 2y = 7$

    Graph each equation by plotting several points for each line. The intercepts are good choices.

    Graph $5x - 3y = 9$.

x	y
$\frac{9}{5}$	0
0	-3
3	2

    Graph $x + 2y = 7$.

x	y
7	0
0	$\frac{7}{2}$
5	1

    (The line $5x - 3y = 9$ is already graphed in the textbook.)

    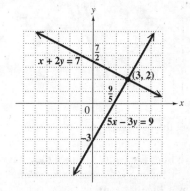

    As suggested by the figure, the solution is $(3, 2)$, the point at which the graphs of the two lines intersect. Check by substituting 3 for x and 2 for y in both equations of the system.

    (b) $x + y = 4$
    $2x - y = -1$

    Graph $x + y = 4$.

x	y
4	0
0	4
3	1

    Graph $2x - y = -1$.

x	y
$-\frac{1}{2}$	0
0	1
2	5

    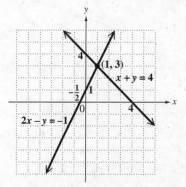

    The solution is $(1, 3)$, the point at which the graphs of the two lines intersect. Check by substituting 1 for x and 3 for y in both equations of the system.

136 Chapter 4 Systems of Linear Equations and Inequalities

3. **(a)** $3x - y = 4$
$6x - 2y = 12$

Graph $3x - y = 4$.

x	y
$\frac{4}{3}$	0
0	-4
1	-1

Graph $6x - 2y = 12$.

x	y
2	0
0	-6
1	3

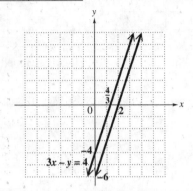

Solving the equations for y gives us $y = 3x - 4$ and $y = 3x - 6$. Since the slope of each line is 3, the two lines are parallel. Since they have no points in common, there is no *solution*. The **system is inconsistent**.

(b) $-x + 3y = 2$
$2x - 6y = -4$

Graph $-x + 3y = 2$.

x	y
-2	0
0	$\frac{2}{3}$
1	1

Graph $2x - 6y = -4$.

x	y
-2	0
0	$\frac{2}{3}$
1	1

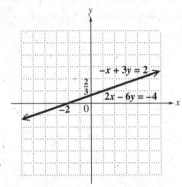

The graphs of these two equations are the same line. There are an *infinite number of solutions*. The **equations are dependent**.

4.1 Section Exercises

1. From the graph, the ordered pair that is a solution of the system is in the second quadrant. Choice **B**, $(-2, 2)$, is the only ordered pair given that is in quadrant II, so it is the only valid choice.

3. There is no way that the sum of two numbers can be both 2 and 4 at the same time.

5. $(2, -3)$

$x + y = -1$
$2x + 5y = 19$

To decide whether $(2, -3)$ is a solution of the system, substitute 2 for x and -3 for y in each equation.

$x + y = -1$
$2 + (-3) = -1$?
$-1 = -1$ True

$2x + 5y = 19$
$2(2) + 5(-3) = 19$?
$4 + (-15) = 19$?
$-11 = 19$ False

The ordered pair $(2, -3)$ satisfies the first equation but not the second. Because it does not satisfy *both* equations, it is not a solution of the system.

7. $(-1, -3)$

$3x + 5y = -18$
$4x + 2y = -10$

Substitute -1 for x and -3 for y in each equation.

$3x + 5y = -18$
$3(-1) + 5(-3) = -18$?
$-3 - 15 = -18$?
$-18 = -18$ True

$4x + 2y = -10$
$4(-1) + 2(-3) = -10$?
$-4 - 6 = -10$?
$-10 = -10$ True

Since $(-1, -3)$ satisfies both equations, it is a solution of the system.

9. $(7, -2)$

$4x = 26 - y$
$3x = 29 + 4y$

Substitute 7 for x and -2 for y in each equation.

$4x = 26 - y$
$4(7) = 26 - (-2)$?
$28 = 26 + 2$?
$28 = 28$ True

$3x = 29 + 4y$
$3(7) = 29 + 4(-2)$?
$21 = 29 - 8$?
$21 = 21$ True

Since $(7, -2)$ satisfies both equations, it is a solution of the system.

11. $(6, -8)$

$$-2y = x + 10$$
$$3y = 2x + 30$$

Substitute 6 for x and -8 for y in each equation.

$$-2y = x + 10$$
$$-2(-8) = 6 + 10 \quad ?$$
$$16 = 16 \quad \quad True$$
$$3y = 2x + 30$$
$$3(-8) = 2(6) + 30 \quad ?$$
$$-24 = 12 + 30 \quad ?$$
$$-24 = 42 \quad \quad False$$

The ordered pair $(6, -8)$ satisfies the first equation but not the second. Because it does not satisfy *both* equations, it is not a solution of the system.

13. $x - y = 2$
$x + y = 6$

To graph the equations, find the intercepts.

$x - y = 2$: Let $y = 0$; then $x = 2$.
Let $x = 0$; then $y = -2$.

Plot the intercepts, $(2, 0)$ and $(0, -2)$, and draw the line through them.

$x + y = 6$: Let $y = 0$; then $x = 6$.
Let $x = 0$; then $y = 6$.

Plot the intercepts, $(6, 0)$ and $(0, 6)$, and draw the line through them.

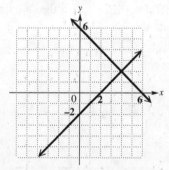

It appears that the lines intersect at the point $(4, 2)$. Check this by substituting 4 for x and 2 for y in both equations. Since $(4, 2)$ satisfies both equations, the solution of this system is $(4, 2)$.

15. $x + y = 4$
$y - x = 4$

To graph the equations, find the intercepts.

$x + y = 4$: Let $y = 0$; then $x = 4$.
Let $x = 0$; then $y = 4$.

Plot the intercepts, $(0, 4)$ and $(4, 0)$, and draw the line through them.

$y - x = 4$: Let $y = 0$; then $x = -4$.
Let $x = 0$; then $y = 4$.

Plot the intercepts, $(-4, 0)$ and $(0, 4)$, and draw the line through them.

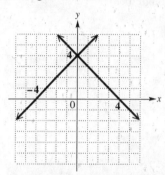

The lines intersect at their common y-intercept, $(0, 4)$, so $(0, 4)$ is the solution of the system.

17. $x - 2y = 6$
$x + 2y = 2$

To graph the equations, find the intercepts.

$x - 2y = 6$: Let $y = 0$; then $x = 6$.
Let $x = 0$; then $y = -3$.

Plot the intercepts, $(6, 0)$ and $(0, -3)$, and draw the line through them.

$x + 2y = 2$: Let $y = 0$; then $x = 2$.
Let $x = 0$; then $y = 1$.

Plot the intercepts, $(2, 0)$ and $(0, 1)$, and draw the line through them.

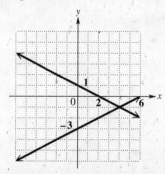

It appears that the lines intersect at the point $(4, -1)$. Since $(4, -1)$ satisfies both equations, the solution of this system is $(4, -1)$.

19. $3x - 2y = -3$
    $-3x - y = -6$

To graph the equations, find the intercepts.

$3x - 2y = -3$: Let $y = 0$; then $x = -1$.
          Let $x = 0$; then $y = \frac{3}{2}$.

Plot the intercepts, $(-1, 0)$ and $(0, \frac{3}{2})$ and draw the line through them.

$-3x - y = -6$: Let $y = 0$; then $x = 2$.
          Let $x = 0$; then $y = 6$.

Plot the intercepts, $(2, 0)$ and $(0, 6)$, and draw the line through them.

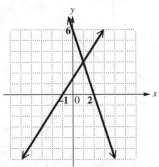

It appears that the lines intersect at the point $(1, 3)$. Since $(1, 3)$ satisfies both equations, the solution of this system is $(1, 3)$.

21. $2x - 3y = -6$
    $y = -3x + 2$

To graph the first line, find the intercepts.

$2x - 3y = -6$: Let $y = 0$; then $x = -3$.
          Let $x = 0$; then $y = 2$.

Plot the intercepts, $(-3, 0)$ and $(0, 2)$, and draw the line through them.

To graph the second line, start by plotting the y-intercept, $(0, 2)$. From this point, go 3 units down and 1 unit to the right (because the slope is -3) to reach the point $(1, -1)$. Draw the line through $(0, 2)$ and $(1, -1)$.

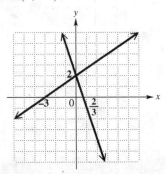

The lines intersect at their common y-intercept, $(0, 2)$, so $(0, 2)$ is the solution of the system.

23. $x + 2y = 6$
    $2x + 4y = 8$

To graph the lines, find the intercepts.

$x + 2y = 6$: Let $y = 0$; then $x = 6$.
         Let $x = 0$; then $y = 3$.

Draw the line through $(6, 0)$ and $(0, 3)$.

$2x + 4y = 8$: Let $y = 0$; then $x = 4$.
         Let $x = 0$; then $y = 2$.

Draw the line through $(4, 0)$ and $(0, 2)$.

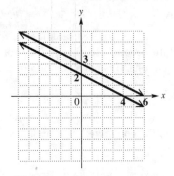

Each line has slope $-\frac{1}{2}$, so the lines are parallel, and the system has no solution.

25. $2x - y = 4$
    $4x = 2y + 8$

Graph the line $2x - y = 4$ through its intercepts, $(2, 0)$ and $(0, -4)$.

Graph the line $4x = 2y + 8$ through its intercepts, $(2, 0)$ and $(0, -4)$.

Since both equations have the same intercepts, they are equations of the same line.

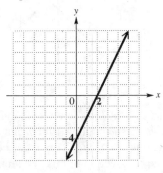

There are an infinite number of solutions. The equations are dependent equations and the solution contains an infinite number of ordered pairs.

27. $3x - 4y = 24$
    $y = -\frac{3}{2}x + 3$

Graph the line $3x - 4y = 24$ through its intercepts, $(8, 0)$ and $(0, -6)$.

To graph the line $y = -\frac{3}{2}x + 3$, plot the y-intercept $(0, 3)$ and then go 3 units down and 2 units to the right (because the slope is $-\frac{3}{2}$) to reach the point $(2, 0)$. Draw the line through $(0, 3)$ and $(2, 0)$.

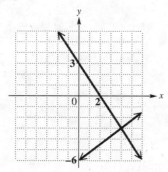

It appears that the lines intersect at the point $(4, -3)$. Since $(4, -3)$ satisfies both equations, the solution of this system is $(4, -3)$.

29. $3x = y + 5$
$6x - 5 = 2y$

To graph the lines, find the intercepts.

$3x = y + 5$: Let $y = 0$; then $x = \frac{5}{3}$.
Let $x = 0$; then $y = -5$.

Draw the line through $\left(\frac{5}{3}, 0\right)$ and $(0, -5)$.

$6x - 5 = 2y$: Let $y = 0$; then $x = \frac{5}{6}$.
Let $x = 0$; then $y = -\frac{5}{2}$.

Draw the line through $\left(\frac{5}{6}, 0\right)$ and $\left(0, -\frac{5}{2}\right)$.

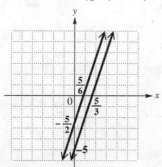

Each line has slope 3, so the lines are parallel, and the system has no solution.

31. $3x + 2y = 6$
$-2y = 3x - 5$

Solve each equation for y.

$3x + 2y = 6$
$2y = -3x + 6$ Subtract $3x$.
$y = -\frac{3}{2}x + 3$ Divide by 2.

$-2y = 3x - 5$
$y = -\frac{3}{2}x + \frac{5}{2}$ Divide by -2.

Since the slopes are the same and the y-intercepts are different, the graphs of these two equations are parallel lines.

32. $2x - y = 4$
$x = .5y + 2$

Solve each equation for y.

$2x - y = 4$
$-y = -2x + 4$ Subtract $2x$.
$y = 2x - 4$ Multiply by -1.

$x = .5y + 2$
$2x = y + 4$ Multiply by 2.
$2x - 4 = y$ Subtract 4.

Since the slopes and the y-intercepts are the same, the graphs of these two equations are the same line.

33. $x - 3y = 5$
$2x + y = 8$

Solve each equation for y.

$x - 3y = 5$
$-3y = -x + 5$ Subtract x.
$y = \frac{1}{3}x - \frac{5}{3}$ Divide by -3.

$2x + y = 8$
$y = -2x + 8$ Subtract $2x$.

Since the slopes are different, the graphs of these two equations are two lines that intersect in exactly one point.

34. If the lines have the same slope and different y-intercepts, they are parallel and the system has no solution (Exercise 31). If the slopes are the same and the y-intercepts are the same, the lines coincide and the system has infinitely many solutions (Exercise 32). If the slopes are unequal, the lines intersect in only one point and the system has only one solution (Exercise 33).

35. The graph for ABC is above the others for years 89–97. Therefore, ABC dominated between 1989 and 1997.

37. The graphs of ABC and CBS intersect at $(1989, 20)$ and $(1998, 16)$. In 1989, with a share of 20% and in 1998, with a share of 16%, ABC and CBS had equal shares.

39. If the coordinates of the point of intersection are not integers, the solution will be difficult to determine from a graph.

140 Chapter 4 Systems of Linear Equations and Inequalities

41. Since $(-2, 3)$ is a solution, we will substitute -2 for x and 3 for y, and choose two different values for m in the slope-intercept form. Then solve each equation for b.

First Equation
$y = mx + b$
$3 = -1(-2) + b \quad m = -1$
$1 = b$
$y = -x + 1$

Second Equation
$y = mx + b$
$3 = 1(-2) + b \quad m = 1$
$5 = b$
$y = x + 5$

So a system of equations is
$y = -x + 1$
$y = x + 5$

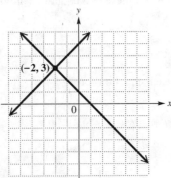

4.2 Solving Systems of Linear Equations by Substitution

4.2 Margin Exercises

1. $3x + 5y = 69$
    $y = 4x$

    Substitute $4x$ for y in the first equation, and solve for x.

    $3x + 5y = 69$
    $3x + 5(\underline{4x}) = 69 \quad$ Let $y = 4x$.
    $3x + 20x = 69 \quad$ Multiply.
    $\underline{23x} = 69 \quad$ Combine terms.
    $x = \underline{3} \quad$ Divide by 23.

    To find y, substitute 3 for x in $y = 4x$.

    $y = 4x = 4(\underline{3}) = \underline{12}$

    The solution is $\underline{(3, 12)}$.

2. $2x + 7y = -12$
    $x = 3 - 2y$

    Substitute $3 - 2y$ for x in the first equation, and solve for y.

    $2x + 7y = -12$
    $2(3 - 2y) + 7y = -12 \quad$ Let $x = 3 - 2y$.
    $6 - 4y + 7y = -12$
    $6 + 3y = -12$
    $3y = -18$
    $y = -6$

    To find x, use $x = 3 - 2y$ and $y = -6$.

    $x = 3 - 2y = 3 - 2(-6) = 3 + 12 = 15$

    The solution is $(15, -6)$.

3. **(a)** $x + 4y = -1 \quad (1)$
    $2x - 5y = 11 \quad (2)$

    Solve (1) for x.

    $x + 4y = -1$
    $x = -1 - \underline{4y} \quad (3)$

    Substitute $-1 - 4y$ for x in equation (2) and solve for y.

    $2(\underline{-1 - 4y}) - 5y = 11 \quad$ Let $x = -1 - 4y$.
    $-2 - 8y - 5y = 11$
    $-2 - \underline{13}y = 11$
    $\underline{-13}y = 13$
    $y = \underline{-1}$

    To find x, let $y = -1$ in equation (3).

    $x = -1 - 4y = -1 - 4(-1)$
    $= -1 - (\underline{-4}) = -1 + 4 = \underline{3}$

    Check that $(3, -1)$ is the solution.

    $x + 4y = -1 \quad (1) \quad | \quad 2x - 5y = 11 \quad (2)$
    $3 + 4(-1) = -1 \, ? \quad | \quad 2(3) - 5(-1) = 11 \, ?$
    $-1 = -1 \quad True \quad | \quad 11 = 11 \quad True$

    The solution of the system is $\underline{(3, -1)}$.

    (b) $2x + 5y = 4 \quad (1)$
    $x + y = -1 \quad (2)$

    Solve equation (2) for x.

    $x + y = -1$
    $x = -1 - y \quad (3)$

    Substitute $-1 - y$ for x in equation (1) and solve for y.

$$2x + 5y = 4$$
$$2(-1 - y) + 5y = 4 \quad \text{Let } x = -1 - y.$$
$$-2 - 2y + 5y = 4$$
$$-2 + 3y = 4$$
$$3y = 6$$
$$y = 2$$

To find x, let $y = 2$ in equation (3).
$$x = -1 - y = -1 - 2 = -3$$

The solution is $(-3, 2)$. Check this solution in both of the original equations.

4. $3x + 2y = 1$ (1)
$3x - 4y = -11$ (2)

Solve equation (1) for x.
$$3x + 2y = 1$$
$$3x = 1 - 2y$$
$$x = \tfrac{1}{3} - \tfrac{2}{3}y \quad (3)$$

Substitute $\tfrac{1}{3} - \tfrac{2}{3}y$ for x in equation (2) and solve for y.
$$3(\tfrac{1}{3} - \tfrac{2}{3}y) - 4y = -11$$
$$1 - 2y - 4y = -11$$
$$1 - 6y = -11$$
$$-6y = -12$$
$$y = 2$$

To find x, let $y = 2$ in equation (3).
$$x = \tfrac{1}{3} - \tfrac{2}{3}y = \tfrac{1}{3} - \tfrac{2}{3}(2) = \tfrac{1}{3} - \tfrac{4}{3} = -1$$

The solution is $(-1, 2)$. Check this solution in both of the original equations.

5. (a) $8x - y = 4$ (1)
$y = 8x + 4$ (2)

Substitute $8x + 4$ for y in equation (1).
$$8x - y = 4$$
$$8x - (8x + 4) = 4 \quad \text{Let } y = 8x + 4.$$
$$8x - 8x - 4 = 4$$
$$-4 = 4 \quad \text{False}$$

This false result indicates that the system has no solution.

(b) $7x - 6y = 10$ (1)
$-14x + 20 = -12y$ (2)

Solve equation (1) for x.
$$7x - 6y = 10$$
$$7x = 6y + 10$$
$$x = \frac{6y + 10}{7}$$

Substitute $\dfrac{6y + 10}{7}$ for x in equation (2).
$$-14x + 20 = -12y$$
$$-14\left(\frac{6y + 10}{7}\right) + 20 = -12y \quad \text{Let } x = \frac{6y + 10}{7}.$$
$$-2(6y + 10) + 20 = -12y$$
$$-12y - 20 + 20 = -12y$$
$$-12y = -12y$$
$$0 = 0 \quad \text{True}$$

This true result indicates that the system has an infinite number of solutions.

6. $\tfrac{2}{3}x + \tfrac{1}{2}y = 6$ (1)
$\tfrac{1}{2}x - \tfrac{3}{4}y = 0$ (2)

Multiply equation (1) by its LCD, 6, and equation (2) by its LCD, 4, to eliminate the fractions. Then solve the resulting equations by the substitution method.

$$4x + 3y = 36 \quad (3)$$
$$2x - 3y = 0 \quad (4)$$

Solve equation (4) for x.
$$2x = 3y$$
$$x = \tfrac{3}{2}y \quad (5)$$

Substitute $\tfrac{3}{2}y$ for x in equation (3).
$$4x + 3y = 36$$
$$4(\tfrac{3}{2}y) + 3y = 36 \quad \text{Let } x = \tfrac{3}{2}y.$$
$$6y + 3y = 36$$
$$9y = 36$$
$$y = 4$$

To find x, let $y = 4$ in equation (5).
$$x = \tfrac{3}{2}y = \tfrac{3}{2}(4) = 6$$

The solution is $(6, 4)$. Check in both of the original equations.

4.2 Section Exercises

1. To give the solution "$x = 3$" for the system
$$5x - y = 15$$
$$7x + y = 21,$$
is not correct. Although 3 is the correct x-value, the solution to a system of equations is an ordered pair. The y-value must also be determined. Substitute $x = 3$ into the second equation.

continued

$7x + y = 21$
$7(3) + y = 21$ Let $x = 3$.
$21 + y = 21$ Multiply.
$y = 0$ Subtract 21.

The solution is the ordered pair $(3, 0)$.

In this section, all solutions should be checked by substituting the proposed solution in *both* equations of the original system. Checks will not be shown here.

3. $x + y = 12$ (1)
       $y = 3x$ (2)

Equation (2) is already solved for y. Substitute $3x$ for y in equation (1) and solve the resulting equation for x.

$x + y = 12$
$x + 3x = 12$ Let $y = 3x$.
$4x = 12$
$x = 3$

To find the y-value of the solution, substitute 3 for x in equation (2).

$y = 3x$
$y = 3(3)$ Let $x = 3$.
$= 9$

The solution is $(3, 9)$.

To check this solution, substitute 3 for x and 9 for y in both equations of the given system.

5. $3x + 2y = 27$ (1)
       $x = y + 4$ (2)

Equation (2) is already solved for x. Substitute $y + 4$ for x in equation (1).

$3x + 2y = 27$
$3(y + 4) + 2y = 27$
$3y + 12 + 2y = 27$
$5y = 15$
$y = 3$

To find x, substitute 3 for y in equation (2).

$x = y + 4 = 3 + 4 = 7$

The solution is $(7, 3)$.

7. $3x + 5y = 14$ (1)
       $x - 2y = -10$ (2)

Solve equation (2) for x since its coefficient is 1.

$x - 2y = -10$
$x = 2y - 10$ (3)

Substitute $2y - 10$ for x in equation (1) and solve for y.

$3x + 5y = 14$
$3(2y - 10) + 5y = 14$
$6y - 30 + 5y = 14$
$11y = 44$
$y = 4$

To find x, substitute 4 for y in equation (3).

$x = 2y - 10 = 2(4) - 10 = -2$

The solution is $(-2, 4)$.

9. $3x + 4 = -y$ (1)
       $2x + y = 0$ (2)

Solve equation (1) for y.

$3x + 4 = -y$
$y = -3x - 4$ (3)

Substitute $-3x - 4$ for y in equation (2) and solve for x.

$2x + y = 0$
$2x + (-3x - 4) = 0$
$-x - 4 = 0$
$-x = 4$
$x = -4$

To find y, substitute -4 for x in equation (3).

$y = -3x - 4 = -3(-4) - 4 = 8$

The solution is $(-4, 8)$.

11. $7x + 4y = 13$ (1)
        $x + y = 1$ (2)

Solve equation (2) for y.

$x + y = 1$
$y = 1 - x$ (3)

Substitute $1 - x$ for y in equation (1).

$7x + 4y = 13$
$7x + 4(1 - x) = 13$
$7x + 4 - 4x = 13$
$3x + 4 = 13$
$3x = 9$
$x = 3$

To find y, substitute 3 for x in equation (3).

$y = 1 - x = 1 - 3 = -2$

The solution is $(3, -2)$.

4.2 Solving Systems of Linear Equations by Substitution

13. $3x - y = 5$ (1)
     $y = 3x - 5$ (2)

Equation (2) is already solved for y, so we substitute $3x - 5$ for y in equation (1).

$$3x - (3x - 5) = 5$$
$$3x - 3x + 5 = 5$$
$$5 = 5 \quad \text{True}$$

This true result means that every solution of one equation is also a solution of the other, so the system has an infinite number of solutions.

15. $6x - 8y = 6$ (1)
     $2y = -2 + 3x$ (2)

Solve equation (2) for y.

$$2y = -2 + 3x$$
$$y = \frac{3x - 2}{2} \quad (3)$$

Substitute $\frac{3x-2}{2}$ for y in equation (1).

$$6x - 8y = 6$$
$$6x - 8\left(\frac{3x-2}{2}\right) = 6$$
$$6x - 4(3x - 2) = 6$$
$$6x - 12x + 8 = 6$$
$$-6x + 8 = 6$$
$$-6x = -2$$
$$x = \frac{-2}{-6} = \frac{1}{3}$$

To find y, let $x = \frac{1}{3}$ in equation (3).

$$y = \frac{3x-2}{2} = \frac{3\left(\frac{1}{3}\right) - 2}{2} = \frac{1-2}{2} = -\frac{1}{2}$$

The solution is $\left(\frac{1}{3}, -\frac{1}{2}\right)$.

17. $2x + 8y = 3$ (1)
     $x = 8 - 4y$ (2)

Equation (2) is already solved for x, so substitute $8 - 4y$ for x in equation (1).

$$2(8 - 4y) + 8y = 3$$
$$16 - 8y + 8y = 3$$
$$16 = 3 \quad \text{False}$$

This false statement means that the system has no solution. The equations of the system represent parallel lines.

19. $12x - 16y = 8$ (1)
     $3x = 4y + 2$ (2)

Solve equation (2) for x.

$$3x = 4y + 2$$
$$x = \frac{4y+2}{3} \quad (3)$$

Substitute $\frac{4y+2}{3}$ for x in equation (1).

$$12x - 16y = 8$$
$$12\left(\frac{4y+2}{3}\right) - 16y = 8$$
$$4(4y + 2) - 16y = 8$$
$$16y + 8 - 16y = 8$$
$$8 = 8 \quad \text{True}$$

This true result means that every solution of one equation is also a solution of the other, so the system has an infinite number of solutions.

21. $5x + 4y = 40$ (1)
     $x + y = 1$ (2)

Solve equation (2) for x.

$$x + y = 1$$
$$x = 1 - y \quad (3)$$

Substitute $1 - y$ for x in equation (1).

$$5x + 4y = 40$$
$$5(1 - y) + 4y = 40 \quad \text{Let } x = 1 - y.$$
$$5 - 5y + 4y = 40$$
$$5 - y = 40$$
$$-y = 35$$
$$y = -35$$

To find x, let $y = -35$ in equation (3).

$$x = 1 - y = 1 - (-35) = 1 + 35 = 36$$

The solution is $(36, -35)$.

23. $3x = 6 - 4y$ (1)
     $9x + 12y = 10$ (2)

Solve equation (1) for x.

$$3x = 6 - 4y$$
$$x = \frac{6-4y}{3} \quad (3)$$

Substitute $\frac{6-4y}{3}$ for x in equation (2).

$$9x + 12y = 10$$
$$9\left(\frac{6-4y}{3}\right) + 12y = 10 \quad \text{Let } x = \frac{6-4y}{3}.$$
$$3(6 - 4y) + 12y = 10$$
$$18 - 12y + 12y = 10$$
$$18 = 10 \quad \text{False}$$

This false statement means that the system has no solution. The equations of the system represent parallel lines.

144 Chapter 4 Systems of Linear Equations and Inequalities

25. $\dfrac{1}{5}x + \dfrac{2}{3}y = -\dfrac{8}{5}$ (1)

$3x - y = 9$ (2)

Multiply each side of equation (1) by 15 to clear fractions.

$$15\left(\tfrac{1}{5}x + \tfrac{2}{3}y\right) = 15\left(-\tfrac{8}{5}\right)$$
$$15\left(\tfrac{1}{5}x\right) + 15\left(\tfrac{2}{3}y\right) = -24$$
$$3x + 10y = -24 \quad (3)$$

Solve equation (2) for y.

$3x - y = 9$
$-y = -3x + 9$ Subtract $3x$.
$y = 3x - 9$ (4)   Divide by -1.

Substitute $3x - 9$ for y in equation (3).

$$3x + 10y = -24$$
$$3x + 10(3x - 9) = -24$$
$$3x + 30x - 90 = -24$$
$$33x = 66 \quad \text{Add 90.}$$
$$x = 2 \quad \text{Divide by 33.}$$

To find y, let $x = 2$ in equation (4).

$$y = 3(2) - 9 = 6 - 9 = -3$$

The solution is $(2, -3)$.

27. $\dfrac{x}{2} - \dfrac{y}{3} = 9$ (1)

$\dfrac{x}{5} - \dfrac{y}{4} = 5$ (2)

To clear fractions, multiply each side of equation (1) by the LCD, 6.

$$6\left(\tfrac{x}{2} - \tfrac{y}{3}\right) = 6(9)$$
$$3x - 2y = 54 \quad (3)$$

To clear fractions, multiply each side of equation (2) by the LCD, 20.

$$20\left(\tfrac{x}{5} - \tfrac{y}{4}\right) = 20(5)$$
$$4x - 5y = 100 \quad (4)$$

We now have the simplified system

$3x - 2y = 54$ (3)
$4x - 5y = 100.$ (4)

Solve equation (3) for x.

$3x - 2y = 54$
$3x = 2y + 54$
$x = \dfrac{2y + 54}{3}$ (5)

Substitute $\dfrac{2y + 54}{3}$ for x in equation (4).

$4x - 5y = 100$
$4\left(\dfrac{2y + 54}{3}\right) - 5y = 100$
$4(2y + 54) - 3(5y) = 3(100)$ Multiply by 3.
$8y + 216 - 15y = 300$
$-7y = 84$
$y = -12$

To find x, let $y = -12$ in equation (5).

$$x = \dfrac{2(-12) + 54}{3} = \dfrac{-24 + 54}{3} = \dfrac{30}{3} = 10$$

The solution is $(10, -12)$.

29. $\dfrac{x}{5} + 2y = \dfrac{16}{5}$ (1)

$\dfrac{3x}{5} + \dfrac{y}{2} = -\dfrac{7}{5}$ (2)

Multiply each side of equation (1) by 5.

$$5\left(\tfrac{x}{5} + 2y\right) = 5\left(\tfrac{16}{5}\right)$$
$$x + 10y = 16 \quad (3)$$

Multiply each side of equation (2) by 10.

$$10\left(\tfrac{3x}{5} + \tfrac{y}{2}\right) = 10\left(-\tfrac{7}{5}\right)$$
$$6x + 5y = -14 \quad (4)$$

We now have the simplified system

$x + 10y = 16$ (3)
$6x + 5y = -14.$ (4)

To solve this system by the substitution method, solve equation (3) for x.

$$x = 16 - 10y \quad (5)$$

Substitute $16 - 10y$ for x in equation (4).

$6x + 5y = -14$
$6(16 - 10y) + 5y = -14$
$96 - 60y + 5y = -14$
$-55y = -110$
$y = 2$

To find x, let $y = 2$ in equation (5).

$x = 16 - 10y$
$= 16 - 10(2) = -4$

The solution is $(-4, 2)$.

30. To find the total cost, multiply the number of bicycles (x) by the cost per bicycle ($\$400$), and add the fixed cost ($\$5000$). Thus,

$y_1 = 400x + 5000$ gives the total cost (in dollars).

31. Since each bicycle sells for $600, the total revenue for selling x bicycles is $600x$ (in dollars). Thus, $y_2 = 600x$ gives the total revenue.

32. $y_1 = 400x + 5000$ (1)
$y_2 = 600x$ (2)

To solve this system by the substitution method, substitute $600x$ for y_1 in equation (1).

$$600x = 400x + 5000$$
$$200x = 5000$$
$$x = 25$$

If $x = 25$, $y_2 = 600(25) = 15{,}000$.

The solution is $(25, 15{,}000)$.

33. The value of x from Exercise 32 is the number of bikes it takes to break even. When 25 bikes are sold, the break-even point is reached. At that point, you have spent 15,000 dollars and taken in 15,000 dollars.

4.3 Solving Systems of Linear Equations by Elimination

4.3 Margin Exercises

1. **(a)** $x + y = 5$
$ 4 + y = 5 \quad$ *Let x = 4.*
$ y = 5 - 4$
$ y = 1$

(b) If $x = 4$, then $y = 1$. The solution is $(4, 1)$.

2. **(a)** $x + y = 8 \quad (1)$
$\underline{x - y = 2} \quad (2)$
$2x + 0 = 10 \quad$ *Add (1) and (2).*
$x = 5 \quad$ *Divide by 2.*

To find y, substitute 5 for x in either of the original equations.

$x - y = 2 \quad$ *Equation (2)*
$\underline{5} - y = 2 \quad$ *Let x = 5.*
$-y = \underline{-3}$
$y = \underline{3}$

Check by substituting 5 for x and 3 for y in both equations of the system.

$x + y = 8 \quad (1) \quad | \quad x - y = 2 \quad (2)$
$5 + 3 = 8 \quad ? \quad | \quad 5 - 3 = 2 \quad ?$
$8 = 8 \quad$ *True* $ | \quad 2 = 2 \quad$ *True*

The solution is $(5, 3)$.

(b) $3x - y = 7 \quad (1)$
$\underline{2x + y = 3} \quad (2)$
$5x = 10 \quad$ *Add (1) and (2).*
$x = 2 \quad$ *Divide by 5.*

To find y, substitute 2 for x in either of the original equations.

$2x + y = 3 \quad$ *Equation (2)*
$2(2) + y = 3 \quad$ *Let x = 2.*
$4 + y = 3$
$y = -1$

Check $x = 2$, $y = -1$: $7 = 7$, $3 = 3$

The solution is $(2, -1)$.

3. **(a)** $2x - y = 2 \quad (1)$
$\underline{4x + y = 10} \quad (2)$
$6x = 12 \quad$ *Add (1) and (2).*
$x = 2 \quad$ *Divide by 6.*

To find y, substitute 2 for x in either of the original equations.

$4x + y = 10 \quad$ *Equation (2)*
$4(2) + y = 10 \quad$ *Let x = 2.*
$8 + y = 10$
$y = 2$

Check $x = 2$, $y = 2$: $2 = 2$, $10 = 10$

The solution is $(2, 2)$.

(b) $8x - 5y = 32 \quad (1)$
$\underline{4x + 5y = 4} \quad (2)$
$12x = 36 \quad$ *Add (1) and (2).*
$x = 3 \quad$ *Divide by 12.*

To find y, substitute 3 for x in either of the original equations.

$4x + 5y = 4 \quad$ *Equation (2)*
$4(3) + 5y = 4 \quad$ *Let x = 3.*
$12 + 5y = 4$
$5y = -8$
$y = -\tfrac{8}{5}$

Check $x = 3$, $y = -\tfrac{8}{5}$: $32 = 32$, $4 = 4$

The solution is $\left(3, -\tfrac{8}{5}\right)$.

4. (a) $2x + 3y = -15$ (1)
$5x + 2y = 1$ (2)

To eliminate y, multiply equation (1) by 2 and equation (2) by -3. Notice that we multiplied by numbers that will cause the absolute value of the coefficients of y to be the *least* common positive multiple of 3 and 2, that is, 6.

$4x + 6y = -30$ (3)
$-15x - 6y = -3$ (4)
$-11x = -33$ Add (3) and (4).
$x = 3$ Divide by -11.

Substitute 3 for x in equation (1).

$2x + 3y = -15$
$2(3) + 3y = -15$
$6 + 3y = -15$
$3y = -21$
$y = -7$

Check $x = 3$, $y = -7$: $-15 = -15$, $1 = 1$

The solution is $(3, -7)$.

(b) $6x + 7y = 4$ (1)
$5x + 8y = -1$ (2)

To eliminate x, multiply equation (1) by -5 and equation (2) by 6.

$-30x - 35y = -20$ (3)
$30x + 48y = -6$ (4)
$13y = -26$ Add (3) and (4).
$y = -2$ Divide by 13.

To find x, substitute -2 for y in equation (1).

$6x + 7y = 4$
$6x + 7(-2) = 4$ Let $y = -2$.
$6x - 14 = 4$
$6x = 18$
$x = 3$

Check $x = 3$, $y = -2$: $4 = 4$, $-1 = -1$

The solution is $(3, -2)$.

5. (a) $5x = 7 + 2y$ (1)
$5y = 5 - 3x$ (2)

Rearrange the terms in both equations as follows so that like terms can be aligned.

$5x - 2y = 7$ (3)
$3x + 5y = 5$ (4)

To eliminate y, multiply equation (3) by 5 and equation (4) by 2.

$25x - 10y = 35$ (5)
$6x + 10y = 10$ (6)
$31x = 45$ Add (5) and (6).
$x = \frac{45}{31}$ Divide by 31.

Substituting $\frac{45}{31}$ for x to find y in one of the original equations would be messy. Instead, solve for y by starting with equations (3) and (4) and eliminating x. Multiply equation (3) by -3 and equation (4) by 5.

$-15x + 6y = -21$ (7)
$15x + 25y = 25$ (8)
$31y = 4$ Add (7) and (8).
$y = \frac{4}{31}$ Divide by 31.

(See the calculator note at the end of the solution for Exercise 31 in this section.)

Check $x = \frac{45}{31}$, $y = \frac{4}{31}$: $\frac{225}{31} = \frac{225}{31}$, $\frac{20}{31} = \frac{20}{31}$

The solution is $\left(\frac{45}{31}, \frac{4}{31}\right)$.

(b) $3y = 8 + 4x$ (1)
$6x = 9 - 2y$ (2)

Rearrange the terms in both equations.

$-4x + 3y = 8$ (3)
$6x + 2y = 9$ (4)

To eliminate x, multiply equation (3) by 3 and equation (4) by 2.

$-12x + 9y = 24$ (5)
$12x + 4y = 18$ (6)
$13y = 42$ Add (5) and (6).
$y = \frac{42}{13}$ Divide by 13.

To eliminate y, multiply equation (3) by 2 and equation (4) by -3.

$-8x + 6y = 16$ (7)
$-18x - 6y = -27$ (8)
$-26x = -11$ Add (7) and (8).
$x = \frac{11}{26}$ Divide by -26.

Check $x = \frac{11}{26}$, $y = \frac{42}{13}$: $\frac{126}{13} = \frac{126}{13}$, $\frac{33}{13} = \frac{33}{13}$

The solution is $\left(\frac{11}{26}, \frac{42}{13}\right)$.

6. (a) $4x + 3y = 10$ (1)
$2x + \frac{3}{2}y = 12$ (2)

To eliminate x, multiply equation (2) by -2.

$4x + 3y = 10$ (1)
$-4x - 3y = -24$ (3)
$0 = -14$ False

A false statement results. The graphs of these equations are parallel lines, so there is no solution.

(b) $\quad 4x - 6y = 10 \quad (1)$
$\quad\quad -10x + 15y = -25 \quad (2)$

To eliminate x, multiply equation (1) by 5 and equation (2) by 2.

$$\begin{aligned} 20x - 30y &= 50 \quad (3) \\ -20x + 30y &= -50 \quad (4) \\ \hline 0 &= 0 \quad \text{True} \end{aligned}$$

The system has an infinite number of solutions.

4.3 Section Exercises

1. The statement is *true*. Both lines would have x- and y-intercepts at $(0, 0)$. Since the point $(0, 0)$ satisfies both equations, it is a solution of the system.

3. It is impossible to have two numbers whose sum is both 1 and 2, so the given statement is *true*.

For all systems in this section, solutions should be checked by substituting the proposed solution in *both* equations of the original system. A detailed check will only be shown here for Exercise 7.

5. $\quad\begin{aligned} x + y &= 2 \quad (1) \\ 2x - y &= -5 \quad (2) \\ \hline 3x &= -3 \quad \text{Add (1) and (2).} \\ x &= -1 \quad \text{Divide by 3.} \end{aligned}$

Replace x with -1 in either equation. If we use equation (1), $x + y = 2$, we get $-1 + y = 2$ or $y = 3$. The solution is $(-1, 3)$.

7. $\quad\begin{aligned} 2x + y &= -5 \quad (1) \\ x - y &= 2 \quad (2) \\ \hline 3x &= -3 \quad \text{Add (1) and (2).} \\ x &= -1 \end{aligned}$

Substitute -1 for x in equation (1) to find the y-value of the solution.

$$\begin{aligned} 2x + y &= -5 \\ 2(-1) + y &= -5 \quad \text{Let } x = -1. \\ -2 + y &= -5 \\ y &= -3 \end{aligned}$$

Check $x = -1, y = -3$:

$$\begin{aligned} 2x + y &= -5 \quad (1) \\ 2(-1) + (-3) &= -5 \quad ? \\ -2 + (-3) &= -5 \quad ? \\ -5 &= -5 \quad \text{True} \end{aligned}$$

$$\begin{aligned} x - y &= 2 \quad (2) \\ -1 - (-3) &= 2 \quad ? \\ -1 + 3 &= 2 \quad ? \\ 2 &= 2 \quad \text{True} \end{aligned}$$

The solution is $(-1, -3)$.

9. $\quad\begin{aligned} 3x + 2y &= 0 \quad (1) \\ -3x - y &= 3 \quad (2) \\ \hline y &= 3 \quad \text{Add (1) and (2).} \end{aligned}$

Substitute 3 for y in equation (1).

$$\begin{aligned} 3x + 2y &= 0 \\ 3x + 2(3) &= 0 \\ 3x + 6 &= 0 \\ 3x &= -6 \\ x &= -2 \end{aligned}$$

The solution is $(-2, 3)$.

11. $\quad 6x - y = -1$
$\quad\quad 5y = 17 + 6x$

Rewrite in standard form.

$$\begin{aligned} 6x - y &= -1 \quad (1) \\ -6x + 5y &= 17 \quad (2) \\ \hline 4y &= 16 \quad \text{Add (1) and (2).} \\ y &= 4 \quad \text{Solve for } y. \end{aligned}$$

Substitute 4 for y in equation (1).

$$\begin{aligned} 6x - y &= -1 \\ 6x - 4 &= -1 \\ 6x &= 3 \\ x &= \tfrac{3}{6} = \tfrac{1}{2} \end{aligned}$$

The solution is $\left(\tfrac{1}{2}, 4\right)$.

13. $\quad 2x - y = 12 \quad (1)$
$\quad\quad 3x + 2y = -3 \quad (2)$

If we simply add the equations, we will not eliminate either variable. To eliminate y, multiply equation (1) by 2 and add the result to equation (2).

$$\begin{aligned} 4x - 2y &= 24 \quad (3) \\ 3x + 2y &= -3 \quad (2) \\ \hline 7x &= 21 \quad \text{Add (3) and (2).} \\ x &= 3 \end{aligned}$$

Substitute 3 for x in equation (1).

$$\begin{aligned} 2x - y &= 12 \\ 2(3) - y &= 12 \\ -y &= 6 \\ y &= -6 \end{aligned}$$

The solution is $(3, -6)$.

15. $x + 3y = 19$ (1)
$2x - y = 10$ (2)

If we simply add the equations, we will not eliminate either variable. To eliminate y, multiply equation (2) by 3 and add the result to equation (1).

$$\begin{aligned} x + 3y &= 19 \quad (1)\\ 6x - 3y &= 30 \quad (3)\\ \hline 7x &= 49 \quad \text{Add (1) and (3).}\\ x &= 7 \end{aligned}$$

Substitute 7 for x in equation (1) to find the y-value of the solution.

$$\begin{aligned} x + 3y &= 19\\ 7 + 3y &= 19\\ 3y &= 12\\ y &= 4 \end{aligned}$$

The solution is $(7, 4)$.

17. $x + 4y = 16$ (1)
$3x + 5y = 20$ (2)

To eliminate x, multiply equation (1) by -3 and add the result to equation (2).

$$\begin{aligned} -3x - 12y &= -48 \quad (3)\\ 3x + 5y &= 20 \quad (2)\\ \hline -7y &= -28 \quad \text{Add (3) and (2).}\\ y &= 4 \end{aligned}$$

Substitute 4 for y in equation (1).

$$\begin{aligned} x + 4y &= 16\\ x + 4(4) &= 16\\ x + 16 &= 16\\ x &= 0 \end{aligned}$$

The solution is $(0, 4)$.

19. $5x - 3y = -20$ (1)
$-3x + 6y = 12$ (2)

To eliminate y, multiply equation (1) by 2 and add the result to equation (2).

$$\begin{aligned} 10x - 6y &= -40 \quad (3)\\ -3x + 6y &= 12 \quad (2)\\ \hline 7x &= -28 \quad \text{Add (3) and (2).}\\ x &= -4 \end{aligned}$$

Substitute -4 for x in equation (2).

$$\begin{aligned} -3x + 6y &= 12\\ -3(-4) + 6y &= 12\\ 12 + 6y &= 12\\ 6y &= 0\\ y &= 0 \end{aligned}$$

The solution is $(-4, 0)$.

21. $2x - 8y = 0$ (1)
$4x + 5y = 0$ (2)

To eliminate x, multiply equation (1) by -2 and add the result to equation (2).

$$\begin{aligned} -4x + 16y &= 0 \quad (3)\\ 4x + 5y &= 0 \quad (2)\\ \hline 21y &= 0 \quad \text{Add (3) and (2).}\\ y &= 0 \end{aligned}$$

Substitute 0 for y in equation (1).

$$\begin{aligned} 2x - 8y &= 0\\ 2x - 8(0) &= 0\\ 2x &= 0\\ x &= 0 \end{aligned}$$

The solution is $(0, 0)$.

23. $x + y = 7$ (1)
$x + y = -3$ (2)

Multiply equation (2) by -1 and add the result to equation (1).

$$\begin{aligned} x + y &= 7 \quad (1)\\ -x - y &= 3 \quad (3)\\ \hline 0 &= 10 \quad \text{Add (1) and (3).} \end{aligned}$$

The false statement $0 = 10$ shows that the given system has no solution.

25. $-x + 3y = 4$ (1)
$-2x + 6y = 8$ (2)

$$\begin{aligned} 2x - 6y &= -8 \quad (3) \quad -2 \times \text{Eq.(1)}\\ -2x + 6y &= 8 \quad (2)\\ \hline 0 &= 0 \quad \text{Add (3) and (2).} \end{aligned}$$

Since $0 = 0$ is a *true* statement, the equations are equivalent. This result indicates that every solution of one equation is also a solution of the other; there are an *infinite number of solutions*.

27. $4x - 3y = -19$ (1)
$3x + 2y = 24$ (2)

$$\begin{aligned} 8x - 6y &= -38 \quad (3) \quad 2 \times \text{Eq.(1)}\\ 9x + 6y &= 72 \quad (4) \quad 3 \times \text{Eq.(2)}\\ \hline 17x &= 34 \quad \text{Add (3) and (4).}\\ x &= 2 \end{aligned}$$

Substitute 2 for x in equation (2).

$$\begin{aligned} 3x + 2y &= 24\\ 3(2) + 2y &= 24\\ 6 + 2y &= 24\\ 2y &= 18\\ y &= 9 \end{aligned}$$

The solution is $(2, 9)$.

4.3 Solving Systems of Linear Equations by Elimination

29. $3x - 7 = -5y$
$5x + 4y = -10$

Rewrite in standard form.

$3x + 5y = 7$ (1)
$5x + 4y = -10$ (2)

To eliminate x, multiply equation (1) by 5 and equation (2) by -3.

$$\begin{array}{rcl} 15x + 25y &=& 35 \quad (3) \\ -15x - 12y &=& 30 \quad (4) \\ \hline 13y &=& 65 \quad \text{Add (3) and (4).} \\ y &=& 5 \end{array}$$

Substitute 5 for y in equation (1).

$3x + 5y = 7$
$3x + 5(5) = 7$
$3x + 25 = 7$
$3x = -18$
$x = -6$

The solution is $(-6, 5)$.

31. $2x + 3y = 0$
$4x + 12 = 9y$

Rewrite in standard form.

$2x + 3y = 0$ (1)
$4x - 9y = -12$ (2)

$$\begin{array}{rcl} 6x + 9y &=& 0 \quad\;\; (3) \quad 3 \times \text{Eq.(1)} \\ 4x - 9y &=& -12 \quad (2) \\ \hline 10x &=& -12 \quad \text{Add (3) and (2).} \\ x &=& \tfrac{-12}{10} = -\tfrac{6}{5} \end{array}$$

$$\begin{array}{rcl} -4x - 6y &=& 0 \quad\;\; (4) \quad -2 \times \text{Eq.(1)} \\ 4x - 9y &=& -12 \quad (2) \\ \hline -15y &=& -12 \quad \text{Add (4) and (2).} \\ y &=& \tfrac{-12}{-15} = \tfrac{4}{5} \end{array}$$

The solution is $\left(-\tfrac{6}{5}, \tfrac{4}{5}\right)$.

When you get a solution that has non-integer components, it is sometimes more difficult to check the problem than it was to solve it. A graphing calculator can be very helpful in this case. Just store the values for x and y in their respective memory locations, and then type the expressions as shown in the following screen. The results 0 and -12 (the right sides of the equations) indicate that we have found the correct solution.

```
-6/5→X:4/5→Y
                .8
2X+3Y
                 0
4X-9Y
               -12
■
```

33. $24x + 12y = -7$
$16x - 17 = 18y$

Rewrite in standard form.

$24x + 12y = -7$ (1)
$16x - 18y = 17$ (2)

$$\begin{array}{rcl} 48x + 24y &=& -14 \quad (3) \quad 2 \times \text{Eq.(1)} \\ -48x + 54y &=& -51 \quad (4) \quad -3 \times \text{Eq.(2)} \\ \hline 78y &=& -65 \quad \text{Add (3) and (4).} \\ y &=& \tfrac{-65}{78} = -\tfrac{5}{6} \end{array}$$

$$\begin{array}{rcl} 72x + 36y &=& -21 \quad (5) \quad 3 \times \text{Eq.(1)} \\ 32x - 36y &=& 34 \quad\;\; (6) \quad 2 \times \text{Eq.(2)} \\ \hline 104x &=& 13 \quad\;\; \text{Add (5) and (6).} \\ x &=& \tfrac{13}{104} = \tfrac{1}{8} \end{array}$$

The solution is $\left(\tfrac{1}{8}, -\tfrac{5}{6}\right)$.

35. $3x = 3 + 2y$ (1)
$-\tfrac{4}{3}x + y = \tfrac{1}{3}$ (2)

Rewrite equation (1) in standard form and multiply equation (2) by 3 to clear fractions.

$3x - 2y = 3$ (3)
$-4x + 3y = 1$ (4)

$$\begin{array}{rcl} 12x - 8y &=& 12 \quad (5) \quad 4 \times \text{Eq.(3)} \\ -12x + 9y &=& 3 \quad\;\; (6) \quad 3 \times \text{Eq.(4)} \\ \hline y &=& 15 \quad \text{Add (5) and (6).} \end{array}$$

$$\begin{array}{rcl} 9x - 6y &=& 9 \quad (7) \quad 3 \times \text{Eq.(3)} \\ -8x + 6y &=& 2 \quad (8) \quad 2 \times \text{Eq.(4)} \\ \hline x &=& 11 \quad \text{Add (7) and (8).} \end{array}$$

The solution is $(11, 15)$.

37. $5x - 2y = 3$ (1)
$10x - 4y = 5$ (2)

$$\begin{array}{rcl} -10x + 4y &=& -6 \quad (3) \quad -2 \times \text{Eq.(1)} \\ 10x - 4y &=& 5 \quad\;\; (2) \\ \hline 0 &=& -1 \quad \text{Add (3) and (2).} \end{array}$$

Since $0 = -1$ is a *false* statement, there are no solutions of the system.

150 Chapter 4 Systems of Linear Equations and Inequalities

39. $6x + 3y = 0$ (1)
$-18x - 9y = 0$ (2)

Multiply equation (1) by 3 and add the result to equation (2).

$$\begin{array}{rcl} 18x + 9y &=& 0 \quad (3) \\ -18x - 9y &=& 0 \quad (2) \\ \hline 0 &=& 0 \quad \text{Add (3) and (2).} \end{array}$$

This true result, $0 = 0$, means that the system has an infinite number of solutions. The equations of the system represent the same line.

41. $y = ax + b$
$1141 = a(1991) + b$ *Let x = 1991, y = 1141.*
$1141 = 1991a + b$

42. As in Exercise 41,
$$1465 = 1999a + b.$$

43. $1991a + b = 1141$ (1)
$1999a + b = 1465$ (2)

Multiply equation (1) by -1 and add the result to equation (2).

$$\begin{array}{rcl} -1991a - b &=& -1141 \\ 1999a + b &=& 1465 \\ \hline 8a &=& 324 \\ a &=& 40.5 \end{array}$$

Substitute 40.5 for a in equation (1).

$$1991(40.5) + b = 1141$$
$$80{,}635.5 + b = 1141$$
$$b = -79{,}494.5$$

The solution is $(40.5, -79{,}494.5)$.

44. An equation of the segment PQ is
$$y = 40.5x - 79{,}494.5$$
for $1991 \le x \le 1999$.

45. $y = 40.5x - 79{,}494.5$
$y = 40.5(1998) - 79{,}494.5$ *Let x = 1998.*
$= 80{,}919 - 79{,}494.5$
$= 1424.5$ (million)

This is less than the actual figure of 1481 million.

46. Since the data do not lie in a perfectly straight line, the quantity obtained from an equation determined in this way will probably be "off" a bit. We cannot put too much faith in models such as this one, because not all sets of data points are linear in nature.

Summary Exercises on Solving Systems of Linear Equations

1. **(a)** $3x + 2y = 18$
$y = 3x$

Use substitution since the second equation is solved for y.

(b) $3x + y = -7$
$x - y = -5$

Use elimination since the coefficients of the y-terms are opposites.

(c) $3x - 2y = 0$
$9x + 8y = 7$

Use elimination since the equations are in standard form with no coefficients of 1 or -1. Solving by substitution would involve fractions.

3. $4x - 3y = -8$ (1)
$x + 3y = 13$ (2)

(a) Solve the system by the elimination method.

$$\begin{array}{rcl} 4x - 3y &=& -8 \quad (1) \\ x + 3y &=& 13 \quad (2) \\ \hline 5x &=& 5 \quad \text{Add (1) and (2).} \\ x &=& 1 \end{array}$$

To find y, let $x = 1$ in equation (2).

$$x + 3y = 13$$
$$1 + 3y = 13$$
$$3y = 12$$
$$y = 4$$

The solution is $(1, 4)$.

(b) To solve this system by the substitution method, begin by solving equation (2) for x.

$$x + 3y = 13$$
$$x = -3y + 13$$

Substitute $-3y + 13$ for x in equation (1).

$$4(-3y + 13) - 3y = -8$$
$$-12y + 52 - 3y = -8$$
$$-15y = -60$$
$$y = 4$$

To find x, let $y = 4$ in equation (2).

$$x + 3y = 13$$
$$x + 3(4) = 13$$
$$x + 12 = 13$$
$$x = 1$$

The solution is $(1, 4)$.

(c) For this particular system, the elimination method is preferable because both equations are already written in the form $Ax + By = C$, and the equations can be added without multiplying either by a constant. Comparing the two methods, we see that the elimination method requires fewer steps than the substitution method for this system.

5. $3x + 2y = 18 \quad (1)$
$y = 3x \quad (2)$

Equation (2) is already solved for y, so we will use the substitution method. Substitute $3x$ for y in equation (1) and solve the resulting equation for x.

$$3x + 2y = 18$$
$$3x + 2(3x) = 18 \quad \text{Let } y = 3x.$$
$$3x + 6x = 18$$
$$9x = 18$$
$$x = \tfrac{18}{9} = 2$$

To find the y-value of the solution, substitute 2 for x in equation (2).

$$y = 3x = 3(2) = 6$$

The solution is $(2, 6)$.

7. $3x - 2y = 0 \quad (1)$
$9x + 8y = 7 \quad (2)$

$$\begin{array}{rcl}
12x - 8y & = & 0 \quad (3) \; 4 \times \text{Eq. (1)} \\
9x + 8y & = & 7 \quad (2) \\ \hline
21x & = & 7 \quad \text{Add (3) and (2).} \\
x & = & \tfrac{7}{21} = \tfrac{1}{3}
\end{array}$$

$$\begin{array}{rcl}
-9x + 6y & = & 0 \quad (4) \; -3 \times \text{Eq.(1)} \\
9x + 8y & = & 7 \quad (2) \\ \hline
14y & = & 7 \quad \text{Add (4) and (2).} \\
y & = & \tfrac{7}{14} = \tfrac{1}{2}
\end{array}$$

The solution is $\left(\tfrac{1}{3}, \tfrac{1}{2}\right)$.

9. $5x - 4y = 15 \quad (1)$
$-3x + 6y = -9 \quad (2)$

$$\begin{array}{rcl}
15x - 12y & = & 45 \quad (3) \; 3 \times \text{Eq.(1)} \\
-15x + 30y & = & -45 \quad (4) \; 5 \times \text{Eq.(2)} \\ \hline
18y & = & 0 \quad \text{Add (3) and (4).} \\
y & = & 0
\end{array}$$

Substitute 0 for y in equation (1).

$$5x - 4y = 15$$
$$5x - 4(0) = 15$$
$$5x = 15$$
$$x = 3$$

The solution is $(3, 0)$.

11. $3x = 7 - y \quad (1)$
$2y = 14 - 6x \quad (2)$

Solve equation (1) for y.

$$3x + y = 7$$
$$y = 7 - 3x$$

Substitute $7 - 3x$ for y in equation (2).

$$2(7 - 3x) = 14 - 6x$$
$$14 - 6x = 14 - 6x$$
$$14 = 14$$

The last equation is true, so there are an infinite number of solutions.

13. $3y = 4x + 2$
$5x - 2y = -3$

Rewrite in standard form.

$-4x + 3y = 2 \quad (1)$
$5x - 2y = -3 \quad (2)$

$$\begin{array}{rcl}
-8x + 6y & = & 4 \quad (3) \; 2 \times \text{Eq.(1)} \\
15x - 6y & = & -9 \quad (4) \; 3 \times \text{Eq.(2)} \\ \hline
7x & = & -5 \quad \text{Add (3) and (4).} \\
x & = & -\tfrac{5}{7}
\end{array}$$

$$\begin{array}{rcl}
-20x + 15y & = & 10 \quad (5) \; 5 \times \text{Eq.(1)} \\
20x - 8y & = & -12 \quad (6) \; 4 \times \text{Eq.(2)} \\ \hline
7y & = & -2 \quad \text{Add (5) and (6).} \\
y & = & -\tfrac{2}{7}
\end{array}$$

The solution is $\left(-\tfrac{5}{7}, -\tfrac{2}{7}\right)$.

15. $2x - 3y = 7 \quad (1)$
$-4x + 6y = 14 \quad (2)$

$$\begin{array}{rcl}
4x - 6y & = & 14 \quad (3) \; 2 \times \text{Eq.(1)} \\
-4x + 6y & = & 14 \quad (2) \\ \hline
0 & = & 28 \quad \text{Add (3) and (2).}
\end{array}$$

Since $0 = 28$ is *false*, there is no solution.

17. $6x + 5y = 13 \quad (1)$
$3x + 3y = 4 \quad (2)$

Multiply equation (2) by -2.

$$\begin{array}{rcl}
6x + 5y & = & 13 \quad (1) \\
-6x - 6y & = & -8 \quad (3) \\ \hline
-y & = & 5 \quad \text{Add (1) and (3).} \\
y & = & -5
\end{array}$$

Substitute -5 for y in equation (2).

continued

$$3x + 3y = 4$$
$$3x + 3(-5) = 4 \quad \text{Let } y = -5.$$
$$3x - 15 = 4$$
$$3x = 19$$
$$x = \tfrac{19}{3}$$

The solution is $\left(\tfrac{19}{3}, -5\right)$.

19. $\dfrac{1}{4}x - \dfrac{1}{5}y = 9 \quad (1)$
    $y = 5x \quad (2)$

First, clear all fractions in equation (1).

$$20\left(\tfrac{1}{4}x - \tfrac{1}{5}y\right) = 20(9) \quad \text{Multiply by the LCD, 20.}$$
$$20\left(\tfrac{1}{4}x\right) - 20\left(\tfrac{1}{5}y\right) = 180 \quad \text{Distributive property}$$
$$5x - 4y = 180 \quad (3)$$

From equation (2), substitute $5x$ for y in equation (3).

$$5x - 4(5x) = 180$$
$$5x - 20x = 180$$
$$-15x = 180$$
$$x = -12$$

To find y, let $x = -12$ in equation (2).

$$y = 5(-12) = -60$$

The solution is $(-12, -60)$.

21. $\dfrac{1}{6}x + \dfrac{1}{6}y = 1 \quad (1)$
    $-\dfrac{1}{2}x - \dfrac{1}{3}y = -5 \quad (2)$

First, clear all fractions.

Equation (1):

$$6\left(\tfrac{1}{6}x + \tfrac{1}{6}y\right) = 6(1) \quad \text{Multiply by 6.}$$
$$6\left(\tfrac{1}{6}x\right) + 6\left(\tfrac{1}{6}y\right) = 6(1) \quad \text{Distributive property}$$
$$x + y = 6 \quad (3)$$

Equation (2):

$$6\left(-\tfrac{1}{2}x - \tfrac{1}{3}y\right) = 6(-5) \quad \text{Multiply by 6.}$$
$$6\left(-\tfrac{1}{2}x\right) - 6\left(\tfrac{1}{3}y\right) = 6(-5) \quad \text{Distributive property}$$
$$-3x - 2y = -30 \quad (4)$$

The system has been simplified to

$$x + y = 6 \quad (3)$$
$$-3x - 2y = -30. \quad (4)$$

Solve equation (3) for y.

$$y = 6 - x \quad (5)$$

Now substitute into equation (4).

$$-3x - 2(6 - x) = -30$$
$$-3x - 12 + 2x = -30$$
$$-x = -18$$
$$x = 18$$

Using $x = 18$ in equation (5), we find

$$y = 6 - 18 = -12.$$

The solution is $(18, -12)$.

23. $\dfrac{x}{5} + y = \dfrac{6}{5} \quad (1)$

$\dfrac{x}{10} + \dfrac{y}{3} = \dfrac{5}{6} \quad (2)$

First, clear all fractions.

Equation (1):

$$5\left(\dfrac{x}{5} + y\right) = 5\left(\dfrac{6}{5}\right) \quad \text{Multiply by 5.}$$
$$x + 5y = 6 \quad (3)$$

Equation (2):

$$30\left(\dfrac{x}{10} + \dfrac{y}{3}\right) = 30\left(\dfrac{5}{6}\right) \quad \text{Multiply by the LCD, 30.}$$
$$3x + 10y = 25 \quad (4)$$

The system has been simplified to

$$x + 5y = 6 \quad (3)$$
$$3x + 10y = 25. \quad (4)$$

Solve this system by the substitution method.

$x = -5y + 6 \quad (5) \quad \text{Solve (3) for } x.$

$3(-5y + 6) + 10y = 25 \quad$ Substitute for x in (4).

$$-15y + 18 + 10y = 25$$
$$-5y + 18 = 25$$
$$-5y = 7$$
$$y = -\tfrac{7}{5}$$

To find x, let $y = -\tfrac{7}{5}$ in equation (5).

$$x = -5\left(-\tfrac{7}{5}\right) + 6 = 7 + 6 = 13$$

The solution is $\left(13, -\tfrac{7}{5}\right)$.

4.4 Applications of Linear Systems

4.4 Margin Exercises

4.4 Applications of Linear Systems

1. $x = 1954 + y$ (1)
   $x + y = 24{,}098$ (2)

   Substitute $1954 + y$ for x in the second equation.

   $$(1954 + y) + y = 24{,}098$$
   $$1954 + 2y = 24{,}098$$
   $$2y = 22{,}144$$
   $$y = 11{,}072$$

   Therefore, $x = 1954 + 11{,}072 = 13{,}026$ and $y = 11{,}072$.

2. Let $x =$ the amount (in millions) that *Lilo and Stitch* grossed, and $y =$ the amount (in millions) that *The Santa Clause 2* grossed. Together they grossed $284.2 million, so

   $$x + y = 284.2.$$

   The Santa Clause 2 grossed $7.4 million less than *Lilo and Stitch*, so

   $$y = x - 7.4.$$

3. (a)

	Number of Tickets Sold	Price (in dollars)	Total Value
Genevans	x	10	$10x$
Parisians	y	8	$8y$
Total	36		298

   (b) From the second column in part (a),

   $$x + y = 36.$$

   From the fourth column in part (a),

   $$10x + 8y = 298.$$

   (c) Multiply the first equation by -8 and add to the second equation.

   $$\begin{array}{rcr} -8x - 8y &=& -288 \\ 10x + 8y &=& 298 \\ \hline 2x &=& 10 \\ x &=& 5 \end{array}$$

   Substitute 5 for x in the first equation.

   $$5 + y = 36$$
   $$y = 31$$

   There were 5 people from Geneva and 31 people from Paris.

   Check: The sum of 5 and 31 is 36, so the number of moviegoers is correct. Since 5 Genevans paid $10 each and 31 Parisians paid $8 each, the total of the admission price is

   $$\$10(5) + \$8(31) = \$298,$$

   which agrees with the total amount stated in the problem.

4. $x + y = 100$
   $.30x + .80y = 50$

   To eliminate the x-terms, multiply the first equation by $-.30$. Then add the result to the second equation.

   $$\begin{array}{rcr} -.30x - .30y &=& -30 \\ .30x + .80y &=& 50 \\ \hline .50y &=& 20 \\ y &=& 40 \end{array}$$

   To find x, substitute 40 for y in the first equation of the original system

   $$x + y = 100$$
   $$x + 40 = 100 \quad \text{Let } y = 40.$$
   $$x = 60$$

   The solution is $x = 60$, $y = 40$.

5. (a)

Percent	Liters	Liters of Pure Alcohol
.25	x	$.25x$
.12	y	$.12y$
.15	13	$.15(13)$

   (b) $x + y = 13$ *From second column*
   $.25x + .12y = .15(13)$ *From third column*

   To eliminate the x-terms, multiply the first equation by $-.25$. Then add the result to the second equation.

   $$\begin{array}{rcr} -.25x - .25y &=& -3.25 \\ .25x + .12y &=& 1.95 \\ \hline -.13y &=& -1.3 \quad \text{Add.} \\ y &=& 10 \end{array}$$

   To find x, substitute 10 for y in the first equation of the original system.

   $$x + y = 13$$
   $$x + 10 = 13 \quad \text{Let } y = 10.$$
   $$x = 3$$

   The solution is $x = 3$, $y = 10$. Mix 3 L of 25% solution with 10 L of 12% solution.

6. Let $x =$ the amount of 10% solution needed, and $y =$ the amount of 25% solution needed.

Make a table.

Percent	Cubic Centimeters	Pure Acid
10	x	$.10x$
25	y	$.25y$
20	100	$.20(100)$

Set up a system of equations.

$$x + y = 100$$
$$.10x + .25y = .20(100)$$

To eliminate the x-terms, multiply the first equation by $-.10$. Then add the result to the second equation.

$$\begin{array}{rcl} -.10x - .10y & = & -10 \\ .10x + .25y & = & 20 \\ \hline .15y & = & 10 \\ y & = & \dfrac{10}{.15} = 66\tfrac{2}{3} \end{array}$$

To find x, substitute $66\tfrac{2}{3}$ for y in the first equation of the original system.

$$x + y = 100$$
$$x + 66\tfrac{2}{3} = 100 \quad \text{Let } y = 66\tfrac{2}{3}.$$
$$x = 33\tfrac{1}{3}$$

$33\tfrac{1}{3}$ cc of 10% solution and $66\tfrac{2}{3}$ cc of 25% solution must be mixed.

7. Use the relationship

$$\text{distance} = \text{rate} \times \text{time}.$$

The rate is 164 mph and the time is 2.5 hr.

$$\text{distance} = (164 \text{ mph}) \times (2.5 \text{ hr})$$

The distance is 410 miles.

8. **(a)** Let $x =$ the speed of the faster car, and $y =$ the speed of the slower car.

	r	t	d
Faster car	x	5	$5x$
Slower car	y	5	$5y$

(b) Write a system of equations.

$$5x + 5y = 450 \quad \text{Total distance}$$
$$x = 2y \quad \text{Faster car is twice as fast.}$$

Substitute $2y$ for x in the first equation, and solve for y.

$$5x + 5y = 450$$
$$5(2y) + 5y = 450 \quad \text{Let } x = 2y.$$
$$10y + 5y = 450$$
$$15y = 450$$
$$y = 30$$

To find x, use $x = 2y$ and $y = 30$.

$$x = 2y = 2(30) = 60$$

The faster car's speed is 60 miles per hour and the slower car's speed is 30 miles per hour.

9. $\begin{array}{rcl} x + y & = & 320 \quad (1) \\ x - y & = & 280 \quad (2) \\ \hline 2x & = & 600 \quad \text{Add (1) and (2).} \\ x & = & 300 \end{array}$

Substitute 300 for x in equation (1).

$$x + y = 320$$
$$300 + y = 320 \quad \text{Let } x = 300.$$
$$y = 20$$

The solution is $(300, 20)$.

10. Let $x =$ the speed of the current, and $y =$ Ann's speed in still water.

When Ann rows against the current, the current works against her, so the rate of the current is subtracted from her speed. When she rows with the current, the rate of the current is added to her speed.

Make a table.

	r	t	d
Against the current	$y - x$	1	2
With the current	$y + x$	1	10

Use $d = rt$ to get each equation of the system.

$$\begin{array}{rcl}(y - x)(1) = 2 & \rightarrow & -x + y = 2 \\ (y + x)(1) = 10 & \rightarrow & x + y = 10 \\ \hline & & 2y = 12 \quad \text{Add.} \\ & & y = 6 \end{array}$$

Substitute y for 6 in the second equation of the system.

$$x + y = 10$$
$$x + 6 = 10 \quad \text{Let } y = 6.$$
$$x = 4$$

The speed of the current is 4 miles per hour. Ann's speed in still water is 6 miles per hour.

4.4 Section Exercises

1. To represent the monetary value of x 20-dollar bills, multiply 20 times x. The answer is **D**, $20x$ dollars.

4.4 Applications of Linear Systems   155

3. The amount of interest earned on d dollars at an interest rate of 2% (.02) is choice **B**, $.02d$ dollars.

5. If a cheetah's rate is 70 mph and it runs at that rate for x hours, then the distance covered is choice **D**, $70x$ miles.

7. Since the plane is traveling *with* the wind, add the rate of the plane, 560 miles per hour, to the rate of the wind, r miles per hour. The answer is **C**, $560 + r$ mph.

9. *Step 2*
   Let $x =$ the first number and
   let $y =$ <u>the second number</u>.

   Step 3
   First equation: $x + y = 98$
   Second equation: $\underline{x - y = 48}$

   Step 4
   Add the two equations.
   $$\begin{aligned} x + y &= 98 \\ x - y &= 48 \\ \hline 2x &= 146 \\ x &= 73 \end{aligned}$$

   Substitute 73 for x in either equation to find $y = 25$.

   Step 5
   The two numbers are 73 and 25.

   Step 6
   The sum of 73 and 25 is 98. The difference between 73 and 25 is 48. The solution satisfies the conditions of the problem.

11. *Step 2*
    Let $x =$ the number of cities visited by the Rolling Stones;
    $y =$ the number of cities visited by Cher.

    Step 3
    The total number of cities visited was 117, so one equation is
    $$x + y = 117. \quad (1)$$
    Cher visited 51 more cities than the Rolling Stones, so another equation is
    $$y = x + 51. \quad (2)$$

    Step 4
    Substitute $x + 51$ for y in equation (1).
    $$\begin{aligned} x + y &= 117 \\ x + (x + 51) &= 117 \\ 2x + 51 &= 117 \\ 2x &= 66 \\ x &= 33 \end{aligned}$$

    Substitute 33 for x in (2) to find
    $y = 33 + 51 = 84$.

    Step 5
    The Rolling Stones visited 33 cities and Cher visited 84 cities.

    Step 6
    The sum of 33 and 84 is 117 and 84 is 51 more than 33.

13. *Step 2*
    Let $x =$ the amount earned by *The Lord of the Rings: The Return of the King*;
    $y =$ the amount earned by *Finding Nemo*.

    Step 3
    Finding Nemo grossed $21.4 million less than *The Lord of the Rings: The Return of the King*, so
    $$y = x - 21.4. \quad (1)$$
    The total earned by these two films was $700.8 million, so
    $$x + y = 700.8 \quad (2)$$

    Step 4
    Substitute $x - 21.4$ for y in equation (2).
    $$\begin{aligned} x + (x - 21.4) &= 700.8 \\ 2x - 21.4 &= 700.8 \\ 2x &= 722.2 \\ x &= 361.1 \end{aligned}$$

    To find y, let $x = 361.1$ in equation (1).
    $$\begin{aligned} y &= x - 21.4 \\ y &= 361.1 - 21.4 = 339.7 \end{aligned}$$

    Step 5
    The Lord of the Rings: The Return of the King earned $361.1 million and *Finding Nemo* earned $339.7 million.

    Step 6
    The sum of 361.1 and 339.7 is 700.8 and 339.7 is 21.4 less than 361.1.

15. *Step 2*
    Let $t =$ the height of the Terminal Tower;
    $k =$ the height of the Key Tower.

    Step 3
    The Terminal Tower is 242 feet shorter than the Key Tower, so
    $$t = k - 242. \quad (1)$$
    The total of the two heights is 1658 feet, so
    $$t + k = 1658. \quad (2)$$

continued

Step 4
Substitute $k - 242$ for t in equation (2).

$$(k - 242) + k = 1658$$
$$2k - 242 = 1658$$
$$2k = 1900$$
$$k = 950$$

To find t, let $k = 950$ in equation (1).

$$t = k - 242$$
$$t = 950 - 242 = 708$$

Step 5
The height of the Terminal Tower is 708 feet and the height of the Key Tower is 950 feet.

Step 6
The sum of 708 and 950 is 1658 and 708 is 242 less than 950.

17. (a) $C = 85x + 900$; $R = 105x$; no more than 38 units can be sold.

To find the break-even quantity, let $C = R$.

$$85x + 900 = 105x$$
$$900 = 20x$$
$$x = \tfrac{900}{20} = 45$$

The break-even quantity is 45 units.

(b) Since no more than 38 units can be sold, do not produce the product (since $38 < 45$). The product will lead to a loss.

19. *Step 2*
Let $x =$ the number of $1 bills;
$y =$ the number of $10 bills.

Complete the table given in the textbook, realizing that the entries in the "total value" column were found by multiplying the denomination of the bill by the number of bills.

Denomination of Bill	Number of Bills	Total Value
$1	x	$1x$
$10	y	$10y$
Totals	74	$326

Step 3
The total number of bills is 74, so

$$x + y = 74. \quad (1)$$

Since the total value is $326, the third column leads to

$$x + 10y = 326. \quad (2)$$

These two equations give the system:

$$x + y = 74 \quad (1)$$
$$x + 10y = 326 \quad (2)$$

Step 4
To solve this system by the elimination method, multiply equation (1) by -1 and add this result to equation (2).

$$\begin{array}{rcr} -x - y &=& -74 \\ x + 10y &=& 326 \\ \hline 9y &=& 252 \\ y &=& 28 \end{array}$$

Substitute 28 for y in equation (2).

$$x + 10y = 326$$
$$x + 10(28) = 326$$
$$x + 280 = 326$$
$$x = 46$$

Step 5
The clerk has 46 ones and 28 tens.

Step 6
46 ones and 28 tens give us 74 bills worth $326.

21. *Step 2*
Let $x =$ the number of *Miracle* DVDs;
$y =$ the number of Linkin Park CDs

Type of Gift	Number Bought	Cost of each (in dollars)	Total Value
DVD	x	14.95	$14.95x$
CD	y	16.88	$16.88y$
Totals	7	—	$114.30

Step 3
From the second and fourth columns of the table, we obtain the system

$$x + y = 7 \quad (1)$$
$$14.95x + 16.88y = 114.30 \quad (2)$$

Step 4
Multiply equation (1) by -14.95 and add the result to equation (2).

$$\begin{array}{rcr} -14.95x - 14.95y &=& -104.65 \\ 14.95x + 16.88y &=& 114.30 \\ \hline 1.93y &=& 9.65 \\ y &=& 5 \end{array}$$

From (1) with $y = 5$, $x = 2$.

Step 5
She bought 2 DVDs and 5 CDs.

Step 6
Two $14.95 DVDs and five $16.88 CDs give us 7 gifts worth $114.30.

4.4 Applications of Linear Systems

23. *Step 2*
Let $x = $ the amount invested at 5%;
$y = $ the amount invested at 4%.

Interest Rate (as a decimal)	Amount Invested	Interest Income (yearly)
.05	x	$.05x$
.04	y	$.04y$
		$350

Step 3
Maria has invested twice as much at 5% as at 4%, so

$$x = 2y. \quad (1)$$

Her total interest income is $350, so

$$.05x + .04y = 350. \quad (2)$$

Step 4
Solve the system by substitution. Substitute $2y$ for x in equation (2).

$$.05(2y) + .04y = 350$$
$$.14y = 350$$
$$y = \frac{350}{.14} = 2500$$

Substitute 2500 for y in equation (1).

$$x = 2y = 2(2500) = 5000$$

Step 5
Maria has $5000 invested at 5% and $2500 invested at 4%.

Step 6
$5000 is twice as much as $2500. 5% of $5000 is $250 and 4% of $2500 is $100, which is a total of $350 in interest each year.

25. *Step 2*
Let $x = $ the average movie ticket cost in Japan;
$y = $ the average movie ticket cost in Switzerland.

Step 3
Three tickets in Japan plus two tickets in Switzerland cost $77.87, so

$$3x + 2y = 77.87. \quad (1)$$

Two tickets in Japan plus three tickets in Switzerland cost $73.83, so

$$2x + 3y = 73.83. \quad (2)$$

Step 4
Multiply (1) by -2, multiply (2) by 3, and add the results.

$$\begin{aligned} -6x - 4y &= -155.74 \\ 6x + 9y &= 221.49 \\ \hline 5y &= 65.75 \\ y &= 13.15 \end{aligned}$$

Substitute $y = 13.15$ in (1).

$$3x + 2(13.15) = 77.87$$
$$3x + 26.30 = 77.87$$
$$3x = 51.57$$
$$x = 17.19$$

Step 5
The average movie ticket cost in Japan is $17.19 and the average movie ticket cost in Switzerland is $13.15.

Step 6
Three tickets (on average) in Japan cost $51.57 and two tickets in Switzerland cost $26.30; a sum of $77.87. Two tickets in Japan cost $34.38 and three tickets in Switzerland cost $39.45; a sum of $73.83.

27. *Step 2*
Let $x = $ the amount of 40% solution;
$y = $ the amount of 70% solution.

Percent (as a decimal)	Liters of Solution	Liters of Pure Dye
.40	x	$.40x$
.70	y	$.70y$
.50	120	$.50(120) = 60$

Step 3
The total number of liters in the final mixture is 120, so

$$x + y = 120. \quad (1)$$

The amount of pure dye in the 40% solution added to the amount of pure dye in the 70% solution is equal to the amount of pure dye in the 50% mixture, so

$$.40x + .70y = 60. \quad (2)$$

Multiply equation (2) by 10 to clear decimals.

$$4x + 7y = 600 \quad (3)$$

We now have the system

$$\begin{aligned} x + y &= 120 & (1) \\ 4x + 7y &= 600. & (3) \end{aligned}$$

Step 4
Solve this system by the elimination method.

$$\begin{aligned} -4x - 4y &= -480 \quad \text{Multiply (1) by } -4. \\ 4x + 7y &= 600 \\ \hline 3y &= 120 \\ y &= 40 \end{aligned}$$

From (1) with $y = 40$, $x = 80$.

continued

158 Chapter 4 Systems of Linear Equations and Inequalities

Step 5
80 liters of 40% solution should be mixed with 40 liters of 70% solution.

Step 6
Since $80 + 40 = 120$ and $.40(80) + .70(40) = 60$, this mixture will give the 120 liters of 50 percent solution, as required in the original problem.

29. *Step 2*
Let $x =$ the number of pounds of coffee worth $6 per pound;
$y =$ the number of pounds of coffee worth $3 per pound.

Complete the table given in the textbook.

Dollars per Pound	Pounds	Cost
6	x	$6x
3	y	$3y
4	90	$90(4) = $360

Step 3
The mixture contains 90 pounds, so
$$x + y = 90. \quad (1)$$

The cost of the mixture is $360, so
$$6x + 3y = 360. \quad (2)$$

Equation (2) may be simplified by dividing each side by 3.
$$2x + y = 120 \quad (3)$$

We now have the system
$$x + y = 90 \quad (1)$$
$$2x + y = 120. \quad (3)$$

Step 4
To solve this system by the elimination method, multiply equation (1) by -1 and add the result to equation (3).

$$\begin{array}{rcr} -x - y &=& -90 \\ 2x + y &=& 120 \\ \hline x &=& 30 \end{array}$$

From (1) with $x = 30$, $y = 60$.

Step 5
The merchant will need to mix 30 pounds of coffee at $6 per pound with 60 pounds at $3 per pound.

Step 6
Since $30 + 60 = 90$ and $$6(30) + $3(60) = 360, this mixture will give the 90 pounds worth $4 per pound, as required in the original problem.

31. *Step 2*
Let $x =$ the number of $40 barrels of pickles;
$y =$ the number of $60 barrels of pickles.

Make a table.

Price per Barrel (in dollars)	Barrels of Pickles	Total Price in Dollars
40	x	$40x$
60	y	$60y$
48	50	$48(50) = 2400$

Step 3
The mixture contains 50 barrels, so
$$x + y = 50. \quad (1)$$

The cost of the mixture is $2400, so
$$40x + 60y = 2400. \quad (2)$$

Step 4
Multiply equation (1) by -40 and add the result to (2).

$$\begin{array}{rcr} -40x - 40y &=& -2000 \\ 40x + 60y &=& 2400 \\ \hline 20y &=& 400 \\ y &=& 20 \end{array}$$

From (1) with $y = 20$, $x = 30$.

Step 5
One should mix 30 barrels at $40 per barrel and 20 barrels at $60 per barrel.

Step 6
Since $30 + 20 = 50$ and $$40(30) + $60(20) = 2400, this mixture will give the 50 barrels worth $48 per barrel, as required in the original problem.

33. *Step 2*
Let $x =$ the average speed of the bicycle;
$y =$ the average speed of the car.

	d	r	t
Bicycle	$7x$	x	7
Car	$7y$	y	7

Step 3
The total distance is 490 miles, so
$$7x + 7y = 490 \quad (1)$$

or, dividing equation (1) by 7,
$$x + y = 70. \quad (2)$$

The car traveled 40 mph faster than the bicycle, so
$$y = x + 40. \quad (3)$$

Step 4
From (3), substitute $x + 40$ for y in (2).

$$x + (x + 40) = 70$$
$$2x + 40 = 70$$
$$2x = 30$$
$$x = 15$$

From (3) with $x = 15$, $y = 15 + 40 = 55$.

Step 5
The average speed of the bicycle was 15 mph and the average speed of the car was 55 mph.

Step 6
55 mph is 40 mph faster than 15 mph. The bicycle traveled $7(15) = 105$ miles and the car traveled $7(55) = 385$ miles. The sum is $105 + 385 = 490$, as required.

35. *Step 2*
Let $x =$ the average speed of the car leaving Cincinnati;
$y =$ the average speed of the car leaving Toledo.

Convert the total time of 1 hour and 36 minutes to hours.

$$1 + \frac{36}{60} = \frac{96}{60} = 1.6$$

	d	r	t
Cincinnati to Toledo	1.6x	x	1.6
Toledo to Cincinnati	1.6y	y	1.6

Step 3
The total distance is 200 miles, so

$$1.6x + 1.6y = 200 \quad (1)$$

or, dividing equation (1) by 1.6,

$$x + y = 125. \quad (2)$$

The car leaving Toledo averages 15 miles per hour more than the other car, so

$$y = x + 15. \quad (3)$$

Step 4
From (3), substitute $x + 15$ for y in (2).

$$x + (x + 15) = 125$$
$$2x = 110$$
$$x = 55$$

From (3) with $x = 55$, $y = 70$.

Step 5
The average speed of the car leaving Cincinnati was 55 miles per hour and the average speed of the car leaving Toledo was 70 miles per hour.

Step 6
70 miles per hour is 15 miles per hour more than 55 miles per hour. The car leaving Cincinnati travels $1.6(55) = 88$ miles and the car leaving Toledo travels $1.6(70) = 112$ miles. The total distance traveled is $88 + 112 = 200$, as required.

37. *Step 2*
Let $x =$ Roberto's speed;
$y =$ Juana's speed.

Step 3
Use the formula $d = rt$ to complete two tables.

Riding in same direction

	d	r	t
Roberto	6x	x	6
Juana	6y	y	6

Roberto rode 30 miles farther than Juana, so

$$6x = 6y + 30$$
$$\text{or} \quad x - y = 5. \quad (1)$$

Riding toward each other

	d	r	t
Roberto	1x	x	1
Juana	1y	y	1

Roberto and Juana rode a total of 30 miles, so

$$x + y = 30. \quad (2)$$

We have the system

$$x - y = 5 \quad (1)$$
$$x + y = 30. \quad (2)$$

Step 4
To solve the system by the elimination method, add equations (1) and (2).

$$\begin{array}{rcr} x - y &=& 5 \\ x + y &=& 30 \\ \hline 2x &=& 35 \\ x &=& 17.5 \end{array}$$

From (2) with $x = 17.5$, $y = 12.5$.

Step 5
Roberto's rate is 17.5 miles per hour and Juana's rate is 12.5 miles per hour.

continued

Step 6
Riding in the same direction, Roberto rides $6(17.5) = 105$ miles and Juana rides $6(12.5) = 75$ miles. The difference is 30 miles, as required.

Riding toward each other, Roberto rides $1(17.5) = 17.5$ miles and Juana rides $1(12.5) = 12.5$ miles. The sum is 30 miles, as required.

39. *Step 2*
Let $x =$ the speed of the boat in still water;
$y =$ the speed of the current.

	d	r	t
Downstream	36	$x+y$	3
Upstream	24	$x-y$	3

Step 3
Use the formula $d = rt$ and the completed table to write the system of equations.

$(x+y)(3) = 36$ (1) *Distance downstream*

$(x-y)(3) = 24$ (2) *Distance upstream*

Step 4
Divide equations (1) and (2) by 3.

$$x + y = 12 \quad (3)$$
$$x - y = 8 \quad (4)$$

Now add equations (3) and (4).

$$\begin{aligned} x + y &= 12 \quad (3) \\ x - y &= 8 \quad (4) \\ \hline 2x &= 20 \quad \text{Add (3) and (4).} \\ x &= 10 \end{aligned}$$

From (3) with $x = 10$, $y = 2$.

Step 5
The speed of the current is 2 miles per hour; the speed of the boat in still water is 10 miles per hour.

Step 6
Traveling downstream for 3 hours at $10 + 2 = 12$ miles per hour gives us a 36-mile trip. Traveling upstream for 3 hours at $10 - 2 = 8$ miles per hour gives us a 24-mile trip, as required.

41. *Step 2*
Let $x =$ the speed of the plane in still air;
$y =$ the speed of the wind.

Step 3
The rate of the plane with the wind is $x + y$, so

$$x + y = 500. \quad (1)$$

The rate of the plane into the wind is $x - y$, so

$$x - y = 440. \quad (2)$$

Step 4
To solve the system by the elimination method, add equations (1) and (2).

$$\begin{aligned} x + y &= 500 \\ x - y &= 440 \\ \hline 2x &= 940 \\ x &= 470 \end{aligned}$$

From (1) with $x = 470$, $y = 30$.

Step 5
The speed of the wind is 30 miles per hour; the speed of the plane in still air is 470 miles per hour.

Step 6
The plane travels $470 + 30 = 500$ miles per hour with the wind and $470 - 30 = 440$ miles per hour into the wind, as required.

4.5 Solving Systems of Linear Inequalities

4.5 Margin Exercises

1. $x - 2y \leq 8$
$3x + y \geq 6$

The graph of $x - 2y = 8$ has intercepts $(8, 0)$ and $(0, -4)$. The graph of $3x + y = 6$ has intercepts $(2, 0)$ and $(0, 6)$. Both are graphed as solid lines because of the $\leq$ and $\geq$ signs. Use $(0, 0)$ as a test point in each case.

$$\begin{aligned} x - 2y &\leq 8 \\ 0 - 2(0) &\leq 8 \quad \text{Let } x = 0, y = 0. \\ 0 &\leq 8 \quad \text{True} \end{aligned}$$

Shade the side of the graph for $x - 2y = 8$ that contains $(0, 0)$.

$$\begin{aligned} 3x + y &\geq 6 \\ 3(0) + 0 &\geq 6 \quad \text{Let } x = 0, y = 0. \\ 0 &\geq 6 \quad \text{False} \end{aligned}$$

Shade the side of the graph for $3x + y = 6$ that does not contain $(0, 0)$. The graph of the solution is the overlap of the two shaded regions.

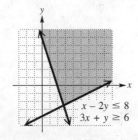

2. **(a)** $x + 2y < 0$
$3x - 4y < 12$

Graph $x + 2y = 0$ as a dashed line through the points $(0,0)$ and $(2,-1)$. Use $(-4,0)$ as a test point.

$$x + 2y < 0$$
$$-4 + 2(0) < 0 \quad \text{Let } x = -4, y = 0.$$
$$-4 < 0 \quad \text{True}$$

The solution is the region that includes $(-4, 0)$.

Then graph $3x - 4y = 12$ as a dashed line through its intercepts, $(4, 0)$ and $(0, -3)$. Use $(0, 0)$ as a test point.

$$3x - 4y < 12$$
$$3(0) - 4(0) < 12 \quad \text{Let } x = 0, y = 0.$$
$$0 < 12 \quad \text{True}$$

The solution is the region that includes $(0, 0)$. The graph of the solution is the overlap of the two shaded regions.

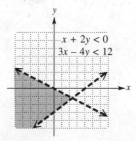

(b) $3x + 2y \leq 12$
$x \leq 2$
$y \leq 4$

Graph $3x + 2y = 12$ as a solid line through its intercepts, $(4, 0)$ and $(0, 6)$. Use $(0, 0)$ as a test point.

$$3x + 2y < 12$$
$$3(0) + 2(0) < 12 \quad \text{Let } x = 0, y = 0.$$
$$0 < 12 \quad \text{True}$$

Shade the region that includes $(0, 0)$. Recall that $x = 2$ is a vertical line through the point $(2, 0)$, and $y = 4$ is a horizontal line through $(0, 4)$. Shade to the left of $x = 2$ and below $y = 4$. The graph of the solution is the overlap of the three shaded regions.

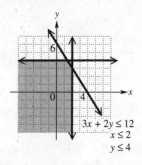

4.5 Section Exercises

1. $x \geq 5$ is the region to the right of the vertical line $x = 5$ and includes the line. $y \leq -3$ is the region below the horizontal line $y = -3$ and includes the line. The correct choice is **C**.

3. $x > 5$ is the region to the right of the vertical line $x = 5$. $y < -3$ is the region below the horizontal line $y = -3$. The correct choice is **B**.

5. $x + y \leq 6$
$x - y \geq 1$

Graph the boundary $x + y = 6$ as a solid line through its intercepts, $(6, 0)$ and $(0, 6)$. Using $(0, 0)$ as a test point will result in the true statement $0 \leq 6$, so shade the region containing the origin.

Graph the boundary $x - y = 1$ as a solid line through its intercepts, $(1, 0)$ and $(0, -1)$. Using $(0, 0)$ as a test point will result in the false statement $0 \geq 1$, so shade the region *not* containing the origin.

The solution of this system is the intersection (overlap) of the two shaded regions, and includes the portions of the boundary lines that bound this region.

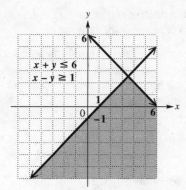

162 Chapter 4 Systems of Linear Equations and Inequalities

7. $4x + 5y \geq 20$
$x - 2y \leq 5$

Graph the boundary $4x + 5y = 20$ as a solid line through its intercepts, $(0, 4)$ and $(5, 0)$. Using $(0, 0)$ as a test point will result in the false statement $0 \geq 20$, so shade the region *not* containing the origin.

Graph the boundary $x - 2y = 5$ as a solid line through $(5, 0)$ and $(1, -2)$. Using $(0, 0)$ as a test point will result in the true statement $0 \leq 5$, so shade the region containing the origin.

The solution of this system is the intersection of the two shaded regions, and includes the portions of the boundary lines that bound the region.

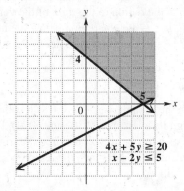

11. $y \leq 2x - 5$
$x < 3y + 2$

Graph $y = 2x - 5$ as a solid line through $(0, -5)$ and $(3, 1)$. Using $(0, 0)$ as a test point will result in the false statement $0 \leq -5$, so shade the region *not* containing the origin.

Now graph $x = 3y + 2$ as a dashed line through $(2, 0)$ and $(-1, -1)$. Using $(0, 0)$ as a test point will result in the true statement $0 < 2$, so shade the region containing the origin.

The solution of the system is the intersection of the two shaded regions. It includes the portion of the line $y = 2x - 5$ that bounds the region, but not the portion of the line $x = 3y + 2$.

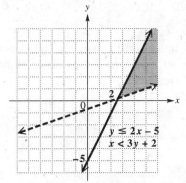

9. $2x + 3y < 6$
$x - y < 5$

Graph $2x + 3y = 6$ as a dashed line through $(3, 0)$ and $(0, 2)$. Using $(0, 0)$ as a test point will result in the true statement $0 < 6$, so shade the region containing the origin.

Now graph $x - y = 5$ as a dashed line through $(5, 0)$ and $(0, -5)$. Using $(0, 0)$ as a test point will result in the true statement $0 < 5$, so shade the region containing the origin.

The solution of the system is the intersection of the two shaded regions. Because the inequality signs are both $<$, the solution does not include the boundary lines.

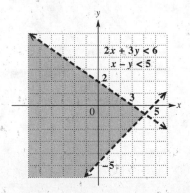

13. $4x + 3y < 6$
$x - 2y > 4$

Graph $4x + 3y = 6$ as a dashed line through $\left(\frac{3}{2}, 0\right)$ and $(0, 2)$. Using $(0, 0)$ as a test point will result in the true statement $0 < 6$, so shade the region containing the origin.

Now graph $x - 2y = 4$ as a dashed line through $(4, 0)$ and $(0, -2)$. Using $(0, 0)$ as a test point will result in the false statement $0 > 4$, so shade the region *not* containing the origin.

The solution of the system is the intersection of the two shaded regions. It does not include the boundary lines.

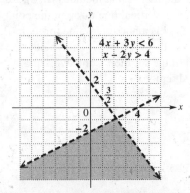

15. $x \leq 2y + 3$
$x + y < 0$

Graph $x = 2y + 3$ as a solid line through $(3, 0)$ and $(5, 1)$. Using $(0, 0)$ as a test point will result in the true statement $0 \leq 3$, so shade the region containing the origin.

Now graph $x + y = 0$ as a dashed line through $(0, 0)$ and $(1, -1)$. Using $(1, 0)$ as a test point will result in the false statement $1 < 0$, so shade the region *not* containing $(1, 0)$.

The solution of the system is the intersection of the two shaded regions. It includes the portion of the line $x = 2y + 3$ that bounds the region, but not the portion of the line $x + y = 0$.

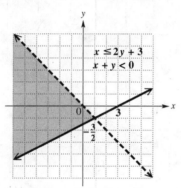

17. $4x + 5y < 8$
$y > -2$
$x > -4$

Graph $4x + 5y = 8$, $y = -2$, and $x = -4$ as dashed lines. All three inequalities are true for $(0, 0)$. Shade the region bounded by the three lines, which contains the test point $(0, 0)$. The solution is the shaded region.

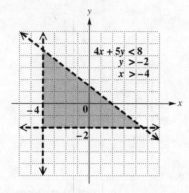

19. $3x - 2y \geq 6$
$x + y \leq 4$
$x \geq 0$
$y \geq -4$

Graph $3x - 2y = 6$, $x + y = 4$, $x = 0$, and $y = -4$ as solid lines. All four inequalities are true for $(2, -2)$. Shade the region bounded by the four lines, which contains the test point $(2, -2)$. The solution is the shaded region.

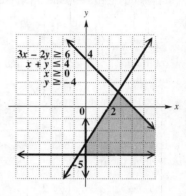

Chapter 4 Review Exercises

1. $(3, 4)$

$4x - 2y = 4$
$5x + y = 19$

To decide whether $(3, 4)$ is a solution of the system, substitute 3 for x and 4 for y in each equation.

$$4x - 2y = 4$$
$$4(3) - 2(4) = 4 \quad ?$$
$$12 - 8 = 4 \quad ?$$
$$4 = 4 \quad \text{True}$$

$$5x + y = 19$$
$$5(3) + 4 = 19 \quad ?$$
$$15 + 4 = 19 \quad ?$$
$$19 = 19 \quad \text{True}$$

Since $(3, 4)$ satisfies both equations, it is a solution of the system.

2. $(-5, 2)$

$x - 4y = -13 \quad (1)$
$2x + 3y = 4 \quad (2)$

Substitute -5 for x and 2 for y in equation (2).

$$2x + 3y = 4$$
$$2(-5) + 3(2) = 4 \quad ?$$
$$-10 + 6 = 4 \quad ?$$
$$-4 = 4 \quad \text{False}$$

Since $(-5, 2)$ is not a solution of the second equation, it cannot be a solution of the system.

3. $x + y = 4$
$2x - y = 5$

To graph the equations, find the intercepts.

$x + y = 4$: Let $y = 0$; then $x = 4$.
Let $x = 0$; then $y = 4$.

Plot the intercepts, $(4, 0)$ and $(0, 4)$, and draw the line through them.

$2x - y = 5$: Let $y = 0$; then $x = \frac{5}{2}$.
Let $x = 0$; then $y = -5$.

Plot the intercepts, $\left(\frac{5}{2}, 0\right)$ and $(0, -5)$, and draw the line through them.

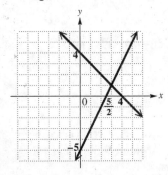

It appears that the lines intersect at the point $(3, 1)$. Check this by substituting 3 for x and 1 for y in both equations. Since $(3, 1)$ satisfies both equations, the solution of this system is $(3, 1)$.

4. $x - 2y = 4$
$2x + y = -2$

To graph the equations, find the intercepts.

$x - 2y = 4$: Let $y = 0$; then $x = 4$.
Let $x = 0$; then $y = -2$.

Plot the intercepts, $(4, 0)$ and $(0, -2)$, and draw the line through them.

$2x + y = -2$: Let $y = 0$; then $x = -1$.
Let $x = 0$; then $y = -2$.

Plot the intercepts, $(-1, 0)$ and $(0, -2)$, and draw the line through them.

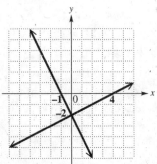

The lines intersect at their common y-intercept, $(0, -2)$, so $(0, -2)$ is the solution of the system.

5. $x - 2 = 2y$
$2x - 4y = 4$

The first equation may be written as $x - 2y = 2$. Find the intercepts for this line.

$x - 2y = 2$: Let $y = 0$; then $x = 2$.
Let $x = 0$; then $y = -1$.

Draw the line through $(2, 0)$ and $(0, -1)$.

Now find the intercepts for the second line.

$2x - 4y = 4$: Let $y = 0$; then $x = 2$.
Let $x = 0$; then $y = -1$.

Again, the intercepts are $(2, 0)$ and $(0, -1)$, so the graph is the same line.

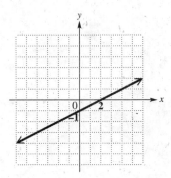

Because the graphs of both equations are the same line, the system has an infinite number of solutions.

6. $2x + 4 = 2y$
$y - x = -3$

Graph the line $2x + 4 = 2y$ through its intercepts, $(-2, 0)$ and $(0, 2)$. Graph the line $y - x = -3$ through its intercepts $(3, 0)$ and $(0, -3)$.

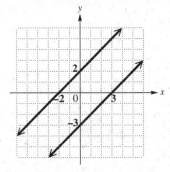

The lines are parallel (both lines have slope 1). Since they do not intersect, there is no solution.

7. It is not a solution of the system because it is not a solution of the second equation, $2x + y = 4$.

8. $3x + y = 7$ (1)
$x = 2y$ (2)

Substitute $2y$ for x in equation (1) and solve the resulting equation for y.

$$3x + y = 7$$
$$3(2y) + y = 7$$
$$6y + y = 7$$
$$7y = 7$$
$$y = 1$$

To find x, let $y = 1$ in equation (2).

$$x = 2y = 2(1) = 2$$

The solution is $(2, 1)$.

9. $2x - 5y = -19$ (1)
$y = x + 2$ (2)

Substitute $x + 2$ for y in equation (1).

$$2x - 5y = -19$$
$$2x - 5(x + 2) = -19$$
$$2x - 5x - 10 = -19$$
$$-3x - 10 = -19$$
$$-3x = -9$$
$$x = 3$$

To find y, let $x = 3$ in equation (2).

$$y = x + 2 = 3 + 2 = 5$$

The solution is $(3, 5)$.

10. $4x + 5y = 44$ (1)
$x + 2 = 2y$ (2)

Solve equation (2) for x.

$$x + 2 = 2y$$
$$x = 2y - 2 \quad (3)$$

Substitute $2y - 2$ for x in equation (1).

$$4x + 5y = 44$$
$$4(2y - 2) + 5y = 44$$
$$8y - 8 + 5y = 44$$
$$13y - 8 = 44$$
$$13y = 52$$
$$y = 4$$

To find x, let $y = 4$ in equation (3).

$$x = 2y - 2 = 2(4) - 2 = 8 - 2 = 6$$

The solution is $(6, 4)$.

11. $5x + 15y = 3$ (1)
$x + 3y = 6$ (2)

Solve equation (2) for x.

$$x + 3y = 6$$
$$x = 6 - 3y \quad (3)$$

Substitute $6 - 3y$ for x in equation (1).

$$5x + 15y = 3$$
$$5(6 - 3y) + 15y = 3$$
$$30 - 15y + 15y = 3$$
$$30 = 3 \quad \text{False}$$

This false statement means that the system has no solution. The equations of the system represent parallel lines.

12. $2x - y = 13$ (1)
$\underline{x + y = 8} $ (2)
$3x = 21$ Add (1) and (2).
$x = 7$

From (2) with $x = 7$, $y = 1$.

The solution is $(7, 1)$.

13. $3x - y = -13$ (1)
$x - 2y = -1$ (2)

Multiply equation (2) by -3 and add the result to equation (1).

$3x - y = -13$ (1)
$\underline{-3x + 6y = 3}$ (3)
$5y = -10$ Add (1) and (3).
$y = -2$

To find x, let $y = -2$ in equation (2).

$$x - 2y = -1$$
$$x - 2(-2) = -1 \quad \text{Let } y = -2.$$
$$x + 4 = -1$$
$$x = -5$$

The solution is $(-5, -2)$.

14. $-4x + 3y = 25$ (1)
$6x - 5y = -39$ (2)

Multiply equation (1) by 3 and equation (2) by 2; then add the results.

$-12x + 9y = 75$
$\underline{12x - 10y = -78}$
$-y = -3$
$y = 3$

To find x, let $y = 3$ in equation (1).

$$-4x + 3y = 25$$
$$-4x + 3(3) = 25$$
$$-4x + 9 = 25$$
$$-4x = 16$$
$$x = -4$$

The solution is $(-4, 3)$.

15. $3x - 4y = 9$ (1)
$6x - 8y = 18$ (2)

Multiply equation (1) by -2 and add the result to equation (2).

$$\begin{aligned} -6x + 8y &= -18 \\ 6x - 8y &= 18 \\ \hline 0 &= 0 \quad \text{True} \end{aligned}$$

This result indicates that all solutions of equation (1) are also solutions of equation (2). The given system has an infinite number of solutions.

16. $2x + 12y = 7$
$3x + 4y = 1$

(a) If we multiply the first equation by -3, the first term will become $-6x$. To eliminate x, we need to change the first term on the left side of the second equation from $3x$ to $6x$. In order to do this, we must multiply the second equation by 2.

(b) If we multiply the first equation by -3, the second term will become $-36y$. To eliminate y, we need to change the second term on the left side of the second equation from $4y$ to $36y$. In order to do this, we must multiply the second equation by 9.

17. $x - 2y = 5$ (1)
$y = x - 7$ (2)

From (2), substitute $x - 7$ for y in equation (1).

$$\begin{aligned} x - 2y &= 5 \\ x - 2(x - 7) &= 5 \\ x - 2x + 14 &= 5 \\ -x &= -9 \\ x &= 9 \end{aligned}$$

Let $x = 9$ in equation (2) to find y.

$$y = 9 - 7 = 2$$

The solution is $(9, 2)$.

18. $5x - 3y = 11$ (1)
$2y = x - 4$ (2)

Solve for x in equation (2).

$$\begin{aligned} 2y &= x - 4 \\ 2y + 4 &= x \quad (3) \end{aligned}$$

Substitute $2y + 4$ for x in equation (1) and then solve for y.

$$\begin{aligned} 5(2y + 4) - 3y &= 11 \\ 10y + 20 - 3y &= 11 \\ 7y + 20 &= 11 \\ 7y &= -9 \\ y &= -\tfrac{9}{7} \end{aligned}$$

To find x, let $y = -\tfrac{9}{7}$ in equation (3).

$$x = 2\left(-\tfrac{9}{7}\right) + 4 = -\tfrac{18}{7} + \tfrac{28}{7} = \tfrac{10}{7}$$

The solution is $\left(\tfrac{10}{7}, -\tfrac{9}{7}\right)$.

19. $\dfrac{x}{2} + \dfrac{y}{3} = 7$ (1)

$\dfrac{x}{4} + \dfrac{2y}{3} = 8$ (2)

Multiply equation (1) by 6 to clear fractions.

$$6\left(\frac{x}{2} + \frac{y}{3}\right) = 6(7)$$
$$3x + 2y = 42 \quad (3)$$

Multiply equation (2) by 12 to clear fractions.

$$12\left(\frac{x}{4} + \frac{2y}{3}\right) = 12(8)$$
$$3x + 8y = 96 \quad (4)$$

To solve this system by the elimination method, multiply equation (3) by -1 and add the result to equation (4).

$$\begin{aligned} -3x - 2y &= -42 \\ 3x + 8y &= 96 \\ \hline 6y &= 54 \\ y &= 9 \end{aligned}$$

To find x, let $y = 9$ in equation (3).

$$\begin{aligned} 3x + 2y &= 42 \\ 3x + 2(9) &= 42 \\ 3x + 18 &= 42 \\ 3x &= 24 \\ x &= 8 \end{aligned}$$

The solution is $(8, 9)$.

20. $\dfrac{3x}{4} - \dfrac{y}{3} = \dfrac{7}{6}$ (1)

$\dfrac{x}{2} + \dfrac{2y}{3} = \dfrac{5}{3}$ (2)

Multiply equation (1) by 12 to clear fractions.

$$12\left(\frac{3x}{4}\right) - 12\left(\frac{y}{3}\right) = 12\left(\frac{7}{6}\right)$$
$$9x - 4y = 14 \quad (3)$$

Multiply equation (2) by 6 to clear fractions.

$$6\left(\frac{x}{2}\right) + 6\left(\frac{2y}{3}\right) = 6\left(\frac{5}{3}\right)$$
$$3x + 4y = 10 \quad (4)$$

Add equations (3) and (4) to eliminate y.

$$\begin{aligned} 9x - 4y &= 14 \\ 3x + 4y &= 10 \\ \hline 12x &= 24 \\ x &= 2 \end{aligned}$$

To find y, let $x = 2$ in equation (4).

$$3(2) + 4y = 10$$
$$6 + 4y = 10$$
$$4y = 4$$
$$y = 1$$

The solution is $(2, 1)$.

21. *Step 2*

    Let $x =$ the number of McDonald's restaurants;
    $y =$ the number of Subway restaurants.

    Step 3

    Subway operated 148 more restaurants than McDonald's, so

    $$y = x + 148. \quad (1)$$

    Together, the two chains had 26,346 restaurants, so

    $$x + y = 26{,}346. \quad (2)$$

    Step 4

    Substitute $x + 148$ for y in (2).

    $$x + (x + 148) = 26{,}346$$
    $$2x + 148 = 26{,}346$$
    $$2x = 26{,}198$$
    $$x = 13{,}099$$

    From (1), $y = 13{,}099 + 148 = 13{,}247$.

    Step 5

    At the end of 2001, Subway had 13,247 restaurants and McDonald's had 13,099 restaurants.

    Step 6

    13,247 is 148 more than 13,099 and the sum of 13,247 and 13,099 is 26,346.

22. *Step 2*

    Let $x =$ the circulation figure for *Modern Maturity*;
    $y =$ the circulation figure for *Reader's Digest*.

    Step 3

    The total circulation was 35.6 million, so

    $$x + y = 35.6. \quad (1)$$

    Reader's Digest circulation was 5.4 million less than that of *Modern Maturity*, so

    $$y = x - 5.4. \quad (2)$$

    Step 4

    Substitute $x - 5.4$ for y in (1).

$$x + (x - 5.4) = 35.6$$
$$2x - 5.4 = 35.6$$
$$2x = 41$$
$$x = 20.5$$

From (2), $y = 20.5 - 5.4 = 15.1$.

Step 5

The average circulation for *Modern Maturity* was 20.5 million and for *Reader's Digest* it was 15.1 million.

Step 6

15.1 is 5.4 less than 20.5 and the sum of 15.1 and 20.5 is 35.6.

23. *Step 2*

    Let $x =$ the length of the rectangle;
    $y =$ the width of the rectangle.

    Step 3

    The perimeter is 90 meters, so

    $$2x + 2y = 90. \quad (1)$$

    The length is $1\frac{1}{2}$ (or $\frac{3}{2}$) times the width, so

    $$x = \tfrac{3}{2}y. \quad (2)$$

    Step 4

    Substitute $\frac{3}{2}y$ for x in equation (1).

    $$2x + 2y = 90$$
    $$2(\tfrac{3}{2}y) + 2y = 90$$
    $$3y + 2y = 90$$
    $$5y = 90$$
    $$y = 18$$

    From (2), $x = \tfrac{3}{2}(18) = 27$.

    Step 5

    The length is 27 meters and the width is 18 meters.

    Step 6

    27 is $1\frac{1}{2}$ times 18 and the perimeter is $2(27) + 2(18) = 90$ meters.

24. *Step 2*

    Let $x =$ the number of \$10 bills;
    $y =$ the number of \$20 bills.

Denomination of Bill	Number of Bills	Total Value
\$10	x	\$$10x$
\$20	y	\$$20y$
Totals	20	\$330

continued

Chapter 4 Systems of Linear Equations and Inequalities

Step 3
The total number of bills is 20, so
$$x + y = 20. \quad (1)$$

The total value of the money is $330, so
$$10x + 20y = 330. \quad (2)$$

We may simplify equation (2) by dividing each side by 10.
$$x + 2y = 33 \quad (3)$$

Step 4
To solve this equation by the elimination method, multiply equation (1) by -1 and add the result to equation (3).

$$\begin{array}{rcl} -x - y &=& -20 \\ x + 2y &=& 33 \\ \hline y &=& 13 \end{array}$$

From (1), $x = 20 - 13 = 7$.

Step 5
The cashier has 13 twenties and 7 tens.

Step 6
There are $13 + 7 = 20$ bills worth $13(\$20) + 7(\$10) = \$330$.

25. Step 2
Let $x =$ the number of pounds of $1.30 per pound candy;
$y =$ the number of pounds of $.90 per pound candy.

Cost per Pound (in dollars)	Number of Pounds	Total Value (in dollars)
1.30	x	$1.30x$
.90	y	$.90y$
1.00	100	$100(1) = 100$

Step 3
From the second and third columns of the table, we obtain the system
$$x + y = 100 \quad (1)$$
$$1.30x + .90y = 100. \quad (2)$$

Step 4
Multiply equation (2) by 10 (to clear decimals) and equation (1) by -9.

$$\begin{array}{rcl} -9x - 9y &=& -900 \\ 13x + 9y &=& 1000 \\ \hline 4x &=& 100 \\ x &=& 25 \end{array}$$

From (1), $y = 100 - 25 = 75$.

Step 5
25 pounds of candy at $1.30 per pound and 75 pounds of candy at $.90 per pound should be used.

Step 6
The value of the mixture is $25(1.30) + 75(.90) = 100$, giving us 100 pounds of candy that can sell for $1 per pound.

26. Let $x =$ the speed of the plane in still air;
$y =$ the speed of the wind.

	d	r	t
With Wind	540	$x + y$	2
Against Wind	690	$x - y$	3

Use the formula $d = rt$.

$540 = (x+y)(2)$ With wind
$270 = x + y$ (1) Divide by 2.
$690 = (x-y)(3)$ Against wind
$230 = x - y$ (2) Divide by 3.

Solve the system of equations (1) and (2) by the elimination method.

$$\begin{array}{rcl} 270 &=& x + y \quad (1) \\ 230 &=& x - y \quad (2) \\ \hline 500 &=& 2x \quad \text{Add (1) and (2).} \\ 250 &=& x \end{array}$$

From (1) with $x = 250$, $y = 20$.

The speed of the plane in still air is 250 miles per hour and the speed of the wind is 20 miles per hour.

27. Step 2
Let $x =$ the amount invested at 3%;
$y =$ the amount invested at 4%.

Rate	Amount of Principal	Interest
.03	$\$x$	$\$.03x$
.04	$\$y$	$\$.04y$
Totals	$\$18,000$	$\$650$

Step 3
From the chart, we obtain the equations
$$x + y = 18{,}000 \quad (1)$$
$$.03x + .04y = 650. \quad (2)$$

Step 4
To clear decimals, multiply each side of equation (2) by 100.

$$100(.03x + .04y) = 100(650)$$
$$3x + 4y = 65{,}000 \quad (3)$$

Multiply equation (1) by -3 and add the result to equation (3).

$$\begin{aligned} -3x - 3y &= -54{,}000 \\ 3x + 4y &= 65{,}000 \\ \hline y &= 11{,}000 \end{aligned}$$

From (1) with $y = 11{,}000$, $x = 7000$.

Step 5
She invested $7000 at 3% and $11,000 at 4%.

Step 6
The sum of $7000 and $11,000 is $18,000. 3% of $7000 is $210 and 4% of $11,000 is $440. This gives us $210 + $440 = $650 in interest, as required.

28. Let $x =$ the number of liters of 40% solution;
    $y =$ the number of liters of 70% solution.

Percent	Number of Liters	Amount of Pure Antifreeze
.40	x	$.40x$
.70	y	$.70y$
.50	90	$.50(90) = 45$

From the second and third columns of the table, we obtain the equations

$$x + y = 90 \quad (1)$$
$$.40x + .70y = 45. \quad (2)$$

To clear decimals, multiply both sides of equation (2) by 10.

$$4x + 7y = 450 \quad (3)$$

To solve this system by the elimination method, multiply equation (1) by -4 and add the result to equation (3).

$$\begin{aligned} -4x - 4y &= -360 \quad (4) \\ 4x + 7y &= 450 \quad (3) \\ \hline 3y &= 90 \quad \text{Add (4) and (3).} \\ y &= 30 \end{aligned}$$

To find x, let $y = 30$ in equation (1).

$$x = 90 - y = 90 - 30 = 60$$

To get 90 liters of a 50% antifreeze solution, 60 liters of a 40% solution and 30 liters of a 70% solution will be needed.

29. $x + y \geq 2$
    $x - y \leq 4$

Graph $x + y = 2$ as a solid line through its intercepts, $(2, 0)$ and $(0, 2)$. Using $(0, 0)$ as a test point will result in the false statement $0 \geq 2$, so shade the region *not* containing the origin.

Graph $x - y = 4$ as a solid line through its intercepts, $(4, 0)$ and $(0, -4)$. Using $(0, 0)$ as a test point will result in the true statement $0 \leq 4$, so shade the region containing the origin.

The solution of this system is the intersection of the two shaded regions, and includes the portions of the two lines that bound this region.

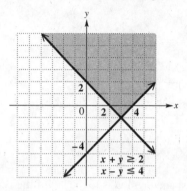

30. $y \geq 2x$
    $2x + 3y \leq 6$

Graph $y = 2x$ as a solid line through $(0, 0)$ and $(1, 2)$. This line goes through the origin, so a different test point must be used. Choosing $(-4, 0)$ as a test point will result in the true statement $0 \geq -8$, so shade the region containing $(-4, 0)$.

Graph $2x + 3y = 6$ as a solid line through its intercepts, $(3, 0)$ and $(0, 2)$. Choosing $(0, 0)$ as a test point will result in the true statement $0 \leq 6$, so shade the region containing the origin.

The solution of this system is the intersection of the two shaded regions, and includes the portions of the two lines that bound this region.

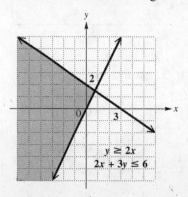

31. $x + y < 3$
$2x > y$

Graph $x + y = 3$ as a dashed line through $(3, 0)$ and $(0, 3)$. Using $(0, 0)$ as a test point will result in the true statement $0 < 3$, so shade the region containing the origin.

Graph $2x = y$ as a dashed line through $(0, 0)$ and $(1, 2)$. Choosing $(0, -3)$ as a test point will result in the true statement $0 \geq -3$, so shade the region containing $(0, -3)$. The solution of this system is the intersection of the two shaded regions. It does not contain the boundary lines.

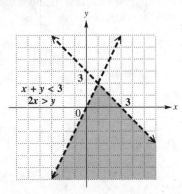

32. The shaded region is to the left of the vertical line $x = 3$, above the horizontal line $y = 1$, and includes both lines. This is the graph of the system

$$x \leq 3$$
$$y \geq 1,$$

so the answer is **B**.

33. It is impossible for the sum of any two numbers to be both greater than 4 and less than 3. Therefore, system **B** has no solution.

34. [4.3] $3x + 4y = 6$ (1)
$4x - 5y = 8$ (2)

To solve this system by the elimination method, multiply equation (1) by -4 and equation (2) by 3; then add the results.

$$-12x - 16y = -24$$
$$\underline{12x - 15y = 24}$$
$$-31y = 0$$
$$y = 0$$

To find x, substitute 0 for y in equation (1).

$$3x + 4(0) = 6$$
$$3x = 6$$
$$x = 2$$

The solution is $(2, 0)$.

35. [4.2] $\dfrac{3x}{2} + \dfrac{y}{5} = -3$ (1)

$4x + \dfrac{y}{3} = -11$ (2)

To clear fractions, multiply each side of equation (1) by 10 and each side of equation (2) by 3.

$$10\left(\dfrac{3x}{2} + \dfrac{y}{5}\right) = 10(-3)$$
$$15x + 2y = -30 \quad (3)$$
$$3\left(4x + \dfrac{y}{3}\right) = 3(-11)$$
$$12x + y = -33 \quad (4)$$

Solve equation (4) for y.

$$y = -33 - 12x \quad (5)$$

Substitute $-33 - 12x$ for y in equation (3) and solve for x.

$$15x + 2y = -30$$
$$15x + 2(-33 - 12x) = -30$$
$$15x - 66 - 24x = -30$$
$$-9x = 36$$
$$x = -4$$

To find y, let $x = -4$ in equation (5).

$$y = -33 - 12(-4)$$
$$= -33 + 48 = 15$$

The solution is $(-4, 15)$.

36. [4.3] $x + 6y = 3$ (1)
$2x + 12y = 2$ (2)

To solve this system by the elimination method, multiply equation (1) by -2 and add the result to equation (2).

$$-2x - 12y = -6 \quad (3)$$
$$\underline{2x + 12y = 2 \quad (2)}$$
$$0 = -4 \quad \text{Add (3) and (2).}$$

The last statement, $0 = -4$, is false, which indicates that the system has no solution.

37. [4.5] $x + y < 5$
$x - y \geq 2$

Graph $x + y = 5$ as a dashed line through its intercepts, $(5, 0)$ and $(0, 5)$. Using $(0, 0)$ as a test point will result in the true statement $0 < 5$, so shade the region containing the origin.

Graph $x - y = 2$ as a solid line through its intercepts, $(2, 0)$ and $(0, -2)$. Using $(0, 0)$ as a test point will result in the false statement $0 \geq 2$, so shade the region *not* containing the origin.

The solution of this system is the intersection of the two shaded regions. It includes the portion of the line $x - y = 2$ that bounds this region, but not the line $x + y = 5$.

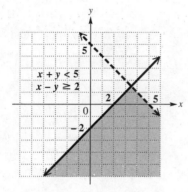

38. [4.5] $y \leq 2x$
$x + 2y > 4$

Graph $y = 2x$ as a solid line through $(0, 0)$ and $(1, 2)$. Shade the region below the line.

Graph $x + 2y = 4$ as a dashed line through its intercepts, $(4, 0)$ and $(0, 2)$. Since $x + 2y > 4$ is equivalent to $y > -\frac{1}{2}x + 2$, shade the region above this line.

The solution of this system is the intersection of the shaded regions. It includes the portion of the line $y = 2x$ that bounds this region, but not the line $x + 2y = 4$.

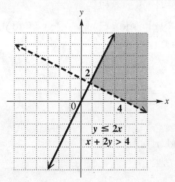

39. [4.5] $y < -4x$
$y < -2$

The graph of $y < -4x$ is the region below the graph of the dashed line $y = -4x$. This line goes through the points $(0, 0)$ and $(1, -4)$.

The graph of $y < -2$ is the region below the graph of the dashed horizontal line $y = -2$. The intersection of these two regions is graphed below. The solution is the shaded region.

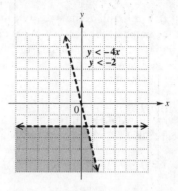

40. [4.4]

Let $x =$ the length of each of the two equal sides;
$y =$ the length of the longer third side.

The perimeter is 29 inches, so

$$x + x + y = 29$$
$$\text{or} \quad 2x + y = 29. \quad (1)$$

The third side is 5 inches longer than each of the two equal sides, so

$$y = x + 5. \quad (2)$$

Substitute $x + 5$ for y in (1).

$$2x + y = 29$$
$$2x + (x + 5) = 29$$
$$3x + 5 = 29$$
$$3x = 24$$
$$x = 8$$

From (2), $y = 8 + 5 = 13$.

The lengths of the sides of the triangle are 8 inches, 8 inches, and 13 inches.

41. [4.4] Let $x =$ the Patriots' score;
$y =$ the Colts' score.

The Patriots beat the Colts by 10, so

$$x = y + 10. \quad (1)$$

The winning score was 4 less than twice the losing score, so

$$x = 2y - 4. \quad (2)$$

Substitute $2y - 4$ for x in equation (1).

$$2y - 4 = y + 10$$
$$2y = y + 14$$
$$y = 14$$

From (1), $x = 14 + 10 = 24$.

The final score of the game was New England 24, Indianapolis 14.

Chapter 4 Test

1. $2x + y = 1$
$3x - y = 9$

The intercepts for the line $2x + y = 1$ are $\left(\frac{1}{2}, 0\right)$ and $(0, 1)$. Because one of the coordinates is a fraction and the intercepts are close together, it may be difficult to graph the line accurately using only these two points. A third point, such as $(-1, 3)$ may be helpful. Graph the line $3x - y = 9$ through its intercepts, $(3, 0)$ and $(0, -9)$.

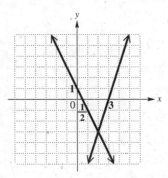

The two lines appear to intersect at the point $(2, -3)$. Checking by substituting 2 for x and -3 for y in the original equations will show that this ordered pair is the solution of the system.

2. Two lines which have the same slope but different y-intercepts are parallel, so the system has no solution.

3. $2x + y = -4 \quad (1)$
$x = y + 7 \quad (2)$

Substitute $y + 7$ for x in equation (1), and solve for y.

$$2x + y = -4$$
$$2(y + 7) + y = -4$$
$$2y + 14 + y = -4$$
$$3y = -18$$
$$y = -6$$

From (2), $x = -6 + 7 = 1$.

The solution is $(1, -6)$.

4. $4x + 3y = -35 \quad (1)$
$x + y = 0 \quad (2)$

Solve equation (2) for y.

$$y = -x \quad (3)$$

Substitute $-x$ for y in equation (1) and solve for x.

$$4x + 3y = -35$$
$$4x + 3(-x) = -35$$
$$4x - 3x = -35$$
$$x = -35$$

From (3), $y = -(-35) = 35$.

The solution is $(-35, 35)$.

5. $2x - y = 4 \quad (1)$
$3x + y = 21 \quad (2)$
$5x = 25 \quad$ *Add (1) and (2).*
$x = 5$

To find y, let $x = 5$ in equation (2).

$$3x + y = 21$$
$$3(5) + y = 21$$
$$15 + y = 21$$
$$y = 6$$

The solution is $(5, 6)$.

6. $4x + 2y = 2 \quad (1)$
$5x + 4y = 7 \quad (2)$

Multiply equation (1) by -2 and add the result to equation (2).

$$-8x - 4y = -4$$
$$5x + 4y = 7$$
$$-3x = 3$$
$$x = -1$$

To find y, let $x = -1$ in equation (1).

$$4x + 2y = 2$$
$$4(-1) + 2y = 2$$
$$-4 + 2y = 2$$
$$2y = 6$$
$$y = 3$$

The solution is $(-1, 3)$.

7. $6x - 5y = 0$ (1)
$-2x + 3y = 0$ (2)

Multiply equation (2) by 3 and add the result to equation (1)

$$\begin{aligned} 6x - 5y &= 0 \\ -6x + 9y &= 0 \\ \hline 4y &= 0 \\ y &= 0 \end{aligned}$$

To find x, let $y = 0$ in equation (1).

$$\begin{aligned} 6x - 5y &= 0 \\ 6x - 5(0) &= 0 \\ 6x &= 0 \\ x &= 0 \end{aligned}$$

The solution is $(0, 0)$.

8. $4x + 5y = 2$ (1)
$-8x - 10y = 6$ (2)

Multiply equation (1) by 2 and add the result to equation (2).

$$\begin{aligned} 8x + 10y &= 4 \\ -8x - 10y &= 6 \\ \hline 0 &= 10 \quad \text{False} \end{aligned}$$

This result indicates that the system has no solution.

9. $3x = 6 + y$ (1)
$6x - 2y = 12$ (2)

Write equation (1) in the form $Ax + By = C$.

$$3x - y = 6 \quad (3)$$
$$6x - 2y = 12 \quad (2)$$

Multiply equation (3) by -2 and add the result to equation (2).

$$\begin{aligned} -6x + 2y &= -12 \quad (4) \\ 6x - 2y &= 12 \quad (2) \\ \hline 0 &= 0 \quad \text{Add (4) and (2).} \end{aligned}$$

This result indicates that the system has an infinite number of solutions.

10. $\dfrac{x}{2} - \dfrac{y}{4} = 7$ (1)
$\dfrac{2x}{3} + \dfrac{5y}{4} = 3$ (2)

To clear fractions, multiply equation (1) by 4 and equation (2) by 12.

$$4\left(\frac{x}{2} - \frac{y}{4}\right) = 4(7)$$
$$2x - y = 28 \quad (3)$$
$$12\left(\frac{2x}{3} + \frac{5y}{4}\right) = 12(3)$$
$$8x + 15y = 36 \quad (4)$$

Multiply equation (3) by -4 and add the result to equation (4).

$$\begin{aligned} -8x + 4y &= -112 \quad (5) \\ 8x + 15y &= 36 \quad (4) \\ \hline 19y &= -76 \quad \text{Add (5) and (4).} \\ y &= -4 \end{aligned}$$

To find x, let $y = -4$ in equation (3).

$$\begin{aligned} 2x - y &= 28 \\ 2x - (-4) &= 28 \\ 2x + 4 &= 28 \\ 2x &= 24 \\ x &= 12 \end{aligned}$$

The solution is $(12, -4)$.

11. *Step 2*
Let $x =$ the distance between Memphis and Atlanta (in miles);
$y =$ the distance between Minneapolis and Houston (in miles).

Step 3
Since the distance between Memphis and Atlanta is 782 miles less than the distance between Minneapolis and Houston,

$$x = y - 782. \quad (1)$$

Together the two distances total 1570 miles, so

$$x + y = 1570. \quad (2)$$

Step 4
Substitute $y - 782$ for x in equation (2).

$$\begin{aligned} (y - 782) + y &= 1570 \\ 2y - 782 &= 1570 \\ 2y &= 2352 \\ y &= 1176 \end{aligned}$$

From (1), $x = 1176 - 782 = 394$.

Step 5
The distance between Memphis and Atlanta is 394 miles, while the distance between Minneapolis and Houston is 1176 miles.

Step 6
394 is 782 less than 1176 and the sum of 394 and 1176 is 1570, as required.

12. *Step 2*

Let $x =$ the number of visitors to the Magic Kingdom (in millions);
$y =$ the number of visitors to Disneyland (in millions).

Step 3
Disneyland had 1.3 million fewer visitors than the Magic Kingdom, so
$$y = x - 1.3 \quad (1)$$
Together they had 26.7 million visitors, so
$$x + y = 26.7 \quad (2)$$

Step 4
From (1), substitute $x - 1.3$ for y in (2).
$$x + (x - 1.3) = 26.7$$
$$2x - 1.3 = 26.7$$
$$2x = 28.0$$
$$x = 14.0$$
From (1), $y = 14.0 - 1.3 = 12.7$.

Step 5
In 2002, the Magic Kingdom had 14.0 million visitors and Disneyland had 12.7 million visitors.

Step 6
12.7 is 1.3 fewer than 14.0 and the sum of 12.7 and 14.0 is 26.7, as required.

13. *Step 2*

Let $x =$ the number of liters of 25% solution;
$y =$ the number of liters of 40% solution.

Percent (as a decimal)	Liters of Solution	Liters of Pure Alcohol
.25	x	$.25x$
.40	y	$.40y$
.30	50	$.30(50) = 15$

Step 3
From the second and third columns of the table, we obtain the equations
$$x + y = 50 \quad (1)$$
$$.25x + .40y = 15. \quad (2)$$
To clear decimals, multiply both sides of equation (2) by 100.
$$25x + 40y = 1500 \quad (3)$$

Step 4
To solve this system by the elimination method, multiply equation (1) by -25 and add the result to equation (3).

$$-25x - 25y = -1250$$
$$\underline{25x + 40y = 1500}$$
$$15y = 250$$
$$y = \tfrac{250}{15} = \tfrac{50}{3} = 16\tfrac{2}{3}$$

From (1), $x = 50 - 16\tfrac{2}{3} = 33\tfrac{1}{3}$.

Step 5
To get 50 liters of a 30% alcohol solution, $33\tfrac{1}{3}$ liters of a 25% solution and $16\tfrac{2}{3}$ liters of a 40% solution should be used.

Step 6
Since $33\tfrac{1}{3} + 16\tfrac{2}{3} = 50$ and
$$.25\left(33\tfrac{1}{3}\right) + .40\left(16\tfrac{2}{3}\right) = \tfrac{1}{4}\left(\tfrac{100}{3}\right) + \tfrac{2}{5}\left(\tfrac{50}{3}\right)$$
$$= \tfrac{25}{3} + \tfrac{20}{3} = \tfrac{45}{3} = 15,$$
this mixture will give the 50 liters of 30% solution, as required.

14. *Step 2*
Let $x =$ the speed of the faster car;
$y =$ the speed of the slower car.

Use $d = rt$ in constructing the table.

	r	t	d
Faster car	x	3	$3x$
Slower car	y	3	$3y$

Step 3
The faster car travels $1\tfrac{1}{3}$ times as fast as the other car, so
$$x = 1\tfrac{1}{3}y$$
or $\quad x = \tfrac{4}{3}y. \quad (1)$

After 3 hours, they are 45 miles apart, that is, the difference between their distances is 45.
$$3x - 3y = 45 \quad (2)$$

Step 4
To solve the system by substitution, substitute $\tfrac{4}{3}y$ for x in equation (2).
$$3\left(\tfrac{4}{3}y\right) - 3y = 45$$
$$4y - 3y = 45$$
$$y = 45$$
From (1), $x = \tfrac{4}{3}(45) = 60$.

Step 5
The speed of the faster car is 60 miles per hour, and the speed of the slower car is 45 miles per hour.

Step 6
60 is one and one-third times 45. In three hours, the faster car travels $3(60) = 180$ miles and the slower car travels $3(45) = 135$ miles. The cars are $180 - 135 = 45$ miles apart, as required.

15. $2x + 7y \leq 14$
$x - y \geq 1$

Graph $2x + 7y = 14$ as a solid line through its intercepts, $(7, 0)$ and $(0, 2)$. Choosing $(0, 0)$ as a test point will result in the true statement $0 \leq 14$, so shade the side of the line containing the origin.

Graph $x - y = 1$ as a solid line through its intercepts, $(1, 0)$ and $(0, -1)$. Choosing $(0, 0)$ as a test point will result in the false statement $0 \geq 1$, so shade the side of the line *not* containing the origin.

The solution of the given system is the intersection of the two shaded regions, and includes the portions of the two lines that bound this region.

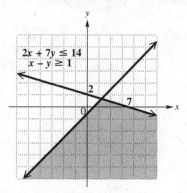

16. $2x - y > 6$
$4y + 12 \geq -3x$

Graph $2x - y = 6$ as a dashed line through its intercepts, $(3, 0)$ and $(0, -6)$. Choosing $(0, 0)$ as a test point will result in the false statement $0 > 6$, so shade the side of the line *not* containing the origin.

Graph $4y + 12 = -3x$ as a solid line through its intercepts, $(-4, 0)$ and $(0, -3)$. Choosing $(0, 0)$ as a test point will result in the true statement $12 \geq 0$, so shade the side of the line containing the origin.

The solution of the given system is the intersection of the two shaded regions. It includes the portion of the line $4y + 12 = -3x$ that bounds this region, but not the line $2x - y = 6$.

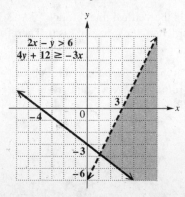

Cumulative Review Exercises (Chapters R–4)

1. The integer factors of 40 are $-1, 1, -2, 2, -4, 4, -5, 5, -8, 8, -10, 10, -20, 20, -40,$ and 40.

2. $\dfrac{3x^2 + 2y^2}{10y + 3} = \dfrac{3 \cdot 1^2 + 2 \cdot 5^2}{10(5) + 3}$ Let $x = 1, y = 5$.

$= \dfrac{3 \cdot 1 + 2 \cdot 25}{50 + 3}$

$= \dfrac{3 + 50}{50 + 3} = \dfrac{53}{53} = 1$

3. $5 + (-4) = (-4) + 5$

The order of the numbers has been changed, so this is an example of the commutative property of addition.

4. $r(s - k) = rs - rk$

This is an example of the distributive property.

5. $-\dfrac{2}{3} + \dfrac{2}{3} = 0$

The numbers $-\dfrac{2}{3}$ and $\dfrac{2}{3}$ are additive inverses (or opposites) of each other. This is an example of the inverse property for addition.

6. $-2 + 6[3 - (4 - 9)] = -2 + 6[3 - (-5)]$
$= -2 + 6(8)$
$= -2 + 48$
$= 46$

7. $2 - 3(6x + 2) = 4(x + 1) + 18$
$2 - 18x - 6 = 4x + 4 + 18$
$-18x - 4 = 4x + 22$
$-22x = 26$
$x = \dfrac{26}{-22} = -\dfrac{13}{11}$

The solution is $-\dfrac{13}{11}$.

8. $\dfrac{3}{2}\left(\dfrac{1}{3}x + 4\right) = 6\left(\dfrac{1}{4} + x\right)$

Multiply each side by 2.

$2\left[\dfrac{3}{2}\left(\dfrac{1}{3}x + 4\right)\right] = 2\left[6\left(\dfrac{1}{4} + x\right)\right]$

Use the associative property and multiply.

$3\left(\dfrac{1}{3}x + 4\right) = 12\left(\dfrac{1}{4} + x\right)$

Use the distributive property to remove parentheses.

$x + 12 = 3 + 12x$
$-11x = -9$
$x = \dfrac{9}{11}$

The solution is $\dfrac{9}{11}$.

176   Chapter 4 Systems of Linear Equations and Inequalities

9. $-\dfrac{5}{6}x < 15$

   Multiply each side by the reciprocal of $-\dfrac{5}{6}$, which is $-\dfrac{6}{5}$, and reverse the direction of the inequality symbol.

   $$-\tfrac{6}{5}\left(-\tfrac{5}{6}x\right) > -\tfrac{6}{5}(15)$$
   $$x > -18$$

   The solution is $x > -18$.

10. $-8 < 2x + 3$
    $-11 < 2x$
    $-\dfrac{11}{2} < x \quad \text{or} \quad x > -\dfrac{11}{2}$

    The solution is $x > -\dfrac{11}{2}$.

11. Let $l =$ the length of the book.
    Then $l - 2.58 =$ the width of the book.

    $$P = 2l + 2w$$
    $$37.8 = 2l + 2(l - 2.58)$$
    $$37.8 = 2l + 2l - 5.16$$
    $$42.96 = 4l$$
    $$10.74 = l$$

    The width is $10.74 - 2.58 = 8.16$ inches and the length is 10.74 inches.

12. $x - y = 4$

    To graph this line, find the intercepts.

    If $y = 0$, $x = 4$, so the x-intercept is $(4, 0)$.
    If $x = 0$, $y = -4$, so the y-intercept is $(0, -4)$.

    Graph the line through these intercepts. A third point, such as $(5, 1)$, may be used as a check.

    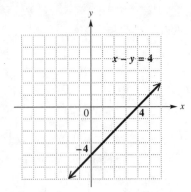

13. $3x + y = 6$

    If $y = 0$, $x = 2$, so the x-intercept is $(2, 0)$.
    If $x = 0$, $y = 6$, so the y-intercept is $(0, 6)$.

    Graph the line through these intercepts. A third point, such as $(1, 3)$, may be used as a check.

    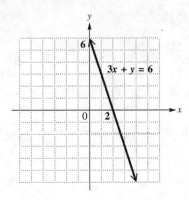

14. The slope m of the line passing through the points $(-5, 6)$ and $(1, -2)$ is

    $$m = \dfrac{y_2 - y_1}{x_2 - x_1} = \dfrac{-2 - 6}{1 - (-5)} = \dfrac{-8}{6} = -\dfrac{4}{3}$$

15. The slope of the line $y = 4x - 3$ is 4. The slope m of the line whose graph is perpendicular to that of $y = 4x - 3$ must satisfy the equation $4 \cdot m = -1$. Thus, $m = -\dfrac{1}{4}$.

16. Through $(-4, 1)$ with slope $\dfrac{1}{2}$
    Use the point-slope form of a line.

    $$y - y_1 = m(x - x_1)$$
    $$y - 1 = \dfrac{1}{2}[x - (-4)]$$
    $$y - 1 = \dfrac{1}{2}(x + 4)$$
    $$y - 1 = \dfrac{1}{2}x + 2$$
    $$y = \dfrac{1}{2}x + 3 \quad \text{Slope-intercept form}$$

17. Through the points $(1, 3)$ and $(-2, -3)$
    Find the slope.

    $$m = \dfrac{y_2 - y_1}{x_2 - x_1} = \dfrac{-3 - 3}{-2 - 1} = \dfrac{-6}{-3} = 2$$

    Use the point-slope form of a line.

    $$y - y_1 = m(x - x_1)$$
    $$y - 3 = 2(x - 1)$$
    $$y - 3 = 2x - 2$$
    $$y = 2x + 1 \quad \text{Slope-intercept form}$$

18. (a) On the vertical line through $(9, -2)$, the x-coordinate of every point is 9. Therefore, an equation of this line is $x = 9$.

    (b) On the horizontal line through $(4, -1)$, the y-coordinate of every point is -1. Therefore, an equation of this line is $y = -1$.

19. $2x - y = -8$ (1)
$x + 2y = 11$ (2)

To solve this system by the elimination method, multiply equation (1) by 2 and add the result to equation (2) to eliminate y.

$$\begin{aligned} 4x - 2y &= -16 \\ x + 2y &= 11 \\ \hline 5x &= -5 \\ x &= -1 \end{aligned}$$

To find y, let $x = -1$ in equation (2).

$$\begin{aligned} -1 + 2y &= 11 \\ 2y &= 12 \\ y &= 6 \end{aligned}$$

The solution is $(-1, 6)$.

20. $4x + 5y = -8$ (1)
$3x + 4y = -7$ (2)

Multiply equation (1) by -3 and equation (2) by 4. Add the resulting equations to eliminate x.

$$\begin{aligned} -12x - 15y &= 24 \\ 12x + 16y &= -28 \\ \hline y &= -4 \end{aligned}$$

To find x, let $y = -4$ in equation (1).

$$\begin{aligned} 4x + 5y &= -8 \\ 4x + 5(-4) &= -8 \\ 4x - 20 &= -8 \\ 4x &= 12 \\ x &= 3 \end{aligned}$$

The solution is $(3, -4)$.

21. $3x + 4y = 2$ (1)
$6x + 8y = 1$ (2)

Multiply equation (1) by -2 and add the result to equation (2).

$$\begin{aligned} -6x - 8y &= -4 \\ 6x + 8y &= 1 \\ \hline 0 &= -3 \quad \text{False} \end{aligned}$$

Since $0 = -3$ is a false statement, there is no solution.

22. *Step 2*
Let x = the number of adults' tickets sold;
y = the number of children's tickets sold.

Kind of Ticket	Number Sold	Cost of Each (in dollars)	Total Value (in dollars)
Adult	x	6	$6x$
Child	y	2	$2y$
Total	454	—	2528

Step 3
The total number of tickets sold was 454, so

$$x + y = 454. \quad (1)$$

Since the total value was $2528, the final column leads to

$$6x + 2y = 2528. \quad (2)$$

Step 4
Multiply both sides of equation (1) by -2 and add this result to equation (2).

$$\begin{aligned} -2x - 2y &= -908 \\ 6x + 2y &= 2528 \\ \hline 4x &= 1620 \\ x &= 405 \end{aligned}$$

From (1), $y = 454 - 405 = 49$.

Step 5
There were 405 adults and 49 children at the game.

Step 6
The total number of tickets sold was $405 + 49 = 454$. Since 405 adults paid $6 each and 49 children paid $2 each, the value of tickets sold should be $405(6) + 49(2) = 2528$, or $2528. The result agrees with the given information.

23. *Step 2*

Let $x =$ the length of each of the equal sides;
$y =$ the length of the shorter third side.

Step 3
The third side measures 4 inches less than each of the equal sides, so
$$y = x - 4. \quad (1)$$
The perimeter of the triangle is 53 inches, so
$$x + x + y = 53$$
$$\text{or} \quad 2x + y = 53. \quad (2)$$

Step 4
Substitute $x - 4$ for y in equation (2).
$$2x + (x - 4) = 53$$
$$3x - 4 = 53$$
$$3x = 57$$
$$x = 19$$
From (1), $y = 19 - 4 = 15$.

Step 5
The lengths of the sides are 19 inches, 19 inches, and 15 inches.

Step 6
15 is 4 less than 19 and the perimeter is $19 + 19 + 15 = 53$ inches, as required.

24. $x + 2y \leq 12$
$2x - y \leq 8$

Graph the boundary with equation $x + 2y = 12$ as a solid line through its intercepts, $(12, 0)$ and $(0, 6)$. Using $(0, 0)$ as a test point results in a true statement, $0 \leq 12$. Shade the region containing the origin.

Graph the boundary with equation $2x - y = 8$ as a solid line through its intercepts, $(4, 0)$ and $(0, -8)$. Using $(0, 0)$ as a test point results in a true statement, $0 \leq 8$. Shade the region containing the origin.

The solution is the intersection of the two shaded regions and includes the portions of the lines that bound this region.

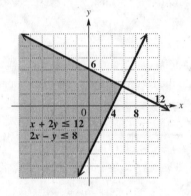

CHAPTER 5 EXPONENTS AND POLYNOMIALS

5.1 Adding and Subtracting Polynomials

5.1 Margin Exercises

1. (a) $5x^4 + 7x^4$
   $= (5+7)x^4$ *Distributive property*
   $= 12x^4$

   (b) $9pq + 3pq - 2pq$
   $= (9+3-2)pq$ *Distributive property*
   $= 10pq$

   (c) $r^2 + 3r + 5r^2$
   $= (1r^2 + 5r^2) + 3r$ *Commutative and Associative properties*
   $= (1+5)r^2 + 3r$ *Distributive property*
   $= 6r^2 + 3r$

   (d) $8t + 6w$ has unlike terms that cannot be added.

2. (a) $3m^5 + 5m^2 - 2m + 1$ is a polynomial written in descending powers, since $m = m^1$ and $1 = m^0$. So **A** and **B** apply.

   (b) $2p^4 + p^6$ is a polynomial, but is not written in descending powers. So **A** applies.

   (c) $\dfrac{1}{x} + 2x^2 + 3$ is not a polynomial, because $\dfrac{1}{x}$ has a variable in the denominator. So **C** applies.

   (d) $x - 3$ is a polynomial written in descending powers, since $x = x^1$ and $3 = 3x^0$. So **A** and **B** apply.

3. (a) $3x^2 + 2x - 4$ cannot be simplified, because there are no like terms. There are three terms, and the largest exponent on the variable x is 2. So the polynomial has degree 2 and is a trinomial.

   (b) $x^3 + 4x^3 = 1x^3 + 4x^3$ simplifies to $(1+4)x^3 = 5x^3$. This polynomial has one term, and the exponent on the variable x is 3. So the polynomial has degree 3 and is a monomial.

   (c) $x^8 - x^7 + 2x^8 = 1x^8 + 2x^8 - x^7$ simplifies to $(1+2)x^8 - x^7 = 3x^8 - x^7$. This polynomial has two terms, and the largest exponent on the variable x is 8. So the polynomial has degree 8 and is a binomial.

4. (a) $2y^3 + 8y - 6$ *Replace y with -1.*
   $= 2(-1)^3 + 8(-1) - 6$
   $= 2(-1) + 8(-1) - 6$
   $= -2 - 8 - 6$
   $= -16$

   (b) $2y^3 + 8y - 6$ *Replace y with 4.*
   $= 2(4)^3 + 8(4) - 6$
   $= 2(64) + 8(4) - 6$
   $= 128 + 32 - 6$
   $= 154$

5. (a) $\begin{array}{r} 4x^3 - 3x^2 + 2x \\ 6x^3 + 2x^2 - 3x \\ \hline 10x^3 - x^2 - x \end{array}$

   (b) $\begin{array}{r} x^2 - 2x + 5 \\ 4x^2 - 2 \\ \hline 5x^2 - 2x + 3 \end{array}$

6. (a) Combine like terms.
   $(2x^4 - 6x^2 + 7) + (-3x^4 + 5x^2 + 2)$
   $= (2x^4 - 3x^4) + (-6x^2 + 5x^2) + (7+2)$
   $= -x^4 - x^2 + 9$

   (b) Combine like terms.
   $(3x^2 + 4x + 2) + (6x^3 - 5x - 7)$
   $= 6x^3 + 3x^2 + (4x - 5x) + (2 - 7)$
   $= 6x^3 + 3x^2 - x - 5$

7. (a) Change subtraction to addition.
   $(14y^3 - 6y^2 + 2y - 5) - (2y^3 - 7y^2 - 4y + 6)$
   $= (14y^3 - 6y^2 + 2y - 5) + (-2y^3 + 7y^2 + 4y - 6)$

   Combine like terms.
   $= (14y^3 - 2y^3) + (-6y^2 + 7y^2) + (2y + 4y)$
   $ + (-5 - 6)$
   $= 12y^3 + y^2 + 6y - 11$

   Check by adding.
   $\begin{array}{r} 2y^3 - 7y^2 - 4y + 6 \\ 12y^3 + y^2 + 6y - 11 \\ \hline 14y^3 - 6y^2 + 2y - 5 \end{array}$ *Second polynomial*
      Answer
      First polynomial

Chapter 5 Exponents and Polynomials

(b) $\left(\dfrac{7}{2}y^2 - \dfrac{11}{3}y + 8\right) - \left(-\dfrac{3}{2}y^2 + \dfrac{4}{3}y + 6\right)$

$= \left(\dfrac{7}{2}y^2 - \dfrac{11}{3}y + 8\right) + \left(\dfrac{3}{2}y^2 - \dfrac{4}{3}y - 6\right)$

Definition of subtraction

$= \left(\dfrac{7}{2}y^2 + \dfrac{3}{2}y^2\right) + \left[\left(-\dfrac{11}{3}y\right) + \left(-\dfrac{4}{3}y\right)\right]$
$+ [8 + (-6)]$ *Group like terms.*

$= \dfrac{10}{2}y^2 - \dfrac{15}{3}y + 2$ *Add like terms.*

$= 5y^2 - 5y + 2$

Check:

$\left(-\dfrac{3}{2}y^2 + \dfrac{4}{3}y + 6\right) + \left(5y^2 - 5y + 2\right)$

$= \left(-\dfrac{3}{2}y^2 + 5y^2\right) + \left(\dfrac{4}{3}y - 5y\right) + (6 + 2)$

$= \left(-\dfrac{3}{2}y^2 + \dfrac{10}{2}y^2\right) + \left(\dfrac{4}{3}y - \dfrac{15}{3}y\right) + (6 + 2)$

$= \dfrac{7}{2}y^2 - \dfrac{11}{3}y + 8$

The answer is correct.

8. $(4y^3 - 16y^2 + 2y) - (12y^3 - 9y^2 + 16)$

Arrange like terms in columns. Insert zeros for missing terms.

$$\begin{array}{r} 4y^3 - 16y^2 + 2y + 0 \\ 12y^3 - 9y^2 + 0y + 16 \end{array}$$

Change all signs in the second row, and then add.

$$\begin{array}{r} 4y^3 - 16y^2 + 2y + 0 \\ -12y^3 + 9y^2 - 0y - 16 \\ \hline -8y^3 - 7y^2 + 2y - 16 \end{array}$$

9. $(6p^4 - 8p^3 + 2p - 1) - (-7p^4 + 6p^2 - 12)$
$\quad + (p^4 - 3p + 8)$

$= (6p^4 - 8p^3 + 2p - 1) + (7p^4 - 6p^2 + 12)$
$\quad + (p^4 - 3p + 8)$

$= (13p^4 - 8p^3 - 6p^2 + 2p + 11) + (p^4 - 3p + 8)$

$= 14p^4 - 8p^3 - 6p^2 - p + 19$

10. (a) $(3mn + 2m - 4n) + (-mn + 4m + n)$
$= [3mn + (-mn)] + (2m + 4m) + [(-4n) + n]$
$= 2mn + 6m - 3n$

(b) $(5p^2q^2 - 4p^2 + 2q) - (2p^2q^2 - p^2 - 3q)$
$= (5p^2q^2 - 4p^2 + 2q) + (-2p^2q^2 + p^2 + 3q)$
$= [5p^2q^2 + (-2p^2q^2)] + [(-4p^2) + p^2]$
$\quad + (2q + 3q)$
$= 3p^2q^2 - 3p^2 + 5q$

5.1 Section Exercises

1. In the term $7x^5$, the coefficient is $\underline{7}$ and the exponent is $\underline{5}$.

3. The degree of the term $-4x^8$ is $\underline{8}$, the exponent.

5. When $x^2 + 10$ is evaluated for $x = 4$, the result is
$$4^2 + 10 = 16 + 10 = \underline{26}.$$

7. The polynomial $6x^4$ has one term. The coefficient of this term is 6.

9. The polynomial t^4 has one term. Since $t^4 = 1 \cdot t^4$, the coefficient of this term is 1.

11. The polynomial $-19r^2 - r$ has two terms. The coefficient of r^2 is -19 and the coefficient of r is -1.

13. $x + 8x^2$ has two terms. The coefficient of x is 1 and the coefficient of x^2 is 8.

In Exercises 15–24, use the distributive property to add like terms.

15. $-3m^5 + 5m^5 = (-3 + 5)m^5 = 2m^5$

17. $2r^5 + (-3r^5) = [2 + (-3)]r^5$
$\qquad = -1r^5 = -r^5$

19. The polynomial $.2m^5 - .5m^2$ cannot be simplified. The two terms are unlike because the exponents on the variables are different, so they cannot be combined.

21. $-3x^5 + 2x^5 - 4x^5 = (-3 + 2 - 4)x^5$
$\qquad = -5x^5$

23. $-4p^7 + 8p^7 + 5p^9 = (-4 + 8)p^7 + 5p^9$
$\qquad = 4p^7 + 5p^9$

In descending powers of the variable, this polynomial is written $5p^9 + 4p^7$.

25. $-4y^2 + 3y^2 - 2y^2 + y^2$
$= (-4 + 3 - 2 + 1)y^2$
$= -2y^2$

27. $6x^4 - 9x$

This polynomial has no like terms, so it is already simplified. It is already written in descending powers of the variable x. The highest degree of any nonzero term is 4, so the degree of the polynomial is 4. There are two terms, so this is a *binomial*.

29. $5m^4 - 3m^2 + 6m^5 - 7m^3$

The polynomial has no like terms, so it is already simplified. We can write it in descending powers of the variable m as

$$6m^5 + 5m^4 - 7m^3 - 3m^2.$$

The highest degree of any nonzero term is 5, so the degree of the polynomial is 5. The polynomial is not a monomial, binomial, or trinomial since it has 4 terms.

31. $\dfrac{5}{3}x^4 - \dfrac{2}{3}x^4 + \dfrac{1}{3}x^2 - 4 = \dfrac{3}{3}x^4 + \dfrac{1}{3}x^2 - 4$

$$= x^4 + \dfrac{1}{3}x^2 - 4$$

The resulting polynomial is a *trinomial* of degree 4.

33. $.8x^4 - .3x^4 - .5x^4 + 7$

$= (.8 - .3 - .5)x^4 + 7$

$= 0x^4 + 7 = 7$

Since 7 can be written as $7x^0$, the degree of the polynomial is 0. The simplified polynomial has one term, so it is a *monomial*.

35. (a) $-2x + 3 = -2(2) + 3$ *Let x = 2.*
$= -4 + 3$
$= -1$

(b) $-2x + 3 = -2(-1) + 3$ *Let x = -1.*
$= 2 + 3$
$= 5$

37. (a) $2x^2 + 5x + 1$
$= 2(2)^2 + 5(2) + 1$ *Let x = 2.*
$= 2(4) + 10 + 1$
$= 8 + 10 + 1$
$= 18 + 1$
$= 19$

(b) $2x^2 + 5x + 1$
$= 2(-1)^2 + 5(-1) + 1$ *Let x = -1.*
$= 2(1) - 5 + 1$
$= 2 - 5 + 1$
$= -3 + 1$
$= -2$

39. (a) $2x^5 - 4x^4 + 5x^3 - x^2$
$= 2(2)^5 - 4(2)^4 + 5(2)^3 - (2)^2$ *Let x = 2.*
$= 2(32) - 4(16) + 5(8) - 4$
$= 64 - 64 + 40 - 4$
$= 36$

(b) $2x^5 - 4x^4 + 5x^3 - x^2$
$= 2(-1)^5 - 4(-1)^4 + 5(-1)^3 - (-1)^2$
  Let x = -1.
$= 2(-1) - 4(1) + 5(-1) - 1$
$= -2 - 4 - 5 - 1$
$= -12$

41. (a) $-4x^5 + x^2$
$= -4(2)^5 + (2)^2$ *Let x = 2.*
$= -4(32) + 4$
$= -128 + 4$
$= -124$

(b) $-4x^5 + x^2$
$= -4(-1)^5 + (-1)^2$ *Let x = -1.*
$= -4(-1) + 1$
$= 4 + 1$
$= 5$

43. $D = 100t - 13t^2$
$= 100(5) - 13(5)^2$ *Let t = 5.*
$= 500 - 13(25)$
$= 500 - 325$
$= 175$

The skidding distance is 175 feet for 5 seconds.

44. $D = 100t - 13t^2$
$= 100(1) - 13(1)^2$ *Let t = 1.*
$= 100 - 13$
$= 87$

The skidding distance is 87 feet for 1 second. The ordered pair is $(t, D) = (1, 87)$.

45. If $x = 4$,

$$1.80x = 1.80(4) = 7.20.$$

If 4 gallons are purchased, the cost is $7.20.

46. If $x = 6$,

$$2x + 15 = 2(6) + 15 = 12 + 15 = 27.$$

If the saw is rented for 6 days, the cost is $27.

47. $\quad 3m^2 + 5m$
$\quad \underline{2m^2 - 2m}$
$\quad 5m^2 + 3m$

49. Subtract.

$$12x^4 - x^2$$
$$\underline{8x^4 + 3x^2}$$

Change all signs in the second row, and then add.

$$12x^4 - x^2$$
$$\underline{-8x^4 - 3x^2}$$
$$4x^4 - 4x^2$$

51. Add.

$$\frac{2}{3}x^2 + \frac{1}{5}x + \frac{1}{6}$$
$$\underline{\frac{1}{2}x^2 - \frac{1}{3}x + \frac{2}{3}}$$

Rewrite the fractions so that the fractions in each column have a common denominator; then add column by column.

$$\frac{4}{6}x^2 + \frac{3}{15}x + \frac{1}{6}$$
$$\underline{\frac{3}{6}x^2 - \frac{5}{15}x + \frac{4}{6}}$$
$$\frac{7}{6}x^2 - \frac{2}{15}x + \frac{5}{6}$$

53. Subtract.

$$12m^3 - 8m^2 + 6m + 7$$
$$\underline{ 5m^2 - 4}$$

Change all signs in the second row, and then add.

$$12m^3 - 8m^2 + 6m + 7$$
$$\underline{ - 5m^2 + 4}$$
$$12m^3 - 13m^2 + 6m + 11$$

55. $(2r^2 + 3r - 12) + (6r^2 + 2r)$
$= (2r^2 + 6r^2) + (3r + 2r) - 12$
$= 8r^2 + 5r - 12$

57. $(8m^2 - 7m) - (3m^2 + 7m - 6)$
$= (8m^2 - 7m) + (-3m^2 - 7m + 6)$
$= (8 - 3)m^2 + (-7 - 7)m + 6$
$= 5m^2 - 14m + 6$

59. $(16x^3 - x^2 + 3x) + (-12x^3 + 3x^2 + 2x)$
$= 16x^3 - x^2 + 3x - 12x^3 + 3x^2 + 2x$
$= (16 - 12)x^3 + (-1 + 3)x^2 + (3 + 2)x$
$= 4x^3 + 2x^2 + 5x$

61. $(7y^4 + 3y^2 + 2y) - (18y^5 - 5y^3 + y)$
$= (7y^4 + 3y^2 + 2y) + (-18y^5 + 5y^3 - y)$
$= (-18y^5) + 7y^4 + 5y^3 + 3y^2 + [2y + (-y)]$
$= -18y^5 + 7y^4 + 5y^3 + 3y^2 + y$

63. $[(8m^2 + 4m - 7) - (2m^3 - 5m + 2)]$
$ - (m^2 + m)$
$= (8m^2 + 4m - 7) + (-2m^3 + 5m - 2)$
$ + (-m^2 - m)$
$= 8m^2 + 4m - 7 - 2m^3 + 5m - 2 - m^2 - m$
$= -2m^3 + (8m^2 - m^2) + (4m + 5m - m)$
$ + (-7 - 2)$
$= -2m^3 + 7m^2 + 8m - 9$

65. Use the formula for the perimeter of a rectangle, $P = 2L + 2W$, with length $L = 4x^2 + 3x + 1$ and width $W = x + 2$.

$P = 2L + 2W$
$= 2(4x^2 + 3x + 1) + 2(x + 2)$
$= 8x^2 + 6x + 2 + 2x + 4$
$= 8x^2 + 8x + 6$

The perimeter of the rectangle is $8x^2 + 8x + 6$.

67. Use the formula for the perimeter of a triangle, $P = a + b + c$, with $a = 3t^2 + 2t + 7$, $b = 5t^2 + 2$, and $c = 6t + 4$.

$P = (3t^2 + 2t + 7) + (5t^2 + 2) + (6t + 4)$
$= (3t^2 + 5t^2) + (2t + 6t) + (7 + 2 + 4)$
$= 8t^2 + 8t + 13$

The perimeter of the triangle is $8t^2 + 8t + 13$.

69. $(-2x^2 - 6x + 4) - (9x^2 - 3x + 7)$
$= (-2x^2 - 6x + 4) + (-9x^2 + 3x - 7)$
$= (-2x^2 - 9x^2) + (-6x + 3x) + (4 - 7)$
$= -11x^2 - 3x - 3$

71. The degree of a term is determined by the exponents on the variables, but 3 is not a variable. The degree of $3^4 = 3^4 x^0$ is 0.

73. $(9a^2b - 3a^2 + 2b) + (4a^2b - 4a^2 - 3b)$
$= (9a^2b + 4a^2b) + [(-3a^2) + (-4a^2)]$
$ + [2b + (-3b)]$
$= 13a^2b + (-7a^2) + (-b)$
$= 13a^2b - 7a^2 - b$

75. $(2c^4d + 3c^2d^2 - 4d^2) - (c^4d + 8c^2d^2 - 5d^2)$
$= (2c^4d + 3c^2d^2 - 4d^2) + (-c^4d - 8c^2d^2 + 5d^2)$
$= (2c^4d - c^4d) + (3c^2d^2 - 8c^2d^2)$
$ + (-4d^2 + 5d^2)$
$= c^4d - 5c^2d^2 + d^2$

77. Subtract.

$$9m^3n - 5m^2n^2 + 4mn^2$$
$$\underline{-3m^3n + 6m^2n^2 + 8mn^2}$$

Change all signs in the second row, and then add.

$$9m^3n - 5m^2n^2 + 4mn^2$$
$$\underline{3m^3n - 6m^2n^2 - 8mn^2}$$
$$12m^3n - 11m^2n^2 - 4mn^2$$

5.2 The Product Rule and Power Rules for Exponents

5.2 Margin Exercises

1. $2 \cdot 2 \cdot 2 \cdot 2 = 2^4$ 2 occurs as a factor 4 times.
   $= 16$

2. (a) $(-2)^5 = (-2)(-2)(-2)(-2)(-2)$
       $= -32$

   Base: -2; exponent: 5

   (b) $-2^5 = -(2 \cdot 2 \cdot 2 \cdot 2 \cdot 2)$
       $= -32$

   Base: 2; exponent: 5

   (c) $-4^2 = -(4 \cdot 4)$
       $= -16$

   Base: 4; exponent: 2

   (d) $(-4)^2 = (-4)(-4)$
       $= 16$

   Base: -4; exponent: 2

3. (a) $8^2 \cdot 8^5 = 8^{2+5}$ Product rule
       $= 8^7$

   (b) $(-7)^5 \cdot (-7)^3 = (-7)^{5+3}$ Product rule
       $= (-7)^8$

   (c) $y^3 \cdot y = y^3 \cdot y^1 = y^{3+1}$ Product rule
       $= y^4$

   (d) $4^2 \cdot 3^5$: Cannot use the product rule because the bases are different.

   (e) $6^4 + 6^2$: Cannot use the product rule because it is a sum, not a product.

4. (a) $5m^2 \cdot 2m^6 = 5 \cdot 2 \cdot m^2 \cdot m^6 = 10m^{2+6}$
       $= 10m^8$

   (b) $3p^5 \cdot 9p^4 = 3 \cdot 9 \cdot p^5 \cdot p^4 = 27p^{5+4}$
       $= 27p^9$

   (c) $-7p^5 \cdot (3p^8) = -7 \cdot 3 \cdot p^5 \cdot p^8 = -21p^{5+8}$
       $= -21p^{13}$

5. (a) $(5^3)^4 = 5^{3 \cdot 4}$ Power rule (a)
       $= 5^{12}$

   (b) $(6^2)^5 = 6^{2 \cdot 5}$ Power rule (a)
       $= 6^{10}$

   (c) $(3^2)^4 = 3^{2 \cdot 4}$ Power rule (a)
       $= 3^8$

   (d) $(a^6)^5 = a^{6 \cdot 5}$ Power rule (a)
       $= a^{30}$

6. (a) $5(mn)^3 = 5(m^3n^3)$ Power rule (b)
       $= 5m^3n^3$

   (b) $(3a^2b^4)^5 = 3^5(a^2)^5(b^4)^5$ Power rule (b)
       $= 3^5 a^{10} b^{20}$ Power rule (a)
       $= 243a^{10}b^{20}$

   (c) $(-5m^2)^3$
       $= (-1 \cdot 5 \cdot m^2)^3$ $-a = -1 \cdot a$
       $= (-1)^3 \cdot 5^3 \cdot (m^2)^3$ Power rule (b)
       $= -1 \cdot 5^3 \cdot m^{2 \cdot 3}$ Power rule (a)
       $= -5^3 m^6$ or $-125m^6$

7. (a) $\left(\dfrac{5}{2}\right)^4 = \dfrac{5^4}{2^4}$ Power rule (c)
       $= \dfrac{625}{16}$

   (b) $\left(\dfrac{p}{q}\right)^2 = \dfrac{p^2}{q^2}$ Power rule (c)

   (c) $\left(\dfrac{r}{t}\right)^3 = \dfrac{r^3}{t^3}$ Power rule (c)

8. (a) $(2m)^3(2m)^4$
       $= (2m)^{3+4}$ Product rule
       $= (2m)^7$
       $= 2^7 m^7$ or $128m^7$ Power rule (b)

   (b)
   $\left(\dfrac{5k^3}{3}\right)^2 = \dfrac{(5k^3)^2}{3^2}$ Power rule (c)
   $= \dfrac{5^2(k^3)^2}{3^2}$ Power rule (b)
   $= \dfrac{5^2 k^{3 \cdot 2}}{3^2}$ Power rule (a)
   $= \dfrac{5^2 k^6}{3^2}$ or $\dfrac{25k^6}{9}$

184 Chapter 5 Exponents and Polynomials

(c) $\left(\dfrac{1}{5}\right)^4 (2x)^2 = \dfrac{1^4}{5^4}(2x)^2$ Power rule (c)

$= \dfrac{1}{5^4}(2^2 x^2)$ Power rule (b)

$= \dfrac{2^2 x^2}{5^4}$ or $\dfrac{4x^2}{625}$

(d) $(-3xy^2)^3 (x^2 y)^4 = (-1)^3 3^3 x^3 (y^2)^3 (x^2)^4 y^4$

Power rule (b)

$= (-1)3^3 x^3 y^{2 \cdot 3} x^{2 \cdot 4} y^4$

Power rule (a)

$= -3^3 x^3 y^6 x^8 y^4$

$= -3^3 x^{3+8} y^{6+4}$

Product rule

$= -3^3 x^{11} y^{10}$ or $-27 x^{11} y^{10}$

9. Use the formula for the area of a rectangle, $A = LW$.

$A = (8x^4)(4x^2)$ Area formula

$= 8 \cdot 4 \cdot x^{4+2}$ Product rule

$= 32 x^6$

5.2 Section Exercises

1. xy^2 can be written as $x^1 y^2$. The exponent on the base x is understood to be 1.

3. $3^3 = 3 \cdot 3 \cdot 3 = 27$, so the statement $3^3 = 9$ is false.

5. $(a^2)^3 = a^{2(3)} = a^6$, so the statement $(a^2)^3 = a^5$ is false.

7. Since -2 occurs as a factor five times, the base is -2 and the exponent is 5. The exponential expression is $(-2)^5$.

9. $\left(\dfrac{1}{2}\right)\left(\dfrac{1}{2}\right)\left(\dfrac{1}{2}\right)\left(\dfrac{1}{2}\right)\left(\dfrac{1}{2}\right)\left(\dfrac{1}{2}\right) = \left(\dfrac{1}{2}\right)^6$

11. $(-8p)(-8p) = (-8p)^2$

13. In $(-3)^4$, -3 is the base.

$(-3)^4 = (-3)(-3)(-3)(-3) = 81$

In -3^4, 3 is the base.

$-3^4 = -(3 \cdot 3 \cdot 3 \cdot 3) = -81$

15. In the exponential expression 3^5, the base is 3 and the exponent is 5.

$3^5 = 3 \cdot 3 \cdot 3 \cdot 3 \cdot 3 = 243$

17. In the expression $(-3)^5$, the base is -3 and the exponent is 5.

$(-3)^5 = (-3)(-3)(-3)(-3)(-3) = -243$

19. In the expression $(-6x)^4$, the base is $-6x$ and the exponent is 4.

21. In the expression $-6x^4$, -6 is not part of the base. The base is x and the exponent is 4.

23. The product rule does not apply to $5^2 + 5^3$ because the expression is a sum, not a product. The product rule would apply if we had $5^2 \cdot 5^3$.

$5^2 + 5^3 = 25 + 125 = 150$

25. $5^2 \cdot 5^6 = 5^{2+6} = 5^8$

27. $4^2 \cdot 4^7 \cdot 4^3 = 4^{2+7+3} = 4^{12}$

29. $(-7)^3 (-7)^6 = (-7)^{3+6} = (-7)^9$

31. $t^3 \cdot t^8 \cdot t^{13} = t^{3+8+13} = t^{24}$

33. $(-8r^4)(7r^3) = -8 \cdot 7 \cdot r^4 \cdot r^3$

$= -56 r^{4+3}$

$= -56 r^7$

35. $(-6p^5)(-7p^5) = (-6)(-7) p^5 \cdot p^5$

$= 42 p^{5+5}$

$= 42 p^{10}$

37. $(4^3)^2 = 4^{3 \cdot 2}$ Power rule (a)

$= 4^6$

39. $(t^4)^5 = t^{4 \cdot 5} = t^{20}$ Power rule (a)

41. $(7r)^3 = 7^3 r^3$ Power rule (b)

43. $(5xy)^5 = 5^5 x^5 y^5$ Power rule (b)

45. $8(qr)^3 = 8 q^3 r^3$ Power rule (b)

47. $\left(\dfrac{1}{2}\right)^3 = \dfrac{1^3}{2^3} = \dfrac{1}{2^3}$ Power rule (c)

49. $\left(\dfrac{a}{b}\right)^3 (b \neq 0) = \dfrac{a^3}{b^3}$ Power rule (c)

51. $\left(\dfrac{9}{5}\right)^8 = \dfrac{9^8}{5^8}$ Power rule (c)

53. $(-2x^2 y)^3 = (-2)^3 \cdot (x^2)^3 \cdot y^3$

$= (-2)^3 \cdot x^{2 \cdot 3} \cdot y^3$

$= (-2)^3 x^6 y^3$

55. $(3a^3 b^2)^2 = 3^2 \cdot (a^3)^2 \cdot (b^2)^2$

$= 3^2 \cdot a^{3 \cdot 2} \cdot b^{2 \cdot 2}$

$= 3^2 a^6 b^4$

57. $\left(\dfrac{5}{2}\right)^3 \cdot \left(\dfrac{5}{2}\right)^2 = \left(\dfrac{5}{2}\right)^{3+2}$ Product rule

$= \left(\dfrac{5}{2}\right)^5$

$= \dfrac{5^5}{2^5}$ Power rule (c)

59. $\left(\dfrac{9}{8}\right)^3 \cdot 9^2 = \dfrac{9^3}{8^3} \cdot \dfrac{9^2}{1}$ Power rule (c)

$= \dfrac{9^3 \cdot 9^2}{8^3 \cdot 1}$ Multiply fractions

$= \dfrac{9^{3+2}}{8^3}$ Product rule

$= \dfrac{9^5}{8^3}$

61. $(2x)^9 (2x)^3 = (2x)^{9+3}$ Product rule

$= (2x)^{12}$

$= 2^{12} x^{12}$ Power rule (b)

63. $(-6p)^4(-6p)$

$= (-6p)^4(-6p)^1$

$= (-6p)^5$ Product rule

$= (-6)^5 p^5$ Power rule (b)

65. $(6x^2 y^3)^5 = 6^5 (x^2)^5 (y^3)^5$ Power rule (b)

$= 6^5 x^{2 \cdot 5} y^{3 \cdot 5}$ Power rule (a)

$= 6^5 x^{10} y^{15}$

67. $(x^2)^3 (x^3)^5 = x^6 \cdot x^{15}$ Power rule (a)

$= x^{21}$ Product rule

69. $(2w^2 x^3 y)^2 (x^4 y)^5$

$= \left[2^2 (w^2)^2 (x^3)^2 y^2\right] \left[(x^4)^5 y^5\right]$ Power rule (b)

$= (2^2 w^4 x^6 y^2)(x^{20} y^5)$ Power rule (a)

$= 2^2 w^4 (x^6 x^{20})(y^2 y^5)$

Commutative and associative properties

$= 2^2 w^4 x^{26} y^7 \text{ or } 4w^4 x^{26} y^7$

71. $(-r^4 s)^2 (-r^2 s^3)^5$

$= \left[(-1) r^4 s\right]^2 \left[(-1) r^2 s^3\right]^5$

$= \left[(-1)^2 (r^4)^2 s^2\right] \left[(-1)^5 (r^2)^5 (s^3)^5\right]$

Power rule (b)

$= \left[(-1)^2 r^8 s^2\right] \left[(-1)^5 r^{10} s^{15}\right]$ Power rule (a)

$= (-1)^7 r^{18} s^{17}$ Product rule

$= -r^{18} s^{17}$

73. $\left(\dfrac{5a^2 b^5}{c^6}\right)^3$ $(c \neq 0)$

$= \dfrac{(5a^2 b^5)^3}{(c^6)^3}$ Power rule (c)

$= \dfrac{5^3 (a^2)^3 (b^5)^3}{(c^6)^3}$ Power rule (b)

$= \dfrac{5^3 a^6 b^{15}}{c^{18}}$ or $\dfrac{125 a^6 b^{15}}{c^{18}}$ Power rule (a)

75. $(-5m^3 p^4 q)^2 (p^2 q)^3$

$= (-1 \cdot 5 m^3 p^4 q)^2 (p^2 q)^3$

$= (-1)^2 \cdot 5^2 \cdot (m^3)^2 \cdot (p^4)^2 \cdot q^2 \cdot (p^2)^3 \cdot q^3$

$= 1 \cdot 25 \cdot m^{3 \cdot 2} \cdot p^{4 \cdot 2} \cdot q^2 \cdot p^{2 \cdot 3} \cdot q^3$

$= 25 m^6 p^8 q^2 p^6 q^3$

$= 25 m^6 p^{8+6} q^{2+3}$

$= 25 m^6 p^{14} q^5$

77. $(2x^2 y^3 z)^4 (xy^2 z^3)^2$

$= 2^4 \cdot (x^2)^4 \cdot (y^3)^4 \cdot z^4 \cdot x^2 \cdot (y^2)^2 \cdot (z^3)^2$

$= 16 x^{2 \cdot 4} \cdot y^{3 \cdot 4} \cdot z^4 \cdot x^2 \cdot y^{2 \cdot 2} \cdot z^{3 \cdot 2}$

$= 16 x^8 y^{12} z^4 x^2 y^4 z^6$

$= 16 x^{8+2} y^{12+4} z^{4+6}$

$= 16 x^{10} y^{16} z^{10}$

79. Use the formula for the area of a rectangle, $A = LW$, with $L = 10x^5$ and $W = 3x^2$.

$A = (10x^5)(3x^2)$

$= 10 \cdot 3 \cdot x^5 \cdot x^2$

$= 30 x^7$

5.3 Multiplying Polynomials

5.3 Margin Exercises

1. **(a)** $5m^3 (2m + 7)$

$= 5m^3 (2m) + 5m^3 (7)$ Distributive property

$= 10m^4 + 35m^3$ Multiply monomials.

(b) $2x^4 (3x^2 + 2x - 5)$

$= 2x^4 (3x^2) + 2x^4 (2x) + 2x^4 (-5)$

Distributive property

$= 6x^6 + 4x^5 + (-10x^4)$

Multiply monomials.

$= 6x^6 + 4x^5 - 10x^4$

186 Chapter 5 Exponents and Polynomials

(c) $-4y^2(3y^3 + 2y^2 - 4y + 8)$
$= -4y^2(3y^3) + (-4y^2)(2y^2)$
$\quad + (-4y^2)(-4y) + (-4y^2)(8)$
  Distributive property
$= -12y^5 + (-8y^4) + (16y^3) + (-32y^2)$
  Multiply monomials.
$= -12y^5 - 8y^4 + 16y^3 - 32y^2$

2. (a) $(m^3 - 2m + 1)(2m^2 + 4m + 3)$
$= m^3(2m^2) + m^3(4m) + m^3(3)$
$\quad - 2m(2m^2) - 2m(4m) - 2m(3)$
$\quad + 1(2m^2) + 1(4m) + 1(3)$
$= 2m^5 + 4m^4 + 3m^3 - 4m^3 - 8m^2 - 6m$
$\quad + 2m^2 + 4m + 3$
$= 2m^5 + 4m^4 - m^3 - 6m^2 - 2m + 3$

(b) $(6p^2 + 2p - 4)(3p^2 - 5)$
$= 6p^2(3p^2) + 6p^2(-5) + (2p)(3p^2)$
$\quad + (2p)(-5) + (-4)(3p^2) + (-4)(-5)$
$= 18p^4 - 30p^2 + 6p^3 - 10p - 12p^2 + 20$
$= 18p^4 + 6p^3 - 42p^2 - 10p + 20$

3.
$\quad\quad\quad\quad 3x^2 + 4x - 5$
$\quad\quad\quad\quad\quad\quad\quad x + 4$
$\overline{\quad\quad\quad 12x^2 + 16x - 20 \leftarrow 4(3x^2 + 4x - 5)}$
$\quad 3x^3 + 4x^2 - 5x \quad\quad \leftarrow x(3x^2 + 4x - 5)$
$\overline{3x^3 + 16x^2 + 11x - 20 \leftarrow \text{Add like terms.}}$

4. (a)
$\quad\quad\quad\quad\quad k^3 - k^2 + k + 1$
$\quad\quad\quad\quad\quad\quad\quad\quad\quad \frac{2}{3}k - \frac{1}{3}$
$\overline{\quad\quad\quad -\frac{1}{3}k^3 + \frac{1}{3}k^2 - \frac{1}{3}k - \frac{1}{3}}$
$\quad \frac{2}{3}k^4 - \frac{2}{3}k^3 + \frac{2}{3}k^2 + \frac{2}{3}k$
$\overline{\frac{2}{3}k^4 - k^3 + k^2 + \frac{1}{3}k - \frac{1}{3}}$

(b)
$\quad\quad\quad\quad\quad a^3 \quad\quad\quad + 3a - 4$
$\quad\quad\quad\quad\quad\quad\quad 2a^2 + 6a + 5$
$\overline{\quad\quad\quad\quad 5a^3 \quad\quad\quad + 15a - 20}$
$\quad\quad\quad\quad 6a^4 \quad\quad\quad + 18a^2 - 24a$
$\quad 2a^5 \quad\quad\quad + 6a^3 - 8a^2$
$\overline{2a^5 + 6a^4 + 11a^3 + 10a^2 - 9a - 20}$

5. (a)

	x	2
$4x$	$4x^2$	$8x$
3	$3x$	6

$(4x + 3)(x + 2) = 4x^2 + 8x + 3x + 6$
$\quad\quad\quad\quad\quad\quad\quad = 4x^2 + 11x + 6$

(b)

	x^2	$3x$	1
x	x^3	$3x^2$	x
5	$5x^2$	$15x$	5

$(x + 5)(x^2 + 3x + 1)$
$= x^3 + 3x^2 + x + 5x^2 + 15x + 5$
$= x^3 + 8x^2 + 16x + 5$

6. $(2p - 5)(3p + 7)$

(a) The product of the first terms is $2p(3p) = 6p^2$.

(b) The outer product is $2p(7) = 14p$.

(c) The inner product is $-5(3p) = -15p$.

(d) The product of the last terms is $-5(7) = -35$.

(e) The complete product is
$6p^2 + 14p + (-15p) + (-35) = 6p^2 - p - 35$.

7. (a) $(m + 4)(m - 3)$
  F: $\quad m(m) = m^2$
  O: $\quad m(-3) = -3m$
  I: $\quad 4(m) = 4m$
  L: $\quad 4(-3) = -12$
  $(m + 4)(m - 3) = m^2 - 3m + 4m - 12$
  $\quad\quad\quad\quad\quad\quad\quad = m^2 + m - 12$

(b) $(y + 7)(y + 2)$
  F: $\quad y(y) = y^2$
  O: $\quad y(2) = 2y$
  I: $\quad 7(y) = 7y$
  L: $\quad 7(2) = 14$
  $(y + 7)(y + 2) = y^2 + 2y + 7y + 14$
  $\quad\quad\quad\quad\quad\quad = y^2 + 9y + 14$

(c) $(r - 8)(r - 5)$
  F: $\quad r(r) = r^2$
  O: $\quad r(-5) = -5r$
  I: $\quad -8(r) = -8r$
  L: $\quad -8(-5) = 40$
  $(r - 8)(r - 5) = r^2 - 5r - 8r + 40$
  $\quad\quad\quad\quad\quad\quad = r^2 - 13r + 40$

8. $(4x - 3)(2y + 5)$
  F: $\quad 4x(2y) = 8xy$
  O: $\quad 4x(5) = 20x$
  I: $\quad -3(2y) = -6y$
  L: $\quad -3(5) = -15$
  $(4x - 3)(2y + 5) = 8xy + 20x - 6y - 15$

9. **(a)** $(6m+5)(m-4)$

F: $\quad 6m(m) = 6m^2$
O: $\quad 6m(-4) = -24m$
I: $\quad 5(m) = 5m$
L: $\quad 5(-4) = -20$

$$(6m+5)(m-4) = 6m^2 - 24m + 5m - 20$$
$$= 6m^2 - 19m - 20$$

(b) $(3r+2t)(3r+4t)$

F: $\quad 3r(3r) = 9r^2$
O: $\quad 3r(4t) = 12rt$
I: $\quad 2t(3r) = 6rt$
L: $\quad 2t(4t) = 8t^2$

$$(3r+2t)(3r+4t) = 9r^2 + 12rt + 6rt + 8t^2$$
$$= 9r^2 + 18rt + 8t^2$$

(c) $y^2(8y+3)(2y+1)$

First multiply the binomials and then multiply that result by y^2.

F: $\quad 8y(2y) = 16y^2$
O: $\quad 8y(1) = 8y$
I: $\quad 3(2y) = 6y$
L: $\quad 3(1) = 3$

$$y^2(8y+3)(2y+1) = y^2(16y^2 + 8y + 6y + 3)$$
$$= y^2(16y^2 + 14y + 3)$$
$$= 16y^4 + 14y^3 + 3y^2$$

5.3 Section Exercises

1. $(x+3)(x+4)$

	x	4
x	x^2	$4x$
3	$3x$	12

$$(x+3)(x+4) = x^2 + 4x + 3x + 12$$
$$= x^2 + 7x + 12$$

3. $(2x+1)(x^2+3x+2)$

	x^2	$3x$	2
$2x$	$2x^3$	$6x^2$	$4x$
1	x^2	$3x$	2

$$(2x+1)(x^2+3x+2)$$
$$= 2x^3 + 6x^2 + 4x + x^2 + 3x + 2$$
$$= 2x^3 + 7x^2 + 7x + 2$$

5. The first property that is used is the *distributive* property. $4x$ is being distributed over $(3x^2 + 7x^3)$.

7. $-2m(3m+2)$
$= -2m(3m) + (-2m)(2) \quad$ Distributive property
$= -6m^2 - 4m \quad$ Multiply monomials.

9. $\dfrac{3}{4}p(8 - 6p + 12p^3)$
$= \dfrac{3}{4}p(8) + \dfrac{3}{4}p(-6p) + \dfrac{3}{4}p(12p^3)$
$= 6p - \dfrac{9}{2}p^2 + 9p^4$

11. $2y^5(3 + 2y + 5y^4)$
$= 2y^5(3) + 2y^5(2y) + 2y^5(5y^4)$
$= 6y^5 + 4y^6 + 10y^9$

In Exercises 13–20, we can multiply the polynomials horizontally or vertically. The following solutions illustrate these two methods.

13. $(6x+1)(2x^2 + 4x + 1)$
$= (6x)(2x^2) + (6x)(4x) + (6x)(1)$
$\quad + (1)(2x^2) + (1)(4x) + (1)(1)$
$= 12x^3 + 24x^2 + 6x + 2x^2 + 4x + 1$
$= 12x^3 + 26x^2 + 10x + 1$

15. $(4m+3)(5m^3 - 4m^2 + m - 5)$

Multiply vertically.

$$\begin{array}{r}
5m^3 - 4m^2 + m - 5 \\
4m + 3 \\
\hline
15m^3 - 12m^2 + 3m - 15 \\
20m^4 - 16m^3 + 4m^2 - 20m \quad\quad \\
\hline
20m^4 - m^3 - 8m^2 - 17m - 15
\end{array}$$

17. $(2x-1)(3x^5 - 2x^3 + x^2 - 2x + 3)$

Multiply vertically.

$$\begin{array}{r}
3x^5 \quad\quad - 2x^3 + x^2 - 2x + 3 \\
2x - 1 \\
\hline
-3x^5 \quad\quad + 2x^3 - x^2 + 2x - 3 \\
6x^6 \quad\quad - 4x^4 + 2x^3 - 4x^2 + 6x \quad\quad \\
\hline
6x^6 - 3x^5 - 4x^4 + 4x^3 - 5x^2 + 8x - 3
\end{array}$$

188 Chapter 5 Exponents and Polynomials

19. $(5x^2 + 2x + 1)(x^2 - 3x + 5)$

Multiply vertically.

$$
\begin{array}{r}
5x^2 + 2x + 1 \\
x^2 - 3x + 5 \\
\hline
25x^2 + 10x + 5 \\
-15x^3 - 6x^2 - 3x \\
5x^4 + 2x^3 + x^2 \\
\hline
5x^4 - 13x^3 + 20x^2 + 7x + 5
\end{array}
$$

21. Multiply each term of the second polynomial by each term of the first. Then combine like terms.

$(n - 2)(n + 3)$

 F O I L
$= n(n) + n(3) + (-2)(n) + (-2)(3)$
$= n^2 + 3n + (-2n) + (-6)$
$= n^2 + n - 6$

23. Multiply each term of the second polynomial by each term of the first. Then combine like terms.

$(4r + 1)(2r - 3)$

 F O I L
$= 4r(2r) + 4r(-3) + 1(2r) + 1(-3)$
$= 8r^2 + (-12r) + 2r + (-3)$
$= 8r^2 - 10r - 3$

25. $(3x + 2)(3x - 2)$

 F O I L
$= 3x(3x) + 3x(-2) + 2(3x) + 2(-2)$
$= 9x^2 - 6x + 6x - 4$
$= 9x^2 - 4$

27. $(3q + 1)(3q + 1)$

 F O I L
$= 3q(3q) + 3q(1) + 1(3q) + 1(1)$
$= 9q^2 + 3q + 3q + 1$
$= 9q^2 + 6q + 1$

29. $(3t + 4s)(2t + 5s)$

 F O I L
$= 3t(2t) + 3t(5s) + 4s(2t) + 4s(5s)$
$= 6t^2 + 15st + 8st + 20s^2$
$= 6t^2 + 23st + 20s^2$

31. $(-.3t + .4)(t + .6)$

 F O I L
$= -.3t(t) + (-.3t)(.6) + .4(t) + .4(.6)$
$= -.3t^2 - .18t + .4t + .24$
$= -.3t^2 + .22t + .24$

33. $\left(x - \frac{2}{3}\right)\left(x + \frac{1}{4}\right)$

 F O I L
$= (x)(x) + (x)\left(\frac{1}{4}\right) + \left(-\frac{2}{3}\right)(x) + \left(-\frac{2}{3}\right)\left(\frac{1}{4}\right)$
$= x^2 + \frac{1}{4}x - \frac{2}{3}x - \frac{1}{6}$
$= x^2 + \left(\frac{3}{12}x - \frac{8}{12}x\right) - \frac{1}{6}$
$= x^2 - \frac{5}{12}x - \frac{1}{6}$

35. $\left(-\frac{5}{4} + 2r\right)\left(-\frac{3}{4} - r\right)$

 F O
$= \left(-\frac{5}{4}\right)\left(-\frac{3}{4}\right) + \left(-\frac{5}{4}\right)(-r)$
 I L
$+ (2r)\left(-\frac{3}{4}\right) + (2r)(-r)$
$= \frac{15}{16} + \frac{5}{4}r - \frac{6}{4}r - 2r^2$
$= \frac{15}{16} - \frac{1}{4}r - 2r^2$

37. $3y^3(2y + 3)(y - 5)$

Use FOIL to multiply the binomials.

$(2y + 3)(y - 5)$

 F O I L
$= 2y(y) + 2y(-5) + 3y + 3(-5)$
$= 2y^2 - 10y + 3y - 15$
$= 2y^2 - 7y - 15$

Now multiply this result by $3y^3$.

$3y^3(2y^2 - 7y - 15)$
$= 3y^3(2y^2) + 3y^3(-7y) + 3y^3(-15)$
$= 6y^5 - 21y^4 - 45y^3$

39. Area = Width · length
$= 10(3x + 6)$
$= 10(3x) + 10(6)$
$= (30x + 60) \text{ yd}^2$

40. Area = 600
$30x + 60 = 600 \quad$ Subtract 60.
$30x = 540 \quad$ Divide by 30.
$\frac{30x}{30} = \frac{540}{30}$
$x = 18$

41. With $x = 18$, $3x + 6 = 3(18) + 6 = 60$.

The rectangle measures 10 yd by 60 yd.

42. $(\$3.50 \text{ per yd}^2)(600 \text{ yd}^2) = \$2100$

43. Perimeter = 2(length) + 2(width)
$= 2(60) + 2(10)$
$= 120 + 20$
$= 140 \text{ yd}$

44. $(\$9.00 \text{ per yd})(140 \text{ yd}) = \$1260$

45. $(x+4)(x-4) = x(x) + x(-4) + 4(x) + 4(-4)$
$= x^2 + (-4x) + 4x + (-16)$
$= x^2 - 16$

$(y+2)(y-2) = y(y) + y(-2) + 2(y) + 2(-2)$
$= y^2 + (-2y) + 2y + (-4)$
$= y^2 - 4$

$(r+7)(r-7) = r(r) + r(-7) + 7(r) + 7(-7)$
$= r^2 + (-7r) + 7r + (-49)$
$= r^2 - 49$

The answers are $x^2 - 16$, $y^2 - 4$, and $r^2 - 49$. Each product is the difference of the square of the first term and the square of the last term of the binomials.

5.4 Special Products

5.4 Margin Exercises

1. (a) x is the first term of the binomial $x + 4$.
$(x)^2 = x^2$

(b) 4 is the last term of the binomial.
$4^2 = 16$

(c) $2 \cdot x \cdot 4 = 8x$

(d) Adding the terms from parts (a), (b), and (c) gives us
$(x + 4)^2 = x^2 + 8x + 16$.

2. (a) $(t + u)^2 = t^2 + 2(t)(u) + u^2$
$= t^2 + 2tu + u^2$

(b) $(2m - p)^2 = (2m)^2 - 2(2m)(p) + p^2$
$= 4m^2 - 4mp + p^2$

(c) $(4p + 3q)^2 = (4p)^2 + 2(4p)(3q) + (3q)^2$
$= 16p^2 + 24pq + 9q^2$

(d) $(5r - 6s)^2 = (5r)^2 - 2(5r)(6s) + (6s)^2$
$= 25r^2 - 60rs + 36s^2$

(e) $\left(3k - \frac{1}{2}\right)^2 = (3k)^2 - 2(3k)\left(\frac{1}{2}\right) + \left(\frac{1}{2}\right)^2$
$= 9k^2 - 3k + \frac{1}{4}$

3. (a) $(6a + 3)(6a - 3) = (6a)^2 - 3^2$
$= 36a^2 - 9$

(b) $(10m + 7)(10m - 7) = (10m)^2 - 7^2$
$= 100m^2 - 49$

(c) $(7p + 2q)(7p - 2q) = (7p)^2 - (2q)^2$
$= 49p^2 - 4q^2$

(d) $\left(3r - \frac{1}{2}\right)\left(3r + \frac{1}{2}\right) = (3r)^2 - \left(\frac{1}{2}\right)^2$
$= 9r^2 - \frac{1}{4}$

(e) $3x(x^3 - 4)(x^3 + 4)$
First use the rule for the product of the sum and difference of two terms.
$(x^3 - 4)(x^3 + 4) = (x^3)^2 - (4)^2$
$= x^6 - 16$

Now multiply this result by $3x$.
$3x(x^6 - 16) = 3x(x^6) + 3x(-16)$
$= 3x^7 - 48x$

4. (a) $(m + 1)^3$
Since $(m+1)^3 = (m+1)^2(m+1)$, the first step is to find the product $(m+1)^2$.
$(m+1)^2 = m^2 + 2 \cdot m \cdot 1 + 1$
$= m^2 + 2m + 1$

Now multiply this result by $m + 1$.
$(m+1)^3 = (m+1)(m^2 + 2m + 1)$
$= m^3 + 2m^2 + m + m^2 + 2m + 1$
$= m^3 + 3m^2 + 3m + 1$

(b) Note that $(3k - 2)^4 = (3k - 2)^2(3k - 2)^2$.
Since $(3k - 2)^2 = (3k)^2 - 2(3k)(2) + 2^2$
$= 9k^2 - 12k + 4,$
we have $(3k - 2)^4 = (9k^2 - 12k + 4)^2$.
Multiply vertically.

$$\begin{array}{r} 9k^2 - 12k + 4 \\ 9k^2 - 12k + 4 \\ \hline 36k^2 - 48k + 16 \\ -108k^3 + 144k^2 - 48k \\ 81k^4 - 108k^3 + 36k^2 \\ \hline 81k^4 - 216k^3 + 216k^2 - 96k + 16 \end{array}$$

(c)
$-3x(x - 4)^3$
$= -3x(x - 4)(x - 4)^2$
$= -3x(x - 4)(x^2 - 8x + 16)$
$= -3x(x^3 - 8x^2 + 16x - 4x^2 + 32x - 64)$
$= -3x(x^3 - 12x^2 + 48x - 64)$
$= -3x^4 + 36x^3 - 144x^2 + 192x$

5.4 Section Exercises

1. $(2x + 3)^2$

(a) The square of the first term is
$(2x)^2 = (2x)(2x) = 4x^2.$

(b) Twice the product of the two terms is
$$2(2x)(3) = 12x.$$

(c) The square of the last term is
$$3^2 = 9.$$

(d) The final product is the trinomial
$$4x^2 + 12x + 9.$$

In Exercises 3–16, use one of the following formulas for the square of a binomial:
$$(a+b)^2 = a^2 + 2ab + b^2$$
$$(a-b)^2 = a^2 - 2ab + b^2$$

3. $(a-c)^2 = (a)^2 - 2(a)(c) + (c)^2$
$= a^2 - 2ac + c^2$

5. $(p+2)^2 = p^2 + 2(p)(2) + 2^2$
$= p^2 + 4p + 4$

7. $(4x-3)^2 = (4x)^2 - 2(4x)(3) + 3^2$
$= 16x^2 - 24x + 9$

9. $(.8t + .7s)^2 = (.8t)^2 + 2(.8t)(.7s) + (.7s)^2$
$= .64t^2 + 1.12ts + .49s^2$

11. $\left(5x + \dfrac{2}{5}y\right)^2 = (5x)^2 + 2(5x)\left(\dfrac{2}{5}y\right) + \left(\dfrac{2}{5}y\right)^2$
$\phantom{\left(5x + \dfrac{2}{5}y\right)^2}= 25x^2 + 4xy + \tfrac{4}{25}y^2$

13. $\left(4a - \dfrac{3}{2}b\right)^2 = (4a)^2 - 2(4a)\left(\dfrac{3}{2}b\right) + \left(\dfrac{3}{2}b\right)^2$
$\phantom{\left(4a - \dfrac{3}{2}b\right)^2}= 16a^2 - 12ab + \tfrac{9}{4}b^2$

15. $-(4r - 2)^2$

First square the binomial.
$(4r-2)^2 = (4r)^2 - 2(4r)(2) + 2^2$
$= 16r^2 - 16r + 4$

Now multiply by -1.
$-1(16r^2 - 16r + 4) = -16r^2 + 16r - 4$

17. **(a)** $(7x)(7x) = 49x^2$

(b) $(7x)(-3y) + (3y)(7x)$
$= -21xy + 21xy = 0$

(c) $(3y)(-3y) = -9y^2$

(d) $49x^2 - 9y^2$

The sum found in part (b) is omitted because it is 0. Adding 0, the identity element for addition, would not change the answer.

In Exercises 19–30, use the formula for the product of the sum and the difference of two terms.
$$(a+b)(a-b) = a^2 - b^2$$

19. $(q+2)(q-2) = q^2 - 2^2$
$= q^2 - 4$

21. $(2w+5)(2w-5) = (2w)^2 - 5^2$
$= 4w^2 - 25$

23. $(10x + 3y)(10x - 3y) = (10x)^2 - (3y)^2$
$= 100x^2 - 9y^2$

25. $(2x^2 - 5)(2x^2 + 5) = (2x^2)^2 - 5^2$
$= 4x^4 - 25$

27. $\left(7x + \dfrac{3}{7}\right)\left(7x - \dfrac{3}{7}\right) = (7x)^2 - \left(\dfrac{3}{7}\right)^2$
$\phantom{\left(7x + \dfrac{3}{7}\right)\left(7x - \dfrac{3}{7}\right)}= 49x^2 - \tfrac{9}{49}$

29. $p(3p+7)(3p-7)$

First use the rule for the product of the sum and difference of two terms.
$(3p+7)(3p-7) = (3p)^2 - 7^2$
$= 9p^2 - 49$

Now multiply by p.
$$p(9p^2 - 49) = 9p^3 - 49p$$

31. The large square has sides of length $a+b$, so its area is $(a+b)^2$.

32. The red square has sides of length a, so its area is a^2.

33. Each blue rectangle has length a and width b, so each has an area of ab. Thus, the sum of the areas of the blue rectangles is
$$ab + ab = 2ab.$$

34. The yellow square has sides of length b, so its area is b^2.

35. Sum $= a^2 + 2ab + b^2$

36. The area of the largest square equals the sum of the areas of the two smaller squares and the two rectangles. Therefore, $(a+b)^2$ must equal $a^2 + 2ab + b^2$.

37. $35^2 = (35)(35)$

$$\begin{array}{r}35\\35\\\hline 175\\105\\\hline 1225\end{array}$$

38. $(a+b)^2 = a^2 + 2ab + b^2$
$(30+5)^2 = 30^2 + 2(30)(5) + 5^2$

39. $30^2 + 2(30)(5) + 5^2$
$= 900 + 60(5) + 25$
$= 900 + 300 + 25$
$= 1225$

40. The answers are equal.

41. $(m-5)^3$
$= (m-5)^2(m-5)$ $a^3 = a^2 \cdot a$
$= (m^2 - 10m + 25)(m-5)$ Square the binomial.
$= m^3 - 10m^2 + 25m$
$\quad - 5m^2 + 50m - 125$ Multiply polynomials.
$= m^3 - 15m^2 + 75m - 125$ Combine like terms.

43. $(2a+1)^3$
$= (2a+1)^2(2a+1)$ $a^3 = a^2 \cdot a$
$= (4a^2 + 4a + 1)(2a+1)$ Square the binomial.
$= 8a^3 + 8a^2 + 2a$
$\quad + 4a^2 + 4a + 1$ Multiply polynomials.
$= 8a^3 + 12a^2 + 6a + 1$ Combine like terms.

45. $(3r-2t)^4$
$= (3r-2t)^2(3r-2t)^2$ $a^4 = a^2 \cdot a^2$
$= (9r^2 - 12rt + 4t^2)(9r^2 - 12rt + 4t^2)$
   Square each binomial.
$= 81r^4 - 108r^3t + 36r^2t^2 - 108r^3t$
$\quad + 144r^2t^2 - 48rt^3 + 36r^2t^2 - 48rt^3 + 16t^4$
   Multiply polynomials.
$= 81r^4 - 216r^3t + 216r^2t^2 - 96rt^3 + 16t^4$
   Combine like terms.

47. $3x^2(x-3)^3$
$= 3x^2(x-3)(x-3)^2$
$= 3x^2(x-3)(x^2 - 6x + 9)$
$= 3x^2(x^3 - 6x^2 + 9x - 3x^2 + 18x - 27)$
$= 3x^2(x^3 - 9x^2 + 27x - 27)$
$= 3x^5 - 27x^4 + 81x^3 - 81x^2$

49. $-8x^2y(x+y)^4$

First expand $(x+y)^4$.

$(x+y)^4$
$= (x+y)^2(x+y)^2$
$= (x^2 + 2xy + y^2)(x^2 + 2xy + y^2)$
$= x^2(x^2 + 2xy + y^2)$
$\quad + 2xy(x^2 + 2xy + y^2)$
$\quad + y^2(x^2 + 2xy + y^2)$
$= x^4 + 2x^3y + x^2y^2$
$\quad + 2x^3y + 4x^2y^2 + 2xy^3$
$\quad + x^2y^2 + 2xy^3 + y^4$
$= x^4 + 4x^3y + 6x^2y^2 + 4xy^3 + y^4$

Now multiply this result by $-8x^2y$.

$-8x^2y(x^4 + 4x^3y + 6x^2y^2 + 4xy^3 + y^4)$
$= -8x^6y - 32x^5y^2 - 48x^4y^3 - 32x^3y^4 - 8x^2y^5$

51. If $x = 6$, $V = (6+2)^3 = 8^3 = 512$.

If the value of x is 6, the volume of the cube is 512 cubic units.

5.5 Integer Exponents and the Quotient Rule

5.5 Margin Exercises

1. $2^{-2} = \frac{1}{4}$ $\left(\frac{1}{2} \div 2 = \frac{1}{4}\right)$

$2^{-3} = \frac{1}{8}$ $\left(\frac{1}{4} \div 2 = \frac{1}{8}\right)$

$2^{-4} = \frac{1}{16}$ $\left(\frac{1}{8} \div 2 = \frac{1}{16}\right)$

2. **(a)** $28^0 = 1$

(b) $(-16)^0 = 1$

(c) $-7^0 = -(7^0) = -1$

(d) $m^0 = 1$ when $m \neq 0$

(e) $-p^0 = -(p^0) = -(1) = -1$ when $p \neq 0$

3. **(a)** $4^{-3} = \frac{1}{4^3}$ Definition of negative exponent

(b) $6^{-2} = \frac{1}{6^2}$ Definition of negative exponent

(c) $\left(\frac{2}{3}\right)^{-2} = \left(\frac{3}{2}\right)^2$ $\frac{2}{3}$ and $\frac{3}{2}$ are reciprocals.

(d) $2^{-1} + 5^{-1} = \frac{1}{2^1} + \frac{1}{5^1}$
$= \frac{1}{2} + \frac{1}{5}$
$= \frac{5}{10} + \frac{2}{10} = \frac{7}{10}$

192 Chapter 5 Exponents and Polynomials

(e) $m^{-5} = \dfrac{1}{m^5}, m \neq 0$

(f) $\dfrac{1}{z^{-4}} = \dfrac{1^{-4}}{z^{-4}}$ $1^{-4} = 1, z \neq 0$

$= \left(\dfrac{1}{z}\right)^{-4}$ Power rule (c)

$= z^4$ $\dfrac{1}{z}$ and z are reciprocals.

4. (a) $\dfrac{7^{-1}}{5^{-4}} = \dfrac{5^4}{7^1} = \dfrac{5^4}{7}$

(b) $\dfrac{x^{-3}}{y^{-2}} = \dfrac{y^2}{x^3}$

(c) $\dfrac{4h^{-5}}{m^{-2}k} = \dfrac{4}{k} \cdot \dfrac{h^{-5}}{m^{-2}} = \dfrac{4}{k} \cdot \dfrac{m^2}{h^5} = \dfrac{4m^2}{h^5 k}$

(d) $p^2 q^{-5} = p^2 \left(\dfrac{1}{q^5}\right) = \dfrac{p^2}{q^5}$

(e) $\left(\dfrac{3m}{p}\right)^{-2} = \left(\dfrac{p}{3m}\right)^2 = \dfrac{p^2}{3^2 m^2}$

5. (a) $\dfrac{5^{11}}{5^8} = 5^{11-8} = 5^3$

(b) $\dfrac{4^7}{4^{10}} = 4^{7-10} = 4^{-3} = \dfrac{1}{4^3}$

(c) $\dfrac{6^{-5}}{6^{-2}} = 6^{-5-(-2)} = 6^{-5+2} = 6^{-3} = \dfrac{1}{6^3}$

(d) $\dfrac{8^4 m^9}{8^5 m^{10}} = \dfrac{8^4}{8^5} \cdot \dfrac{m^9}{m^{10}}$

$= 8^{4-5} \cdot m^{9-10}$

$= 8^{-1} \cdot m^{-1}$

$= \dfrac{1}{8} \cdot \dfrac{1}{m}$

$= \dfrac{1}{8m}$ $(m \neq 0)$

(e) $\dfrac{3^{-1}(x+y)^{-3}}{2^{-2}(x+y)^{-4}} = \dfrac{2^2}{3^1}(x+y)^{-3-(-4)}$

$= \dfrac{4}{3}(x+y)^{-3+4}$

$= \dfrac{4}{3}(x+y)$ $(x \neq -y)$

Note that x cannot equal $-y$ because then $x+y$ would equal 0, and the original expression would be undefined.

6. (a) $12^5 \cdot 12^{-7} \cdot 12^6 = 12^{5+(-7)+6}$ Product rule

$= 12^{-2+6}$

$= 12^4$ or $20{,}736$

(b) $y^{-2} \cdot y^5 \cdot y^{-8} = y^{-2+5+(-8)}$ Product rule

$= y^{3+(-8)}$

$= y^{-5}$

$= \dfrac{1}{y^5}$

(c) $\dfrac{(6x)^{-1}}{(3x^2)^{-2}} \cdot \dfrac{(3x^2)^2}{(6x)^1}$

$= \dfrac{9x^4}{6x}$

$= \dfrac{3}{2}\left(x^{4-1}\right)$

$= \dfrac{3x^3}{2}$

(d) $\dfrac{3^9 \cdot (x^2 y)^{-2}}{3^3 \cdot x^{-4} y} = \dfrac{3^9}{3^3} \cdot \dfrac{(x^2 y)^{-2}}{x^{-4}} \cdot \dfrac{1}{y}$

$= 3^{9-3} \cdot \dfrac{x^4}{(x^2 y)^2} \cdot \dfrac{1}{y}$

$= 3^6 \cdot \dfrac{x^4}{x^4 y^2} \cdot \dfrac{1}{y}$

$= 3^6 x^{4-4} \left(\dfrac{1}{y^{2+1}}\right)$

$= \dfrac{3^6 x^0}{y^3}$

$= \dfrac{3^6}{y^3}$ or $\dfrac{729}{y^3}$

5.5 Section Exercises

1. $(-2)^{-3} = \dfrac{1}{(-2)^3}$ is negative, because $(-2)^3$ is a negative number raised to an odd exponent, which is negative, and the quotient of a positive number and a negative number is a negative number.

3. $-2^4 = -(2^4)$ is negative, because 2^4 is positive.

5. $\left(\dfrac{1}{4}\right)^{-2}$ is positive. A positive base to any power will have a positive result.

7. $1 - 5^0 = 1 - 1 = 0$

9. $(-4)^0 = 1$ Definition of zero exponent

11. $-9^0 = -(9^0) = -(1) = -1$

13. $(-2)^0 - 2^0 = 1 - 1 = 0$

15. $\dfrac{0^{10}}{10^0} = \dfrac{0}{1} = 0$

17. $7^0 + 9^0 = 1 + 1 = 2$

19. $4^{-3} = \dfrac{1}{4^3}$ Definition of negative exponent

$= \dfrac{1}{64}$

21. When we evaluate a fraction raised to a negative exponent, we can use a shortcut. Note that

$$\left(\frac{a}{b}\right)^{-n} = \frac{1}{\left(\frac{a}{b}\right)^n} = \frac{1}{\frac{a^n}{b^n}} = \frac{b^n}{a^n} = \left(\frac{b}{a}\right)^n.$$

In words, a fraction raised to the negative of a number is equal to its reciprocal raised to the number. We will use the simple phrase "$\frac{a}{b}$ and $\frac{b}{a}$ are reciprocals" to indicate our use of this evaluation shortcut.

$$\left(\frac{1}{2}\right)^{-4} = 2^4 = 16 \quad \text{$\frac{1}{2}$ and 2 are reciprocals.}$$

23. $\left(\frac{6}{7}\right)^{-2} = \left(\frac{7}{6}\right)^2$ $\frac{6}{7}$ and $\frac{7}{6}$ are reciprocals.

$\quad = \frac{7^2}{6^2}$ Power rule (c)

$\quad = \frac{49}{36}$

25. $5^{-1} + 3^{-1} = \frac{1}{5^1} + \frac{1}{3^1}$

$\quad = \frac{1}{5} + \frac{1}{3}$

$\quad = \frac{3}{15} + \frac{5}{15} = \frac{8}{15}$

27. $-2^{-1} + 3^{-2} = -(2^{-1}) + 3^{-2}$

$\quad = -\frac{1}{2^1} + \frac{1}{3^2}$

$\quad = -\frac{1}{2} + \frac{1}{9}$

$\quad = -\frac{9}{18} + \frac{2}{18}$

$\quad = -\frac{7}{18}$

29. In simplest form,

$$\frac{25}{25} = 1.$$

30. $\frac{25}{25} = \frac{5^2}{5^2}$

31. $\frac{5^2}{5^2} = 5^{2-2} = 5^0$

32. $1 = 5^0$; This supports the definition of 0 as an exponent.

33. $\frac{9^4}{9^5} = 9^{4-5}$

$\quad = 9^{-1}$

$\quad = \frac{1}{9}$

35. $\frac{6^{-3}}{6^2} = 6^{-3-2}$

$\quad = 6^{-5}$

$\quad = \frac{1}{6^5}$

37. $\frac{1}{6^{-3}} = 6^3$ Changing from negative to positive exponents

39. $\frac{2}{r^{-4}} = 2r^4$ Changing from negative to positive exponents

41. $\frac{4^{-3}}{5^{-2}} = \frac{5^2}{4^3}$ Changing from negative to positive exponents

43. $p^5 q^{-8} = \frac{p^5}{q^8}$ Changing from negative to positive exponents

45. $\frac{r^5}{r^{-4}} = r^5 \cdot r^4 = r^{5+4} = r^9$

Or we can use the quotient rule:

$\frac{r^5}{r^{-4}} = r^{5-(-4)} = r^{5+4} = r^9$

47. $\frac{6^4 x^8}{6^5 x^3} = 6^{4-5} \cdot x^{8-3}$

$\quad = 6^{-1} x^5$

$\quad = \frac{x^5}{6^1} \text{ or } \frac{x^5}{6}$

49. $\frac{6y^3}{2y} = \frac{6}{2} y^{3-1} = 3y^2$

51. $\frac{3x^5}{3x^2} = 3^{1-1} x^{5-2} = 3^0 x^3 = 1x^3 = x^3$

53. $\frac{(7^4)^3}{7^9} = \frac{7^{4 \cdot 3}}{7^9}$ Power rule (a)

$\quad = \frac{7^{12}}{7^9}$

$\quad = 7^{12-9}$ Quotient rule

$\quad = 7^3 \text{ or } 343$

55. $x^{-3} \cdot x^5 \cdot x^{-4}$

$\quad = x^{-3+5+(-4)}$ Product rule

$\quad = x^{-2}$

$\quad = \frac{1}{x^2}$ Definition of negative exponent

Chapter 5 Exponents and Polynomials

57. $\dfrac{(3x)^{-2}}{(4x)^{-3}} = \dfrac{(4x)^3}{(3x)^2}$ *Changing from negative to positive exponents.*

$= \dfrac{4^3 x^3}{3^2 x^2}$ *Power rule (b)*

$= \dfrac{4^3 x^{3-2}}{3^2}$ *Quotient rule*

$= \dfrac{4^3 x}{3^2} = \dfrac{64x}{9}$

59. $\left(\dfrac{x^{-1}y}{z^2}\right)^{-2} = \dfrac{(x^{-1}y)^{-2}}{(z^2)^{-2}}$ *Power rule (c)*

$= \dfrac{(x^{-1})^{-2} y^{-2}}{(z^2)^{-2}}$ *Power rule (b)*

$= \dfrac{x^2 y^{-2}}{z^{-4}}$ *Power rule (a)*

$= \dfrac{x^2 z^4}{y^2}$ *Definition of negative exponent*

61. $(6x)^4 (6x)^{-3} = (6x)^{4+(-3)}$ *Product rule*

$= (6x)^1 = 6x$

63. $\dfrac{(m^7 n)^{-2}}{m^{-4} n^3} = \dfrac{(m^7)^{-2} n^{-2}}{m^{-4} n^3}$

$= \dfrac{m^{7(-2)} n^{-2}}{m^{-4} n^3}$

$= \dfrac{m^{-14} n^{-2}}{m^{-4} n^3}$

$= m^{-14-(-4)} n^{-2-3}$

$= m^{-10} n^{-5}$

$= \dfrac{1}{m^{10} n^5}$

65. $\dfrac{5x^{-3}}{(4x)^2} = \dfrac{5x^{-3}}{4^2 x^2}$

$= \dfrac{5}{16 x^2 x^3}$

$= \dfrac{5}{16 x^5}$

67. $\left(\dfrac{2p^{-1}q}{3^{-1} m^2}\right)^2 = \dfrac{2^2 (p^{-1})^2 q^2}{(3^{-1})^2 (m^2)^2}$

$= \dfrac{2^2 p^{-2} q^2}{3^{-2} m^4}$

$= \dfrac{2^2 \cdot 3^2 q^2}{m^4 p^2}$

$= \dfrac{4 \cdot 9 q^2}{m^4 p^2}$

$= \dfrac{36 q^2}{m^4 p^2}$

Summary Exercises on the Rules for Exponents

1. $\left(\dfrac{6x^2}{5}\right)^{12} = \dfrac{(6x^2)^{12}}{5^{12}}$

$= \dfrac{6^{12} (x^2)^{12}}{5^{12}}$

$= \dfrac{6^{12} x^{24}}{5^{12}}$

3. $(10 x^2 y^4)^2 (10 x y^2)^3$

$= 10^2 (x^2)^2 (y^4)^2 \cdot 10^3 x^3 (y^2)^3$

$= 10^2 x^4 y^8 10^3 x^3 y^6$

$= 10^5 x^7 y^{14}$

5. $\left(\dfrac{9 w x^3}{y^4}\right)^3 = \dfrac{(9 w x^3)^3}{(y^4)^3}$

$= \dfrac{9^3 w^3 (x^3)^3}{y^{12}}$

$= \dfrac{729 w^3 x^9}{y^{12}}$

7. $\dfrac{c^{11} (c^2)^4}{(c^3)^3 (c^2)^{-6}} = \dfrac{c^{11} c^8}{c^9 c^{-12}}$

$= \dfrac{c^{19}}{c^{-3}}$

$= c^{19-(-3)} = c^{22}$

9. $5^{-1} + 6^{-1} = \dfrac{1}{5^1} + \dfrac{1}{6^1}$

$= \dfrac{6}{30} + \dfrac{5}{30}$

$= \dfrac{11}{30}$

11. $\dfrac{(2 x y^{-1})^3}{2^3 x^{-3} y^2} = \dfrac{2^3 x^3 y^{-3}}{2^3 x^{-3} y^2}$

$= x^6 y^{-5}$

$= \dfrac{x^6}{y^5}$

13. $(z^4)^{-3} (z^{-2})^{-5}$

$= z^{-12} z^{10} = z^{-2} = \dfrac{1}{z^2}$

Summary Exercises on the Rules for Exponents

15. $\dfrac{(3^{-1}x^{-3}y)^{-1}(2x^2y^{-3})^2}{(5x^{-2}y^2)^{-2}}$

$= \dfrac{(5x^{-2}y^2)^2(2x^2y^{-3})^2}{(3^{-1}x^{-3}y)^1}$

$= \dfrac{5^2 x^{-4} y^4 2^2 x^4 y^{-6}}{3^{-1} x^{-3} y}$

$= \dfrac{25 x^0 y^{-2} \cdot 4}{3^{-1} x^{-3} y}$

$= 100 x^3 y^{-3} \cdot 3$

$= \dfrac{300 x^3}{y^3}$

17. $\left(\dfrac{-2x^{-2}}{2x^2}\right)^{-2} = \left(\dfrac{2x^2}{-2x^{-2}}\right)^2$

$= \left(\dfrac{x^4}{-1}\right)^2 = \dfrac{(x^4)^2}{(-1)^2}$

$= \dfrac{x^8}{1} = x^8$

19. $\dfrac{(a^{-2}b^3)^{-4}}{(a^{-3}b^2)^{-2}(ab)^{-4}}$

$= \dfrac{a^8 b^{-12}}{a^6 b^{-4} a^{-4} b^{-4}}$

$= \dfrac{a^8 b^{-12}}{a^2 b^{-8}}$

$= a^6 b^{-4}$

$= \dfrac{a^6}{b^4}$

21. $5^{-2} + 6^{-2} = \dfrac{1}{5^2} + \dfrac{1}{6^2}$

$= \dfrac{1}{25} + \dfrac{1}{36}$

$= \dfrac{36}{25 \cdot 36} + \dfrac{25}{25 \cdot 36}$

$= \dfrac{36 + 25}{900} = \dfrac{61}{900}$

23. $\left(\dfrac{7a^2b^3}{2}\right)^3 = \dfrac{(7a^2b^3)^3}{2^3}$

$= \dfrac{7^3 a^6 b^9}{8} = \dfrac{343 a^6 b^9}{8}$

25. $-(-12)^0 = -1$

27. $\dfrac{(2xy^{-3})^{-2}}{(3x^{-2}y^4)^{-3}}$

$= \dfrac{(3x^{-2}y^4)^3}{(2xy^{-3})^2}$

$= \dfrac{3^3 x^{-6} y^{12}}{2^2 x^2 y^{-6}}$

$= \dfrac{27 x^{-8} y^{18}}{4}$

$= \dfrac{27 y^{18}}{4 x^8}$

29. $(6x^{-5}z^3)^{-3}$

$= 6^{-3} x^{15} z^{-9}$

$= \dfrac{x^{15}}{6^3 z^9}$

$= \dfrac{x^{15}}{216 z^9}$

31. $\dfrac{(xy)^{-3}(xy)^5}{(xy)^{-4}} = (xy)^{-3+5-(-4)}$

$= (xy)^6 = x^6 y^6$

33. $\dfrac{(7^{-1}x^{-3})^{-2}(x^4)^{-6}}{7^{-1}x^{-3}}$

$= \dfrac{7^2 x^6 x^{-24}}{7^{-1} x^{-3}}$

$= 7^{2-(-1)} x^{6-24-(-3)}$

$= 7^3 x^{-15} = \dfrac{343}{x^{15}}$

35. $(5p^{-2}q)^{-3}(5pq^3)^4$

$= 5^{-3} p^6 q^{-3} 5^4 p^4 q^{12}$

$= 5 p^{10} q^9$

37. $\left(\dfrac{4r^{-6}s^{-2}t}{2r^8 s^{-4} t^2}\right)^{-1}$

$= (2r^{-6-8} s^{-2-(-4)} t^{1-2})^{-1}$

$= (2r^{-14} s^2 t^{-1})^{-1}$

$= \left(\dfrac{2s^2}{r^{14}t}\right)^{-1}$

$= \dfrac{r^{14} t}{2 s^2}$

39. $\dfrac{(8pq^{-2})^4}{(8p^{-2}q^{-3})^3}$

$= \dfrac{8^4 p^4 q^{-8}}{8^3 p^{-6} q^{-9}}$

$= 8^{4-3} p^{4-(-6)} q^{-8-(-9)}$

$= 8 p^{10} q$

Chapter 5 Exponents and Polynomials

41. $-\left(-3^0\right)^0 = -(1) = -1$

5.6 Dividing a Polynomial by a Monomial

5.6 Margin Exercises

1. (a) $\dfrac{6p^4 + 18p^7}{3p^2} = \dfrac{6p^4}{3p^2} + \dfrac{18p^7}{3p^2}$
$= 2p^2 + 6p^5$

   (b) $\dfrac{12m^6 + 18m^5 + 30m^4}{6m^2}$
$= \dfrac{12m^6}{6m^2} + \dfrac{18m^5}{6m^2} + \dfrac{30m^4}{6m^2}$
$= 2m^4 + 3m^3 + 5m^2$

   (c) $(18r^7 - 9r^2) \div (3r) = \dfrac{18r^7 - 9r^2}{3r}$
$= \dfrac{18r^7}{3r} - \dfrac{9r^2}{3r}$
$= 6r^6 - 3r$

2. (a) $\dfrac{20x^4 - 25x^3 + 5x}{5x^2} = \dfrac{20x^4}{5x^2} - \dfrac{25x^3}{5x^2} + \dfrac{5x}{5x^2}$
$= 4x^2 - 5x + \dfrac{1}{x}$

   (b) $\dfrac{50m^4 - 30m^3 + 20m}{10m^3}$
$= \dfrac{50m^4}{10m^3} - \dfrac{30m^3}{10m^3} + \dfrac{20m}{10m^3}$
$= 5m - 3 + \dfrac{2}{m^2}$

3. (a) $\dfrac{8y^7 - 9y^6 - 11y - 4}{y^2}$
$= \dfrac{8y^7}{y^2} - \dfrac{9y^6}{y^2} - \dfrac{11y}{y^2} - \dfrac{4}{y^2}$
$= 8y^5 - 9y^4 - \dfrac{11}{y} - \dfrac{4}{y^2}$

   (b) $\dfrac{12p^5 + 8p^4 + 3p^3 - 5p^2}{3p^3}$
$= \dfrac{12p^5}{3p^3} + \dfrac{8p^4}{3p^3} + \dfrac{3p^3}{3p^3} - \dfrac{5p^2}{3p^3}$
$= 4p^2 + \dfrac{8p}{3} + 1 - \dfrac{5}{3p}$

4. $\dfrac{45x^4y^3 + 30x^3y^2 - 60x^2y}{-15x^2y}$
$= \dfrac{45x^4y^3}{-15x^2y} + \dfrac{30x^3y^2}{-15x^2y} - \dfrac{60x^2y}{-15x^2y}$
$= -3x^2y^2 - 2xy + 4$

5.6 Section Exercises

1. In the statement $\dfrac{6x^2 + 8}{2} = 3x^2 + 4$, $\underline{6x^2 + 8}$ is the dividend, $\underline{2}$ is the divisor, and $\underline{3x^2 + 4}$ is the quotient.

3. To check the division shown in Exercise 1, multiply $\underline{3x^2 + 4}$ by $\underline{2}$ (or 2 by $3x^2 + 4$) and show that the product is $\underline{6x^2 + 8}$.

5. To use the method of this section, the denominator must be a monomial (one term). This is true of $\dfrac{16m^3 - 12m^2}{4m}$, but not of $\dfrac{4m}{16m^3 - 12m^2}$.

7. $\dfrac{60x^4 - 20x^2 + 10x}{2x}$
$= \dfrac{60x^4}{2x} - \dfrac{20x^2}{2x} + \dfrac{10x}{2x}$
$= \dfrac{60}{2}x^{4-1} - \dfrac{20}{2}x^{2-1} + \dfrac{10}{2}$
$= 30x^3 - 10x + 5$

9. $\dfrac{20m^5 - 10m^4 + 5m^2}{-5m^2}$
$= \dfrac{20m^5}{-5m^2} - \dfrac{10m^4}{-5m^2} + \dfrac{5m^2}{-5m^2}$
$= -4m^3 + 2m^2 - 1$

11. $\dfrac{8t^5 - 4t^3 + 4t^2}{2t}$
$= \dfrac{8t^5}{2t} - \dfrac{4t^3}{2t} + \dfrac{4t^2}{2t}$
$= 4t^4 - 2t^2 + 2t$

13. $\dfrac{4a^5 - 4a^2 + 8}{4a}$
$= \dfrac{4a^5}{4a} - \dfrac{4a^2}{4a} + \dfrac{8}{4a}$
$= a^4 - a + \dfrac{2}{a}$

15. $\dfrac{12x^5 - 4x^4 + 6x^3}{-6x^2}$
$= \dfrac{12x^5}{-6x^2} - \dfrac{4x^4}{-6x^2} + \dfrac{6x^3}{-6x^2}$
$= -2x^3 + \dfrac{2x^2}{3} - x$

17. $\dfrac{4x^2 + 20x^3 - 36x^4}{4x^2}$
$= \dfrac{4x^2}{4x^2} + \dfrac{20x^3}{4x^2} - \dfrac{36x^4}{4x^2}$
$= 1 + 5x - 9x^2$

19. $\dfrac{4x^4 + 3x^3 + 2x}{3x^2}$

$= \dfrac{4x^4}{3x^2} + \dfrac{3x^3}{3x^2} + \dfrac{2x}{3x^2}$

$= \dfrac{4x^2}{3} + x + \dfrac{2}{3x}$

21. $\dfrac{27r^4 - 36r^3 - 6r^2 + 3r - 2}{3r}$

$= \dfrac{27r^4}{3r} - \dfrac{36r^3}{3r} - \dfrac{6r^2}{3r} + \dfrac{3r}{3r} - \dfrac{2}{3r}$

$= 9r^3 - 12r^2 - 2r + 1 - \dfrac{2}{3r}$

23. $\dfrac{2m^5 - 6m^4 + 8m^2}{-2m^3}$

$= \dfrac{2m^5}{-2m^3} - \dfrac{6m^4}{-2m^3} + \dfrac{8m^2}{-2m^3}$

$= -m^2 + 3m - \dfrac{4}{m}$

25. $(20a^4b^3 - 15a^5b^2 + 25a^3b) \div (-5a^4b)$

$= \dfrac{20a^4b^3}{-5a^4b} - \dfrac{15a^5b^2}{-5a^4b} + \dfrac{25a^3b}{-5a^4b}$

$= -4b^2 + 3ab - \dfrac{5}{a}$

27. $(120x^{11} - 60x^{10} + 140x^9 - 100x^8) \div (10x^{12})$

$= \dfrac{120x^{11} - 60x^{10} + 140x^9 - 100x^8}{10x^{12}}$

$= \dfrac{120x^{11}}{10x^{12}} - \dfrac{60x^{10}}{10x^{12}} + \dfrac{140x^9}{10x^{12}} - \dfrac{100x^8}{10x^{12}}$

$= \dfrac{12}{x} - \dfrac{6}{x^2} + \dfrac{14}{x^3} - \dfrac{10}{x^4}$

29. No, $\dfrac{2}{3}x$ means $\dfrac{2x}{3}$, which is not the same as $\dfrac{2}{3x}$. In the first case we multiply by x, in the second case we divide by x. Yes, $\dfrac{4}{3}x^2 = \dfrac{4x^2}{3}$. In both cases we are multiplying by x^2.

31. $\dfrac{(\ \ )}{5x^3} = 3x^2 - 7x + 7$

Multiply the quotient and the divisor to find the missing polynomial.

$(5x^3)(3x^2 - 7x + 7)$
$= 5x^3(3x^2) + 5x^3(-7x) + 5x^3(7)$
$= 15x^5 - 35x^4 + 35x^3$

33. $2\overline{\smash{)}2846}$ with 1423 above

34. $1423 = (1 \times 10^3) + (4 \times 10^2)$
$\qquad + (2 \times 10^1) + (3 \times 10^0)$

35. $\dfrac{2x^3 + 8x^2 + 4x + 6}{2}$

$= \dfrac{2x^3}{2} + \dfrac{8x^2}{2} + \dfrac{4x}{2} + \dfrac{6}{2}$

$= x^3 + 4x^2 + 2x + 3$

36. They are similar in that the coefficients of the powers of ten are equal to the coefficients of the powers of x. They are different in that one is a constant while the other is a polynomial. They are equal if $x = 10$ (the base of our decimal system).

5.7 Dividing a Polynomial by a Polynomial

5.7 Margin Exercises

1. **(a)** $(x^3 + x^2 + 4x - 6) \div (x - 1)$

$$\begin{array}{r} x^2 + 2x + 6 \\ x - 1 \overline{\smash{)}x^3 + x^2 + 4x - 6} \\ \underline{x^3 - x^2} \\ 2x^2 + 4x \\ \underline{2x^2 - 2x} \\ 6x - 6 \\ \underline{6x - 6} \\ 0 \end{array}$$

$(x^3 + x^2 + 4x - 6) \div (x - 1) = x^2 + 2x + 6$

(b) $\dfrac{p^3 - 2p^2 - 5p + 9}{p + 2}$

$$\begin{array}{r} p^2 - 4p + 3 \\ p + 2 \overline{\smash{)}p^3 - 2p^2 - 5p + 9} \\ \underline{p^3 + 2p^2} \\ -4p^2 - 5p \\ \underline{-4p^2 - 8p} \\ 3p + 9 \\ \underline{3p + 6} \\ 3 \end{array}$$

Write the remainder, 3, in the numerator of a fraction with the divisor as the denominator. Add this fraction to the quotient to get the answer.

$\dfrac{p^3 - 2p^2 - 5p + 9}{p + 2} = p^2 - 4p + 3 + \dfrac{3}{p + 2}$

Chapter 5 Exponents and Polynomials

2. (a) $\dfrac{r^2 - 5}{r + 4}$

   Use 0 as the coefficient of the missing r-term.

   $$\begin{array}{r} r - 4 \\ r + 4 \overline{\smash{)}r^2 + 0r - 5} \\ \underline{r^2 + 4r} \\ -4r - 5 \\ \underline{-4r - 16} \\ 11 \leftarrow \text{Remainder} \end{array}$$

   $\dfrac{r^2 - 5}{r + 4} = r - 4 + \dfrac{11}{r + 4}$

   (b) $(x^3 - 8) \div (x - 2)$

   Use 0 as the coefficient of the missing x^2- and x-terms.

   $$\begin{array}{r} x^2 + 2x + 4 \\ x - 2 \overline{\smash{)}x^3 + 0x^2 + 0x - 8} \\ \underline{x^3 - 2x^2} \\ 2x^2 + 0x \\ \underline{2x^2 - 4x} \\ 4x - 8 \\ \underline{4x - 8} \\ 0 \end{array}$$

   $(x^3 - 8) \div (x - 2) = x^2 + 2x + 4$

3. (a) $(2x^4 + 3x^3 - x^2 + 6x + 5) \div (x^2 - 1)$

   $$\begin{array}{r} 2x^2 + 3x + 1 \\ x^2 + 0x - 1 \overline{\smash{)}2x^4 + 3x^3 - x^2 + 6x + 5} \\ \underline{2x^4 + 0x^3 - 2x^2} \\ 3x^3 + x^2 + 6x \\ \underline{3x^3 + 0x^2 - 3x} \\ x^2 + 9x + 5 \\ \underline{x^2 + 0x - 1} \\ 9x + 6 \end{array}$$

   The remainder is $9x + 6$.

   The answer is $2x^2 + 3x + 1 + \dfrac{9x + 6}{x^2 - 1}$.

   (b) $\dfrac{2m^5 + m^4 + 6m^3 - 3m^2 - 18}{m^2 + 3}$

   $$\begin{array}{r} 2m^3 + m^2 - 6 \\ m^2 + 0m + 3 \overline{\smash{)}2m^5 + m^4 + 6m^3 - 3m^2 + 0m - 18} \\ \underline{2m^5 + 0m^4 + 6m^3} \\ m^4 + 0m^3 - 3m^2 \\ \underline{m^4 + 0m^3 + 3m^2} \\ -6m^2 + 0m - 18 \\ \underline{-6m^2 + 0m - 18} \\ 0 \end{array}$$

   $\dfrac{2m^5 + m^4 + 6m^3 - 3m^2 - 18}{m^2 + 3} = 2m^3 + m^2 - 6$

4. $$\begin{array}{r} x^2 + \tfrac{1}{3}x + \tfrac{5}{3} \\ 3x + 6 \overline{\smash{)}3x^3 + 7x^2 + 7x + 10} \\ \underline{3x^3 + 6x^2} \\ x^2 + 7x \quad \tfrac{x^2}{3x} = \tfrac{1}{3}x \\ \underline{x^2 + 2x} \\ 5x + 10 \quad \tfrac{5x}{3x} = \tfrac{5}{3} \\ \underline{5x + 10} \\ 0 \end{array}$$

   $(3x^3 + 7x^2 + 7x + 10)$ divided by $(3x + 6)$ is $x^2 + \tfrac{1}{3}x + \tfrac{5}{3}$.

5. $$\begin{array}{r} x^2 + 2x + 4 \\ x + 2 \overline{\smash{)}x^3 + 4x^2 + 8x + 8} \\ \underline{x^3 + 2x^2} \\ 2x^2 + 8x \\ \underline{2x^2 + 4x} \\ 4x + 8 \\ \underline{4x + 8} \\ 0 \end{array}$$

   $(x^3 + 4x^2 + 8x + 8)$ divided by $(x + 2)$ is $x^2 + 2x + 4$.

5.7 Section Exercises

1. In the division problem, the divisor is $2x + 5$ and the quotient is $2x^3 - 4x^2 + 3x + 2$.

3. In dividing $12m^2 - 20m + 3$ by $2m - 3$, the first step is to divide $12m^2$ by $2m$ to get $6m$.

5. $\dfrac{x^2 - x - 6}{x - 3}$

   $$\begin{array}{r} x + 2 \\ x - 3 \overline{\smash{)}x^2 - x - 6} \\ \underline{x^2 - 3x} \\ 2x - 6 \\ \underline{2x - 6} \\ 0 \end{array}$$

   The remainder is 0. The answer is the quotient, $x + 2$.

7. $\dfrac{2y^2 + 9y - 35}{y + 7}$

   $$\begin{array}{r} 2y - 5 \\ y + 7 \overline{\smash{)}2y^2 + 9y - 35} \\ \underline{2y^2 + 14y} \\ -5y - 35 \\ \underline{-5y - 35} \\ 0 \end{array}$$

   The remainder is 0. The answer is the quotient, $2y - 5$.

5.7 Dividing a Polynomial by a Polynomial

9. $\dfrac{p^2 + 2p + 20}{p + 6}$

$$\begin{array}{r} p - 4 \\ p+6 \overline{\smash{\big)}\, p^2 + 2p + 20} \\ \underline{p^2 + 6p } \\ -4p + 20 \\ \underline{-4p - 24} \\ 44 \end{array}$$

The remainder is 44. Write the remainder as the numerator of a fraction that has the divisor $p + 6$ as its denominator. The answer is

$$p - 4 + \dfrac{44}{p+6}.$$

11. $(r^2 - 8r + 15) \div (r - 3)$

$$\begin{array}{r} r - 5 \\ r-3 \overline{\smash{\big)}\, r^2 - 8r + 15} \\ \underline{r^2 - 3r } \\ -5r + 15 \\ \underline{-5r + 15} \\ 0 \end{array}$$

The remainder is 0. The answer is the quotient, $r - 5$.

13. $\dfrac{4a^2 - 22a + 32}{2a + 3}$

$$\begin{array}{r} 2a - 14 \\ 2a+3 \overline{\smash{\big)}\, 4a^2 - 22a + 32} \\ \underline{4a^2 + 6a } \\ -28a + 32 \\ \underline{-28a - 42} \\ 74 \end{array}$$

The remainder is 74. The answer is

$$2a - 14 + \dfrac{74}{2a+3}.$$

15. $\dfrac{8x^3 - 10x^2 - x + 3}{2x + 1}$

$$\begin{array}{r} 4x^2 - 7x + 3 \\ 2x+1 \overline{\smash{\big)}\, 8x^3 - 10x^2 - x + 3} \\ \underline{8x^3 + 4x^2 } \\ -14x^2 - x \\ \underline{-14x^2 - 7x } \\ 6x + 3 \\ \underline{6x + 3} \\ 0 \end{array}$$

The remainder is 0. The answer is the quotient,

$$4x^2 - 7x + 3.$$

17. $\dfrac{3y^3 + y^2 + 2}{y + 1}$

Use 0 as the coefficient of the missing y-term.

$$\begin{array}{r} 3y^2 - 2y + 2 \\ y+1 \overline{\smash{\big)}\, 3y^3 + y^2 + 0y + 2} \\ \underline{3y^3 + 3y^2 } \\ -2y^2 + 0y \\ \underline{-2y^2 - 2y } \\ 2y + 2 \\ \underline{2y + 2} \\ 0 \end{array}$$

The remainder is 0.

The answer is the quotient, $3y^2 - 2y + 2$.

19. $\dfrac{3k^3 - 4k^2 - 6k + 10}{k^2 - 2}$

Use 0 as the coefficient of the missing k-term in the divisor.

$$\begin{array}{r} 3k - 4 \\ k^2+0k-2 \overline{\smash{\big)}\, 3k^3 - 4k^2 - 6k + 10} \\ \underline{3k^3 + 0k^2 - 6k } \\ -4k^2 + 0k + 10 \\ \underline{-4k^2 + 0k + 8} \\ 2 \end{array}$$

The remainder is 2.

The quotient is $3k - 4$.

The answer is $3k - 4 + \dfrac{2}{k^2 - 2}.$

21. $(x^4 - x^2 - 2) \div (x^2 - 2)$

Use 0 as the coefficient of the missing x^3- and x-terms.

$$\begin{array}{r} x^2 + 1 \\ x^2+0x-2 \overline{\smash{\big)}\, x^4 + 0x^3 - x^2 + 0x - 2} \\ \underline{x^4 + 0x^3 - 2x^2 } \\ x^2 + 0x - 2 \\ \underline{x^2 + 0x - 2} \\ 0 \end{array}$$

The remainder is 0. The answer is the quotient, $x^2 + 1$.

23. $\dfrac{6p^4 - 15p^3 + 14p^2 - 5p + 10}{3p^2 + 1}$

$$
\begin{array}{r}
2p^2 - 5p + 4 \\
3p^2 + 0p + 1 \overline{\smash{\big)}\, 6p^4 - 15p^3 + 14p^2 - 5p + 10} \\
\underline{6p^4 + 0p^3 + 2p^2} \\
-15p^3 + 12p^2 - 5p \\
\underline{-15p^3 + 0p^2 - 5p} \\
12p^2 + 0p + 10 \\
\underline{12p^2 + 0p + 4} \\
6
\end{array}
$$

The remainder is 6.
The quotient is $2p^2 - 5p + 4$.
The answer is $2p^2 - 5p + 4 + \dfrac{6}{3p^2 + 1}$.

25. $\dfrac{2x^5 + x^4 + 11x^3 - 8x^2 - 13x + 7}{2x^2 + x - 1}$

$$
\begin{array}{r}
x^3 + 6x - 7 \\
2x^2 + x - 1 \overline{\smash{\big)}\, 2x^5 + x^4 + 11x^3 - 8x^2 - 13x + 7} \\
\underline{2x^5 + x^4 - x^3} \\
12x^3 - 8x^2 - 13x \\
\underline{12x^3 + 6x^2 - 6x} \\
-14x^2 - 7x + 7 \\
\underline{-14x^2 - 7x + 7} \\
0
\end{array}
$$

The remainder is 0. The answer is the quotient,
$x^3 + 6x - 7$.

27. $\dfrac{x^4 - 1}{x^2 - 1}$

$$
\begin{array}{r}
x^2 + 1 \\
x^2 + 0x - 1 \overline{\smash{\big)}\, x^4 + 0x^3 + 0x^2 + 0x - 1} \\
\underline{x^4 + 0x^3 - x^2} \\
x^2 + 0x - 1 \\
\underline{x^2 + 0x - 1} \\
0
\end{array}
$$

The remainder is 0. The answer is the quotient,
$x^2 + 1$.

29. $(10x^3 + 13x^2 + 4x + 1) \div (5x + 5)$

$$
\begin{array}{r}
2x^2 + \tfrac{3}{5}x + \tfrac{1}{5} \\
5x + 5 \overline{\smash{\big)}\, 10x^3 + 13x^2 + 4x + 1} \\
\underline{10x^3 + 10x^2} \\
3x^2 + 4x \\
\underline{3x^2 + 3x} \\
x + 1 \\
\underline{x + 1} \\
0
\end{array}
$$

The remainder is 0. The answer is the quotient,
$2x^2 + \tfrac{3}{5}x + \tfrac{1}{5}$.

31. Use $A = LW$ with
$$A = 5x^3 + 7x^2 - 13x - 6$$
and $W = 5x + 2$.
$$5x^3 + 7x^2 - 13x - 6 = L(5x + 2)$$
$$\dfrac{5x^3 + 7x^2 - 13x - 6}{5x + 2} = L$$

$$
\begin{array}{r}
x^2 + x - 3 \\
5x + 2 \overline{\smash{\big)}\, 5x^3 + 7x^2 - 13x - 6} \\
\underline{5x^3 + 2x^2} \\
5x^2 - 13x \\
\underline{5x^2 + 2x} \\
-15x - 6 \\
\underline{-15x - 6} \\
0
\end{array}
$$

The length is $x^2 + x - 3$ units.

33. $2x^2 - 4x + 3 = 2(-3)^2 - 4(-3) + 3$ Let $x = -3$.
$= 18 + 12 + 3$
$= 33$

34. $$
\begin{array}{r}
2x - 10 \\
x + 3 \overline{\smash{\big)}\, 2x^2 - 4x + 3} \\
\underline{2x^2 + 6x} \\
-10x + 3 \\
\underline{-10x - 30} \\
33
\end{array}
$$

The remainder is 33.

35. The answers to Exercises 33 and 34 are the same.

36. We will choose the polynomial
$5x^3 - 12x^2 - 25x - 20$ and the value $x = 4$.

Substitution:

$5x^3 - 12x^2 - 25x - 20$
$= 5(4)^3 - 12(4)^2 - 25(4) - 20$ Let $x = 4$.
$= 320 - 192 - 100 - 20$
$= 8$

Division:

$$
\begin{array}{r}
5x^2 + 8x + 7 \\
x - 4 \overline{\smash{\big)}\, 5x^3 - 12x^2 - 25x - 20} \\
\underline{5x^3 - 20x^2} \\
8x^2 - 25x \\
\underline{8x^2 - 32x} \\
7x - 20 \\
\underline{7x - 28} \\
8
\end{array}
$$

The remainder is 8. The answers agree.

5.8 An Application of Exponents: Scientific Notation

5.8 Margin Exercises

1. (a) $63,000 = 6.3 \times 10^4$

    The decimal point has been moved 4 places to put it after the first nonzero digit. Since 6.3 is smaller than the original number, it must be multiplied by a number larger than 1 to get 63,000. Thus, the exponent on 10 must be positive.

    (b) $5,870,000 = 5.87 \times 10^6$

    (c) $.0571 = 5.71 \times 10^{-2}$

    The decimal point has been moved 2 places to put it after the first nonzero digit. Since 5.71 is larger than the original number, it must be multiplied by a number smaller than 1 to get .0571. Thus, the exponent on 10 must be negative.

    (d) $-.000062 = -6.2 \times 10^{-5}$

2. (a) $4.2 \times 10^3 = 4200$

    Since the exponent is positive, move the decimal point 3 places to the *right*.

    (b) $8.7 \times 10^5 = 870,000$

    Since the exponent is positive, move the decimal point 5 places to the *right*.

    (c) $6.42 \times 10^{-3} = 0.00642$

    Since the exponent is negative, move the decimal point 3 places to the *left*.

3. (a) $(2.6 \times 10^4)(2 \times 10^{-6})$
    $= (2.6 \times 2)(10^4 \times 10^{-6})$
    $= 5.2 \times 10^{4+(-6)}$
    $= 5.2 \times 10^{-2}$
    $= .052$

    (b) $\dfrac{4.8 \times 10^2}{2.4 \times 10^{-3}} = \dfrac{4.8}{2.4} \times \dfrac{10^2}{10^{-3}}$
    $= 2 \times 10^{2-(-3)}$
    $= 2 \times 10^{2+3}$
    $= 2 \times 10^5$
    $= 200,000$

4. (a) $\left(3.0 \times 10^5 \, \dfrac{\text{km}}{\text{sec}}\right)(6.0 \times 10^1 \text{ sec})$
    $= (3.0 \times 6.0) \times (10^5 \times 10^1)$
    $= 18.0 \times 10^{5+1}$
    $= 18.0 \times 10^6$
    $= 18,000,000$

    Light travels 18,000,000 km in 6.0×10^1 sec.

    (b) $\dfrac{1.5 \times 10^8}{3.0 \times 10^5} = \dfrac{1.5}{3.0} \times \dfrac{10^8}{10^5}$
    $= .5 \times 10^{8-5}$
    $= .5 \times 10^3$
    $= 500$

    It takes light 500 sec to travel approximately 1.5×10^8 km from the sun to Earth.

5.8 Section Exercises

1. 6,130,900,000

    Move the decimal point to the right of the first nonzero digit and count the number of places the decimal point was moved.

    $6\,1\,3\,0\,9\,0\,0\,0\,0\,0$ *9 places*

    Because moving the decimal point to the *left* made the number *smaller*, we must multiply by a *positive* power of 10 so that the product 6.1309×10^n will equal the larger number. Thus, $n = 9$, and

    $6,130,900,000 = 6.1309 \times 10^9.$

    Similarly,

    $5,868,900,000 = 5.8689 \times 10^9.$

3. 69,627,000,000

    Move the decimal point left 10 places so it is to the right of the first nonzero digit.

    $6.962\,700\,000\,0$ *10 places*

    Since the number got smaller, multiply by a positive power of 10.

    $69,627,000,000 = 6.9627 \times 10^{10}$

    Similarly,

    $181,040,000,000 = 1.8104 \times 10^{11}.$

5. 4.56×10^3 is written in scientific notation because 4.56 is between 1 and 10, and 10^3 is a power of 10.

7. 5,600,000 is not in scientific notation. It can be written in scientific notation as 5.6×10^6.

9. .004 is not in scientific notation because $|.004| = .004$ is not between 1 and 10. It can be written in scientific notation as 4×10^{-3}.

11. $.8 \times 10^2$ is not in scientific notation because $|.8| = .8$ is not greater than or equal to 1 and less than 10. It can be written in scientific notation as 8×10^1.

13. A number is written in scientific notation if it is the product of a number whose absolute value is between 1 and 10 (inclusive of 1) and a power of 10.

15. 5,876,000,000

    Move the decimal point to the right of the first nonzero digit and count the number of places the decimal point was moved.

    $$5.876{,}000{,}000 \quad 9 \text{ places}$$

    Because moving the decimal point to the *left* made the number *smaller*, we must multiply by a *positive* power of 10 so that the product 5.876×10^n will equal the larger number. Thus, $n = 9$, and

    $$5{,}876{,}000{,}000 = 5.876 \times 10^9.$$

17. 82,350

    Move the decimal point left 4 places so it is to the right of the first nonzero digit.

    $$8.2350 \quad 4 \text{ places}$$

    Since the number got smaller, multiply by a positive power of 10.

    $$82{,}350 = 8.2350 \times 10^4 = 8.235 \times 10^4$$

    (Note that the final zero need not be written.)

19. .000007

    Move the decimal point to the right of the first nonzero digit.

    $$0\,0\,0\,0\,0\,7. \quad 6 \text{ places}$$

    Since moving the decimal point to the *right* made the number *larger*, we must multiply by a *negative* power of 10 so that the product 7×10^n will equal the smaller number. Thus, $n = -6$, and

    $$.000007 = 7 \times 10^{-6}.$$

21. $-.00203$

    To move the decimal point to the right of the first nonzero digit, we move it 3 places. Since 2.03 is larger than .00203, the exponent on 10 must be negative.

    $$-.00203 = -2.03 \times 10^{-3}$$

23. 7.5×10^5

    Since the exponent is positive, make 7.5 larger by moving the decimal point 5 places to the right.

    $$7.5 \times 10^5 = 750{,}000$$

25. 5.677×10^{12}

    Since the exponent is positive, make 5.677 larger by moving the decimal point 12 places to the right. We need to add 9 zeros.

    $$5.677 \times 10^{12} = 5{,}677{,}000{,}000{,}000$$

27. -6.21×10^0

    Because the exponent is 0, the decimal point should not be moved.

    $$-6.21 \times 10^0 = -6.21$$

    We know this result is correct because $10^0 = 1$.

29. 7.8×10^{-4}

    Since the exponent is negative, make 7.8 smaller by moving the decimal point 4 places to the left.

    $$7.8 \times 10^{-4} = .00078$$

31. 5.134×10^{-9}

    Since the exponent is negative, make 5.134 smaller by moving the decimal point 9 places to the left.

    $$5.134 \times 10^{-9} = .000\,000\,005\,134$$

33. $(2 \times 10^8) \times (3 \times 10^3)$

    $ = (2 \times 3)(10^8 \times 10^3)$ *Commutative and associative properties*

    $= 6 \times 10^{11}$ *Product rule for exponents; Scientific notation*

    $= 600{,}000{,}000{,}000$ *Without exponents*

35. $(5 \times 10^4) \times (3 \times 10^2)$

    $= (5 \times 3)(10^4 \times 10^2)$

    $= 15 \times 10^6$

    $= 1.5 \times 10^7$ *Scientific notation*

    $= 15{,}000{,}000$ *Without exponents*

37. $(3.15 \times 10^{-4}) \times (2.04 \times 10^8)$
    $= (3.15 \times 2.04)(10^{-4} \times 10^8)$
    $= 6.426 \times 10^4$ *Scientific notation*
    $= 64{,}260$ *Without exponents*

39. $\dfrac{9 \times 10^{-5}}{3 \times 10^{-1}} = \dfrac{9}{3} \times \dfrac{10^{-5}}{10^{-1}}$
    $= 3 \times 10^{-5-(-1)}$
    $= 3 \times 10^{-4}$

41. $\dfrac{8 \times 10^3}{2 \times 10^2} = \dfrac{8}{2} \times \dfrac{10^3}{10^2}$
    $= 4 \times 10^1$

43. $\dfrac{2.6 \times 10^{-3} \times 7.0 \times 10^{-1}}{2 \times 10^2 \times 3.5 \times 10^{-3}} = \dfrac{2.6}{2} \times \dfrac{7.0}{3.5} \times \dfrac{10^{-4}}{10^{-1}}$
    $= 1.3 \times 2 \times 10^{-4-(-1)}$
    $= 2.6 \times 10^{-3}$

45. $\dfrac{10^9}{3 \times 10^8} = \dfrac{1 \times 10^9}{3 \times 10^8}$
    $= \dfrac{1}{3} \times 10^{9-8}$
    $\approx .33 \times 10^1$
    $= 3.3$

    There are about 3.3 social security numbers available for each person.

47. Let $x =$ the total receipts.
    Dividing the receipts for *Titanic* by the total receipts gives us the fraction. An equation for this is
    $$\dfrac{6 \times 10^8}{x} = 9.5 \times 10^{-3}$$
    Solving for x we get
    $$6 \times 10^8 = (9.5 \times 10^{-3})x$$
    and so $x = \dfrac{6 \times 10^8}{9.5 \times 10^{-3}}.$
    $\dfrac{6 \times 10^8}{9.5 \times 10^{-3}} = \dfrac{6}{9.5} \times \dfrac{10^8}{10^{-3}}$
    $\approx .63 \times 10^{8-(-3)}$
    $= .63 \times 10^{8+3}$
    $= .63 \times 10^{11}$
    $= 63{,}000{,}000{,}000$

    The total receipts were about $63,000,000,000.

49. $1200(2.3 \times 10^{-4}) = (1.2 \times 10^3)(2.3 \times 10^{-4})$
    $= (1.2 \times 2.3) \times (10^3 \times 10^{-4})$
    $= 2.76 \times 10^{3+(-4)}$
    $= 2.76 \times 10^{-1}$
    $= .276$

    There is about .276 lb of copper in 1200 such people.

Chapter 5 Review Exercises

1. $9m^2 + 11m^2 + 2m^2 = (9 + 11 + 2)m^2$
   $= 22m^2$

   The degree is 2.

   To determine if the polynomial is a monomial, binomial, or trinomial, count the number of terms in the final expression.

   There is one term, so this is a *monomial*.

2. $-4p + p^3 - p^2 + 8p + 2$
   $= p^3 - p^2 + (-4 + 8)p + 2$
   $= p^3 - p^2 + 4p + 2$

   The degree is 3.

   To determine if the polynomial is a monomial, binomial, or trinomial, count the number of terms in the final expression. Since there are four terms, it is none of these.

3. $12a^5 - 9a^4 + 8a^3 + 2a^2 - a + 3$ cannot be simplified further and is already written in descending powers of the variable.

   The degree is 5.

   This polynomial has 6 terms, so it is none of the names listed.

4. $-7y^5 - 8y^4 - y^5 + y^4 + 9y$
   $= -7y^5 - 1y^5 - 8y^4 + 1y^4 + 9y$
   $= (-7 - 1)y^5 + (-8 + 1)y^4 + 9y$
   $= -8y^5 - 7y^4 + 9y$

   The degree is 5.

   There are three terms, so the polynomial is a *trinomial*.

5. Add.
   $$\begin{array}{r} -2a^3 + 5a^2 \\ -3a^3 - a^2 \\ \hline -5a^3 + 4a^2 \end{array}$$

6. Add.

$$4r^3 - 8r^2 + 6r$$
$$-2r^3 + 5r^2 + 3r$$
$$\overline{2r^3 - 3r^2 + 9r}$$

7. Subtract.

$$6y^2 - 8y + 2$$
$$-5y^2 + 2y - 7$$

Change all signs in the second row and then add.

$$6y^2 - 8y + 2$$
$$5y^2 - 2y + 7$$
$$\overline{11y^2 - 10y + 9}$$

8. Subtract.

$$-12k^4 - 8k^2 + 7k - 5$$
$$k^4 + 7k^2 + 11k + 1$$

Change all signs in the second row and then add.

$$-12k^4 - 8k^2 + 7k - 5$$
$$-k^4 - 7k^2 - 11k - 1$$
$$\overline{-13k^4 - 15k^2 - 4k - 6}$$

9. $(2m^3 - 8m^2 + 4) + (8m^3 + 2m^2 - 7)$
   $= (2m^3 + 8m^3) + (-8m^2 + 2m^2) + (4 - 7)$
   $= 10m^3 - 6m^2 - 3$

10. $(-5y^2 + 3y + 11) + (4y^2 - 7y + 15)$
    $= (-5y^2 + 4y^2) + (3y - 7y) + (11 + 15)$
    $= -y^2 - 4y + 26$

11. $(6p^2 - p - 8) - (-4p^2 + 2p + 3)$
    $= (6p^2 - p - 8) + (4p^2 - 2p - 3)$
    $= (6p^2 + 4p^2) + (-p - 2p) + (-8 - 3)$
    $= 10p^2 - 3p - 11$

12. $(12r^4 - 7r^3 + 2r^2) - (5r^4 - 3r^3 + 2r^2 + 1)$
    $= (12r^4 - 7r^3 + 2r^2) + (-5r^4 + 3r^3 - 2r^2 - 1)$
    $= (12r^4 - 5r^4) + (-7r^3 + 3r^3) + (2r^2 - 2r^2) - 1$
    $= 7r^4 - 4r^3 - 1$

13. $4^3 \cdot 4^8 = 4^{3+8} = 4^{11}$

14. $(-5)^6(-5)^5 = (-5)^{6+5} = (-5)^{11}$

15. $(-8x^4)(9x^3) = (-8)(9)(x^4)(x^3)$
    $= -72x^{4+3} = -72x^7$

16. $(2x^2)(5x^3)(x^9) = (2)(5)(x^2)(x^3)(x^9)$
    $= 10x^{2+3+9} = 10x^{14}$

17. $(19x)^5 = 19^5 x^5$

18. $(-4y)^7 = (-4)^7 y^7$

19. $5(pt)^4 = 5p^4 t^4$

20. $\left(\dfrac{7}{5}\right)^6 = \dfrac{7^6}{5^6}$

21. $(3x^2 y^3)^3$
    $= 3^3 (x^2)^3 (y^3)^3$
    $= 3^3 x^{2 \cdot 3} y^{3 \cdot 3}$
    $= 3^3 x^6 y^9$

22. $(t^4)^8 (t^2)^5 = t^{4 \cdot 8} \cdot t^{2 \cdot 5}$
    $= t^{32} \cdot t^{10}$
    $= t^{32+10}$
    $= t^{42}$

23. $(6x^2 z^4)^2 (x^3 y z^2)^4$
    $= 6^2 (x^2)^2 (z^4)^2 (x^3)^4 (y)^4 (z^2)^4$
    $= 6^2 x^4 z^8 x^{12} y^4 z^8$
    $= 6^2 x^{4+12} y^4 z^{8+8}$
    $= 6^2 x^{16} y^4 z^{16}$

24. The product rule does not apply to $7^2 + 7^4$ because you are adding powers of 7, not multiplying them.

25. $5x(2x + 14)$
    $= 5x(2x) + 5x(14)$
    $= 10x^2 + 70x$

26. $-3p^3(2p^2 - 5p)$
    $= -3p^3(2p^2) + (-3p^3)(-5p)$
    $= -6p^5 + 15p^4$

27. $(3r - 2)(2r^2 + 4r - 3)$
    Multiply vertically.

    $$\begin{array}{r} 2r^2 + 4r - 3 \\ 3r - 2 \\ \hline -4r^2 - 8r + 6 \\ 6r^3 + 12r^2 - 9r \\ \hline 6r^3 + 8r^2 - 17r + 6 \end{array}$$

28. $(2y + 3)(4y^2 - 6y + 9)$
    Multiply vertically.

    $$\begin{array}{r} 4y^2 - 6y + 9 \\ 2y + 3 \\ \hline 12y^2 - 18y + 27 \\ 8y^3 - 12y^2 + 18y \\ \hline 8y^3 + 0y^2 + 0y + 27 \end{array} = 8y^3 + 27$$

29. $(5p^2 + 3p)(p^3 - p^2 + 5)$
$= 5p^2(p^3) + 5p^2(-p^2) + 5p^2(5)$
$\quad + 3p(p^3) + 3p(-p^2) + 3p(5)$
$= 5p^5 - 5p^4 + 25p^2 + 3p^4 - 3p^3 + 15p$
$= 5p^5 - 2p^4 - 3p^3 + 25p^2 + 15p$

30. $(3k - 6)(2k + 1)$
$= (3k)(2k) + (3k)(1) + (-6)(2k) + (-6)(1)$
$= 6k^2 + 3k - 12k - 6$
$= 6k^2 - 9k - 6$

31. $(6p - 3q)(2p - 7q)$
$= 6p(2p) + 6p(-7q) + (-3q)(2p) + (-3q)(-7q)$
$= 12p^2 + (-42pq) + (-6pq) + (21q^2)$
$= 12p^2 - 48pq + 21q^2$

32.
$\quad\quad\quad\quad m^2 + m - 9$
$\quad\quad\quad\quad 2m^2 + 3m - 1$
$\quad\quad\quad\overline{\quad -m^2 - m + 9}$
$\quad\quad 3m^3 + 3m^2 - 27m$
$\underline{2m^4 + 2m^3 - 18m^2\quad\quad\quad}$
$2m^4 + 5m^3 - 16m^2 - 28m + 9$

33. $(a + 4)^2 = a^2 + 2(a)(4) + 4^2$
$= a^2 + 8a + 16$

34. $(3p - 2)^2 = (3p)^2 - 2(3p)(2) + (2)^2$
$= 9p^2 - 12p + 4$

35. $(2r + 5s)^2 = (2r)^2 + 2(2r)(5s) + (5s)^2$
$= 4r^2 + 20rs + 25s^2$

36. $(r + 2)^3 = (r + 2)^2(r + 2)$
$= (r^2 + 4r + 4)(r + 2)$
$= r^3 + 4r^2 + 4r + 2r^2 + 8r + 8$
$= r^3 + 6r^2 + 12r + 8$

37. $(2x - 1)^3 = (2x - 1)(2x - 1)^2$
$= (2x - 1)(4x^2 - 4x + 1)$
$= 8x^3 - 8x^2 + 2x - 4x^2 + 4x - 1$
$= 8x^3 - 12x^2 + 6x - 1$

38. $(6m - 5)(6m + 5) = (6m)^2 - 5^2$
$= 36m^2 - 25$

39. $(2z + 7)(2z - 7) = (2z)^2 - 7^2$
$= 4z^2 - 49$

40. $(5a + 6b)(5a - 6b) = (5a)^2 - (6b)^2$
$= 25a^2 - 36b^2$

41. $(2x^2 + 5)(2x^2 - 5) = (2x^2)^2 - 5^2$
$= 4x^4 - 25$

42. $(a + b)^2 = (a + b)(a + b) = a^2 + 2ab + b^2$. The term $2ab$ is not in $a^2 + b^2$.

43. $5^0 + 8^0 = 1 + 1 = 2$

44. $2^{-5} = \dfrac{1}{2^5} = \dfrac{1}{32}$

45. $\left(\dfrac{6}{5}\right)^{-2} = \left(\dfrac{5}{6}\right)^2$
$= \dfrac{5^2}{6^2}$ or $\dfrac{25}{36}$

46. $4^{-2} - 4^{-1} = \dfrac{1}{4^2} - \dfrac{1}{4^1}$
$= \dfrac{1}{16} - \dfrac{1}{4}$
$= \dfrac{1}{16} - \dfrac{4}{16} = -\dfrac{3}{16}$

47. $\dfrac{6^{-3}}{6^{-5}} = 6^{-3-(-5)} = 6^{-3+5} = 6^2$

48. $\dfrac{x^{-7}}{x^{-9}} = x^{-7-(-9)} = x^{-7+9} = x^2$

49. $\dfrac{p^{-8}}{p^4} = p^{-8-4} = p^{-12} = \dfrac{1}{p^{12}}$

50. $\dfrac{r^{-2}}{r^{-6}} = r^{-2-(-6)} = r^{-2+6} = r^4$

51. $(2^4)^2 = 2^{4 \cdot 2} = 2^8$

52. $(9^3)^{-2} = 9^{(3)(-2)} = 9^{-6} = \dfrac{1}{9^6}$

53. $(5^{-2})^{-4} = 5^{(-2)(-4)} = 5^8$

54. $(8^{-3})^4 = 8^{(-3)(4)} = 8^{-12} = \dfrac{1}{8^{12}}$

55. $\dfrac{(m^2)^3}{(m^4)^2} = \dfrac{m^6}{m^8}$
$= m^{6-8}$
$= m^{-2}$
$= \dfrac{1}{m^2}$

56. $\dfrac{y^4 \cdot y^{-2}}{y^{-5}} = \dfrac{y^4 \cdot y^5}{y^2}$
$= \dfrac{y^9}{y^2} = y^7$

57. $\dfrac{r^9 \cdot r^{-5}}{r^{-2} \cdot r^{-7}} = \dfrac{r^{9+(-5)}}{r^{-2+(-7)}}$
$= \dfrac{r^4}{r^{-9}}$
$= r^{4-(-9)}$
$= r^{13}$

58. $(-5m^3)^2 = (-5)^2(m^3)^2$
$= (-5)^2 m^{3 \cdot 2}$
$= (-5)^2 m^6$

59. $(2y^{-4})^{-3} = 2^{-3}(y^{-4})^{-3}$
$= 2^{-3} y^{(-4)(-3)}$
$= 2^{-3} y^{12}$
$= \dfrac{1}{2^3} \cdot y^{12}$
$= \dfrac{y^{12}}{2^3}$

60. $\dfrac{ab^{-3}}{a^4 b^2} = \dfrac{a}{a^4 b^2 b^3} = \dfrac{1}{a^3 b^5}$

61. $\dfrac{(6r^{-1})^2 (2r^{-4})}{r^{-5}(r^2)^{-3}} = \dfrac{(6^2 r^{-2})(2r^{-4})}{r^{-5} r^{-6}}$
$= \dfrac{2 \cdot 6^2 \cdot r^{-6}}{r^{-11}}$
$= \dfrac{2 \cdot 6^2 \cdot r^{11}}{r^6}$
$= 2 \cdot 6^2 \cdot r^5$

62. $\dfrac{(2m^{-5}n^2)^3 (3m^2)^{-1}}{m^{-2}n^{-4}(m^{-1})^2}$
$= \dfrac{2^3 (m^{-5})^3 (n^2)^3 3^{-1}(m^2)^{-1}}{m^{-2}n^{-4}(m^{-1})^2}$
$= \dfrac{2^3 m^{(-5)(3)} n^{2 \cdot 3} 3^{-1} m^{2(-1)}}{m^{-2} n^{-4} m^{(-1)(2)}}$
$= \dfrac{2^3 m^{-15} n^6 \cdot 3^{-1} m^{-2}}{m^{-2} n^{-4} m^{-2}}$
$= \dfrac{2^3 \cdot 3^{-1} m^{-15+(-2)} n^6}{m^{-2+(-2)} n^{-4}}$
$= \dfrac{2^3 \cdot 3^{-1} m^{-17} n^6}{m^{-4} n^{-4}}$
$= 2^3 \cdot 3^{-1} m^{-17-(-4)} n^{6-(-4)}$
$= 2^3 \cdot 3^{-1} m^{-13} n^{10}$
$= \dfrac{2^3 n^{10}}{3 m^{13}}$

63. $\dfrac{-15 y^4}{-9 y^2} = \dfrac{-15}{-9} \cdot \dfrac{y^4}{y^2} = \dfrac{5}{3} \cdot y^{4-2} = \dfrac{5y^2}{3}$

64. $\dfrac{-12 x^3 y^2}{6xy} = \dfrac{-12}{6} \cdot \dfrac{x^3}{x} \cdot \dfrac{y^2}{y}$
$= -2 x^{3-1} y^{2-1}$
$= -2 x^2 y$

65. $\dfrac{6y^4 - 12y^2 + 18y}{-6y} = \dfrac{6y^4}{-6y} - \dfrac{12y^2}{-6y} + \dfrac{18y}{-6y}$
$= -y^3 + 2y - 3$

66. $\dfrac{2p^3 - 6p^2 + 5p}{2p^2} = \dfrac{2p^3}{2p^2} - \dfrac{6p^2}{2p^2} + \dfrac{5p}{2p^2}$
$= p - 3 + \dfrac{5}{2p}$

67. $(5x^{13} - 10x^{12} + 20x^7 - 35x^5) \div (-5x^4)$
$= \dfrac{5x^{13}}{-5x^4} - \dfrac{10x^{12}}{-5x^4} + \dfrac{20x^7}{-5x^4} - \dfrac{35x^5}{-5x^4}$
$= -x^9 + 2x^8 - 4x^3 + 7x$

68. $(-10 m^4 n^2 + 5 m^3 n^3 + 6 m^2 n^4) \div (5 m^2 n)$
$= \dfrac{-10 m^4 n^2 + 5 m^3 n^3 + 6 m^2 n^4}{5 m^2 n}$
$= \dfrac{-10 m^4 n^2}{5 m^2 n} + \dfrac{5 m^3 n^3}{5 m^2 n} + \dfrac{6 m^2 n^4}{5 m^2 n}$
$= -2 m^2 n + m n^2 + \dfrac{6 n^3}{5}$

69. $(2r^2 + 3r - 14) \div (r - 2)$

$$
\begin{array}{r}
2r + 7 \\
r - 2 \overline{\smash{)} 2r^2 + 3r - 14} \\
\underline{2r^2 - 4r} \\
7r - 14 \\
\underline{7r - 14} \\
0
\end{array}
$$

Answer: $2r + 7$

70. $\dfrac{12m^2 - 11m - 10}{3m - 5}$

$$
\begin{array}{r}
4m + 3 \\
3m - 5 \overline{\smash{)} 12m^2 - 11m - 10} \\
\underline{12m^2 - 20m} \\
9m - 10 \\
\underline{9m - 15} \\
5
\end{array}
$$

Answer: $4m + 3 + \dfrac{5}{3m - 5}$

71. $\dfrac{10 a^3 + 5 a^2 - 14 a + 9}{5 a^2 - 3}$

$$
\begin{array}{r}
2a + 1 \\
5a^2 + 0a - 3 \overline{\smash{)} 10a^3 + 5a^2 - 14a + 9} \\
\underline{10a^3 + 0a^2 - 6a} \\
5a^2 - 8a + 9 \\
\underline{5a^2 + 0a - 3} \\
-8a + 12
\end{array}
$$

Answer: $2a + 1 + \dfrac{-8a + 12}{5a^2 - 3}$

72. $\dfrac{2k^4 + 4k^3 + 9k^2 - 8}{2k^2 + 1}$

$$
\begin{array}{r}
k^2 + 2k + 4 \\
2k^2 + 0k + 1 \overline{) 2k^4 + 4k^3 + 9k^2 + 0k - 8} \\
\underline{2k^4 + 0k^3 + k^2} \\
4k^3 + 8k^2 + 0k \\
\underline{4k^3 + 0k^2 + 2k} \\
8k^2 - 2k - 8 \\
\underline{8k^2 + 0k + 4} \\
-2k - 12
\end{array}
$$

Answer: $k^2 + 2k + 4 + \dfrac{-2k - 12}{2k^2 + 1}$

73. $48{,}000{,}000 = 4.8 \times 10^7$

Move the decimal point left 7 places so it is to the right of the first nonzero digit. 48,000,000 is *larger* than 4.8, so the power is *positive*.

74. $28{,}988{,}000{,}000 = 2.8988 \times 10^{10}$

Move the decimal point left 10 places so it is to the right of the first nonzero digit. 28,988,000,000 is *larger* than 2.8988, so the power is *positive*.

75. $0.000065 = 6.5 \times 10^{-5}$

Move the decimal point right 5 places so it is to the right of the first nonzero digit. The *original* number was "small" (between 0 and 1), so the exponent is *negative* 5.

76. $.000\,000\,082\,4 = 8.24 \times 10^{-8}$

Move the decimal point right 8 places so it is to the right of the first nonzero digit. .0000000824 is *smaller* than 8.24, so the power is *negative*.

77. $2.4 \times 10^4 = 24{,}000$

Move the decimal point 4 places to the right.

78. $7.83 \times 10^7 = 78{,}300{,}000$

Move the decimal point 7 places to the right.

79. $8.97 \times 10^{-7} = .000\,000\,897$

Move the decimal point 7 places to the left.

80. $9.95 \times 10^{-12} = 0.000\,000\,000\,009\,95$

Since the exponent is negative, move the decimal point 12 places to the *left*.

81. $(2 \times 10^{-3}) \times (4 \times 10^5)$
$= (2 \times 4)(10^{-3} \times 10^5)$
$= 8 \times 10^{-3+5} = 8 \times 10^2$
$= 800$

82. $\dfrac{8 \times 10^4}{2 \times 10^{-2}} = \dfrac{8}{2} \times \dfrac{10^4}{10^{-2}} = 4 \times 10^{4-(-2)}$
$= 4 \times 10^6 = 4{,}000{,}000$

83. $\dfrac{12 \times 10^{-5} \times 5 \times 10^4}{4 \times 10^3 \times 6 \times 10^{-2}}$

$= \dfrac{12 \times 5}{4 \times 6} \times \dfrac{10^{-5} \times 10^4}{10^3 \times 10^{-2}}$

$= \dfrac{60}{24} \times \dfrac{10^{-1}}{10^1}$

$= \dfrac{5}{2} \times 10^{-1-1}$

$= 2.5 \times 10^{-2}$

$= .025$

84. $\dfrac{2.5 \times 10^5 \times 4.8 \times 10^{-4}}{7.5 \times 10^8 \times 1.6 \times 10^{-5}}$

$= \dfrac{2.5 \times 4.8 \times 10^5 \times 10^{-4}}{7.5 \times 1.6 \times 10^8 \times 10^{-5}}$

$= \dfrac{2.5}{7.5} \times \dfrac{4.8}{1.6} \times \dfrac{10^{5+(-4)}}{10^{8+(-5)}}$

$= \dfrac{1}{3} \times \dfrac{3}{1} \times \dfrac{10^1}{10^3}$

$= 1 \times 10^{1-3}$

$= 1 \times 10^{-2}$

$= .01$

85. $1.6 \times 10^{-12} = .000\,000\,000\,001\,6$

86. (a) $1000 = 1 \times 10^3$

(b) $2000 = 2 \times 10^3$

(c) $50{,}000 = 5 \times 10^4$

(d) $100{,}000 = 1 \times 10^5$

87. [5.5] $19^0 - 3^0 = 1 - 1 = 0$

88. [5.5] $(3p)^4(3p^{-7}) = (3^4)(p^4)(3p^{-7})$
$= (3^4)(3)(p^4)(p^{-7})$
$= 3^{4+1} p^{4+(-7)}$
$= 3^5 p^{-3}$
$= \dfrac{3^5}{p^3}$

89. [5.5] $7^{-2} = \dfrac{1}{7^2}$ or $\dfrac{1}{49}$

90. [5.4] $(-7 + 2k)^2 = (-7)^2 + 2(-7)(2k) + (2k)^2$
$= 49 - 28k + 4k^2$

208 Chapter 5 Exponents and Polynomials

91. [5.7] $\dfrac{2y^3 + 17y^2 + 37y + 7}{2y + 7}$

$$\begin{array}{r} y^2 + 5y + 1 \\ 2y+7\overline{\smash{)}2y^3 + 17y^2 + 37y + 7} \\ \underline{2y^3 + 7y^2} \\ 10y^2 + 37y \\ \underline{10y^2 + 35y} \\ 2y + 7 \\ \underline{2y + 7} \\ 0 \end{array}$$

Answer: $y^2 + 5y + 1$

92. [5.2] $\left(\dfrac{6r^2 s}{5}\right)^4 = \dfrac{(6r^2 s)^4}{5^4}$

$= \dfrac{6^4 (r^2)^4 s^4}{5^4}$

$= \dfrac{6^4 r^8 s^4}{5^4}$

93. [5.3]

$-m^5(8m^2 + 10m + 6)$
$= -m^5(8m^2) + (-m^5)(10m) + (-m^5)(6)$
$= -8m^7 + (-10m^6) + (-6m^5)$
$= -8m^7 - 10m^6 - 6m^5$

94. [5.5] $\left(\dfrac{1}{2}\right)^{-5} = \left(\dfrac{2}{1}\right)^5 = 2^5$

95. [5.6] $(25x^2 y^3 - 8xy^2 + 15x^3 y) \div (5x)$

$= \dfrac{25x^2 y^3}{5x} - \dfrac{8xy^2}{5x} + \dfrac{15x^3 y}{5x}$

$= 5xy^3 - \dfrac{8y^2}{5} + 3x^2 y$

96. [5.5] $(6r^{-2})^{-1} = 6^{-1}(r^{-2})^{-1}$

$= 6^{-1} r^2$

$= \dfrac{r^2}{6}$

97. [5.4] $(2x + y)^3 = (2x + y)(2x + y)(2x + y)$

First find $(2x + y)(2x + y) = (2x + y)^2$.

$(2x + y)^2 = (2x)^2 + 2(2x)(y) + y^2$
$= 4x^2 + 4xy + y^2$

Now multiply this result by $2x + y$.

$$\begin{array}{r} 4x^2 + 4xy + y^2 \\ \times 2x + y \\ \hline 4x^2 y + 4xy^2 + y^3 \\ 8x^3 + 8x^2 y + 2xy^2 \\ \hline 8x^3 + 12x^2 y + 6xy^2 + y^3 \end{array}$$

Thus,

$(2x + y)^3 = 8x^3 + 12x^2 y + 6xy^2 + y^3$.

98. [5.5] $2^{-1} + 4^{-1} = \dfrac{1}{2} + \dfrac{1}{4} = \dfrac{2}{4} + \dfrac{1}{4} = \dfrac{3}{4}$

99. [5.3] $(a + 2)(a^2 - 4a + 1)$
$= a(a^2 - 4a + 1) + 2(a^2 - 4a + 1)$
$= a^3 - 4a^2 + a + 2a^2 - 8a + 2$
$= a^3 - 2a^2 - 7a + 2$

100. [5.1] $(5y^3 - 8y^2 + 7) - (-3y^3 + y^2 + 2)$
$= (5y^3 - 8y^2 + 7) + (3y^3 - y^2 - 2)$
$= (5y^3 + 3y^3) + (-8y^2 - y^2) + (7 - 2)$
$= 8y^3 - 9y^2 + 5$

101. [5.3] $(2r + 5)(5r - 2)$

  F O I L
$= 2r(5r) + 2r(-2) + 5(5r) + 5(-2)$
$= 10r^2 - 4r + 25r - 10$
$= 10r^2 + 21r - 10$

102. [5.4] $(12a + 1)(12a - 1) = (12a)^2 - (1)^2$
$= 144a^2 - 1$

103. [5.3] Use the formula for the area of a rectangle, $A = LW$, with $L = 2x - 3$ and $W = x + 2$.

$A = (2x - 3)(x + 2)$
$= (2x)(x) + (2x)(2) + (-3)(x) + (-3)(2)$
$= 2x^2 + 4x - 3x - 6$
$= 2x^2 + x - 6$

The area of the rectangle is

$2x^2 + x - 6$ square units.

104. [5.4] Use the formula for the area of a square, $A = s^2$, with $s = 5x^4 + 2x^2$.

$A = (5x^4 + 2x^2)^2$
$= (5x^4)^2 + 2(5x^4)(2x^2) + (2x^2)^2$
$= 25x^8 + 20x^6 + 4x^4$

The area of the square is

$25x^8 + 20x^6 + 4x^4$

square units.

Chapter 5 Test

1. $(5t^4 - 3t^2 + 7t + 3) - (t^4 - t^3 + 3t^2 + 8t + 3)$
$= (5t^4 - 3t^2 + 7t + 3)$
$ + (-t^4 + t^3 - 3t^2 - 8t - 3)$
$= (5t^4 - t^4) + t^3 + (-3t^2 - 3t^2)$
$ + (7t - 8t) + (3 - 3)$
$= 4t^4 + t^3 - 6t^2 - t$

2. $(2y^2 - 8y + 8) + (-3y^2 + 2y + 3)$
$\quad - (y^2 + 3y - 6)$
$= (2y^2 - 8y + 8) + (-3y^2 + 2y + 3)$
$\quad + (-y^2 - 3y + 6)$
$= 2y^2 - 3y^2 - y^2 - 8y + 2y - 3y + 8 + 3 + 6$
$= -2y^2 - 9y + 17$

3. Subtract.

$\quad 9t^3 - 4t^2 + 2t + 2$
$\quad \underline{9t^3 + 8t^2 - 3t - 6}$

Change all signs in the second row and then add.

$\quad 9t^3 - 4t^2 + 2t + 2$
$\quad \underline{-9t^3 - 8t^2 + 3t + 6}$
$\quad \quad \quad -12t^2 + 5t + 8$

4. $(-2)^3(-2)^2 = (-2)^{3+2} = (-2)^5$ or -2^5

5. $\left(\dfrac{6}{m^2}\right)^3 = \dfrac{6^3}{(m^2)^3} = \dfrac{6^3}{m^{2\cdot 3}} = \dfrac{6^3}{m^6}, m \neq 0$

6. $3x^2(-9x^3 + 6x^2 - 2x + 1)$
$= (3x^2)(-9x^3) + (3x^2)(6x^2)$
$\quad + (3x^2)(-2x) + (3x^2)(1)$
$= -27x^5 + 18x^4 - 6x^3 + 3x^2$

7. $(2r - 3)(r^2 + 2r - 5)$

Multiply vertically.

$\quad \quad \quad r^2 + 2r - 5$
$\quad \quad \quad \quad \quad \quad 2r - 3$
$\quad \quad \overline{\quad -3r^2 - 6r + 15}$
$\quad \quad 2r^3 + 4r^2 - 10r$
$\quad \overline{2r^3 + r^2 - 16r + 15}$

8. $\quad \quad \quad \text{F O I L}$
$(t - 8)(t + 3) = t^2 + 3t - 8t - 24$
$\quad \quad \quad \quad \quad = t^2 - 5t - 24$

9. $(4x + 3y)(2x - y)$
$\quad \text{F O I L}$
$= 8x^2 - 4xy + 6xy - 3y^2$
$= 8x^2 + 2xy - 3y^2$

10. $(5x - 2y)^2 = (5x)^2 - 2(5x)(2y) + (2y)^2$
$\quad \quad \quad \quad = 25x^2 - 20xy + 4y^2$

11. $(10v + 3w)(10v - 3w) = (10v)^2 - (3w)^2$
$\quad \quad \quad \quad \quad \quad \quad \quad = 100v^2 - 9w^2$

12. $(x+1)^3$
$= (x+1)(x+1)^2$
$= (x+1)(x^2 + 2x + 1)$
$= x(x^2 + 2x + 1) + 1(x^2 + 2x + 1)$
$= x^3 + 2x^2 + x + x^2 + 2x + 1$
$= x^3 + 3x^2 + 3x + 1$

13. $5^{-4} = \dfrac{1}{5^4} = \dfrac{1}{625}$

14. $(-3)^0 + 4^0 = 1 + 1 = 2$

15. $4^{-1} + 3^{-1} = \dfrac{1}{4^1} + \dfrac{1}{3^1} = \dfrac{3}{12} + \dfrac{4}{12} = \dfrac{7}{12}$

16. $\dfrac{8^{-1} \cdot 8^4}{8^{-2}} = \dfrac{8^{(-1)+4}}{8^{-2}} = \dfrac{8^3}{8^{-2}} = 8^{3-(-2)} = 8^5$

17. $\dfrac{(x^{-3})^{-2}(x^{-1}y)^2}{(xy^{-2})^2} = \dfrac{(x^{-3})^{-2}(x^{-1})^2(y)^2}{(x)^2(y^{-2})^2}$
$= \dfrac{x^6 x^{-2} y^2}{x^2 y^{-4}}$
$= \dfrac{x^4 y^2}{x^2 y^{-4}}$
$= x^{4-2} y^{2-(-4)}$
$= x^2 y^6$

18. $\dfrac{8y^3 - 6y^2 + 4y + 10}{2y}$
$= \dfrac{8y^3}{2y} - \dfrac{6y^2}{2y} + \dfrac{4y}{2y} + \dfrac{10}{2y}$
$= 4y^2 - 3y + 2 + \dfrac{5}{y}$

19. $(-9x^2y^3 + 6x^4y^3 + 12xy^3) \div (3xy)$
$= \dfrac{-9x^2y^3 + 6x^4y^3 + 12xy^3}{3xy}$
$= \dfrac{-9x^2y^3}{3xy} + \dfrac{6x^4y^3}{3xy} + \dfrac{12xy^3}{3xy}$
$= -3xy^2 + 2x^3y^2 + 4y^2$

20. $\dfrac{2x^2 + x - 36}{x - 4}$

$\quad \quad \quad \quad \quad 2x + 9$
$x - 4 \overline{\smash{)}\, 2x^2 + x - 36}$
$\quad \quad \quad \underline{2x^2 - 8x}$
$\quad \quad \quad \quad \quad \quad 9x - 36$
$\quad \quad \quad \quad \quad \underline{9x - 36}$
$\quad \quad \quad \quad \quad \quad \quad \quad 0$

Answer: $2x + 9$

Chapter 5 Exponents and Polynomials

21. $(3x^3 - x + 4) \div (x - 2)$

$$\begin{array}{r}
3x^2 + 6x + 11 \\
x - 2 \overline{\smash{\big)}\, 3x^3 + 0x^2 - x + 4} \\
\underline{3x^3 - 6x^2 } \\
6x^2 - x \\
\underline{6x^2 - 12x } \\
11x + 4 \\
\underline{11x - 22} \\
26
\end{array}$$

The answer is

$$3x^2 + 6x + 11 + \frac{26}{x-2}.$$

22. **(a)** $344{,}000{,}000{,}000 = 3.44 \times 10^{11}$

Move the decimal point left 11 places so it is to the right of the first nonzero digit.
344,000,000,000 is *larger* than 3.44, so the power is *positive*.

(b) $.00000557 = 5.57 \times 10^{-6}$

Move the decimal point right 6 places so it is to the right of the first nonzero digit. .00000557 is *smaller* than 5.57, so the power is *negative*.

23. **(a)** $2.96 \times 10^7 = 29{,}600{,}000$

Move the decimal point 7 places to the right.

(b) $6.07 \times 10^{-8} = .0000000607$

Move the decimal point 8 places to the left.

24. Use the formula for the area of a square, $A = s^2$, with $s = 3x + 9$.

$$\begin{aligned}
A &= (3x+9)^2 \\
&= (3x)^2 + 2(3x)(9) + 9^2 \\
&= 9x^2 + 54x + 81
\end{aligned}$$

25. For the sum of two fourth degree polynomials in x to be a third degree polynomial in x, the degree 4 terms would have to have a sum of 0 and the sum of the degree 3 terms would have to be nonzero. Therefore, the coefficients of the fourth degree terms must be opposites of each other.
For example,

$$\begin{aligned}
&\left(-4x^4 + 3x^3 + 2x + 1\right) + \left(4x^4 - 8x^3 + 2x + 7\right) \\
&= -5x^3 + 4x + 8.
\end{aligned}$$

Notice that the degree 4 terms, $-4x^4$ and $4x^4$, are opposites of each other.

Cumulative Review Exercises (Chapters R–5)

1. $\dfrac{2}{3} + \dfrac{1}{8} = \dfrac{16}{24} + \dfrac{3}{24} = \dfrac{19}{24}$

2. $\dfrac{7}{4} - \dfrac{9}{5} = \dfrac{35}{20} - \dfrac{36}{20} = -\dfrac{1}{20}$

3. $8.32 - 4.6$

$$\begin{array}{r}
\overset{7\;13}{8.\cancel{3}\,2} \\
- 4.6\,0 \\
\hline
3.7\,2
\end{array}$$

4. 7.21×8.6

$$\begin{array}{r}
7.2\,1 \\
\times\;\;8.6 \\
\hline
4\,3\,2\,6 \\
5\,7\,6\,8 \\
\hline
62.0\,0\,6
\end{array}$$

5. Use the formula for simple interest, $I = Prt$, with $P = \$34{,}000$, $r = 5.4\%$, and $t = 1$.

$$\begin{aligned}
I &= Prt \\
&= (34{,}000)(.054)(1) \\
&= 1836
\end{aligned}$$

She earned $1836 in interest.

6. $\dfrac{4x - 2y}{x + y} = \dfrac{4(-2) - 2(4)}{(-2) + 4}$ Let $x = -2$, $y = 4$.

$$= \dfrac{-8 - 8}{2} = \dfrac{-16}{2} = -8$$

7. $x^3 - 4xy = (-2)^3 - 4(-2)(4)$ Let $x = -2$, $y = 4$.

$$= -8 + 32$$
$$= 24$$

8. $\dfrac{(-13 + 15) - (3 + 2)}{6 - 12} = \dfrac{2 - 5}{-6} = \dfrac{-3}{-6} = \dfrac{1}{2}$

9. $-7 - 3[2 + (5 - 8)] = -7 - 3[2 + (-3)]$
$$= -7 - 3[-1]$$
$$= -7 + 3 = -4$$

10. $(9 + 2) + 3 = 9 + (2 + 3)$

The numbers are in the same order but grouped differently, so this is an example of the associative property of addition.

11. $-7 + 7 = 0$

The sum of the two numbers is 0, so they are additive inverses (or opposites) of each other. This is an example of the additive inverse property.

12. $6(4 + 2) = 6(4) + 6(2)$

The number 6 outside the parentheses is "distributed" over the 4 and the 2. This is an example of the distributive property.

13. $2x - 7x + 8x = 30$
    $3x = 30$
    $x = 10$

    The solution is 10.

14. $2 - 3(t - 5) = 4 + t$
    $2 - 3t + 15 = 4 + t$
    $-3t + 17 = 4 + t$
    $-4t + 17 = 4$
    $-4t = -13$
    $t = \frac{-13}{-4} = \frac{13}{4}$

    The solution is $\frac{13}{4}$.

15. $2(5h + 1) = 10h + 4$
    $10h + 2 = 10h + 4$
    $2 = 4$ False

    The false statement indicates that the equation has no solution.

16. Solve $d = rt$ for r.

    $\frac{d}{t} = \frac{rt}{t}$ Divide by t.

    $\frac{d}{t} = r$

17. $\frac{x}{5} = \frac{x-2}{7}$

    $7x = 5(x - 2)$ *Cross products are equal*

    $7x = 5x - 10$
    $2x = -10$
    $x = -5$

    The solution is -5.

18. $\frac{1}{3}p - \frac{1}{6}p = -2$

    To clear fractions, multiply both sides of the equation by the least common denominator, which is 6.

    $6\left(\frac{1}{3}p - \frac{1}{6}p\right) = (6)(-2)$
    $6\left(\frac{1}{3}p\right) - 6\left(\frac{1}{6}p\right) = -12$
    $2p - p = -12$
    $p = -12$

    The solution is -12.

19. $.05x + .15(50 - x) = 5.50$

    To clear decimals, multiply both sides of the equation by 100.

    $100[.05x + .15(50 - x)] = 100(5.50)$
    $100(.05x) + 100[.15(50 - x)] = 100(5.50)$
    $5x + 15(50 - x) = 550$
    $5x + 750 - 15x = 550$
    $-10x + 750 = 550$
    $-10x = -200$
    $x = 20$

    The solution is 20.

20. $4 - (3x + 12) = (2x - 9) - (5x - 1)$
    $4 - 3x - 12 = 2x - 9 - 5x + 1$
    $-3x - 8 = -3x - 8$
    $-8 = -8$ True

    The true statement indicates that the solution is all real numbers.

21. Let $x =$ the number of breaths per minute taken by the elephant.
    Then $16x =$ the number for the mouse.
    The total number of breaths per minute is 170.

    $x + 16x = 170$
    $17x = 170$
    $x = 10$

    The elephant takes 10 breaths per minute and the mouse takes $16(10) = 160$ breaths per minute.

22. Let $x =$ the unknown number.

    $3(8 - x) = 3x$
    $24 - 3x = 3x$
    $24 = 6x$
    $x = 4$

    The unknown number is 4.

23. $-8x \leq -80$
    $\frac{-8x}{-8} \geq \frac{-80}{-8}$ Divide by -8; reverse the symbol.
    $x \geq 10$

24. $-2(x + 4) > 3x + 6$
    $-2x - 8 > 3x + 6$
    $-2x > 3x + 14$
    $-5x > 14$
    $\frac{-5x}{-5} < \frac{14}{-5}$ Divide by -5; reverse the symbol.
    $x < -\frac{14}{5}$

Chapter 5 Exponents and Polynomials

25. $-3 \leq 2x + 5 < 9$
$-8 \leq 2x < 4$ Subtract 5.
$\dfrac{-8}{2} \leq \dfrac{2x}{2} < \dfrac{4}{2}$ Divide by 2.
$-4 \leq x < 2$

26. $2x - 3y = -6$
Let $x = 0$.
$$2(0) - 3y = -6$$
$$0 - 3y = -6$$
$$\dfrac{-3y}{-3} = \dfrac{-6}{-3}$$
$$y = 2$$

The y-intercept is $(0, 2)$.

Let $y = 0$.
$$2x - 3(0) = -6$$
$$2x - 0 = -6$$
$$\dfrac{2x}{2} = \dfrac{-6}{2}$$
$$x = -3$$

The x-intercept is $(-3, 0)$.

27. $2x - 3y = -6$

Graph the y-intercept, $(0, 2)$, and the x-intercept, $(-3, 0)$, and draw a line through them.

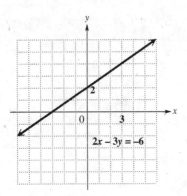

28. Solve the equation for y.
$$2x - 3y = -6$$
$$-3y = -2x - 6 \quad \text{Subtract } 2x.$$
$$y = \tfrac{2}{3}x + 2 \quad \text{Divide by } -3.$$

The slope is the coefficient of x, $\tfrac{2}{3}$.

29. $y = 2x + 5$ (1)
$x + y = -4$ (2)

To solve the system by the substitution method, let $y = 2x + 5$ in equation (2).
$$x + y = -4$$
$$x + (2x + 5) = -4$$
$$3x + 5 = -4$$
$$3x = -9$$
$$x = -3$$

From (1), $y = 2(-3) + 5 = -1$.
The solution is $(-3, -1)$.

30. $3x + 2y = 2$ (1)
$2x + 3y = -7$ (2)

We solve this system by the elimination method. To eliminate x, multiply equation (1) by 2, equation (2) by -3, and then add.

$$\begin{array}{rcr} 6x + 4y &=& 4 \\ -6x - 9y &=& 21 \\ \hline -5y &=& 25 \\ y &=& -5 \end{array}$$

To eliminate y, multiply equation (1) by 3, equation (2) by -2, and then add.

$$\begin{array}{rcr} 9x + 6y &=& 6 \\ -4x - 6y &=& 14 \\ \hline 5x &=& 20 \\ x &=& 4 \end{array}$$

The solution is $(4, -5)$.

31. $4^{-1} + 3^0 = \dfrac{1}{4^1} + 1 = 1\dfrac{1}{4}$ or $\dfrac{5}{4}$

32. $2^{-4} \cdot 2^5 = 2^{-4+5} = 2^1 = 2$

33. $\dfrac{8^{-5} \cdot 8^7}{8^2} = \dfrac{8^{-5+7}}{8^2} = \dfrac{8^2}{8^2} = 1$

34. $\dfrac{(a^{-3}b^2)^2}{(2a^{-4}b^{-3})^{-1}} = \dfrac{(a^{-3})^2(b^2)^2}{2^{-1}(a^{-4})^{-1}(b^{-3})^{-1}}$
$= \dfrac{a^{-6}b^4}{2^{-1}a^4b^3}$
$= \dfrac{2b^4}{a^6a^4b^3} = \dfrac{2b}{a^{10}}$

35. $34{,}500 = 3.45 \times 10^4$

Move the decimal point left 4 places so it is to the right of the first nonzero digit. 34,500 is *larger* than 3.45, so the power is *positive*.

36. $(7x^3 - 12x^2 - 3x + 8) + (6x^2 + 4)$
$\quad - (-4x^3 + 8x^2 - 2x - 2)$
$= (7x^3 - 12x^2 - 3x + 8) + (6x^2 + 4)$
$\quad + (4x^3 - 8x^2 + 2x + 2)$
$= (7 + 4)x^3 + (-12 + 6 - 8)x^2$
$\quad + (-3 + 2)x + (8 + 4 + 2)$
$= 11x^3 - 14x^2 - x + 14$

37. $6x^5(3x^2 - 9x + 10)$
$= (6x^5)(3x^2) + (6x^5)(-9x)$
$\quad + (6x^5)(10)$
$= 18x^7 - 54x^6 + 60x^5$

38. $(7x + 4)(9x + 3)$
$= 63x^2 + 21x + 36x + 12 \quad FOIL$
$= 63x^2 + 57x + 12$

39. $(5x + 8)^2 = (5x)^2 + 2(5x)(8) + (8)^2$
$\quad = 25x^2 + 80x + 64$

40. $\dfrac{y^3 - 3y^2 + 8y - 6}{y - 1}$

$$\begin{array}{r}
y^2 - 2y + 6 \\
y - 1 \overline{\smash{)}\, y^3 - 3y^2 + 8y - 6} \\
\underline{y^3 - y^2 } \\
-2y^2 + 8y \\
\underline{-2y^2 + 2y } \\
6y - 6 \\
\underline{6y - 6} \\
0
\end{array}$$

The remainder is 0. The answer is the quotient,
$$y^2 - 2y + 6.$$

CHAPTER 6 FACTORING AND APPLICATIONS

6.1 Factors; The Greatest Common Factor

6.1 Margin Exercises

1. **(a)** 30, 20, 15

   Write each number in prime factored form.
   $$30 = 2 \cdot 3 \cdot 5, \quad 20 = 2^2 \cdot 5, \quad 15 = 3 \cdot 5$$
   Use each prime the *least* number of times it appears in *all* the factored forms. The primes 2 and 3 do not appear in *all* the factored forms, so they will not appear in the greatest common factor. The greatest common factor (**GCF**) is 5.

   (b) 42, 28, 35

   Write each number in prime factored form.
   $$42 = 2 \cdot 3 \cdot 7, \quad 28 = 2^2 \cdot 7, \quad 35 = 5 \cdot 7$$
   Use each prime the least number of times it appears. The greatest common factor is 7.

   (c) 12, 18, 26, 32

   Write each number in prime factored form.
   $$12 = 2^2 \cdot 3, \quad 18 = 2 \cdot 3^2,$$
   $$26 = 2 \cdot 13, \quad 32 = 2^5$$
   Use each prime the least number of times it appears. The greatest common factor is 2.

   (d) 10, 15, 21

   Write each number in prime factored form.
   $$10 = 2 \cdot 5, \quad 15 = 3 \cdot 5, \quad 21 = 3 \cdot 7$$
   Use each prime the least number of times it appears. Since no prime appears in all the factored forms, the GCF is 1.

2. **(a)** $6m^4$, $9m^2$, $12m^5$

   Write each number in prime factored form.
   $$6m^4 = 2 \cdot 3 \cdot m^4, \quad 9m^2 = 3^2 \cdot m^2,$$
   $$12m^5 = 2^2 \cdot 3 \cdot m^5$$
   The greatest common factor of the coefficients 6, 9, and 12 is 3. The greatest common factor of the terms m^4, m^2, and m^5 is m^2 since 2 is the least exponent on m. Thus, the GCF of these terms is the product of 3 and m^2, that is, $3m^2$.

   (b) $-12p^5$, $-18q^4$

   The greatest common factor of -12 and -18 is 6. Since these terms do not contain the same variables, no variable is common to them. The GCF is 6.

   (c) y^4z^2, y^6z^8, z^9

   The least exponent on z is 2, and y does not occur in z^9. Thus, the GCF is z^2.

   (d) $12p^{11}$, $17q^5$

   Since $12p^{11}$ and $17q^5$ do not contain the same variables and 1 is the greatest common factor of 12 and 17, the GCF is 1.

3. **(a)** $4x^2 + 6x$

   2 is the greatest common factor of 4 and 6, and 1 is the least exponent on x. Hence, $2x$ is the GCF.
   $$4x^2 + 6x = 2x(2x) + 2x(3)$$
   $$= 2x(2x + 3)$$

   (b) $10y^5 - 8y^4 + 6y^2$

   2 is the greatest common factor of 10, 8, and 6, so 2 can be factored out. Since y occurs in every term and the least exponent on y is 2, y^2 can be factored out. Hence, $2y^2$ is the GCF.
   $$10y^5 - 8y^4 + 6y^2$$
   $$= (2y^2)(5y^3) - (2y^2)(4y^2) + (2y^2)(3)$$
   $$= 2y^2(5y^3 - 4y^2 + 3)$$

   (c) $m^7 + m^9$

   m occurs in every term and 7 is the least exponent on m, so m^7 is the GCF.
   $$m^7 + m^9 = m^7 \cdot 1 + m^7 \cdot m^2$$
   $$= m^7(1 + m^2)$$

   (d) $8p^5q^2 + 16p^6q^3 - 12p^4q^7$

   4 is the greatest common factor of 8, 16, and 12. The least exponents on p and q are 4 and 2, respectively. Hence, $4p^4q^2$ is the GCF.
   $$8p^5q^2 + 16p^6q^3 - 12p^4q^7$$
   $$= (4p^4q^2)(2p) + (4p^4q^2)(4p^2q) - (4p^4q^2)(3q^5)$$
   $$= 4p^4q^2(2p + 4p^2q - 3q^5)$$

   (e) $\frac{1}{3}b^2 - \frac{2}{3}b$

   $\frac{1}{3}$ is the greatest common factor of $\frac{1}{3}$ and $-\frac{2}{3}$, and 1 is the least exponent on b. Hence, $\frac{1}{3}b$ is the GCF.
   $$\tfrac{1}{3}b^2 - \tfrac{2}{3}b = \tfrac{1}{3}b(b) + \tfrac{1}{3}b(-2)$$
   $$= \tfrac{1}{3}b(b - 2)$$

(f) $13x^2 - 27$

The greatest common factor of 13 and 27 is 1, and x does not appear in the second term. Thus, there is no common factor (except 1).

4. (a) $r(t-4) + 5(t-4)$

The binomial $t - 4$ is the greatest common factor here.

$$r(t-4) + 5(t-4) = (t-4)(r) + (t-4)(5)$$
$$= (t-4)(r+5)$$

(b) $y^2(y+2) - 3(y+2)$

The binomial $y + 2$ is the greatest common factor.

$$y^2(y+2) - 3(y+2) = (y+2)(y^2 - 3)$$

(c) $x(x-1) - 5(x-1)$

The binomial $x - 1$ is the greatest common factor.

$$x(x-1) - 5(x-1) = (x-1)(x-5)$$

5. (a) $pq + 5q + 2p + 10$
$= (pq + 5q) + (2p + 10)$ *Group terms.*
$= q(p+5) + 2(p+5)$ *Factor each group.*
$= (p+5)(q+2)$ *Factor out $p + 5$.*

(b) $2xy + 3y + 2x + 3$
$= (2xy + 3y) + (2x + 3)$ *Group terms.*
$= y(2x+3) + 1(2x+3)$ *Factor each group; remember the 1.*
$= (2x+3)(y+1)$ *Factor out $2x + 3$.*

(c) $2a^2 - 4a + 3ab - 6b$
$= (2a^2 - 4a) + (3ab - 6b)$ *Group terms.*
$= 2a(a-2) + 3b(a-2)$ *Factor each group.*
$= (a-2)(2a+3b)$ *Factor out $a - 2$.*

(d) $x^3 + 3x^2 - 5x - 15$
$= (x^3 + 3x^2) + (-5x - 15)$ *Group terms.*
$= x^2(x+3) - 5(x+3)$ *Factor each group.*
$= (x+3)(x^2 - 5)$ *Factor out $x + 3$.*

6. (a) $6y^2 - 20w + 15y - 8yw$

Factoring out the common factor 2 from the first two terms and the common factor y from the last two terms gives

$$6y^2 - 20w + 15y - 8yw$$
$$= 2(3y - 10w) + y(15 - 8w).$$

This does not lead to a common factor, so we try rearranging the terms.

$= (6y^2 + 15y) + (-20w - 8yw)$ *Rearrange.*
$= 3y(2y+5) - 4w(5+2y)$ *Factor each group.*
$= (2y+5)(3y-4w)$ *Factor out $2y + 5$.*

Here's another rearrangement.

$= (6y^2 - 8yw) + (15y - 20w)$
$= 2y(3y - 4w) + 5(3y - 4w)$
$= (3y - 4w)(2y + 5)$

This is an equivalent answer.

(b) $9mn - 4 + 12m - 3n$
$= (9mn + 12m) + (-4 - 3n)$
$= 3m(3n+4) - 1(4+3n)$
$= (3n+4)(3m-1)$

6.1 Section Exercises

1. 12, 16

Write each number in prime factored form.

$$12 = 2^2 \cdot 3, \ 16 = 2^4$$

Use each prime the *least* number of times it appears in both factored forms. 2 is the only prime that appears in both forms and it appears at least twice in each form, so the greatest common factor is $2^2 = 4$.

3. Find the prime factored form of each number.

$$40 = 2 \cdot 2 \cdot 2 \cdot 5$$
$$20 = 2 \cdot 2 \cdot 5$$
$$4 = 2 \cdot 2$$

The least number of times 2 appears in all the factored forms is 2. There is no 5 in the prime factored form of 4, so the

$$\text{GCF} = 2^2 = 4.$$

5. Find the prime factored form of each number.
$$18 = 2 \cdot 3 \cdot 3$$
$$24 = 2 \cdot 2 \cdot 2 \cdot 3$$
$$36 = 2 \cdot 2 \cdot 3 \cdot 3$$
$$48 = 2 \cdot 2 \cdot 2 \cdot 2 \cdot 3$$

The least number of times the primes 2 and 3 appear in all four factored forms is once, so
$$\text{GCF} = 2 \cdot 3 = 6.$$

7. Write each number in prime factored form.
$$4 = 2^2,\ 9 = 3^2,\ 12 = 2^2 \cdot 3$$

There are no prime factors common to all three numbers, so the greatest common factor is 1.

9. Write each term in prime factored form.
$$16y = 2^4 \cdot y$$
$$24 = 2^3 \cdot 3$$

There is no y in the second term, so y will not appear in the GCF. Thus, the GCF of $16y$ and 24 is
$$2^3 = 8.$$

11.
$$30x^3 = 2 \cdot 3 \cdot 5 \cdot x^3$$
$$40x^6 = 2^3 \cdot 5 \cdot x^6$$
$$50x^7 = 2 \cdot 5^2 \cdot x^7$$

The GCF of the coefficients, 30, 40, and 50, is $2^1 \cdot 5^1 = 10$. The smallest exponent on the variable x is 3. Thus the GCF of the given terms is $10x^3$.

13.
$$-x^4 y^3 = -1 \cdot x^4 \cdot y^3$$
$$-xy^2 = -1 \cdot x \cdot y^2$$

The GCF is xy^2.

15.
$$42ab^3 = 2 \cdot 3 \cdot 7 \cdot a \cdot b^3$$
$$-36a = -1 \cdot 2^2 \cdot 3^2 \cdot a$$
$$90b = 2 \cdot 3^2 \cdot 5 \cdot b$$
$$-48ab = -1 \cdot 2^4 \cdot 3 \cdot a \cdot b$$

The GCF is $2 \cdot 3 = 6$.

17. $9m^4 = 3m^2(3m^2)$

Factor out $3m^2$ from $9m^4$ to obtain $3m^2$.

19. $-8z^9 = -4z^5(2z^4)$

Factor out $-4z^5$ from $-8z^9$ to obtain $2z^4$.

21. $6m^4 n^5 = 3m^3 n(2mn^4)$

Factor out $3m^3n$ from $6m^4n^5$ to obtain $2mn^4$.

23. $12y + 24 = 12 \cdot y + 12 \cdot 2$
$$= 12(y+2)$$

25. $10a^2 - 20a = 10a(a) - 10a(2)$
$$= 10a(a-2)$$

27. $8x^2 y + 12x^3 y^2 = 4x^2 y(2) + 4x^2 y(3xy)$
$$= 4x^2 y(2 + 3xy)$$

29. The greatest common factor for $x^2 - 4x$ is x.
$$x^2 - 4x = x(x) + x(-4)$$
$$= x(x-4)$$

31. The greatest common factor for $6t^2 + 15t$ is $3t$.
$$6t^2 + 15t = 3t(2t) + 3t(5)$$
$$= 3t(2t+5)$$

33. $\frac{1}{4}d^2 - \frac{3}{4}d$

As in Example 3(e), we factor out $\frac{1}{4}d$.
$$\tfrac{1}{4}d^2 - \tfrac{3}{4}d = \tfrac{1}{4}d(d) + \tfrac{1}{4}d(-3)$$
$$= \tfrac{1}{4}d(d-3)$$

35. $12x^3 + 6x^2$
The GCF is $6x^2$.
$$12x^3 + 6x^2 = 6x^2(2x) + 6x^2(1)$$
$$= 6x^2(2x+1)$$

37. $65y^{10} + 35y^6$
The GCF is $5y^6$.
$$65y^{10} + 35y^6 = (5y^6)(13y^4) + (5y^6)(7)$$
$$= 5y^6(13y^4 + 7)$$

39. $11w^3 - 100$

The two terms of this expression have no common factor (except 1).

41. $8m^2 n^3 + 24m^2 n^2$
The GCF is $8m^2 n^2$.
$$8m^2 n^3 + 24m^2 n^2$$
$$= (8m^2 n^2)(n) + (8m^2 n^2)(3)$$
$$= 8m^2 n^2(n+3)$$

43. The greatest common factor for $4x^3 - 10x^2 + 6x$ is $2x$.
$$4x^3 - 10x^2 + 6x$$
$$= (2x)(2x^2) + (2x)(-5x) + (2x)(3)$$
$$= 2x(2x^2 - 5x + 3)$$

45. $13y^8 + 26y^4 - 39y^2$
The GCF is $13y^2$.
$$13y^8 + 26y^4 - 39y^2$$
$$= 13y^2(y^6) + 13y^2(2y^2) + 13y^2(-3)$$
$$= 13y^2(y^6 + 2y^2 - 3)$$

47. $45q^4p^5 + 36qp^6 + 81q^2p^3$
The GCF is $9qp^3$.

$$45q^4p^5 + 36qp^6 + 81q^2p^3$$
$$= 9qp^3(5q^3p^2) + 9qp^3(4p^3)$$
$$+ 9qp^3(9q)$$
$$= 9qp^3(5q^3p^2 + 4p^3 + 9q)$$

49. The GCF of the terms of $c(x+2) + d(x+2)$ is the binomial $x+2$.

$$c(x+2) + d(x+2)$$
$$= (x+2)(c) + (x+2)(d)$$
$$= (x+2)(c+d)$$

51. The greatest common factor for $a^2(2a+b) - b(2a+b)$ is $2a+b$.

$$a^2(2a+b) - b(2a+b) = (2a+b)(a^2 - b)$$

53. The GCF for $q(p+4) - 1(p+4)$ is $p+4$.

$$q(p+4) - 1(p+4) = (p+4)(q-1)$$

55. $5m + mn + 20 + 4n$
$= (5m + mn) + (20 + 4n)$
$= m(5 + n) + 4(5 + n)$
$= (5 + n)(m + 4)$

57. $6xy - 21x + 8y - 28$
$= (6xy - 21x) + (8y - 28)$
$= 3x(2y - 7) + 4(2y - 7)$
$= (2y - 7)(3x + 4)$

59. $3xy + 9x + y + 3$
$= (3xy + 9x) + (y + 3)$
$= 3x(y + 3) + 1(y + 3)$
$= (y + 3)(3x + 1)$

61. $7z^2 + 14z - az - 2a$
$= (7z^2 + 14z) + (-az - 2a)$ *Group the terms.*
$= 7z(z + 2) - a(z + 2)$ *Factor each group.*
$= (z + 2)(7z - a)$ *Factor out $z+2$.*

63. $18r^2 + 12ry - 3xr - 2xy$
$= (18r^2 + 12ry) + (-3xr - 2xy)$ *Group the terms.*
$= 6r(3r + 2y) - x(3r + 2y)$ *Factor each group.*
$= (3r + 2y)(6r - x)$ *Factor out $3r+2y$.*

65. $w^3 + w^2 + 9w + 9$
$= (w^3 + w^2) + (9w + 9)$
$= w^2(w + 1) + 9(w + 1)$
$= (w + 1)(w^2 + 9)$

67. $3a^3 + 6a^2 - 2a - 4$
$= (3a^3 + 6a^2) + (-2a - 4)$
$= 3a^2(a + 2) - 2(a + 2)$
$= (a + 2)(3a^2 - 2)$

69. $16m^3 - 4m^2p^2 - 4mp + p^3$
$= (16m^3 - 4m^2p^2) + (-4mp + p^3)$
$= 4m^2(4m - p^2) - p(4m - p^2)$
$= (4m - p^2)(4m^2 - p)$

71. $y^2 + 3x + 3y + xy$
$= y^2 + 3y + xy + 3x$ *Rearrange.*
$= (y^2 + 3y) + (xy + 3x)$
$= y(y + 3) + x(y + 3)$
$= (y + 3)(y + x)$

73. $2z^2 + 6w - 4z - 3wz$
$= 2z^2 - 4z - 3wz + 6w$ *Rearrange.*
$= (2z^2 - 4z) + (-3wz + 6w)$
$= 2z(z - 2) - 3w(z - 2)$
$= (z - 2)(2z - 3w)$

75. In order to rewrite
$$2xy + 12 - 3y - 8x$$
as
$$2xy - 8x - 3y + 12,$$
we must change the order of the terms. The property that allows us to do this is the commutative property of addition.

76. After we group both pairs of terms in the rearranged polynomial, we have
$$(2xy - 8x) + (-3y + 12).$$
The greatest common factor for the first pair of terms is $2x$. The GCF for the second pair is -3. Factoring each group gives us
$$2x(y - 4) - 3(y - 4).$$

77. The expression obtained in Exercise 76 is the *difference* between two terms, $2x(y-4)$ and $3(y-4)$, so it is *not* in factored form.

78. $2x(y-4) - 3(y-4)$
$= (y-4)(2x-3)$
or $(2x-3)(y-4)$

Yes, this is the same result as the one shown in Example 6(b), even though the terms were grouped in a different way.

6.2 Factoring Trinomials

6.2 Margin Exercises

1. **(a)** All pairs of positive integers whose product is 6 are 1, 6 and 2, 3.

(b) The pair 2, 3 has a sum of 5.

2. **(a)** $y^2 + 12y + 20$

Factors of 20	Sums of factors
20, 1	$20 + 1 = 21$
10, 2	$10 + 2 = 12$
5, 4	$5 + 4 = 9$

The pair of integers whose product is 20 and whose sum is 12 is 10 and 2. Thus,

$$y^2 + 12y + 20 = (y + 10)(y + 2).$$

(b) $x^2 + 9x + 18$

Factors of 18	Sums of factors
18, 1	$18 + 1 = 19$
9, 2	$9 + 2 = 11$
6, 3	$6 + 3 = 9$

The pair of integers whose product is 18 and whose sum is 9 is 6 and 3. Thus,

$$x^2 + 9x + 18 = (x + 6)(x + 3).$$

3. **(a)** $t^2 - 12t + 32$

Factors of 32	Sums of factors
$-32, -1$	$-32 + (-1) = -33$
$-16, -2$	$-16 + (-2) = -18$
$-8, -4$	$-8 + (-4) = -12$

The pair of integers whose product is 32 and whose sum is -12 is -8 and -4. Thus,

$$t^2 - 12t + 32 = (t - 8)(t - 4).$$

(b) $y^2 - 10y + 24$

Factors of 24	Sums of factors
$-24, -1$	$-24 + (-1) = -25$
$-12, -2$	$-12 + (-2) = -14$
$-8, -3$	$-8 + (-3) = -11$
$-6, -4$	$-6 + (-4) = -10$

The pair of integers whose product is 24 and whose sum is -10 is -6 and -4. Thus,

$$y^2 - 10y + 24 = (y - 6)(y - 4).$$

4. **(a)** $a^2 - 9a - 22$

Find the two integers whose product is -22 and whose sum is -9. Because the last term is negative, the pair must include one positive and one negative integer.

Factors of -22	Sums of factors
$22, -1$	$22 + (-1) = 21$
$11, -2$	$11 + (-2) = 9$
$-22, 1$	$-22 + 1 = -21$
$-11, 2$	$-11 + 2 = -9$

The required integers are -11 and 2, so

$$a^2 - 9a - 22 = (a - 11)(a + 2).$$

(b) $r^2 - 6r - 16$

Find two integers whose product is -16 and whose sum is -6.

Factors of -16	Sums of factors
$16, -1$	$16 + (-1) = 15$
$8, -2$	$8 + (-2) = 6$
$4, -4$	$4 + (-4) = 0$
$-16, 1$	$-16 + 1 = -15$
$-8, 2$	$-8 + 2 = -6$

We can stop here since we have found the required integers, -8 and 2. Thus,

$$r^2 - 6r - 16 = (r - 8)(r + 2).$$

5. **(a)** $r^2 - 3r - 4$

Find two integers whose product is -4 and whose sum is -3.

Factors of -4	Sums of factors
$2, -2$	$2 + (-2) = 0$
$4, -1$	$4 + (-1) = 3$
$-4, 1$	$-4 + 1 = -3$

The required integers are -4 and 1, so

$$r^2 - 3r - 4 = (r - 4)(r + 1).$$

(b) $m^2 - 2m + 5$

There is no pair of integers whose product is 5 and whose sum is -2, so $m^2 - 2m + 5$ is a prime polynomial.

6. **(a)** $b^2 - 3ab - 4a^2$

Two expressions whose product is $-4a^2$ and whose sum is $-3a$ are $-4a$ and a, so

$$b^2 - 3ab - 4a^2 = (b - 4a)(b + a).$$

(b) $r^2 - 6rs + 8s^2$

Two expressions whose product is $8s^2$ and whose sum is $-6s$ are $-4s$ and $-2s$, so

$$r^2 - 6rs + 8s^2 = (r - 4s)(r - 2s).$$

7. (a) $2p^3 + 6p^2 - 8p$

First, factor out the greatest common factor, $2p$.

$$2p^3 + 6p^2 - 8p = 2p(p^2 + 3p - 4)$$

Now factor $p^2 + 3p - 4$.

The integers 4 and -1 have a product of -4 and a sum of 3, so

$$p^2 + 3p - 4 = (p + 4)(p - 1).$$

The complete factored form is

$$2p^3 + 6p^2 - 8p = 2p(p + 4)(p - 1).$$

(b) $3x^4 - 15x^3 + 18x^2 = 3x^2(x^2 - 5x + 6)$

Factor $x^2 - 5x + 6$. The integers -3 and -2 have a product of 6 and a sum of -5, so

$$x^2 - 5x + 6 = (x-3)(x - 2).$$

The complete factored form is

$$3x^4 - 15x^3 + 18x^2 = 3x^2(x-3)(x-2).$$

6.2 Section Exercises

1. If the coefficient of the last term of the trinomial is negative, then a and b must have different signs, one positive and one negative.

3. A *prime polynomial* is one that cannot be factored using only integers in the factors.

5. Product: 12   Sum: 7

List all pairs of integers whose product is 12, and then find the sum of each pair.

Factors of 12	Sums of factors
1, 12	$12 + 1 = 13$
$-1, -12$	$-12 + (-1) = -13$
2, 6	$6 + 2 = 8$
$-2, -6$	$-6 + (-2) = -8$
3, 4	$4 + 3 = 7$ ←
$-3, -4$	$-4 + (-3) = -7$

The pair of integers whose product is 12 and whose sum is 7 is 4 and 3.

7. Product: -24 Sum: -5

Factors of -24	Sums of factors
1, -24	$1 + (-24) = -23$
$-1, 24$	$-1 + 24 = 23$
2, -12	$2 + (-12) = -10$
$-2, 12$	$-2 + 12 = 10$
3, -8	$3 + (-8) = -5$ ←
$-3, 8$	$-3 + 8 = 5$
4, -6	$4 + (-6) = -2$
$-4, 6$	$-4 + 6 = 2$

The pair of integers whose product is -24 and whose sum is -5 is 3 and -8.

9. $x^2 - 12x + 32$

Multiply each of the given pairs of factors to determine which one gives the required product.

A. $(x - 8)(x + 4) = x^2 - 4x - 32$

B. $(x + 8)(x - 4) = x^2 + 4x - 32$

C. $(x - 8)(x - 4) = x^2 - 12x + 32$

D. $(x + 8)(x + 4) = x^2 + 12x + 32$

Choice **C** is the correct factored form.

11. $x^2 + 15x + 44 = (x + 4)(\quad)$

Look for an integer whose product with 4 is 44 and whose sum with 4 is 15. That integer is 11.

$$x^2 + 15x + 44 = (x + 4)(x + 11)$$

13. $x^2 - 9x + 8 = (x - 1)(\quad)$

Look for an integer whose product with -1 is 8 and whose sum with -1 is -9. That integer is -8.

$$x^2 - 9x + 8 = (x - 1)(x - 8)$$

15. $y^2 - 2y - 15 = (y + 3)(\quad)$

Look for an integer whose product with 3 is -15 and whose sum with 3 is -2. That integer is -5.

$$y^2 - 2y - 15 = (y + 3)(y - 5)$$

17. $x^2 + 9x - 22 = (x - 2)(\quad)$

Look for an integer whose product with -2 is -22 and whose sum with -2 is 9. That integer is 11.

$$x^2 + 9x - 22 = (x - 2)(x + 11)$$

19. $y^2 - 7y - 18 = (y + 2)(\quad)$

Look for an integer whose product with 2 is -18 and whose sum with 2 is -7. That integer is -9.

$$y^2 - 7y - 18 = (y + 2)(y - 9)$$

6.2 Factoring Trinomials

21. $y^2 + 9y + 8$

Look for two integers whose product is 8 and whose sum is 9. Both integers must be positive because b and c are both positive.

Factors of 8	Sums of factors
1, 8	9 ←
2, 4	6

Thus,
$$y^2 + 9y + 8 = (y + 8)(y + 1).$$

23. $b^2 + 8b + 15$

Look for two integers whose product is 15 and whose sum is 8. Both integers must be positive because b and c are both positive.

Factors of 15	Sums of factors
1, 15	16
3, 5	8 ←

Thus,
$$b^2 + 8b + 15 = (b + 3)(b + 5).$$

25. $m^2 + m - 20$

Look for two integers whose product is -20 and whose sum is 1. Since c is negative, one integer must be positive and one must be negative.

Factors of -20	Sums of factors
$-1, 20$	19
$1, -20$	-19
$-2, 10$	8
$2, -10$	-8
$-4, 5$	1 ←
$4, -5$	-1

Thus,
$$m^2 + m - 20 = (m - 4)(m + 5).$$

27. $x^2 + 3x - 40$

Look for two integers whose product is -40 and whose sum is 3.

Factors of 40	Sums of factors
$-1, 40$	$-1 + 40 = 39$
$1, -40$	$1 + (-40) = -39$
$-2, 20$	$-2 + 20 = 18$
$2, -20$	$2 + (-20) = -18$
$-4, 10$	$-4 + 10 = 6$
$4, -10$	$4 + (-10) = -6$
$-5, 8$	$-5 + 8 = 3$ ←
$5, -8$	$5 + (-8) = -3$

Thus,
$$x^2 + 3x - 40 = (x - 5)(x + 8).$$

29. $y^2 - 8y + 15$

Find two integers whose product is 15 and whose sum is -8. Since c is positive and b is negative, both integers must be negative.

Factors of 15	Sums of factors
$-1, -15$	-16
$-3, -5$	-8 ←

Thus,
$$y^2 - 8y + 15 = (y - 5)(y - 3).$$

31. $z^2 - 15z + 56$

Find two integers whose product is 56 and whose sum is -15. Since c is positive and b is negative, both integers must be negative.

Factors of 56	Sums of factors
$-1, -56$	-57
$-2, -28$	-30
$-4, -14$	-18
$-7, -8$	-15 ←

Thus,
$$z^2 - 15z + 56 = (z - 7)(z - 8).$$

33. $r^2 - r - 30$

Look for two integers whose product is -30 and whose sum is -1. Because c is negative, one integer must be positive and the other must be negative.

Factors of -30	Sums of factors
$-1, 30$	29
$1, -30$	-29
$-2, 15$	13
$2, -15$	-13
$-3, 10$	7
$3, -10$	-7
$-5, 6$	1
$5, -6$	-1 ←

Thus,
$$r^2 - r - 30 = (r + 5)(r - 6).$$

35. $a^2 - 8a - 48$

Find two integers whose product is -48 and whose sum is -8. Since c is negative, one integer must be positive and one must be negative.

Factors of -48	Sums of factors
$-1, 48$	47
$1, -48$	-47
$-2, 24$	22
$2, -24$	-22
$-3, 16$	13
$3, -16$	-13
$-4, 12$	8
$4, -12$	-8 ←
$-6, 8$	2
$6, -8$	-2

Thus,
$$a^2 - 8a - 48 = (a + 4)(a - 12).$$

37. $x^2 + 4x + 5$

Look for two integers whose product is 5 and whose sum is 4. Both integers must be positive since b and c are both positive.

Product	Sum
$5 \cdot 1 = 5$	$5 + 1 = 6$

There is no other pair of positive integers whose product is 5. Since there is no pair of integers whose product is 5 and whose sum is 4, $x^2 + 4x + 5$ is a *prime* polynomial.

39. $r^2 + 3ra + 2a^2$

Look for two expressions whose product is $2a^2$ and whose sum is $3a$. They are $2a$ and a, so
$$r^2 + 3ra + 2a^2 = (r + 2a)(r + a).$$

41. $x^2 + 4xy + 3y^2$

Look for two expressions whose product is $3y^2$ and whose sum is $4y$. The expressions are $3y$ and y, so
$$x^2 + 4xy + 3y^2 = (x + 3y)(x + y).$$

43. $t^2 - tz - 6z^2$

Look for two expressions whose product is $-6z^2$ and whose sum is $-z$. They are $2z$ and $-3z$, so
$$t^2 - tz - 6z^2 = (t + 2z)(t - 3z).$$

45. $v^2 - 11vw + 30w^2$

Factors of $30w^2$	Sums of factors
$-30w, -w$	$-31w$
$-15w, -2w$	$-17w$
$-10w, -3w$	$-13w$
$-5w, -6w$	$-11w$ ←

The complete factored form is
$$v^2 - 11vw + 30w^2 = (v - 5w)(v - 6w).$$

47. $4x^2 + 12x - 40$

First, factor out the GCF, 4.
$$4x^2 + 12x - 40 = 4(x^2 + 3x - 10)$$

Now factor $x^2 + 3x - 10$.

Factors of -10	Sums of factors
$-1, 10$	9
$1, -10$	-9
$2, -5$	-3
$-2, 5$	3 ←

Thus,
$$x^2 + 3x - 10 = (x - 2)(x + 5).$$

The complete factored form is
$$4x^2 + 12x - 40 = 4(x - 2)(x + 5).$$

49. $2t^3 + 8t^2 + 6t$

First, factor out the GCF, $2t$.
$$2t^3 + 8t^2 + 6t = 2t(t^2 + 4t + 3)$$

Then factor $t^2 + 4t + 3$.
$$t^2 + 4t + 3 = (t + 1)(t + 3)$$

The complete factored form is
$$2t^3 + 8t^2 + 6t = 2t(t + 1)(t + 3).$$

51. $2x^6 + 8x^5 - 42x^4$

First, factor out the GCF, $2x^4$.
$$2x^6 + 8x^5 - 42x^4 = 2x^4(x^2 + 4x - 21)$$

Now factor $x^2 + 4x - 21$.

Factors of -21	Sums of factors
$1, -21$	-20
$-1, 21$	20
$3, -7$	-4
$-3, 7$	4 ←

Thus,
$$x^2 + 4x - 21 = (x-3)(x+7).$$

The complete factored form is
$$2x^6 + 8x^5 - 42x^4 = 2x^4(x-3)(x+7).$$

53. $a^5 + 3a^4b - 4a^3b^2$

The GCF is a^3, so
$$a^5 + 3a^4b - 4a^3b^2$$
$$= a^3(a^2 + 3ab - 4b^2).$$

Now factor $a^2 + 3ab - 4b^2$. The expressions $4b$ and $-b$ have a product of $-4b^2$ and a sum of $3b$. The complete factored form is
$$a^5 + 3a^4b - 4a^3b^2 = a^3(a+4b)(a-b).$$

55. $m^3n - 10m^2n^2 + 24mn^3$

First, factor out the GCF, mn.
$$m^3n - 10m^2n^2 + 24mn^3$$
$$= mn(m^2 - 10mn + 24n^2)$$

The expressions $-6n$ and $-4n$ have a product of $24n^2$ and a sum of $-10n$. The complete factored form is
$$m^3n - 10m^2n^2 + 24mn^3$$
$$= mn(m-6n)(m-4n).$$

57. $(2x+4)(x-3)$

$$\begin{aligned}&\quad\;\,\mathbf{F}\qquad\;\;\mathbf{O}\qquad\;\;\mathbf{I}\qquad\;\;\mathbf{L}\\&= (2x)(x) + (2x)(-3) + (4)(x) + (4)(-3)\\&= 2x^2 - 6x + 4x - 12\\&= 2x^2 - 2x - 12\end{aligned}$$

It is incorrect to completely factor $2x^2 - 2x - 12$ as $(2x+4)(x-3)$ because $2x+4$ can be factored further as $2(x+2)$. The first step should be to factor out the GCF, 2. The correct factorization is
$$2x^2 - 2x - 12 = 2(x^2 - x - 6)$$
$$= 2(x+2)(x-3).$$

6.3 Factoring Trinomials by Grouping

6.3 Margin Exercises

1. **(a)** $2x^2 + 7x + 6$
$$= 2x^2 + 4x + 3x + 6$$
$$= (2x^2 + \underline{4x}\,) + (3x + \underline{6}\,)$$
$$= 2x(x + \underline{2}\,) + 3(x + \underline{2}\,)$$
$$= \underline{(x+2)}\,(2x+3)$$

(b) Yes, the answer is the same.

2. **(a)** $2m^2 + 7m + 3$

Find the two integers whose product is $2(3) = 6$ and whose sum is 7. The integers are 1 and 6. Write the middle term, $7m$, as $1m + 6m$.
$$\begin{aligned}2m^2 + 7m + 3 &= 2m^2 + 1m + 6m + 3\\&= (2m^2 + 1m) + (6m + 3)\\&= m(2m+1) + 3(2m+1)\\&= (2m+1)(m+3)\end{aligned}$$

(b) $5p^2 - 2p - 3$

Find two integers whose product is $5(-3) = -15$ and whose sum is -2. The integers are -5 and 3.
$$\begin{aligned}5p^2 - 2p - 3 &= 5p^2 - 5p + 3p - 3\\&= (5p^2 - 5p) + (3p - 3)\\&= 5p(p-1) + 3(p-1)\\&= (p-1)(5p+3)\end{aligned}$$

(c) $15k^2 - km - 2m^2$

Find two integers whose product is $15(-2) = -30$ and whose sum is -1. The integers are -6 and 5.
$$15k^2 - km - 2m^2$$
$$= 15k^2 - 6km + 5km - 2m^2$$
$$= (15k^2 - 6km) + (5km - 2m^2)$$
$$= 3k(5k - 2m) + m(5k - 2m)$$
$$= (5k - 2m)(3k + m)$$

3. **(a)** $4x^2 - 2x - 30$

First factor out the greatest common factor, 2.
$$4x^2 - 2x - 30 = 2(2x^2 - x - 15)$$

Find two integers whose product is $2(-15) = -30$ and whose sum is -1. The integers are -6 and 5.
$$\begin{aligned}2x^2 - x - 15 &= 2x^2 - 6x + 5x - 15\\&= 2x(x-3) + 5(x-3)\\&= (2x+5)(x-3)\end{aligned}$$

The complete factored form is
$$4x^2 - 2x - 30 = 2(2x+5)(x-3).$$

(b) $18p^4 + 63p^3 + 27p^2 = 9p^2(2p^2 + 7p + 3)$

Find two integers whose product is $2(3) = 6$ and whose sum is 7. The integers are 6 and 1.
$$\begin{aligned}2p^2 + 7p + 3 &= 2p^2 + 6p + 1p + 3\\&= 2p(p+3) + 1(p+3)\\&= (2p+1)(p+3)\end{aligned}$$

The complete factored form is
$$18p^4 + 63p^3 + 27p^2 = 9p^2(2p+1)(p+3).$$

(c) $6a^2 + 3ab - 18b^2$

First factor out the greatest common factor, 3.

$$6a^2 + 3ab - 18b^2 = 3(2a^2 + ab - 6b^2)$$

Find two integers whose product is $2(-6) = -12$ and whose sum is 1. The integers are 4 and -3.

$$\begin{aligned}2a^2 + ab - 6b^2 &= 2a^2 + 4ab - 3ab - 6b^2 \\ &= 2a(a + 2b) - 3b(a + 2b) \\ &= (2a - 3b)(a + 2b)\end{aligned}$$

The complete factored form is

$$6a^2 + 3ab - 18b^2 = 3(2a - 3b)(a + 2b).$$

6.3 Section Exercises

1. $m^2 + 6m + 2m + 12$
$= (m^2 + 6m) + (2m + 12)$ Group terms.
$= m(m + 6) + 2(m + 6)$ Factor each group.
$= (m + 6)(m + 2)$ Factor out $m + 6$.

3. $a^2 + 5a - 2a - 10$
$= (a^2 + 5a) + (-2a - 10)$ Group terms.
$= a(a + 5) - 2(a + 5)$ Factor each group.
$= (a + 5)(a - 2)$ Factor out $a + 5$.

5. $10t^2 + 5t + 4t + 2$
$= (10t^2 + 5t) + (4t + 2)$ Group terms.
$= 5t(2t + 1) + 2(2t + 1)$ Factor each group.
$= (2t + 1)(5t + 2)$ Factor out $2t + 1$.

7. $15z^2 - 10z - 9z + 6$
$= (15z^2 - 10z) + (-9z + 6)$ Group terms.
$= 5z(3z - 2) - 3(3z - 2)$ Factor each group.
$= (3z - 2)(5z - 3)$ Factor out $3z - 2$.

9. $8s^2 - 4st + 6st - 3t^2$
$= (8s^2 - 4st) + (6st - 3t^2)$ Group terms.
$= 4s(2s - t) + 3t(2s - t)$ Factor each group.
$= (2s - t)(4s + 3t)$ Factor out $2s - t$.

11. $15a^2 + 10ab + 12ab + 8b^2$
$= (15a^2 + 10ab) + (12ab + 8b^2)$ Group terms.
$= 5a(3a + 2b) + 4b(3a + 2b)$ Factor each group.
$= (3a + 2b)(5a + 4b)$ Factor out $3a + 2b$.

13. To factor $12y^2 + 5y - 2$, we must find two integers with a product of $12(-2) = -24$ and a sum of 5. The only pair of integers satisfying those conditions is 8 and -3, choice **B**.

15. $2m^2 + 11m + 12$

(a) Find two integers whose product is $\underline{2} \cdot \underline{12} = \underline{24}$ and whose sum is $\underline{11}$.

(b) The required integers are $\underline{3}$ and $\underline{8}$. (Order is irrelevant.)

(c) Write the middle term $11m$ as $\underline{3m} + \underline{8m}$.

(d) Rewrite the given trinomial as $\underline{2m^2 + 3m + 8m + 12}$.

(e) $(2m^2 + 3m) + (8m + 12)$ Group terms.
$= m(2m + 3) + 4(2m + 3)$ Factor each group.
$= (2m + 3)(m + 4)$ Factor out $2m + 3$.

(f) $(2m + 3)(m + 4)$
      F   O   I   L
$= 2m(m) + 2m(4) + 3(m) + 3(4)$
$= 2m^2 + 8m + 3m + 12$
$= 2m^2 + 11m + 12$

17. $2x^2 + 7x + 3$

Look for two integers whose product is $2(3) = 6$ and whose sum is 7. The integers are 1 and 6.

$$\begin{aligned}2x^2 + 7x + 3 &= 2x^2 + x + 6x + 3 \\ &= (2x^2 + x) + (6x + 3) \\ &= x(2x + 1) + 3(2x + 1) \\ &= (2x + 1)(x + 3)\end{aligned}$$

19. $4r^2 + r - 3$

Look for two integers whose product is $4(-3) = -12$ and whose sum is 1. The integers are -3 and 4.

$$\begin{aligned}4r^2 + r - 3 &= 4r^2 - 3r + 4r - 3 \\ &= (4r^2 - 3r) + (4r - 3) \\ &= r(4r - 3) + 1(4r - 3) \\ &= (4r - 3)(r + 1)\end{aligned}$$

21. $8m^2 - 10m - 3$

Look for two integers whose product is $8(-3) = -24$ and whose sum is -10. The integers are -12 and 2.

$$\begin{aligned}8m^2 - 10m - 3 &= 8m^2 - 12m + 2m - 3 \\ &= 4m(2m - 3) + 1(2m - 3) \\ &= (2m - 3)(4m + 1)\end{aligned}$$

6.4 Factoring Trinomials Using FOIL

23. $21m^2 + 13m + 2$

Look for two integers whose product is $21(2) = 42$ and whose sum is 13. The integers are 7 and 6.

$$21m^2 + 13m + 2$$
$$= 21m^2 + 7m + 6m + 2$$
$$= 7m(3m+1) + 2(3m+1)$$
$$= (3m+1)(7m+2)$$

25. $6b^2 + 7b + 2$

Find two integers whose product is $6(2) = 12$ and whose sum is 7. The integers are 3 and 4.

$$6b^2 + 7b + 2 = 6b^2 + 3b + 4b + 2$$
$$= 3b(2b+1) + 2(2b+1)$$
$$= (2b+1)(3b+2)$$

27. $12y^2 - 13y + 3$

Look for two integers whose product is $12(3) = 36$ and whose sum is -13. The integers are -9 and -4.

$$12y^2 - 13y + 3 = 12y^2 - 9y - 4y + 3$$
$$= 3y(4y-3) - 1(4y-3)$$
$$= (4y-3)(3y-1)$$

29. $24x^2 - 42x + 9$

Factor out the greatest common factor, 3.

$$24x^2 - 42x + 9 = 3(8x^2 - 14x + 3)$$

Look for two integers whose product is $8(3) = 24$ and whose sum is -14. The integers are -2 and -12.

$$24x^2 - 42x + 9$$
$$= 3(8x^2 - 2x - 12x + 3)$$
$$= 3[2x(4x-1) - 3(4x-1)]$$
$$= 3(4x-1)(2x-3)$$

31. $2m^3 + 2m^2 - 40m$

Factor out the greatest common factor, $2m$.

$$2m^3 + 2m^2 - 40m = 2m(m^2 + m - 20)$$

Find two integers whose product is $1(-20) = -20$ and whose sum is 1. The integers are -4 and 5.

$$2m^3 + 2m^2 - 40m$$
$$= 2m(m^2 - 4m + 5m - 20)$$
$$= 2m[m(m-4) + 5(m-4)]$$
$$= 2m(m-4)(m+5)$$

33. $32z^5 - 20z^4 - 12z^3$

Factor out the greatest common factor, $4z^3$.

$$32z^5 - 20z^4 - 12z^3 = 4z^3(8z^2 - 5z - 3)$$

Find two integers whose product is $8(-3) = -24$ and whose sum is -5. The integers are 3 and -8.

$$32z^5 - 20z^4 - 12z^3$$
$$= 4z^3(8z^2 + 3z - 8z - 3)$$
$$= 4z^3[z(8z+3) - 1(8z+3)]$$
$$= 4z^3(8z+3)(z-1)$$

35. $12p^2 + 7pq - 12q^2$

Find two integers whose product is $12(-12) = -144$ and whose sum is 7. The integers are 16 and -9.

$$12p^2 + 7pq - 12q^2$$
$$= 12p^2 + 16pq - 9pq - 12q^2$$
$$= 4p(3p + 4q) - 3q(3p + 4q)$$
$$= (3p + 4q)(4p - 3q)$$

37. $6a^2 - 7ab - 5b^2$

Find two integers whose product is $6(-5) = -30$ and whose sum is -7. The integers are -10 and 3.

$$6a^2 - 7ab - 5b^2$$
$$= 6a^2 - 10ab + 3ab - 5b^2$$
$$= 2a(3a - 5b) + b(3a - 5b)$$
$$= (3a - 5b)(2a + b)$$

39. $5 - 6x + x^2$

Find two integers whose product is $5(1) = 5$ and whose sum is -6. The integers are -1 and -5.

$$5 - 6x + x^2 = 5 - x - 5x + x^2$$
$$= 1(5-x) - x(5-x)$$
$$= (5-x)(1-x)$$

41. The student stopped too soon.
He needs to factor out the common factor $4x - 1$ to get $(4x-1)(4x-5)$ as the correct answer.

6.4 Factoring Trinomials Using FOIL

6.4 Margin Exercises

1. **(a)** Multiply using FOIL.

$$(2x+1)(x+6) = 2x^2 + 12x + x + 6$$
$$= 2x^2 + 13x + 6$$

Thus, $(2x+1)(x+6)$ is an *incorrect* factoring of $2x^2 + 7x + 6$.

(b) Multiply using FOIL.

$$(2x+6)(x+1) = 2x^2 + 2x + 6x + 6$$
$$= 2x^2 + 8x + 6$$

Thus, $(2x+6)(x+1)$ is an *incorrect* factoring of $2x^2 + 7x + 6$.

(c) Multiply using FOIL.

$$(2x+2)(x+3) = 2x^2 + 6x + 2x + 6$$
$$= 2x^2 + 8x + 6$$

Thus, $(2x+2)(x+3)$ is an *incorrect* factoring of $2x^2 + 7x + 6$.

2. (a) $2p^2 + 9p + 9$

The only factors of $2p^2$ are $2p$ and p. The only factors of 9 are 3 and 3, or 9 and 1. Try various combinations, checking to see if the middle term is $9p$ in each case.

$(2p+9)(p+1) = 2p^2 + 11p + 9$ *Incorrect*
$(2p+3)(p+3) = 2p^2 + 9p + 9$ *Correct*

(b) $6p^2 + 19p + 10$

The factors of $6p^2$ are $2p$ and $3p$, or $6p$ and p. The factors of 10 are 10 and 1, or 5 and 2. Try various combinations, checking to see if the middle term is $19p$ in each case.

$(3p+5)(2p+2) = 6p^2 + 16p + 10$ *Incorrect*
$(3p+2)(2p+5) = 6p^2 + 19p + 10$ *Correct*

(c) $8x^2 + 14x + 3$

The factors of $8x^2$ are $8x$ and x, or $4x$ and $2x$. The factors of 3 are 3 and 1. Try various combinations, checking to see if the middle term is $14x$ in each case.

$(8x+3)(x+1) = 8x^2 + 11x + 3$ *Incorrect*
$(8x+1)(x+3) = 8x^2 + 25x + 3$ *Incorrect*
$(4x+1)(2x+3) = 8x^2 + 14x + 3$ *Correct*

3. (a) $4y^2 - 11y + 6$

The factors of $4y^2$ are $4y$ and y, or $2y$ and $2y$. Try $4y$ and y. Since the last term is positive and the coefficient of the middle term is negative, only negative factors of 6 should be considered. The factors of 6 are -6 and -1, or -3 and -2.

Try -6 and -1.

$(4y-6)(y-1) = 4y^2 - 10y + 6$ *Incorrect*

Try -3 and -2.

$(4y-3)(y-2) = 4y^2 - 11y + 6$ *Correct*

Thus,

$$4y^2 - 11y + 6 = (4y-3)(y-2).$$

(b) $9x^2 - 21x + 10$

The factors of $9x^2$ are $9x$ and x, or $3x$ and $3x$. Try $3x$ and $3x$. Since the last term is positive and the coefficient of the middle term is negative, only negative factors of 10 should be considered. The factors of 10 are -10 and -1, or -5 and -2.

Try -10 and -1.

$(3x-10)(3x-1) = 9x^2 - 33x + 10$ *Incorrect*

Try -5 and -2.

$(3x-5)(3x-2) = 9x^2 - 21x + 10$ *Correct*

Thus,

$$9x^2 - 21x + 10 = (3x-5)(3x-2).$$

4. (a) $6x^2 + 5x - 4$

6 and -4 each have several factors. Since the middle coefficient, 5, is not large, we try $3x$ and $2x$ as factors of $6x$, rather than $6x$ and x. The last term, -4, has factors -4 and 1, 4 and -1, -2 and 2.

Try -4 and 1.

$(3x-4)(2x+1) = 6x^2 - 5x - 4$ *Incorrect*

Try 4 and -1.

$(3x+4)(2x-1) = 6x^2 + 5x - 4$ *Correct*

Thus,

$$6x^2 + 5x - 4 = (3x+4)(2x-1).$$

(b) $6m^2 - 11m - 10$

The factors of $6m^2$ are $6m$ and m, or $3m$ and $2m$. Start with $3m$ and $2m$. Some factors of -10 are 10 and -1, or 2 and -5. Try various possibilities.

$(3m-1)(2m+10) = 6m^2 - 28m - 10$
   Incorrect
$(3m+2)(2m-5) = 6m^2 - 11m - 10$
   Correct

Thus,

$$6m^2 - 11m - 10 = (3m+2)(2m-5).$$

(c) $4x^2 - 3x - 7$

The factors of $4x^2$ are $4x$ and x, or $2x$ and $2x$. Start with $2x$ and $2x$. The factors of -7 are -7 and 1. Try various possibilities.

$$(2x-7)(2x+1) = 4x^2 - 12x - 7$$
   Incorrect

It is obvious that $2x$ and $2x$ won't work, so we will try $4x$ and x.

$$(4x - 7)(x + 1) = 4x^2 - 3x - 7$$
Correct

Thus,

$$4x^2 - 3x - 7 = (4x - 7)(x + 1).$$

(d) $3y^2 + 8y - 6$

The factors of $3y^2$ are $3y$ and y.
The factors of -6 are -6 and 1, 6 and -1, 2 and -3, or -2 and 3.

1) $(3y - 6)(y + 1) = 3y^2 - 3y - 6$ Incorrect
2) $(3y + 1)(y - 6) = 3y^2 - 17y - 6$ Incorrect
3) $(3y - 1)(y + 6) = 3y^2 + 17y - 6$ Incorrect
4) $(3y + 6)(y - 1) = 3y^2 + 3y - 6$ Incorrect
5) $(3y + 2)(y - 3) = 3y^2 - 7y - 6$ Incorrect
6) $(3y - 3)(y + 2) = 3y^2 + 3y - 6$ Incorrect
7) $(3y - 2)(y + 3) = 3y^2 + 7y - 6$ Incorrect
8) $(3y + 3)(y - 2) = 3y^2 - 3y - 6$ Incorrect

All 8 attempts failed. Thus, the polynomial is *prime*. Notice in lines 1, 4, 6, and 8, that 3 is a factor of the first factor, so those factorings cannot be correct since 3 is not a factor of the original polynomial.

5. **(a)** $2x^2 - 5xy - 3y^2$

Try various possibilities.

$$(2x - y)(x + 3y) = 2x^2 + 5xy - 3y^2$$
Incorrect

The middle terms differ only in sign, so reverse the signs of the two factors.

$$(2x + y)(x - 3y) = 2x^2 - 5xy - 3y^2$$
Correct

Thus,

$$2x^2 - 5xy - 3y^2 = (2x + y)(x - 3y).$$

(b) $8a^2 + 2ab - 3b^2$

Try various possibilities.

$$(8a + 3b)(a - b) = 8a^2 - 5ab - 3b^2$$
Incorrect

$$(4a + 3b)(2a - b) = 8a^2 + 2ab - 3b^2$$
Correct

Thus,

$$8a^2 + 2ab - 3b^2 = (4a + 3b)(2a - b).$$

6. **(a)** $36z^3 - 6z^2 - 72z$

Factor out the common factor, $6z$.

$$36z^3 - 6z^2 - 72z = 6z(6z^2 - z - 12)$$

Factor $6z^2 - z - 12$ by trial and error to obtain

$$6z^2 - z - 12 = (2z - 3)(3z + 4).$$

The complete factored form is

$$36z^3 - 6z^2 - 72z = 6z(2z - 3)(3z + 4).$$

(b) $-24x^3 + 32x^2y + 6xy^2$

The common factor could be $2x$ or $-2x$. If we factor out $-2x$, the first term of the trinomial factor will be positive, which makes it easier to factor.

$$-24x^3 + 32x^2y + 6xy^2$$
$$= -2x(12x^2 - 16xy - 3y^2)$$

Factor $12x^2 - 16xy - 3y^2$ by trial and error to obtain

$$12x^2 - 16xy - 3y^2 = (6x + y)(2x - 3y).$$

The complete factored form is

$$-24x^3 + 32x^2y + 6xy^2$$
$$= -2x(6x + y)(2x - 3y).$$

6.4 Section Exercises

For Exercises 1–6, multiply the factors in the choices together to see which ones give the correct product.

1. $2x^2 - x - 1$

  A. $(2x - 1)(x + 1) = 2x^2 + x - 1$

  B. $(2x + 1)(x - 1) = 2x^2 - x - 1$

B is the correct factored form.

3. $4y^2 + 17y - 15$

  A. $(y + 5)(4y - 3) = 4y^2 + 17y - 15$

  B. $(2y - 5)(2y + 3) = 4y^2 - 4y - 15$

A is the correct factored form.

5. $4k^2 + 13mk + 3m^2$

  A. $(4k + m)(k + 3m) = 4k^2 + 13mk + 3m^2$

  B. $(4k + 3m)(k + m) = 4k^2 + 7mk + 3m^2$

A is the correct factored form.

7. $6a^2 + 7ab - 20b^2 = (3a - 4b)(\quad)$

The first term in the missing expression must be $2a$ since
$$(3a)(2a) = 6a^2.$$

The second term in the missing expression must be $5b$ since
$$(-4b)(5b) = -20b^2.$$

Checking our answer by multiplying, we see that $(3a - 4b)(2a + 5b) = 6a^2 + 7ab - 20b^2$, as desired.

9. $2x^2 + 6x - 8 = 2(x^2 + 3x - 4)$

To factor $x^2 + 3x - 4$, we look for two integers whose product is -4 and whose sum is 3. The integers are 4 and -1. Thus,
$$2x^2 + 6x - 8 = 2(x + 4)(x - 1).$$

11. $4z^3 - 10z^2 - 6z = 2z(2z^2 - 5z - 3)$
$\qquad\qquad\qquad\quad = 2z(2z + 1)(z - 3)$

13. Since 2 is not a factor of $12x^2 + 7x - 12$, it cannot be a factor of any factor of $12x^2 + 7x - 12$. Since 2 is a factor of $2x - 6$, this means that $2x - 6$ cannot be a factor of $12x^2 + 7x - 12$.

15. $3a^2 + 10a + 7$

Possible factors of $3a^2$ are $3a$ and a.

Possible factors of 7 are 7 and 1.

$(3a + 1)(a + 7) = 3a^2 + 22a + 7\quad$ Incorrect
$(3a + 7)(a + 1) = 3a^2 + 10a + 7\quad$ Correct

17. $2y^2 + 7y + 6$

Possible factors of $2y^2$ are $2y$ and y.

Possible factors of 6 are 6 and 1, or 3 and 2.

$(2y + 6)(y + 1) = 2y^2 + 8y + 6\quad$ Incorrect
$(2y + 1)(y + 6) = 2y^2 + 13y + 6\quad$ Incorrect
$(2y + 3)(y + 2) = 2y^2 + 7y + 6\quad$ Correct

19. $15m^2 + m - 2$

Possible factors of $15m^2$ are $15m$ and m, or $3m$ and $5m$.

Possible factors of -2 are -2 and 1, or 2 and -1.

$(3m - 2)(5m + 1) = 15m^2 - 7m - 2$
$\qquad\qquad\qquad\qquad$ Incorrect
$(3m + 2)(5m - 1) = 15m^2 + 7m - 2$
$\qquad\qquad\qquad\qquad$ Incorrect
$(3m - 1)(5m + 2) = 15m^2 + m - 2$
$\qquad\qquad\qquad\qquad$ Correct

21. $12s^2 + 11s - 5$

Possible factors of $12s^2$ are s and $12s$, $2s$ and $6s$, or $3s$ and $4s$.

Factors of -5 are -1 and 5 or -5 and 1.

$(2s - 1)(6s + 5) = 12s^2 + 4s - 5\quad$ Incorrect
$(2s + 1)(6s - 5) = 12s^2 - 4s - 5\quad$ Incorrect
$(3s - 1)(4s + 5) = 12s^2 + 11s - 5\quad$ Correct

23. $10m^2 - 23m + 12$

Possible factors of $10m^2$ are m and $10m$, or $2m$ and $5m$.

Possible factors of 12 are -12 and -1, -2 and -6, or -3 and -4.

$(2m - 12)(5m - 1) = 10m^2 - 62m + 12$
$\qquad\qquad\qquad\qquad$ Incorrect
$(2m - 1)(5m - 12) = 10m^2 - 29m + 12$
$\qquad\qquad\qquad\qquad$ Incorrect
$(2m - 3)(5m - 4) = 10m^2 - 23m + 12$
$\qquad\qquad\qquad\qquad$ Correct

25. $8w^2 - 14w + 3$

Possible factors of $8w^2$ are w and $8w$ or $2w$ and $4w$.

Factors of 3 are -1 and -3 (since $b = -14$ is negative).

$(4w - 3)(2w - 1) = 8w^2 - 10w + 3\quad$ Incorrect
$(4w - 1)(2w - 3) = 8w^2 - 14w + 3\quad$ Correct

27. $20y^2 - 39y - 11$

Possible factors of $20y^2$ are $20y$ and y, $10y$ and $2y$, $5y$ and $4y$.

Possible factors of -11 are -11 and 1, or -1 and 11.

$(4y - 1)(5y + 11) = 20y^2 + 39y - 11$
$\qquad\qquad\qquad\qquad$ Incorrect
$(4y + 1)(5y - 11) = 20y^2 - 39y - 11$
$\qquad\qquad\qquad\qquad$ Correct

29. $3x^2 - 15x + 16$

Possible factors of $3x^2$ are $3x$ and x.

Possible factors of 16 are -16 and -1, -8 and -2, or -4 and -4.

$(3x - 16)(x - 1) = 3x^2 - 19x + 16\quad$ Incorrect
$(3x - 1)(x - 16) = 3x^2 - 49x + 16\quad$ Incorrect
$(3x - 8)(x - 2) = 3x^2 - 14x + 16\quad$ Incorrect
$(3x - 2)(x - 8) = 3x^2 - 26x + 16\quad$ Incorrect
$(3x - 4)(x - 4) = 3x^2 - 16x + 16\quad$ Incorrect

All are incorrect, so the polynomial is *prime*.

31. $20x^2 + 22x + 6$

   First, factor out the greatest common factor, 2.
   $$20x^2 + 22x + 6 = 2(10x^2 + 11x + 3)$$
   Now factor $10x^2 + 11x + 3$ by trial and error to obtain
   $$10x^2 + 11x + 3 = (5x + 3)(2x + 1).$$
   The complete factorization is
   $$20x^2 + 22x + 6 = 2(5x + 3)(2x + 1).$$

33. $40m^2q + mq - 6q$

   First, factor out the greatest common factor, q.
   $$40m^2q + mq - 6q = q(40m^2 + m - 6)$$
   Now factor $40m^2 + m - 6$ by trial and error to obtain
   $$40m^2 + m - 6 = (5m + 2)(8m - 3).$$
   The complete factorization is
   $$40m^2q + mq - 6q = q(5m + 2)(8m - 3).$$

35. $15n^4 - 39n^3 + 18n^2$

   Factor out the greatest common factor, $3n^2$.
   $$15n^4 - 39n^3 + 18n^2 = 3n^2(5n^2 - 13n + 6)$$
   Factor $5n^2 - 13n + 6$ by the trial and error method. Possible factors of $5n^2$ are $5n$ and n.
   Possible factors of 6 are -6 and -1, -3 and -2, -2 and -3, or -1 and -6.
   $(5n - 6)(n - 1) = 5n^2 - 11n + 6$ Incorrect
   $(5n - 3)(n - 2) = 5n^2 - 13n + 6$ Correct
   The complete factored form is
   $$15n^4 - 39n^3 + 18n^2 = 3n^2(5n - 3)(n - 2).$$

37. $15x^2y^2 - 7xy^2 - 4y^2$

   Factor out the greatest common factor, y^2.
   $$15x^2y^2 - 7xy^2 - 4y^2 = y^2(15x^2 - 7x - 4)$$
   Factor $15x^2 - 7x - 4$ by trial and error to obtain
   $$15x^2 - 7x - 4 = (5x - 4)(3x + 1).$$
   Thus,
   $$15x^2y^2 - 7xy^2 - 4y^2 = y^2(5x - 4)(3x + 1).$$

39. $5a^2 - 7ab - 6b^2$

   The possible factors of $5a^2$ are $5a$ and a.

   The possible factors of $-6b^2$ are $-6b$ and b, $6b$ and $-b$, $-3b$ and $2b$, or $3b$ and $-2b$.

$(5a - 3b)(a + 2b) = 5a^2 + 7ab - 6b^2$
   Incorrect
$(5a + 3b)(a - 2b) = 5a^2 - 7ab - 6b^2$
   Correct

41. $12s^2 + 11st - 5t^2$

   The possible factors of $12s^2$ are $12s$ and s, $6s$ and $2s$, or $4s$ and $3s$.

   The possible factors of $-5t^2$ are $-5t$ and t, or $5t$ and $-t$.

$(4s - 5t)(3s + t) = 12s^2 - 11st - 5t^2$
   Incorrect
$(4s + 5t)(3s - t) = 12s^2 + 11st - 5t^2$
   Correct

43. $6m^6n + 7m^5n^2 + 2m^4n^3$

   Factor out the GCF, m^4n.
   $$6m^6n + 7m^5n^2 + 2m^4n^3 = m^4n(6m^2 + 7mn + 2n^2)$$
   Now factor $6m^2 + 7mn + 2n^2$ by trial and error.

   Possible factors of $6m^2$ are $6m$ and m or $3m$ and $2m$. Possible factors of $2n^2$ are $2n$ and n.
   $$(3m + 2n)(2m + n) = 6m^2 + 7mn + 2n^2$$
   Correct

   The complete factored form is
   $$6m^6n + 7m^5n^2 + 2m^4n^3 = m^4n(3m + 2n)(2m + n).$$

45. $-x^2 - 4x + 21 = -1(x^2 + 4x - 21)$
   $= -1(x + 7)(x - 3)$

47. $-3x^2 - x + 4 = -1(3x^2 + x - 4)$
   $= -1(3x + 4)(x - 1)$

49. $-2a^2 - 5ab - 2b^2$
   $= -1(2a^2 + 5ab + 2b^2)$
   Factor out -1.
   $= -1(2a^2 + 4ab + ab + 2b^2)$
   $5ab = 4ab + ab$
   $= -1[(2a^2 + 4ab) + (ab + 2b^2)]$
   Group the terms.
   $= -1[2a(a + 2b) + b(a + 2b)]$
   Factor each group.
   $= -1(a + 2b)(2a + b)$
   Factor out $a + 2b$.

51. $35 = 5 \cdot 7$

52. $35 = (-5)(-7)$

230 Chapter 6 Factoring and Applications

53. Verify the given factorization by multiplying the two binomial factors by FOIL.

$$(3x-4)(2x-1) = 6x^2 - 3x - 8x + 4$$
$$= 6x^2 - 11x + 4$$

Thus, the product of $3x-4$ and $2x-1$ is indeed $6x^2 - 11x + 4$.

54. Verify the given factorization by multiplying the two binomial factors by FOIL.

$$(4-3x)(1-2x) = 4 - 8x - 3x + 6x^2$$
$$= 6x^2 - 11x + 4$$

Thus, the product of $4-3x$ and $1-2x$ is indeed $6x^2 - 11x + 4$.

55. The factors of Exercise 53 are the opposites of the factors of Exercise 54.

56. We can form another valid factorization by taking the opposites of the two given factors. The opposite of $7t-3$ is

$$-1(7t-3) = -7t + 3 = 3 - 7t,$$

and the opposite of $2t-5$ is $5-2t$, so we obtain the factorization $(3-7t)(5-2t)$.

6.5 Special Factoring Techniques

6.5 Margin Exercises

1. (a) $p^2 - 100 = p^2 - 10^2$
$$= (p+10)(p-10)$$

(b) $x^2 - \dfrac{25}{36} = x^2 - \left(\dfrac{5}{6}\right)^2$
$$= \left(x + \dfrac{5}{6}\right)\left(x - \dfrac{5}{6}\right)$$

(c) $x^2 + y^2$

This binomial is the sum of two squares and the terms have no common factor. Unlike the difference of two squares, it cannot be factored. It is a *prime* polynomial.

(d) $9m^2 - 49 = (3m)^2 - 7^2$
$$= (3m+7)(3m-7)$$

(e) $64a^2 - 25 = (8a)^2 - 5^2$
$$= (8a+5)(8a-5)$$

2. (a) $50r^2 - 32$
$$= 2(25r^2 - 16) \quad \text{Factor out 2.}$$
$$= 2[(5r)^2 - 4^2]$$
$$= 2(5r+4)(5r-4) \quad \text{Difference of squares}$$

(b) $27y^2 - 75$
$$= 3(9y^2 - 25) \quad \text{Factor out 3.}$$
$$= 3[(3y)^2 - 5^2]$$
$$= 3(3y+5)(3y-5) \quad \text{Difference of squares}$$

(c) $25a^2 - 64b^2 = (5a)^2 - (8b)^2$
$$= (5a+8b)(5a-8b)$$

(d) $k^4 - 49 = (k^2)^2 - 7^2$
$$= (k^2+7)(k^2-7)$$

(e) $81r^4 - 16 = (9r^2)^2 - 4^2$
$$= (9r^2+4)(9r^2-4)$$
$$= (9r^2+4)[(3r)^2 - 2^2]$$
$$= (9r^2+4)(3r+2)(3r-2)$$

3. (a) $p^2 + 14p + 49 = (p)^2 + 2 \cdot p \cdot 7 + (7)^2$
$$= (p+7)^2$$

(b) $m^2 + 8m + 16 = (m)^2 + 2 \cdot m \cdot 4 + (4)^2$
$$= (m+4)^2$$

(c) $x^2 + 2x + 1 = (x)^2 + 2 \cdot x \cdot 1 + (1)^2$
$$= (x+1)^2$$

4. (a) $p^2 - 18p + 81 \stackrel{?}{=} (p-q)^2$

$2 \cdot p \cdot 9 = 18p$, so this is a perfect square trinomial, and

$$p^2 - 18p + 81 = (p-9)^2.$$

(b) $16a^2 + 56a + 49 \stackrel{?}{=} (4a+7)^2$

$2 \cdot 4a \cdot 7 = 56a$, so this is a perfect square trinomial, and

$$16a^2 + 56a + 49 = (4a+7)^2.$$

(c) $121p^2 + 110p + 100 \stackrel{?}{=} (11p+10)^2$

$2 \cdot 11p \cdot 10 = 220p \neq 110p$, so this is not a perfect square trinomial. It is *prime*.

(d) $64x^2 - 48x + 9 \stackrel{?}{=} (8x-3)^2$

$2 \cdot 8x \cdot 3 = 48x$, so this is a perfect square trinomial, and

$$64x^2 - 48x + 9 = (8x-3)^2.$$

(e) $27y^3 + 72y^2 + 48y$

Factor out the greatest common factor, $3y$.

$$27y^3 + 72y^2 + 48y = 3y(9y^2 + 24y + 16)$$
$$\stackrel{?}{=} 3y(3y+4)^4$$

$2 \cdot 3y \cdot 4 = 24y$, so this is a perfect square trinomial, and

$$3y(9y^2 + 24y + 16) = 3y(3y+4)^2.$$

6.5 Section Exercises

1.
$1^2 = \underline{1}$ $2^2 = \underline{4}$ $3^2 = \underline{9}$
$4^2 = \underline{16}$ $5^2 = \underline{25}$ $6^2 = \underline{36}$
$7^2 = \underline{49}$ $8^2 = \underline{64}$ $9^2 = \underline{81}$
$10^2 = \underline{100}$ $11^2 = \underline{121}$ $12^2 = \underline{144}$
$13^2 = \underline{169}$ $14^2 = \underline{196}$ $15^2 = \underline{225}$
$16^2 = \underline{256}$ $17^2 = \underline{289}$ $18^2 = \underline{324}$
$19^2 = \underline{361}$ $20^2 = \underline{400}$

3. The following powers of x are all perfect squares: x^2, x^4, x^6, x^8, x^{10}. Based on this observation, we may make a conjecture (an educated guess) that if the power of a variable is divisible by $\underline{2}$ (with 0 remainder), then we have a perfect square.

5. $y^2 - 25 = y^2 - 5^2$
$= (y+5)(y-5)$

7. $p^2 - \dfrac{1}{9} = p^2 - \left(\dfrac{1}{3}\right)^2$
$= \left(p + \dfrac{1}{3}\right)\left(p - \dfrac{1}{3}\right)$

9. $m^2 - 12$

Because 12 is not the square of an integer, this binomial is not a difference of squares. It is a *prime* polynomial.

11. $9r^2 - 4 = (3r)^2 - 2^2$
$= (3r+2)(3r-2)$

13. $36m^2 - \dfrac{16}{25} = (6m)^2 - \left(\dfrac{4}{5}\right)^2$
$= \left(6m + \dfrac{4}{5}\right)\left(6m - \dfrac{4}{5}\right)$

15. $36x^2 - 16$

First factor out the GCF, 4; then use the rule for factoring the difference of squares.

$36x^2 - 16 = 4(9x^2 - 4)$
$= 4\left[(3x)^2 - 2^2\right]$
$= 4(3x+2)(3x-2)$

17. $196p^2 - 225 = (14p)^2 - 15^2$
$= (14p+15)(14p-15)$

19. $16r^2 - 25a^2 = (4r)^2 - (5a)^2$
$= (4r+5a)(4r-5a)$

21. $100x^2 + 49$

This binomial is the *sum* of squares and the terms have no common factor. Unlike the *difference* of squares, it cannot be factored. It is a *prime* polynomial.

23. $p^4 - 49 = \left(p^2\right)^2 - 7^2$
$= \left(p^2 + 7\right)\left(p^2 - 7\right)$

25. $x^4 - 1$

To factor this binomial completely, factor the difference of squares twice.

$x^4 - 1 = \left(x^2\right)^2 - 1^2$
$= \left(x^2 + 1\right)\left(x^2 - 1\right)$
$= \left(x^2 + 1\right)\left(x^2 - 1^2\right)$
$= \left(x^2 + 1\right)(x+1)(x-1)$

27. $p^4 - 256$

To factor this binomial completely, factor the difference of squares twice.

$p^4 - 256 = \left(p^2\right)^2 - 16^2$
$= \left(p^2 + 16\right)\left(p^2 - 16\right)$
$= \left(p^2 + 16\right)\left(p^2 - 4^2\right)$
$= \left(p^2 + 16\right)(p+4)(p-4)$

29. The student's answer is not a complete factorization because $x^2 - 9$ can be factored further. The correct complete factorization is

$x^4 - 81 = \left(x^2 + 9\right)(x+3)(x-3)$.

Because the teacher had directed the student to factor the polynomial *completely*, she was justified in her grading of the item.

31. Since $9y^2 = (3y)^2$ and $25 = 5^2$, if a trinomial with $9y^2$ as its first term and 25 as its last term were a perfect square trinomial, the middle term would be

$2(3y)(5) = 30y$
or $-2(3y)(5) = -30y$.

Because $9y^2 + 14y + 25$ does not have either of these middle terms, it is not a perfect square trinomial. In fact, it is a prime polynomial.

33. $w^2 + 2w + 1$

The first and last terms are perfect squares, w^2 and 1^2. This trinomial is a perfect square, since the middle term is twice the product of w and 1, or

$2 \cdot w \cdot 1 = 2w.$

Therefore,

$w^2 + 2w + 1 = (w+1)^2.$

35. $x^2 - 8x + 16$

The first and last terms are perfect squares, x^2 and $(-4)^2$. This trinomial is a perfect square, since the middle term is twice the product of x and -4, or
$$2 \cdot x \cdot (-4) = -8x.$$
Therefore,
$$x^2 - 8x + 16 = (x - 4)^2.$$

37. $t^2 + t + \dfrac{1}{4}$

t^2 is a perfect square, and $\frac{1}{4}$ is a perfect square since $\frac{1}{2} \cdot \frac{1}{2} = \frac{1}{4}$. The middle term is twice the product of t and $\frac{1}{2}$, or
$$t = 2(t)\left(\tfrac{1}{2}\right).$$
Therefore,
$$t^2 + t + \frac{1}{4} = \left(t + \frac{1}{2}\right)^2.$$

39. $x^2 - 1.0x + .25$

The first and last terms are perfect squares, x^2 and $(-.5)^2$. The trinomial is a perfect square, since the middle term is
$$2 \cdot x \cdot (-.5) = -1.0x.$$
Therefore,
$$x^2 - 1.0x + .25 = (x - .5)^2.$$

41. $2x^2 + 24x + 72$

First, factor out the GCF, 2.
$$2x^2 + 24x + 72 = 2(x^2 + 12x + 36)$$
Now factor $x^2 + 12x + 36$ as a perfect square trinomial.
$$x^2 + 12x + 36 = (x + 6)^2$$
The final factored form is
$$2x^2 + 24x + 72 = 2(x + 6)^2.$$

43. $16x^2 - 40x + 25$

The first and last terms are perfect squares, $(4x)^2$ and $(-5)^2$. The middle term is
$$2(4x)(-5) = -40x.$$
Therefore,
$$16x^2 - 40x + 25 = (4x - 5)^2.$$

45. $49x^2 - 28xy + 4y^2$

The first and last terms are perfect squares, $(7x)^2$ and $(-2y)^2$. The middle term is
$$2(7x)(-2y) = -28xy.$$
Therefore,
$$49x^2 - 28xy + 4y^2 = (7x - 2y)^2.$$

47. $64x^2 + 48xy + 9y^2$
$$= (8x)^2 + 2(8x)(3y) + (3y)^2$$
$$= (8x + 3y)^2$$

49. $50h^3 - 40h^2y + 8hy^2$
$$= 2h(25h^2 - 20hy + 4y^2)$$
$$= 2h\left[(5h)^2 - 2(5h)(2y) + (2y)^2\right]$$
$$= 2h(5h - 2y)^2$$

51. Factor by trial and error to obtain
$$10x^2 + 11x - 6 = (5x - 2)(2x + 3).$$

52.
$$\begin{array}{r} 5x - 2 \\ 2x+3 \overline{\smash{\big)}\, 10x^2 + 11x - 6} \\ \underline{10x^2 + 15x} \\ -4x - 6 \\ \underline{-4x - 6} \\ 0 \end{array}$$

Thus,
$$\frac{10x^2 + 11x - 6}{2x + 3} = 5x - 2.$$

53. Yes. If $10x^2 + 11x - 6$ factors as $(5x - 2)(2x + 3)$, then when $10x^2 + 11x - 6$ is divided by $2x + 3$, the quotient should be $5x - 2$.

54.
$$\begin{array}{r} x^2 + x + 1 \\ x-1 \overline{\smash{\big)}\, x^3 + 0x^2 + 0x - 1} \\ \underline{x^3 - x^2} \\ x^2 + 0x \\ \underline{x^2 - x} \\ x - 1 \\ \underline{x - 1} \\ 0 \end{array}$$

The quotient is $x^2 + x + 1$, so
$$x^3 - 1 = (x - 1)(x^2 + x + 1).$$

Summary Exercises on Factoring

1. $32m^9 + 16m^5 + 24m^3$
$$= 8m^3(4m^6 + 2m^2 + 3)$$

3. $14k^3 + 7k^2 - 70k$
$$= 7k(2k^2 + k - 10)$$
$$= 7k(2k + 5)(k - 2)$$

Summary Exercises on Factoring 233

5. $6z^2 + 31z + 5 = 6z^2 + 30z + z + 5$
$= (6z^2 + 30z) + (z + 5)$
$= 6z(z + 5) + 1(z + 5)$
$= (z + 5)(6z + 1)$

7. $49z^2 - 16y^2 = (7z)^2 - (4y)^2$
$= (7z + 4y)(7z - 4y)$

9. $16x^2 + 20x = 4x(4x) + 4x(5)$
$= 4x(4x + 5)$

11. $10y^2 - 7yz - 6z^2$
$= 10y^2 - 12yz + 5yz - 6z^2$
$= 2y(5y - 6z) + z(5y - 6z)$ *Factor each group.*
$= (5y - 6z)(2y + z)$ *Factor out $5y - 6z$.*

13. $m^2 + 2m - 15$

    Look for a pair of numbers whose product is -15 and whose sum is 2. The numbers are -3 and 5, so
    $$m^2 + 2m - 15 = (m - 3)(m + 5).$$

15. $32z^3 + 56z^2 - 16z = 8z(4z^2 + 7z - 2)$
$= 8z(4z^2 + 8z - z - 2)$
$= 8z[4z(z + 2) - 1(z + 2)]$
$= 8z(z + 2)(4z - 1)$

17. $z^2 - 12z + 36$
$= z^2 - 2 \cdot 6z + 6^2$
$= (z - 6)^2$

19. $y^2 - 4yk - 12k^2 = (y - 6k)(y + 2k)$

21. $6y^2 - 6y - 12 = 6(y^2 - y - 2)$
$= 6(y - 2)(y + 1)$

23. $p^2 - 17p + 66 = (p - 11)(p - 6)$

25. $k^2 + 9$ cannot be factored because it is the sum of squares with no GCF. The expression is *prime*.

27. $z^2 - 3za - 10a^2 = (z - 5a)(z + 2a)$

29. $4k^2 - 12k + 9$
$= (2k)^2 - 2 \cdot 2k \cdot 3 + 3^2$
$= (2k - 3)^2$ *Perfect square trinomial*

31. $16r^2 + 24rm + 9m^2$
$= (4r)^2 + 2 \cdot 4r \cdot 3m + (3m)^2$
$= (4r + 3m)^2$ *Perfect square trinomial*

33. $n^2 - 12n - 35$ is *prime*.

35. $16k^2 - 48k + 36$
$= 4(4k^2 - 12k + 9)$
$= 4[(2k)^2 - 2(2k)(3) + 3^2]$
$= 4(2k - 3)^2$

37. $36y^6 - 42y^5 - 120y^4$
$= 6y^4(6y^2 - 7y - 20)$
$= 6y^4(6y^2 - 15y + 8y - 20)$
$= 6y^4[3y(2y - 5) + 4(2y - 5)]$
$= 6y^4(2y - 5)(3y + 4)$

39. $8p^2 + 23p - 3$
$= 8p^2 + 24p - p - 3$ $24 \cdot (-1) = -24;$
$= 8p(p + 3) - 1(p + 3)$ $24 + (-1) = 23$
$= (p + 3)(8p - 1)$

41. $54m^2 - 24z^2$
$= 6(9m^2 - 4z^2)$
$= 6[(3m)^2 - (2z)^2]$
$= 6(3m + 2z)(3m - 2z)$

43. $6a^2 + 10a - 4$
$= 2(3a^2 + 5a - 2)$
$= 2(3a - 1)(a + 2)$

45. $m^2 - 81 = m^2 - 9^2$
$= (m + 9)(m - 9)$

47. $125m^4 - 400m^3n + 195m^2n^2$
$= 5m^2(25m^2 - 80mn + 39n^2)$
$= 5m^2(25m^2 - 15mn - 65mn + 39n^2)$
$= 5m^2[5m(5m - 3n) - 13n(5m - 3n)]$
$= 5m^2(5m - 3n)(5m - 13n)$

49. $m^2 - 4m + 4$
$= m^2 - 2 \cdot 2m + 2^2$
$= (m - 2)^2$ *Perfect square trinomial*

51. $27p^{10} - 45p^9 - 252p^8$
$= 9p^8(3p^2 - 5p - 28)$
$= 9p^8(3p^2 - 12p + 7p - 28)$
$= 9p^8[3p(p - 4) + 7(p - 4)]$
$= 9p^8(p - 4)(3p + 7)$

53. $4 - 2q - 6p + 3pq$
$= 2(2 - q) - 3p(2 - q)$
$= (2 - q)(2 - 3p)$

55. $64p^2 - 100m^2$
$= 4(16p^2 - 25m^2)$
$= 4[(4p)^2 - (5m)^2]$
$= 4(4p + 5m)(4p - 5m)$

57. $100a^2 - 81y^2$
$= (10a)^2 - (9y)^2$
$= (10a + 9y)(10a - 9y)$

59. $a^2 + 8a + 16 = (a + 4)^2$
Perfect square trinomial

6.6 Solving Quadratic Equations by Factoring

6.6 Margin Exercises

1. **A.** $y^2 - 4y - 5 = 0$ is in the form $ax^2 + bx + c = 0$.

B. $x^3 - x^2 + 16 = 0$ is not a quadratic equation because of the x^3-term.

C. $2z^2 + 7z = -3$, or $2z^2 + 7z + 3 = 0$, is in the form $ax^2 + bx + c = 0$.

D. $x + 2y = -4$ is not a quadratic equation because there is no x^2-term.

A and **C** are quadratic equations.

2. **(a)** $x^2 - 3x = 4$

To write this equation in standard form, subtract 4 from both sides of the equation.

$$x^2 - 3x - 4 = 0$$

(b) $y^2 = 9y - 8$

To write this equation in standard form, subtract $9y$ from both sides and add 8 to both sides of the equation.

$$y^2 - 9y + 8 = 0$$

3. **(a)** $(x - 5)(x + 2) = 0$

By the zero-factor property, either $x - 5 = 0$ or $x + 2 = 0$. The equation $x - 5 = 0$ gives the solution $x = 5$; $x + 2 = 0$ gives $x = -2$. The solutions are 5 and -2. Check both solutions by substituting first 5 and then -2 for x in the original equation.

(b) $(3x - 2)(x + 6) = 0$

By the zero-factor property, either $3x - 2 = 0$ or $x + 6 = 0$.
Solve each equation.

$3x - 2 = 0$ or $x + 6 = 0$
$3x = 2$
$x = \frac{2}{3}$ or $x = -6$

The solutions are $\frac{2}{3}$ and -6. Check by substituting each solution for x in the original equation.

(c) $z(2z + 5) = 0$

By the zero-factor property, either $z = 0$ or $2z + 5 = 0$. Since the first equation is solved, we only need to solve the second.

$2z + 5 = 0$
$2z = -5$
$z = -\frac{5}{2}$

The solutions are 0 and $-\frac{5}{2}$. Check by substituting each solution for z in the original equation.

4. **(a)** $m^2 - 3m - 10 = 0$

First factor the equation to get

$$(m-5)(m+2) = 0.$$

Then set each factor equal to zero and solve.

$m - 5 = 0$ or $m + 2 = 0$
$m = 5$ or $m = -2$

The solutions are -2 and 5.

(b) $r^2 + 2r = 8$

Bring all nonzero terms to the same side of the equals sign by subtracting 8 from each side.

$r^2 + 2r - 8 = 0$
$(r - 2)(r + 4) = 0$
$r - 2 = 0$ or $r + 4 = 0$
$r = 2$ or $r = -4$

The solutions are -4 and 2.

5. **(a)** $10a^2 - 5a - 15 = 0$
$5(2a^2 - a - 3) = 0$ *Factor out 5.*
$2a^2 - a - 3 = 0$ *Divide by 5.*
$(2a - 3)(a + 1) = 0$
$2a - 3 = 0$ or $a + 1 = 0$
$2a = 3$
$a = \frac{3}{2}$ or $a = -1$

The solutions are -1 and $\frac{3}{2}$. Check them by substituting in the original equation.

(b) $4x^2 - 2x = 42$
$4x^2 - 2x - 42 = 0$
$2(2x^2 - x - 21) = 0$
$2x^2 - x - 21 = 0$
$(2x-7)(x + 3) = 0$
$2x - 7 = 0$ or $x + 3 = 0$
$2x = 7$
$x = \frac{7}{2}$ or $x = -3$

The solutions are -3 and $\frac{7}{2}$. Check them by substituting in the original equation.

6. (a)
$$49m^2 - 9 = 0$$
$$(7m+3)(7m-3) = 0$$
$$7m+3 = 0 \quad \text{or} \quad 7m-3 = 0$$
$$7m = -3 \qquad\qquad 7m = 3$$
$$m = -\tfrac{3}{7} \quad \text{or} \quad m = \tfrac{3}{7}$$

The solutions are $-\tfrac{3}{7}$ and $\tfrac{3}{7}$.

(b) $p(4p+7) = 2$

Get all terms on one side of the equals sign, with 0 on the other side.
$$4p^2 + 7p = 2$$
$$4p^2 + 7p - 2 = 0$$
$$(4p-1)(p+2) = 0$$
$$4p - 1 = 0 \quad \text{or} \quad p+2 = 0$$
$$4p = 1$$
$$p = \tfrac{1}{4} \quad \text{or} \quad p = -2$$

The solutions are -2 and $\tfrac{1}{4}$.

(c) $m^2 = 3m$

Get all terms on one side of the equals sign, with 0 on the other side.
$$m^2 - 3m = 0$$

Factor the binomial.
$$m(m-3) = 0$$

Set each factor equal to 0 and solve the resulting equations.
$$m = 0 \quad \text{or} \quad m - 3 = 0$$
$$m = 0 \quad \text{or} \quad m = 3$$

The solutions are 0 and 3.

7. (a)
$$r^3 - 16r = 0$$
$$r(r^2 - 16) = 0$$
$$r(r-4)(r+4) = 0$$
$$r = 0 \quad \text{or} \quad r-4 = 0 \quad \text{or} \quad r+4 = 0$$
$$r = 0 \quad \text{or} \quad r = 4 \quad \text{or} \quad r = -4$$

The solutions are -4, 0, and 4.
Check each solution.

(b)
$$x^3 - 3x^2 - 18x = 0$$
$$x(x^2 - 3x - 18) = 0$$
$$x(x-6)(x+3) = 0$$
$$x = 0 \quad \text{or} \quad x-6 = 0 \quad \text{or} \quad x+3 = 0$$
$$x = 0 \quad \text{or} \quad x = 6 \quad \text{or} \quad x = -3$$

The solutions are -3, 0, and 6.
Check each solution.

8. (a)
$$(m+3)(m^2 - 11m + 10) = 0$$
$$(m+3)(m-10)(m-1) = 0$$
$$m+3 = 0 \quad \text{or} \quad m-10 = 0 \quad \text{or} \quad m-1 = 0$$
$$m = -3 \quad \text{or} \quad m = 10 \quad \text{or} \quad m = 1$$

The solutions are -3, 1, and 10.
Check each solution.

(b)
$$(2x+5)(4x^2 - 9) = 0$$
$$(2x+5)(2x+3)(2x-3) = 0$$
$$2x+5 = 0 \quad \text{or} \quad 2x+3 = 0 \quad \text{or} \quad 2x-3 = 0$$
$$2x = -5 \quad \text{or} \quad 2x = -3 \quad \text{or} \quad 2x = 3$$
$$x = -\tfrac{5}{2} \quad \text{or} \quad x = -\tfrac{3}{2} \quad \text{or} \quad x = \tfrac{3}{2}$$

The solutions are $-\tfrac{5}{2}$, $-\tfrac{3}{2}$, and $\tfrac{3}{2}$.
Check each solution.

6.6 Section Exercises

For all equations in this section, answers should be checked by substituting into the original equation. These checks will be shown here for only a few of the exercises.

1. $(x+5)(x-2) = 0$

By the zero-factor property, the only way that the product of these two factors can be zero is if at least one of the factors is zero.
$$x+5 = 0 \quad \text{or} \quad x-2 = 0$$

Solve each of these linear equations.
$$x = -5 \quad \text{or} \quad x = 2$$

The solutions are -5 and 2.

3. $(2m-7)(m-3) = 0$

Set each factor equal to zero and solve the resulting linear equations.
$$2m - 7 = 0 \quad \text{or} \quad m - 3 = 0$$
$$2m = 7$$
$$m = \tfrac{7}{2} \quad \text{or} \quad m = 3$$

The solutions are 3 and $\tfrac{7}{2}$.

5. $t(6t+5) = 0$

Set each factor equal to zero and solve the resulting linear equations.
$$t = 0 \quad \text{or} \quad 6t+5 = 0$$
$$6t = -5$$
$$t = -\tfrac{5}{6}$$

The solutions are $-\tfrac{5}{6}$ and 0.

236 Chapter 6 Factoring and Applications

7. $2x(3x - 4) = 0$

Set each factor equal to zero and solve the resulting linear equations.

$$2x = 0 \quad \text{or} \quad 3x - 4 = 0$$
$$x = 0 \quad \text{or} \quad 3x = 4$$
$$x = \tfrac{4}{3}$$

The solutions are 0 and $\tfrac{4}{3}$.

9. $\left(x + \tfrac{1}{2}\right)\left(2x - \tfrac{1}{3}\right) = 0$

Set each factor equal to zero and solve the resulting linear equations.

$$x + \tfrac{1}{2} = 0 \quad \text{or} \quad 2x - \tfrac{1}{3} = 0$$
$$x = -\tfrac{1}{2} \quad \text{or} \quad 2x = \tfrac{1}{3}$$
$$x = \tfrac{1}{6}$$

The solutions are $-\tfrac{1}{2}$ and $\tfrac{1}{6}$.

11. $(.5z - 1)(2.5z + 2) = 0$

Set each factor equal to zero and solve the resulting linear equations.

$$.5z - 1 = 0 \quad \text{or} \quad 2.5z + 2 = 0$$
$$.5z = 1 \quad \text{or} \quad 2.5z = -2$$
$$z = \tfrac{1}{.5} \quad \text{or} \quad z = -\tfrac{2}{2.5}$$
$$z = 2 \quad \text{or} \quad z = -.8$$

The solutions are $-.8$ and 2.

13. 9 is called a *double solution* for $(x - 9)^2 = 0$ because it occurs twice when the equation is solved.

$$(x - 9)(x - 9) = 0$$
$$x - 9 = 0 \quad \text{or} \quad x - 9 = 0$$
$$x = 9 \quad \text{or} \quad x = 9$$

The solution is 9.

15. We can consider the factored form as $2 \cdot x(3x - 4)$, the product of three factors, 2, x, and $3x - 4$. Applying the zero-factor property yields three equations,

$$2 = 0 \quad \text{or} \quad x = 0 \quad \text{or} \quad 3x - 4 = 0.$$

Since "$2 = 0$" is impossible, it has no solution, so we end up with the two solutions, $x = 0$ and $x = \tfrac{4}{3}$.

We conclude that multiplying a polynomial by a constant does not affect the solutions of the corresponding equation.

17. $y^2 + 3y + 2 = 0$

Factor the polynomial.

$$(y + 2)(y + 1) = 0$$

Set each factor equal to 0.

$$y + 2 = 0 \quad \text{or} \quad y + 1 = 0$$

Solve each equation.

$$y = -2 \quad \text{or} \quad y = -1$$

Check these solutions by substituting -2 for y and then -1 for y in the original equation.

$$y^2 + 3y + 2 = 0$$
$$(-2)^2 + 3(-2) + 2 = 0 \text{ ? } \textit{Let } y = -2.$$
$$4 - 6 + 2 = 0 \text{ ?}$$
$$-2 + 2 = 0 \quad \textit{True}$$
$$y^2 + 3y + 2 = 0$$
$$(-1)^2 + 3(-1) + 2 = 0 \text{ ? } \textit{Let } y = -1.$$
$$1 - 3 + 2 = 0 \text{ ?}$$
$$-2 + 2 = 0 \quad \textit{True}$$

The solutions are -2 and -1.

19. $y^2 - 3y + 2 = 0$

Factor the polynomial.

$$(y - 1)(y - 2) = 0$$

Set each factor equal to 0.

$$y - 1 = 0 \quad \text{or} \quad y - 2 = 0$$

Solve each equation.

$$y = 1 \quad \text{or} \quad y = 2$$

The solutions are 1 and 2.

21. $x^2 = 24 - 5x$

Write the equation in standard form.

$$x^2 + 5x - 24 = 0$$

Factor the polynomial.

$$(x + 8)(x - 3) = 0$$

Set each factor equal to 0.

$$x + 8 = 0 \quad \text{or} \quad x - 3 = 0$$

Solve each equation.

$$x = -8 \quad \text{or} \quad x = 3$$

The solutions are -8 and 3.

23. $x^2 = 3 + 2x$

   Write the equation in standard form.
   $$x^2 - 2x - 3 = 0$$
   $$(x+1)(x-3) = 0$$
   $x + 1 = 0$ or $x - 3 = 0$
   $x = -1$ or $x = 3$

   The solutions are -1 and 3.

25. $z^2 + 3z = -2$

   Write the equation in standard form.
   $$z^2 + 3z + 2 = 0$$
   Factor the polynomial.
   $$(z+2)(z+1) = 0$$
   Set each factor equal to 0.
   $z + 2 = 0$ or $z + 1 = 0$
   $z = -2$ or $z = -1$

   The solutions are -2 and -1.

27. $m^2 + 8m + 16 = 0$

   Factor $m^2 + 8m + 16$ as a perfect square trinomial.
   $$(m+4)^2 = 0$$
   Set the factor $m + 4$ equal to 0 and solve.
   $$m + 4 = 0$$
   $$m = -4$$

   The solution is -4.

29. $3x^2 + 5x - 2 = 0$

   Factor the polynomial.
   $$(3x - 1)(x + 2) = 0$$
   Set each factor equal to 0.
   $3x - 1 = 0$ or $x + 2 = 0$
   $3x = 1$
   $x = \frac{1}{3}$ or $x = -2$

   The solutions are -2 and $\frac{1}{3}$.

31. $6p^2 = 4 - 5p$
   $$6p^2 + 5p - 4 = 0$$
   $$(3p + 4)(2p - 1) = 0$$
   $3p + 4 = 0$ or $2p - 1 = 0$
   $3p = -4$ $2p = 1$
   $p = -\frac{4}{3}$ or $p = \frac{1}{2}$

   The solutions are $-\frac{4}{3}$ and $\frac{1}{2}$.

33. $9s^2 + 12s = -4$
   $$9s^2 + 12s + 4 = 0$$
   $$(3s + 2)^2 = 0$$
   Set the factor $3s + 2$ equal to 0 and solve.
   $$3s + 2 = 0$$
   $$3s = -2$$
   $$s = -\frac{2}{3}$$

   The solution is $-\frac{2}{3}$.

35. $y^2 - 9 = 0$
   $$(y + 3)(y - 3) = 0$$
   $y + 3 = 0$ or $y - 3 = 0$
   $y = -3$ or $y = 3$

   The solutions are -3 and 3.

37. $16k^2 - 49 = 0$
   $$(4k + 7)(4k - 7) = 0$$
   $4k + 7 = 0$ or $4k - 7 = 0$
   $4k = -7$ $4k = 7$
   $k = -\frac{7}{4}$ or $k = \frac{7}{4}$

   The solutions are $-\frac{7}{4}$ and $\frac{7}{4}$.

39. $n^2 = 121$
   $$n^2 - 121 = 0$$
   $$(n + 11)(n - 11) = 0$$
   $n + 11 = 0$ or $n - 11 = 0$
   $n = -11$ or $n = 11$

   The solutions are -11 and 11.

41. $x^2 = 7x$
   $$x^2 - 7x = 0$$
   $$x(x - 7) = 0$$
   $x = 0$ or $x - 7 = 0$
   $x = 0$ or $x = 7$

   The solutions are 0 and 7.

43. $6r^2 = 3r$
   $$6r^2 - 3r = 0$$
   $$3r(2r - 1) = 0$$
   $3r = 0$ or $2r - 1 = 0$
                $2r = 1$
   $r = 0$ or $r = \frac{1}{2}$

   The solutions are 0 and $\frac{1}{2}$.

45.
$$g(g-7) = -10$$
$$g^2 - 7g = -10$$
$$g^2 - 7g + 10 = 0$$
$$(g-2)(g-5) = 0$$
$g - 2 = 0$ or $g - 5 = 0$
$g = 2$ or $g = 5$

The solutions are 2 and 5.

47.
$$z(2z+7) = 4$$
$$2z^2 + 7z = 4$$
$$2z^2 + 7z - 4 = 0$$
$$(2z-1)(z+4) = 0$$
$2z - 1 = 0$ or $z + 4 = 0$
$2z = 1$
$z = \frac{1}{2}$ or $z = -4$

The solutions are -4 and $\frac{1}{2}$.

49.
$$2(y^2 - 66) = -13y$$
$$2y^2 - 132 = -13y$$
$$2y^2 + 13y - 132 = 0$$
$$(2y - 11)(y + 12) = 0$$
$2y - 11 = 0$ or $y + 12 = 0$
$2y = 11$
$y = \frac{11}{2}$ or $y = -12$

The solutions are -12 and $\frac{11}{2}$.

51.
$$5x^3 - 20x = 0$$
$$5x(x^2 - 4) = 0$$
$$5x(x+2)(x-2) = 0$$
$5x = 0$ or $x + 2 = 0$ or $x - 2 = 0$
$x = 0$ or $x = -2$ or $x = 2$

The solutions are 0, -2, and 2.

53. $9y^3 - 49y = 0$

To factor the polynomial, begin by factoring out the greatest common factor.

$$y(9y^2 - 49) = 0$$

Now factor $9y^2 - 49$ as the difference of two squares.

$$y(3y+7)(3y-7) = 0$$

Set each of the three factors equal to 0 and solve.

$y = 0$ or $3y + 7 = 0$ or $3y - 7 = 0$
$3y = -7$ $3y = 7$
$y = 0$ or $y = -\frac{7}{3}$ or $y = \frac{7}{3}$

The solutions are $-\frac{7}{3}$, 0, and $\frac{7}{3}$.

55. $(2r+5)(3r^2 - 16r + 5) = 0$

Begin by factoring $3r^2 - 16r + 5$.

$$(2r+5)(3r-1)(r-5) = 0$$

Set each of the three factors equal to 0 and solve the resulting equations.

$2r + 5 = 0$ or $3r - 1 = 0$ or $r - 5 = 0$
$2r = -5$ $3r = 1$
$r = -\frac{5}{2}$ or $r = \frac{1}{3}$ or $r = 5$

The solutions are $-\frac{5}{2}$, $\frac{1}{3}$, and 5.

57. $(2x+7)(x^2 + 2x - 3) = 0$
$$(2x+7)(x+3)(x-1) = 0$$
$2x + 7 = 0$ or $x + 3 = 0$ or $x - 1 = 0$
$2x = -7$
$x = -\frac{7}{2}$ or $x = -3$ or $x = 1$

The solutions are $-\frac{7}{2}$, -3, and 1.

59. (a) $d = 16t^2$

$t = 2: d = 16(2)^2 = 16(4) = 64$
$t = 3: d = 16(3)^2 = 16(9) = 144$
$d = 256: 256 = 16t^2; 16 = t^2; t = 4$
$d = 576: 576 = 16t^2; 36 = t^2; t = 6$

t in seconds	0	1	2	3	4	6
d in feet	0	16	64	144	256	576

(b) When $t = 0$, $d = 0$, no time has elapsed, so the object hasn't fallen (been released) yet.

(c) Time cannot be negative.

6.7 Applications of Quadratic Equations

6.7 Margin Exercises

1. (a) Let $x = $ the width of the room.
Then $x + 2 = $ the length.

Use the formula for the area of a rectangle, $A = LW$.

Substitute 48 for A, $x + 2$ for L, and x for W.

$A = LW$
$48 = (x+2)x$
$48 = x^2 + 2x$
$0 = x^2 + 2x - 48$ *Subtract 48.*
$0 = (x-6)(x+8)$ *Factor.*

$x - 6 = 0$ or $x + 8 = 0$
$x = 6$ or $x = -8$

Discard the solution -8 since the width cannot be negative. The width is 6 meters and the length is $x + 2 = 6 + 2 = 8$ meters.

6.7 Applications of Quadratic Equations   239

(b) Let $x =$ the length of a side of the original square. Then $x + 4 =$ the length of a side of the larger square.

Use the formula for the area of a square, $A = s^2$. The area of the original square is x^2, and the area of the larger square is $(x + 4)^2$. The sum of the areas of the two squares is 106 square inches, so

$$x^2 + (x + 4)^2 = 106.$$

Solve this equation.

$$x^2 + x^2 + 8x + 16 = 106$$
$$2x^2 + 8x - 90 = 0$$
$$x^2 + 4x - 45 = 0 \quad \text{Divide by 2.}$$
$$(x - 5)(x + 9) = 0$$
$$x - 5 = 0 \quad \text{or} \quad x + 9 = 0$$
$$x = 5 \quad \text{or} \quad x = -9$$

Discard the solution -9 since the length of a side cannot be negative. The length of a side of the original square is 5 inches.

2. Let $x =$ the first locker number.
Then $x + 1 =$ the second locker number.
The product is 132.

$$x(x + 1) = 132$$
$$x^2 + x = 132$$
$$x^2 + x - 132 = 0$$
$$(x - 11)(x + 12) = 0$$
$$x - 11 = 0 \quad \text{or} \quad x + 12 = 0$$
$$x = 11 \quad \text{or} \quad x = -12$$

A locker number is positive, so discard -12. Therefore, the locker numbers are 11 and $x + 1 = 11 + 1 = 12$.

3. **(a)** Let $x =$ the smaller integer.
Then $x + 2 =$ the next larger even integer.

The product is 4 more than two times their sum.

$$x(x + 2) = 4 + 2 \cdot [x + (x + 2)]$$

Solve this equation.

$$x^2 + 2x = 4 + 2(2x + 2)$$
$$x^2 + 2x = 4 + 4x + 4$$
$$x^2 - 2x - 8 = 0$$
$$(x - 4)(x + 2) = 0$$
$$x - 4 = 0 \quad \text{or} \quad x + 2 = 0$$
$$x = 4 \quad \text{or} \quad x = -2$$

We need to find two consecutive even integers. If $x = 4$ is the smaller, then the larger is

$$x + 2 = 4 + 2 = 6.$$

If $x = -2$ is the smaller, then the larger is

$$x + 2 = -2 + 2 = 0$$

The two integers are 4 and 6, or they are -2 and 0.

(b) Let $x =$ the smallest integer.
Then $x + 2 =$ the middle integer
and $x + 4 =$ the largest integer.

From the given information, we write the equation

$$x(x + 4) = (x + 2) + 16.$$

Solve this equation.

$$x^2 + 4x = x + 18$$
$$x^2 + 3x - 18 = 0$$
$$(x + 6)(x - 3) = 0$$
$$x + 6 = 0 \quad \text{or} \quad x - 3 = 0$$
$$x = -6 \quad \text{or} \quad x = 3$$

Reject -6 since it is not an odd integer.
If $x = 3$, then $x + 2 = 5$ and $x + 4 = 7$. The integers are 3, 5, and 7.

4. Let $x =$ the length of the longer leg of the right triangle.
Then $x + 3 =$ the length of the hypotenuse and $x - 3 =$ the length of the shorter leg.

Use the Pythagorean formula, substituting $x - 3$ for a, x for b, and $x + 3$ for c.

$$a^2 + b^2 = c^2$$
$$(x - 3)^2 + x^2 = (x + 3)^2$$
$$x^2 - 6x + 9 + x^2 = x^2 + 6x + 9$$
$$2x^2 - 6x + 9 = x^2 + 6x + 9$$
$$x^2 - 12x = 0$$
$$x(x - 12) = 0$$
$$x = 0 \quad \text{or} \quad x - 12 = 0$$
$$x = 0 \quad \text{or} \quad x = 12$$

Discard 0 since the length of a side of a triangle cannot be zero. The length of the longer leg is 12 inches, the length of the hypotenuse is $x + 3 = 12 + 3 = 15$ inches, and the length of the shorter leg is $x - 3 = 12 - 3 = 9$ inches.

5. $I = -x^2 + 2x + 60$

To find the time when 45 impulses occur, let $I = 45$ and solve the resulting equation. (For convenience, we reverse the sides of the equation.)

$$-x^2 + 2x + 60 = 45$$
$$-x^2 + 2x + 15 = 0 \quad \text{Subtract 45.}$$
$$-1(x^2 - 2x - 15) = 0 \quad \text{Factor out } -1.$$
$$-1(x + 3)(x - 5) = 0 \quad \text{Factor.}$$
$$x + 3 = 0 \quad \text{or} \quad x - 5 = 0$$
$$x = -3 \quad \text{or} \quad x = 5$$

There will be 45 impulses after 5 milliseconds. There are two solutions to the equation, -3 and 5. Only one answer makes sense here; a negative answer is not appropriate.

6. In 1995, $x = 1995 - 1990 = 5$.

$$y = .37x^2 - 4.1x + 12$$
$$y = .37(5)^2 - 4.1(5) + 12 \quad \text{Let } x = 5.$$
$$= .75$$

To the nearest tenth, pharmacies paid about .8% more for drugs in 1995. The actual data for 1995 is .8%, so our answer using the model is the same.

6.7 Section Exercises

1. Read; variable; equation; Solve; answer; Check; original

3. $A = bh$; $A = 45$, $b = 2x + 1$, $h = x + 1$

Step 3
$$A = bh$$
$$45 = (2x + 1)(x + 1)$$

Step 4
$$45 = 2x^2 + 3x + 1$$
$$0 = 2x^2 + 3x - 44$$
$$0 = (2x + 11)(x - 4)$$
$$2x + 11 = 0 \quad \text{or} \quad x - 4 = 0$$
$$2x = -11$$
$$x = -\tfrac{11}{2} \quad \text{or} \quad x = 4$$

Step 5
Substitute these values for x in the expressions $2x + 1$ and $x + 1$ to find the values of b and h.

$$b = 2x + 1 = 2\left(-\tfrac{11}{2}\right) + 1$$
$$= -11 + 1 = -10$$
or $b = 2x + 1 = 2(4) + 1$
$$= 8 + 1 = 9$$

We must discard the first solution because the base of a parallelogram cannot have a negative length. Since $x = -\tfrac{11}{2}$ will not give a realistic answer for the base, we only need to substitute 4 for x to compute the height.

$$h = x + 1 = 4 + 1 = 5$$

The base is 9 units and the height is 5 units.

Step 6
$bh = 9 \cdot 5 = 45$, the desired value of A.

5. $A = LW$; $A = 80$, $L = x + 8$, $W = x - 8$

Step 3
$$A = LW$$
$$80 = (x + 8)(x - 8)$$

Step 4
$$80 = x^2 - 64$$
$$0 = x^2 - 144$$
$$0 = (x + 12)(x - 12)$$
$$x + 12 = 0 \quad \text{or} \quad x - 12 = 0$$
$$x = -12 \quad \text{or} \quad x = 12$$

Step 5
The solution cannot be $x = -12$ since, when substituted, $-12 + 8$ and $-12 - 8$ are negative numbers and length and width cannot be negative. Thus, $x = 12$ and

$$L = x + 8 = 12 + 8 = 20$$
$$W = x - 8 = 12 - 8 = 4.$$

The length is 20 units, and the width is 4 units.

Step 6
$LW = 20 \cdot 4 = 80$, the desired value of A.

7. Let $x =$ the width of the shell.
Then $x + 3 =$ the length of the shell.

Substitute 28 for the area, x for the width, and $x + 3$ for the length in the formula for the area of a rectangle.

$$A = LW$$
$$28 = (x + 3)x$$
$$28 = x^2 + 3x$$
$$0 = x^2 + 3x - 28$$
$$0 = (x + 7)(x - 4)$$
$$x + 7 = 0 \quad \text{or} \quad x - 4 = 0$$
$$x = -7 \quad \text{or} \quad x = 4$$

The width of a rectangle cannot be negative, so we reject -7. The width of the shell is 4 inches, and the length is $4 + 3 = 7$ inches.

9. Let $x =$ the width of the original screen.
   Then $x + 3 =$ the length of the original screen
   and $x + 4 =$ the length of the new screen.

   Length × width   length × width
   (area) of new   equals   (area) of
   screen   original screen
   ↓   ↓   ↓
   $x(x + 4)$ $=$ $x(x + 3)$

   increased
   by   10.
   ↓   ↓
   $+$ 10

   Simplify this equation and solve it.
   $$x^2 + 4x = x^2 + 3x + 10$$
   $$x = 10$$

   The width of the original screen is 10 inches, and the length is $x + 3 = 10 + 3 = 13$ inches.

11. Let $x =$ the width of the aquarium.
    Then $x + 3 =$ the height of the aquarium.

    Use the formula for the volume of a rectangular box.
    $$V = LWH$$
    $$2730 = 21x(x + 3)$$
    $$130 = x(x + 3) \quad \text{Divide by 21.}$$
    $$130 = x^2 + 3x$$
    $$0 = x^2 + 3x - 130$$
    $$0 = (x + 13)(x - 10)$$
    $$x + 13 = 0 \quad \text{or} \quad x - 10 = 0$$
    $$x = -13 \quad \text{or} \quad x = 10$$

    We discard -13 because the width cannot be negative. The width is 10 inches. The height is $10 + 3 = 13$ inches.

13. Let $x =$ the length of a side of the square painting.
    Then $x - 2 =$ the length of a side of the square mirror.

    Since the formula for the area of a square is $A = s^2$, the area of the painting is x^2, and the area of the mirror is $(x - 2)^2$. The difference between their areas is 32, so
    $$x^2 - (x - 2)^2 = 32$$
    $$x^2 - (x^2 - 4x + 4) = 32$$
    $$x^2 - x^2 + 4x - 4 = 32$$
    $$4x - 4 = 32$$
    $$4x = 36$$
    $$x = 9.$$

The length of a side of the painting is 9 feet.
The length of a side of the mirror is $9 - 2 = 7$ feet.

Check: $9^2 - 7^2 = 81 - 49 = 32$

15. Let $x =$ the first volume number.
    Then $x + 1 =$ the second volume number.
    The product of the numbers is 420.
    $$x(x + 1) = 420$$
    $$x^2 + x - 420 = 0$$
    $$(x - 20)(x + 21) = 0$$
    $$x - 20 = 0 \quad \text{or} \quad x + 21 = 0$$
    $$x = 20 \quad \text{or} \quad x = -21$$

    The volume number cannot be negative, so we reject -21. The volume numbers are 20 and $x + 1 = 20 + 1 = 21$.

17. Let $n =$ the first integer.
    Then $n + 1 =$ the next integer.

    The product is 11 more than their sum.
    ↓   ↓   ↓
    $n(n + 1) = 11 + [n + (n + 1)]$
    $$n^2 + n = 11 + n + n + 1$$
    $$n^2 + n = 2n + 12$$
    $$n^2 - n - 12 = 0$$
    $$(n - 4)(n + 3) = 0$$
    $$n - 4 = 0 \quad \text{or} \quad n + 3 = 0$$
    $$n = 4 \quad \text{or} \quad n = -3$$

    If $n = 4$, then $n + 1 = 5$.
    If $n = -3$, then $n + 1 = -2$.
    The two integers are 4 and 5, or -3 and -2.

19. Let $x =$ the smaller odd integer.
    Then $x + 2 =$ the larger odd integer.
    Their product is 15 more than three times their sum.
    $$x(x + 2) = 3(x + x + 2) + 15$$
    $$x^2 + 2x = 6x + 6 + 15$$
    $$x^2 - 4x - 21 = 0$$
    $$(x - 7)(x + 3) = 0$$
    $$x - 7 = 0 \quad \text{or} \quad x + 3 = 0$$
    $$x = 7 \quad \text{or} \quad x = -3$$

    The two integers are 7 and $7 + 2 = 9$ or -3 and $-3 + 2 = -1$.

    Check: $7(9) = 3(7 + 9) + 15; \; 63 = 63$
    $-3(-1) = 3(-3 - 1) + 15; \; 3 = 3$

Chapter 6 Factoring and Applications

21. Let n = the smallest even integer. Then $n+2$ and $n+4$ are the next two even integers.

$$n^2 + (n+2)^2 = (n+4)^2$$
$$n^2 + n^2 + 4n + 4 = n^2 + 8n + 16$$
$$n^2 - 4n - 12 = 0$$
$$(n-6)(n+2) = 0$$

$$n - 6 = 0 \quad \text{or} \quad n + 2 = 0$$
$$n = 6 \quad \text{or} \quad n = -2$$

If $n = 6$, $n+2 = 8$, and $n+4 = 10$.
If $n = -2$, $n+2 = 0$, and $n+4 = 2$.

The three integers are 6, 8, and 10 or -2, 0, and 2.

23. Let x = the length of the longer leg of the right triangle.
Then $x+1$ = the length of the hypotenuse and $x-7$ = the length of the shorter leg.

Refer to the figure in the text. Use the Pythagorean formula with $a = x$, $b = x-7$, and $c = x+1$.

$$a^2 + b^2 = c^2$$
$$x^2 + (x-7)^2 = (x+1)^2$$
$$x^2 + (x^2 - 14x + 49) = x^2 + 2x + 1$$
$$2x^2 - 14x + 49 = x^2 + 2x + 1$$
$$x^2 - 16x + 48 = 0$$
$$(x-12)(x-4) = 0$$

$$x - 12 = 0 \quad \text{or} \quad x - 4 = 0$$
$$x = 12 \quad \text{or} \quad x = 4$$

Discard 4 because if the length of the longer leg is 4 centimeters, by the conditions of the problem, the length of the shorter leg would be $4 - 7 = -3$ centimeters, which is impossible. The length of the longer leg is 12 centimeters.

Check: $12^2 + 5^2 = 13^2$; $169 = 169$

25. Let x = Denny's distance from home. Then $x+1$ = the distance between Terri and Denny.

Refer to the diagram in the textbook.
Use the Pythagorean formula.

$$a^2 + b^2 = c^2$$
$$x^2 + 5^2 = (x+1)^2$$
$$x^2 + 25 = x^2 + 2x + 1$$
$$24 = 2x$$
$$12 = x$$

Denny is 12 miles from home.

Check: $12^2 + 5^2 = 13^2$; $169 = 169$

27. Let x = the length of the ladder. Then
$x - 4$ = the distance from the bottom of the ladder to the building and
$x - 2$ = the distance on the side of the building to the top of the ladder.

Substitute into the Pythagorean formula.

$$a^2 + b^2 = c^2$$
$$(x-2)^2 + (x-4)^2 = x^2$$
$$x^2 - 4x + 4 + x^2 - 8x + 16 = x^2$$
$$x^2 - 12x + 20 = 0$$
$$(x-10)(x-2) = 0$$

$$x - 10 = 0 \quad \text{or} \quad x - 2 = 0$$
$$x = 10 \quad \text{or} \quad x = 2$$

The solution cannot be 2 because then a negative distance results. Thus, $x = 10$ and the top of the ladder reaches $x - 2 = 10 - 2 = 8$ feet up the side of the building.

Check: $8^2 + 6^2 = 10^2$; $100 = 100$

29. **(a)** Let $h = 64$ in the given formula and solve for t.

$$h = -16t^2 + 32t + 48$$
$$64 = -16t^2 + 32t + 48$$
$$16t^2 - 32t + 16 = 0$$
$$16(t^2 - 2t + 1) = 0$$
$$16(t-1)^2 = 0$$
$$t - 1 = 0$$
$$t = 1$$

The height of the object will be 64 feet after 1 second.

(b) To find the time when the height is 60 feet, let $h = 60$ in the given equation and solve for t.

$$h = -16t^2 + 32t + 48$$
$$60 = -16t^2 + 32t + 48$$
$$16t^2 - 32t + 12 = 0$$
$$4(4t^2 - 8t + 3) = 0$$
$$4(2t-1)(2t-3) = 0$$

$$2t - 1 = 0 \quad \text{or} \quad 2t - 3 = 0$$
$$2t = 1 \quad \quad \quad 2t = 3$$
$$t = \tfrac{1}{2} \quad \text{or} \quad t = \tfrac{3}{2}$$

The height of the object is 60 feet after $\tfrac{1}{2}$ second (on the way up) and after $\tfrac{3}{2}$ or $1\tfrac{1}{2}$ seconds (on the way down).

(c) To find the time when the object hits the ground, let $h = 0$ and solve for t.

$$h = -16t^2 + 32t + 48$$
$$0 = -16t^2 + 32t + 48$$
$$16t^2 - 32t - 48 = 0$$
$$16(t^2 - 2t - 3) = 0$$
$$16(t + 1)(t - 3) = 0$$

$t + 1 = 0$ or $t - 3 = 0$
$t = -1$ or $t = 3$

We discard -1 because time cannot be negative. The object will hit the ground after 3 seconds.

(d) The negative solution, -1, does not make sense, since t represents time, which cannot be negative.

31. (a) $x = 6$ in 1996.
$$y = .866x^2 + 1.02x + 5.29$$
$$y = .866(6)^2 + 1.02(6) + 5.29$$
$$y \approx 42.6$$

In 1996, the model predicts there were about 42.6 million cellular phones. The result using the model is a little less than 44 million, the actual number for 1996.

(b) $x = 2002 - 1990 = 12$
$x = 12$ corresponds to 2002.

(c) $y = .866(12)^2 + 1.02(12) + 5.29$
$y \approx 142.2$

In 2002, the model predicts there were 142.2 million cellular phones.

The result is a little more than 140 million, the actual number for 2002.

(d) $x = 2005 - 1990 = 15$
$y = .866(15)^2 + 1.02(15) + 5.29$
$y \approx 215.4$

In 2005, the model predicts there will be 215.4 million cellular phones.

32. $378.7 - 271.3 = 107.4$

The trade deficit increased 107.4 billion dollars from 1999 to 2000.

107.4 is what percent of 271.3?

$$\frac{a}{b} = \frac{p}{100}$$
$$\frac{107.4}{271.3} = \frac{p}{100}$$
$$271.3p = 10,740$$
$$p \approx 39.59$$

This is about a 40% increase.

33. $y = 40.8x + 66.9$

In 1997, $x = 2$.
$$y = 40.8(2) + 66.9 = 148.5$$

In 1999, $x = 4$.
$$y = 40.8(4) + 66.9 = 230.1$$

In 2000, $x = 5$.
$$y = 40.8(5) + 66.9 = 270.9$$

The deficits in billions of dollars for 1997, 1999, and 2000 are 148.5, 230.1, and 270.9, respectively.

34. The answers using the linear equation are not at all close to the actual data.

35. $y = 18.5x^2 - 33.4x + 104$

In 1997, $x = 2$.
$$y = 18.5(2)^2 - 33.4(2) + 104 = 111.2$$

In 1999, $x = 4$.
$$y = 18.5(4)^2 - 33.4(4) + 104 = 266.4$$

In 2000, $x = 5$.
$$y = 18.5(5)^2 - 33.4(5) + 104 = 399.5$$

The trade deficit in billions of dollars for 1997, 1999, and 2000 is 111.2, 266.4, and 399.5, respectively.

36. The answers in Exercise 35 are fairly close to the actual data. The quadratic equation models the data better.

37. The x-coordinates are $0, 1, 2, 3, 4$, and 5. The y-coordinates are the deficits given in the table. The ordered pairs are: $(0, 97.5)$, $(1, 104.3)$, $(2, 104.7)$, $(3, 164.3)$, $(4, 271.3)$, $(5, 378.7)$

38.

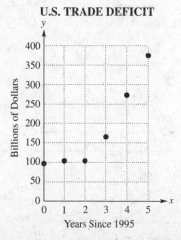

U.S. TRADE DEFICIT

No, the ordered pairs do not lie in a linear pattern.

39. In 2002, $x = 7$.

$$y = 18.5(7)^2 - 33.4(7) + 104 = 776.7$$

In 2002, the model predicts the trade deficit to be 776.7 billion dollars.

40. (a) $776.7 - 417.9 = 358.8$

The actual deficit is quite a bit less than the prediction.

(b) No, the equation is based on data for the years 1995–2000 and is valid only for those years. Data for later years might not follow the same pattern.

Chapter 6 Review Exercises

1. $7t + 14 = 7 \cdot t + 7 \cdot 2 = 7(t+2)$

2. $60z^3 + 30z = 30z \cdot 2z^2 + 30z \cdot 1$
   $= 30z(2z^2 + 1)$

3. $35x^3 + 70x^2 = 35x^2(x) + 35x^2(2)$
   $= 35x^2(x+2)$

4. $100m^2n^3 - 50m^3n^4 + 150m^2n^2$
   $= 50m^2n^2(2n) + 50m^2n^2(-mn^2)$
   $\quad + 50m^2n^2(3)$
   $= 50m^2n^2(2n - mn^2 + 3)$

5. $2xy - 8y + 3x - 12$
   $= (2xy - 8y) + (3x - 12)$ *Group terms.*
   $= 2y(x-4) + 3(x-4)$ *Factor each group.*
   $= (x-4)(2y+3)$ *Factor out x − 4.*

6. $6y^2 + 9y + 4xy + 6x$
   $= (6y^2 + 9y) + (4xy + 6x)$
   $= 3y(2y+3) + 2x(2y+3)$
   $= (2y+3)(3y+2x)$

7. $x^2 + 5x + 6$

   Find two integers whose product is 6 and whose sum is 5. The integers are 3 and 2. Thus,

   $$x^2 + 5x + 6 = (x+3)(x+2).$$

8. $y^2 - 13y + 40$

   Find two integers whose product is 40 and whose sum is −13.

Factors of 40	Sums of factors
−1, −40	−41
−2, −20	−22
−4, −10	−14
−5, −8	−13

   The integers are −5 and −8, so

   $$y^2 - 13y + 40 = (y-5)(y-8).$$

9. $q^2 + 6q - 27$

   Find two integers whose product is −27 and whose sum is 6. The integers are −3 and 9, so

   $$q^2 + 6q - 27 = (q-3)(q+9).$$

10. $r^2 - r - 56$

    Find two integers whose product is −56 and whose sum is −1. The integers are 7 and −8, so

    $$r^2 - r - 56 = (r+7)(r-8).$$

11. $r^2 - 4rs - 96s^2$

    Find two expressions whose product is $-96s^2$ and whose sum is $-4s$. The expressions are $8s$ and $-12s$, so

    $$r^2 - 4rs - 96s^2 = (r+8s)(r-12s).$$

12. $p^2 + 2pq - 120q^2$

    Find two expressions whose product is $-120q^2$ and whose sum is $2q$. The expressions are $12q$ and $-10q$, so

    $$p^2 + 2pq - 120q^2 = (p+12q)(p-10q).$$

13. $8p^3 - 24p^2 - 80p$

    First, factor out the GCF, $8p$.

    $$8p^3 - 24p^2 - 80p = 8p(p^2 - 3p - 10)$$

    Now factor $p^2 - 3p - 10$.

    $$p^2 - 3p - 10 = (p+2)(p-5)$$

    The complete factored form is

    $$8p^3 - 24p^2 - 80p = 8p(p+2)(p-5).$$

14. $3x^4 + 30x^3 + 48x^2$
    $= 3x^2(x^2 + 10x + 16)$
    $= 3x^2(x+2)(x+8)$

15. Find two expressions whose product is $-18n^2$ and whose sum is $-3n$. The expressions are $3n$ and $-6n$, so

    $$m^2 - 3mn - 18n^2 = (m+3n)(m-6n).$$

16. Find two expressions whose product is $15z^2$ and whose sum is $-8z$. The expressions are $-3z$ and $-5z$, so

    $$y^2 - 8yz + 15z^2 = (y-3z)(y-5z).$$

17. $p^7 - p^6q - 2p^5q^2 = p^5(p^2 - pq - 2q^2)$
    $\quad\quad\quad\quad\quad\quad\quad = p^5(p+q)(p-2q)$

18. $3r^5 - 6r^4s - 45r^3s^2$
    $= 3r^3(r^2 - 2rs - 15s^2)$
    $= 3r^3(r+3s)(r-5s)$

19. There is no pair of integers whose product is 1 and whose sum is 1, so $x^2 + x + 1$ is *prime*.

20. $3x^2 + 6x + 6 = 3(x^2 + 2x + 2)$

    There is no pair of integers whose product is 2 and whose sum is 2, so $x^2 + 2x + 2$ is prime. Thus, the complete factored form is
    $$3x^2 + 6x + 6 = 3(x^2 + 2x + 2).$$

21. To begin factoring $6r^2 - 5r - 6$, the possible first terms of the two binomial factors are r and $6r$, or $2r$ and $3r$, if we consider only positive integer coefficients.

22. When factoring $2z^3 + 9z^2 - 5z$, the first step is to factor out the GCF, z.

In Exercises 23–30, either the trial and error method or the grouping method can be used to factor each polynomial.

23. Factor $2k^2 - 5k + 2$ by trial and error.
    $$2k^2 - 5k + 2 = (2k - 1)(k - 2)$$

24. Factor $3r^2 + 11r - 4$ by grouping. Look for two integers whose product is $3(-4) = -12$ and whose sum is 11. The integers are 12 and -1.
    $$\begin{aligned}3r^2 + 11r - 4 &= 3r^2 + 12r - r - 4\\ &= (3r^2 + 12r) + (-r - 4)\\ &= 3r(r + 4) - 1(r + 4)\\ &= (r + 4)(3r - 1)\end{aligned}$$

25. Factor $6r^2 - 5r - 6$ by grouping. Find two integers whose product is $6(-6) = -36$ and whose sum is -5. The integers are -9 and 4.
    $$\begin{aligned}6r^2 - 5r - 6 &= 6r^2 - 9r + 4r - 6\\ &= (6r^2 - 9r) + (4r - 6)\\ &= 3r(2r - 3) + 2(2r - 3)\\ &= (2r - 3)(3r + 2)\end{aligned}$$

26. Factor $10z^2 - 3z - 1$ by trial and error.
    $$10z^2 - 3z - 1 = (5z + 1)(2z - 1)$$

27. $5t^2 - 11t + 12$ is *prime*.

28. $24x^5 - 20x^4 + 4x^3$

    Factor out the GCF, $4x^3$. Then complete the factoring by trial and error.
    $$\begin{aligned}24x^5 - 20x^4 + 4x^3 &\\ &= 4x^3(6x^2 - 5x + 1)\\ &= 4x^3(3x - 1)(2x - 1)\end{aligned}$$

29. $\begin{aligned}-6x^2 + 3x + 30 &= -3(2x^2 - x - 10)\\ &= -3(2x - 5)(x + 2)\end{aligned}$

30. $\begin{aligned}10r^3s &+ 17r^2s^2 + 6rs^3\\ &= rs(10r^2 + 17rs + 6s^2)\\ &= rs(5r + 6s)(2r + s)\end{aligned}$

31. Only choice **B**, $4x^2y^2 - 25z^2$, is the difference of squares. In **A**, 32 is not a perfect square. In **C**, we have a sum, not a difference. In **D**, y^3 is not a square. The correct choice is **B**.

32. Only choice **D**, $x^2 - 20x + 100$, is a perfect square trinomial because $x^2 = x \cdot x$, $100 = 10 \cdot 10$, and $-20x = -2(x)(10)$.

In Exercises 33–38, use the rule for factoring a difference of two squares, if possible.

33. $n^2 - 64 = n^2 - 8^2 = (n + 8)(n - 8)$

34. $\begin{aligned}25b^2 - 121 &= (5b)^2 - 11^2\\ &= (5b + 11)(5b - 11)\end{aligned}$

35. $\begin{aligned}49y^2 - 25w^2 &= (7y)^2 - (5w)^2\\ &= (7y + 5w)(7y - 5w)\end{aligned}$

36. $\begin{aligned}144p^2 - 36q^2 &= 36(4p^2 - q^2)\\ &= 36[(2p)^2 - q^2]\\ &= 36(2p + q)(2p - q)\end{aligned}$

37. $x^2 + 100$

    This polynomial is *prime* because it is the sum of squares and the two terms have no common factor.

38. $\begin{aligned}x^2 - \frac{49}{100} &= x^2 - \left(\frac{7}{10}\right)^2\\ &= \left(x + \frac{7}{10}\right)\left(x - \frac{7}{10}\right)\end{aligned}$

In Exercises 39–44, use the rules for factoring a perfect square trinomial.

39. $\begin{aligned}z^2 + 10z + 25 &= z^2 + 2(5)(z) + 5^2\\ &= (z + 5)^2\end{aligned}$

40. $\begin{aligned}r^2 - 12r + 36 &= r^2 - 2(6)(r) + 6^2\\ &= (r - 6)^2\end{aligned}$

41. $\begin{aligned}9t^2 - 42t + 49 &= (3t)^2 - 2(3t)(7) + 7^2\\ &= (3t - 7)^2\end{aligned}$

42. $\begin{aligned}16m^2 &+ 40mn + 25n^2\\ &= (4m)^2 + 2(4m)(5n) + (5n)^2\\ &= (4m + 5n)^2\end{aligned}$

43. $\begin{aligned}54x^3 &- 72x^2 + 24x\\ &= 6x(9x^2 - 12x + 4)\\ &= 6x[(3x)^2 - 2(3x)(2) + 2^2]\\ &= 6x(3x - 2)^2\end{aligned}$

44. $x^2 + \dfrac{2}{3}x + \dfrac{1}{9} = x^2 + 2(x)\left(\dfrac{1}{3}\right) + \left(\dfrac{1}{3}\right)^2$

$\phantom{x^2 + \dfrac{2}{3}x + \dfrac{1}{9}} = \left(x + \dfrac{1}{3}\right)^2$

In Exercises 45–58, all solutions should be checked by substituting in the original equations. The checks will not be shown here.

45. $(4t + 3)(t - 1) = 0$

$\quad 4t + 3 = 0 \quad$ or $\quad t - 1 = 0$
$\quad 4t = -3$
$\quad t = -\dfrac{3}{4} \quad$ or $\quad t = 1$

The solutions are $-\dfrac{3}{4}$ and 1.

46. $(x + 7)(x - 4)(x + 3) = 0$

$x + 7 = 0 \quad$ or $\quad x - 4 = 0 \quad$ or $\quad x + 3 = 0$
$x = -7 \quad$ or $\quad x = 4 \quad$ or $\quad x = -3$

The solutions are -7, -3, and 4.

47. $x(2x - 5) = 0$

$\quad x = 0 \quad$ or $\quad 2x - 5 = 0$
$\phantom{x = 0 \quad \text{or} \quad} 2x = 5$
$\phantom{x = 0 \quad \text{or} \quad} x = \dfrac{5}{2}$

The solutions are 0 and $\dfrac{5}{2}$.

48. $z^2 + 4z + 3 = 0$

$(z + 3)(z + 1) = 0$
$z + 3 = 0 \quad$ or $\quad z + 1 = 0$
$z = -3 \quad$ or $\quad z = -1$

The solutions are -3 and -1.

49. $m^2 - 5m + 4 = 0$

$(m - 1)(m - 4) = 0$
$m - 1 = 0 \quad$ or $\quad m - 4 = 0$
$m = 1 \quad$ or $\quad m = 4$

The solutions are 1 and 4.

50. $x^2 = -15 + 8x$

$x^2 - 8x + 15 = 0$
$(x - 3)(x - 5) = 0$
$x - 3 = 0 \quad$ or $\quad x - 5 = 0$
$x = 3 \quad$ or $\quad x = 5$

The solutions are 3 and 5.

51. $3z^2 - 11z - 20 = 0$

$(3z + 4)(z - 5) = 0$
$3z + 4 = 0 \quad$ or $\quad z - 5 = 0$
$3z = -4$
$z = -\dfrac{4}{3} \quad$ or $\quad z = 5$

The solutions are $-\dfrac{4}{3}$ and 5.

52. $81t^2 - 64 = 0$

$(9t + 8)(9t - 8) = 0$
$9t + 8 = 0 \quad$ or $\quad 9t - 8 = 0$
$9t = -8 9t = 8$
$t = -\dfrac{8}{9} \quad$ or $\quad t = \dfrac{8}{9}$

The solutions are $-\dfrac{8}{9}$ and $\dfrac{8}{9}$.

53. $y^2 = 8y$

$y^2 - 8y = 0$
$y(y - 8) = 0$
$y = 0 \quad$ or $\quad y - 8 = 0$
$y = 0 \quad$ or $\quad y = 8$

The solutions are 0 and 8.

54. $n(n - 5) = 6$

$n^2 - 5n = 6$
$n^2 - 5n - 6 = 0$
$(n + 1)(n - 6) = 0$
$n + 1 = 0 \quad$ or $\quad n - 6 = 0$
$n = -1 \quad$ or $\quad n = 6$

The solutions are -1 and 6.

55. $t^2 - 14t + 49 = 0$

$(t - 7)^2 = 0$
$t - 7 = 0$
$t = 7$

The solution is 7.

56. $t^2 = 12(t - 3)$

$t^2 = 12t - 36$
$t^2 - 12t + 36 = 0$
$(t - 6)^2 = 0$
$t - 6 = 0$
$t = 6$

The solution is 6.

57. $(5z + 2)(z^2 + 3z + 2) = 0$

$(5z + 2)(z + 2)(z + 1) = 0$
$5z + 2 = 0 \quad$ or $\quad z + 2 = 0 \quad$ or $\quad z + 1 = 0$
$5z = -2$
$z = -\dfrac{2}{5} \quad$ or $\quad z = -2 \quad$ or $\quad z = -1$

The solutions are $-\dfrac{2}{5}$, -2, and -1.

58. $x^2 = 9$

$x^2 - 9 = 0$
$(x + 3)(x - 3) = 0$
$x + 3 = 0 \quad$ or $\quad x - 3 = 0$
$x = -3 \quad$ or $\quad x = 3$

The solutions are -3 and 3.

59. Let $x =$ the width of the rug.
Then $x + 6 =$ the length of the rug.

$$A = LW$$
$$40 = (x+6)x$$
$$40 = x^2 + 6x$$
$$0 = x^2 + 6x - 40$$
$$0 = (x+10)(x-4)$$

$x + 10 = 0 \quad$ or $\quad x - 4 = 0$
$x = -10 \quad$ or $\quad x = 4$

Reject -10 since the width cannot be negative. The width of the rug is 4 feet and the length is $4 + 6$ or 10 feet.

60. From the figure, we have $L = 20$, $W = x$, and $H = x + 4$.

$$S = 2WH + 2WL + 2LH$$
$$650 = 2x(x+4) + 2x(20) + 2(20)(x+4)$$
$$650 = 2x^2 + 8x + 40x + 40(x+4)$$
$$650 = 2x^2 + 48x + 40x + 160$$
$$0 = 2x^2 + 88x - 490$$
$$0 = 2(x^2 + 44x - 245)$$
$$0 = 2(x+49)(x-5)$$

$x + 49 = 0 \quad$ or $\quad x - 5 = 0$
$x = -49 \quad$ or $\quad x = 5$

Reject -49 because the width cannot be negative. The width of the chest is 5 feet.

61. Let $x =$ the width of the rectangle.
Then $3x =$ the length. Use $A = LW$.

The width increased by 3	times	the same length	would be	an area of 30.
↓	↓	↓	↓	↓
$(x+3)$	·	$3x$	$=$	30

Solve the equation.

$$(x+3)(3x) = 30$$
$$3x^2 + 9x = 30$$
$$3x^2 + 9x - 30 = 0$$
$$3(x^2 + 3x - 10) = 0$$
$$3(x+5)(x-2) = 0$$

$x + 5 = 0 \quad$ or $\quad x - 2 = 0$
$x = -5 \quad$ or $\quad x = 2$

Reject -5. The width of the original rectangle is 2 meters and the length is $3(2)$ or 6 meters.

62. Let $x =$ the length of the box.
Then $x - 1 =$ the height of the box.

$$V = LWH$$
$$120 = x(4)(x-1)$$
$$120 = 4x^2 - 4x$$
$$4x^2 - 4x - 120 = 0$$
$$4(x^2 - x - 30) = 0$$
$$4(x-6)(x+5) = 0$$

$x - 6 = 0 \quad$ or $\quad x + 5 = 0$
$x = 6 \quad$ or $\quad x = -5$

Reject -5. The length of the box is 6 meters and the height is $6 - 1 = 5$ meters.

63. Let $x =$ the first integer.
Then $x + 1 =$ the next integer.

The product of the integers is 29 more than their sum, so

$$x(x+1) = 29 + [x + (x+1)].$$

Solve this equation.

$$x^2 + x = 29 + 2x + 1$$
$$x^2 - x - 30 = 0$$
$$(x-6)(x+5) = 0$$

$x - 6 = 0 \quad$ or $\quad x + 5 = 0$
$x = 6 \quad$ or $\quad x = -5$

If $x = 6$, $x + 1 = 6 + 1 = 7$.
If $x = -5$, $x + 1 = -5 + 1 = -4$.

The consecutive integers are 6 and 7 or -5 and -4.

64. Let $x =$ the distance traveled west.
Then $x - 14 =$ the distance traveled south, and $(x - 14) + 16 = x + 2 =$ the distance between the cars.

These three distances form a right triangle with x and $x - 14$ representing the lengths of the legs and $x + 2$ representing the length of the hypotenuse. Use the Pythagorean formula.

$$a^2 + b^2 = c^2$$
$$x^2 + (x-14)^2 = (x+2)^2$$
$$x^2 + x^2 - 28x + 196 = x^2 + 4x + 4$$
$$x^2 - 32x + 192 = 0$$
$$(x-8)(x-24) = 0$$

$x - 8 = 0 \quad$ or $\quad x - 24 = 0$
$x = 8 \quad$ or $\quad x = 24$

If $x = 8$, then $x - 14 = -6$, which is not possible because a distance cannot be negative.

If $x = 24$, then $x - 14 = 10$ and $x + 2 = 26$. The cars were 26 miles apart.

65. $h = 128t - 16t^2$
$h = 128(1) - 16(1)^2$ *Let t = 1.*
$= 128 - 16$
$= 112$

After 1 second, the height is 112 feet.

66. $h = 128t - 16t^2$
$h = 128(2) - 16(2)^2$ *Let t = 2.*
$= 256 - 16(4)$
$= 256 - 64$
$= 192$

After 2 seconds, the height is 192 feet.

67. $h = 128t - 16t^2$
$h = 128(4) - 16(4)^2$ *Let t = 4.*
$= 512 - 256$
$= 256$

After 4 seconds, the height is 256 feet.

68. The object hits the ground when $h = 0$.
$h = 128t - 16t^2$
$0 = 128t - 16t^2$ *Let h = 0.*
$0 = 16t(8 - t)$

$16t = 0$ or $8 - t = 0$
$t = 0$ or $8 = t$

The solution $t = 0$ represents the time before the object is propelled upward. The object returns to the ground after 8 seconds.

69. **(a)** In 2001, $x = 2001 - 1997 = 4$.
$y = 47.8x^2 - .135x + 7.65$
$y = 47.8(4)^2 - .135(4) + 7.65 = 771.91$

In 2001, the model predicts about $771.9 million in annual revenue for eBay. The answer using the model is a little high.

(b) In 2003, $x = 2003 - 1997 = 6$.
$y = 47.8(6)^2 - .135(6) + 7.65 = 1727.64$

In 2003, the model predicts about $1727.6 million in annual revenue for eBay. The answer using the model is a little high.

(c) No, the prediction seems low. If eBay revenues were $1516.7 million through three quarters, they were approximately $500 million per quarter, which would lead to annual revenue in 2003 of about $2000 million.

70. [6.1] **D** is not factored completely.
$3(7t + 4) + x(7t + 4) = (7t + 4)(3 + x)$

71. [6.1] The factor $2x + 8$ has a common factor of 2. The complete factored form is $2(x + 4)(3x - 4)$.

72. [6.4] $z^2 - 11zx + 10x^2 = (z - x)(z - 10x)$

73. [6.2] $3k^2 + 11k + 10$

Two integers with product $3(10) = 30$ and sum 11 are 5 and 6.

$3k^2 + 11k + 10$
$= 3k^2 + 5k + 6k + 10$
$= (3k^2 + 5k) + (6k + 10)$
$= k(3k + 5) + 2(3k + 5)$
$= (3k + 5)(k + 2)$

74. [6.3] $15m^2 + 20mp - 12m - 16p$
$= (15m^2 + 20mp) + (-12m - 16p)$
$= 5m(3m + 4p) - 4(3m + 4p)$
$= (3m + 4p)(5m - 4)$

75. [6.5]
$y^4 - 625$
$= (y^2)^2 - 25^2$
$= (y^2 + 25)(y^2 - 25)$ *Difference of squares*
$= (y^2 + 25)(y + 5)(y - 5)$ *Difference of squares*

76. [6.3]
$6m^3 - 21m^2 - 45m$
$= 3m(2m^2 - 7m - 15)$
$= 3m[(2m^2 - 10m) + (3m - 15)]$ *Factor by grouping.*
$= 3m[2m(m - 5) + 3(m - 5)]$
$= 3m(m - 5)(2m + 3)$

77. [6.1] $24ab^3c^2 - 56a^2bc^3 + 72a^2b^2c$
$= 8abc(3b^2c - 7ac^2 + 9ab)$

78. [6.2] $25a^2 + 15ab + 9b^2$ is a *prime* polynomial.

79. [6.1] $12x^2yz^3 + 12xy^2z - 30x^3y^2z^4$
$= 6xyz(2xz^2 + 2y - 5x^2yz^3)$

80. [6.4] $2a^5 - 8a^4 - 24a^3$
$= 2a^3(a^2 - 4a - 12)$
$= 2a^3(a - 6)(a + 2)$

81. [6.3] $12r^2 + 8rq - 15q^2$
$= 12r^2 + 18rq - 10rq - 15q^2$
$= (12r^2 + 18rq) + (-10rq - 15q^2)$
$= 6r(2r + 3q) - 5q(2r + 3q)$
$= (2r + 3q)(6r - 5q)$

82. [6.5] $100a^2 - 9 = (10a)^2 - 3^2$
$= (10a + 3)(10a - 3)$

83. [6.5] $49t^2 + 56t + 16$
$= (7t)^2 + 2(7t)(4) + 4^2$
$= (7t + 4)^2$

Chapter 6 Review Exercises 249

84. [6.6] $t(t-7) = 0$

$t = 0$ or $t - 7 = 0$
$t = 0$ or $t = 7$

The solutions are 0 and 7.

85. [6.6] $\quad x^2 + 3x = 10$
$x^2 + 3x - 10 = 0$
$(x+5)(x-2) = 0$

$x + 5 = 0$ or $x - 2 = 0$
$x = -5$ or $x = 2$

The solutions are -5 and 2.

86. [6.6] $\quad 25x^2 + 20x + 4 = 0$
$(5x)^2 + 2(5x)(2) + 2^2 = 0$
$(5x + 2)^2 = 0$
$5x + 2 = 0$
$5x = -2$
$x = -\frac{2}{5}$

The solution is $-\frac{2}{5}$.

87. [6.7] Let x = the length of the shorter leg. Then $2x + 6$ = the length of the longer leg and $(2x + 6) + 3 = 2x + 9$ = the length of the hypotenuse.

Use the Pythagorean formula, $a^2 + b^2 = c^2$.

$x^2 + (2x+6)^2 = (2x+9)^2$
$x^2 + 4x^2 + 24x + 36 = 4x^2 + 36x + 81$
$x^2 - 12x - 45 = 0$
$(x - 15)(x + 3) = 0$

$x - 15 = 0$ or $x + 3 = 0$
$x = 15$ or $x = -3$

Reject -3 because a length cannot be negative. The sides of the lot are 15 meters, $2(15) + 6 = 36$ meters, and $36 + 3 = 39$ meters.

88. [6.7] Let x = the width of the base. Then $x + 2$ = the length of the base.

The area of the base, B, is given by LW, so
$B = x(x+2)$.

Use the formula for the volume of a pyramid,
$V = \frac{1}{3} \cdot B \cdot h$.
$48 = \frac{1}{3}x(x+2)(6)$
$48 = 2x(x+2)$
$24 = x^2 + 2x$
$x^2 + 2x - 24 = 0$
$(x + 6)(x - 4) = 0$

$x + 6 = 0$ or $x - 4 = 0$
$x = -6$ or $x = 4$

Reject -6. The width of the base is 4 meters and the length is $4 + 2$ or 6 meters.

89. [6.7] Let x = the smallest integer. Then $x + 1$ and $x + 2$ are the next two larger integers.

The product of the smaller two of three consecutive integers is equal to 23 plus the largest.

$x(x+1) = 23 + (x+2)$
$x^2 + x = 23 + x + 2$
$x^2 - 25 = 0$
$(x+5)(x-5) = 0$

$x + 5 = 0$ or $x - 5 = 0$
$x = -5$ or $x = 5$

If $x = -5$, then $x + 1 = -4$ and $x + 2 = -3$.
If $x = 5$, then $x + 1 = 6$ and $x + 2 = 7$.
The integers are $-5, -4,$ and -3, or $5, 6,$ and 7.

90. [6.7] $d = 16t^2$

(a) $t = 4: d = 16(4)^2 = 256$

In 4 seconds, the object would fall 256 feet.

(b) $t = 8: d = 16(8)^2 = 1024$

In 8 seconds, the object would fall 1024 feet.

91. [6.7] Let x = the width of the house. Then $x + 7$ = the length of the house.

Use $A = LW$ with 170 for A, $x + 7$ for L, and x for W.

$170 = (x+7)(x)$
$170 = x^2 + 7x$
$0 = x^2 + 7x - 170$
$0 = (x+17)(x-10)$

$x + 17 = 0$ or $x - 10 = 0$
$x = -17$ or $x = 10$

Discard -17 because the width cannot be negative. If $x = 10$, $x + 7 = 10 + 7 = 17$.

The width is 10 meters and the length is 17 meters.

92. [6.7] Let b = the base of the sail. Then $b + 4$ = the height of the sail.

Use the formula for the area of a triangle.

$A = \frac{1}{2}bh$
$30 = \frac{1}{2}(b)(b+4) \quad \text{Let } A = 30$
$60 = b^2 + 4b$
$0 = b^2 + 4b - 60$
$0 = (b+10)(b-6)$

continued

$$b + 10 = 0 \quad \text{or} \quad b - 6 = 0$$
$$b = -10 \quad \text{or} \quad b = 6$$

Discard -10 since the base of a triangle cannot be negative. The base of the triangular sail is 6 meters.

93. [6.7] $y = .25x^2 - 25.65x + 496.6$

(a) If $x = 102$ (for 2002), then $y = 481.3$ (about 481,000 vehicles).

(b) The estimate may be unreliable because the conditions that prevailed in the years 1998–2001 may have changed, causing either a greater increase or a greater decrease in the numbers of alternative-fueled vehicles.

Chapter 6 Test

1. $2x^2 - 2x - 24 = 2(x^2 - x - 12)$
$= 2(x + 3)(x - 4)$

The correct completely factored form is choice **D**. Note that the factored forms **A**, $(2x + 6)(x - 4)$, and **B**, $(x + 3)(2x - 8)$, also can be multiplied to give a product of $2x^2 - 2x - 24$, but neither of these is completely factored because $2x + 6$ and $2x - 8$ both contain a common factor of 2.

2. $12x^2 - 30x = 6x(2x - 5)$

3. $2m^3n^2 + 3m^3n - 5m^2n^2$
$= m^2n(2mn + 3m - 5n)$

4. $2ax - 2bx + ay - by$
$= 2x(a - b) + y(a - b)$
$= (a - b)(2x + y)$

5. Find two integers whose product is 14 and whose sum is -9. The integers are -7 and -2.
$x^2 - 9x + 14 = (x - 7)(x - 2)$

6. Factor $2x^2 + x - 3$ by trial and error.
$2x^2 + x - 3 = (2x + 3)(x - 1)$

7. $6x^2 - 19x - 7$

To factor by grouping, find two integers whose product is $6(-7) = -42$ and whose sum is -19. The integers are -21 and 2.
$6x^2 - 19x - 7 = 6x^2 - 21x + 2x - 7$
$= 3x(2x - 7) + 1(2x - 7)$
$= (2x - 7)(3x + 1)$

8. $3x^2 - 12x - 15 = 3(x^2 - 4x - 5)$
$= 3(x + 1)(x - 5)$

9. Factor $10z^2 - 17z + 3$ by trial and error.
$10z^2 - 17z + 3 = (2z - 3)(5z - 1)$

10. $t^2 + 2t + 3$

We cannot find two integers whose product is 3 and whose sum is 2. This polynomial is *prime*.

11. $x^2 + \frac{1}{36}$

This polynomial is *prime* because the sum of squares cannot be factored and the two terms have no common factor.

12. $y^2 - 49 = y^2 - 7^2 = (y + 7)(y - 7)$

13. $9y^2 - 64 = (3y)^2 - 8^2$
$= (3y + 8)(3y - 8)$

14. $x^2 + 16x + 64 = x^2 + 2(8)(x) + 8^2$
$= (x + 8)^2$

15. $4x^2 - 28xy + 49y^2$
$= (2x)^2 - 2(2x)(7y) + (7y)^2$
$= (2x - 7y)^2$

16. $-2x^2 - 4x - 2$
$= -2(x^2 + 2x + 1)$
$= -2(x^2 + 2 \cdot x \cdot 1 + 1^2)$
$= -2(x + 1)^2$

17. $6t^4 + 3t^3 - 108t^2$
$= 3t^2(2t^2 + t - 36)$
$= 3t^2(2t + 9)(t - 4)$

18. Note that $(2r + 5t)^2 = 4r^2 + 20rt + 25t^2$, not $4r^2 + 10rt + 25t^2$, so the given polynomial is not a perfect square trinomial. This polynomial is *prime* because there are no two integers whose product is $4(25) = 100$ and whose sum is 10.

19. $4t^3 + 32t^2 + 64t = 4t(t^2 + 8t + 16)$
$= 4t(t + 4)^2$

20. $x^4 - 81 = (x^2)^2 - 9^2$
$= (x^2 + 9)(x^2 - 9)$
$= (x^2 + 9)(x + 3)(x - 3)$

21. $(p + 3)(p + 3)$ is not the correct factored form of $p^2 + 9$ because
$(p + 3)(p + 3) = p^2 + 6p + 9$
$\neq p^2 + 9.$

The binomial $p^2 + 9$ is a *prime* polynomial.

22. $(x + 3)(x - 9) = 0$
$x + 3 = 0 \quad \text{or} \quad x - 9 = 0$
$x = -3 \quad \text{or} \quad x = 9$

The solutions are -3 and 9.

23. $2r^2 - 13r + 6 = 0$
    $(2r - 1)(r - 6) = 0$
    $2r - 1 = 0$ or $r - 6 = 0$
    $2r = 1$
    $r = \frac{1}{2}$ or $r = 6$

    The solutions are $\frac{1}{2}$ and 6.

24. $25x^2 - 4 = 0$
    $(5x + 2)(5x - 2) = 0$
    $5x + 2 = 0$ or $5x - 2 = 0$
    $5x = -2$ $5x = 2$
    $x = -\frac{2}{5}$ or $x = \frac{2}{5}$

    The solutions are $-\frac{2}{5}$ and $\frac{2}{5}$.

25. $x(x - 20) = -100$
    $x^2 - 20x = -100$
    $x^2 - 20x + 100 = 0$
    $(x - 10)^2 = 0$
    $x - 10 = 0$
    $x = 10$

    The solution is 10.

26. $t^2 = 3t$
    $t^2 - 3t = 0$
    $t(t - 3) = 0$
    $t = 0$ or $t - 3 = 0$
    $t = 0$ or $t = 3$

    The solutions are 0 and 3.

27. Let $x =$ the width of the flower bed.
    Then $2x - 3 =$ the length of the flower bed.
    Use the formula $A = LW$.

    $x(2x - 3) = 54$
    $2x^2 - 3x = 54$
    $2x^2 - 3x - 54 = 0$
    $(2x + 9)(x - 6) = 0$
    $2x + 9 = 0$ or $x - 6 = 0$
    $2x = -9$
    $x = -\frac{9}{2}$ or $x = 6$

    Reject $-\frac{9}{2}$. If $x = 6, 2x - 3 = 2(6) - 3 = 9$.
    The dimensions of the flower bed are 6 feet by 9 feet.

28. Let $x =$ the first integer.
    Then $x + 1 =$ the second integer.

    The square of the sum of the two integers is 11 more than the smaller integer.

    $[x + (x + 1)]^2 = x + 11$
    $(2x + 1)^2 = x + 11$
    $4x^2 + 4x + 1 = x + 11$
    $4x^2 + 3x - 10 = 0$
    $(4x - 5)(x + 2) = 0$
    $4x - 5 = 0$ or $x + 2 = 0$
    $4x = 5$
    $x = \frac{5}{4}$ or $x = -2$

    Reject $\frac{5}{4}$ because it is not an integer. If $x = -2$, $x + 1 = -1$. The integers are -2 and -1.

29. Let $x =$ the length of the stud.
    Then $3x - 7 =$ the length of the brace.

    The figure shows that a right triangle is formed with the brace as the hypotenuse. Use the Pythagorean formula, $a^2 + b^2 = c^2$.

    $x^2 + 15^2 = (3x - 7)^2$
    $x^2 + 225 = 9x^2 - 42x + 49$
    $0 = 8x^2 - 42x - 176$
    $0 = 2(4x^2 - 21x - 88)$
    $0 = 2(4x + 11)(x - 8)$
    $4x + 11 = 0$ or $x - 8 = 0$
    $4x = -11$
    $x = -\frac{11}{4}$ or $x = 8$

    Reject $-\frac{11}{4}$. If $x = 8, 3x - 7 = 24 - 7 = 17$, so the brace should be 17 feet long.

30. For 2000, $x = 2000 - 1984 = 16$.
    $y = 1.06x^2 - 4.77x + 47.9$
    $y = 1.06(16)^2 - 4.77(16) + 47.9 = 242.94$

    In 2000, the model estimates that the number of cable TV channels is 243.

Cumulative Review Exercises (Chapters R–6)

1. $3x + 2(x - 4) = 4(x - 2)$
    $3x + 2x - 8 = 4x - 8$
    $5x - 8 = 4x - 8$
    $x - 8 = -8$
    $x = 0$

    The solution is 0.

2. $.3x + .9x = .06$

   Multiply both sides by 100 to clear decimals.

   $$100(.3x + .9x) = 100(.06)$$
   $$30x + 90x = 6$$
   $$120x = 6$$
   $$x = \frac{6}{120} = \frac{1}{20} = .05$$

   The solution is .05.

3. $\frac{2}{3}n - \frac{1}{2}(n - 4) = 3$

   To clear fractions, multiply both sides by the least common denominator, which is 6.

   $$6\left[\frac{2}{3}n - \frac{1}{2}(n - 4)\right] = 6(3)$$
   $$4n - 3(n - 4) = 18$$
   $$4n - 3n + 12 = 18$$
   $$n + 12 = 18$$
   $$n = 6$$

   The solution is 6.

4. $A = P + Prt$
   $A = P(1 + rt)$ Factor out P.
   $\frac{A}{1 + rt} = \frac{P(1 + rt)}{1 + rt}$ Divide by $1 + rt$.
   $\frac{A}{1 + rt} = P$ or $P = \frac{A}{1 + rt}$

5. 69% of 500 is what number?

   $$\frac{a}{b} = \frac{p}{100}$$
   $$\frac{a}{500} = \frac{69}{100}$$
   $$100a = 69(500)$$
   $$a = 345$$

   42% of 500 is what number?

   $$\frac{a}{b} = \frac{p}{100}$$
   $$\frac{a}{500} = \frac{42}{100}$$
   $$100a = 500(42)$$
   $$a = 210$$

   What percent of 500 is 190?

   $$\frac{a}{b} = \frac{p}{100}$$
   $$\frac{190}{500} = \frac{p}{100}$$
   $$500p = 190(100)$$
   $$p = 38 \quad (38\%)$$

   What percent of 500 is 75?

   $$\frac{a}{b} = \frac{p}{100}$$
   $$\frac{75}{500} = \frac{p}{100}$$
   $$500p = 75(100)$$
   $$p = 15 \quad (15\%)$$

Item	Percent	Number
Toilet paper	69%	345
Zipper	42%	210
Frozen foods	38%	190
Self-stick note pads	15%	75

6. Let x = number of bronze medals.
   Then $x + 9$ = number of silver medals,
   and $(x + 9) - 4 = x + 5$ = the number of gold medals.

   The total number of medals was 35.

   $$x + (x + 9) + (x + 5) = 35$$
   $$3x + 14 = 35$$
   $$3x = 21$$
   $$x = 7$$

   Since $x = 7$, $x + 9 = 16$, and $x + 5 = 12$.
   Germany won 12 gold medals, 16 silver medals, and 7 bronze medals.

7. "77.5 cents for every dollar" indicates that we want 77.5% of $45,000.

   $$.775(\$45,000) = \$34,875$$

8. The angles are supplementary, so the sum of the angles is 180°.

   $$(2x + 16) + (x + 23) = 180$$
   $$3x + 39 = 180$$
   $$3x = 141$$
   $$x = 47$$

   Since $x = 47$, $2x + 16 = 2(47) + 16 = 110$ and $x + 23 = 47 + 23 = 70$.
   The angles are 110° and 70°.

9. The point with coordinates (a, b) is in

   (a) quadrant II if a is *negative* and b is *positive*.

   (b) quadrant III if a is *negative* and b is *negative*.

10. The equation $y = 12x + 3$ is in slope-intercept form, so the y-intercept is $(0, 3)$.

    Let $y = 0$ to find the x-intercept.

    $$0 = 12x + 3$$
    $$-3 = 12x$$
    $$-\tfrac{1}{4} = x$$

    The x-intercept is $\left(-\tfrac{1}{4}, 0\right)$.

11. The equation $y = 12x + 3$ is in slope-intercept form, so the slope is the coefficient of x, that is, 12.

12.
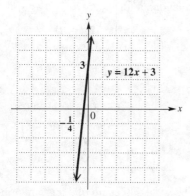

13. **(a)** $(1995, 9889), (2002, 10{,}569)$

$$m = \frac{y_2 - y_1}{x_2 - x_1}$$

$$= \frac{10{,}569 - 9889}{2002 - 1995}$$

$$= \frac{680}{7} = 97 \text{ (to the nearest whole number)}$$

A slope of 97 means that the number of radio stations increased by about 97 stations per year.

(b) The graph of the line intersects the vertical grid line over 2000 about halfway between 10,250 and 10,500. The average of these numbers is

$$\frac{10{,}250 + 10{,}500}{2} = \frac{20{,}750}{2} = 10{,}375.$$

The ordered pair is $(2000, 10{,}375)$.

14. $4x - y = -6$ (1)
$2x + 3y = 4$ (2)

$12x - 3y = -18$ (3) $3 \times$ Eq. (1)
$\underline{2x + 3y = 4}$ (2)
$14x = -14$ Add (3) and (2).
$x = -1$

To find y, substitute -1 for x in equation (1).

$$4(-1) - y = -6$$
$$-4 - y = -6$$
$$-y = -2$$
$$y = 2$$

The solution of the system is $(-1, 2)$.

15. $5x + 3y = 10$ (1)
$2x + \dfrac{6}{5}y = 5$ (2)

$-10x - 6y = -20$ (3) $-2 \times$ Eq. (1)
$\underline{10x + 6y = 25}$ (4) $5 \times$ Eq. (2)
$0 = 5$ Add (3) and (4).

The system of equations has *no solution*.

16. $2^{-3} \cdot 2^5 = 2^{-3+5} = 2^2 = 4$

17. $\left(\dfrac{3}{4}\right)^{-2} = \left(\dfrac{4}{3}\right)^2 = \dfrac{16}{9}$

18. $\dfrac{6^5 \cdot 6^{-2}}{6^3} = \dfrac{6^{5+(-2)}}{6^3}$

$$= \dfrac{6^3}{6^3} = 1$$

19. $\left(\dfrac{4^{-3} \cdot 4^4}{4^5}\right)^{-1} = \left(\dfrac{4^5}{4^{-3} \cdot 4^4}\right)^1$

$$= \dfrac{4^5}{4^1} = 4^4 = 256$$

20. $\dfrac{(p^2)^3 p^{-4}}{(p^{-3})^{-1} p} = \dfrac{p^{2 \cdot 3} p^{-4}}{p^{(-3)(-1)} p}$

$$= \dfrac{p^6 p^{-4}}{p^3 p^1}$$

$$= \dfrac{p^{6-4}}{p^{3+1}}$$

$$= \dfrac{p^2}{p^4} = \dfrac{1}{p^2}$$

21. $\dfrac{(m^{-2})^3 m}{m^5 m^{-4}} = \dfrac{m^{-2(3)} m^1}{m^{5+(-4)}}$

$$= \dfrac{m^{-6+1}}{m^1}$$

$$= \dfrac{m^{-5}}{m^1} = \dfrac{1}{m^6}$$

22. $(2k^2 + 4k) - (5k^2 - 2) - (k^2 + 8k - 6)$
$= (2k^2 + 4k) + (-5k^2 + 2)$
   $+ (-k^2 - 8k + 6)$
$= 2k^2 + 4k - 5k^2 + 2 - k^2 - 8k + 6$
$= -4k^2 - 4k + 8$

23. $(9x + 6)(5x - 3)$
    F   O   I   L
$= (9x)(5x) + (9x)(-3) + (6)(5x) + (6)(-3)$
$= 45x^2 - 27x + 30x - 18$
$= 45x^2 + 3x - 18$

24. $(3p + 2)^2 = (3p)^2 + 2 \cdot 3p \cdot 2 + 2^2$
$= 9p^2 + 12p + 4$

25. $\dfrac{8x^4 + 12x^3 - 6x^2 + 20x}{2x}$

$= \dfrac{8x^4}{2x} + \dfrac{12x^3}{2x} - \dfrac{6x^2}{2x} + \dfrac{20x}{2x}$

$= 4x^3 + 6x^2 - 3x + 10$

26. Factor $2a^2 + 7a - 4$ by trial and error.

$2a^2 + 7a - 4 = (a + 4)(2a - 1)$

27. $10m^2 + 19m + 6$

To factor by grouping, find two integers whose product is $10(6) = 60$ and whose sum is 19. The integers are 15 and 4.

$10m^2 + 19m + 6 = 10m^2 + 15m + 4m + 6$
$= 5m(2m + 3) + 2(2m + 3)$
$= (2m + 3)(5m + 2)$

28. Factor $8t^2 + 10tv + 3v^2$ by trial and error.

$8t^2 + 10tv + 3v^2 = (4t + 3v)(2t + v)$

29. $4p^2 - 12p + 9 = (2p - 3)(2p - 3)$
$= (2p - 3)^2$

30. $25r^2 - 81t^2 = (5r)^2 - (9t)^2$
$= (5r + 9t)(5r - 9t)$

31. $2pq + 6p^3q + 8p^2q$
$= 2pq(1 + 3p^2 + 4p)$
$= 2pq(3p^2 + 4p + 1)$
$= 2pq(3p + 1)(p + 1)$

32. $6m^2 + m - 2 = 0$
$(3m + 2)(2m - 1) = 0$

$3m + 2 = 0$ or $2m - 1 = 0$
$3m = -2$ $\qquad$ $2m = 1$
$m = -\tfrac{2}{3}$ or $m = \tfrac{1}{2}$

The solutions are $-\tfrac{2}{3}$ and $\tfrac{1}{2}$.

33. $8x^2 = 64x$
$8x^2 - 64x = 0$
$8x(x - 8) = 0$

$8x = 0$ or $x - 8 = 0$
$x = 0$ or $x = 8$

The solutions are 0 and 8.

34. Let $x =$ the length of the shorter leg.
Then $x + 7 =$ the length of the longer leg, and $2x + 3 =$ the length of the hypotenuse.

Use the Pythagorean formula.

$x^2 + (x + 7)^2 = (2x + 3)^2$
$x^2 + (x^2 + 14x + 49) = 4x^2 + 12x + 9$
$2x^2 + 14x + 49 = 4x^2 + 12x + 9$
$0 = 2x^2 - 2x - 40$
$0 = 2(x^2 - x - 20)$
$0 = (x - 5)(x + 4)$

$x - 5 = 0$ or $x + 4 = 0$
$x = 5$ or $x = -4$

Reject -4 because the length of a leg cannot be negative. Since $x = 5$, $x + 7 = 12$, and $2x + 3 = 2(5) + 3 = 13$. The length of the sides are 5 meters, 12 meters, and 13 meters.

CHAPTER 7 RATIONAL EXPRESSIONS AND APPLICATIONS

7.1 The Fundamental Property of Rational Expressions

7.1 Margin Exercises

1. (a) $\dfrac{x+2}{x-5}$

   Solve $x - 5 = 0$.
   Since $x = 5$ will make the denominator zero, the expression is undefined for 5.

   (b) $\dfrac{3r}{r^2 + 6r + 8}$

   Solve $r^2 + 6r + 8 = 0$.
   $(r+4)(r+2) = 0$, so the denominator is zero when either $r + 4 = 0$ or $r + 2 = 0$, that is, when $r = -4$ or $r = -2$. The expression is undefined for -4 and -2.

   (c) $\dfrac{-5m}{m^2 + 4}$

   Since $m^2 + 4$ cannot equal 0, there are no values of the variable that make the expression undefined. The expression is never undefined.

2. (a) $\dfrac{x}{2x+1} = \dfrac{3}{2(3)+1}$ Let $x = 3$.
   $= \dfrac{3}{6+1}$
   $= \dfrac{3}{7}$

   (b) $\dfrac{2x+6}{x-3} = \dfrac{2(3)+6}{3-3}$ Let $x = 3$.
   $= \dfrac{12}{0}$

   If we substitute 3 for x, the denominator is zero, so the quotient is undefined when $x = 3$.

3. (a) $\dfrac{5x^4}{15x^2} = \dfrac{5 \cdot x \cdot x \cdot x \cdot x}{3 \cdot 5 \cdot x \cdot x}$ Factor.
   $= \dfrac{x \cdot x(5 \cdot x \cdot x)}{3(5 \cdot x \cdot x)}$ Group.
   $= \dfrac{x^2}{3}$ Fundamental property

   (b) $\dfrac{6p^3}{2p^2} = \dfrac{2 \cdot 3 \cdot p \cdot p \cdot p}{2 \cdot p \cdot p}$ Factor.
   $= \dfrac{3 \cdot p(2 \cdot p \cdot p)}{1(2 \cdot p \cdot p)}$ Group.
   $= \dfrac{3p}{1}$ Fundamental property
   $= 3p$

4. (a) $\dfrac{4y+2}{6y+3} = \dfrac{2(2y+1)}{3(2y+1)}$ Factor.
   $= \dfrac{2}{3}$ Fundamental property

   (b) $\dfrac{8p+8q}{5p+5q} = \dfrac{8(p+q)}{5(p+q)}$ Factor.
   $= \dfrac{8}{5}$ Fundamental property

   (c) $\dfrac{x^2+4x+4}{4x+8}$
   $= \dfrac{(x+2)(x+2)}{4(x+2)}$ Factor.
   $= \dfrac{x+2}{4}$ Fundamental property

   (d) $\dfrac{a^2 - b^2}{a^2 + 2ab + b^2}$
   $= \dfrac{(a-b)(a+b)}{(a+b)(a+b)}$ Factor.
   $= \dfrac{a-b}{a+b}$ Fundamental property

5. (a) $\dfrac{5-y}{y-5}$

   Since $y - 5 = -1(-y + 5)$
   $= -1(5 - y)$,

   $5 - y$ and $y - 5$ are negatives (or opposites) of each other. Therefore,
   $$\dfrac{5-y}{y-5} = -1.$$

   (b) $\dfrac{m-n}{n-m}$

   $n - m = -1(-n + m)$
   $= -1(m - n)$

   $m - n$ and $n - m$ are negatives of each other. Therefore,
   $$\dfrac{m-n}{n-m} = -1.$$

Chapter 7 Rational Expressions and Applications

(c) $\dfrac{25x^2 - 16}{12 - 15x} = \dfrac{(5x)^2 - 4^2}{3(4 - 5x)}$

$= \dfrac{(5x + 4)(5x - 4)}{3(-1)(5x - 4)}$

$= \dfrac{5x + 4}{3(-1)}$

$= \dfrac{5x + 4}{-3}$ or $-\dfrac{5x + 4}{3}$

(d) $\dfrac{9 - k}{9 + k}$

$9 - k = -1(-9 + k) \neq -1(9 + k)$

The expressions $9 - k$ and $9 + k$ are not negatives of each other. They do not have any common factors (other than 1), so the rational expression is already in lowest terms.

6. (a) $\dfrac{-(2x - 6)}{x + 3} = -\dfrac{2x - 6}{x + 3}$

This expression is equivalent to $-\dfrac{2x - 6}{x + 3}$.

(b) $\dfrac{-2x + 6}{x + 3} = \dfrac{-(2x - 6)}{x + 3} = -\dfrac{2x - 6}{x + 3}$

This expression is equivalent to $-\dfrac{2x - 6}{x + 3}$.

(c) $\dfrac{-2x - 6}{x + 3} = \dfrac{-(2x + 6)}{x + 3} = -\dfrac{2x + 6}{x + 3}$

This expression is *not* equivalent to $-\dfrac{2x - 6}{x + 3}$.

(d) $\dfrac{2x - 6}{-(x + 3)} = -\dfrac{2x - 6}{x + 3}$

This expression is equivalent to $-\dfrac{2x - 6}{x + 3}$.

(e) $\dfrac{2x - 6}{-x - 3} = \dfrac{2x - 6}{-(x + 3)} = -\dfrac{2x - 6}{x + 3}$

This expression is equivalent to $-\dfrac{2x - 6}{x + 3}$.

(f) $\dfrac{2x - 6}{x - 3} = \dfrac{2x - 6}{-(-x + 3)} = -\dfrac{2x - 6}{-x + 3}$

This expression is *not* equivalent to $-\dfrac{2x - 6}{x + 3}$.

7.1 Section Exercises

1. (a) The rational expression $\dfrac{x + 5}{x - 3}$ is undefined when its denominator is equal to 0; that is, when $x - 3 = 0$, or $x = \underline{3}$, so $x \neq \underline{3}$. It is equal to 0 when its numerator is equal to 0; that is, when $x + 5 = 0$, or $x = \underline{-5}$.

(b) The denominator is 0 when $q - p = 0$, or $q = p$. The expression can be simplified when $q - p \neq 0$:

$\dfrac{p - q}{q - p} = \dfrac{p - q}{-1(-q + p)} = \dfrac{p - q}{-1(p - q)} = -1$

The rational expression $\dfrac{p - q}{q - p}$ is undefined when $p = \underline{q}$, and in all other cases when written in lowest terms is equal to $\underline{-1}$.

3. A rational expression is a quotient of two polynomials, such as $\dfrac{x + 3}{x^2 - 4}$. One can think of this as an algebraic fraction.

5. $\dfrac{2}{5y}$

The denominator $5y$ will be zero when $y = 0$, so the given expression is undefined for $y = 0$.

7. $\dfrac{4x^2}{3x - 5}$

To find the values for which this expression is undefined, set the denominator equal to zero and solve for x.

$3x - 5 = 0$
$3x = 5$
$x = \tfrac{5}{3}$

Because $x = \tfrac{5}{3}$ will make the denominator zero, the given expression is undefined for $\tfrac{5}{3}$.

9. $\dfrac{m + 2}{m^2 + m - 6}$

To find the numbers that make the denominator 0, we must solve

$m^2 + m - 6 = 0$
$(m + 3)(m - 2) = 0$

$m + 3 = 0$ or $m - 2 = 0$
$m = -3$ or $m = 2$

The given expression is undefined for $m = -3$ and for $m = 2$.

11. $\dfrac{3x}{x^2 + 2}$

This denominator cannot equal zero for any value of x because x^2 is always greater than or equal to zero, and adding 2 makes the sum greater than zero. Thus, the given rational expression is never undefined.

7.1 The Fundamental Property of Rational Expressions

13. (a) $\dfrac{5x-2}{4x} = \dfrac{5\cdot 2 - 2}{4\cdot 2}$ Let $x = 2$.

    $= \dfrac{10-2}{8}$

    $= \dfrac{8}{8} = 1$

    (b) $\dfrac{5x-2}{4x} = \dfrac{5(-3)-2}{4(-3)}$ Let $x = -3$.

    $= \dfrac{-15-2}{-12}$

    $= \dfrac{-17}{-12} = \dfrac{17}{12}$

15. (a) $\dfrac{2x^2 - 4x}{3x} = \dfrac{2(2)^2 - 4(2)}{3(2)}$ Let $x = 2$.

    $= \dfrac{2(4) - 4(2)}{3(2)}$

    $= \dfrac{8-8}{6}$

    $= \dfrac{0}{6} = 0$

    (b) $\dfrac{2x^2 - 4x}{3x} = \dfrac{2(-3)^2 - 4(-3)}{3(-3)}$ Let $x = -3$.

    $= \dfrac{2(9) - (-12)}{-9}$

    $= \dfrac{18+12}{-9}$

    $= \dfrac{30}{-9} = -\dfrac{10}{3}$

17. (a) $\dfrac{(-3x)^2}{4x+12} = \dfrac{(-3\cdot 2)^2}{4\cdot 2 + 12}$ Let $x = 2$.

    $= \dfrac{(-6)^2}{8+12}$

    $= \dfrac{36}{20} = \dfrac{9}{5}$

    (b) $\dfrac{(-3x)^2}{4x+12} = \dfrac{[-3(-3)]^2}{4(-3)+12}$ Let $x = -3$.

    $= \dfrac{9^2}{-12+12}$

    $= \dfrac{81}{0}$

    Since substituting -3 for x makes the denominator zero, the given rational expression is *undefined* when $x = -3$.

19. (a) $\dfrac{5x+2}{2x^2 + 11x + 12}$

    $= \dfrac{5(2)+2}{2(2)^2 + 11(2) + 12}$ Let $x = 2$.

    $= \dfrac{10+2}{2(4) + 11(2) + 12}$

    $= \dfrac{12}{8 + 22 + 12}$

    $= \dfrac{12}{42} = \dfrac{2}{7}$

    (b) $\dfrac{5x+2}{2x^2 + 11x + 12}$

    $= \dfrac{5(-3)+2}{2(-3)^2 + 11(-3) + 12}$ Let $x = -3$.

    $= \dfrac{-15+2}{2(9) - 33 + 12}$

    $= \dfrac{-13}{18 - 33 + 12}$

    $= \dfrac{-13}{-3} = \dfrac{13}{3}$

21. Any number divided by itself is 1, *provided the number is not 0*. This expression is equal to

    $\dfrac{1}{x+2}$ for all values of x except -2 and 2, since

    $\dfrac{x-2}{x^2 - 4} = \dfrac{x-2}{(x+2)(x-2)}.$

23. $\dfrac{18r^3}{6r} = \dfrac{3r^2(6r)}{1(6r)}$ Factor.

    $= 3r^2$ Fundamental property

25. $\dfrac{4(y-2)}{10(y-2)} = \dfrac{2\cdot 2(y-2)}{5\cdot 2(y-2)}$ Factor.

    $= \dfrac{2}{5}$ Fundamental property

27. $\dfrac{(x+1)(x-1)}{(x+1)^2} = \dfrac{(x+1)(x-1)}{(x+1)(x+1)}$

    $= \dfrac{x-1}{x+1}$ Fundamental property

29. $\dfrac{7m+14}{5m+10} = \dfrac{7(m+2)}{5(m+2)}$ Factor.

    $= \dfrac{7}{5}$ Fundamental property

31. $\dfrac{m^2 - n^2}{m+n} = \dfrac{(m+n)(m-n)}{m+n}$

    $= m - n$

33. $\dfrac{12m^2 - 3}{8m - 4} = \dfrac{3(4m^2 - 1)}{4(2m - 1)}$

$= \dfrac{3(2m + 1)(2m - 1)}{4(2m - 1)}$

$= \dfrac{3(2m + 1)}{4}$

35. $\dfrac{3m^2 - 3m}{5m - 5} = \dfrac{3m(m - 1)}{5(m - 1)}$

$= \dfrac{3m}{5}$

37. $\dfrac{9r^2 - 4s^2}{9r + 6s} = \dfrac{(3r + 2s)(3r - 2s)}{3(3r + 2s)}$

$= \dfrac{3r - 2s}{3}$

39. $\dfrac{2x^2 - 3x - 5}{2x^2 - 7x + 5} = \dfrac{(2x - 5)(x + 1)}{(2x - 5)(x - 1)}$

$= \dfrac{x + 1}{x - 1}$

41. Factor the numerator and denominator by grouping.

$\dfrac{zw + 4z - 3w - 12}{zw + 4z + 5w + 20}$

$= \dfrac{z(w + 4) - 3(w + 4)}{z(w + 4) + 5(w + 4)}$

$= \dfrac{(w + 4)(z - 3)}{(w + 4)(z + 5)}$

$= \dfrac{z - 3}{z + 5}$

43. $\dfrac{6 - t}{t - 6} = \dfrac{-1(t - 6)}{1(t - 6)} = \dfrac{-1}{1} = -1$

Note that $6 - t$ and $t - 6$ are opposites, so we know that their quotient will be -1.

45. $\dfrac{m^2 - 1}{1 - m} = \dfrac{(m + 1)(m - 1)}{-1(m - 1)}$

$= \dfrac{m + 1}{-1}$

$= -(m + 1)$ or $-m - 1$

47. $\dfrac{q^2 - 4q}{4q - q^2} = \dfrac{q(q - 4)}{q(4 - q)}$

$= \dfrac{q - 4}{4 - q} = -1$

$q - 4$ and $4 - q$ are opposites.

49. To write four equivalent expressions for $-\dfrac{x + 4}{x - 3}$, we will follow the outline in Example 7. Applying the negative sign to the numerator we have

$$\dfrac{-(x + 4)}{x - 3}.$$

Distributing the negative sign gives us

$$\dfrac{-x - 4}{x - 3}.$$

Applying the negative sign to the denominator yields

$$\dfrac{x + 4}{-(x - 3)}.$$

Again, we distribute to get

$$\dfrac{x + 4}{-x + 3}.$$

51. $-\dfrac{2x - 3}{x + 3}$ is equivalent to each of the following:

$\dfrac{-(2x - 3)}{x + 3}$, $\dfrac{-2x + 3}{x + 3}$,

$\dfrac{2x - 3}{-(x + 3)}$, $\dfrac{2x - 3}{-x - 3}$

53. First, factor out -1 in the numerator.

$$\dfrac{-3x + 1}{5x - 6} = \dfrac{-(3x - 1)}{5x - 6}$$

The negative sign may be placed in front of the fraction, giving us

$$-\dfrac{3x - 1}{5x - 6}.$$

Applying the negative sign to the denominator gives us

$$\dfrac{3x - 1}{-(5x - 6)}.$$

Distributing the negative sign yields

$$\dfrac{3x - 1}{-5x + 6}.$$

55. $L \cdot W = A$

$$W = \frac{A}{L}$$

$$W = \frac{x^4 + 10x^2 + 21}{x^2 + 7}$$

$$= \frac{(x^2 + 7)(x^2 + 3)}{x^2 + 7}$$

$$= x^2 + 3$$

Note: If it is not apparent that we can factor A as $x^4 + 10x^2 + 21 = (x^2 + 7)(x^2 + 3)$, we may use "long division" to find the quotient $\frac{A}{L}$. Remember to insert zeros for the coefficients of the missing terms in the dividend and divisor.

$$\begin{array}{r}
x^2 + 3 \\
x^2 + 0x + 7 \overline{\smash{\big)}\, x^4 + 0x^3 + 10x^2 + 0x + 21} \\
\underline{x^4 + 0x^3 + 7x^2 } \\
3x^2 + 0x + 21 \\
\underline{3x^2 + 0x + 21} \\
0
\end{array}$$

The width of the rectangle is $x^2 + 3$.

7.2 Multiplying and Dividing Rational Expressions

7.2 Margin Exercises

1. (a) $\dfrac{2}{7} \cdot \dfrac{5}{10} = \dfrac{2 \cdot 5}{7 \cdot 10}$

    $= \dfrac{2 \cdot 5}{7 \cdot 2 \cdot 5}$

    $= \dfrac{1}{7}$

    (b) $\dfrac{3m^2}{2} \cdot \dfrac{10}{m} = \dfrac{3m^2 \cdot 10}{2 \cdot m}$

    $= \dfrac{3 \cdot m \cdot m \cdot 2 \cdot 5}{m \cdot 2}$

    $= \dfrac{3 \cdot m \cdot 5}{1} = 15m$

    (c) $\dfrac{8p^2 q}{3} \cdot \dfrac{9}{q^2 p} = \dfrac{2 \cdot 4 \cdot p \cdot p \cdot q \cdot 3 \cdot 3}{3 \cdot q \cdot q \cdot p}$

    $= \dfrac{2 \cdot 4 \cdot p \cdot 3}{q} = \dfrac{24p}{q}$

2. (a) $\dfrac{a+b}{5} \cdot \dfrac{30}{2(a+b)} = \dfrac{2 \cdot 3 \cdot 5(a+b)}{5 \cdot 2(a+b)}$

    $= \dfrac{3}{1} = 3$

(b) $\dfrac{3(p-q)}{q^2} \cdot \dfrac{q}{2(p-q)^2}$

$= \dfrac{3q(p-q)}{2 \cdot q \cdot q(p-q)(p-q)}$

$= \dfrac{3}{2q(p-q)}$

3. (a) $\dfrac{x^2 + 7x + 10}{3x + 6} \cdot \dfrac{6x - 6}{x^2 + 2x - 15}$

    $= \dfrac{(x+2)(x+5)}{3(x+2)} \cdot \dfrac{6(x-1)}{(x+5)(x-3)}$ Factor.

    $= \dfrac{6(x+2)(x+5)(x-1)}{3(x+2)(x+5)(x-3)}$

    Multiply.

    $= \dfrac{2(x-1)}{x-3}$ Lowest terms

    (b) $\dfrac{m^2 + 4m - 5}{m + 5} \cdot \dfrac{m^2 + 8m + 15}{m - 1}$

    $= \dfrac{(m-1)(m+5)}{m+5} \cdot \dfrac{(m+3)(m+5)}{m-1}$

    Factor.

    $= \dfrac{(m-1)(m+5)(m+3)(m+5)}{(m+5)(m-1)}$

    Multiply.

    $= (m+3)(m+5)$ Lowest terms

4. (a) The reciprocal of a rational expression is found by interchanging the numerator and the denominator, so the reciprocal of

    $$\dfrac{6b^5}{3r^2 b} \text{ is } \dfrac{3r^2 b}{6b^5}.$$

    (b) The reciprocal of

    $$\dfrac{t^2 - 4t}{t^2 + 2t - 3} \text{ is } \dfrac{t^2 + 2t - 3}{t^2 - 4t}.$$

5. (a) $\dfrac{3}{4} \div \dfrac{5}{16} = \dfrac{3}{4} \cdot \dfrac{16}{5}$

    Multiply the first expression by the reciprocal of the second.

    $= \dfrac{3 \cdot 16}{4 \cdot 5}$ Multiply.

    $= \dfrac{3 \cdot 4}{5} = \dfrac{12}{5}$ Lowest terms

    (b) $\dfrac{r}{r-1} \div \dfrac{3r}{r+4} = \dfrac{r}{r-1} \cdot \dfrac{r+4}{3r}$

    Multiply the first expression by the reciprocal of the second.

    $= \dfrac{r(r+4)}{3r(r-1)}$ Multiply.

    $= \dfrac{r+4}{3(r-1)}$ Lowest terms

260 Chapter 7 Rational Expressions and Applications

(c) $\dfrac{6x-4}{3} \div \dfrac{15x-10}{9}$

$= \dfrac{6x-4}{3} \cdot \dfrac{9}{15x-10}$ *Multiply by reciprocal.*

$= \dfrac{2(3x-2)}{3} \cdot \dfrac{3 \cdot 3}{5(3x-2)}$ *Factor.*

$= \dfrac{2 \cdot 3 \cdot 3(3x-2)}{3 \cdot 5(3x-2)}$ *Multiply.*

$= \dfrac{6}{5}$ *Lowest terms*

6. (a) $\dfrac{5a^2b}{2} \div \dfrac{10ab^2}{8}$

$= \dfrac{5a^2b}{2} \cdot \dfrac{8}{10ab^2}$

    Multiply by reciprocal.

$= \dfrac{5 \cdot 8a^2b}{2 \cdot 10ab^2}$ *Multiply.*

$= \dfrac{2a}{b}$ *Lowest terms*

(b) $\dfrac{(3t)^2}{w} \div \dfrac{3t^2}{5w^4}$

$= \dfrac{9t^2}{w} \cdot \dfrac{5w^4}{3t^2}$

    Multiply by reciprocal.

$= \dfrac{9 \cdot 5t^2 w^4}{3t^2 w}$ *Multiply.*

$= 15w^3$ *Lowest terms*

7. (a) $\dfrac{y^2 + 4y + 3}{y+3} \div \dfrac{y^2 - 4y - 5}{y-3}$

$= \dfrac{y^2 + 4y + 3}{y+3} \cdot \dfrac{y-3}{y^2 - 4y - 5}$

    Multiply by reciprocal.

$= \dfrac{(y+3)(y+1)}{y+3} \cdot \dfrac{y-3}{(y-5)(y+1)}$ *Factor.*

$= \dfrac{(y+3)(y+1)(y-3)}{(y+3)(y-5)(y+1)}$ *Multiply.*

$= \dfrac{y-3}{y-5}$ *Lowest terms*

(b) $\dfrac{4x(x+3)}{2x+1} \div \dfrac{-x^2(x+3)}{4x^2-1}$

$= \dfrac{4x(x+3)}{2x+1} \cdot \dfrac{4x^2-1}{-x^2(x+3)}$

    Multiply by reciprocal.

$= \dfrac{4x(x+3)}{2x+1} \cdot \dfrac{(2x+1)(2x-1)}{-x^2(x+3)}$ *Factor.*

$= \dfrac{4x(x+3)(2x+1)(2x-1)}{-x^2(2x+1)(x+3)}$ *Multiply.*

$= -\dfrac{4(2x-1)}{x}$ *Lowest terms*

8. (a) $\dfrac{ab-a^2}{a^2-1} \div \dfrac{a-b}{a-1} = \dfrac{ab-a^2}{a^2-1} \cdot \dfrac{a-1}{a-b}$

$= \dfrac{a(b-a)}{(a+1)(a-1)} \cdot \dfrac{a-1}{a-b}$

$= \dfrac{-a(a-b)}{(a+1)(a-1)} \cdot \dfrac{a-1}{a-b}$

$= \dfrac{-a}{a+1}$

(b) $\dfrac{x^2-9}{2x+6} \div \dfrac{9-x^2}{4x-12} = \dfrac{x^2-9}{2x+6} \cdot \dfrac{4x-12}{9-x^2}$

$= \dfrac{-(9-x^2)}{2(x+3)} \cdot \dfrac{4(x-3)}{9-x^2}$

$= \dfrac{-2(x-3)}{x+3}$

$= \dfrac{-2x+6}{x+3}$

7.2 Section Exercises

1. (a) $\dfrac{5x^3}{10x^4} \cdot \dfrac{10x^7}{2x} = \dfrac{5 \cdot 10 \cdot x^3 \cdot x^7}{10 \cdot 2 \cdot x^4 \cdot x}$

$= \dfrac{5x^{10}}{2x^5}$

$= \dfrac{5x^5}{2}$; **B**

(b) $\dfrac{10x^4}{5x^3} \cdot \dfrac{10x^7}{2x} = \dfrac{10 \cdot 10 \cdot x^4 \cdot x^7}{5 \cdot 2 \cdot x^3 \cdot x}$

$= \dfrac{10x^{11}}{1x^4}$

$= 10x^7$; **D**

(c) $\dfrac{5x^3}{10x^4} \cdot \dfrac{2x}{10x^7} = \dfrac{5 \cdot 2 \cdot x^3 \cdot x}{10 \cdot 10 \cdot x^4 \cdot x^7}$

$= \dfrac{1x^4}{10x^{11}}$

$= \dfrac{1}{10x^7}$; **C**

(d) $\dfrac{10x^4}{5x^3} \cdot \dfrac{2x}{10x^7} = \dfrac{10 \cdot 2 \cdot x^4 \cdot x}{5 \cdot 10 \cdot x^3 \cdot x^7}$

$= \dfrac{2x^5}{5x^{10}}$

$= \dfrac{2}{5x^5}$; **A**

3. $\dfrac{10m^2}{7} \cdot \dfrac{14}{15m} = \dfrac{5 \cdot 2 \cdot m \cdot m \cdot 2 \cdot 7}{7 \cdot 3 \cdot 5 \cdot m}$

$= \dfrac{2 \cdot 2 \cdot m}{3} = \dfrac{4m}{3}$

5. $\dfrac{16y^4}{18y^5} \cdot \dfrac{15y^5}{y^2} = \dfrac{2 \cdot 8 \cdot 3 \cdot 5y^9}{2 \cdot 3 \cdot 3y^7}$

$= \dfrac{8 \cdot 5y^{9-7}}{3} = \dfrac{40y^2}{3}$

7.2 Multiplying and Dividing Rational Expressions

7. $\dfrac{2(c+d)}{3} \cdot \dfrac{18}{6(c+d)^2}$

$= \dfrac{2 \cdot 3 \cdot 6(c+d)}{3 \cdot 6(c+d)(c+d)}$ *Multiply and factor.*

$= \dfrac{2}{c+d}$ *Lowest terms*

9. The reciprocal of a rational expression is found by interchanging the numerator and the denominator, so the reciprocal of

$$\dfrac{3p^3}{16q} \text{ is } \dfrac{16q}{3p^3}.$$

11. The reciprocal of a rational expression is found by interchanging the numerator and the denominator, so the reciprocal of

$$\dfrac{r^2+rp}{7} \text{ is } \dfrac{7}{r^2+rp}.$$

13. The reciprocal of

$$\dfrac{z^2+7z+12}{z^2-9} \text{ is } \dfrac{z^2-9}{z^2+7z+12}.$$

15. $\dfrac{9z^4}{3z^5} \div \dfrac{3z^2}{5z^3} = \dfrac{9z^4}{3z^5} \cdot \dfrac{5z^3}{3z^2}$

$= \dfrac{9 \cdot 5 z^7}{3 \cdot 3 z^7}$

$= 5$

17. $\dfrac{4t^4}{2t^5} \div \dfrac{(2t)^3}{-6} = \dfrac{4t^4}{2t^5} \cdot \dfrac{-6}{(2t)^3}$

$= \dfrac{4t^4}{2t^5} \cdot \dfrac{-6}{8t^3}$

$= \dfrac{-24t^4}{16t^8}$

$= \dfrac{-3(8t^4)}{2t^4(8t^4)}$

$= \dfrac{-3}{2t^4} = -\dfrac{3}{2t^4}$

19. $\dfrac{3}{2y-6} \div \dfrac{6}{y-3} = \dfrac{3}{2y-6} \cdot \dfrac{y-3}{6}$

$= \dfrac{3}{2(y-3)} \cdot \dfrac{y-3}{6}$

$= \dfrac{3(y-3)}{2 \cdot 2 \cdot 3(y-3)}$

$= \dfrac{1}{2 \cdot 2} = \dfrac{1}{4}$

21. To multiply two rational expressions, multiply the numerators and multiply the denominators. Write the answer in lowest terms.

23. $\dfrac{5x-15}{3x+9} \cdot \dfrac{4x+12}{6x-18}$

$= \dfrac{5(x-3)}{3(x+3)} \cdot \dfrac{4(x+3)}{6(x-3)}$

$= \dfrac{5 \cdot 4 \cdot (x-3)(x+3)}{3 \cdot 6 \cdot (x-3)(x+3)}$

$= \dfrac{10}{9}$

25. $\dfrac{2-t}{8} \div \dfrac{t-2}{6} = \dfrac{2-t}{8} \cdot \dfrac{6}{t-2}$ *Multiply by reciprocal.*

$= \dfrac{6(2-t)}{8(t-2)}$ *Multiply numerators; multiply denominators.*

$= \dfrac{6(-1)}{8}$ $\dfrac{2-t}{t-2} = -1$

$= -\dfrac{3}{4}$ *Lowest terms*

27. $\dfrac{5-4x}{5+4x} \cdot \dfrac{4x+5}{4x-5}$

$= \dfrac{(5-4x)(4x+5)}{(5+4x)(4x-5)}$

$= \dfrac{5-4x}{4x-5} \qquad \dfrac{4x+5}{5+4x} = 1$

$= -1$ *Opposites*

29. $\dfrac{6(m-2)^2}{5(m+4)^2} \cdot \dfrac{15(m+4)}{2(2-m)}$

$= \dfrac{2 \cdot 3(m-2)(m-2)}{5(m+4)(m+4)} \cdot \dfrac{3 \cdot 5(m+4)}{2(-1)(m-2)}$

$= \dfrac{3(m-2)(3)}{(m+4)(-1)}$

$= \dfrac{9(m-2)}{-(m+4)}$ or $\dfrac{-9(m-2)}{m+4}$

31. $\dfrac{p^2+4p-5}{p^2+7p+10} \div \dfrac{p-1}{p+4}$

$= \dfrac{p^2+4p-5}{p^2+7p+10} \cdot \dfrac{p+4}{p-1}$

$= \dfrac{(p+5)(p-1) \cdot (p+4)}{(p+5)(p+2) \cdot (p-1)}$

$= \dfrac{p+4}{p+2}$

262 Chapter 7 Rational Expressions and Applications

33. $\dfrac{2k^2 - k - 1}{2k^2 + 5k + 3} \div \dfrac{4k^2 - 1}{2k^2 + k - 3}$

$= \dfrac{2k^2 - k - 1}{2k^2 + 5k + 3} \cdot \dfrac{2k^2 + k - 3}{4k^2 - 1}$

$= \dfrac{(2k + 1)(k - 1)(2k + 3)(k - 1)}{(2k + 3)(k + 1)(2k + 1)(2k - 1)}$

$= \dfrac{(k - 1)(k - 1)}{(k + 1)(2k - 1)}$

$= \dfrac{(k - 1)^2}{(k + 1)(2k - 1)}$

35. $\dfrac{2k^2 + 3k - 2}{6k^2 - 7k + 2} \cdot \dfrac{4k^2 - 5k + 1}{k^2 + k - 2}$

$= \dfrac{(2k - 1)(k + 2)}{(3k - 2)(2k - 1)} \cdot \dfrac{(4k - 1)(k - 1)}{(k + 2)(k - 1)}$

$= \dfrac{(2k - 1)(k + 2)(4k - 1)(k - 1)}{(3k - 2)(2k - 1)(k + 2)(k - 1)}$

$= \dfrac{4k - 1}{3k - 2}$

37. $\dfrac{m^2 + 2mp - 3p^2}{m^2 - 3mp + 2p^2} \div \dfrac{m^2 + 4mp + 3p^2}{m^2 + 2mp - 8p^2}$

$= \dfrac{m^2 + 2mp - 3p^2}{m^2 - 3mp + 2p^2} \cdot \dfrac{m^2 + 2mp - 8p^2}{m^2 + 4mp + 3p^2}$

$= \dfrac{(m + 3p)(m - p)(m + 4p)(m - 2p)}{(m - 2p)(m - p)(m + 3p)(m + p)}$

$= \dfrac{m + 4p}{m + p}$

39. $\left(\dfrac{x^2 + 10x + 25}{x^2 + 10x} \cdot \dfrac{10x}{x^2 + 15x + 50}\right) \div \dfrac{x + 5}{x + 10}$

$= \left[\dfrac{(x + 5)^2 \cdot 10x}{x(x + 10)(x + 5)(x + 10)}\right] \div \dfrac{x + 5}{x + 10}$

$= \left[\dfrac{10(x + 5)}{(x + 10)^2}\right] \div \dfrac{x + 5}{x + 10}$

$= \dfrac{10(x + 5)}{(x + 10)^2} \cdot \dfrac{x + 10}{x + 5}$

$= \dfrac{10}{x + 10}$

41. Division requires multiplying by the reciprocal of the second rational expression. In the reciprocal, $x + 7$ is in the denominator, so $x \neq -7$.

7.3 Least Common Denominators

7.3 Margin Exercises

1. **(a)** $\dfrac{7}{10}, \dfrac{1}{25}$

   Factor each denominator.

   $10 = 2 \cdot 5, \quad 25 = 5 \cdot 5 = 5^2$

   Take each different factor the *greatest* number of times it appears as a factor in any of the denominators, and use it to form the least common denominator (LCD).

   $LCD = 2 \cdot 5^2 = 50$

   (b) $\dfrac{7}{20p}, \dfrac{11}{30p}$

   Factor each denominator.

   $20p = 2 \cdot 2 \cdot 5 \cdot p = 2^2 \cdot 5 \cdot p$
   $30p = 2 \cdot 3 \cdot 5 \cdot p$

   Take each factor the greatest number of times it appears in any denominator; then multiply.

   $LCD = 2^2 \cdot 3 \cdot 5 \cdot p = 60p$

   (c) $\dfrac{4}{5x}, \dfrac{12}{10x}$

   Factor each denominator.

   $5x = 5 \cdot x, \quad 10x = 2 \cdot 5 \cdot x$

   Take each factor the greatest number of times it appears in any denominator; then multiply.

   $LCD = 2 \cdot 5 \cdot x = 10x$

2. **(a)** $\dfrac{4}{16m^3n}, \dfrac{5}{9m^5}$

   Factor each denominator.

   $16m^3n = 2^4 \cdot m^3 \cdot n$
   $9m^5 = 3^2 \cdot m^5$

   Take each factor the greatest number of times it appears in any denominator; then multiply.

   $LCD = 2^4 \cdot 3^2 \cdot m^5 \cdot n = 144m^5n$

   (b) $\dfrac{3}{25a^2}, \dfrac{2}{10a^3b}$

   Factor each denominator.

   $25a^2 = 5^2 \cdot a^2$
   $10a^3b = 2 \cdot 5 \cdot a^3 \cdot b$

   Take each factor the greatest number of times it appears in any denominator; then multiply.

   $LCD = 2 \cdot 5^2 \cdot a^3 \cdot b = 50a^3b$

7.3 Least Common Denominators

3. (a) $\dfrac{7}{3a}, \dfrac{11}{a^2 - 4a}$

   Factor each denominator.
   $$3a = 3 \cdot a$$
   $$a^2 - 4a = a(a - 4)$$

   Take each factor the greatest number of times it appears in any denominator; then multiply.
   $$\text{LCD} = 3 \cdot a(a - 4) = 3a(a - 4)$$

   (b) $\dfrac{2m}{m^2 - 3m + 2}, \dfrac{5m - 3}{m^2 + 3m - 10}, \dfrac{4m + 7}{m^2 + 4m - 5}$

   Factor each denominator.
   $$m^2 - 3m + 2 = (m - 1)(m - 2)$$
   $$m^2 + 3m - 10 = (m + 5)(m - 2)$$
   $$m^2 + 4m - 5 = (m - 1)(m + 5)$$

   Take each factor the greatest number of times it appears in any denominator; then multiply.
   $$\text{LCD} = (m - 1)(m - 2)(m + 5)$$

   (c) $\dfrac{6}{x - 4}, \dfrac{3x - 1}{4 - x}$

   The expressions $x - 4$ and $4 - x$ are opposites of each other because
   $$-(x - 4) = -x + 4 = 4 - x.$$

   Therefore, either $x - 4$ or $4 - x$ can be used as the LCD.

4. (a) $\dfrac{3}{4} = \dfrac{}{36}$

   First factor the denominator on the right. Then compare the denominator on the left with the one on the right to decide what factors are missing.
   $$\dfrac{3}{4} = \dfrac{}{4 \cdot 9}$$

   A factor of 9 is missing, so multiply $\tfrac{3}{4}$ by $\tfrac{9}{9}$, which is equal to 1.
   $$\dfrac{3}{4} = \dfrac{3}{4} \cdot \dfrac{9}{9} = \dfrac{27}{36}$$

   (b) $\dfrac{7k}{5} = \dfrac{}{30p}$

   Factor the denominator on the right; then compare it to the denominator on the left.
   $$\dfrac{7k}{5} = \dfrac{}{5 \cdot 6 \cdot p}$$

   The factors 6 and p are missing on the left, so multiply by $\tfrac{6p}{6p}$.
   $$\dfrac{7k}{5} = \dfrac{7k}{5} \cdot \dfrac{6p}{6p} = \dfrac{42kp}{30p}$$

5. (a) $\dfrac{9}{2a + 5} = \dfrac{}{6a + 15}$

   Factor the denominator on the right.
   $$\dfrac{9}{2a + 5} = \dfrac{}{3(2a + 5)}$$

   The factor 3 is missing on the left, so multiply by $\tfrac{3}{3}$.
   $$\dfrac{9}{2a + 5} = \dfrac{9}{2a + 5} \cdot \dfrac{3}{3} = \dfrac{27}{6a + 15}$$

   (b) $\dfrac{5k + 1}{k^2 + 2k} = \dfrac{}{k^3 + k^2 - 2k}$

   Factor and compare the denominators.
   $$\dfrac{5k + 1}{k(k + 2)} = \dfrac{}{k(k + 2)(k - 1)}$$

   The factor $k - 1$ is missing on the left, so multiply by $\tfrac{k-1}{k-1}$.
   $$\dfrac{5k + 1}{k^2 + 2k} = \dfrac{5k + 1}{k(k + 2)} \cdot \dfrac{k - 1}{k - 1}$$
   $$= \dfrac{(5k + 1)(k - 1)}{(k^2 + 2k)(k - 1)}$$
   $$= \dfrac{(5k + 1)(k - 1)}{k^3 + k^2 - 2k}$$

7.3 Section Exercises

1. The factor a appears at most one time in any denominator as does the factor b. Thus, the LCD is the product of the two factors, ab. The correct response is **C**.

3. Since $20 = 2^2 \cdot 5$, the LCD of $\tfrac{11}{20}$ and $\tfrac{1}{2}$ must have 5 as a factor and 2^2 as a factor. Because 2 appears twice in $2^2 \cdot 5$, we don't have to include another 2 in the LCD for the number $\tfrac{1}{2}$. Thus, the LCD is just $2^2 \cdot 5 = 20$. Note that this is a specific case of Exercise 2 since 2 is a factor of 20. The correct response is **C**.

5. $\dfrac{2}{15}, \dfrac{3}{10}, \dfrac{7}{30}$

   Write each denominator in prime factored form.
   $$15 = 3 \cdot 5, \quad 10 = 2 \cdot 5, \quad 30 = 2 \cdot 3 \cdot 5$$

   To find the LCD, take each different factor the *greatest* number of times it appears as a factor in *any* denominator; then multiply.
   $$\text{LCD} = 3 \cdot 5 \cdot 2 = 30$$

7. $\dfrac{3}{x^4}, \dfrac{5}{x^7}$

   The only factor in any denominator is x. The greatest number of times it appears is 7, so the least common denominator is x^7.

9. $\dfrac{5}{36q}, \dfrac{17}{24q}$

Factor each denominator.
$$36q = 2^2 \cdot 3^2 \cdot q$$
$$24q = 2^3 \cdot 3 \cdot q$$

Take each factor the greatest number of times it appears, and use the greatest exponent on q. The least common denominator is $2^3 \cdot 3^2 \cdot q = 72q$.

11. $\dfrac{6}{21r^3}, \dfrac{8}{12r^5}$

Factor each denominator.
$$21r^3 = 3 \cdot 7 \cdot r^3$$
$$12r^5 = 2^2 \cdot 3 \cdot r^5$$

Take each factor the greatest number of times it appears; then multiply.
$$\text{LCD} = 2^2 \cdot 3 \cdot 7 \cdot r^5 = 84r^5$$

13. Take each factor the greatest number of times it appears. The factored form of their LCD is $2^3 \cdot 3 \cdot 5$.

15. If two denominators have a greatest common factor equal to 1, their least common denominator is the product of the two denominators.

17. $\dfrac{9}{28m^2}, \dfrac{3}{12m - 20}$

Factor each denominator.
$$28m^2 = 2^2 \cdot 7 \cdot m^2$$
$$12m - 20 = 4(3m - 5) = 2^2(3m - 5)$$

Take each factor the greatest number of times it appears; then multiply.
$$\text{LCD} = 2^2 \cdot 7m^2(3m - 5) = 28m^2(3m - 5)$$

19. $\dfrac{7}{5b - 10}, \dfrac{11}{6b - 12}$

Factor each denominator.
$$5b - 10 = 5(b - 2)$$
$$6b - 12 = 6(b - 2) = 2 \cdot 3(b - 2)$$

Take each factor the greatest number of times it appears; then multiply.
$$\text{LCD} = 2 \cdot 3 \cdot 5(b - 2) = 30(b - 2)$$

21. $\dfrac{5}{c - d}, \dfrac{8}{d - c}$

The denominators, $c - d$ and $d - c$, are opposites of each other since
$$-(c - d) = -c + d = d - c.$$

Therefore, either $c - d$ or $d - c$ can be used as the LCD.

23. $\dfrac{3}{k^2 + 5k}, \dfrac{2}{k^2 + 3k - 10}$

Factor each denominator.
$$k^2 + 5k = k(k + 5)$$
$$k^2 + 3k - 10 = (k + 5)(k - 2)$$

$$\text{LCD} = k(k + 5)(k - 2)$$

25. $\dfrac{5}{p^2 + 8p + 15}, \dfrac{3}{p^2 - 3p - 18}, \dfrac{2}{p^2 - p - 30}$

Factor each denominator.
$$p^2 + 8p + 15 = (p + 5)(p + 3)$$
$$p^2 - 3p - 18 = (p - 6)(p + 3)$$
$$p^2 - p - 30 = (p - 6)(p + 5)$$

$$\text{LCD} = (p + 3)(p + 5)(p - 6)$$

27. $\dfrac{4}{11} = \dfrac{}{55}$

First factor the denominator on the right. Then compare the denominator on the left with the one on the right to decide what factors are missing.

$$\dfrac{4}{11} = \dfrac{}{11 \cdot 5}$$

A factor of 5 is missing, so multiply $\dfrac{4}{11}$ by $\dfrac{5}{5}$, which is equal to 1.

$$\dfrac{4}{11} \cdot \dfrac{5}{5} = \dfrac{20}{55}$$

29. $\dfrac{-5}{k} = \dfrac{}{9k}$

A factor of 9 is missing in the first fraction, so multiply numerator and denominator by 9.

$$\dfrac{-5}{k} \cdot \dfrac{9}{9} = \dfrac{-45}{9k}$$

31. $\dfrac{13}{40y} = \dfrac{}{80y^3}$

$80y^3 = (40y)(2y^2)$, so we must multiply the numerator and denominator by $2y^2$.

$$\dfrac{13}{40y} = \dfrac{13}{40y} \cdot \dfrac{2y^2}{2y^2} = \dfrac{26y^2}{80y^3}$$

33. $\dfrac{5t^2}{6r} = \dfrac{}{42r^4}$

$42r^4 = (6r)(7r^3)$, so we must multiply the numerator and denominator by $7r^3$.

$$\dfrac{5t^2}{6r} = \dfrac{5t^2}{6r} \cdot \dfrac{7r^3}{7r^3} = \dfrac{35t^2r^3}{42r^4}$$

35. $\dfrac{5}{2(m+3)} = \dfrac{}{8(m+3)}$

$8(m+3) = 2(m+3) \cdot 4$, so we must multiply the numerator and denominator by 4.

$$\dfrac{5}{2(m+3)} = \dfrac{5}{2(m+3)} \cdot \dfrac{4}{4} = \dfrac{20}{8(m+3)}$$

37. $\dfrac{-4t}{3t-6} = \dfrac{}{12-6t}$

Factor the denominators.

$$3t - 6 = 3(t-2)$$
$$12 - 6t = -6(-2+t) = -6(t-2)$$

$-6(t-2) = 3(t-2) \cdot (-2)$, so we must multiply the numerator and denominator by -2.

$$\dfrac{-2(-4t)}{-2 \cdot 3(t-2)} = \dfrac{8t}{-6(t-2)} \text{ or } \dfrac{8t}{12-6t}$$

39. $\dfrac{14}{z^2 - 3z} = \dfrac{}{z(z-3)(z-2)}$

Compared to the second denominator, $z(z-3)(z-2)$, the first denominator, $z(z-3)$, is missing a factor of $z - 2$. Multiply the numerator and denominator by $z - 2$.

$$\dfrac{14}{z^2 - 3z} = \dfrac{14}{z(z-3)} \cdot \dfrac{z-2}{z-2}$$
$$= \dfrac{14(z-2)}{z(z-3)(z-2)}$$

41. $\dfrac{2(b-1)}{b^2 + b} = \dfrac{}{b^3 + 3b^2 + 2b}$

Compared to the second denominator,

$$b^3 + 3b^2 + 2b = b(b^2 + 3b + 2)$$
$$= b(b+1)(b+2),$$

the first denominator, $b(b+1)$, is missing a factor of $b + 2$. Multiply the numerator and denominator by $b + 2$.

$$\dfrac{2(b-1)}{b(b+1)} \cdot \dfrac{b+2}{b+2} = \dfrac{2(b-1)(b+2)}{b^3 + 3b^2 + 2b}$$

7.4 Adding and Subtracting Rational Expressions

7.4 Margin Exercises

1. (a) $\dfrac{7}{15} + \dfrac{3}{15} = \dfrac{7+3}{15}$
$= \dfrac{10}{15} = \dfrac{2}{3}$

(b) $\dfrac{3}{y+4} + \dfrac{2}{y+4} = \dfrac{3+2}{y+4}$
$= \dfrac{5}{y+4}$

(c) $\dfrac{x}{x+y} + \dfrac{1}{x+y} = \dfrac{x+1}{x+y}$

(d) $\dfrac{a}{a+b} + \dfrac{b}{a+b} = \dfrac{a+b}{a+b} = 1$

(e) $\dfrac{x^2}{x+1} + \dfrac{x}{x+1} = \dfrac{x^2 + x}{x+1}$
$= \dfrac{x(x+1)}{x+1} = x$

2. (a) $\dfrac{1}{10} + \dfrac{1}{15}$

Step 1
LCD $= 2 \cdot 3 \cdot 5 = 30$

Step 2
$\dfrac{1}{10} = \dfrac{1(3)}{10(3)} = \dfrac{3}{30}$
$\dfrac{1}{15} = \dfrac{1(2)}{15(2)} = \dfrac{2}{30}$

Step 3
$\dfrac{3}{30} + \dfrac{2}{30} = \dfrac{3+2}{30}$
$= \dfrac{5}{30}$

Step 4
$\dfrac{5}{30} = \dfrac{1(5)}{6(5)} = \dfrac{1}{6}$

(b) $\dfrac{6}{5x} + \dfrac{9}{2x}$

Step 1
LCD $= 5 \cdot 2 \cdot x = 10x$

Step 2
$\dfrac{6}{5x} = \dfrac{6(2)}{5x(2)} = \dfrac{12}{10x}$
$\dfrac{9}{2x} = \dfrac{9(5)}{2x(5)} = \dfrac{45}{10x}$

Step 3
$\dfrac{12}{10x} + \dfrac{45}{10x} = \dfrac{12+45}{10x}$
$= \dfrac{57}{10x}$

(c) $\dfrac{m}{3n} + \dfrac{2}{7n}$

Step 1
LCD $= 3 \cdot 7 \cdot n = 21n$

Step 2
$\dfrac{m}{3n} = \dfrac{m(7)}{3n(7)} = \dfrac{7m}{21n}$
$\dfrac{2}{7n} = \dfrac{2(3)}{7n(3)} = \dfrac{6}{21n}$

Step 3
$\dfrac{7m}{21n} + \dfrac{6}{21n} = \dfrac{7m+6}{21n}$

3. **(a)** $\dfrac{2p}{3p+3} + \dfrac{5p}{2p+2}$

Since $3p + 3 = 3(p+1)$ and $2p + 2 = 2(p+1)$, the LCD is $2 \cdot 3(p+1) = 6(p+1)$.

$$\dfrac{2p}{3p+3} + \dfrac{5p}{2p+2} = \dfrac{2(2p)}{2(3p+3)} + \dfrac{3(5p)}{3(2p+2)}$$

$$= \dfrac{4p}{6p+6} + \dfrac{15p}{6p+6}$$

$$= \dfrac{4p + 15p}{6p+6}$$

$$= \dfrac{19p}{6p+6}$$

$$= \dfrac{19p}{6(p+1)}$$

(b) $\dfrac{4}{y^2-1} + \dfrac{6}{y+1}$

$$= \dfrac{4}{(y+1)(y-1)} + \dfrac{6}{y+1}$$

$\qquad\qquad$ Factor denominator.

$$= \dfrac{4}{(y+1)(y-1)} + \dfrac{6(y-1)}{(y+1)(y-1)}$$

$\qquad\qquad$ LCD $= (y+1)(y-1)$

$$= \dfrac{4 + 6(y-1)}{(y+1)(y-1)} \quad \text{Add numerators.}$$

$$= \dfrac{4 + 6y - 6}{(y+1)(y-1)} \quad \text{Distributive property}$$

$$= \dfrac{6y - 2}{(y+1)(y-1)} \quad \text{Combine terms.}$$

$$= \dfrac{2(3y-1)}{(y+1)(y-1)} \quad \text{Factor numerator.}$$

(c) $\dfrac{-2}{p+1} + \dfrac{4p}{p^2-1}$

$$= \dfrac{-2}{p+1} + \dfrac{4p}{(p+1)(p-1)} \quad \text{Factor denominator.}$$

$$= \dfrac{-2(p-1)}{(p+1)(p-1)} + \dfrac{4p}{(p+1)(p-1)}$$

$\qquad\qquad$ LCD $= (p+1)(p-1)$

$$= \dfrac{-2(p-1) + 4p}{(p+1)(p-1)} \quad \text{Add numerators.}$$

$$= \dfrac{-2p + 2 + 4p}{(p+1)(p-1)} \quad \text{Distributive property}$$

$$= \dfrac{2p + 2}{(p+1)(p-1)} \quad \text{Combine terms.}$$

$$= \dfrac{2(p+1)}{(p+1)(p-1)} \quad \text{Factor numerator.}$$

$$= \dfrac{2}{p-1} \quad \text{Lowest terms}$$

4. **(a)** $\dfrac{2k}{k^2 - 5k + 4} + \dfrac{3}{k^2 - 1}$

$$= \dfrac{2k}{(k-4)(k-1)} + \dfrac{3}{(k+1)(k-1)}$$

$\qquad\qquad$ Factor.

$$= \dfrac{2k(k+1)}{(k-4)(k-1)(k+1)}$$

$$+ \dfrac{3(k-4)}{(k-4)(k-1)(k+1)}$$

$\qquad\qquad$ LCD $= (k-4)(k-1)(k+1)$

$$= \dfrac{2k(k+1) + 3(k-4)}{(k-4)(k-1)(k+1)} \quad \text{Add numerators.}$$

$$= \dfrac{2k^2 + 2k + 3k - 12}{(k-4)(k-1)(k+1)}$$

$\qquad\qquad$ Distributive property

$$= \dfrac{2k^2 + 5k - 12}{(k-4)(k-1)(k+1)} \quad \text{Combine terms.}$$

$$= \dfrac{(2k-3)(k+4)}{(k-4)(k-1)(k+1)} \quad \text{Factor numerator.}$$

(b) $\dfrac{4m}{m^2 + 3m + 2} + \dfrac{2m-1}{m^2 + 6m + 5}$

$$= \dfrac{4m}{(m+2)(m+1)} + \dfrac{2m-1}{(m+5)(m+1)}$$

$\qquad\qquad$ Factor.

$$= \dfrac{4m(m+5)}{(m+2)(m+1)(m+5)}$$

$$+ \dfrac{(2m-1)(m+2)}{(m+5)(m+1)(m+2)}$$

$\qquad\qquad$ LCD $= (m+2)(m+1)(m+5)$

$$= \dfrac{4m^2 + 20m + 2m^2 + 3m - 2}{(m+2)(m+1)(m+5)}$$

$$= \dfrac{6m^2 + 23m - 2}{(m+2)(m+1)(m+5)}$$

5. $\dfrac{m}{2m - 3n} + \dfrac{n}{3n - 2m}$

$$= \dfrac{m}{2m - 3n} + \dfrac{n(-1)}{(3n - 2m)(-1)}$$

$$= \dfrac{m}{2m - 3n} + \dfrac{-n}{2m - 3n}$$

$$= \dfrac{m - n}{2m - 3n} \quad \text{or} \quad \dfrac{n - m}{3n - 2m}$$

6. **(a)** $\dfrac{3}{m^2} - \dfrac{2}{m^2} = \dfrac{3-2}{m^2} = \dfrac{1}{m^2}$

7.4 Adding and Subtracting Rational Expressions

(b) $\dfrac{x}{2x+3} - \dfrac{3x+4}{2x+3}$

$= \dfrac{x-(3x+4)}{2x+3}$ Use parentheses.

$= \dfrac{x-3x-4}{2x+3}$

$= \dfrac{-2x-4}{2x+3}$

$= \dfrac{-2(x+2)}{2x+3}$

7. (a) $\dfrac{1}{k+4} - \dfrac{2}{k} = \dfrac{k \cdot 1}{k(k+4)} - \dfrac{2(k+4)}{k(k+4)}$

$\qquad\qquad\qquad$ LCD $= k(k+4)$

$= \dfrac{k-2(k+4)}{k(k+4)}$

$= \dfrac{k-2k-8}{k(k+4)}$

$= \dfrac{-k-8}{k(k+4)}$

(b) $\dfrac{6}{a+2} - \dfrac{1}{a-3}$

$= \dfrac{6(a-3)}{(a+2)(a-3)} - \dfrac{1(a+2)}{(a-3)(a+2)}$

$\qquad\qquad$ LCD $= (a+2)(a-3)$

$= \dfrac{6(a-3) - 1(a+2)}{(a+2)(a-3)}$

$= \dfrac{6a - 18 - a - 2}{(a+2)(a-3)}$

$= \dfrac{5a - 20}{(a+2)(a-3)}$

$= \dfrac{5(a-4)}{(a+2)(a-3)}$

8. (a) $\dfrac{5}{x-1} - \dfrac{3x}{1-x}$

The denominators are opposites, so either may be used as the common denominator. We will choose $x-1$.

$\dfrac{5}{x-1} - \dfrac{3x}{1-x} = \dfrac{5}{x-1} - \dfrac{3x}{1-x} \cdot \dfrac{-1}{-1}$

$= \dfrac{5}{x-1} - \dfrac{-3x}{x-1}$

$= \dfrac{5 - (-3x)}{x-1}$

$= \dfrac{5 + 3x}{x-1}$

(b) $\dfrac{2y}{y-2} - \dfrac{1+y}{2-y}$

The denominators are opposites, so either may be used as the common denominator. We will choose $y - 2$.

$\dfrac{2y}{y-2} - \dfrac{1+y}{2-y} = \dfrac{2y}{y-2} - \dfrac{1+y}{2-y} \cdot \dfrac{-1}{-1}$

$= \dfrac{2y}{y-2} - \dfrac{-1-y}{y-2}$

$= \dfrac{2y - (-1 - y)}{y-2}$

$= \dfrac{2y + 1 + y}{y-2}$

$= \dfrac{3y + 1}{y-2}$

9. (a) $\dfrac{4y}{y^2-1} - \dfrac{5}{y^2+2y+1}$

$= \dfrac{4y}{(y-1)(y+1)} - \dfrac{5}{(y+1)^2}$ Factor.

$= \dfrac{4y(y+1)}{(y-1)(y+1)^2} - \dfrac{5(y-1)}{(y+1)^2(y-1)}$

$\qquad\qquad$ LCD $= (y-1)(y+1)^2$

$= \dfrac{4y(y+1) - 5(y-1)}{(y-1)(y+1)^2}$

$= \dfrac{4y^2 + 4y - 5y + 5}{(y-1)(y+1)^2}$

$= \dfrac{4y^2 - y + 5}{(y-1)(y+1)^2}$

(b) $\dfrac{3r}{r^2-5r} - \dfrac{4}{r^2-10r+25}$

$= \dfrac{3r}{r(r-5)} - \dfrac{4}{(r-5)^2}$ Factor.

$= \dfrac{3}{r-5} - \dfrac{4}{(r-5)^2}$ Reduce.

$= \dfrac{3(r-5)}{(r-5)^2} - \dfrac{4}{(r-5)^2}$

$\qquad\qquad$ LCD $= (r-5)^2$

$= \dfrac{3(r-5) - 4}{(r-5)(r-5)}$

$= \dfrac{3r - 15 - 4}{(r-5)(r-5)}$

$= \dfrac{3r - 19}{(r-5)^2}$

Chapter 7 Rational Expressions and Applications

10. (a) $\dfrac{2}{p^2-5p+4} - \dfrac{3}{p^2-1}$

$= \dfrac{2}{(p-4)(p-1)} - \dfrac{3}{(p+1)(p-1)}$

$= \dfrac{2(p+1)}{(p-4)(p-1)(p+1)}$

$\quad - \dfrac{3(p-4)}{(p+1)(p-1)(p-4)}$

$LCD = (p-4)(p-1)(p+1)$

$= \dfrac{(2p+2) - (3p-12)}{(p-4)(p-1)(p+1)}$

$= \dfrac{2p+2-3p+12}{(p-4)(p-1)(p+1)}$

$= \dfrac{14-p}{(p-4)(p-1)(p+1)}$

(b) $\dfrac{q}{2q^2+5q-3} - \dfrac{3q+4}{3q^2+10q+3}$

$= \dfrac{q}{(2q+1)(q+3)} - \dfrac{3q+4}{(q+3)(3q+1)}$

$= \dfrac{q(3q+1)}{(2q-1)(q+3)(3q+1)}$

$\quad - \dfrac{(2q-1)((3q+4)}{(2q-1)(q+3)(3q+1)}$

$LCD = (2q-1)(q+3)(3q+1)$

$= \dfrac{(3q^2+q) - (6q^2+5q-4)}{(2q-1)(q+3)(3q+1)}$

$= \dfrac{3q^2+q-6q^2-5q+4}{(2q-1)(q+3)(3q+1)}$

$= \dfrac{-3q^2-4q+4}{(2q-1)(q+3)(3q+1)}$ or

$\dfrac{(-3q+2)(q+2)}{(2q-1)(q+3)(3q+1)}$

7.4 Section Exercises

1. $\dfrac{x}{x+6} + \dfrac{6}{x+6}$

The denominators are the same, so the sum is found by adding the two numerators and keeping the same (common) denominator.

$\dfrac{x}{x+6} + \dfrac{6}{x+6} = \dfrac{x+6}{x+6} = 1$

Choice **E** is correct.

3. $\dfrac{6}{x-6} - \dfrac{x}{x-6}$

The denominators are the same, so the difference is found by subtracting the two numerators and keeping the same (common) denominator.

$\dfrac{6}{x-6} - \dfrac{x}{x-6} = \dfrac{6-x}{x-6}$

$= \dfrac{-1(x-6)}{x-6} = -1$

Choice **C** is correct.

5. $\dfrac{x}{x+6} - \dfrac{6}{x+6}$

The denominators are the same, so the difference is found by subtracting the two numerators and keeping the same (common) denominator.

$\dfrac{x}{x+6} - \dfrac{6}{x+6} = \dfrac{x-6}{x+6}$

Choice **B** is correct.

7. $\dfrac{1}{6} - \dfrac{1}{x}$

The LCD is $6x$. Now rewrite each rational expression as a fraction with the LCD as its denominator.

$\dfrac{1}{6} \cdot \dfrac{x}{x} = \dfrac{x}{6x}$

$\dfrac{1}{x} \cdot \dfrac{6}{6} = \dfrac{6}{6x}$

Since the fractions now have a common denominator, subtract the numerators and use the LCD as the denominator of the sum.

$\dfrac{1}{6} - \dfrac{1}{x} = \dfrac{x}{6x} - \dfrac{6}{6x} = \dfrac{x-6}{6x}$

Choice **G** is correct.

9. $\dfrac{4}{m} + \dfrac{7}{m}$

The denominators are the same, so the sum is found by adding the two numerators and keeping the same (common) denominator.

$\dfrac{4}{m} + \dfrac{7}{m} = \dfrac{4+7}{m} = \dfrac{11}{m}$

11. $\dfrac{a+b}{2} - \dfrac{a-b}{2}$

The denominators are the same, so the difference is found by subtracting the two numerators and keeping the same (common) denominator. Don't forget the parentheses on the second numerator.

$\dfrac{a+b}{2} - \dfrac{a-b}{2} = \dfrac{(a+b)-(a-b)}{2}$

$= \dfrac{a+b-a+b}{2}$

$= \dfrac{2b}{2} = b$

7.4 Adding and Subtracting Rational Expressions

13. $\dfrac{x^2}{x+5} + \dfrac{5x}{x+5} = \dfrac{x^2+5x}{x+5}$ *Add numerators.*

    $= \dfrac{x(x+5)}{x+5}$ *Factor numerator.*

    $= x$ *Lowest terms*

15. $\dfrac{y^2-3y}{y+3} + \dfrac{-18}{y+3} = \dfrac{y^2-3y-18}{y+3}$

    $= \dfrac{(y-6)(y+3)}{y+3}$

    $= y - 6$

17. To add or subtract rational expressions with the same denominators, combine the numerators and keep the same denominator. For example,

    $\dfrac{3x+2}{x-6} + \dfrac{-2x-8}{x-6} = \dfrac{x-6}{x-6}.$

    Then write in lowest terms. In this example, the sum simplifies to 1.

19. $\dfrac{z}{5} + \dfrac{1}{3}$

    The LCD is 15. Now rewrite each rational expression as a fraction with the LCD as its denominator.

    $\dfrac{z}{5} \cdot \dfrac{3}{3} = \dfrac{3z}{15}$

    $\dfrac{1}{3} \cdot \dfrac{5}{5} = \dfrac{5}{15}$

    Since the fractions now have a common denominator, add the numerators and use the LCD as the denominator of the sum.

    $\dfrac{z}{5} + \dfrac{1}{3} = \dfrac{3z}{15} + \dfrac{5}{15} = \dfrac{3z+5}{15}$

21. $\dfrac{5}{7} - \dfrac{r}{2} = \dfrac{5}{7} \cdot \dfrac{2}{2} - \dfrac{r}{2} \cdot \dfrac{7}{7}$ *LCD = 14*

    $= \dfrac{10}{14} - \dfrac{7r}{14}$

    $= \dfrac{10 - 7r}{14}$

23. $-\dfrac{3}{4} - \dfrac{1}{2x} = -\dfrac{3 \cdot x}{4 \cdot x} - \dfrac{1 \cdot 2}{2x \cdot 2}$ *LCD = 4x*

    $= \dfrac{-3x - 2}{4x}$

25. $\dfrac{x+1}{6} + \dfrac{3x+3}{9}$

    First reduce the second fraction.

    $\dfrac{3x+3}{9} = \dfrac{3(x+1)}{9} = \dfrac{x+1}{3}$

    Now the LCD of $\dfrac{x+1}{6}$ and $\dfrac{x+1}{3}$ is 6. Thus,

    $\dfrac{x+1}{6} + \dfrac{x+1}{3} = \dfrac{x+1}{6} + \dfrac{x+1}{3} \cdot \dfrac{2}{2}$

    $= \dfrac{x+1+2x+2}{6}$

    $= \dfrac{3x+3}{6}$

    $= \dfrac{3(x+1)}{6} = \dfrac{x+1}{2}.$

27. $\dfrac{x+3}{3x} + \dfrac{2x+2}{4x} = \dfrac{x+3}{3x} + \dfrac{2(x+1)}{4x}$

    $= \dfrac{x+3}{3x} + \dfrac{x+1}{2x}$ *Reduce.*

    $= \dfrac{x+3}{3x} \cdot \dfrac{2}{2} + \dfrac{x+1}{2x} \cdot \dfrac{3}{3}$

     *LCD = 6x*

    $= \dfrac{2x+6+3x+3}{6x}$

    $= \dfrac{5x+9}{6x}$

29. $\dfrac{2}{x+3} + \dfrac{1}{x}$

    $= \dfrac{2(x)}{(x+3)(x)} + \dfrac{1(x+3)}{x(x+3)}$ *LCD = x(x + 3)*

    $= \dfrac{2x}{x(x+3)} + \dfrac{x+3}{x(x+3)}$

    $= \dfrac{2x+x+3}{x(x+3)}$

    $= \dfrac{3x+3}{x(x+3)}$ or $\dfrac{3(x+1)}{x(x+3)}$

31. $\dfrac{x}{x-2} + \dfrac{4}{x+2}$

    $= \dfrac{x(x+2)}{(x-2)(x+2)} + \dfrac{4(x-2)}{(x+2)(x-2)}$

     *LCD = (x + 2)(x − 2)*

    $= \dfrac{x^2+2x}{(x-2)(x+2)} + \dfrac{4x-8}{(x+2)(x-2)}$

    $= \dfrac{x^2+2x+4x-8}{(x+2)(x-2)}$

    $= \dfrac{x^2+6x-8}{(x+2)(x-2)}$

33. $\dfrac{t}{t+2} + \dfrac{5-t}{t} - \dfrac{4}{t^2+2t}$

$= \dfrac{t}{t+2} + \dfrac{5-t}{t} - \dfrac{4}{t(t+2)}$

$= \dfrac{t}{t+2} \cdot \dfrac{t}{t} + \dfrac{5-t}{t} \cdot \dfrac{t+2}{t+2}$

$\quad - \dfrac{4}{t(t+2)} \quad LCD = t(t+2)$

$= \dfrac{t \cdot t + (5-t)(t+2) - 4}{t(t+2)}$

$= \dfrac{t^2 + 5t + 10 - t^2 - 2t - 4}{t(t+2)}$

$= \dfrac{3t+6}{t(t+2)}$

$= \dfrac{3(t+2)}{t(t+2)} = \dfrac{3}{t}$

35. $\dfrac{10}{m-2} + \dfrac{5}{2-m}$

Since

$$2 - m = -1(m-2),$$

either $m-2$ or $2-m$ could be used as the LCD.

37. $\dfrac{4}{x-5} + \dfrac{6}{5-x}$

The two denominators, $x-5$ and $5-x$, are opposites of each other, so either one may be used as the common denominator. We will work the exercise both ways and compare the answers.

$\dfrac{4}{x-5} + \dfrac{6}{5-x} = \dfrac{4}{x-5} + \dfrac{6(-1)}{(5-x)(-1)}$

$\qquad\qquad\qquad\qquad LCD = x-5$

$= \dfrac{4}{x-5} + \dfrac{-6}{x-5}$

$= \dfrac{-2}{x-5}$

$\dfrac{4}{x-5} + \dfrac{6}{5-x} = \dfrac{4(-1)}{(x-5)(-1)} + \dfrac{6}{5-x}$

$\qquad\qquad\qquad\qquad LCD = 5-x$

$= \dfrac{-4}{5-x} + \dfrac{6}{5-x}$

$= \dfrac{2}{5-x}$

The two answers are equivalent, since

$$\dfrac{-2}{x-5} \cdot \dfrac{-1}{-1} = \dfrac{2}{5-x}.$$

39. $\dfrac{-1}{1-y} + \dfrac{3-4y}{y-1}$

$= \dfrac{-1}{1-y} + \dfrac{(3-4y)(-1)}{(y-1)(-1)} \quad LCD = 1-y$

$= \dfrac{-1}{1-y} + \dfrac{-3+4y}{1-y}$

$= \dfrac{-1-3+4y}{1-y}$

$= \dfrac{-4+4y}{1-y}$

$= \dfrac{-4(1-y)}{1-y} = -4$

41. $\dfrac{2}{x-y^2} + \dfrac{7}{y^2-x}$

$LCD = x - y^2$ or $y^2 - x$

We will use $x - y^2$.

$\dfrac{2}{x-y^2} + \dfrac{7}{y^2-x}$

$= \dfrac{2}{x-y^2} + \dfrac{-1(7)}{-1(y^2-x)}$

$= \dfrac{2}{x-y^2} + \dfrac{-7}{-y^2+x}$

$= \dfrac{2}{x-y^2} + \dfrac{-7}{x-y^2}$

$= \dfrac{2+(-7)}{x-y^2} = \dfrac{-5}{x-y^2}$

If $y^2 - x$ is used as the LCD, we will obtain the equivalent answer

$$\dfrac{5}{y^2-x}.$$

43. $\dfrac{x}{5x-3y} - \dfrac{y}{3y-5x}$

$LCD = 5x - 3y$ or $3y - 5x$

We will use $5x - 3y$.

$\dfrac{x}{5x-3y} - \dfrac{y}{3y-5x}$

$= \dfrac{x}{5x-3y} - \dfrac{-1(y)}{-1(3y-5x)}$

$= \dfrac{x}{5x-3y} - \dfrac{-y}{-3y+5x}$

$= \dfrac{x}{5x-3y} - \dfrac{-y}{5x-3y}$

$= \dfrac{x-(-y)}{5x-3y} = \dfrac{x+y}{5x-3y}$

If $3y - 5x$ is used as the LCD, we will obtain the equivalent answer

$$\dfrac{-x-y}{3y-5x}.$$

45. $\dfrac{3}{4p-5} + \dfrac{9}{5-4p}$

LCD $= 4p - 5$ or $5 - 4p$

We will use $4p - 5$.

$$\dfrac{3}{4p-5} + \dfrac{9}{5-4p}$$

$$= \dfrac{3}{4p-5} + \dfrac{-1(9)}{-1(5-4p)}$$

$$= \dfrac{3}{4p-5} + \dfrac{-9}{-5+4p}$$

$$= \dfrac{3}{4p-5} + \dfrac{-9}{4p-5}$$

$$= \dfrac{3 + (-9)}{4p-5} = \dfrac{-6}{4p-5}$$

If $5 - 4p$ is used as the LCD, we will obtain the equivalent answer

$$\dfrac{6}{5-4p}.$$

47. $\dfrac{2m}{m-n} - \dfrac{5m+n}{2m-2n}$

$$= \dfrac{2m}{m-n} - \dfrac{5m+n}{2(m-n)} \quad \textit{Factor second denominator.}$$

$$= \dfrac{2m}{m-n} \cdot \dfrac{2}{2} - \dfrac{5m+n}{2(m-n)} \quad \textit{LCD = 2(m-n)}$$

$$= \dfrac{4m - (5m+n)}{2(m-n)}$$

$$= \dfrac{4m - 5m - n}{2(m-n)}$$

$$= \dfrac{-m - n}{2(m-n)}$$

$$= \dfrac{-(m+n)}{2(m-n)}$$

49. $\dfrac{5}{x^2-9} - \dfrac{x+2}{x^2+4x+3}$

To find the LCD, factor the denominators.

$$x^2 - 9 = (x+3)(x-3)$$
$$x^2 + 4x + 3 = (x+3)(x+1)$$

The LCD is $(x+3)(x-3)(x+1)$.

$$\dfrac{5}{x^2-9} - \dfrac{x+2}{x^2+4x+3}$$

$$= \dfrac{5 \cdot (x+1)}{(x+3)(x-3) \cdot (x+1)}$$

$$- \dfrac{(x+2) \cdot (x-3)}{(x+3)(x+1) \cdot (x-3)}$$

$$= \dfrac{5x+5}{(x+3)(x-3)(x+1)}$$

$$- \dfrac{x^2 - x - 6}{(x+3)(x+1)(x-3)}$$

$$= \dfrac{(5x+5) - (x^2 - x - 6)}{(x+3)(x-3)(x+1)}$$

$$= \dfrac{5x + 5 - x^2 + x + 6}{(x+3)(x-3)(x+1)}$$

$$= \dfrac{-x^2 + 6x + 11}{(x+3)(x-3)(x+1)}$$

51. $\dfrac{2q+1}{3q^2+10q-8} - \dfrac{3q+5}{2q^2+5q-12}$

$$= \dfrac{2q+1}{(3q-2)(q+4)} - \dfrac{3q+5}{(2q-3)(q+4)}$$

$$= \dfrac{(2q+1) \cdot (2q-3)}{(3q-2)(q+4) \cdot (2q-3)}$$

$$- \dfrac{(3q+5) \cdot (3q-2)}{(2q-3)(q+4) \cdot (3q-2)}$$

LCD $= (3q-2)(q+4)(2q-3)$

$$= \dfrac{(4q^2 - 4q - 3) - (9q^2 + 9q - 10)}{(3q-2)(q+4)(2q-3)}$$

$$= \dfrac{4q^2 - 4q - 3 - 9q^2 - 9q + 10}{(3q-2)(q+4)(2q-3)}$$

$$= \dfrac{-5q^2 - 13q + 7}{(3q-2)(q+4)(2q-3)}$$

53. $\dfrac{4}{r^2-r} + \dfrac{6}{r^2+2r} - \dfrac{1}{r^2+r-2}$

$$= \dfrac{4}{r(r-1)} + \dfrac{6}{r(r+2)} - \dfrac{1}{(r+2)(r-1)}$$

$$= \dfrac{4 \cdot (r+2)}{r(r-1) \cdot (r+2)} + \dfrac{6 \cdot (r-1)}{r(r+2) \cdot (r-1)}$$

$$- \dfrac{1 \cdot r}{r \cdot (r+2)(r-1)}$$

LCD $= r(r+2)(r-1)$

$$= \dfrac{4r + 8 + 6r - 6 - r}{r(r+2)(r-1)}$$

$$= \dfrac{9r + 2}{r(r+2)(r-1)}$$

55. $\dfrac{x+3y}{x^2+2xy+y^2} + \dfrac{x-y}{x^2+4xy+3y^2}$

$= \dfrac{x+3y}{(x+y)(x+y)} + \dfrac{x-y}{(x+3y)(x+y)}$

$= \dfrac{(x+3y)\cdot(x+3y)}{(x+y)(x+y)\cdot(x+3y)}$

$\quad + \dfrac{(x-y)\cdot(x+y)}{(x+3y)(x+y)\cdot(x+y)}$

$\quad\quad LCD = (x+y)(x+y)(x+3y)$

$= \dfrac{(x^2+6xy+9y^2)+(x^2-y^2)}{(x+y)(x+y)(x+3y)}$

$= \dfrac{2x^2+6xy+8y^2}{(x+y)(x+y)(x+3y)}$

$= \dfrac{2(x^2+3xy+4y^2)}{(x+y)(x+y)(x+3y)}$

or $\dfrac{2(x^2+3xy+4y^2)}{(x+y)^2(x+3y)}$

57. $\dfrac{r+y}{18r^2+9ry-2y^2} + \dfrac{3r-y}{36r^2-y^2}$

$= \dfrac{r+y}{(3r+2y)(6r-y)} + \dfrac{3r-y}{(6r-y)(6r+y)}$

Rewrite fractions with the LCD,
$(3r+2y)(6r-y)(6r+y)$.

$= \dfrac{(r+y)\cdot(6r+y)}{(3r+2y)(6r-y)\cdot(6r+y)}$

$\quad + \dfrac{(3r-y)\cdot(3r+2y)}{(6r-y)(6r+y)\cdot(3r+2y)}$

$= \dfrac{6r^2+7ry+y^2}{(3r+2y)(6r-y)(6r+y)}$

$\quad + \dfrac{9r^2+3ry-2y^2}{(3r+2y)(6r-y)(6r+y)}$

$= \dfrac{6r^2+7ry+y^2+9r^2+3ry-2y^2}{(3r+2y)(6r-y)(6r+y)}$

$= \dfrac{15r^2+10ry-y^2}{(3r+2y)(6r-y)(6r+y)}$

59. (a) $P = 2L + 2W$

$= 2\left(\dfrac{3k+1}{10}\right) + 2\left(\dfrac{5}{6k+2}\right)$

$= 2\left(\dfrac{3k+1}{2\cdot 5}\right) + 2\left(\dfrac{5}{2(3k+1)}\right)$

$= \dfrac{3k+1}{5} + \dfrac{5}{3k+1}$

To add the two fractions on the right, use $5(3k+1)$ as the LCD.

$P = \dfrac{(3k+1)(3k+1)}{5(3k+1)} + \dfrac{(5)(5)}{5(3k+1)}$

$= \dfrac{(3k+1)(3k+1)+(5)(5)}{5(3k+1)}$

$= \dfrac{9k^2+6k+1+25}{5(3k+1)}$

$= \dfrac{9k^2+6k+26}{5(3k+1)}$

(b) $A = L \cdot W$

$A = \dfrac{3k+1}{10} \cdot \dfrac{5}{6k+2}$

$= \dfrac{3k+1}{5\cdot 2} \cdot \dfrac{5}{2(3k+1)}$

$= \dfrac{1}{2\cdot 2} = \dfrac{1}{4}$

7.5 Complex Fractions

7.5 Margin Exercises

1. (a) $\dfrac{\dfrac{2}{5}+\dfrac{1}{4}}{\dfrac{1}{2}+\dfrac{1}{3}} = \dfrac{\dfrac{2(4)}{5(4)}+\dfrac{1(5)}{4(5)}}{\dfrac{1(3)}{2(3)}+\dfrac{1(2)}{3(2)}}$

$= \dfrac{\dfrac{8+5}{20}}{\dfrac{3+2}{6}}$

$= \dfrac{13}{20} \div \dfrac{5}{6}$

$= \dfrac{13}{20} \cdot \dfrac{6}{5}$ *Multiply by reciprocal.*

$= \dfrac{13\cdot 2\cdot 3}{10\cdot 2\cdot 5}$

$= \dfrac{39}{50}$

(b) $\dfrac{6+\dfrac{1}{x}}{5-\dfrac{2}{x}} = \dfrac{\dfrac{6x}{x}+\dfrac{1}{x}}{\dfrac{5x}{x}-\dfrac{2}{x}}$

$= \dfrac{\dfrac{6x+1}{x}}{\dfrac{5x-2}{x}}$

$= \dfrac{6x+1}{x} \div \dfrac{5x-2}{x}$

$= \dfrac{6x+1}{x} \cdot \dfrac{x}{5x-2}$ *Multiply by reciprocal.*

$= \dfrac{6x+1}{5x-2}$

7.5 Complex Fractions 273

(c) $\dfrac{9 - \dfrac{4}{p}}{\dfrac{2}{p} + 1} = \dfrac{\dfrac{9p}{p} - \dfrac{4}{p}}{\dfrac{2}{p} + \dfrac{p}{p}}$

$= \dfrac{\dfrac{9p - 4}{p}}{\dfrac{2 + p}{p}}$

$= \dfrac{9p - 4}{p} \div \dfrac{2 + p}{p}$

$= \dfrac{9p - 4}{p} \cdot \dfrac{p}{2 + p}$ *Multiply by reciprocal.*

$= \dfrac{9p - 4}{2 + p}$

2. (a) $\dfrac{\dfrac{rs^2}{t}}{\dfrac{r^2 s}{t^2}} = \dfrac{rs^2}{t} \div \dfrac{r^2 s}{t^2} = \dfrac{rs^2}{t} \cdot \dfrac{t^2}{r^2 s} = \dfrac{st}{r}$

(b) $\dfrac{\dfrac{m^2 n^3}{p}}{\dfrac{m^4 n}{p^2}} = \dfrac{m^2 n^3}{p} \div \dfrac{m^4 n}{p^2}$

$= \dfrac{m^2 n^3}{p} \cdot \dfrac{p^2}{m^4 n} = \dfrac{n^2 p}{m^2}$

3. $\dfrac{\dfrac{2}{x-1} + \dfrac{1}{x+1}}{\dfrac{3}{x-1} - \dfrac{4}{x+1}}$

$= \dfrac{\dfrac{2(x+1)}{(x-1)(x+1)} + \dfrac{1(x-1)}{(x+1)(x-1)}}{\dfrac{3(x+1)}{(x-1)(x+1)} - \dfrac{4(x-1)}{(x+1)(x-1)}}$

$= \dfrac{\dfrac{2x+2}{(x-1)(x+1)} + \dfrac{x-1}{(x+1)(x-1)}}{\dfrac{3x+3}{(x-1)(x+1)} - \dfrac{4x-4}{(x+1)(x-1)}}$

$= \dfrac{\dfrac{(2x+2) + (x-1)}{(x+1)(x-1)}}{\dfrac{(3x+3) - (4x-4)}{(x-1)(x+1)}}$

$= \dfrac{\dfrac{2x+2+x-1}{(x+1)(x-1)}}{\dfrac{3x+3-4x+4}{(x-1)(x+1)}}$

$= \dfrac{3x+1}{\dfrac{(x+1)(x-1)}{\dfrac{-x+7}{(x-1)(x+1)}}}$

$= \dfrac{3x+1}{(x+1)(x-1)} \cdot \dfrac{(x-1)(x+1)}{-x+7}$

$= \dfrac{3x+1}{-x+7}$

4. (a) $\dfrac{\dfrac{2}{3} - \dfrac{1}{4}}{\dfrac{4}{9} + \dfrac{1}{2}} = \dfrac{36\left(\dfrac{2}{3} - \dfrac{1}{4}\right)}{36\left(\dfrac{4}{9} + \dfrac{1}{2}\right)}$ $LCD = 2^2 \cdot 3^2 = 36$

$= \dfrac{36\left(\dfrac{2}{3}\right) - 36\left(\dfrac{1}{4}\right)}{36\left(\dfrac{4}{9}\right) + 36\left(\dfrac{1}{2}\right)}$

$= \dfrac{24 - 9}{16 + 18}$

$= \dfrac{15}{34}$

(b) $\dfrac{2 - \dfrac{6}{a}}{3 + \dfrac{4}{a}}$ $LCD = a$

$= \dfrac{a\left(2 - \dfrac{6}{a}\right)}{a\left(3 + \dfrac{4}{a}\right)}$ *Multiply numerator and denominator by a.*

$= \dfrac{a(2) - a\left(\dfrac{6}{a}\right)}{a(3) + a\left(\dfrac{4}{a}\right)}$

$= \dfrac{2a - 6}{3a + 4}$

(c) $\dfrac{\dfrac{p}{5-p}}{\dfrac{4p}{2p+1}}$ $LCD = (5-p)(2p+1)$

$= \dfrac{(5-p)(2p+1)\left(\dfrac{p}{5-p}\right)}{(5-p)(2p+1)\left(\dfrac{4p}{2p+1}\right)}$

$= \dfrac{(2p+1)(p)}{(5-p)(4p)}$

$= \dfrac{2p+1}{4(5-p)}$

274 Chapter 7 Rational Expressions and Applications

5. $\dfrac{\dfrac{2}{5x} - \dfrac{3}{x^2}}{\dfrac{7}{4x} + \dfrac{1}{2x^2}}$ LCD $= 20x^2$

$= \dfrac{20x^2\left(\dfrac{2}{5x} - \dfrac{3}{x^2}\right)}{20x^2\left(\dfrac{7}{4x} + \dfrac{1}{2x^2}\right)}$

$= \dfrac{20x^2\left(\dfrac{2}{5x}\right) - 20x^2\left(\dfrac{3}{x^2}\right)}{20x^2\left(\dfrac{7}{4x}\right) + 20x^2\left(\dfrac{1}{2x^2}\right)}$

$= \dfrac{8x - 60}{35x + 10}$

6. **(a)** $\dfrac{\dfrac{1}{x} + \dfrac{2}{x-1}}{\dfrac{2}{x} - \dfrac{4}{x-1}}$ LCD $= x(x-1)$

$= \dfrac{x(x-1)\left(\dfrac{1}{x} + \dfrac{2}{x-1}\right)}{x(x-1)\left(\dfrac{2}{x} - \dfrac{4}{x-1}\right)}$ Use Method 2.

$= \dfrac{(x-1)(1) + (x)(2)}{(x-1)(2) - (x)(4)}$

$= \dfrac{x - 1 + 2x}{2x - 2 - 4x}$

$= \dfrac{3x - 1}{-2x - 2}$

(b) $\dfrac{1 - \dfrac{2}{x} - \dfrac{15}{x^2}}{1 + \dfrac{5}{x} + \dfrac{6}{x^2}}$ LCD $= x^2$

$= \dfrac{x^2\left(1 - \dfrac{2}{x} - \dfrac{15}{x^2}\right)}{x^2\left(1 + \dfrac{5}{x} + \dfrac{6}{x^2}\right)}$ Use Method 2.

$= \dfrac{x^2 - 2x - 15}{x^2 + 5x + 6}$

$= \dfrac{(x-5)(x+3)}{(x+2)(x+3)}$

$= \dfrac{x-5}{x+2}$

(c) $\dfrac{\dfrac{2x+3}{x-4}}{\dfrac{4x^2-9}{x^2-16}} = \dfrac{2x+3}{x-4} \div \dfrac{4x^2-9}{x^2-16}$

$= \dfrac{2x+3}{x-4} \cdot \dfrac{x^2-16}{4x^2-9}$

$= \dfrac{2x+3}{x-4} \cdot \dfrac{(x+4)(x-4)}{(2x+3)(2x-3)}$

$= \dfrac{x+4}{2x-3}$

7.5 Section Exercises

1. **(a)** The LCD of $\tfrac{1}{2}$ and $\tfrac{1}{3}$ is $2 \cdot 3 = 6$. The simplified form of the numerator is

$\dfrac{1}{2} - \dfrac{1}{3} = \dfrac{3}{6} - \dfrac{2}{6} = \dfrac{1}{6}.$

(b) The LCD of $\tfrac{5}{6}$ and $\tfrac{1}{12}$ is 12 since 12 is a multiple of 6. The simplified form of the denominator is

$\dfrac{5}{6} - \dfrac{1}{12} = \dfrac{10}{12} - \dfrac{1}{12} = \dfrac{9}{12} = \dfrac{3}{4}.$

(c) $\dfrac{\dfrac{1}{6}}{\dfrac{3}{4}} = \dfrac{1}{6} \div \dfrac{3}{4}$

(d) $\dfrac{1}{6} \div \dfrac{3}{4} = \dfrac{1}{6} \cdot \dfrac{4}{3}$

$= \dfrac{2 \cdot 2}{2 \cdot 3 \cdot 3} = \dfrac{2}{9}$

In Exercises 3–28, either Method 1 or Method 2 can be used to simplify each complex fraction. Only one method will be shown for each exercise.

3. To use Method 1, divide the numerator of the complex fraction by the denominator.

$\dfrac{-\dfrac{4}{3}}{\dfrac{2}{9}} = -\dfrac{4}{3} \div \dfrac{2}{9} = -\dfrac{4}{3} \cdot \dfrac{9}{2}$

$= -\dfrac{36}{6} = -6$

5. $\dfrac{\dfrac{p}{q^2}}{\dfrac{p^2}{q}} = \dfrac{p}{q^2} \div \dfrac{p^2}{q} = \dfrac{p}{q^2} \cdot \dfrac{q}{p^2} = \dfrac{1}{pq}$

7. To use Method 2, multiply the numerator and denominator of the complex fraction by the LCD, y^2.

$\dfrac{\dfrac{x}{y^2}}{\dfrac{x^2}{y}} = \dfrac{y^2\left(\dfrac{x}{y^2}\right)}{y^2\left(\dfrac{x^2}{y}\right)}$

$= \dfrac{x}{yx^2} = \dfrac{1}{xy}$

7.5 Complex Fractions

9. $\dfrac{\frac{4a^4b^3}{3a}}{\frac{2ab^4}{b^2}} = \dfrac{4a^4b^3}{3a} \div \dfrac{2ab^4}{b^2}$ Method 1

$= \dfrac{4a^4b^3}{3a} \cdot \dfrac{b^2}{2ab^4}$

$= \dfrac{4a^4b^3 \cdot b^2}{3a \cdot 2ab^4}$

$= \dfrac{4a^4b^5}{6a^2b^4}$

$= \dfrac{2a^2b}{3}$

11. To use Method 2, multiply the numerator and denominator of the complex fraction by the LCD, $3m$.

$\dfrac{\frac{m+2}{3}}{\frac{m-4}{m}} = \dfrac{3m\left(\frac{m+2}{3}\right)}{3m\left(\frac{m-4}{m}\right)}$

$= \dfrac{m(m+2)}{3(m-4)}$

13. $\dfrac{\frac{2}{x} - 3}{\frac{2-3x}{2}} = \dfrac{2x\left(\frac{2}{x} - 3\right)}{2x\left(\frac{2-3x}{2}\right)}$ Method 2; LCD = $2x$

$= \dfrac{2x\left(\frac{2}{x}\right) - 2x(3)}{x(2-3x)}$

$= \dfrac{4 - 6x}{x(2-3x)}$

$= \dfrac{2(2-3x)}{x(2-3x)}$ Factor.

$= \dfrac{2}{x}$ Lowest terms

15. $\dfrac{\frac{1}{x} + x}{\frac{x^2+1}{8}} = \dfrac{8x\left(\frac{1}{x} + x\right)}{8x\left(\frac{x^2+1}{8}\right)}$ Method 2; LCD = $8x$

$= \dfrac{8 + 8x^2}{x(x^2+1)}$ Distributive property

$= \dfrac{8(1+x^2)}{x(x^2+1)}$ Factor.

$= \dfrac{8}{x}$ Lowest terms

17. $\dfrac{a - \frac{5}{a}}{a + \frac{1}{a}} = \dfrac{a\left(a - \frac{5}{a}\right)}{a\left(a + \frac{1}{a}\right)}$ Method 2; LCD = a

$= \dfrac{a^2 - 5}{a^2 + 1}$

19. $\dfrac{\frac{1}{2} + \frac{1}{p}}{\frac{2}{3} + \frac{1}{p}} = \dfrac{6p\left(\frac{1}{2} + \frac{1}{p}\right)}{6p\left(\frac{2}{3} + \frac{1}{p}\right)}$ Method 2; LCD = $6p$

$= \dfrac{6p\left(\frac{1}{2}\right) + 6p\left(\frac{1}{p}\right)}{6p\left(\frac{2}{3}\right) + 6p\left(\frac{1}{p}\right)}$

$= \dfrac{3p + 6}{4p + 6}$

$= \dfrac{3(p+2)}{2(2p+3)}$

21. $\dfrac{\frac{t}{t+2}}{\frac{4}{t^2-4}} = \dfrac{t}{t+2} \div \dfrac{4}{t^2-4}$ Method 1

$= \dfrac{t}{t+2} \cdot \dfrac{t^2-4}{4}$

$= \dfrac{t \cdot (t+2)(t-2)}{(t+2) \cdot 4}$

$= \dfrac{t(t-2)}{4}$

23. $\dfrac{\frac{1}{k+1} - 1}{\frac{1}{k+1} + 1}$

$= \dfrac{(k+1)\left(\frac{1}{k+1} - 1\right)}{(k+1)\left(\frac{1}{k+1} + 1\right)}$ Method 2; LCD = $k+1$

$= \dfrac{1 - 1(k+1)}{1 + 1(k+1)}$ Distributive property

$= \dfrac{1 - k - 1}{1 + k + 1}$ Distributive property

$= \dfrac{-k}{k+2}$

276 Chapter 7 Rational Expressions and Applications

25. $\dfrac{2 + \dfrac{1}{x} - \dfrac{28}{x^2}}{3 + \dfrac{13}{x} + \dfrac{4}{x^2}}$ $LCD = x^2$

$= \dfrac{x^2\left(2 + \dfrac{1}{x} - \dfrac{28}{x^2}\right)}{x^2\left(3 + \dfrac{13}{x} + \dfrac{4}{x^2}\right)}$ *Method 2.*

$= \dfrac{2x^2 + x - 28}{3x^2 + 13x + 4}$

$= \dfrac{(2x - 7)(x + 4)}{(3x + 1)(x + 4)}$

$= \dfrac{2x - 7}{3x + 1}$

27. $\dfrac{\dfrac{1}{m-1} + \dfrac{2}{m+2}}{\dfrac{2}{m+2} - \dfrac{1}{m-3}}$

$= \dfrac{(m-1)(m+2)(m-3)\left(\dfrac{1}{m-1} + \dfrac{2}{m+2}\right)}{(m-1)(m+2)(m-3)\left(\dfrac{2}{m+2} - \dfrac{1}{m-3}\right)}$

Method 2;
$LCD = (m-1)(m+2)(m-3)$

$= \dfrac{(m+2)(m-3) + 2(m-1)(m-3)}{2(m-1)(m-3) - (m-1)(m+2)}$

Distributive property

$= \dfrac{(m-3)[(m+2) + 2(m-1)]}{(m-1)[2(m-3) - (m+2)]}$

Factor out m − 3 in numerator and m − 1 in denominator.

$= \dfrac{(m-3)[m + 2 + 2m - 2]}{(m-1)[2m - 6 - m - 2]}$

Distributive property

$= \dfrac{3m(m-3)}{(m-1)(m-8)}$ *Combine like terms.*

29. $2 - \dfrac{2}{2 + \dfrac{2}{2+2}} = 2 - \dfrac{2}{2 + \dfrac{2}{4}}$

$= 2 - \dfrac{2}{\frac{5}{2}}$

$= 2 - 2 \cdot \dfrac{2}{5}$

$= 2 - \dfrac{4}{5}$

$= \dfrac{10}{5} - \dfrac{4}{5}$

$= \dfrac{6}{5}$

7.6 Solving Equations with Rational Expressions

7.6 Margin Exercises

1. **(a)** $\dfrac{x}{3} + \dfrac{x}{5} = 7 + x$ has an equals sign, so this is an *equation* to be solved. Use the multiplication property of equality to clear fractions. The LCD is 15.

$\dfrac{x}{3} + \dfrac{x}{5} = 7 + x$

$15\left(\dfrac{x}{3} + \dfrac{x}{5}\right) = 15(7 + x)$ *Multiply by 15.*

$15\left(\dfrac{x}{3}\right) + 15\left(\dfrac{x}{5}\right) = 15(7) + 15(x)$ *Distributive property*

$5x + 3x = 105 + 15x$ *Multiply.*

$8x = 105 + 15x$ *Combine like terms.*

$-7x = 105$ *Subtract 15x.*

$x = -15$ *Divide by −7.*

Check $x = -15$: $-8 = -8$

The solution is -15.

(b) $\dfrac{2x}{3} - \dfrac{4x}{9}$ is the sum of two terms, so it is an *expression* to be simplified. Simplify by finding the LCD, writing each coefficient with this LCD, and combining like terms.

$\dfrac{2x}{3} - \dfrac{4x}{9} = \dfrac{2x \cdot 3}{3 \cdot 3} - \dfrac{4x}{9}$ $LCD = 9$

$= \dfrac{6x - 4x}{9}$

$= \dfrac{2x}{9}$

2. **(a)** $\dfrac{x}{5} + 3 = \dfrac{3}{5}$ has an equals sign, so this is an *equation* to be solved. Use the multiplication property of equality to clear fractions. The LCD is 5.

$\dfrac{x}{5} + 3 = \dfrac{3}{5}$

$5\left(\dfrac{x}{5} + 3\right) = 5\left(\dfrac{3}{5}\right)$ *Multiply by 5.*

$5\left(\dfrac{x}{5}\right) + 5(3) = 5\left(\dfrac{3}{5}\right)$ *Distributive property*

$x + 15 = 3$

$x = -12$ *Subtract 15.*

Check $x = -12$: $\dfrac{3}{5} = \dfrac{3}{5}$

The solution is -12.

7.6 Solving Equations with Rational Expressions

(b) $\dfrac{x}{2} - \dfrac{x}{3} = \dfrac{5}{6}$ has an equals sign, so this is an *equation* to be solved. Use the multiplication property of equality to clear fractions. The LCD is 6.

$$\dfrac{x}{2} - \dfrac{x}{3} = \dfrac{5}{6}$$

$$6\left(\dfrac{x}{2} - \dfrac{x}{3}\right) = 6\left(\dfrac{5}{6}\right) \quad \text{Multiply by 6.}$$

$$6\left(\dfrac{x}{2}\right) - 6\left(\dfrac{x}{3}\right) = 6\left(\dfrac{5}{6}\right) \quad \text{Distributive property}$$

$$3x - 2x = 5$$

$$x = 5 \quad \text{Combine terms.}$$

Check $x = 5$: $\dfrac{5}{6} = \dfrac{5}{6}$

The solution is 5.

3. (a) $\dfrac{k}{6} - \dfrac{k+1}{4} = -\dfrac{1}{2}$ has an equals sign, so this is an *equation* to be solved. Use the multiplication property of equality to clear fractions. The LCD is 12.

$$\dfrac{k}{6} - \dfrac{k+1}{4} = -\dfrac{1}{2}$$

$$12\left(\dfrac{k}{6} - \dfrac{k+1}{4}\right) = 12\left(-\dfrac{1}{2}\right) \quad \text{Multiply by 12.}$$

$$12\left(\dfrac{k}{6}\right) - 12\left(\dfrac{k+1}{4}\right) = 12\left(-\dfrac{1}{2}\right) \quad \text{Distributive property}$$

$$2k - 3(k+1) = -6 \quad \text{Multiply.}$$

$$2k - 3k - 3 = -6 \quad \text{Distributive property}$$

$$-k - 3 = -6 \quad \text{Combine like terms.}$$

$$-k = -3 \quad \text{Add 3.}$$

$$k = 3 \quad \text{Divide by } -1.$$

Check $k = 3$: $-\dfrac{1}{2} = -\dfrac{1}{2}$

The solution is 3.

(b) $\dfrac{2m-3}{5} - \dfrac{m}{3} = -\dfrac{6}{5}$

$$15\left(\dfrac{2m-3}{5} - \dfrac{m}{3}\right) = 15\left(-\dfrac{6}{5}\right)$$

Multiply by the LCD, 15.

$$3(2m - 3) - 5m = 3(-6)$$

$$6m - 9 - 5m = -18$$

$$m - 9 = -18 \quad \text{Combine like terms.}$$

$$m = -9 \quad \text{Add 9.}$$

Check $m = -9$: $-\dfrac{6}{5} = -\dfrac{6}{5}$

The solution is -9.

4. $1 - \dfrac{2}{x+1} = \dfrac{2x}{x+1}$

$$(x+1)\left(1 - \dfrac{2}{x+1}\right) = (x+1)\dfrac{2x}{x+1}$$

Multiply by the LCD, $x + 1$.

$$(x + 1) - 2 = 2x$$

$$x - 1 = 2x$$

$$-1 = x \quad \text{Subtract } x.$$

Check $x = -1$: $1 - \dfrac{2}{0} = -\dfrac{2}{0}$

The fractions are undefined, so the equation has no solution.

5. (a) $\dfrac{4}{x^2 - 3x} = \dfrac{1}{x^2 - 9}$

$$\dfrac{4}{x(x-3)} = \dfrac{1}{(x-3)(x+3)} \quad \text{Factor.}$$

Multiply by the LCD, $x(x+3)(x-3)$.

$$x(x+3)(x-3)\left(\dfrac{4}{x(x-3)}\right)$$

$$= x(x+3)(x-3)\left(\dfrac{1}{(x-3)(x+3)}\right)$$

$$4(x+3) = 1x$$

$$4x + 12 = x$$

$$3x = -12 \quad \text{Subtract } x.$$

$$x = -4 \quad \text{Divide by 3.}$$

Check $x = -4$: $\dfrac{1}{7} = \dfrac{1}{7}$

The solution is -4.

(b) $\dfrac{2}{p^2 - 2p} = \dfrac{3}{p^2 - p}$

$$\dfrac{2}{p(p-2)} = \dfrac{3}{p(p-1)} \quad \text{Factor.}$$

Multiply by the LCD, $p(p-2)(p-1)$.

$$p(p-2)(p-1)\left(\dfrac{2}{p(p-2)}\right)$$

$$= p(p-1)(p-2)\left(\dfrac{3}{p(p-1)}\right)$$

$$2(p-1) = 3(p-2)$$

$$2p - 2 = 3p - 6$$

$$2p + 4 = 3p \quad \text{Add 6.}$$

$$4 = p \quad \text{Subtract } 2p.$$

Check $p = 4$: $\dfrac{1}{4} = \dfrac{1}{4}$

The solution is 4.

278 Chapter 7 Rational Expressions and Applications

6. (a) $\dfrac{2p}{p^2-1} = \dfrac{2}{p+1} - \dfrac{1}{p-1}$

$\dfrac{2p}{(p+1)(p-1)} = \dfrac{2}{p+1} - \dfrac{1}{p-1}$ *Factor.*

Multiply by the LCD, $(p+1)(p-1)$.

$(p+1)(p-1)\left(\dfrac{2p}{(p+1)(p-1)}\right)$

$= (p+1)(p-1)\left(\dfrac{2}{p+1} - \dfrac{1}{p-1}\right)$

$2p = 2(p-1) - (p+1)$
$2p = 2p - 2 - p - 1$
$2p = p - 3$
$p = -3$ *Subtract p.*

Check $p = -3$: $-\tfrac{3}{4} = -\tfrac{3}{4}$

The solution is -3.

(b) $\dfrac{8r}{4r^2-1} = \dfrac{3}{2r+1} + \dfrac{3}{2r-1}$

$\dfrac{8r}{(2r+1)(2r-1)} = \dfrac{3}{2r+1} + \dfrac{3}{2r-1}$ *Factor.*

Multiply by the LCD, $(2r+1)(2r-1)$.

$(2r+1)(2r-1)\left(\dfrac{8r}{(2r+1)(2r-1)}\right)$

$= (2r+1)(2r-1)\left(\dfrac{3}{2r+1} + \dfrac{3}{2r-1}\right)$

$8r = 3(2r-1) + 3(2r+1)$
$8r = 6r - 3 + 6r + 3$
$8r = 12r$
$0 = 4r$ *Subtract 8r.*
$0 = r$ *Divide by 4.*

Check $r = 0$: $0 = 0$

The solution is 0.

7. $\dfrac{2}{3x+1} - \dfrac{1}{x} = \dfrac{-6x}{3x+1}$

Multiply by the LCD, $x(3x+1)$.

$x(3x+1)\left(\dfrac{2}{3x+1} - \dfrac{1}{x}\right) = x(3x+1)\left(\dfrac{-6x}{3x+1}\right)$

$2x - (3x+1) = -6x^2$
$2x - 3x - 1 = -6x^2$
$-x - 1 = -6x^2$
$6x^2 - x - 1 = 0$
$(2x-1)(3x+1) = 0$
$x = \tfrac{1}{2}$ or $x = -\tfrac{1}{3}$

Check $x = \tfrac{1}{2}$: $-\tfrac{6}{5} = -\tfrac{6}{5}$

Since $-\tfrac{1}{3}$ makes a denominator of the original equation equal 0, $-\tfrac{1}{3}$ is not a solution. The solution is $\tfrac{1}{2}$.

8. (a) $\dfrac{1}{x-2} + \dfrac{1}{5} = \dfrac{2}{5(x^2-4)}$

$\dfrac{1}{x-2} + \dfrac{1}{5} = \dfrac{2}{5(x-2)(x+2)}$ *Factor.*

Multiply by the LCD, $5(x-2)(x+2)$.

$5(x-2)(x+2)\left(\dfrac{1}{x-2} + \dfrac{1}{5}\right)$

$= 5(x-2)(x+2)\left(\dfrac{2}{5(x-2)(x+2)}\right)$

$5(x+2) + (x+2)(x-2) = 2$
$5x + 10 + x^2 - 4 = 2$
$x^2 + 5x + 6 = 2$
$x^2 + 5x + 4 = 0$
$(x+4)(x+1) = 0$
$x+4 = 0$ or $x+1 = 0$
$x = -4$ or $x = -1$

Check $x = -4$: $\tfrac{1}{30} = \tfrac{1}{30}$

Check $x = -1$: $-\tfrac{2}{15} = -\tfrac{2}{15}$

The solutions are -4 and -1.

(b) $\dfrac{6}{5a+10} - \dfrac{1}{a-5} = \dfrac{4}{a^2-3a-10}$

$\dfrac{6}{5(a+2)} - \dfrac{1}{a-5} = \dfrac{4}{(a-5)(a+2)}$ *Factor.*

Multiply by the LCD, $5(a+2)(a-5)$.

$5(a+2)(a-5)\left(\dfrac{6}{5(a+2)} - \dfrac{1}{a-5}\right)$

$= 5(a+2)(a-5)\left(\dfrac{4}{(a-5)(a+2)}\right)$

$6(a-5) - 5(a+2) = 5(4)$
$6a - 30 - 5a - 10 = 20$
$a - 40 = 20$
$a = 60$

Check $a = 60$: $\tfrac{2}{1705} = \tfrac{2}{1705}$

The solution is 60.

9. Solve $z = \dfrac{x}{x+y}$ for y.

Multiply by the LCD, $x+y$.

$z(x+y) = \dfrac{x}{x+y}(x+y)$

$zx + zy = x$
$zy = x - zx$ *Isolate zy.*
$y = \dfrac{x - zx}{z}$ *Divide by z.*

7.6 Solving Equations with Rational Expressions

10. Solve $\dfrac{2}{x} = \dfrac{1}{y} + \dfrac{1}{z}$ for z.

Multiply by the LCD, xyz.

$$xyz\left(\dfrac{2}{x}\right) = xyz\left(\dfrac{1}{y} + \dfrac{1}{z}\right)$$

$$xyz\left(\dfrac{2}{x}\right) = xyz\left(\dfrac{1}{y}\right) + xyz\left(\dfrac{1}{z}\right)$$

$$2yz = xz + xy$$

Get the z-terms on one side.

$$2yz - xz = xy$$
$$z(2y - x) = xy \quad \text{Factor.}$$
$$z = \dfrac{xy}{2y - x} \quad \text{or} \quad z = \dfrac{-xy}{x - 2y}$$

7.6 Section Exercises

1. $\dfrac{7}{8}x + \dfrac{1}{5}x$ is the sum of two terms, so it is an *expression* to be simplified. Simplify by finding the LCD, writing each coefficient with this LCD, and combining like terms.

$$\dfrac{7}{8}x + \dfrac{1}{5}x = \dfrac{35}{40}x + \dfrac{8}{40}x \quad \text{LCD} = 40$$
$$= \dfrac{43}{40}x \quad \text{Combine like terms.}$$

3. $\dfrac{7}{8}x + \dfrac{1}{5}x = 1$ has an equals sign, so this is an *equation* to be solved. Use the multiplication property of equality to clear fractions. The LCD is 40.

$$\dfrac{7}{8}x + \dfrac{1}{5}x = 1$$

$$40\left(\dfrac{7}{8}x + \dfrac{1}{5}x\right) = 40 \cdot 1 \quad \text{Multiply by 40.}$$

$$40\left(\dfrac{7}{8}x\right) + 40\left(\dfrac{1}{5}x\right) = 40 \cdot 1 \quad \text{Distributive property}$$

$$35x + 8x = 40 \quad \text{Multiply.}$$
$$43x = 40 \quad \text{Combine like terms.}$$
$$x = \dfrac{40}{43} \quad \text{Divide by 43.}$$

The solution is $\dfrac{40}{43}$.

5. $\dfrac{3}{5}y - \dfrac{7}{10}y$ is the difference of two terms, so it is an *expression* to be simplified.

$$\dfrac{3}{5}y - \dfrac{7}{10}y = \dfrac{6}{10}y - \dfrac{7}{10}y \quad \text{LCD} = 10$$

$$= -\dfrac{1}{10}y \quad \text{Combine like terms.}$$

7. When solving equations, the LCD is used as a multiplier for every term in the equation. As a result, the fractions are removed from the equation.

When adding and subtracting rational expressions, the LCD is used to make it possible to combine several separate rational expressions into one rational expression. This does not necessarily eliminate fractions.

Note: In Exercises 9–28 and 31–48, all proposed solutions should be checked by substituting in the original equation.

9.
$$\dfrac{2}{3}x + \dfrac{1}{2}x = -7$$

$$6\left(\dfrac{2}{3}x + \dfrac{1}{2}x\right) = 6(-7) \quad \text{Multiply by LCD, 6.}$$

$$6\left(\dfrac{2}{3}x\right) + 6\left(\dfrac{1}{2}x\right) = -42 \quad \text{Distributive property}$$

$$4x + 3x = -42 \quad \text{Multiply.}$$

$$7x = -42 \quad \text{Combine like terms.}$$

$$x = -6 \quad \text{Divide by 7.}$$

Check $x = -6$: $-7 = -7$

The solution is -6.

11.
$$\dfrac{p}{3} - \dfrac{p}{6} = 4$$

$$6\left(\dfrac{p}{3} - \dfrac{p}{6}\right) = 6(4) \quad \text{Multiply by LCD, 6.}$$

$$6\left(\dfrac{p}{3}\right) - 6\left(\dfrac{p}{6}\right) = 24 \quad \text{Distributive property}$$

$$2p - p = 24$$
$$p = 24$$

Check $p = 24$: $4 = 4$

The solution is 24.

13.
$$\dfrac{3x}{5} - 6 = x$$

$$5\left(\dfrac{3x}{5} - 6\right) = 5(x) \quad \text{Multiply by LCD, 5.}$$

$$5\left(\dfrac{3x}{5}\right) - 5(6) = 5x \quad \text{Distributive property}$$

$$3x - 30 = 5x$$
$$-30 = 2x$$
$$-15 = x$$

Check $x = -15$: $-15 = -15$

The solution is -15.

15.
$$\frac{4m}{7} + m = 11$$
$$7\left(\frac{4m}{7} + m\right) = 7(11) \quad \text{Multiply by LCD, 7.}$$
$$7\left(\frac{4m}{7}\right) + 7(m) = 77 \quad \text{Distributive property}$$
$$4m + 7m = 77$$
$$11m = 77$$
$$m = 7$$

Check $m = 7$: $11 = 11$

The solution is 7.

17.
$$\frac{z-1}{4} = \frac{z+3}{3}$$
$$12\left(\frac{z-1}{4}\right) = 12\left(\frac{z+3}{3}\right) \quad \text{Multiply by LCD, 12.}$$
$$3(z-1) = 4(z+3)$$
$$3z - 3 = 4z + 12 \quad \text{Distributive property}$$
$$-15 = z$$

Check $z = -15$: $-4 = -4$

The solution is -15.

19.
$$\frac{3p+6}{8} = \frac{3p-3}{16}$$
$$16\left(\frac{3p+6}{8}\right) = 16\left(\frac{3p-3}{16}\right) \quad \text{Multiply by LCD, 16.}$$
$$2(3p+6) = 3p - 3$$
$$6p + 12 = 3p - 3 \quad \text{Distributive property}$$
$$3p = -15$$
$$p = -5$$

Check $p = -5$: $-\frac{9}{8} = -\frac{9}{8}$

The solution is -5.

21.
$$\frac{2x+3}{-6} = \frac{3}{2}$$
$$-6\left(\frac{2x+3}{-6}\right) = -6\left(\frac{3}{2}\right) \quad \text{Multiply by LCD, }-6.$$
$$2x + 3 = -9$$
$$2x = -12$$
$$x = -6$$

Check $x = -6$: $\frac{3}{2} = \frac{3}{2}$

The solution is -6.

23.
$$\frac{q+2}{3} + \frac{q-5}{5} = \frac{7}{3}$$
$$15\left(\frac{q+2}{3} + \frac{q-5}{5}\right) = 15\left(\frac{7}{3}\right) \quad \text{Multiply by LCD, 15.}$$
$$15\left(\frac{q+2}{3}\right) + 15\left(\frac{q-5}{5}\right) = 5 \cdot 7$$
$$5(q+2) + 3(q-5) = 35$$
$$5q + 10 + 3q - 15 = 35$$
$$8q - 5 = 35$$
$$8q = 40$$
$$q = 5$$

Check $q = 5$: $\frac{7}{3} = \frac{7}{3}$

The solution is 5.

25.
$$\frac{t}{6} + \frac{4}{3} = \frac{t-2}{3}$$
$$6\left(\frac{t}{6} + \frac{4}{3}\right) = 6\left(\frac{t-2}{3}\right) \quad \text{Multiply by LCD, 6.}$$
$$6\left(\frac{t}{6}\right) + 6\left(\frac{4}{3}\right) = 2(t-2)$$
$$t + 8 = 2t - 4$$
$$12 = t$$

Check $t = 12$: $\frac{10}{3} = \frac{10}{3}$

The solution is 12.

27.
$$\frac{3m}{5} - \frac{3m-2}{4} = \frac{1}{5}$$
$$20\left(\frac{3m}{5} - \frac{3m-2}{4}\right) = 20\left(\frac{1}{5}\right) \quad \text{Multiply by LCD, 20.}$$
$$4(3m) - 5(3m-2) = 4$$
$$12m - 15m + 10 = 4$$
$$-3m + 10 = 4$$
$$-3m = -6$$
$$m = 2$$

Check $m = 2$: $\frac{1}{5} = \frac{1}{5}$

The solution is 2.

29. $$\frac{1}{x-4} = \frac{3}{2x}$$

x cannot take values which would make $x - 4$ or $2x$ equal to 0.

$$x - 4 = 0 \quad \text{or} \quad 2x = 0$$
$$x = 4 \quad \text{or} \quad x = 0$$

Thus, 0 and 4 cannot be solutions of the given equation.

7.6 Solving Equations with Rational Expressions

31.
$$\frac{2x+3}{x} = \frac{3}{2}$$
$$2x\left(\frac{2x+3}{x}\right) = 2x\left(\frac{3}{2}\right) \quad \text{Multiply by LCD, } 2x.$$
$$2(2x+3) = 3x$$
$$4x + 6 = 3x \quad \text{Distributive property}$$
$$x = -6$$

Check $x = -6$: $\frac{3}{2} = \frac{3}{2}$

The solution is -6.

33.
$$\frac{k}{k-4} - 5 = \frac{4}{k-4}$$
$$(k-4)\left(\frac{k}{k-4} - 5\right) = (k-4)\left(\frac{4}{k-4}\right)$$
Multiply by LCD, $k-4$.
$$(k-4)\left(\frac{k}{k-4}\right) - 5(k-4) = 4 \quad \text{Distributive property}$$
$$k - 5k + 20 = 4$$
$$-4k = -16$$
$$k = 4$$

The proposed solution is 4. However, 4 cannot be a solution because it makes the denominator $k-4$ equal 0. Therefore, the given equation has *no solution*.

35.
$$\frac{3}{x-1} + \frac{2}{4x-4} = \frac{7}{4}$$
$$\frac{3}{x-1} + \frac{2}{4(x-1)} = \frac{7}{4}$$
$$4(x-1)\left(\frac{3}{x-1} + \frac{2}{4(x-1)}\right) = 4(x-1)\left(\frac{7}{4}\right)$$
Multiply by LCD, $4(x-1)$.
$$4(3) + 2 = (x-1)(7)$$
$$14 = 7x - 7$$
$$21 = 7x$$
$$3 = x$$

Check $x = 3$: $\frac{7}{4} = \frac{7}{4}$

The solution is 3.

37.
$$\frac{y}{3y+3} = \frac{2y-3}{y+1} - \frac{2y}{3y+3}$$
$$\frac{y}{3(y+1)} = \frac{2y-3}{y+1} - \frac{2y}{3(y+1)}$$
$$3(y+1)\left(\frac{y}{3(y+1)}\right) =$$
$$3(y+1)\left[\frac{2y-3}{y+1} - \frac{2y}{3(y+1)}\right]$$
Multiply by LCD, $3(y+1)$.

$$y = 3(y+1)\left(\frac{2y-3}{y+1}\right)$$
$$- 3(y+1)\left(\frac{2y}{3(y+1)}\right)$$
$$y = 3(2y-3) - 2y$$
$$y = 6y - 9 - 2y$$
$$y = 4y - 9$$
$$-3y = -9$$
$$y = 3$$

Check $y = 3$: $\frac{1}{4} = \frac{1}{4}$

The solution is 3.

39.
$$\frac{2}{m} = \frac{m}{5m+12}$$
$$m(5m+12)\left(\frac{2}{m}\right) = m(5m+12)\left(\frac{m}{5m+12}\right)$$
$$(5m+12)(2) = m(m)$$
$$10m + 24 = m^2$$
$$0 = m^2 - 10m - 24$$
$$0 = (m-12)(m+2)$$
$$m - 12 = 0 \quad \text{or} \quad m + 2 = 0$$
$$m = 12 \quad \text{or} \quad m = -2$$

Check $m = -2$: $-1 = -1$

Check $m = 12$: $\frac{1}{6} = \frac{1}{6}$

The solutions are -2 and 12.

41.
$$\frac{-2}{z+5} + \frac{3}{z-5} = \frac{20}{z^2 - 25}$$
$$\frac{-2}{z+5} + \frac{3}{z-5} = \frac{20}{(z+5)(z-5)} \quad \text{Factor.}$$
$$(z+5)(z-5)\left(\frac{-2}{z+5} + \frac{3}{z-5}\right)$$
$$= (z+5)(z-5)\left(\frac{20}{(z+5)(z-5)}\right)$$
Multiply by LCD, $(z+5)(z-5)$.
$$-2(z-5) + 3(z+5) = 20$$
$$-2z + 10 + 3z + 15 = 20$$
$$z + 25 = 20$$
$$z = -5$$

The proposed solution, -5, cannot be a solution because it makes the denominators $z + 5$ and $z^2 - 25$ equal 0 and the corresponding fractions undefined. Therefore, the equation has *no solution*.

43. $\dfrac{3y}{y^2+5y+6}$

$= \dfrac{5y}{y^2+2y-3} - \dfrac{2}{y^2+y-2}$

$\dfrac{3y}{(y+2)(y+3)}$

$= \dfrac{5y}{(y+3)(y-1)} - \dfrac{2}{(y-1)(y+2)}$

$(y+2)(y+3)(y-1) \cdot \left[\dfrac{3y}{(y+2)(y+3)}\right]$

$= (y+2)(y+3)(y-1) \cdot \left[\dfrac{5y}{(y+3)(y-1)}\right]$

$\quad - (y+2)(y+3)(y-1) \cdot \left[\dfrac{2}{(y-1)(y+2)}\right]$

Multiply by LCD, (y + 2)(y + 3)(y − 1).

$3y(y-1) = 5y(y+2) - 2(y+3)$
$3y^2 - 3y = 5y^2 + 10y - 2y - 6$
$0 = 2y^2 + 11y - 6$
$0 = (2y-1)(y+6)$

Note to reader: We may skip writing out the zero-factor property since this step can be easily performed mentally.

$\qquad y = \tfrac{1}{2} \quad \text{or} \quad y = -6$

Check $y = \tfrac{1}{2}$: $\tfrac{6}{35} = -\tfrac{10}{7} - \left(-\tfrac{8}{5}\right)$

Check $y = -6$: $-\tfrac{3}{2} = -\tfrac{10}{7} - \tfrac{1}{14}$

The solutions are -6 and $\tfrac{1}{2}$.

45. $\dfrac{5x}{14x+3} = \dfrac{1}{x}$

$x(14x+3)\left(\dfrac{5x}{14x+3}\right) = x(14x+3)\left(\dfrac{1}{x}\right)$

Multiply by LCD, x(14x + 3).

$x(5x) = (14x+3)(1)$
$5x^2 = 14x + 3$
$5x^2 - 14x - 3 = 0$
$(5x+1)(x-3) = 0$
$\qquad x = -\tfrac{1}{5} \quad \text{or} \quad x = 3$

Check $x = -\tfrac{1}{5}$: $-5 = -5$

Check $x = 3$: $\tfrac{1}{3} = \tfrac{1}{3}$

The solutions are $-\tfrac{1}{5}$ and 3.

47. $\dfrac{2}{z-1} - \dfrac{5}{4} = \dfrac{-1}{z+1}$

$4(z+1)(z-1)\left(\dfrac{2}{z-1} - \dfrac{5}{4}\right)$

$\qquad = 4(z+1)(z-1)\left(\dfrac{-1}{z+1}\right)$

Multiply by LCD, 4(z + 1)(z − 1).

$8(z+1) - 5(z^2-1) = -4(z-1)$
$8z + 8 - 5z^2 + 5 = -4z + 4$
$-5z^2 + 12z + 9 = 0$
$5z^2 - 12z - 9 = 0$
$(5z+3)(z-3) = 0$
$\qquad z = -\tfrac{3}{5} \quad \text{or} \quad z = 3$

Check $z = -\tfrac{3}{5}$: $-\tfrac{5}{2} = -\tfrac{5}{2}$

Check $z = 3$: $-\tfrac{1}{4} = -\tfrac{1}{4}$

The solutions are $-\tfrac{3}{5}$ and 3.

49. $kr - mr = km$

If you are solving for k, put both terms with k on one side and the remaining term on the other side.

$\qquad kr - km = mr$

51. $m = \dfrac{kF}{a}$ for F

We need to isolate F on one side of the equation.

$m \cdot a = \left(\dfrac{kF}{a}\right)(a)$ *Multiply by a.*

$ma = kF$

$\dfrac{ma}{k} = \dfrac{kF}{k}$ *Divide by k.*

$\dfrac{ma}{k} = F$

53. $m = \dfrac{kF}{a}$ for a

$m \cdot a = \left(\dfrac{kF}{a}\right)(a)$ *Multiply by a.*

$ma = kF$

$\dfrac{ma}{m} = \dfrac{kF}{m}$ *Divide by m.*

$a = \dfrac{kF}{m}$

55. $I = \dfrac{E}{R+r}$ for R

We need to isolate R on one side of the equation.

$$I(R+r) = \left(\dfrac{E}{R+r}\right)(R+r) \quad \text{Multiply by } R+r.$$

$$IR + Ir = E \quad \text{Distributive property}$$

$$IR = E - Ir \quad \text{Subtract } Ir.$$

$$R = \dfrac{E - Ir}{I} \quad \text{Divide by } I.$$

$$\text{or} \quad R = \dfrac{E}{I} - r$$

57. $h = \dfrac{2A}{B+b}$ for A

$$(B+b)h = (B+b) \cdot \dfrac{2A}{B+b}$$

Multiply by $B + b$.

$$h(B+b) = 2A$$

$$\dfrac{h(B+b)}{2} = A \quad \text{Divide by 2.}$$

59. $d = \dfrac{2S}{n(a+L)}$ for a

We need to isolate a on one side of the equation.

$$d \cdot n(a+L) = \dfrac{2S}{n(a+L)} \cdot n(a+L)$$

Multiply by $n(a+L)$.

$$nd(a+L) = 2S$$

$$nda + ndL = 2S$$

$$nda = 2S - ndL \quad \text{Subtract } ndL.$$

$$a = \dfrac{2S - ndL}{nd} \quad \text{Divide by } nd.$$

$$\text{or} \quad a = \dfrac{2S}{nd} - L$$

61. $\dfrac{2}{r} + \dfrac{3}{s} + \dfrac{1}{t} = 1$ for t

The LCD of all the fractions in the equation is rst, so multiply both sides by rst.

$$rst\left(\dfrac{2}{r} + \dfrac{3}{s} + \dfrac{1}{t}\right) = rst(1)$$

$$rst\left(\dfrac{2}{r}\right) + rst\left(\dfrac{3}{s}\right) + rst\left(\dfrac{1}{t}\right) = rst$$

$$2st + 3rt + rs = rst$$

Since we are solving for t, get all terms with t on one side of the equation.

$$2st + 3rt - rst = -rs$$

Factor out the common factor t on the left.

$$t(2s + 3r - rs) = -rs$$

Finally, divide both sides by the coefficient of t, which is $2s + 3r - rs$.

$$t = \dfrac{-rs}{2s + 3r - rs} \quad \text{or} \quad t = \dfrac{rs}{-2s - 3r + rs}$$

63. $\dfrac{1}{a} - \dfrac{1}{b} - \dfrac{1}{c} = 2$ for c

$$abc\left(\dfrac{1}{a} - \dfrac{1}{b} - \dfrac{1}{c}\right) = abc(2)$$

$$bc - ac - ab = 2abc$$

Get all the terms with c on one side.

$$bc - ac - 2abc = ab$$

$$c(b - a - 2ab) = ab$$

$$c = \dfrac{ab}{b - a - 2ab}$$

$$\text{or} \quad c = \dfrac{-ab}{-b + a + 2ab}$$

65. $9x + \dfrac{3}{z} = \dfrac{5}{y}$ for z

$$yz\left(9x + \dfrac{3}{z}\right) = yz\left(\dfrac{5}{y}\right) \quad \text{Multiply by LCD, } yz.$$

$$yz(9x) + yz\left(\dfrac{3}{z}\right) = yz\left(\dfrac{5}{y}\right) \quad \text{Distributive property}$$

$$9xyz + 3y = 5z$$

$$9xyz - 5z = -3y \quad \text{Get the } z \text{ terms on one side.}$$

$$z(9xy - 5) = -3y \quad \text{Factor out } z.$$

$$z = \dfrac{-3y}{9xy - 5} \quad \text{Divide by } 9xy - 5.$$

$$\text{or} \quad z = \dfrac{3y}{5 - 9xy}$$

Summary Exercises on Rational Expressions and Equations

1. No equals sign appears so this is an *expression*.

$$\dfrac{4}{p} + \dfrac{6}{p} = \dfrac{4+6}{p} = \dfrac{10}{p}$$

3. No equals sign appears so this is an *expression*.

$$\dfrac{1}{x^2 + x - 2} \div \dfrac{4x^2}{2x - 2}$$

$$= \dfrac{1}{x^2 + x - 2} \cdot \dfrac{2x - 2}{4x^2}$$

$$= \dfrac{1}{(x+2)(x-1)} \cdot \dfrac{2(x-1)}{2 \cdot 2x^2}$$

$$= \dfrac{1}{2x^2(x+2)}$$

284 Chapter 7 Rational Expressions and Applications

5. No equals sign appears so this is an *expression*.

$$\frac{2y^2 + y - 6}{2y^2 - 9y + 9} \cdot \frac{y^2 - 2y - 3}{y^2 - 1}$$

$$= \frac{(2y - 3)(y + 2)(y - 3)(y + 1)}{(2y - 3)(y - 3)(y + 1)(y - 1)}$$

$$= \frac{y + 2}{y - 1}$$

7. $\dfrac{x - 4}{5} = \dfrac{x + 3}{6}$

There is an equals sign, so this is an *equation*.

$$30\left(\frac{x-4}{5}\right) = 30\left(\frac{x+3}{6}\right) \quad \text{Multiply by LCD, 30.}$$

$$6(x - 4) = 5(x + 3)$$
$$6x - 24 = 5x + 15$$
$$x = 39$$

Check $x = 39$: $7 = 7$

The solution is 39.

9. No equals sign appears so this is an *expression*.

$$\frac{4}{p+2} + \frac{1}{3p+6} = \frac{4}{p+2} + \frac{1}{3(p+2)}$$

$$= \frac{3 \cdot 4}{3(p+2)} + \frac{1}{3(p+2)} \quad \text{LCD} = 3(p+2)$$

$$= \frac{12 + 1}{3(p+2)}$$

$$= \frac{13}{3(p+2)}$$

11. $\dfrac{3}{t-1} + \dfrac{1}{t} = \dfrac{7}{2}$

There is an equals sign, so this is an *equation*.

$$2t(t-1)\left(\frac{3}{t-1} + \frac{1}{t}\right) = 2t(t-1)\left(\frac{7}{2}\right)$$

Multiply by LCD, $2t(t-1)$.

$$2t(t-1)\left(\frac{3}{t-1}\right) + 2t(t-1)\left(\frac{1}{t}\right) = 7t(t-1)$$

$$2t(3) + 2(t-1) = 7t(t-1)$$
$$6t + 2t - 2 = 7t^2 - 7t$$
$$0 = 7t^2 - 15t + 2$$
$$0 = (7t - 1)(t - 2)$$
$$t = \tfrac{1}{7} \quad \text{or} \quad t = 2$$

Check $t = \tfrac{1}{7}$: $-\tfrac{7}{2} + 7 = \tfrac{7}{2}$

Check $t = 2$: $3 + \tfrac{1}{2} = \tfrac{7}{2}$

The solutions are $\tfrac{1}{7}$ and 2.

13. No equals sign appears so this is an *expression*.

$$\frac{5}{4z} - \frac{2}{3z} = \frac{3 \cdot 5}{3 \cdot 4z} - \frac{4 \cdot 2}{4 \cdot 3z} \quad \text{LCD} = 12z$$

$$= \frac{15}{12z} - \frac{8}{12z}$$

$$= \frac{15 - 8}{12z} = \frac{7}{12z}$$

15. No equals sign appears so this is an *expression*.

$$\frac{1}{m^2 + 5m + 6} + \frac{2}{m^2 + 4m + 3}$$

$$= \frac{1}{(m+2)(m+3)} + \frac{2}{(m+1)(m+3)}$$

$$= \frac{1(m+1)}{(m+1)(m+2)(m+3)}$$

$$+ \frac{2(m+2)}{(m+1)(m+2)(m+3)}$$

LCD = $(m + 1)(m + 2)(m + 3)$

$$= \frac{(m+1) + (2m+4)}{(m+1)(m+2)(m+3)}$$

$$= \frac{3m + 5}{(m+1)(m+2)(m+3)}$$

17. $\dfrac{2}{x+1} + \dfrac{5}{x-1} = \dfrac{10}{x^2 - 1}$

There is an equals sign, so this is an *equation*.

$$\frac{2}{x+1} + \frac{5}{x-1} = \frac{10}{(x+1)(x-1)}$$

$$(x+1)(x-1)\left(\frac{2}{x+1} + \frac{5}{x-1}\right)$$

$$= (x+1)(x-1)\left[\frac{10}{(x+1)(x-1)}\right]$$

Multiply by LCD, $(x + 1)(x - 1)$.

$$(x+1)(x-1)\left(\frac{2}{x+1}\right)$$

$$+ (x+1)(x-1)\left(\frac{5}{x-1}\right) = 10$$

Distributive property

$$2(x-1) + 5(x+1) = 10$$
$$2x - 2 + 5x + 5 = 10$$
$$3 + 7x = 10$$
$$7x = 7$$
$$x = 1$$

Replacing x by 1 in the original equation makes the denominators $x - 1$ and $x^2 - 1$ equal to 0, so there is *no solution*.

7.7 Applications of Rational Expressions

7.7 Margin Exercises

1. **(a)** *Step 2*
   Let $x =$ the number added to the numerator and subtracted from the denominator of $\frac{5}{8}$.

   Step 3
   $\frac{5+x}{8-x} =$ the reciprocal of $\frac{5}{8}$.

   Write this as an equation.
   $$\frac{5+x}{8-x} = \frac{8}{5}$$

   Step 4
   Multiply by the LCD $5(8-x)$.
   $$5(8-x)\frac{5+x}{8-x} = 5(8-x)\frac{8}{5}$$
   $$5(5+x) = 8(8-x)$$
   $$25 + 5x = 64 - 8x$$
   $$13x = 39$$
   $$x = 3$$

   Step 5
   The number is 3.

   Step 6
   If 3 is added to the numerator and subtracted from the denominator of $\frac{5}{8}$, the result is $\frac{8}{5}$, which is the reciprocal of $\frac{5}{8}$.

   (b) *Step 2*
   Let $x =$ the numerator of the original fraction. Then $x + 1 =$ the denominator.

   Step 3
   The original fraction is $\frac{x}{x+1}$. Add 6 to the numerator and subtract 6 from the denominator of this fraction to get
   $$\frac{x+6}{x+1-6} = \frac{x+6}{x-5}.$$
   The result is $\frac{15}{4}$, so
   $$\frac{x+6}{x-5} = \frac{15}{4}.$$

   Step 4
   Multiply by the LCD, $4(x-5)$.
   $$4(x-5)\frac{x+6}{x-5} = 4(x-5)\left(\frac{15}{4}\right)$$
   $$4(x+6) = 15(x-5)$$
   $$4x + 24 = 15x - 75$$
   $$24 = 11x - 75$$
   $$99 = 11x$$
   $$9 = x$$

   Step 5
   The original fraction is
   $$\frac{9}{9+1} = \frac{9}{10}.$$

   Step 6
   If we add 6 to the numerator and subtract 6 from the denominator, the fraction $\frac{9}{10}$ becomes $\frac{15}{4}$.

2. **(a)** $r = \frac{d}{t} = \frac{100}{9.79} \approx 10.21$ m per sec

   His average speed was 10.21 meters per second.

   (b) $t = \frac{d}{r} = \frac{3000}{6.173} \approx 486$ sec

   Her winning time was 486 seconds.

   (c) $d = rt = (145)(2) = 290$ mi

   The distance between Chicago and St. Louis is 290 miles.

3. **(a)** Let $t =$ the number of hours until the distance between Lupe and Maria is 55 miles.

	Rate	×	Time	=	Distance
Lupe	10		t		$10t$
Maria	12		t		$12t$

   The sum of the distances is 55 miles.
   $$10t + 12t = 55$$
   $$22t = 55$$
   $$t = \frac{55}{22} = 2.5 \text{ or } 2\frac{1}{2}$$

   They will be 55 miles apart in $2\frac{1}{2}$ hours.

   (b) Let $t =$ the number of hours until the distance between the boats is 35 miles.

	Rate	×	Time	=	Distance
Slower Boat	18		t		$18t$
Faster Boat	25		t		$25t$

   The difference of the distances is 35 miles.
   $$25t - 18t = 35$$
   $$7t = 35$$
   $$t = 5$$

   They will be 35 miles apart in 5 hours.

4. **(a)** Let $x =$ the speed of the boat with no current. Complete a table.

	d	r	t
With the Current	60	$x+4$	$\dfrac{60}{x+4}$
Against the Current	20	$x-4$	$\dfrac{20}{x-4}$

Since the times are equal, we get the following equation.
$$\frac{60}{x+4} = \frac{20}{x-4}$$
Multiply by the LCD, $(x+4)(x-4)$.
$$60(x-4) = 20(x+4)$$
$$60x - 240 = 20x + 80$$
$$40x - 240 = 80$$
$$40x = 320$$
$$x = 8$$

The speed of the boat with no current is 8 miles per hour.

(b) Let $x =$ the speed of the plane in still air. Complete a table.

	d	r	t
With the Wind	450	$x+15$	$\dfrac{450}{x+15}$
Against the Wind	375	$x-15$	$\dfrac{375}{x-15}$

Since the times are equal, we get the following equation.
$$\frac{450}{x+15} = \frac{375}{x-15}$$
Multiply by the LCD, $(x+15)(x-15)$.
$$450(x-15) = 375(x+15)$$
$$18(x-15) = 15(x+15) \quad \text{Divide by 25.}$$
$$18x - 270 = 15x + 225$$
$$3x = 495$$
$$x = 165$$

The plane's speed is 165 miles per hour.

5. **(a)**

	Rate	Time Working Together	Fractional Part of the Job Done When Working Together
Michael	$\frac{1}{8}$	x	$\frac{1}{8}x$
Lindsay	$\frac{1}{6}$	x	$\frac{1}{6}x$

Since together Michael and Lindsay complete 1 whole job, we must add their individual parts and set the sum equal to 1.

$$\tfrac{1}{8}x + \tfrac{1}{6}x = 1$$
$$48\left(\tfrac{1}{8}x + \tfrac{1}{6}x\right) = 48(1)$$
$$48\left(\tfrac{1}{8}x\right) + 48\left(\tfrac{1}{6}x\right) = 48(1) \quad \text{Multiply by the LCD, 48.}$$
$$6x + 8x = 48$$
$$14x = 48$$
$$x = \tfrac{48}{14} = \tfrac{24}{7} \text{ or } 3\tfrac{3}{7}$$

Working together, Michael and Lindsay can paint the room in $3\tfrac{3}{7}$ hours.

(b)

	Rate	Time Working Together	Fractional Part of the Job Done When Working Together
Roberto	$\frac{1}{2}$	x	$\frac{1}{2}x$
Marco	$\frac{1}{3}$	x	$\frac{1}{3}x$

Since together Roberto and Marco complete 1 whole job, we must add their individual parts and set the sum equal to 1.

$$\tfrac{1}{2}x + \tfrac{1}{3}x = 1$$
$$6\left(\tfrac{1}{2}x + \tfrac{1}{3}x\right) = 6(1)$$
$$6\left(\tfrac{1}{2}x\right) + 6\left(\tfrac{1}{3}x\right) = 6 \quad \text{Multiply by the LCD, 6.}$$
$$3x + 2x = 6$$
$$5x = 6$$
$$x = \tfrac{6}{5} \text{ or } 1\tfrac{1}{5}$$

Working together, Roberto and Marco can tune up the Bronco in $1\tfrac{1}{5}$ hours.

7.7 Section Exercises

1. **(a)** Let $x =$ __the amount__.

(b) An expression for "the numerator of the fraction $\tfrac{5}{6}$ is increased by an amount" is __$5+x$__. We could also use $\dfrac{5+x}{6}$.

(c) An equation that can be used to solve the problem is
$$\frac{5+x}{6} = \frac{13}{3}.$$

3. *Step 2*
Let $x =$ the numerator of the original fraction. Then $x - 4 =$ the denominator of the original fraction.

Step 3
If 3 is added to both the numerator and denominator, the resulting fraction is equivalent to $\tfrac{3}{2}$ translates to

$$\frac{x+3}{(x-4)+3} = \frac{3}{2}.$$

Step 4
Since we have a fraction equal to another fraction, we can use cross multiplication.
$$2(x+3) = 3[(x-4)+3]$$
$$2x+6 = 3x-3$$
$$9 = x$$

Step 5
The original fraction is
$$\frac{x}{x-4} = \frac{9}{9-4} = \frac{9}{5}.$$

Step 6
Adding 3 to both the numerator and the denominator gives us
$$\frac{9+3}{5+3} = \frac{12}{8},$$
which is equivalent to $\frac{3}{2}$.

5. *Step 2*
Let $x =$ the numerator of the original fraction. Then $3x =$ the denominator of the original fraction.

Step 3
If 2 is added to the numerator and subtracted from the denominator, the resulting fraction is equivalent to 1 translates to
$$\frac{x+2}{3x-2} = 1.$$

Step 4
$$x+2 = 1(3x-2)$$
$$x+2 = 3x-2$$
$$4 = 2x$$
$$2 = x$$

Step 5 The original fraction is
$$\frac{x}{3x} = \frac{2}{3(2)} = \frac{2}{6}.$$

Step 6
$$\frac{2+2}{3(2)-2} = \frac{4}{4} = 1$$

7. *Step 2*
Let $x =$ the number.

Step 3
One-sixth of a number is 5 more than the same number translates to
$$\frac{1}{6}x = 5 + x.$$

Step 4
Multiply both sides by the LCD, 6.
$$6\left(\frac{1}{6}x\right) = 6(5+x)$$
$$x = 30 + 6x$$
$$-30 = 5x$$
$$-6 = x$$

Step 5 The number is -6.

Step 6
One-sixth of -6 is -1 and -1 is 5 more than -6.

9. *Step 2*
Let $x =$ the quantity.
Then its $\frac{3}{4}$, its $\frac{1}{2}$, and its $\frac{1}{3}$ are
$$\tfrac{3}{4}x, \tfrac{1}{2}x, \text{ and } \tfrac{1}{3}x.$$

Step 3
Their sum becomes 93 translates to
$$x + \tfrac{3}{4}x + \tfrac{1}{2}x + \tfrac{1}{3}x = 93.$$

Step 4
Multiply both sides by the LCD of 4, 2, and 3, which is 12.
$$12\left(x + \tfrac{3}{4}x + \tfrac{1}{2}x + \tfrac{1}{3}x\right) = 12(93)$$
$$12x + 12\left(\tfrac{3}{4}x\right) + 12\left(\tfrac{1}{2}x\right) + 12\left(\tfrac{1}{3}x\right) = 12(93)$$
$$12x + 9x + 6x + 4x = 1116$$
$$31x = 1116$$
$$x = 36$$

Step 5
The quantity is 36.

Step 6
Check 36 in the original problem.
$$x = 36, \tfrac{3}{4}x = 27, \tfrac{1}{2}x = 18, \tfrac{1}{3}x = 12$$
Adding gives us
$$36 + 27 + 18 + 12 = 93, \text{ as desired.}$$

11. We are asked to find the *rate*, so we'll use the distance, rate, and time relationship
$$r = \frac{d}{t}.$$
$$r = \frac{60 \text{ meters}}{7.10 \text{ seconds}}$$
$$\approx 8.45 \text{ meters per second}$$

13. We are asked to find the *time*, so we'll use the distance, rate, and time relationship
$$t = \frac{d}{r}.$$
$$t = \frac{500 \text{ miles}}{133.870 \text{ miles per hour}}$$
$$\approx 3.735 \text{ hours}$$

15. We are asked to find the average *rate*, so we'll use the distance, rate, and time relationship

$$r = \frac{d}{t}.$$

$$r = \frac{5000 \text{ meters}}{14.761 \text{ minutes}}$$

$$\approx 338.730 \text{ meters per minute}$$

17. Let $x =$ speed of the current. Then the rate is $4 - x$ upstream (against the current) and $4 + x$ downstream (with the current). The time to row upstream is

$$t = \frac{d}{r} = \frac{8}{4 - x},$$

and the time to row downstream is

$$t = \frac{d}{r} = \frac{24}{4 + x}.$$

Now complete the chart.

	d	r	t
Upstream	8	$4 - x$	$\frac{8}{4-x}$
Downstream	24	$4 + x$	$\frac{24}{4+x}$

Since the problem states that the two times are equal, we have

$$\frac{8}{4 - x} = \frac{24}{4 + x}.$$

We would use this equation to solve the problem.

19. When the hawk flies *into a headwind*, the wind works against the hawk, so the rate of the hawk is 5 miles per hour *less* than if it traveled in still air, or $m - 5$ miles per hour.

When the hawk flies *with a tailwind*, the wind pushes the hawk, so the rate of the hawk is 5 miles per hour *more* than if it traveled in still air, or $m + 5$ miles per hour.

21. Let $x =$ the speed of the wind.
Then $165 + x =$ the speed of the plane flying with the wind, and $165 - x =$ the speed of the plane flying against the wind.

Use $t = \frac{d}{r}$ to complete the table.

	d	r	t
With the Wind	350	$165 + x$	$\frac{350}{165+x}$
Against the Wind	310	$165 - x$	$\frac{310}{165-x}$

Since the times are equal, we get the following equation.

$$\frac{350}{165 + x} = \frac{310}{165 - x}$$

$$(165 + x)(165 - x)\left(\frac{350}{165 + x}\right)$$
$$= (165 + x)(165 - x)\left(\frac{310}{165 - x}\right)$$

$$350(165 - x) = 310(165 + x)$$
$$57{,}750 - 350x = 51{,}150 + 310x$$
$$6600 = 660x$$
$$10 = x$$

The speed of the wind is 10 miles per hour.

23. Let $x =$ the average speed of the ferry.

Use the formula $t = \frac{d}{r}$ to make a chart.

	d	r	t
Seattle-Victoria	148	x	$\frac{148}{x}$
Victoria-Vancouver	74	x	$\frac{74}{x}$

Since the time for the Victoria-Vancouver trip is 4 hours less than the time for the Seattle-Victoria trip, solve the equation

$$\frac{74}{x} = \frac{148}{x} - 4.$$

$$x\left(\frac{74}{x}\right) = x\left(\frac{148}{x} - 4\right) \quad \text{Multiply by LCD, } x.$$

$$74 = 148 - 4x$$
$$4x = 74$$
$$x = \frac{74}{4} = \frac{37}{2} \text{ or } 18\tfrac{1}{2}$$

The average speed of the ferry is $\frac{37}{2}$ or $18\tfrac{1}{2}$ miles per hour.

25. Let $t =$ the number of hours it would take Edwin and his son to tune up the Chevy, working together.

	Rate	Time Working Together	Fractional Part of the Job Done When Working Together
Edwin	$\frac{1}{2}$	t	$\frac{1}{2}t$
Son	$\frac{1}{3}$	t	$\frac{1}{3}t$

Working together, they complete 1 whole job, so add their individual fractional parts and set the sum equal to 1.

fractional part done by Edwin	+	fractional part done by his son	=	1 whole job
↓	↓	↓	↓	↓
$\frac{1}{2}t$	+	$\frac{1}{3}t$	=	1

An equation that can be used to solve this problem is

$$\frac{1}{2}t + \frac{1}{3}t = 1.$$

As in the previous solution, another equation is

$$\frac{1}{2} + \frac{1}{3} = \frac{1}{t}.$$

27. Let x = the number of hours it will take Lea and Tran to groom the horses, working together.

	Rate	Time Working Together	Fractional Part of the Job Done When Working Together
Lea	$\frac{1}{5}$	x	$\frac{1}{5}x$
Tran	$\frac{1}{4}$	x	$\frac{1}{4}x$

Since together Lea and Tran complete 1 whole job, we must add their individual fractional parts and set the sum equal to 1.

$$\frac{1}{5}x + \frac{1}{4}x = 1$$
$$20\left(\frac{1}{5}x + \frac{1}{4}x\right) = 20(1)$$
$$20\left(\frac{1}{5}x\right) + 20\left(\frac{1}{4}x\right) = 20 \quad \text{Multiply by LCD, 20.}$$
$$4x + 5x = 20$$
$$9x = 20$$
$$x = \frac{20}{9} \text{ or } 2\frac{2}{9}$$

It will take Lea and Tran $\frac{20}{9}$ or $2\frac{2}{9}$ hours to groom the horses.

29. Let x = the time required for the printing job with the copiers working together.

	Rate	Time Working Together	Fractional Part of the Job Done When Working Together
First Copier	$\frac{1}{7}$	x	$\frac{1}{7}x$
Second Copier	$\frac{1}{12}$	x	$\frac{1}{12}x$

Since together the two copiers complete 1 whole job, we must add their individual fractional parts and set the sum equal to 1.

$$\frac{1}{7}x + \frac{1}{12}x = 1$$
$$84\left(\frac{1}{7}x + \frac{1}{12}x\right) = 84(1) \quad \text{Multiply by LCD, 84.}$$
$$84\left(\frac{1}{7}x\right) + 84\left(\frac{1}{12}x\right) = 84$$
$$12x + 7x = 84$$
$$19x = 84$$
$$x = \frac{84}{19} \text{ or } 4\frac{8}{19}$$

The copiers together can do the printing job in $\frac{84}{19}$ or $4\frac{8}{19}$ hours.

31. Let x = represent the number of hours it would take Brenda to paint the room by herself. Complete a table.

	Rate	Time Working Together	Fractional Part of the Job Done When Working Together
Hilda	$\frac{1}{6}$	$3\frac{3}{4} = \frac{15}{4}$	$\frac{1}{6} \cdot \frac{15}{4} = \frac{5}{8}$
Brenda	$\frac{1}{x}$	$3\frac{3}{4} = \frac{15}{4}$	$\frac{1}{x} \cdot \frac{15}{4} = \frac{15}{4x}$

Since together Hilda and Brenda complete 1 whole job, we must add their individual fractional parts and set the sum equal to 1.

$$\frac{5}{8} + \frac{15}{4x} = 1$$
$$8x\left(\frac{5}{8} + \frac{15}{4x}\right) = 8x(1)$$
$$8x\left(\frac{5}{8}\right) + 8x\left(\frac{15}{4x}\right) = 8x \quad \text{Multiply by LCD, 8x.}$$
$$5x + 30 = 8x$$
$$30 = 3x$$
$$10 = x$$

It will take Brenda 10 hours to paint the room by herself.

33. Let x = the number of hours to fill the pool with both pipes left open.

	Rate	Time to Fill the Pool	Part Done by Each Pipe
Inlet Pipe	$\frac{1}{9}$	x	$\frac{1}{9}x$
Outlet Pipe	$\frac{1}{12}$	x	$\frac{1}{12}x$

Part done by inlet pipe	−	Part done by outlet pipe	=	Full pool
↓	↓	↓	↓	↓
$\frac{1}{9}x$	−	$\frac{1}{12}x$	=	1

continued

$$36\left(\tfrac{1}{9}x - \tfrac{1}{12}x\right) = 36(1) \quad \text{Multiply by LCD, 36.}$$
$$36\left(\tfrac{1}{9}x\right) - 36\left(\tfrac{1}{12}x\right) = 36$$
$$4x - 3x = 36$$
$$x = 36$$

It takes 36 hours to fill the pool.

35. After both pipes have been running for 3 hours, the pool is $3 \cdot \tfrac{1}{36}$ or $\tfrac{1}{12}$ full. When the outlet pipe is closed, the inlet pipe can completely fill the pool in 9 hours. But only $\tfrac{11}{12}$ of the pool still has to be filled, so the required time (in hours) is

$$\tfrac{11}{12}(9) = \tfrac{99}{12} = \tfrac{33}{4} = 8\tfrac{1}{4}.$$

$\tfrac{33}{4}$ or $8\tfrac{1}{4}$ hours more are needed to fill the pool.

37. The equation would be erroneously written $\dfrac{8}{4+x} = \dfrac{24}{4-x}$. Solving for x gives $x = -2$. Because x represents the speed of the current, it cannot be negative. Therefore, the student should realize that there is a problem in the setup.

7.8 Variation

7.8 Margin Exercises

1. **(a)** $z = kt$ z varies directly as t.
  $11 = 4k$ Substitute $z = 11$ and $t = 4$.
  $\tfrac{11}{4} = k$ Solve for k.

Since $z = kt$ and $k = \tfrac{11}{4}$,

  $z = \tfrac{11}{4}t$.   Substitute $k = \tfrac{11}{4}$ in $z = kt$.
  $z = \tfrac{11}{4}(32)$ Substitute $t = 32$.
  $z = 88$ Multiply.

(b) $C = kr$ Circumference varies directly as radius.
  $43.96 = 7k$ Substitute $r = 7$ and $C = 43.96$.
  $6.28 = k$ Solve for k.

Since $C = kr$ and $k = 6.28$,

  $C = 6.28r$.
  $C = 6.28(11)$ Substitute $r = 11$.
  $C = 69.08$

The circumference is 69.08 centimeters when the radius is 11 centimeters.

2. $z = \dfrac{k}{t}$ z varies inversely as t.
  $8 = \dfrac{k}{2}$ Substitute $z = 8$ and $t = 2$.
  $16 = k$ Solve for k.

Since $z = \dfrac{k}{t}$ and $k = 16$,

  $z = \dfrac{16}{t}$.   Substitute $k = 16$ in $z = \dfrac{k}{t}$.
  $z = \dfrac{16}{32}$ Substitute $t = 32$.
  $z = \dfrac{1}{2}$ Reduce.

3. $c = \dfrac{k}{r}$ Current varies inversely as resistance.
  $80 = \dfrac{k}{10}$ Substitute $c = 80$ and $r = 10$.
  $800 = k$ Solve for k.

Since $c = \dfrac{k}{r}$ and $k = 800$,

  $c = \dfrac{800}{r}$.   Substitute $k = 800$ in $c = \dfrac{k}{r}$.
  $c = \dfrac{800}{16}$ Substitute $r = 16$.
  $c = 50$ Reduce.

If the resistance is 16 ohms, the current is 50 amps.

7.8 Section Exercises

1. **(a)** If the constant of variation is positive and y varies directly as x, then as x increases, y <u>increases</u>.

(b) If the constant of variation is positive and y varies inversely as x, then as x increases, y <u>decreases</u>.

3. Since z varies directly as x, there is a constant k such that $z = kx$. First find the value of k.

$$30 = k(8) \quad \text{Let } z = 30, x = 8.$$
$$k = \tfrac{30}{8} = \tfrac{15}{4}$$

When $k = \tfrac{15}{4}$, $z = kx$ becomes

$$z = \tfrac{15}{4}x.$$

Now find z when $x = 4$.

$$z = \tfrac{15}{4}(4) \quad \text{Let } x = 4.$$
$$= 15$$

5. Since d varies directly as r, there is a constant k such that $d = kr$. First find the value of k.

$$200 = k(40) \quad \text{Let } d = 200, r = 40.$$
$$k = \tfrac{200}{40} = 5$$

When $k = 5$, $d = kr$ becomes
$$d = 5r.$$

Now find d when $r = 60$.
$$d = 5(60) \quad \text{Let } r = 60.$$
$$= 300$$

7. Since z varies inversely as x, there is a constant k such that $z = \dfrac{k}{x}$. First find the value of k.

$$50 = \tfrac{k}{2} \quad \text{Let } z = 50, x = 2.$$
$$k = 50(2) = 100$$

When $k = 100$, $z = \dfrac{k}{x}$ becomes
$$z = \tfrac{100}{x}.$$

Now find z when $x = 25$.
$$z = \tfrac{100}{25} = 4$$

9. Since m varies inversely as r, there is a constant k such that $m = \dfrac{k}{r}$. First find the value of k.

$$12 = \tfrac{k}{8} \quad \text{Let } m = 12, r = 8.$$
$$k = 12(8) = 96$$

When $k = 96$, $m = \dfrac{k}{r}$ becomes
$$m = \tfrac{96}{r}.$$

Now find m when $r = 16$.
$$m = \tfrac{96}{16} = 6$$

11. For a given base, the area A of a triangle varies directly as its height h, so there is a constant k such that $A = kh$. Find the value of k.

$$10 = k(4) \quad \text{Let } A = 10, h = 4.$$
$$k = \tfrac{10}{4} = 2.5$$

When $k = 2.5$, $A = kh$ becomes
$$A = 2.5h.$$

Now find A when $h = 6$.
$$A = 2.5(6) = 15$$

When the height of the triangle is 6 inches, the area of the triangle is 15 square inches.

13. The distance d that a spring stretches varies directly with the force F applied, so
$$d = kF.$$
$$16 = k(75) \quad \text{Let } d = 16, F = 75.$$
$$\tfrac{16}{75} = k$$

So $d = \tfrac{16}{75}F$ and when $F = 200$,
$$d = \tfrac{16}{75}(200) = \tfrac{16(8)}{3} = \tfrac{128}{3}.$$

A force of 200 pounds stretches the spring $42\tfrac{2}{3}$ inches.

15. For a constant area, the length L of a rectangle varies inversely as the width W, so
$$L = \tfrac{k}{W}.$$
$$27 = \tfrac{k}{10} \quad \text{Let } L = 27, W = 10.$$
$$k = 27 \cdot 10 = 270$$

So $L = \dfrac{270}{W}$ and when $L = 18$,
$$18 = \tfrac{270}{W}$$
$$18W = 270$$
$$W = \tfrac{270}{18} = 15.$$

When the length is 18 feet, the width is 15 feet.

17. If the temperature is constant, the pressure P of a gas in a container varies inversely as the volume V of the container, so $P = \dfrac{k}{V}$.

$$10 = \tfrac{k}{3} \quad \text{Let } P = 10, V = 3.$$
$$k = 3 \cdot 10 = 30$$

So $P = \dfrac{30}{V}$ and when $V = 1.5$,
$$P = \tfrac{30}{1.5} = 20.$$

The pressure is 20 pounds per square foot.

292 Chapter 7 Rational Expressions and Applications

19. The rate of change r of the amount of raw sugar varies directly as the amount a of raw sugar remaining, so there is a constant k such that $r = ka$. Find the value of k.

$$200 = k(800) \quad \text{Let } r = 200,\ a = 800.$$
$$k = \frac{200}{800} = \frac{1}{4}$$

When $k = \frac{1}{4}$, $r = ka$ becomes

$$r = \frac{1}{4}a.$$

Now find r when $a = 100$.

$$r = \frac{1}{4}(100) = 25$$

When only 100 kilograms of raw sugar are left, the rate of change is 25 kilograms per hour.

21. For a constant time of 3 hours, if the rate of the pickup truck *increases*, then the distance traveled *increases*. Thus, the variation between the quantities is *direct*.

23. As the number of days from now until December 25 *decreases*, the magnitude of the frenzy of Christmas shopping *increases*. Thus, the variation between the quantities is *inverse*.

25. As the amount of gasoline that you pump *increases*, the amount of empty space left in your tank *decreases*. Thus, the variation between the quantities is *inverse*.

27. As the amount of gasoline you pump *increases*, the amount you will pay *increases*. Thus, the variation between the quantities is *direct*.

29. Ratios of corresponding sides are equal, so since $\frac{12}{3} = 4$, we must have

$$\frac{x}{2} = 4, \text{ and hence } x = 8.$$

31. Ratios of corresponding sides are equal, so $\frac{2}{3} = \frac{x}{3}$, and we must have $x = 2$.

33. Let $x = $ the height of the candle.

$$\frac{5}{2} = \frac{x}{32}$$
$$32\left(\frac{5}{2}\right) = 32\left(\frac{x}{32}\right)$$
$$80 = x$$

The candle is 80 ft tall.

Chapter 7 Review Exercises

1. $\dfrac{4}{x-3}$

To find the values for which this expression is undefined, set the denominator equal to zero and solve for x.

$$x - 3 = 0$$
$$x = 3$$

Because $x = 3$ will make the denominator zero, the given expression is undefined for 3.

2. $\dfrac{y+3}{2y}$

Set the denominator equal to zero and solve for y.

$$2y = 0$$
$$y = 0$$

The given expression is undefined for 0.

3. $\dfrac{m-2}{m^2 - 2m - 3}$

Set the denominator equal to zero and solve for m.

$$m^2 - 2m - 3 = 0$$
$$(m-3)(m+1) = 0$$
$$m - 3 = 0 \quad \text{or} \quad m + 1 = 0$$
$$m = 3 \quad \text{or} \quad m = -1$$

The given expression is undefined for -1 and 3.

4. $\dfrac{2k+1}{3k^2 + 17k + 10}$

Set the denominator equal to zero and solve for k.

$$3k^2 + 17k + 10 = 0$$
$$(3k+2)(k+5) = 0$$
$$k = -\tfrac{2}{3} \quad \text{or} \quad k = -5$$

The given expression is undefined for -5 and $-\frac{2}{3}$.

5. **(a)** $\dfrac{x^2}{x-5} = \dfrac{(-2)^2}{-2-5} \quad \text{Let } x = -2.$

$$= \dfrac{4}{-7} = -\dfrac{4}{7}$$

(b) $\dfrac{x^2}{x-5} = \dfrac{4^2}{4-5} \quad \text{Let } x = 4.$

$$= \dfrac{16}{-1} = -16$$

6. **(a)** $\dfrac{4x-3}{5x+2} = \dfrac{4(-2)-3}{5(-2)+2} \quad \text{Let } x = -2.$

$$= \dfrac{-8-3}{-10+2} = \dfrac{-11}{-8} = \dfrac{11}{8}$$

(b) $\dfrac{4x-3}{5x+2} = \dfrac{4(4)-3}{5(4)+2}$ Let x = 4.

$= \dfrac{16-3}{20+2} = \dfrac{13}{22}$

7. (a) $\dfrac{3x}{x^2-4} = \dfrac{3(-2)}{(-2)^2-4}$ Let x = –2.

$= \dfrac{-6}{4-4} = \dfrac{-6}{0}$

Substituting –2 for x makes the denominator zero, so the given expression is *undefined* when $x = -2$.

(b) $\dfrac{3x}{x^2-4} = \dfrac{3(4)}{(4)^2-4}$ Let x = 4.

$= \dfrac{12}{16-4} = \dfrac{12}{12} = 1$

8. (a) $\dfrac{x-1}{x+2} = \dfrac{-2-1}{-2+2}$ Let x = –2.

$= \dfrac{-3}{0}$

Substituting –2 for x makes the denominator zero, so the given expression is *undefined* when $x = -2$.

(b) $\dfrac{x-1}{x+2} = \dfrac{4-1}{4+2}$ Let x = 4.

$= \dfrac{3}{6} = \dfrac{1}{2}$

9. $\dfrac{5a^3b^3}{15a^4b^2} = \dfrac{b \cdot 5a^3b^2}{3a \cdot 5a^3b^2} = \dfrac{b}{3a}$

10. $\dfrac{m-4}{4-m} = \dfrac{-1(4-m)}{4-m} = -1$

11. $\dfrac{4x^2-9}{6-4x} = \dfrac{(2x+3)(2x-3)}{-2(2x-3)}$

$= \dfrac{2x+3}{-2} = \dfrac{-1(2x+3)}{2}$

$= \dfrac{-(2x+3)}{2}$

12. $\dfrac{4p^2+8pq-5q^2}{10p^2-3pq-q^2} = \dfrac{(2p-q)(2p+5q)}{(5p+q)(2p-q)}$

$= \dfrac{2p+5q}{5p+q}$

13. $-\dfrac{4x-9}{2x+3}$

Apply the negative sign to the numerator:

$\dfrac{-(4x-9)}{2x+3}$

Now distribute the negative sign:

$\dfrac{-4x+9}{2x+3}$

Apply the negative sign to the denominator:

$\dfrac{4x-9}{-(2x+3)}$

Again, distribute the negative sign:

$\dfrac{4x-9}{-2x-3}$

14. $\dfrac{8-3x}{3+6x}$

Four equivalent forms are:

$\dfrac{-8+3x}{-3-6x}$, $\dfrac{-(-8+3x)}{3+6x}$,

$\dfrac{8-3x}{-(-3-6x)}$, $-\dfrac{-8+3x}{3+6x}$

15. $\dfrac{8x^2}{12x^5} \cdot \dfrac{6x^4}{2x} = \dfrac{2 \cdot 4}{3 \cdot 4x^3} \cdot \dfrac{3x^3}{1} = 2$

16. $\dfrac{9m^2}{(3m)^4} \div \dfrac{6m^5}{36m} = \dfrac{9m^2}{(3m)^4} \cdot \dfrac{36m}{6m^5}$

$= \dfrac{9m^2}{81m^4} \cdot \dfrac{36m}{6m^5}$

$= \dfrac{6m^3}{9m^9}$

$= \dfrac{2}{3m^6}$

17. $\dfrac{x-3}{4} \cdot \dfrac{5}{2x-6} = \dfrac{x-3}{4} \cdot \dfrac{5}{2(x-3)} = \dfrac{5}{8}$

18. $\dfrac{2r+3}{r-4} \cdot \dfrac{r^2-16}{6r+9}$

$= \dfrac{2r+3}{r-4} \cdot \dfrac{(r+4)(r-4)}{3(2r+3)}$

$= \dfrac{r+4}{3}$

19. $\dfrac{3q+3}{5-6q} \div \dfrac{4q+4}{2(5-6q)}$

$= \dfrac{3(q+1)}{5-6q} \cdot \dfrac{2(5-6q)}{4(q+1)}$

$= \dfrac{3 \cdot 2}{1 \cdot 4} = \dfrac{3}{2}$

20. $\dfrac{y^2-6y+8}{y^2+3y-18} \div \dfrac{y-4}{y+6}$

$= \dfrac{y^2-6y+8}{y^2+3y-18} \cdot \dfrac{y+6}{y-4}$

$= \dfrac{(y-4)(y-2)}{(y+6)(y-3)} \cdot \dfrac{y+6}{y-4}$

$= \dfrac{y-2}{y-3}$

21. $\dfrac{2p^2+13p+20}{p^2+p-12} \cdot \dfrac{p^2+2p-15}{2p^2+7p+5}$

$= \dfrac{(2p+5)(p+4)}{(p+4)(p-3)} \cdot \dfrac{(p+5)(p-3)}{(2p+5)(p+1)}$

$= \dfrac{p+5}{p+1}$

22. $\dfrac{3z^2+5z-2}{9z^2-1} \cdot \dfrac{9z^2+6z+1}{z^2+5z+6}$

$= \dfrac{(3z-1)(z+2)}{(3z-1)(3z+1)} \cdot \dfrac{(3z+1)^2}{(z+3)(z+2)}$

$= \dfrac{3z+1}{z+3}$

23. $\dfrac{1}{8}, \dfrac{5}{12}, \dfrac{7}{32}$

Factor the denominators.

$8 = 2^3, 12 = 2^2 \cdot 3, 32 = 2^5$

LCD $= 2^5 \cdot 3 = 96$

24. $\dfrac{4}{9y}, \dfrac{7}{12y^2}, \dfrac{5}{27y^4}$

Factor each denominator.

$9y = 3^2 y$
$12y^2 = 2^2 \cdot 3 \cdot y^2$
$27y^4 = 3^3 \cdot y^4$

LCD $= 2^2 \cdot 3^3 \cdot y^4 = 108y^4$

25. $\dfrac{1}{m^2+2m}, \dfrac{4}{m^2+7m+10}$

Factor the denominators.

$m^2 + 2m = m(m+2)$
$m^2 + 7m + 10 = (m+2)(m+5)$

LCD $= m(m+2)(m+5)$

26. $\dfrac{3}{x^2+4x+3}, \dfrac{5}{x^2+5x+4}, \dfrac{2}{x^2+7x+12}$

Factor each denominator.

$x^2 + 4x + 3 = (x+3)(x+1)$
$x^2 + 5x + 4 = (x+1)(x+4)$
$x^2 + 7x + 12 = (x+3)(x+4)$

LCD $= (x+3)(x+1)(x+4)$

27. $\dfrac{5}{8} = \dfrac{}{56}$

A factor of 7 is missing in the first fraction, so multiply numerator and denominator by 7.

$\dfrac{5}{8} = \dfrac{5}{8} \cdot \dfrac{7}{7} = \dfrac{35}{56}$

28. $\dfrac{10}{k} = \dfrac{}{4k}$

Multiply the numerator and the denominator by 4.

$\dfrac{10}{k} = \dfrac{10}{k} \cdot \dfrac{4}{4} = \dfrac{40}{4k}$

29. $\dfrac{3}{2a^3} = \dfrac{}{10a^4}$

$\dfrac{3}{2a^3} = \dfrac{3}{2a^3} \cdot \dfrac{5a}{5a} = \dfrac{15a}{10a^4}$

30. $\dfrac{9}{x-3} = \dfrac{}{18-6x} = \dfrac{}{-6(-3+x)}$

$\dfrac{9}{x-3} = \dfrac{9}{x-3} \cdot \dfrac{-6}{-6}$

$= \dfrac{-54}{-6x+18}$

$= \dfrac{-54}{18-6x}$

31. $\dfrac{-3y}{2y-10} = \dfrac{}{50-10y} = \dfrac{}{-5(2y-10)}$

$\dfrac{-3y}{2y-10} = \dfrac{-3y}{2y-10} \cdot \dfrac{-5}{-5}$

$= \dfrac{15y}{-10y+50}$

$= \dfrac{15y}{50-10y}$

32. $\dfrac{4b}{b^2+2b-3} = \dfrac{}{(b+3)(b-1)(b+2)}$

$\dfrac{4b}{b^2+2b-3} = \dfrac{4b}{(b+3)(b-1)}$

$= \dfrac{4b}{(b+3)(b-1)} \cdot \dfrac{b+2}{b+2}$

$= \dfrac{4b(b+2)}{(b+3)(b-1)(b+2)}$

33. $\dfrac{10}{x} + \dfrac{5}{x} = \dfrac{10+5}{x} = \dfrac{15}{x}$

34. $\dfrac{6}{3p} - \dfrac{12}{3p} = \dfrac{6-12}{3p} = \dfrac{-6}{3p} = -\dfrac{2}{p}$

35. $\dfrac{9}{k} - \dfrac{5}{k-5} = \dfrac{9(k-5)}{k(k-5)} - \dfrac{5 \cdot k}{(k-5)k}$

LCD $= k(k-5)$

$= \dfrac{9(k-5) - 5k}{k(k-5)}$

$= \dfrac{9k - 45 - 5k}{k(k-5)}$

$= \dfrac{4k - 45}{k(k-5)}$

36. $\dfrac{4}{y} + \dfrac{7}{7+y} = \dfrac{4(7+y)}{y(7+y)} + \dfrac{7 \cdot y}{(7+y)y}$
$\hspace{5em} LCD = y(7+y)$
$= \dfrac{28 + 4y + 7y}{y(7+y)}$
$= \dfrac{28 + 11y}{y(7+y)}$

37. $\dfrac{m}{3} - \dfrac{2+5m}{6} = \dfrac{m \cdot 2}{3 \cdot 2} - \dfrac{2+5m}{6}$ $\quad LCD = 6$
$= \dfrac{2m - (2+5m)}{6}$
$= \dfrac{2m - 2 - 5m}{6}$
$= \dfrac{-2 - 3m}{6}$

38. $\dfrac{12}{x^2} - \dfrac{3}{4x} = \dfrac{12 \cdot 4}{x^2 \cdot 4} - \dfrac{3 \cdot x}{4x \cdot x}$ $\quad LCD = 4x^2$
$= \dfrac{48 - 3x}{4x^2}$
$= \dfrac{3(16 - x)}{4x^2}$

39. $\dfrac{5}{a-2b} + \dfrac{2}{a+2b}$
$= \dfrac{5(a+2b)}{(a-2b)(a+2b)} + \dfrac{2(a-2b)}{(a+2b)(a-2b)}$
$\hspace{5em} LCD = (a-2b)(a+2b)$
$= \dfrac{5(a+2b) + 2(a-2b)}{(a-2b)(a+2b)}$
$= \dfrac{5a + 10b + 2a - 4b}{(a-2b)(a+2b)}$
$= \dfrac{7a + 6b}{(a-2b)(a+2b)}$

40. $\dfrac{4}{k^2 - 9} - \dfrac{k+3}{3k - 9}$
$= \dfrac{4}{(k+3)(k-3)} - \dfrac{k+3}{3(k-3)}$
$\hspace{5em} LCD = 3(k+3)(k-3)$
$= \dfrac{4 \cdot 3}{(k+3)(k-3) \cdot 3} - \dfrac{(k+3)(k+3)}{3(k-3)(k+3)}$
$= \dfrac{12 - (k+3)(k+3)}{3(k+3)(k-3)}$
$= \dfrac{12 - (k^2 + 6k + 9)}{3(k+3)(k-3)}$
$= \dfrac{12 - k^2 - 6k - 9}{3(k+3)(k-3)}$
$= \dfrac{-k^2 - 6k + 3}{3(k+3)(k-3)}$

41. $\dfrac{8}{z^2 + 6z} - \dfrac{3}{z^2 + 4z - 12}$
$= \dfrac{8}{z(z+6)} - \dfrac{3}{(z+6)(z-2)}$
$\hspace{3em} LCD = z(z+6)(z-2)$
$= \dfrac{8(z-2)}{z(z+6)(z-2)} - \dfrac{3 \cdot z}{(z+6)(z-2) \cdot z}$
$= \dfrac{8(z-2) - 3z}{z(z+6)(z-2)}$
$= \dfrac{8z - 16 - 3z}{z(z+6)(z-2)}$
$= \dfrac{5z - 16}{z(z+6)(z-2)}$

42. $\dfrac{11}{2p - p^2} - \dfrac{2}{p^2 - 5p + 6}$
$= \dfrac{11}{p(2-p)} - \dfrac{2}{(p-3)(p-2)}$
$\hspace{3em} LCD = p(p-3)(p-2)$
$= \dfrac{11(-1)(p-3)}{p(2-p)(-1)(p-3)}$
$\quad - \dfrac{2 \cdot p}{(p-3)(p-2)p}$
$= \dfrac{-11(p-3) - 2p}{p(p-2)(p-3)}$
$= \dfrac{-11p + 33 - 2p}{p(p-2)(p-3)}$
$= \dfrac{-13p + 33}{p(p-2)(p-3)}$

43. $\dfrac{\dfrac{a^4}{b^2}}{\dfrac{a^3}{b}} = \dfrac{a^4}{b^2} \div \dfrac{a^3}{b} = \dfrac{a^4}{b^2} \cdot \dfrac{b}{a^3} = \dfrac{a^4 b}{a^3 b^2} = \dfrac{a}{b}$

44. $\dfrac{\dfrac{y-3}{y}}{\dfrac{y+3}{4y}} = \dfrac{y-3}{y} \cdot \dfrac{4y}{y+3} = \dfrac{4(y-3)}{y+3}$

45. $\dfrac{\dfrac{3m+2}{m}}{\dfrac{2m-5}{6m}} = \dfrac{3m+2}{m} \div \dfrac{2m-5}{6m}$
$= \dfrac{3m+2}{m} \cdot \dfrac{6m}{2m-5} = \dfrac{6(3m+2)}{2m-5}$

46.
$$\frac{\frac{1}{p} - \frac{1}{q}}{\frac{1}{q-p}} = \frac{\left(\frac{1}{p} - \frac{1}{q}\right)pq(q-p)}{\left(\frac{1}{q-p}\right)pq(q-p)}$$

Multiply by LCD, $pq(q-p)$.

$$= \frac{\frac{1}{p}[pq(q-p)] - \frac{1}{q}[pq(q-p)]}{pq}$$

$$= \frac{q(q-p) - p(q-p)}{pq}$$

$$= \frac{q^2 - pq - pq + p^2}{pq}$$

$$= \frac{q^2 - 2pq + p^2}{pq}$$

$$= \frac{(q-p)^2}{pq}$$

47.
$$\frac{x + \frac{1}{w}}{x - \frac{1}{w}}$$

$$= \frac{\left(x + \frac{1}{w}\right) \cdot w}{\left(x - \frac{1}{w}\right) \cdot w} \quad \text{Multiply by LCD, } w.$$

$$= \frac{xw + \left(\frac{1}{w}\right)w}{xw - \left(\frac{1}{w}\right)w}$$

$$= \frac{xw + 1}{xw - 1}$$

48.
$$\frac{\frac{1}{r+t} - 1}{\frac{1}{r+t} + 1}$$

$$= \frac{\left(\frac{1}{r+t} - 1\right)(r+t)}{\left(\frac{1}{r+t} + 1\right)(r+t)} \quad \text{Multiply by LCD, } r+t.$$

$$= \frac{\frac{1}{r+t}(r+t) - 1(r+t)}{\frac{1}{r+t}(r+t) + 1(r+t)}$$

$$= \frac{1 - r - t}{1 + r + t}$$

49. $\frac{k}{5} - \frac{2}{3} = \frac{1}{2}$

Multiply by the LCD, 30.

$$30\left(\frac{k}{5} - \frac{2}{3}\right) = 30\left(\frac{1}{2}\right)$$
$$30\left(\frac{k}{5}\right) - 30\left(\frac{2}{3}\right) = 15$$
$$6k - 20 = 15$$
$$6k = 35$$
$$k = \frac{35}{6}$$

Check $k = \frac{35}{6}$: $\frac{1}{2} = \frac{1}{2}$

The solution is $\frac{35}{6}$.

50. $\frac{4-z}{z} + \frac{3}{2} = \frac{-4}{z}$

Multiply each side by the LCD, $2z$.

$$2z\left(\frac{4-z}{z} + \frac{3}{2}\right) = 2z\left(-\frac{4}{z}\right)$$
$$2z\left(\frac{4-z}{z}\right) + 2z\left(\frac{3}{2}\right) = -8$$
$$2(4-z) + 3z = -8$$
$$8 - 2z + 3z = -8$$
$$8 + z = -8$$
$$z = -16$$

Check $z = -16$: $-\frac{5}{4} + \frac{3}{2} = \frac{1}{4}$

The solution is -16.

51. $\frac{x}{2} - \frac{x-3}{7} = -1$

Multiply by the LCD, 14.

$$14\left(\frac{x}{2} - \frac{x-3}{7}\right) = 14(-1)$$
$$14\left(\frac{x}{2}\right) - 14\left(\frac{x-3}{7}\right) = -14$$
$$7x - 2(x-3) = -14$$
$$7x - 2x + 6 = -14$$
$$5x + 6 = -14$$
$$5x = -20$$
$$x = -4$$

Check $x = -4$: $-1 = -1$

The solution is -4.

52.
$$\frac{3y-1}{y-2} = \frac{5}{y-2} + 1$$
$$(y-2)\left(\frac{3y-1}{y-2}\right) = (y-2)\left(\frac{5}{y-2} + 1\right)$$
Multiply by LCD, $y-2$.
$$(y-2)\left(\frac{3y-1}{y-2}\right) = (y-2)\left(\frac{5}{y-2}\right) + (y-2)(1)$$
Distributive property
$$3y - 1 = 5 + y - 2$$
$$3y - 1 = 3 + y$$
$$2y = 4$$
$$y = 2$$

There is *no solution* because $y = 2$ makes the original denominators equal to zero.

53.
$$\frac{3}{m-2} + \frac{1}{m-1} = \frac{7}{m^2 - 3m + 2}$$
$$\frac{3}{m-2} + \frac{1}{m-1} = \frac{7}{(m-2)(m-1)}$$
$$(m-2)(m-1)\left(\frac{3}{m-2} + \frac{1}{m-1}\right)$$
$$= (m-2)(m-1) \cdot \frac{7}{(m-2)(m-1)}$$
Multiply by LCD, $(m-2)(m-1)$.
$$3(m-1) + 1(m-2) = 7$$
$$3m - 3 + m - 2 = 7$$
$$4m - 5 = 7$$
$$4m = 12$$
$$m = 3$$

Check $m = 3$: $3 + \frac{1}{2} = \frac{7}{2}$ ✓

The solution is 3.

54. $m = \dfrac{Ry}{t}$ for t
$$t \cdot m = t\left(\frac{Ry}{t}\right) \quad \text{Multiply by } t.$$
$$tm = Ry$$
$$t = \frac{Ry}{m} \quad \text{Divide by } m.$$

55. $x = \dfrac{3y-5}{4}$ for y
$$4x = 4\left(\frac{3y-5}{4}\right)$$
$$4x = 3y - 5$$
$$4x + 5 = 3y$$
$$\frac{4x+5}{3} = y$$

56. $\dfrac{1}{r} - \dfrac{1}{s} = \dfrac{1}{t}$ for t
$$rst\left(\frac{1}{r} - \frac{1}{s}\right) = rst\left(\frac{1}{t}\right)$$
$$st - rt = rs$$
$$t(s - r) = rs$$
$$t = \frac{rs}{s - r}$$

57. Let x = the number.

One-fourth of the number	is	9 less than the number.
↓	↓	↓
$\frac{1}{4}x$	$=$	$x - 9$

$$4\left(\tfrac{1}{4}x\right) = 4(x-9)$$
$$x = 4x - 36$$
$$36 = 3x$$
$$12 = x$$

The number is 12.

58. Let x = the numerator. Then $x - 5$ = the denominator. Adding 5 to both the numerator and the denominator gives us a fraction that is equivalent to $\frac{5}{4}$.

$$\frac{x+5}{x-5+5} = \frac{5}{4}$$
$$\frac{x+5}{x} = \frac{5}{4}$$
$$4x\left(\frac{x+5}{x}\right) = 4x\left(\frac{5}{4}\right)$$
$$4(x+5) = x(5)$$
$$4x + 20 = 5x$$
$$20 = x$$

The numerator is 20 and the denominator is $20 - 5 = 15$, so the original fraction is $\frac{20}{15}$.

59. Let x = the numerator. Then $6x$ = the denominator. Adding 3 to the numerator and subtracting 3 from the denominator gives us a fraction equivalent to $\frac{2}{5}$.

$$\frac{x+3}{6x-3} = \frac{2}{5}$$
$$5(6x-3)\left(\frac{x+3}{6x-3}\right) = 5(6x-3)\left(\frac{2}{5}\right)$$
$$5(x+3) = 2(6x-3)$$
$$5x + 15 = 12x - 6$$
$$21 = 7x$$
$$3 = x$$

The numerator is 3 and the denominator is $6 \cdot 3 = 18$, so the original fraction is $\frac{3}{18}$.

60. $t = \dfrac{d}{r}$

$= \dfrac{200}{130.934} \approx 1.527$ hr

His time was about 1.527 hours.

61. *Step 2*
Let $x =$ the number of hours it takes them to do the job working together.

	Rate	Time Working Together	Fractional Part of the Job Done When Working Together
Man	$\frac{1}{5}$	x	$\frac{1}{5}x$
Daughter	$\frac{1}{8}$	x	$\frac{1}{8}x$

Step 3
Working together, they do 1 whole job, so

$\frac{1}{5}x + \frac{1}{8}x = 1$.

Step 4
Solve this equation by multiplying both sides by the LCD, 40.

$40\left(\frac{1}{5}x + \frac{1}{8}x\right) = 40(1)$
$8x + 5x = 40$
$13x = 40$
$x = \frac{40}{13}$ or $3\frac{1}{13}$

Step 5
Working together, it takes them $\frac{40}{13}$ or $3\frac{1}{13}$ hours.

Step 6
The man does $\frac{1}{5}$ of the job per hour for $\frac{40}{13}$ hours:

$\dfrac{1}{5} \cdot \dfrac{40}{13} = \dfrac{8}{13}$ of the job

His daughter does $\frac{1}{8}$ of the job per hour for $\frac{40}{13}$ hours:

$\dfrac{1}{8} \cdot \dfrac{40}{13} = \dfrac{5}{13}$ of the job

Together, they have done

$\dfrac{8}{13} + \dfrac{5}{13} = \dfrac{13}{13} = 1$ total job.

62. Let $t =$ the number of hours until the distance between the steamboats is 70 miles.

	Rate	×	Time	=	Distance
Slower Steamboat	18		t		$18t$
Faster Steamboat	25		t		$25t$

The difference of the distances is 70 miles.

$25t - 18t = 70$
$7t = 70$
$t = 10$

They will be 70 miles apart in 10 hours.

63. Let $h =$ the height of the parallelogram and
$b =$ the length of the base of the parallelogram.

The height varies inversely as the base, so

$h = \dfrac{k}{b}$.

Find k by replacing h with 8 and b with 12.

$8 = \dfrac{k}{12}$
$k = 8 \cdot 12 = 96$

So $h = \dfrac{96}{b}$ and when $b = 24$,

$h = \dfrac{96}{24} = 4$.

The height of the parallelogram is 4 centimeters.

64. Since y varies directly as x, there is a constant k such that $y = kx$. First find the value of k.

$5 = k(12)$ *Let x = 12, y = 5.*
$k = \dfrac{5}{12}$

When $k = \dfrac{5}{12}$, $y = kx$ becomes

$y = \dfrac{5}{12}x$.

Now find x when $y = 3$.

$3 = \dfrac{5}{12}x$ *Let y = 3.*
$x = 3 \cdot \dfrac{12}{5} = \dfrac{36}{5}$

65. [7.4] $\dfrac{4}{m-1} - \dfrac{3}{m+1}$

To perform the indicated subtraction, use $(m-1)(m+1)$ as the LCD.

$\dfrac{4}{m-1} - \dfrac{3}{m+1}$

$= \dfrac{4(m+1)}{(m-1)(m+1)} - \dfrac{3(m-1)}{(m+1)(m-1)}$

$= \dfrac{4(m+1) - 3(m-1)}{(m-1)(m+1)}$

$= \dfrac{4m + 4 - 3m + 3}{(m-1)(m+1)}$

$= \dfrac{m+7}{(m-1)(m+1)}$

66. [7.2] $\dfrac{8p^5}{5} \div \dfrac{2p^3}{10}$

To perform the indicated division, multiply the first rational expression by the reciprocal of the second.

$$\dfrac{8p^5}{5} \div \dfrac{2p^3}{10} = \dfrac{8p^5}{5} \cdot \dfrac{10}{2p^3}$$

$$= \dfrac{80p^5}{10p^3}$$

$$= 8p^2$$

67. [7.2] $\dfrac{r-3}{8} \div \dfrac{3r-9}{4} = \dfrac{r-3}{8} \cdot \dfrac{4}{3r-9}$

$$= \dfrac{r-3}{8} \cdot \dfrac{4}{3(r-3)}$$

$$= \dfrac{4}{24} = \dfrac{1}{6}$$

68. [7.5] $\dfrac{\dfrac{5}{x}-1}{\dfrac{5-x}{3x}} = \dfrac{\left(\dfrac{5}{x}-1\right)3x}{\left(\dfrac{5-x}{3x}\right)3x}$ *Multiply by LCD, $3x$.*

$$= \dfrac{\dfrac{5}{x}(3x) - 1(3x)}{5-x}$$

$$= \dfrac{15 - 3x}{5 - x}$$

$$= \dfrac{3(5-x)}{5-x} = 3$$

69. [7.4] $\dfrac{4}{z^2 - 2z + 1} - \dfrac{3}{z^2 - 1}$

$$= \dfrac{4}{(z-1)^2} - \dfrac{3}{(z+1)(z-1)}$$

$\quad$ LCD $= (z+1)(z-1)^2$

$$= \dfrac{4(z+1)}{(z-1)^2(z+1)}$$

$$\quad - \dfrac{3(z-1)}{(z+1)(z-1)(z-1)}$$

$$= \dfrac{4(z+1) - 3(z-1)}{(z+1)(z-1)^2}$$

$$= \dfrac{4z + 4 - 3z + 3}{(z+1)(z-1)^2}$$

$$= \dfrac{z+7}{(z+1)(z-1)^2}$$

70. [7.6] $F = \dfrac{k}{d-D}$ for d

$$F(d-D) = \left(\dfrac{k}{d-D}\right)(d-D)$$

$$Fd - FD = k$$

$$Fd = k + FD$$

$$d = \dfrac{k + FD}{F}$$

or $\quad d = \dfrac{k}{F} + \dfrac{FD}{F} = \dfrac{k}{F} + D$

71. [7.6] $\dfrac{2}{z} - \dfrac{z}{z+3} = \dfrac{1}{z+3}$

Multiply each side of the equation by the LCD, $z(z+3)$.

$$z(z+3)\left(\dfrac{2}{z} - \dfrac{z}{z+3}\right) = z(z+3)\left(\dfrac{1}{z+3}\right)$$

$$z(z+3)\left(\dfrac{2}{z}\right) - z(z+3)\left(\dfrac{z}{z+3}\right) = z(1)$$

$$2(z+3) - z^2 = z$$

$$2z + 6 - z^2 = z$$

$$0 = z^2 - z - 6$$

$$0 = (z-3)(z+2)$$

$z - 3 = 0 \quad$ or $\quad z + 2 = 0$
$\quad z = 3 \quad$ or $\quad z = -2$

Check $z = -2$: $-1 - (-2) = 1$

Check $z = 3$: $\tfrac{2}{3} - \tfrac{1}{2} = \tfrac{1}{6}$

The solutions are -2 and 3.

72. [7.7] Let $x =$ the speed of the plane in still air. Then the speed of the plane with the wind is $x + 50$, and the speed of the plane against the wind is $x - 50$. Use $t = \dfrac{d}{r}$ to complete the chart.

	d	r	t
With the Wind	400	$x + 50$	$\dfrac{400}{x+50}$
Against the Wind	200	$x - 50$	$\dfrac{200}{x-50}$

The times are the same, so

$$\dfrac{400}{x+50} = \dfrac{200}{x-50}.$$

To solve this equation, multiply both sides by the LCD, $(x+50)(x-50)$.

continued

$$(x+50)(x-50) \cdot \frac{400}{x+50}$$
$$= (x+50)(x-50) \cdot \frac{200}{x-50}$$
$$400(x-50) = 200(x+50)$$
$$400x - 20{,}000 = 200x + 10{,}000$$
$$200x = 30{,}000$$
$$x = 150$$

The speed of the plane is 150 kilometers per hour.

73. [7.7] Let $x =$ the number of hours it takes them to do the job working together.

	Rate	Time Working Together	Fractional Part of the Job Done When Working Together
Joe	$\frac{1}{3}$	x	$\frac{1}{3}x$
Sam	$\frac{1}{5}$	x	$\frac{1}{5}x$

Working together, they do 1 whole job, so
$$\tfrac{1}{3}x + \tfrac{1}{5}x = 1.$$

To clear fractions, multiply both sides by the LCD, 15.
$$15\left(\tfrac{1}{3}x + \tfrac{1}{5}x\right) = 15(1)$$
$$5x + 3x = 15$$
$$8x = 15$$
$$x = \tfrac{15}{8} \text{ or } 1\tfrac{7}{8}$$

Working together, they can paint the house in $1\tfrac{7}{8}$ hours.

74. [7.8] For a constant area, the length L of a rectangle varies inversely as the width W, so
$$L = \frac{k}{W}.$$
$$24 = \frac{k}{2} \quad \text{Let } L = 24, W = 2.$$
$$k = 24 \cdot 2 = 48$$

So $L = \dfrac{48}{W}$ and when $L = 12$,
$$12 = \frac{48}{W}$$
$$12W = 48$$
$$W = \tfrac{48}{12} = 4.$$

When the length is 12, the width is 4.

75. [7.8] Since w varies inversely as z, there is a constant k such that $w = \dfrac{k}{z}$. First find the value of k.
$$16 = \frac{k}{3} \quad \text{Let } w = 16, z = 3.$$
$$k = 16(3) = 48$$

When $k = 48$, $w = \dfrac{k}{z}$ becomes
$$w = \frac{48}{z}.$$

Now find w when $z = 2$.
$$w = \tfrac{48}{2} = 24$$

Chapter 7 Test

1. $\dfrac{3x-1}{x^2 - 2x - 8}$

Set the denominator equal to zero and solve for x.
$$x^2 - 2x - 8 = 0$$
$$(x+2)(x-4) = 0$$
$$x + 2 = 0 \quad \text{or} \quad x - 4 = 0$$
$$x = -2 \quad \text{or} \quad x = 4$$

The expression is undefined for -2 and 4.

2. (a) $\dfrac{6r+1}{2r^2 - 3r - 20}$
$$= \frac{6(-2)+1}{2(-2)^2 - 3(-2) - 20} \quad \text{Let } r = -2.$$
$$= \frac{-12+1}{2 \cdot 4 + 6 - 20}$$
$$= \frac{-11}{8 + 6 - 20}$$
$$= \frac{-11}{-6} = \frac{11}{6}$$

(b) $\dfrac{6r+1}{2r^2 - 3r - 20}$
$$= \frac{6(4)+1}{2(4)^2 - 3(4) - 20} \quad \text{Let } r = 4.$$
$$= \frac{24+1}{2 \cdot 16 - 12 - 20}$$
$$= \frac{25}{32 - 12 - 20}$$
$$= \frac{25}{20 - 20} = \frac{25}{0}$$

The expression is *undefined* when $r = 4$ because the denominator is 0.

3. $-\dfrac{6x-5}{2x+3}$

   Apply the negative sign to the numerator:
   $$\dfrac{-(6x-5)}{2x+3}$$
   Now distribute the negative sign:
   $$\dfrac{-6x+5}{2x+3}$$
   Apply the negative sign to the denominator:
   $$\dfrac{6x-5}{-(2x+3)}$$
   Again, distribute the negative sign:
   $$\dfrac{6x-5}{-2x-3}$$

4. $\dfrac{-15x^6y^4}{5x^4y} = \dfrac{(5x^4y)(-3x^2y^3)}{(5x^4y)(1)}$
   $= \dfrac{5x^4y}{5x^4y} \cdot \dfrac{-3x^2y^3}{1} = -3x^2y^3$

5. $\dfrac{6a^2+a-2}{2a^2-3a+1} = \dfrac{(3a+2)(2a-1)}{(2a-1)(a-1)}$
   $= \dfrac{3a+2}{a-1}$

6. $\dfrac{5(d-2)}{9} \div \dfrac{3(d-2)}{5}$
   $= \dfrac{5(d-2)}{9} \cdot \dfrac{5}{3(d-2)}$
   $= \dfrac{5 \cdot 5}{9 \cdot 3} = \dfrac{25}{27}$

7. $\dfrac{6k^2-k-2}{8k^2+10k+3} \cdot \dfrac{4k^2+7k+3}{3k^2+5k+2}$
   $= \dfrac{(3k-2)(2k+1)}{(4k+3)(2k+1)} \cdot \dfrac{(4k+3)(k+1)}{(3k+2)(k+1)}$
   $= \dfrac{3k-2}{3k+2}$

8. $\dfrac{4a^2+9a+2}{3a^2+11a+10} \div \dfrac{4a^2+17a+4}{3a^2+2a-5}$
   $= \dfrac{4a^2+9a+2}{3a^2+11a+10} \cdot \dfrac{3a^2+2a-5}{4a^2+17a+4}$
   $= \dfrac{(4a+1)(a+2)}{(3a+5)(a+2)} \cdot \dfrac{(3a+5)(a-1)}{(4a+1)(a+4)}$
   $= \dfrac{a-1}{a+4}$

9. $\dfrac{-3}{10p^2}, \dfrac{21}{25p^3}, \dfrac{-7}{30p^5}$

   Factor each denominator.
   $$10p^2 = 2 \cdot 5 \cdot p^2$$
   $$25p^3 = 5^2 \cdot p^3$$
   $$30p^5 = 2 \cdot 3 \cdot 5 \cdot p^5$$
   $\text{LCD} = 2 \cdot 3 \cdot 5^2 \cdot p^5 = 150p^5$

10. $\dfrac{r+1}{2r^2+7r+6}, \dfrac{-2r+1}{2r^2-7r-15}$

    Factor each denominator.
    $$2r^2+7r+6 = (2r+3)(r+2)$$
    $$2r^2-7r-15 = (2r+3)(r-5)$$
    $\text{LCD} = (2r+3)(r+2)(r-5)$

11. $\dfrac{15}{4p} = \dfrac{}{64p^3} = \dfrac{}{4p \cdot 16p^2}$
    $\dfrac{15}{4p} = \dfrac{15 \cdot 16p^2}{4p \cdot 16p^2} = \dfrac{240p^2}{64p^3}$

12. $\dfrac{3}{6m-12} = \dfrac{}{42m-84} = \dfrac{}{7(6m-12)}$
    $\dfrac{3}{6m-12} = \dfrac{3 \cdot 7}{(6m-12)7} = \dfrac{21}{42m-84}$

13. $\dfrac{4x+2}{x+5} + \dfrac{-2x+8}{x+5}$
    $= \dfrac{(4x+2)+(-2x+8)}{x+5}$
    $= \dfrac{2x+10}{x+5}$
    $= \dfrac{2(x+5)}{x+5} = 2$

14. $\dfrac{-4}{y+2} + \dfrac{6}{5y+10}$
    $= \dfrac{-4}{y+2} + \dfrac{6}{5(y+2)}$ $\text{LCD} = 5(y+2)$
    $= \dfrac{-4 \cdot 5}{(y+2) \cdot 5} + \dfrac{6}{5(y+2)}$
    $= \dfrac{-20+6}{5(y+2)} = \dfrac{-14}{5(y+2)}$

15. Using LCD $= 3 - x$,

$$\frac{x+1}{3-x} - \frac{x^2}{x-3} = \frac{x+1}{3-x} + \frac{x^2}{-1(x-3)}$$

$$= \frac{x+1}{3-x} + \frac{x^2}{-x+3}$$

$$= \frac{x+1}{3-x} + \frac{x^2}{3-x}$$

$$= \frac{(x+1) + x^2}{3-x}$$

$$= \frac{x^2 + x + 1}{3-x}.$$

If we use $x - 3$ for the LCD, we obtain the equivalent answer

$$\frac{-x^2 - x - 1}{x - 3}.$$

16. $\dfrac{3}{2m^2 - 9m - 5} - \dfrac{m+1}{2m^2 - m - 1}$

$= \dfrac{3}{(2m+1)(m-5)} - \dfrac{m+1}{(2m+1)(m-1)}$

LCD $= (2m + 1)(m - 5)(m - 1)$

$= \dfrac{3(m-1)}{(2m+1)(m-5)(m-1)}$

$- \dfrac{(m+1)(m-5)}{(2m+1)(m-1)(m-5)}$

$= \dfrac{3(m-1) - (m+1)(m-5)}{(2m+1)(m-5)(m-1)}$

$= \dfrac{(3m-3) - (m^2 - 4m - 5)}{(2m+1)(m-5)(m-1)}$

$= \dfrac{3m - 3 - m^2 + 4m + 5}{(2m+1)(m-5)(m-1)}$

$= \dfrac{-m^2 + 7m + 2}{(2m+1)(m-5)(m-1)}$

17. $\dfrac{\frac{2p}{k^2}}{\frac{3p^2}{k^3}} = \dfrac{2p}{k^2} \div \dfrac{3p^2}{k^3}$

$= \dfrac{2p}{k^2} \cdot \dfrac{k^3}{3p^2}$

$= \dfrac{2k^3 p}{3k^2 p^2} = \dfrac{2k}{3p}$

18. $\dfrac{\frac{1}{x+3} - 1}{1 + \frac{1}{x+3}}$

$= \dfrac{(x+3)\left(\frac{1}{x+3} - 1\right)}{(x+3)\left(1 + \frac{1}{x+3}\right)}$ Multiply by LCD, $x + 3$.

$= \dfrac{(x+3)\left(\frac{1}{x+3}\right) - (x+3)(1)}{(x+3)(1) + (x+3)\left(\frac{1}{x+3}\right)}$

$= \dfrac{1 - (x+3)}{(x+3) + 1}$

$= \dfrac{1 - x - 3}{x + 4}$

$= \dfrac{-2 - x}{x + 4}$

19. $\dfrac{2x}{x-3} + \dfrac{1}{x+3} = \dfrac{-6}{x^2 - 9}$

$\dfrac{2x}{x-3} + \dfrac{1}{x+3} = \dfrac{-6}{(x+3)(x-3)}$

$(x+3)(x-3)\left(\dfrac{2x}{x-3} + \dfrac{1}{x+3}\right)$

$= (x+3)(x-3)\left(\dfrac{-6}{(x+3)(x-3)}\right)$

Multiply by LCD, $(x + 3)(x - 3)$.

$2x(x+3) + 1(x-3) = -6$

$2x^2 + 6x + x - 3 = -6$

$2x^2 + 7x + 3 = 0$

$(2x+1)(x+3) = 0$

$x = -\tfrac{1}{2}$ or $x = -3$

x cannot equal -3 because the denominator $x + 3$ would equal 0.

Check $x = -\tfrac{1}{2}$: $\tfrac{2}{7} + \tfrac{2}{5} = \tfrac{24}{35}$ True

The solution is $-\tfrac{1}{2}$.

20. Solve $F = \dfrac{k}{d - D}$ for D.

$(d - D)(F) = (d - D)\left(\dfrac{k}{d - D}\right)$ Multiply by LCD, $d - D$.

$(d - D)(F) = k$

$dF - DF = k$

$-DF = k - dF$

$D = \dfrac{k - dF}{-F}$ or $D = \dfrac{dF - k}{F}$

21. Let $x =$ the number.
If x is added to the numerator and subtracted from the denominator of $\frac{5}{6}$, the resulting fraction is equivalent to $\frac{1}{10}$.

$$\frac{5+x}{6-x} = \frac{1}{10}$$

$$10(6-x)\left(\frac{5+x}{6-x}\right) = 10(6-x)\left(\frac{1}{10}\right)$$

$$10(5+x) = (6-x)(1)$$

$$50 + 10x = 6 - x$$

$$11x = -44$$

$$x = -4$$

The number is -4.

22. Let $x =$ the speed of the current.

	d	r	t
Upstream	20	$7-x$	$\frac{20}{7-x}$
Downstream	50	$7+x$	$\frac{50}{7+x}$

The times are equal, so

$$\frac{20}{7-x} = \frac{50}{7+x}.$$

$$(7-x)(7+x)\left(\frac{20}{7-x}\right) = (7-x)(7+x)\left(\frac{50}{7+x}\right)$$

Multiply by LCD, $(7-x)(7+x)$.

$$20(7+x) = 50(7-x)$$

$$140 + 20x = 350 - 50x$$

$$70x = 210$$

$$x = 3$$

The speed of the current is 3 miles per hour.

23. Let $x =$ the time required for the couple to paint the room working together.

	Rate	Time Working Together	Fractional Part of the Job Done When Working Together
Husband	$\frac{1}{5}$	x	$\frac{1}{5}x$
Wife	$\frac{1}{4}$	x	$\frac{1}{4}x$

Working together, they do 1 whole job, so

$$\tfrac{1}{5}x + \tfrac{1}{4}x = 1.$$

$$20\left(\tfrac{1}{5}x + \tfrac{1}{4}x\right) = 20(1) \quad \text{Multiply by LCD, 20.}$$

$$20\left(\tfrac{1}{5}x\right) + 20\left(\tfrac{1}{4}x\right) = 20$$

$$4x + 5x = 20$$

$$9x = 20$$

$$x = \tfrac{20}{9} \text{ or } 2\tfrac{2}{9}$$

The couple can paint the room in $\frac{20}{9}$ or $2\frac{2}{9}$ hours.

24. Since x varies directly as y, there is a constant k such that $x = ky$. First find the value of k.

$$x = ky$$
$$12 = k \cdot 4 \quad \text{Let } x = 12, y = 4.$$
$$k = \tfrac{12}{4} = 3$$

Since $x = ky$ and $k = 3$,

$$x = 3y.$$

Now find x when $y = 9$.

$$x = 3(9) = 27$$

25. The length of time L that it takes for fruit to ripen during the growing season varies inversely as the average maximum temperature T during the season, so

$$L = \frac{k}{T}.$$

$$25 = \frac{k}{80} \quad \text{Let } L = 25, T = 80.$$

$$k = 25 \cdot 80 = 2000$$

So $L = \dfrac{2000}{T}$ and when $T = 75$,

$$L = \tfrac{2000}{75} = \tfrac{80}{3} \text{ or } 26\tfrac{2}{3}.$$

To the nearest whole number, L is 27 days.

Cumulative Review Exercises (Chapters R–7)

1. $3 + 4\left(\dfrac{1}{2} - \dfrac{3}{4}\right)$

$= 3 + 4\left(\dfrac{2}{4} - \dfrac{3}{4}\right)$

$= 3 + 4\left(-\dfrac{1}{4}\right)$ Parentheses

$= 3 + (-1)$ Multiplication

$= 2$ Addition

2. $3(2y - 5) = 2 + 5y$

$6y - 15 = 2 + 5y$

$y = 17$

The solution is 17.

3. Solve $A = \dfrac{1}{2}bh$ for b.

$2 \cdot A = 2 \cdot \dfrac{1}{2}bh$

$2A = bh$

$\dfrac{2A}{h} = b$

304 Chapter 7 Rational Expressions and Applications

4. $\dfrac{2+m}{2-m} = \dfrac{3}{4}$

$4(2+m) = 3(2-m)$ *Cross multiply.*

$8 + 4m = 6 - 3m$

$7m = -2$

$m = -\dfrac{2}{7}$

The solution is $-\dfrac{2}{7}$.

5. $5y \leq 6y + 8$

$-y \leq 8$

$y \geq -8$ *Reverse the inequality symbol.*

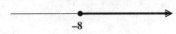

6. $5m - 9 > 2m + 3$

$3m > 12$

$m > 4$

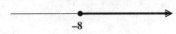

7. $y = -3x + 2$

This is an equation of a line.

If $x = 0$, $y = 2$; so the y-intercept is $(0, 2)$.
If $x = 1$, $y = -1$; and if $x = 2, y = -4$.

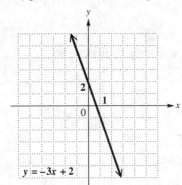

8. Graph the solid line $y = 2x + 3$ because the inequality is $\geq$.

Find the x- and y-intercepts. If $x = 0$, then $y = 3$, so the y-intercept is $(0, 3)$.

If $y = 0$, then $\quad 0 = 2x + 3$

$-3 = 2x$

$x = -\dfrac{3}{2}$,

so the x-intercept is $\left(-\dfrac{3}{2}, 0\right)$.

Use $(0, 0)$ for the test point.

$y \geq 2x + 3$

$0 \geq 2(0) + 3$?

$0 \geq 3 \qquad$ *False*

Shade the half-plane *not* containing $(0, 0)$.

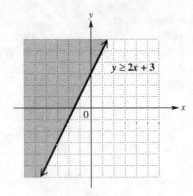

9. $3x + 4y = 5 \quad (1)$
$6x + 7y = 8 \quad (2)$

$\begin{array}{rll} -6x - 8y = & -10 & (3) \quad -2 \times \text{Eq.}(1) \\ 6x + 7y = & 8 & (2) \\ \hline -y = & -2 & \text{Add } (3) \text{ and } (2). \\ y = & 2 \end{array}$

Substitute 2 for y in equation (1).

$3x + 4(2) = 5$

$3x + 8 = 5$

$3x = -3$

$x = -1$

The solution is $(-1, 2)$.

10. $y = -3x + 1 \quad (1)$
$x + 2y = -3 \quad (2)$

Substitute $-3x + 1$ for y in equation (2).

$x + 2(-3x + 1) = -3$

$x - 6x + 2 = -3$

$-5x = -5$

$x = 1$

From (1), $y = -3(1) + 1 = -2$.

The solution is $(1, -2)$.

11. $\dfrac{(2x^3)^{-1} \cdot x}{2^3 x^5} = \dfrac{2^{-1}(x^3)^{-1} \cdot x}{2^3 x^5} = \dfrac{2^{-1} x^{-3} x}{2^3 x^5}$

$= \dfrac{2^{-1} x^{-2}}{2^3 x^5} = \dfrac{1}{2^1 \cdot 2^3 \cdot x^2 \cdot x^5}$

$= \dfrac{1}{2^4 x^7}$

12. $\dfrac{(m^{-2})^3 m}{m^5 m^{-4}} = \dfrac{m^{-6} m}{m^5 m^{-4}} = \dfrac{m \cdot m^4}{m^5 \cdot m^6}$

$= \dfrac{m^5}{m^{11}} = \dfrac{1}{m^6}$

13. $\dfrac{2p^3 q^4}{8p^5 q^3} = \dfrac{q \cdot 2p^3 q^3}{4p^2 \cdot 2p^3 q^3} = \dfrac{q}{4p^2}$

14. $(2k^2 + 3k) - (k^2 + k - 1)$
    $= 2k^2 + 3k - k^2 - k + 1$
    $= k^2 + 2k + 1$

15. $8x^2y^2(9x^4y^5) = 72x^6y^7$

16. $(2a - b)^2 = (2a)^2 - 2(2a)(b) + (b)^2$
    $= 4a^2 - 4ab + b^2$

17. $(y^2 + 3y + 5)(3y - 1)$

    Multiply vertically.

    $$\begin{array}{r} y^2 + 3y + 5 \\ 3y - 1 \\ \hline -y^2 - 3y - 5 \\ 3y^3 + 9y^2 + 15y \\ \hline 3y^3 + 8y^2 + 12y - 5 \end{array}$$

18. $\dfrac{12p^3 + 2p^2 - 12p + 4}{2p - 2}$

    $$\begin{array}{r} 6p^2 + 7p + 1 \\ 2p-2 \overline{\smash{\big)}\, 12p^3 + 2p^2 - 12p + 4} \\ \underline{12p^3 - 12p^2} \\ 14p^2 - 12p \\ \underline{14p^2 - 14p} \\ 2p + 4 \\ \underline{2p - 2} \\ 6 \end{array}$$

    The result is

    $6p^2 + 7p + 1 + \dfrac{6}{2p - 2}$

    $= 6p^2 + 7p + 1 + \dfrac{2 \cdot 3}{2(p - 1)}$

    $= 6p^2 + 7p + 1 + \dfrac{3}{p - 1}$.

19. $8t^2 + 10tv + 3v^2$

    $= 8t^2 + 6tv + 4tv + 3v^2 \quad \begin{array}{l} 6 \cdot 4 = 24; \\ 6 + 4 = 10 \end{array}$

    $= (8t^2 + 6tv) + (4tv + 3v^2)$
    $= 2t(4t + 3v) + v(4t + 3v)$
    $= (4t + 3v)(2t + v)$

20. $8r^2 - 9rs + 12s^2$

    To factor this polynomial by the grouping method, we must find two integers whose product is $(8)(12) = 96$ and whose sum is -9. There is no pair of integers that satisfies both of these conditions, so the polynomial is *prime*.

21. $16x^4 - 1$
    $= (4x^2)^2 - (1)^2$
    $= (4x^2 + 1)(4x^2 - 1)$
    $= (4x^2 + 1)[(2x)^2 - (1)^2]$
    $= (4x^2 + 1)(2x + 1)(2x - 1)$

22. $ r^2 = 2r + 15$
    $r^2 - 2r - 15 = 0$
    $(r + 3)(r - 5) = 0$
    $r + 3 = 0 \quad \text{or} \quad r - 5 = 0$
    $ r = -3 \quad \text{or} \quad r = 5$

    The solutions are -3 and 5.

23. $(r - 5)(2r + 1)(3r - 2) = 0$
    $r = 5 \quad \text{or} \quad r = -\tfrac{1}{2} \quad \text{or} \quad r = \tfrac{2}{3}$

    The solutions are $5, -\tfrac{1}{2},$ and $\tfrac{2}{3}$.

24. Let $x =$ the smaller number
    Then $x + 4 =$ the larger number.

    The product of the numbers is 2 less than the smaller number translates to

    $x(x + 4) = x - 2$.
    $x^2 + 4x = x - 2$
    $x^2 + 3x + 2 = 0$
    $(x + 2)(x + 1) = 0$
    $x = -2 \quad \text{or} \quad x = -1$

    The smaller number can be either -2 or -1.

25. Let $w =$ the width of the rectangle.
    Then $2w - 2 =$ the length of the rectangle.

    Use the formula $A = LW$ with the area $= 60$.

    $60 = (2w - 2)w$
    $60 = 2w^2 - 2w$
    $0 = 2w^2 - 2w - 60$
    $0 = 2(w^2 - w - 30)$
    $0 = 2(w - 6)(w + 5)$
    $w = 6 \quad \text{or} \quad w = -5$

    Discard -5 because the width cannot be negative. The width of the rectangle is 6 meters.

26. $\dfrac{2 + t}{t^2 - 4}$

    Set the denominator equal to 0.

    $t^2 - 4 = 0$
    $(t - 2)(t + 2) = 0$
    $t - 2 = 0 \quad \text{or} \quad t + 2 = 0$
    $ t = 2 \quad \text{or} \quad t = -2$

    The expression is undefined for -2 and 2.

Chapter 7 Rational Expressions and Applications

27. All of the given expressions are equal to 1 for all real numbers for which they are defined. However, expressions **B**, **C**, and **D** all have one or more values for which the expression is undefined and therefore cannot be equal to 1. Since $k^2 + 2$ is *always* positive, the denominator in expression **A** is never equal to zero. This expression is defined and equal to 1 for all real numbers, so the correct choice is **A**.

28. The appropriate choice is **D** since
$$\frac{-(3x+4)}{7} = \frac{-3x-4}{7}$$
$$\neq \frac{4-3x}{7}.$$

29. $\dfrac{5}{q} - \dfrac{1}{q} = \dfrac{5-1}{q} = \dfrac{4}{q}$

30. $\dfrac{3}{7} + \dfrac{4}{r} = \dfrac{3 \cdot r}{7 \cdot r} + \dfrac{4 \cdot 7}{r \cdot 7}$ *LCD = 7r*

$= \dfrac{3r + 28}{7r}$

31. $\dfrac{4}{5q - 20} - \dfrac{1}{3q - 12}$

$= \dfrac{4}{5(q-4)} - \dfrac{1}{3(q-4)}$

$= \dfrac{4 \cdot 3}{5(q-4) \cdot 3} - \dfrac{1 \cdot 5}{3(q-4) \cdot 5}$

                   $LCD = 5 \cdot 3 \cdot (q-4) = 15(q-4)$

$= \dfrac{12 - 5}{15(q-4)} = \dfrac{7}{15(q-4)}$

32. $\dfrac{2}{k^2 + k} - \dfrac{3}{k^2 - k}$

$= \dfrac{2}{k(k+1)} - \dfrac{3}{k(k-1)}$

$= \dfrac{2(k-1)}{k(k+1)(k-1)} - \dfrac{3(k+1)}{k(k-1)(k+1)}$

                   $LCD = k(k+1)(k-1)$

$= \dfrac{2(k-1) - 3(k+1)}{k(k+1)(k-1)}$

$= \dfrac{2k - 2 - 3k - 3}{k(k+1)(k-1)}$

$= \dfrac{-k - 5}{k(k+1)(k-1)}$

33. $\dfrac{7z^2 + 49z + 70}{16z^2 + 72z - 40} \div \dfrac{3z + 6}{4z^2 - 1}$

$= \dfrac{7z^2 + 49z + 70}{16z^2 + 72z - 40} \cdot \dfrac{4z^2 - 1}{3z + 6}$

$= \dfrac{7(z^2 + 7z + 10)}{8(2z^2 + 9z - 5)} \cdot \dfrac{(2z+1)(2z-1)}{3(z+2)}$

$= \dfrac{7(z+5)(z+2)}{8(2z-1)(z+5)} \cdot \dfrac{(2z+1)(2z-1)}{3(z+2)}$

$= \dfrac{7(2z+1)}{8 \cdot 3} = \dfrac{7(2z+1)}{24}$

34. $\dfrac{\dfrac{4}{a} + \dfrac{5}{2a}}{\dfrac{7}{6a} - \dfrac{1}{5a}}$

$= \dfrac{\left(\dfrac{4}{a} + \dfrac{5}{2a}\right) \cdot 30a}{\left(\dfrac{7}{6a} - \dfrac{1}{5a}\right) \cdot 30a}$ *Multiply by LCD, 30a.*

$= \dfrac{\dfrac{4}{a}(30a) + \dfrac{5}{2a}(30a)}{\dfrac{7}{6a}(30a) - \dfrac{1}{5a}(30a)}$

$= \dfrac{4 \cdot 30 + 5 \cdot 15}{7 \cdot 5 - 1 \cdot 6}$

$= \dfrac{120 + 75}{35 - 6} = \dfrac{195}{29}$

35. $\dfrac{r+2}{5} = \dfrac{r-3}{3}$

$15\left(\dfrac{r+2}{5}\right) = 15\left(\dfrac{r-3}{3}\right)$ *Multiply by LCD, 15.*

$3(r+2) = 5(r-3)$

$3r + 6 = 5r - 15$

$21 = 2r$

$\dfrac{21}{2} = r$

Check $r = \dfrac{21}{2}$: $\dfrac{5}{2} = \dfrac{5}{2}$

The solution is $\dfrac{21}{2}$.

36.
$$\frac{1}{x} = \frac{1}{x+1} + \frac{1}{2}$$
$$2x(x+1)\left(\frac{1}{x}\right) = 2x(x+1)\left(\frac{1}{x+1} + \frac{1}{2}\right)$$
Multiply by LCD, $2x(x+1)$.
$$2(x+1) = 2x(x+1)\left(\frac{1}{x+1}\right)$$
$$+ 2x(x+1)\left(\frac{1}{2}\right)$$
$$2(x+1) = 2x + x(x+1)$$
$$2x + 2 = 2x + x^2 + x$$
$$0 = x^2 + x - 2$$
$$0 = (x+2)(x-1)$$
$$x = -2 \quad \text{or} \quad x = 1$$

Check $x = -2$: $-\frac{1}{2} = -1 + \frac{1}{2}$ *True*
Check $x = 1$: $1 = \frac{1}{2} + \frac{1}{2}$ *True*

The solutions are -2 and 1.

37.

	Rate ×	Time =	Distance
To Business	60	x	$60x$
Coming Home	50	$x + \frac{1}{2}$	$50\left(x + \frac{1}{2}\right)$

The distances are equal, so we get the following equation.
$$60x = 50\left(x + \frac{1}{2}\right)$$
$$60x = 50x + 25$$
$$10x = 25$$
$$x = \frac{25}{10} = \frac{5}{2}$$

Monica's one-way trip was $60 \cdot \frac{5}{2} = 150$ miles.

38. Let $x = $ the number of hours it will take Chandler and Ross to weed the yard working together.

	Rate	Time Working Together	Fractional Part of the Job Done When Working Together
Chandler	$\frac{1}{3}$	x	$\frac{1}{3}x$
Ross	$\frac{1}{2}$	x	$\frac{1}{2}x$

Working together, they can do 1 whole job, so
$$\frac{1}{3}x + \frac{1}{2}x = 1.$$
$$6\left(\frac{1}{3}x + \frac{1}{2}x\right) = 6 \cdot 1 \quad \text{\textit{Multiply by LCD, 6.}}$$
$$6\left(\frac{1}{3}x\right) + 6\left(\frac{1}{2}x\right) = 6$$
$$2x + 3x = 6$$
$$5x = 6$$
$$x = \frac{6}{5}$$

If Chandler and Ross worked together, it would take them $\frac{6}{5}$ or $1\frac{1}{5}$ hours to weed the yard.

CHAPTER 8 ROOTS AND RADICALS

8.1 Evaluating Roots

8.1 Margin Exercises

1. **(a)** The square roots of 100 are 10 and -10 because $10 \cdot 10 = 100$ and $(-10)(-10) = 100$.

   (b) The square roots of 25 are 5 and -5 because $5 \cdot 5 = 25$ and $(-5)(-5) = 25$.

   (c) The square roots of 36 are 6 and -6 because $6 \cdot 6 = 36$ and $(-6)(-6) = 36$.

   (d) The square roots of $\frac{25}{36}$ are $\frac{5}{6}$ and $-\frac{5}{6}$ because $\frac{5}{6} \cdot \frac{5}{6} = \frac{25}{36}$ and $\left(-\frac{5}{6}\right)\left(-\frac{5}{6}\right) = \frac{25}{36}$.

2. **(a)** $\sqrt{16}$ is the positive square root of 16.
   $$\sqrt{16} = 4$$

   (b) $-\sqrt{169}$ is the negative square root of 169.
   $$-\sqrt{169} = -13$$

   (c) $-\sqrt{225}$ is the negative square root of 225.
   $$-\sqrt{225} = -15$$

   (d) $\sqrt{729}$ is the positive square root of 729.
   $$\sqrt{729} = 27$$

   (e) $\sqrt{\frac{36}{25}}$ is the positive square root of $\frac{36}{25}$.
   $$\sqrt{\frac{36}{25}} = \frac{6}{5}$$

   (f) $\sqrt{.49}$ is the positive square root of .49.
   $$\sqrt{.49} = .7$$

3. **(a)** The square of $\sqrt{41}$ is
   $$\left(\sqrt{41}\right)^2 = 41.$$

   (b) The square of $-\sqrt{39}$ is
   $$\left(\sqrt{39}\right)^2 = 39.$$

   (c) The square of $\sqrt{2x^2 + 3}$ is
   $$\left(\sqrt{2x^2 + 3}\right)^2 = 2x^2 + 3.$$

4. **(a)** 9 is a perfect square, so $\sqrt{9} = 3$ is rational.

   (b) 7 is not a perfect square, so $\sqrt{7}$ is irrational.

   (c) $\frac{4}{9}$ is a perfect square, so $\sqrt{\frac{4}{9}} = \frac{2}{3}$ is rational.

   (d) 72 is not a perfect square, so $\sqrt{72}$ is irrational.

   (e) There is no real number whose square is -43, so $\sqrt{-43}$ is not a real number.

5. Use the square root key of a calculator to find a decimal approximation for each root. Answers are given to the nearest thousandth.

   (a) $\sqrt{28} \approx 5.292$

   (b) $\sqrt{63} \approx 7.937$

   (c) $-\sqrt{190} \approx -13.784$

   (d) $\sqrt{1000} \approx 31.623$

6. Substitute the given values in the Pythagorean formula, $c^2 = a^2 + b^2$. Then solve for the variable that is not given.

   (a) $c^2 = a^2 + b^2$
   $c^2 = 7^2 + 24^2$ Let $a = 7$, $b = 24$.
   $c^2 = 49 + 576$ Square.
   $c^2 = 625$ Add.
   $c = \sqrt{625}$ Take square root.
   $c = 25$ $25^2 = 625$

   (b) $c^2 = a^2 + b^2$
   $15^2 = a^2 + 13^2$ Let $c = 15$, $b = 24$.
   $225 = a^2 + 169$
   $56 = a^2$ Subtract 169.
   $a = \sqrt{56}$ Take square root.
   $a \approx 7.483$ Use a calculator.

   (c) $c^2 = a^2 + b^2$
   $11^2 = 8^2 + b^2$ Let $c = 11$, $a = 8$.
   $121 = 64 + b^2$
   $57 = b^2$ Subtract 64.
   $b = \sqrt{57}$ Take square root.
   $b \approx 7.550$ Use a calculator.

7. Let $c = $ the length of the diagonal. Use the Pythagorean formula, since the width, length, and diagonal of a rectangle form a right triangle.

   $c^2 = a^2 + b^2$
   $c^2 = 5^2 + 12^2$ Let $a = 5$, $b = 12$.
   $c^2 = 25 + 144$
   $c^2 = 169$
   $c = \sqrt{169}$
   $c = 13$

   The diagonal is 13 feet long.

8.

Perfect Cubes	Perfect Fourth Powers
$1^3 = 1$	$1^4 = 1$
$2^3 = 8$	$2^4 = 16$
$3^3 = 27$	$3^4 = 81$
$4^3 = \underline{64}$	$4^4 = \underline{256}$
$5^3 = \underline{125}$	$5^4 = \underline{625}$
$6^3 = \underline{216}$	$6^4 = \underline{1296}$
$7^3 = \underline{343}$	$7^4 = \underline{2401}$
$8^3 = \underline{512}$	$8^4 = \underline{4096}$
$9^3 = \underline{729}$	$9^4 = \underline{6561}$
$10^3 = \underline{1000}$	$10^4 = \underline{10{,}000}$

9. (a) $\sqrt[3]{27} = 3$ because $3^3 = 27$.

   (b) $\sqrt[3]{64} = 4$ because $4^3 = 64$.

   (c) $\sqrt[3]{-125} = -5$ because $(-5)^3 = -125$.

10. (a) $\sqrt[4]{81} = 3$ because $3^4 = 81$.

    (b) $\sqrt[4]{-81}$ is not a real number because a fourth power of a real number cannot be negative.

    (c) $-\sqrt[4]{625}$

    First find $\sqrt[4]{625}$. Because $5^4 = 625$, $\sqrt[4]{625} = 5$. Thus, $-\sqrt[4]{625} = -5$.

    (d) $\sqrt[5]{243} = 3$ because $3^5 = 243$.

    (e) $\sqrt[5]{-243} = -3$ because $(-3)^5 = -243$.

8.1 Section Exercises

1. Every positive number has two real square roots. This statement is *true*. One of the real square roots is a positive number and the other is its opposite.

3. Every nonnegative number has two real square roots. This statement is *false* since zero is a nonnegative number that has only one square root, namely 0.

5. The cube root of every real number has the same sign as the number itself. This statement is *true*. The cube root of a positive real number is positive and the cube root of a negative real number is negative. The cube root of 0 is 0.

7. The square roots of 9 are -3 and 3 because $(-3)(-3) = 9$ and $3 \cdot 3 = 9$.

9. The square roots of 64 are -8 and 8 because $(-8)(-8) = 64$ and $8 \cdot 8 = 64$.

11. The square roots of 169 are -13 and 13 because $(-13)(-13) = 169$ and $13 \cdot 13 = 169$.

13. The square roots of $\frac{25}{196}$ are $-\frac{5}{14}$ and $\frac{5}{14}$ because
$$\left(-\tfrac{5}{14}\right)\left(-\tfrac{5}{14}\right) = \tfrac{25}{196}$$
and
$$\tfrac{5}{14} \cdot \tfrac{5}{14} = \tfrac{25}{196}.$$

15. The square roots of 900 are -30 and 30 because $(-30)(-30) = 900$ and $30 \cdot 30 = 900$.

17. $\sqrt{1}$ represents the positive square root of 1. Since $1 \cdot 1 = 1$,
$$\sqrt{1} = 1.$$

19. $\sqrt{49}$ represents the positive square root of 49. Since $7 \cdot 7 = 49$,
$$\sqrt{49} = 7.$$

21. $-\sqrt{256}$ represents the negative square root of 256. Since $16 \cdot 16 = 256$,
$$-\sqrt{256} = -16.$$

23. $-\sqrt{\dfrac{144}{121}}$ represents the negative square root of $\dfrac{144}{121}$. Since $\dfrac{12}{11} \cdot \dfrac{12}{11} = \dfrac{144}{121}$,
$$-\sqrt{\dfrac{144}{121}} = -\dfrac{12}{11}.$$

25. $\sqrt{.64}$ represents the positive square root of $.64$. Since $(.8)(.8) = .64$,
$$\sqrt{.64} = .8.$$

27. $\sqrt{-121}$ is not a real number because there is no real number whose square is -121.

29. $\sqrt{-49}$ is not a real number because there is no real number whose square is -49. Thus, $-\sqrt{-49}$ is not a real number.

31. The square of $\sqrt{100}$ is
$$\left(\sqrt{100}\right)^2 = 100,$$
by the definition of square root.

33. The square of $-\sqrt{19}$ is
$$\left(-\sqrt{19}\right)^2 = 19,$$
since the square of a negative number is positive.

35. The square of $\sqrt{\dfrac{2}{3}}$ is
$$\left(\sqrt{\dfrac{2}{3}}\right)^2 = \dfrac{2}{3},$$
by the definition of square root.

37. The square of $\sqrt{3x^2+4}$ is
$$\left(\sqrt{3x^2+4}\right)^2 = 3x^2+4.$$

39. For the statement "$\sqrt{a}$ represents a positive number" to be true, a must be positive because the square root of a negative number is not a real number and $\sqrt{0} = 0$.

41. For the statement "$\sqrt{a}$ is not a real number" to be true, a must be negative.

43. $\sqrt{25}$

The number 25 is a perfect square, 5^2, so $\sqrt{25}$ is a *rational* number.
$$\sqrt{25} = 5$$

45. $\sqrt{29}$

Because 29 is not a perfect square, $\sqrt{29}$ is *irrational*. Using a calculator, we obtain
$$\sqrt{29} \approx 5.385.$$

47. $-\sqrt{64}$

The number 64 is a perfect square, 8^2, so $-\sqrt{64}$ is *rational*.
$$-\sqrt{64} = -8$$

49. $-\sqrt{300}$

The number 300 is not a perfect square, so $-\sqrt{300}$ is *irrational*. Using a calculator, we obtain
$$-\sqrt{300} \approx -17.321.$$

51. $\sqrt{-29}$

There is no real number whose square is -29. Therefore, $\sqrt{-29}$ is *not a real number*.

53. $\sqrt{1200}$

Because 1200 is not a perfect square, $\sqrt{1200}$ is *irrational*. Using a calculator, we obtain
$$\sqrt{1200} \approx 34.641.$$

55. $\sqrt{103} \approx \sqrt{100} = 10$
$\sqrt{48} \approx \sqrt{49} = 7$

The best estimate for the length and width of the rectangle is 10 by 7, choice **C**.

57. $a = 8, b = 15$

Substitute the given values in the Pythagorean formula and then solve for c^2.
$$c^2 = a^2 + b^2$$
$$c^2 = 8^2 + 15^2$$
$$= 64 + 225$$
$$= 289$$

Now find the positive square root of 289 to obtain the length of the hypotenuse, c.
$$c = \sqrt{289} = 17$$

59. $a = 6, c = 10$

Substitute the given values in the Pythagorean formula and then solve for b^2.
$$c^2 = a^2 + b^2$$
$$10^2 = 6^2 + b^2$$
$$100 = 36 + b^2$$
$$64 = b^2$$

Now find the positive square root of 64 to obtain the length of the leg b.
$$b = \sqrt{64} = 8$$

61. $a = 11, b = 4$
$$c^2 = a^2 + b^2$$
$$c^2 = 11^2 + 4^2$$
$$= 121 + 16$$
$$= 137$$
$$c = \sqrt{137} \approx 11.705$$

63. The given information involves a right triangle with hypotenuse 25 centimeters and a leg of length 7 centimeters. Let a represent the length of the other leg, and use the Pythagorean formula.
$$c^2 = a^2 + b^2$$
$$25^2 = a^2 + 7^2$$
$$625 = a^2 + 49$$
$$576 = a^2$$
$$a = \sqrt{576} = 24$$

The length of the rectangle is 24 centimeters.

65. *Step 2*
Let x represent the vertical distance of the kite above Tyler's hand. The kite string forms the hypotenuse of a right triangle.

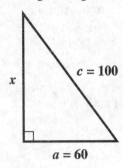

Step 3
Use the Pythagorean formula.
$$a^2 + x^2 = c^2$$
$$60^2 + x^2 = 100^2$$

Step 4
$$3600 + x^2 = 10,000$$
$$x^2 = 6400$$
$$x = \sqrt{6400} = 80$$

Step 5
The kite is 80 feet above his hand.

Step 6
From the figure, we see that we must have
$$60^2 + 80^2 = 100^2 \text{ ?}$$
$$3600 + 6400 = 10,000. \quad \textit{True}$$

67. *Step 2*
Let x represent the distance from R to S.

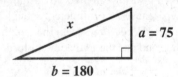

Step 3
Triangle RST is a right triangle, so we can use the Pythagorean formula.
$$x^2 = a^2 + b^2$$
$$x^2 = 75^2 + 180^2$$
$$= 5625 + 32,400$$

Step 4
$$x^2 = 38,025$$
$$x = \sqrt{38,025} = 195$$

Step 5
The distance across the lake is 195 feet.

Step 6
From the figure, we see that we must have
$$75^2 + 180^2 = 195^2 \text{ ?}$$
$$5625 + 32,400 = 38,025. \quad \textit{True}$$

69. Use the Pythagorean formula with $a = 5$, $b = 8$, and $c = x$.
$$c^2 = a^2 + b^2$$
$$x^2 = 5^2 + 8^2$$
$$= 25 + 64$$
$$= 89$$
$$x = \sqrt{89} \approx 9.434$$

71. Let $a = 4.5$ and $c = 12$. Use the Pythagorean formula.
$$c^2 = a^2 + b^2$$
$$12^2 = (4.5)^2 + b^2$$
$$144 = 20.25 + b^2$$
$$123.75 = b^2$$
$$b = \sqrt{123.75} \approx 11.1$$

The distance from the base of the tree to the point where the broken part touches the ground is 11.1 feet (to the nearest tenth).

73. Answers will vary.

For example, if we choose $a = 2$ and $b = 7$,
$$\sqrt{a^2 + b^2} = \sqrt{2^2 + 7^2} = \sqrt{53},$$
while
$$a + b = 2 + 7 = 9.$$
$\sqrt{53} \neq 9$, so we have
$$\sqrt{a^2 + b^2} \neq a + b.$$

75. $\sqrt[3]{1} = 1$ because $1^3 = 1$.

77. $\sqrt[3]{125} = 5$ because $5^3 = 125$.

79. $\sqrt[3]{-27} = -3$ because $(-3)^3 = -27$.

81. $\sqrt[3]{-216} = -6$ because $(-6)^3 = -216$.

83. $\sqrt[3]{8} = 2$ because $2^3 = 8$. Thus,
$$-\sqrt[3]{8} = -2.$$

85. $\sqrt[4]{256} = 4$ because 4 is positive and $4^4 = 256$.

87. $\sqrt[4]{1296} = 6$ because 6 is positive and $6^4 = 1296$.

89. $\sqrt[4]{-1}$ is not a real number because the fourth power of a real number cannot be negative.

91. $\sqrt[4]{81} = 3$ because 3 is positive and $3^4 = 81$. Thus,
$$-\sqrt[4]{81} = -3.$$

93. $\sqrt[5]{-1024} = -4$ because $(-4)^5 = -1024$.

8.2 Multiplying, Dividing, and Simplifying Radicals

8.2 Margin Exercises

1. **(a)** $\sqrt{6} \cdot \sqrt{11} = \sqrt{6 \cdot 11} = \sqrt{66}$ *Product rule*

   (b) $\sqrt{2} \cdot \sqrt{5} = \sqrt{2 \cdot 5} = \sqrt{10}$ *Product rule*

   (c) $\sqrt{10} \cdot \sqrt{r} = \sqrt{10r}$

   Note that this expression is a real number; since $r \geq 0$, $10r$ is nonnegative.

2. **(a)** $\sqrt{8} = \sqrt{4 \cdot 2}$ *4 is a perfect square.*
   $= \sqrt{4} \cdot \sqrt{2}$ *Product rule*
   $= 2\sqrt{2}$

   (b) $\sqrt{27} = \sqrt{9 \cdot 3}$ *9 is a perfect square.*
   $= \sqrt{9} \cdot \sqrt{3}$ *Product rule*
   $= 3\sqrt{3}$

   (c) $\sqrt{50} = \sqrt{25 \cdot 2}$ *25 is a perfect square.*
   $= \sqrt{25} \cdot \sqrt{2}$ *Product rule*
   $= 5\sqrt{2}$

   (d) $\sqrt{60} = \sqrt{4 \cdot 15}$ *4 is a perfect square.*
   $= \sqrt{4} \cdot \sqrt{15}$ *Product rule*
   $= 2\sqrt{15}$

   (e) $\sqrt{30}$ cannot be simplified further because 30 has no perfect square factors (except 1).

3. **(a)** $\sqrt{3} \cdot \sqrt{15} = \sqrt{3 \cdot 15}$
   $= \sqrt{3 \cdot 3 \cdot 5}$
   $= \sqrt{9 \cdot 5}$ *9 is a perfect square.*
   $= \sqrt{9} \cdot \sqrt{5}$ *Product rule*
   $= 3\sqrt{5}$

   (b) $\sqrt{10} \cdot \sqrt{50} = \sqrt{10 \cdot 50}$
   $= \sqrt{10 \cdot 10 \cdot 5}$
   $= \sqrt{100 \cdot 5}$ *100 is a perfect square.*
   $= \sqrt{100} \cdot \sqrt{5}$ *Product rule*
   $= 10\sqrt{5}$

   (c) $\sqrt{12} \cdot \sqrt{2} = \sqrt{12 \cdot 2}$
   $= \sqrt{2 \cdot 6 \cdot 2}$
   $= \sqrt{4 \cdot 6}$ *4 is a perfect square.*
   $= \sqrt{4} \cdot \sqrt{6}$ *Product rule*
   $= 2\sqrt{6}$

   (d) $\sqrt{7} \cdot \sqrt{14} = \sqrt{7 \cdot 14}$
   $= \sqrt{7 \cdot 7 \cdot 2}$
   $= \sqrt{49 \cdot 2}$ *49 is a perfect square.*
   $= \sqrt{49} \cdot \sqrt{2}$ *Product rule*
   $= 7\sqrt{2}$

   (e) $3\sqrt{5} \cdot 4\sqrt{10} = 3 \cdot 4 \cdot \sqrt{5 \cdot 10}$
   $= 12\sqrt{5 \cdot 5 \cdot 2}$
   $= 12\sqrt{25} \cdot \sqrt{2}$
   $= 12 \cdot 5 \cdot \sqrt{2}$
   $= 60\sqrt{2}$

4. **(a)** $\sqrt{\dfrac{81}{16}} = \dfrac{\sqrt{81}}{\sqrt{16}} = \dfrac{9}{4}$ *Quotient rule*

   (b) $\dfrac{\sqrt{192}}{\sqrt{3}} = \sqrt{\dfrac{192}{3}} = \sqrt{64} = 8$

   (c) $\sqrt{\dfrac{10}{49}} = \dfrac{\sqrt{10}}{\sqrt{49}} = \dfrac{\sqrt{10}}{7}$

5. $\dfrac{8\sqrt{50}}{4\sqrt{5}} = \dfrac{8}{4} \cdot \dfrac{\sqrt{50}}{\sqrt{5}}$
   $= 2 \cdot \sqrt{\dfrac{50}{5}}$
   $= 2\sqrt{10}$

6. **(a)** $\sqrt{\dfrac{5}{6}} \cdot \sqrt{120} = \sqrt{\dfrac{5}{6}} \cdot \sqrt{\dfrac{120}{1}}$
   $= \sqrt{\dfrac{5}{6} \cdot \dfrac{120}{1}}$ *Product rule*
   $= \sqrt{\dfrac{600}{6}}$ *Multiply fractions.*
   $= \sqrt{100}$
   $= 10$

   (b) $\sqrt{\dfrac{3}{8}} \cdot \sqrt{\dfrac{7}{2}} = \sqrt{\dfrac{3}{8} \cdot \dfrac{7}{2}}$ *Product rule*
   $= \sqrt{\dfrac{21}{16}}$ *Multiply fractions.*
   $= \dfrac{\sqrt{21}}{\sqrt{16}}$ *Quotient rule*
   $= \dfrac{\sqrt{21}}{4}$ $\sqrt{16} = 4$

7. **(a)** $\sqrt{x^8} = x^4$ since $(x^4)^2 = x^8$.

   (b) $\sqrt{36y^6} = \sqrt{36} \cdot \sqrt{y^6}$ *Product rule*
   $= 6y^3$

   (c) $\sqrt{100p^{12}} = \sqrt{100} \cdot \sqrt{p^{12}} = 10p^6$

314 Chapter 8 Roots and Radicals

(d) $\sqrt{12z^2} = \sqrt{4} \cdot \sqrt{3} \cdot \sqrt{z^2}$
$= 2 \cdot \sqrt{3} \cdot z = 2z\sqrt{3}$

(e) $\sqrt{a^5} = \sqrt{a^4 \cdot a}$ a^4 is a square.
$= \sqrt{a^4} \cdot \sqrt{a} = a^2\sqrt{a}$

(f) $\sqrt{\dfrac{10}{n^4}} = \dfrac{\sqrt{10}}{\sqrt{n^4}} = \dfrac{\sqrt{10}}{n^2}$

8. (a) $\sqrt[3]{108} = \sqrt[3]{27 \cdot 4}$ 27 is a perfect cube.
$= \sqrt[3]{27} \cdot \sqrt[3]{4}$ Product rule
$= 3\sqrt[3]{4}$

(b) $\sqrt[4]{160} = \sqrt[4]{16 \cdot 10}$ 16 is a perfect fourth power.
$= \sqrt[4]{16} \cdot \sqrt[4]{10}$ Product rule
$= 2\sqrt[4]{10}$

(c) $\sqrt[4]{\dfrac{16}{625}} = \dfrac{\sqrt[4]{16}}{\sqrt[4]{625}}$ Quotient rule
$= \dfrac{2}{5}$

9. (a) $\sqrt[3]{z^9} = z^3$ since $(z^3)^3 = z^9$.

(b) $\sqrt[3]{8x^6} = \sqrt[3]{8} \cdot \sqrt[3]{x^6}$ Product rule
$= 2x^2$ $2^3 = 8$; $(x^2)^3 = x^6$

(c) $\sqrt[3]{54t^5} = \sqrt[3]{27t^3 \cdot 2t^2}$ $27t^3$ is a cube.
$= \sqrt[3]{27t^3} \cdot \sqrt[3]{2t^2}$ Product rule
$= 3t\sqrt[3]{2t^2}$ $(3t)^3 = 27t^3$

(d) $\sqrt[3]{\dfrac{a^{15}}{64}} = \dfrac{\sqrt[3]{a^{15}}}{\sqrt[3]{64}}$ Quotient rule
$= \dfrac{a^5}{4}$

8.2 Section Exercises

1. $\sqrt{(-6)^2} = -6$

This statement is *false*.

$\sqrt{(-6)^2} = \sqrt{36} = 6 \neq -6$

In general, $\sqrt{a^2} = a$ only if a is nonnegative.

3. $\sqrt{3} \cdot \sqrt{5}$

Since 3 and 5 are nonnegative real numbers, the *Product Rule for Radicals* applies. Thus,

$\sqrt{3} \cdot \sqrt{5} = \sqrt{3 \cdot 5} = \sqrt{15}$.

5. $\sqrt{2} \cdot \sqrt{11} = \sqrt{2 \cdot 11} = \sqrt{22}$

7. $\sqrt{6} \cdot \sqrt{7} = \sqrt{6 \cdot 7} = \sqrt{42}$

9. $\sqrt{13} \cdot \sqrt{r}$ $(r \geq 0) = \sqrt{13r}$

11. $\sqrt{47}$ is in simplified form since 47 has no perfect square factor (other than 1).

The other three choices could be simplified as follows.

$\sqrt{45} = \sqrt{9 \cdot 5} = 3\sqrt{5}$
$\sqrt{48} = \sqrt{16 \cdot 3} = 4\sqrt{3}$
$\sqrt{44} = \sqrt{4 \cdot 11} = 2\sqrt{11}$

The correct choice is **A**.

13. $\sqrt{45} = \sqrt{9 \cdot 5} = \sqrt{9} \cdot \sqrt{5} = 3\sqrt{5}$

15. $\sqrt{24} = \sqrt{4 \cdot 6} = \sqrt{4} \cdot \sqrt{6} = 2\sqrt{6}$

17. $\sqrt{90} = \sqrt{9 \cdot 10} = \sqrt{9} \cdot \sqrt{10} = 3\sqrt{10}$

19. $\sqrt{75} = \sqrt{25 \cdot 3} = \sqrt{25} \cdot \sqrt{3} = 5\sqrt{3}$

21. $\sqrt{125} = \sqrt{25 \cdot 5} = \sqrt{25} \cdot \sqrt{5} = 5\sqrt{5}$

23. $145 = 5 \cdot 29$, so 145 has no perfect square factors (except 1) and $\sqrt{145}$ cannot be simplified further.

25. $\sqrt{160} = \sqrt{16 \cdot 10} = \sqrt{16} \cdot \sqrt{10} = 4\sqrt{10}$

27. $-\sqrt{700} = -\sqrt{100 \cdot 7}$
$= -\sqrt{100} \cdot \sqrt{7} = -10\sqrt{7}$

29. $\sqrt{3} \cdot \sqrt{18} = \sqrt{3 \cdot 18} = \sqrt{54} = \sqrt{9 \cdot 6}$
$= \sqrt{9} \cdot \sqrt{6} = 3\sqrt{6}$

31. $\sqrt{12} \cdot \sqrt{48} = \sqrt{12 \cdot 48}$
$= \sqrt{12 \cdot 12 \cdot 4}$
$= \sqrt{12 \cdot 12} \cdot \sqrt{4}$
$= 12 \cdot 2$
$= 24$

33. $\sqrt{12} \cdot \sqrt{30} = \sqrt{12 \cdot 30}$
$= \sqrt{360}$
$= \sqrt{36 \cdot 10}$
$= \sqrt{36} \cdot \sqrt{10}$
$= 6\sqrt{10}$

35. $2\sqrt{10} \cdot 3\sqrt{2} = 2 \cdot 3 \cdot \sqrt{10 \cdot 2}$ Product rule
$= 6\sqrt{20}$ Multiply.
$= 6\sqrt{4 \cdot 5}$ Factor; 4 is a perfect square.
$= 6\sqrt{4} \cdot \sqrt{5}$ Product rule
$= 6 \cdot 2 \cdot \sqrt{5}$ $\sqrt{4} = 2$
$= 12\sqrt{5}$ Multiply.

8.2 Multiplying, Dividing, and Simplifying Radicals

37. $5\sqrt{3} \cdot 2\sqrt{15} = 5 \cdot 2 \cdot \sqrt{3 \cdot 15}$ *Product rule*
$= 10\sqrt{45}$ *Multiply.*
$= 10\sqrt{9 \cdot 5}$ *Factor; 9 is a perfect square.*
$= 10\sqrt{9} \cdot \sqrt{5}$ *Product rule*
$= 10 \cdot 3 \cdot \sqrt{5}$ $\sqrt{9} = 3$
$= 30\sqrt{5}$ *Multiply.*

39. $\sqrt{8} \cdot \sqrt{32} = \sqrt{8 \cdot 32} = \sqrt{256} = 16$.
Also, $\sqrt{8} = 2\sqrt{2}$ and $\sqrt{32} = 4\sqrt{2}$, so
$\sqrt{8} \cdot \sqrt{32} = 2\sqrt{2} \cdot 4\sqrt{2} = 8 \cdot 2 = 16$.

Both methods give the same answer, and the correct answer can always be obtained using either method.

41. $\sqrt{\dfrac{16}{225}} = \dfrac{\sqrt{16}}{\sqrt{225}} = \dfrac{4}{15}$

43. $\sqrt{\dfrac{7}{16}} = \dfrac{\sqrt{7}}{\sqrt{16}} = \dfrac{\sqrt{7}}{4}$

45. $\dfrac{\sqrt{75}}{\sqrt{3}} = \sqrt{\dfrac{75}{3}} = \sqrt{25} = 5$

47. $\sqrt{\dfrac{5}{2}} \cdot \sqrt{\dfrac{125}{8}} = \sqrt{\dfrac{5}{2} \cdot \dfrac{125}{8}}$
$= \sqrt{\dfrac{625}{16}}$
$= \dfrac{\sqrt{625}}{\sqrt{16}} = \dfrac{25}{4}$

49. $\dfrac{30\sqrt{10}}{5\sqrt{2}} = \dfrac{30}{5}\sqrt{\dfrac{10}{2}} = 6\sqrt{5}$

51. $\sqrt{m^2} = m$ $(m \geq 0)$

53. $\sqrt{y^4} = \sqrt{(y^2)^2} = y^2$

55. $\sqrt{36z^2} = \sqrt{36} \cdot \sqrt{z^2} = 6z$

57. $\sqrt{400x^6} = \sqrt{20 \cdot 20 \cdot x^3 \cdot x^3} = 20x^3$

59. $\sqrt{18x^8} = \sqrt{9 \cdot 2 \cdot x^8}$
$= \sqrt{9} \cdot \sqrt{2} \cdot \sqrt{x^8}$
$= 3 \cdot \sqrt{2} \cdot x^4$
$= 3x^4\sqrt{2}$

61. $\sqrt{45c^{14}} = \sqrt{9 \cdot 5 \cdot c^{14}}$
$= \sqrt{9} \cdot \sqrt{5} \cdot \sqrt{c^{14}}$
$= 3 \cdot \sqrt{5} \cdot c^7$
$= 3c^7\sqrt{5}$

63. $\sqrt{z^5} = \sqrt{z^4 \cdot z} = \sqrt{z^4} \cdot \sqrt{z} = z^2\sqrt{z}$

65. $\sqrt{a^{13}} = \sqrt{a^{12}} \cdot \sqrt{a}$
$= \sqrt{(a^6)^2} \cdot \sqrt{a} = a^6\sqrt{a}$

67. $\sqrt{64x^7} = \sqrt{64} \cdot \sqrt{x^6}\sqrt{x}$
$= 8x^3\sqrt{x}$

69. $\sqrt{x^6 y^{12}} = \sqrt{(x^3)^2 \cdot (y^6)^2} = x^3 y^6$

71. $\sqrt{81m^4 n^2} = \sqrt{81} \cdot \sqrt{m^4} \cdot \sqrt{n^2}$
$= 9m^2 n$

73. $\sqrt{\dfrac{7}{x^{10}}} = \dfrac{\sqrt{7}}{\sqrt{x^{10}}} = \dfrac{\sqrt{7}}{x^5}$ $(x \neq 0)$

75. $\sqrt{\dfrac{y^4}{100}} = \dfrac{\sqrt{y^4}}{\sqrt{100}} = \dfrac{y^2}{10}$

77. $\sqrt{\dfrac{x^6}{y^8}} = \dfrac{\sqrt{x^6}}{\sqrt{y^8}} = \dfrac{x^3}{y^4}$ $(y \neq 0)$

79. $\sqrt[3]{40}$

8 is a perfect cube that is a factor of 40.

$\sqrt[3]{40} = \sqrt[3]{8 \cdot 5}$
$= \sqrt[3]{8} \cdot \sqrt[3]{5} = 2\sqrt[3]{5}$

81. $\sqrt[3]{54}$

27 is a perfect cube that is a factor of 54.

$\sqrt[3]{54} = \sqrt[3]{27 \cdot 2}$
$= \sqrt[3]{27} \cdot \sqrt[3]{2} = 3\sqrt[3]{2}$

83. $\sqrt[3]{128}$

64 is a perfect cube that is a factor of 128.

$\sqrt[3]{128} = \sqrt[3]{64 \cdot 2}$
$= \sqrt[3]{64} \cdot \sqrt[3]{2} = 4\sqrt[3]{2}$

85. $\sqrt[4]{80}$

16 is a perfect fourth power that is a factor of 80.

$\sqrt[4]{80} = \sqrt[4]{16 \cdot 5}$
$= \sqrt[4]{16} \cdot \sqrt[4]{5} = 2\sqrt[4]{5}$

87. $\sqrt[3]{\dfrac{8}{27}}$

8 and 27 are both perfect cubes.

$\sqrt[3]{\dfrac{8}{27}} = \dfrac{\sqrt[3]{8}}{\sqrt[3]{27}} = \dfrac{2}{3}$

316 Chapter 8 Roots and Radicals

89. $\sqrt[3]{-\dfrac{216}{125}} = -\sqrt[3]{\dfrac{216}{125}} = -\dfrac{\sqrt[3]{216}}{\sqrt[3]{125}} = -\dfrac{6}{5}$

91. $\sqrt[3]{p^3} = p$ because $(p)^3 = p^3$.

93. $\sqrt[3]{x^9} = \sqrt[3]{(x^3)^3} = x^3$

95. $\sqrt[3]{64z^6} = \sqrt[3]{64} \cdot \sqrt[3]{(z^2)^3} = 4z^2$

97. $\sqrt[3]{343a^9b^3} = \sqrt[3]{343} \cdot \sqrt[3]{a^9} \cdot \sqrt[3]{b^3} = 7a^3b$

99. $\sqrt[3]{16t^5} = \sqrt[3]{8t^3} \cdot \sqrt[3]{2t^2} = 2t\sqrt[3]{2t^2}$

101. $\sqrt[3]{\dfrac{m^{12}}{8}} = \sqrt[3]{\dfrac{(m^4)^3}{2^3}} = \dfrac{m^4}{2}$

103. Use the formula for the volume of a cube.

$V = s^3$
$216 = s^3$ Let $V = 216$.
$\sqrt[3]{216} = s$
$6 = s$

The depth of the container is 6 centimeters.

105. Use the formula for the volume of a sphere,

$V = \tfrac{4}{3}\pi r^3$.

Let $V = 288\pi$ and solve for r.

$\tfrac{4}{3}\pi r^3 = 288\pi$
$\tfrac{3}{4}(\tfrac{4}{3}\pi r^3) = \tfrac{3}{4}(288\pi)$
$\pi r^3 = 216\pi$
$r^3 = 216$
$r = \sqrt[3]{216} = 6$

The radius is 6 inches.

107. $2\sqrt{26} \approx 2\sqrt{25} = 2 \cdot 5 = 10$
$\sqrt{83} \approx \sqrt{81} = 9$

Using 10 and 9 as estimates for the length and the width of the rectangle gives us $10 \cdot 9 = 90$ as an estimate for the area. Thus, choice **D** is the best estimate.

8.3 Adding and Subtracting Radicals

8.3 Margin Exercises

1. (a) $5\sqrt{6}$ and $4\sqrt{6}$ are *like* radicals because they are multiples of the same root of the same number.

   (b) $2\sqrt{3}$ and $3\sqrt{2}$ are *unlike* radicals because the radicands, 3 and 2, respectively, are different.

   (c) $\sqrt{10}$ and $\sqrt[3]{10}$ are *unlike* radicals because the indexes, 2 and 3, are different.

   (d) $7\sqrt{2x}$ and $8\sqrt{2x}$ are *like* radicals because they are multiples of the same root of the same number.

   (e) $\sqrt{3y}$ and $\sqrt{6y}$ are *unlike* radicals because the radicands, $3y$ and $6y$, are different.

2. (a) $8\sqrt{5} + 2\sqrt{5} = (8+2)\sqrt{5}$ Distributive property
   $= 10\sqrt{5}$

   (b) $-4\sqrt{3} + 9\sqrt{3} = (-4+9)\sqrt{3}$
   $= 5\sqrt{3}$

   (c) $12\sqrt{11} - 3\sqrt{11} = (12-3)\sqrt{11}$
   $= 9\sqrt{11}$

   (d) $\sqrt{15} + \sqrt{15} = 1\sqrt{15} + 1\sqrt{15}$
   $= (1+1)\sqrt{15}$
   $= 2\sqrt{15}$

   (e) $2\sqrt{7} + 2\sqrt{10}$ cannot be simplified further because $\sqrt{7}$ and $\sqrt{10}$ are unlike radicals.

3. (a) $\sqrt{8} + 4\sqrt{2} = \sqrt{4 \cdot 2} + 4\sqrt{2}$ Factor.
   $= \sqrt{4} \cdot \sqrt{2} + 4\sqrt{2}$ Product rule
   $= 2\sqrt{2} + 4\sqrt{2}$ $\sqrt{4} = 2$
   $= 6\sqrt{2}$ Add like radicals.

   (b) $\sqrt{27} + \sqrt{12} = \sqrt{9 \cdot 3} + \sqrt{4 \cdot 3}$
   $= \sqrt{9} \cdot \sqrt{3} + \sqrt{4} \cdot \sqrt{3}$
   $= 3\sqrt{3} + 2\sqrt{3}$
   $= 5\sqrt{3}$

   (c) $5\sqrt{200} - 6\sqrt{18}$
   $= 5\left(\sqrt{100} \cdot \sqrt{2}\right) - 6\left(\sqrt{9} \cdot \sqrt{2}\right)$
   $= 5\left(10\sqrt{2}\right) - 6\left(3\sqrt{2}\right)$
   $= 50\sqrt{2} - 18\sqrt{2}$
   $= 32\sqrt{2}$

4. (a) $\sqrt{7} \cdot \sqrt{21} + 2\sqrt{27}$
   $= \sqrt{7} \cdot \sqrt{7} \cdot \sqrt{3} + 2\left(\sqrt{9} \cdot \sqrt{3}\right)$
   $= \sqrt{49} \cdot \sqrt{3} + 2\left(3\sqrt{3}\right)$
   $= 7\sqrt{3} + 6\sqrt{3}$
   $= 13\sqrt{3}$

(b) $\sqrt{3r} \cdot \sqrt{6} + \sqrt{8r} = \sqrt{18r} + \sqrt{8r}$
$= \sqrt{9 \cdot 2r} + \sqrt{4 \cdot 2r}$
$= \sqrt{9} \cdot \sqrt{2r} + \sqrt{4} \cdot \sqrt{2r}$
$= 3\sqrt{2r} + 2\sqrt{2r}$
$= 5\sqrt{2r}$

(c) $y\sqrt{72} - \sqrt{18y^2}$
$= y \cdot \sqrt{36 \cdot 2} - \sqrt{9y^2 \cdot 2}$
$= y \cdot \sqrt{36} \cdot \sqrt{2} - \sqrt{9y^2} \cdot \sqrt{2}$
$= y \cdot 6\sqrt{2} - 3y\sqrt{2}$
$= (6y - 3y)\sqrt{2}$
$= 3y\sqrt{2}$

(d) $\sqrt[3]{81x^4} + 5\sqrt[3]{24x^4}$
$= \sqrt[3]{27x^3 \cdot 3x} + 5\sqrt[3]{8x^3 \cdot 3x}$
$= \sqrt[3]{27x^3} \cdot \sqrt[3]{3x} + 5\left(\sqrt[3]{8x^3} \cdot \sqrt[3]{3x}\right)$
$= 3x\sqrt[3]{3x} + 5\left(2x \cdot \sqrt[3]{3x}\right)$
$= 3x\sqrt[3]{3x} + 10x\sqrt[3]{3x}$
$= (3x + 10x)\sqrt[3]{3x}$
$= 13x\sqrt[3]{3x}$

8.3 Section Exercises

1. $5\sqrt{2} + 6\sqrt{2} = (5+6)\sqrt{2} = 11\sqrt{2}$ is an example of the __distributive__ property.

3. $\sqrt{5} + 5\sqrt{3}$ cannot be simplified because the __radicands__ are different.

5. $14\sqrt{7} - 19\sqrt{7} = (14 - 19)\sqrt{7}$
  Distributive property
$= -5\sqrt{7}$

7. $\sqrt{17} + 4\sqrt{17} = 1\sqrt{17} + 4\sqrt{17}$
$= (1 + 4)\sqrt{17}$
$= 5\sqrt{17}$

9. $6\sqrt{7} - \sqrt{7} = 6\sqrt{7} - 1\sqrt{7}$
$= (6-1)\sqrt{7} = 5\sqrt{7}$

11. $\sqrt{45} + 4\sqrt{20} = \sqrt{9 \cdot 5} + 4\sqrt{4 \cdot 5}$
$= \sqrt{9} \cdot \sqrt{5} + 4\left(\sqrt{4} \cdot \sqrt{5}\right)$
$= 3\sqrt{5} + 4\left(2\sqrt{5}\right)$
$= 3\sqrt{5} + 8\sqrt{5}$
$= 11\sqrt{5}$

13. $5\sqrt{72} - 3\sqrt{50} = 5\sqrt{36 \cdot 2} - 3\sqrt{25 \cdot 2}$
$= 5 \cdot \sqrt{36} \cdot \sqrt{2} - 3\sqrt{25} \cdot \sqrt{2}$
$= 5 \cdot 6 \cdot \sqrt{2} - 3 \cdot 5 \cdot \sqrt{2}$
$= 30\sqrt{2} - 15\sqrt{2}$
$= 15\sqrt{2}$

15. $-5\sqrt{32} + 2\sqrt{98}$
$= -5\left(\sqrt{16} \cdot \sqrt{2}\right) + 2\left(\sqrt{49} \cdot \sqrt{2}\right)$
$= -5\left(4\sqrt{2}\right) + 2\left(7\sqrt{2}\right)$
$= -20\sqrt{2} + 14\sqrt{2}$
$= -6\sqrt{2}$

17. $5\sqrt{7} - 3\sqrt{28} + 6\sqrt{63}$
$= 5\sqrt{7} - 3\left(\sqrt{4} \cdot \sqrt{7}\right) + 6\left(\sqrt{9} \cdot \sqrt{7}\right)$
$= 5\sqrt{7} - 3\left(2\sqrt{7}\right) + 6\left(3\sqrt{7}\right)$
$= 5\sqrt{7} - 6\sqrt{7} + 18\sqrt{7}$
$= (5 - 6 + 18)\sqrt{7}$
$= 17\sqrt{7}$

19. $2\sqrt{8} - 5\sqrt{32} - 2\sqrt{48}$
$= 2\left(\sqrt{4} \cdot \sqrt{2}\right) - 5\left(\sqrt{16} \cdot \sqrt{2}\right)$
$\quad - 2\left(\sqrt{16} \cdot \sqrt{3}\right)$
$= 2\left(2\sqrt{2}\right) - 5\left(4\sqrt{2}\right) - 2\left(4\sqrt{3}\right)$
$= 4\sqrt{2} - 20\sqrt{2} - 8\sqrt{3}$
$= -16\sqrt{2} - 8\sqrt{3}$

21. $4\sqrt{50} + 3\sqrt{12} - 5\sqrt{45}$
$= 4\left(\sqrt{25} \cdot \sqrt{2}\right) + 3\left(\sqrt{4} \cdot \sqrt{3}\right) - 5\left(\sqrt{9} \cdot \sqrt{5}\right)$
$= 4\left(5\sqrt{2}\right) + 3\left(2\sqrt{3}\right) - 5\left(3\sqrt{5}\right)$
$= 20\sqrt{2} + 6\sqrt{3} - 15\sqrt{5}$

23. $\frac{1}{4}\sqrt{288} + \frac{1}{6}\sqrt{72}$
$= \frac{1}{4}\left(\sqrt{144} \cdot \sqrt{2}\right) + \frac{1}{6}\left(\sqrt{36} \cdot \sqrt{2}\right)$
$= \frac{1}{4}\left(12\sqrt{2}\right) + \frac{1}{6}\left(6\sqrt{2}\right)$
$= 3\sqrt{2} + 1\sqrt{2}$
$= 4\sqrt{2}$

25. Use the formula for the perimeter of a rectangle.
$P = 2L + 2W$
$= 2\left(7\sqrt{2}\right) + 2\left(4\sqrt{2}\right)$
$= 14\sqrt{2} + 8\sqrt{2}$
$= 22\sqrt{2}$

27. $\sqrt{6} \cdot \sqrt{2} + 9\sqrt{3} = \sqrt{6 \cdot 2} + 9\sqrt{3}$
$= \sqrt{12} + 9\sqrt{3}$
$= \sqrt{4} \cdot \sqrt{3} + 9\sqrt{3}$
$= 2\sqrt{3} + 9\sqrt{3}$
$= 11\sqrt{3}$

29. $\sqrt{9x} + \sqrt{49x} - \sqrt{25x}$
$= \sqrt{9} \cdot \sqrt{x} + \sqrt{49} \cdot \sqrt{x} - \sqrt{25} \cdot \sqrt{x}$
$= 3\sqrt{x} + 7\sqrt{x} - 5\sqrt{x}$
$= (3 + 7 - 5)\sqrt{x}$
$= 5\sqrt{x}$

31. $\sqrt{6x^2} + x\sqrt{24} = \sqrt{x^2 \cdot 6} + x\sqrt{4 \cdot 6}$
$= \sqrt{x^2} \cdot \sqrt{6} + x \cdot \sqrt{4} \cdot \sqrt{6}$
$= x\sqrt{6} + x \cdot 2\sqrt{6}$
$= (x + 2x)\sqrt{6}$
$= 3x\sqrt{6}$

33. $3\sqrt{8x^2} - 4x\sqrt{2} - x\sqrt{8}$
$= 3\sqrt{4x^2 \cdot 2} - 4x\sqrt{2} - x\sqrt{4 \cdot 2}$
$= 3 \cdot \sqrt{4x^2} \cdot \sqrt{2} - 4x\sqrt{2} - x \cdot \sqrt{4} \cdot \sqrt{2}$
$= 3 \cdot 2x \cdot \sqrt{2} - 4x\sqrt{2} - x \cdot 2 \cdot \sqrt{2}$
$= 6x\sqrt{2} - 4x\sqrt{2} - 2x\sqrt{2}$
$= (6x - 4x - 2x)\sqrt{2}$
$= 0 \cdot \sqrt{2}$
$= 0$

35. $-8\sqrt{32k} + 6\sqrt{8k}$
$= -8\left(\sqrt{16 \cdot 2k}\right) + 6\left(\sqrt{4 \cdot 2k}\right)$
$= -8\left(4\sqrt{2k}\right) + 6\left(2\sqrt{2k}\right)$
$= -32\sqrt{2k} + 12\sqrt{2k}$
$= (-32 + 12)\sqrt{2k}$
$= -20\sqrt{2k}$

37. $2\sqrt{125x^2z} + 8x\sqrt{80z}$
$= 2\sqrt{25x^2 \cdot 5z} + 8x\sqrt{16 \cdot 5z}$
$= 2\sqrt{25x^2} \cdot \sqrt{5z} + 8x\sqrt{16} \cdot \sqrt{5z}$
$= 2 \cdot 5x \cdot \sqrt{5z} + 8x \cdot 4 \cdot \sqrt{5z}$
$= 10x\sqrt{5z} + 32x\sqrt{5z}$
$= (10x + 32x)\sqrt{5z}$
$= 42x\sqrt{5z}$

39. $4\sqrt[3]{16} - 3\sqrt[3]{54}$

Recall that 8 and 27 are perfect cubes.
$4\sqrt[3]{16} - 3\sqrt[3]{54}$
$= 4\left(\sqrt[3]{8 \cdot 2}\right) - 3\left(\sqrt[3]{27 \cdot 2}\right)$
$= 4\left(\sqrt[3]{8} \cdot \sqrt[3]{2}\right) - 3\left(\sqrt[3]{27} \cdot \sqrt[3]{2}\right)$
$= 4\left(2\sqrt[3]{2}\right) - 3\left(3\sqrt[3]{2}\right)$
$= 8\sqrt[3]{2} - 9\sqrt[3]{2}$
$= (8 - 9)\sqrt[3]{2} = -1\sqrt[3]{2} = -\sqrt[3]{2}$

41. $6\sqrt[3]{8p^2} - 2\sqrt[3]{27p^2}$
$= 6 \cdot \sqrt[3]{8} \cdot \sqrt[3]{p^2} - 2 \cdot \sqrt[3]{27} \cdot \sqrt[3]{p^2}$
$= 6 \cdot 2 \cdot \sqrt[3]{p^2} - 2 \cdot 3 \cdot \sqrt[3]{p^2}$
$= 12\sqrt[3]{p^2} - 6\sqrt[3]{p^2}$
$= 6\sqrt[3]{p^2}$

43. $5\sqrt[4]{m^3} + 8\sqrt[4]{16m^3}$
$= 5\sqrt[4]{m^3} + 8\sqrt[4]{16}\sqrt[4]{m^3}$
$= 5\sqrt[4]{m^3} + 8 \cdot 2 \cdot \sqrt[4]{m^3}$
$= 5\sqrt[4]{m^3} + 16\sqrt[4]{m^3}$
$= 21\sqrt[4]{m^3}$

45. $5x^2y + 3x^2y - 14x^2y$
$= (5 + 3 - 14)x^2y$
$= -6x^2y$

46. $5(p - 2q)^2(a + b) + 3(p - 2q)^2(a + b)$
$\quad - 14(p - 2q)^2(a + b)$
$= (5 + 3 - 14)(p - 2q)^2(a + b)$
$= -6(p - 2q)^2(a + b)$

47. $5a^2\sqrt{xy} + 3a^2\sqrt{xy} - 14a^2\sqrt{xy}$
$= (5 + 3 - 14)a^2\sqrt{xy}$
$= -6a^2\sqrt{xy}$

48. In Exercises 45–47, the problems are alike since each is of the form
$5A + 3A - 14A = (5 + 3 - 14)A = -6A$.
They are different in that A stands for x^2y in Exercise 45, $(p - 2q)^2 \cdot (a + b)$ in Exercise 46, and $a^2\sqrt{xy}$ in Exercise 47. Also, the first variable factor is raised to the second power, and the second variable factor is raised to the first power. The answers are different because the variables are different: x and y, then $p - 2q$ and $a + b$, and then a and $\sqrt{xy}$.

8.4 Rationalizing the Denominator

8.4 Margin Exercises

1. (a) $\dfrac{3}{\sqrt{5}}$

   We will eliminate the radical in the denominator by multiplying the numerator and the denominator by $\sqrt{5}$.

   $$\dfrac{3}{\sqrt{5}} = \dfrac{3 \cdot \sqrt{5}}{\sqrt{5} \cdot \sqrt{5}} = \dfrac{3\sqrt{5}}{5}$$

   (b) $\dfrac{-6}{\sqrt{11}} = \dfrac{-6 \cdot \sqrt{11}}{\sqrt{11} \cdot \sqrt{11}} = \dfrac{-6\sqrt{11}}{11}$

   (c) $-\dfrac{\sqrt{7}}{\sqrt{2}} = -\dfrac{\sqrt{7} \cdot \sqrt{2}}{\sqrt{2} \cdot \sqrt{2}} = -\dfrac{\sqrt{14}}{2}$

   (d) $\dfrac{20}{\sqrt{18}}$

   Since $\sqrt{18} = \sqrt{9 \cdot 2}$, we will multiply the numerator and denominator by $\sqrt{2}$ to obtain a perfect square in the radicand in the denominator.

   $$\dfrac{20}{\sqrt{18}} = \dfrac{20 \cdot \sqrt{2}}{\sqrt{18} \cdot \sqrt{2}}$$
   $$= \dfrac{20\sqrt{2}}{\sqrt{36}}$$
   $$= \dfrac{20\sqrt{2}}{6}$$
   $$= \dfrac{10\sqrt{2}}{3}$$

2. (a) $\sqrt{\dfrac{16}{11}} = \dfrac{\sqrt{16}}{\sqrt{11}}$
   $$= \dfrac{\sqrt{16} \cdot \sqrt{11}}{\sqrt{11} \cdot \sqrt{11}}$$
   $$= \dfrac{4\sqrt{11}}{11}$$

   (b) $\sqrt{\dfrac{5}{18}} = \dfrac{\sqrt{5}}{\sqrt{18}}$
   $$= \dfrac{\sqrt{5} \cdot \sqrt{2}}{\sqrt{18} \cdot \sqrt{2}}$$
   $$= \dfrac{\sqrt{10}}{\sqrt{36}}$$
   $$= \dfrac{\sqrt{10}}{6}$$

   (c) $\sqrt{\dfrac{8}{32}}$

   When rationalizing the denominator, there are often several ways to approach the problem. Three ways to simplify this radical are shown here.

   $$\sqrt{\dfrac{8}{32}} = \dfrac{\sqrt{8}}{\sqrt{32}} = \dfrac{\sqrt{8} \cdot \sqrt{2}}{\sqrt{32} \cdot \sqrt{2}}$$
   $$= \dfrac{\sqrt{16}}{\sqrt{64}} = \dfrac{4}{8} = \dfrac{1}{2}$$

   OR $\sqrt{\dfrac{8}{32}} = \dfrac{\sqrt{8}}{\sqrt{32}} = \dfrac{\sqrt{4} \cdot \sqrt{2}}{\sqrt{16} \cdot \sqrt{2}}$
   $$= \dfrac{2\sqrt{2}}{4\sqrt{2}} = \dfrac{2}{4} = \dfrac{1}{2}$$

   OR $\sqrt{\dfrac{8}{32}} = \sqrt{\dfrac{1}{4}} = \dfrac{1}{2}$

3. (a) $\sqrt{\dfrac{1}{2}} \cdot \sqrt{\dfrac{5}{6}} = \sqrt{\dfrac{1}{2} \cdot \dfrac{5}{6}} = \sqrt{\dfrac{5}{12}}$
   $$= \dfrac{\sqrt{5}}{\sqrt{12}} = \dfrac{\sqrt{5} \cdot \sqrt{3}}{\sqrt{12} \cdot \sqrt{3}}$$
   $$= \dfrac{\sqrt{15}}{\sqrt{36}} = \dfrac{\sqrt{15}}{6}$$

   (b) $\sqrt{\dfrac{1}{10}} \cdot \sqrt{20} = \sqrt{\dfrac{1}{10} \cdot \dfrac{20}{1}} = \sqrt{2}$

   (c) $\sqrt{\dfrac{5}{8}} \cdot \sqrt{\dfrac{24}{10}} = \sqrt{\dfrac{5}{8} \cdot \dfrac{24}{10}}$
   $$= \sqrt{\dfrac{3}{2}} = \dfrac{\sqrt{3} \cdot \sqrt{2}}{\sqrt{2} \cdot \sqrt{2}}$$
   $$= \dfrac{\sqrt{6}}{\sqrt{4}} = \dfrac{\sqrt{6}}{2}$$

4. $\dfrac{\sqrt{5p}}{\sqrt{q}} = \dfrac{\sqrt{5p} \cdot \sqrt{q}}{\sqrt{q} \cdot \sqrt{q}} = \dfrac{\sqrt{5pq}}{q}$

5. $\sqrt{\dfrac{5r^2t^2}{7}} = \dfrac{\sqrt{5r^2t^2}}{\sqrt{7}}$
   $$= \dfrac{\sqrt{r^2t^2}\sqrt{5}}{\sqrt{7}}$$
   $$= \dfrac{rt\sqrt{5} \cdot \sqrt{7}}{\sqrt{7} \cdot \sqrt{7}}$$
   $$= \dfrac{rt\sqrt{35}}{7}$$

320 Chapter 8 Roots and Radicals

6. (a) $\sqrt[3]{\dfrac{5}{7}} = \dfrac{\sqrt[3]{5}}{\sqrt[3]{7}} = \dfrac{\sqrt[3]{5} \cdot \sqrt[3]{7^2}}{\sqrt[3]{7} \cdot \sqrt[3]{7^2}}$

$= \dfrac{\sqrt[3]{5 \cdot 7^2}}{\sqrt[3]{7^3}} = \dfrac{\sqrt[3]{245}}{7}$

(b) $\dfrac{\sqrt[3]{5}}{\sqrt[3]{9}} = \dfrac{\sqrt[3]{5} \cdot \sqrt[3]{3}}{\sqrt[3]{9} \cdot \sqrt[3]{3}} = \dfrac{\sqrt[3]{15}}{\sqrt[3]{27}} = \dfrac{\sqrt[3]{15}}{3}$

(c) $\dfrac{\sqrt[3]{4}}{\sqrt[3]{25y}} = \dfrac{\sqrt[3]{4} \cdot \sqrt[3]{5y^2}}{\sqrt[3]{25y} \cdot \sqrt[3]{5y^2}} = \dfrac{\sqrt[3]{20y^2}}{5y}$

8.4 Section Exercises

1. $\dfrac{8}{\sqrt{2}} = \dfrac{8 \cdot \sqrt{2}}{\sqrt{2} \cdot \sqrt{2}} = \dfrac{8\sqrt{2}}{2} = 4\sqrt{2}$

3. $\dfrac{-\sqrt{11}}{\sqrt{3}} = \dfrac{-\sqrt{11} \cdot \sqrt{3}}{\sqrt{3} \cdot \sqrt{3}} = \dfrac{-\sqrt{33}}{3}$

5. $\dfrac{7\sqrt{3}}{\sqrt{5}} = \dfrac{7\sqrt{3} \cdot \sqrt{5}}{\sqrt{5} \cdot \sqrt{5}} = \dfrac{7\sqrt{15}}{5}$

7. $\dfrac{24\sqrt{10}}{16\sqrt{3}} = \dfrac{3\sqrt{10}}{2\sqrt{3}}$

$= \dfrac{3\sqrt{10} \cdot \sqrt{3}}{2\sqrt{3} \cdot \sqrt{3}}$

$= \dfrac{3\sqrt{30}}{2 \cdot 3} = \dfrac{\sqrt{30}}{2}$

9. $\dfrac{16}{\sqrt{27}} = \dfrac{16}{\sqrt{9 \cdot 3}} = \dfrac{16}{\sqrt{9} \cdot \sqrt{3}}$

$= \dfrac{16}{3\sqrt{3}} = \dfrac{16 \cdot \sqrt{3}}{3\sqrt{3} \cdot \sqrt{3}} = \dfrac{16\sqrt{3}}{9}$

11. $\dfrac{-3}{\sqrt{50}} = \dfrac{-3}{\sqrt{25 \cdot 2}}$

$= \dfrac{-3}{5\sqrt{2}}$

$= \dfrac{-3 \cdot \sqrt{2}}{5\sqrt{2} \cdot \sqrt{2}}$

$= \dfrac{-3\sqrt{2}}{5 \cdot 2} = \dfrac{-3\sqrt{2}}{10}$

13. $\dfrac{63}{\sqrt{45}} = \dfrac{63}{\sqrt{9} \cdot \sqrt{5}} = \dfrac{63}{3\sqrt{5}} = \dfrac{21}{\sqrt{5}}$

$= \dfrac{21 \cdot \sqrt{5}}{\sqrt{5} \cdot \sqrt{5}} = \dfrac{21\sqrt{5}}{5}$

15. $\dfrac{\sqrt{24}}{\sqrt{8}} = \sqrt{\dfrac{24}{8}} = \sqrt{3}$

17. $\sqrt{\dfrac{1}{2}} = \dfrac{\sqrt{1}}{\sqrt{2}} = \dfrac{1}{\sqrt{2}} = \dfrac{1 \cdot \sqrt{2}}{\sqrt{2} \cdot \sqrt{2}} = \dfrac{\sqrt{2}}{2}$

19. $\sqrt{\dfrac{13}{5}} = \dfrac{\sqrt{13}}{\sqrt{5}} = \dfrac{\sqrt{13} \cdot \sqrt{5}}{\sqrt{5} \cdot \sqrt{5}} = \dfrac{\sqrt{65}}{5}$

21. The given expression is being multiplied by $\dfrac{\sqrt{3}}{\sqrt{3}}$, which is 1. According to the identity property for multiplication, multiplying an expression by 1 does not change the value of the expression.

23. $\sqrt{\dfrac{7}{13}} \cdot \sqrt{\dfrac{13}{3}} = \sqrt{\dfrac{7}{13} \cdot \dfrac{13}{3}}$ Product rule

$= \sqrt{\dfrac{7}{3}} = \dfrac{\sqrt{7}}{\sqrt{3}}$

$= \dfrac{\sqrt{7} \cdot \sqrt{3}}{\sqrt{3} \cdot \sqrt{3}} = \dfrac{\sqrt{21}}{3}$

25. $\sqrt{\dfrac{21}{7}} \cdot \sqrt{\dfrac{21}{8}} = \dfrac{\sqrt{21}}{\sqrt{7}} \cdot \dfrac{\sqrt{21}}{\sqrt{8}} = \dfrac{21}{\sqrt{7 \cdot 2 \cdot 4}}$

$= \dfrac{21}{2\sqrt{14}} = \dfrac{21 \cdot \sqrt{14}}{2 \cdot \sqrt{14} \cdot \sqrt{14}}$

$= \dfrac{21\sqrt{14}}{2 \cdot 14} = \dfrac{3\sqrt{14}}{4}$

27. $\sqrt{\dfrac{1}{12}} \cdot \sqrt{\dfrac{1}{3}} = \sqrt{\dfrac{1}{12} \cdot \dfrac{1}{3}}$

$= \sqrt{\dfrac{1}{36}} = \dfrac{\sqrt{1}}{\sqrt{36}} = \dfrac{1}{6}$

29. $\sqrt{\dfrac{2}{9}} \cdot \sqrt{\dfrac{9}{2}} = \sqrt{\dfrac{2}{9} \cdot \dfrac{9}{2}} = \sqrt{1} = 1$

31. $\dfrac{\sqrt{7}}{\sqrt{x}} = \dfrac{\sqrt{7} \cdot \sqrt{x}}{\sqrt{x} \cdot \sqrt{x}} = \dfrac{\sqrt{7x}}{x}$

33. $\dfrac{\sqrt{4x^3}}{\sqrt{y}} = \dfrac{\sqrt{4x^2} \cdot \sqrt{x}}{\sqrt{y}} = \dfrac{2x\sqrt{x}}{\sqrt{y}}$

$= \dfrac{2x\sqrt{x} \cdot \sqrt{y}}{\sqrt{y} \cdot \sqrt{y}} = \dfrac{2x\sqrt{xy}}{y}$

35. $\sqrt{\dfrac{5x^3z}{6}} = \dfrac{\sqrt{5x^3z}}{\sqrt{6}} = \dfrac{\sqrt{x^2} \cdot \sqrt{5xz}}{\sqrt{6}} = \dfrac{x\sqrt{5xz}}{\sqrt{6}}$

$= \dfrac{x\sqrt{5xz} \cdot \sqrt{6}}{\sqrt{6} \cdot \sqrt{6}} = \dfrac{x\sqrt{30xz}}{6}$

37. $\sqrt{\dfrac{9a^2r^5}{7t}} = \dfrac{\sqrt{9a^2r^5}}{\sqrt{7t}} = \dfrac{\sqrt{9a^2r^4 \cdot r}}{\sqrt{7t}}$

$= \dfrac{\sqrt{9a^2r^4} \cdot \sqrt{r}}{\sqrt{7t}} = \dfrac{3ar^2\sqrt{r}}{\sqrt{7t}}$

$= \dfrac{3ar^2\sqrt{r} \cdot \sqrt{7t}}{\sqrt{7t} \cdot \sqrt{7t}} = \dfrac{3ar^2\sqrt{7rt}}{7t}$

39. We need to multiply the numerator and denominator of $\dfrac{\sqrt[3]{2}}{\sqrt[3]{5}}$ by enough factors of 5 to make the radicand in the denominator a perfect cube. In this case we have one factor of 5, so we need to multiply by two more factors of 5 to make three factors of 5. Thus, the correct choice for a rationalizing factor in this problem is $\sqrt[3]{5^2} = \sqrt[3]{25}$, which corresponds to choice **B**.

41. $\sqrt[3]{\dfrac{3}{2}}$

Multiply the numerator and the denominator by enough factors of 2 to make the radicand in the denominator a perfect cube. This will eliminate the radical in the denominator. Here, we multiply by $\sqrt[3]{2^2}$ or $\sqrt[3]{4}$.

$\sqrt[3]{\dfrac{3}{2}} = \dfrac{\sqrt[3]{3}}{\sqrt[3]{2}} = \dfrac{\sqrt[3]{3} \cdot \sqrt[3]{2^2}}{\sqrt[3]{2} \cdot \sqrt[3]{2^2}}$

$= \dfrac{\sqrt[3]{3 \cdot 2^2}}{\sqrt[3]{2 \cdot 2^2}} = \dfrac{\sqrt[3]{12}}{\sqrt[3]{2^3}} = \dfrac{\sqrt[3]{12}}{2}$

43. $\dfrac{\sqrt[3]{4}}{\sqrt[3]{7}} = \dfrac{\sqrt[3]{4} \cdot \sqrt[3]{7^2}}{\sqrt[3]{7} \cdot \sqrt[3]{7^2}} = \dfrac{\sqrt[3]{4} \cdot \sqrt[3]{49}}{\sqrt[3]{7^3}} = \dfrac{\sqrt[3]{196}}{7}$

45. To make the radicand in the denominator, $4y^2$, into a perfect cube, we must multiply 4 by 2 to get the perfect cube 8 and y^2 by y to get the perfect cube y^3. So we multiply the numerator and denominator by $\sqrt[3]{2y}$.

$\sqrt[3]{\dfrac{3}{4y^2}} = \dfrac{\sqrt[3]{3}}{\sqrt[3]{4y^2}} = \dfrac{\sqrt[3]{3} \cdot \sqrt[3]{2y}}{\sqrt[3]{4y^2} \cdot \sqrt[3]{2y}}$

$= \dfrac{\sqrt[3]{6y}}{\sqrt[3]{8y^3}} = \dfrac{\sqrt[3]{6y}}{2y}$

47. $\dfrac{\sqrt[3]{7m}}{\sqrt[3]{36n}} = \dfrac{\sqrt[3]{7m}}{\sqrt[3]{6^2 n}} = \dfrac{\sqrt[3]{7m} \cdot \sqrt[3]{6n^2}}{\sqrt[3]{6^2 n} \cdot \sqrt[3]{6n^2}}$

$= \dfrac{\sqrt[3]{42mn^2}}{\sqrt[3]{6^3 n^3}} = \dfrac{\sqrt[3]{42mn^2}}{6n}$

49. (a) $p = k \cdot \sqrt{\dfrac{L}{g}}$

$p = 6 \cdot \sqrt{\dfrac{9}{32}}$ Let $k = 6$, $L = 9$, $g = 32$.

$= \dfrac{6\sqrt{9}}{\sqrt{32}} = \dfrac{6 \cdot 3}{\sqrt{16 \cdot 2}}$

$= \dfrac{18}{4\sqrt{2}} = \dfrac{9}{2\sqrt{2}}$

$= \dfrac{9 \cdot \sqrt{2}}{2\sqrt{2} \cdot \sqrt{2}}$ *Rationalize the denominator.*

$= \dfrac{9\sqrt{2}}{4}$

The period of the pendulum is $\dfrac{9\sqrt{2}}{4}$ seconds.

(b) Using a calculator, we obtain

$\dfrac{9\sqrt{2}}{4} \approx 3.182$ seconds.

8.5 More Simplifying and Operations with Radicals

8.5 Margin Exercises

1. (a) $\sqrt{7}\left(\sqrt{2} + \sqrt{5}\right) = \sqrt{7}\left(\sqrt{2}\right) + \sqrt{7}\left(\sqrt{5}\right)$

   $= \sqrt{14} + \sqrt{35}$

   (b) $\sqrt{2}\left(\sqrt{8} + \sqrt{20}\right)$

   $= \sqrt{2}\left(\sqrt{4} \cdot \sqrt{2} + \sqrt{4} \cdot \sqrt{5}\right)$

   $= \sqrt{2}\left(2\sqrt{2} + 2\sqrt{5}\right)$

   $= \sqrt{2} \cdot 2\sqrt{2} + \sqrt{2} \cdot 2\sqrt{5}$

   $= 2 \cdot 2 + 2\sqrt{10}$

   $= 4 + 2\sqrt{10}$

   (c) $\left(\sqrt{2} + 5\sqrt{3}\right)\left(\sqrt{3} - 2\sqrt{2}\right)$

   $= \sqrt{2} \cdot \sqrt{3} - \sqrt{2} \cdot 2\sqrt{2}$
   $\quad + 5\sqrt{3} \cdot \sqrt{3} - 5\sqrt{3} \cdot 2\sqrt{2}$ *FOIL*

   $= \sqrt{6} - 2 \cdot 2 + 5 \cdot 3 - 10\sqrt{6}$

   $= 1\sqrt{6} - 4 + 15 - 10\sqrt{6}$

   $= 11 - 9\sqrt{6}$

(d) $\left(\sqrt{2}-\sqrt{5}\right)\left(\sqrt{10}+\sqrt{2}\right)$
$= \sqrt{2}\cdot\sqrt{10}+\sqrt{2}\cdot\sqrt{2}$
$\quad -\sqrt{5}\cdot\sqrt{10}-\sqrt{5}\cdot\sqrt{2}$
$= \sqrt{20}+2-\sqrt{50}-\sqrt{10}$
$= \sqrt{4\cdot 5}+2-\sqrt{25\cdot 2}-\sqrt{10}$
$= 2\sqrt{5}+2-5\sqrt{2}-\sqrt{10}$

2. (a) Use the special product formula,
$$(a-b)^2 = a^2 - 2ab + b^2.$$
$\left(\sqrt{5}-3\right)^2 = \left(\sqrt{5}\right)^2 - 2\left(\sqrt{5}\right)(3) + 3^2$
$= 5 - 6\sqrt{5} + 9$
$= 14 - 6\sqrt{5}$

(b) Use the special product formula,
$$(a+b)^2 = a^2 + 2ab + b^2.$$
$\left(4\sqrt{2}+5\right)^2 = \left(4\sqrt{2}\right)^2 + 2\left(4\sqrt{2}\right)(5) + 5^2$
$= 16\cdot 2 + 40\sqrt{2} + 25$
$= 32 + 40\sqrt{2} + 25$
$= 57 + 40\sqrt{2}$

(c) $\left(6+\sqrt{m}\right)^2 = 6^2 + 2(6)\left(\sqrt{m}\right) + \left(\sqrt{m}\right)^2$
$= 36 + 12\sqrt{m} + m$

3. (a) Use the rule for the product of the sum and the difference of two terms,
$$(a+b)(a-b) = a^2 - b^2.$$
$\left(3+\sqrt{5}\right)\left(3-\sqrt{5}\right) = 3^2 - \left(\sqrt{5}\right)^2$
$= 9 - 5$
$= 4$

(b) $\left(\sqrt{3}-2\right)\left(\sqrt{3}+2\right) = \left(\sqrt{3}\right)^2 - 2^2$
$= 3 - 4$
$= -1$

(c) $\left(\sqrt{5}+\sqrt{3}\right)\left(\sqrt{5}-\sqrt{3}\right)$
$= \left(\sqrt{5}\right)^2 - \left(\sqrt{3}\right)^2$
$= 5 - 3$
$= 2$

(d) $\left(\sqrt{10}-\sqrt{y}\right)\left(\sqrt{10}+\sqrt{y}\right)$
$= \left(\sqrt{10}\right)^2 - \left(\sqrt{y}\right)^2$
$= 10 - y$

4. (a) $\dfrac{5}{4+\sqrt{2}}$

We can eliminate the radical in the denominator by multiplying both the numerator and denominator by $4-\sqrt{2}$, the conjugate of the denominator.

$\dfrac{5}{4+\sqrt{2}}$

$= \dfrac{5\left(4-\sqrt{2}\right)}{\left(4+\sqrt{2}\right)\left(4-\sqrt{2}\right)}$

$= \dfrac{5\left(4-\sqrt{2}\right)}{4^2 - \left(\sqrt{2}\right)^2} \qquad \begin{array}{l}(a+b)(a-b)\\ = a^2 - b^2\end{array}$

$= \dfrac{5\left(4-\sqrt{2}\right)}{16 - 2}$

$= \dfrac{5\left(4-\sqrt{2}\right)}{14}$

(b) $\dfrac{\sqrt{5}+3}{2-\sqrt{5}} = \dfrac{\left(\sqrt{5}+3\right)\left(2+\sqrt{5}\right)}{\left(2-\sqrt{5}\right)\left(2+\sqrt{5}\right)}$

$= \dfrac{2\sqrt{5}+5+6+3\sqrt{5}}{2^2 - \left(\sqrt{5}\right)^2}$

$= \dfrac{11 + 5\sqrt{5}}{4 - 5}$

$= \dfrac{11 + 5\sqrt{5}}{-1}$

$= -11 - 5\sqrt{5}$

(c) $\dfrac{1}{\sqrt{6}+\sqrt{3}} = \dfrac{1\left(\sqrt{6}-\sqrt{3}\right)}{\left(\sqrt{6}+\sqrt{3}\right)\left(\sqrt{6}-\sqrt{3}\right)}$

$= \dfrac{\sqrt{6}-\sqrt{3}}{\left(\sqrt{6}\right)^2 - \left(\sqrt{3}\right)^2}$

$= \dfrac{\sqrt{6}-\sqrt{3}}{6 - 3}$

$= \dfrac{\sqrt{6}-\sqrt{3}}{3}$

(d) $\dfrac{7}{5-\sqrt{x}} = \dfrac{7\left(5+\sqrt{x}\right)}{\left(5-\sqrt{x}\right)\left(5+\sqrt{x}\right)}$

$= \dfrac{7\left(5+\sqrt{x}\right)}{25 - x}$

5. (a) $\dfrac{5\sqrt{3}-15}{10} = \dfrac{5(\sqrt{3}-3)}{5(2)}$

$= \dfrac{\sqrt{3}-3}{2}$

(b) $\dfrac{12+8\sqrt{5}}{16} = \dfrac{4(3+2\sqrt{5})}{4(4)}$

$= \dfrac{3+2\sqrt{5}}{4}$

8.5 Section Exercises

1. $\sqrt{49} + \sqrt{36} = 13$
$\left(\sqrt{49} + \sqrt{36} = 7 + 6\right)$

3. $\sqrt{2} \cdot \sqrt{8} = 4$
$\left(\sqrt{2} \cdot \sqrt{8} = \sqrt{16}\right)$

5. $\sqrt{5}\left(\sqrt{3} - \sqrt{7}\right) = \sqrt{5} \cdot \sqrt{3} - \sqrt{5} \cdot \sqrt{7}$
$= \sqrt{15} - \sqrt{35}$

7. $2\sqrt{5}\left(\sqrt{2} + 3\sqrt{5}\right)$
$= 2\sqrt{5} \cdot \sqrt{2} + 2\sqrt{5} \cdot 3\sqrt{5}$
$= 2\sqrt{10} + 2 \cdot 3 \cdot \sqrt{5} \cdot \sqrt{5}$
$= 2\sqrt{10} + 6 \cdot 5$
$= 2\sqrt{10} + 30$

9. $3\sqrt{14} \cdot \sqrt{2} - \sqrt{28} = 3\sqrt{14 \cdot 2} - \sqrt{28}$
$= 3\sqrt{28} - 1\sqrt{28}$
$= 2\sqrt{28}$
$= 2\sqrt{4 \cdot 7}$
$= 2 \cdot \sqrt{4} \cdot \sqrt{7}$
$= 2 \cdot 2 \cdot \sqrt{7}$
$= 4\sqrt{7}$

11. $\left(2\sqrt{6} + 3\right)\left(3\sqrt{6} + 7\right)$
$= 2\sqrt{6} \cdot 3\sqrt{6} + 7 \cdot 2\sqrt{6} + 3 \cdot 3\sqrt{6}$
$\quad + 3 \cdot 7 \quad$ FOIL
$= 2 \cdot 3 \cdot \sqrt{6} \cdot \sqrt{6} + 14\sqrt{6} + 9\sqrt{6} + 21$
$= 6 \cdot 6 + 23\sqrt{6} + 21$
$= 36 + 23\sqrt{6} + 21$
$= 57 + 23\sqrt{6}$

13. $\left(5\sqrt{7} - 2\sqrt{3}\right)\left(3\sqrt{7} + 4\sqrt{3}\right)$
$= 5\sqrt{7}\left(3\sqrt{7}\right) + 5\sqrt{7}\left(4\sqrt{3}\right)$
$\quad - 2\sqrt{3}\left(3\sqrt{7}\right) - 2\sqrt{3}\left(4\sqrt{3}\right) \quad$ FOIL
$= 15 \cdot 7 + 20\sqrt{21} - 6\sqrt{21} - 8 \cdot 3$
$= 105 + 14\sqrt{21} - 24$
$= 81 + 14\sqrt{21}$

15. $\left(8 - \sqrt{7}\right)^2$
$= (8)^2 - 2(8)\left(\sqrt{7}\right) + \left(\sqrt{7}\right)^2$
$\quad$ Square of a binomial
$= 64 - 16\sqrt{7} + 7$
$= 71 - 16\sqrt{7}$

17. $\left(2\sqrt{7} + 3\right)^2$
$= \left(2\sqrt{7}\right)^2 + 2\left(2\sqrt{7}\right)(3) + (3)^2$
$\quad$ Square of a binomial
$= 4 \cdot 7 + 12\sqrt{7} + 9$
$= 28 + 12\sqrt{7} + 9$
$= 37 + 12\sqrt{7}$

19. $\left(\sqrt{a} + 1\right)^2 = \left(\sqrt{a}\right)^2 + 2\left(\sqrt{a}\right)(1) + (1)^2$
$\quad$ Square of a binomial
$= a + 2\sqrt{a} + 1$

21. $\left(5 - \sqrt{2}\right)\left(5 + \sqrt{2}\right) = (5)^2 - \left(\sqrt{2}\right)^2$
$\quad$ Product of the sum and
$\quad$ difference of two terms
$= 25 - 2 = 23$

23. $\left(\sqrt{8} - \sqrt{7}\right)\left(\sqrt{8} + \sqrt{7}\right)$
$= \left(\sqrt{8}\right)^2 - \left(\sqrt{7}\right)^2$
$\quad$ Product of the sum and
$\quad$ difference of two terms
$= 8 - 7 = 1$

25. $\left(\sqrt{y} - \sqrt{10}\right)\left(\sqrt{y} + \sqrt{10}\right)$
$= \left(\sqrt{y}\right)^2 - \left(\sqrt{10}\right)^2$
$= y - 10$

27. $(\sqrt{2}+\sqrt{3})(\sqrt{6}-\sqrt{2})$
$= \sqrt{2}(\sqrt{6}) - \sqrt{2}(\sqrt{2}) + \sqrt{3}(\sqrt{6})$
$ - \sqrt{3}(\sqrt{2})$ FOIL
$= \sqrt{12} - 2 + \sqrt{18} - \sqrt{6}$ Product rule
$= \sqrt{4}\cdot\sqrt{3} - 2 + \sqrt{9}\cdot\sqrt{2} - \sqrt{6}$
$= 2\sqrt{3} - 2 + 3\sqrt{2} - \sqrt{6}$

29. $(\sqrt{10}-\sqrt{5})(\sqrt{5}+\sqrt{20})$
$= \sqrt{10}\cdot\sqrt{5} + \sqrt{10}\cdot\sqrt{20} - \sqrt{5}\cdot\sqrt{5}$
$ - \sqrt{5}\cdot\sqrt{20}$ FOIL
$= \sqrt{50} + \sqrt{200} - 5 - \sqrt{100}$
$= \sqrt{25\cdot2} + \sqrt{100\cdot2} - 5 - 10$
$= 5\sqrt{2} + 10\sqrt{2} - 15$
$= 15\sqrt{2} - 15$

31. $(\sqrt{5}+\sqrt{30})(\sqrt{6}+\sqrt{3})$
$= \sqrt{5}\cdot\sqrt{6} + \sqrt{5}\cdot\sqrt{3} + \sqrt{30}\cdot\sqrt{6}$
$ + \sqrt{30}\cdot\sqrt{3}$ FOIL
$= \sqrt{30} + \sqrt{15} + \sqrt{180} + \sqrt{90}$
$= \sqrt{30} + \sqrt{15} + \sqrt{36\cdot5} + \sqrt{9\cdot10}$
$= \sqrt{30} + \sqrt{15} + 6\sqrt{5} + 3\sqrt{10}$

33. $(\sqrt{5}-\sqrt{10})(\sqrt{x}-\sqrt{2})$
$= \sqrt{5}(\sqrt{x}) + \sqrt{5}(-\sqrt{2})$
$ - \sqrt{10}(\sqrt{x}) - \sqrt{10}(-\sqrt{2})$
$= \sqrt{5x} - \sqrt{10} - \sqrt{10x} + \sqrt{20}$
$= \sqrt{5x} - \sqrt{10} - \sqrt{10x} + 2\sqrt{5}$

35. Because multiplication must be performed before addition, it is incorrect to add -37 and -2. Since $-2\sqrt{15}$ cannot be simplified, the expression cannot be written in a simpler form, and the final answer is $-37 - 2\sqrt{15}$.

37. $\dfrac{1}{3+\sqrt{2}} = \dfrac{1(3-\sqrt{2})}{(3+\sqrt{2})(3-\sqrt{2})}$

Multiply numerator and denominator by the conjugate of the denominator.

$= \dfrac{3-\sqrt{2}}{3^2 - (\sqrt{2})^2}$

$= \dfrac{3-\sqrt{2}}{9-2}$

$= \dfrac{3-\sqrt{2}}{7}$

39. $\dfrac{14}{2-\sqrt{11}} = \dfrac{14(2+\sqrt{11})}{(2-\sqrt{11})(2+\sqrt{11})}$

Multiply numerator and denominator by the conjugate of the denominator.

$= \dfrac{14(2+\sqrt{11})}{(2)^2 - (\sqrt{11})^2}$

$(a+b)(a-b) = a^2 - b^2$

$= \dfrac{14(2+\sqrt{11})}{4-11}$

$= \dfrac{14(2+\sqrt{11})}{-7}$

$= -2(2+\sqrt{11})$

$= -4 - 2\sqrt{11}$

41. $\dfrac{\sqrt{2}}{2-\sqrt{2}} = \dfrac{\sqrt{2}(2+\sqrt{2})}{(2-\sqrt{2})(2+\sqrt{2})}$

$= \dfrac{2\sqrt{2}+2}{2^2 - (\sqrt{2})^2}$

$= \dfrac{2\sqrt{2}+2}{4-2} = \dfrac{2\sqrt{2}+2}{2}$

$= \dfrac{2(\sqrt{2}+1)}{2}$

$= \sqrt{2}+1$ or $1+\sqrt{2}$

43. $\dfrac{\sqrt{5}}{\sqrt{2}+\sqrt{3}} = \dfrac{\sqrt{5}(\sqrt{2}-\sqrt{3})}{(\sqrt{2}+\sqrt{3})(\sqrt{2}-\sqrt{3})}$

Multiply by the conjugate.

$= \dfrac{\sqrt{5}\cdot\sqrt{2} - \sqrt{5}\cdot\sqrt{3}}{(\sqrt{2})^2 - (\sqrt{3})^2}$

$= \dfrac{\sqrt{10}-\sqrt{15}}{2-3}$

$= \dfrac{\sqrt{10}-\sqrt{15}}{-1} = -\sqrt{10} + \sqrt{15}$

45. $\dfrac{\sqrt{5}+2}{2-\sqrt{3}}$

$= \dfrac{(\sqrt{5}+2)(2+\sqrt{3})}{(2-\sqrt{3})(2+\sqrt{3})}$

$= \dfrac{2\sqrt{5}+\sqrt{15}+4+2\sqrt{3}}{(2)^2 - (\sqrt{3})^2}$

$$= \frac{2\sqrt{5} + \sqrt{15} + 4 + 2\sqrt{3}}{4 - 3}$$
$$= \frac{2\sqrt{5} + \sqrt{15} + 4 + 2\sqrt{3}}{1}$$
$$= 2\sqrt{5} + \sqrt{15} + 4 + 2\sqrt{3}$$

47. $\dfrac{12}{\sqrt{x}+1} = \dfrac{12(\sqrt{x}-1)}{(\sqrt{x}+1)(\sqrt{x}-1)}$
$$= \dfrac{12(\sqrt{x}-1)}{x-1}$$

49. $\dfrac{3}{7-\sqrt{x}} = \dfrac{3(7+\sqrt{x})}{(7-\sqrt{x})(7+\sqrt{x})}$
$$= \dfrac{3(7+\sqrt{x})}{49 - x}$$

51. $\dfrac{6\sqrt{11} - 12}{6}$
$$= \dfrac{6(\sqrt{11} - 2)}{6} \quad \text{Factor numerator.}$$
$$= \sqrt{11} - 2 \quad \text{Lowest terms}$$

53. $\dfrac{2\sqrt{3} + 10}{16} = \dfrac{2(\sqrt{3}+5)}{2 \cdot 8} = \dfrac{\sqrt{3}+5}{8}$

55. $\dfrac{12 - \sqrt{40}}{4} = \dfrac{12 - \sqrt{4} \cdot \sqrt{10}}{4}$
$$= \dfrac{12 - 2\sqrt{10}}{4}$$
$$= \dfrac{2(6 - \sqrt{10})}{2 \cdot 2}$$
$$= \dfrac{6 - \sqrt{10}}{2}$$

57. $6(5 + 3x) = (6)(5) + (6)(3x)$
$$= 30 + 18x$$

58. 30 and $18x$ cannot be combined because they are not like terms.

59. $\left(2\sqrt{10} + 5\sqrt{2}\right)\left(3\sqrt{10} - 3\sqrt{2}\right)$
$$= 2\sqrt{10}\left(3\sqrt{10}\right) + 2\sqrt{10}\left(-3\sqrt{2}\right)$$
$$\quad + 5\sqrt{2}\left(3\sqrt{10}\right) + 5\sqrt{2}\left(-3\sqrt{2}\right) \quad FOIL$$
$$= 6 \cdot 10 - 6\sqrt{20} + 15\sqrt{20} - 15 \cdot 2$$
$$= 60 + 9\sqrt{20} - 30$$
$$= 30 + 9\sqrt{4} \cdot \sqrt{5}$$
$$= 30 + 9(2\sqrt{5})$$
$$= 30 + 18\sqrt{5}$$

60. 30 and $18\sqrt{5}$ cannot be combined because they are not like radicals.

61. In the expression $30 + 18x$, make the first term $30x$, so that
$$30x + 18x = 48x.$$
In the expression $30 + 18\sqrt{5}$, make the first term $30\sqrt{5}$, so that
$$30\sqrt{5} + 18\sqrt{5} = 48\sqrt{5}.$$

62. When combining like terms, we add (or subtract) the coefficients of the common factors of the terms: $2xy + 5xy = 7xy$. When combining like radicals, we add (or subtract) the coefficients of the common radical terms:
$2\sqrt{ab} + 5\sqrt{ab} = 7\sqrt{ab}.$

63. $r = \dfrac{-h + \sqrt{h^2 + .64S}}{2}$

Substitute 12 for h and 400 for S.
$$r = \dfrac{-12 + \sqrt{12^2 + .64(400)}}{2}$$
$$= \dfrac{-12 + \sqrt{144 + 256}}{2}$$
$$= \dfrac{-12 + \sqrt{400}}{2}$$
$$= \dfrac{-12 + 20}{2}$$
$$= \dfrac{8}{2} = 4$$

The radius should be 4 inches.

Summary Exercises on Operations with Radicals

1. $5\sqrt{10} - 8\sqrt{10} = (5 - 8)\sqrt{10}$
$$= -3\sqrt{10}$$

3. $\left(1 + \sqrt{3}\right)\left(2 - \sqrt{6}\right)$
$$= 1 \cdot 2 - 1 \cdot \sqrt{6} + 2 \cdot \sqrt{3} - \sqrt{3} \cdot \sqrt{6}$$
$$= 2 - \sqrt{6} + 2\sqrt{3} - \sqrt{18}$$
$$= 2 - \sqrt{6} + 2\sqrt{3} - \sqrt{9 \cdot 2}$$
$$= 2 - \sqrt{6} + 2\sqrt{3} - 3\sqrt{2}$$

5. $\left(3\sqrt{5} - 2\sqrt{7}\right)^2$
$$= \left(3\sqrt{5}\right)^2 - 2\left(3\sqrt{5}\right)\left(2\sqrt{7}\right) + \left(2\sqrt{7}\right)^2$$
$$= 3^2\left(\sqrt{5}\right)^2 - 2 \cdot 3 \cdot 2 \cdot \sqrt{5} \cdot \sqrt{7} + 2^2\left(\sqrt{7}\right)^2$$
$$= 9 \cdot 5 - 12\sqrt{35} + 4 \cdot 7$$
$$= 45 - 12\sqrt{35} + 28$$
$$= 73 - 12\sqrt{35}$$

Chapter 8 Roots and Radicals

7. $\sqrt[3]{16t^2} - \sqrt[3]{54t^2} + \sqrt[3]{128t^2}$
$= \sqrt[3]{8} \cdot \sqrt[3]{2t^2} - \sqrt[3]{27} \cdot \sqrt[3]{2t^2} + \sqrt[3]{64} \cdot \sqrt[3]{2t^2}$
$= 2\sqrt[3]{2t^2} - 3\sqrt[3]{2t^2} + 4\sqrt[3]{2t^2}$
$= (2 - 3 + 4)\sqrt[3]{2t^2}$
$= 3\sqrt[3]{2t^2}$

9. $\dfrac{1+\sqrt{2}}{1-\sqrt{2}} = \dfrac{1+\sqrt{2}}{1-\sqrt{2}} \cdot \dfrac{1+\sqrt{2}}{1+\sqrt{2}}$
$= \dfrac{1 + \sqrt{2} + \sqrt{2} + \sqrt{2} \cdot \sqrt{2}}{1^2 - (\sqrt{2})^2}$
$= \dfrac{1 + 2\sqrt{2} + 2}{1 - 2}$
$= \dfrac{3 + 2\sqrt{2}}{-1} = -3 - 2\sqrt{2}$

11. $(\sqrt{3}+6)(\sqrt{3}-6) = (\sqrt{3})^2 - 6^2$
$= 3 - 36$
$= -33$

13. $\sqrt[3]{8x^3y^5z^6} = \sqrt[3]{8x^3y^3z^6} \cdot \sqrt[3]{y^2}$
$= 2xyz^2\sqrt[3]{y^2}$

15. $\dfrac{5}{\sqrt{6}-1} = \dfrac{5}{\sqrt{6}-1} \cdot \dfrac{\sqrt{6}+1}{\sqrt{6}+1}$
$= \dfrac{5(\sqrt{6}+1)}{(\sqrt{6})^2 - 1^2}$
$= \dfrac{5(\sqrt{6}+1)}{6-1}$
$= \dfrac{5(\sqrt{6}+1)}{5} = \sqrt{6}+1$

17. $\dfrac{6\sqrt{3}}{5\sqrt{12}} = \dfrac{6\sqrt{3}}{5\sqrt{4}\cdot\sqrt{3}} = \dfrac{6}{5\cdot 2} = \dfrac{3}{5}$

19. $\dfrac{-4}{\sqrt[3]{4}} = \dfrac{-4 \cdot \sqrt[3]{2}}{\sqrt[3]{4} \cdot \sqrt[3]{2}}$
$= \dfrac{-4\sqrt[3]{2}}{\sqrt[3]{8}}$
$= \dfrac{-4\sqrt[3]{2}}{2} = -2\sqrt[3]{2}$

21. $\sqrt{75x} - \sqrt{12x} = \sqrt{25\cdot 3x} - \sqrt{4\cdot 3x}$
$= 5\sqrt{3x} - 2\sqrt{3x}$
$= (5-2)\sqrt{3x}$
$= 3\sqrt{3x}$

23. $(\sqrt{7}-\sqrt{6})(\sqrt{7}+\sqrt{6})$
$= (\sqrt{7})^2 - (\sqrt{6})^2$
$= 7 - 6 = 1$

25. $x\sqrt[4]{x^5} - 3\sqrt[4]{x^9} + x^2\sqrt[4]{x}$
$= x\sqrt[4]{x^4}\cdot\sqrt[4]{x} - 3\sqrt[4]{x^8}\cdot\sqrt[4]{x} + x^2\sqrt[4]{x}$
$= x\cdot x\cdot\sqrt[4]{x} - 3\cdot x^2\cdot\sqrt[4]{x} + x^2\sqrt[4]{x}$
$= (x^2 - 3x^2 + x^2)\sqrt[4]{x}$
$= -x^2\sqrt[4]{x}$

27. $\sqrt{14} + \sqrt{17}$ cannot be added using the distributive property.

29. $\sqrt{\dfrac{3}{4}} \cdot \sqrt{\dfrac{1}{5}} = \dfrac{\sqrt{3}}{\sqrt{4}} \cdot \dfrac{\sqrt{1}}{\sqrt{5}}$
$= \dfrac{\sqrt{3}\cdot\sqrt{5}}{2\cdot\sqrt{5}\cdot\sqrt{5}}$
$= \dfrac{\sqrt{15}}{2\cdot 5}$
$= \dfrac{\sqrt{15}}{10}$

31. $\sqrt[3]{24} + 6\sqrt[3]{81}$
$= \sqrt[3]{8}\cdot\sqrt[3]{3} + 6\sqrt[3]{27}\cdot\sqrt[3]{3}$
$= 2\sqrt[3]{3} + 6(3\sqrt[3]{3})$
$= 2\sqrt[3]{3} + 18\sqrt[3]{3}$
$= 20\sqrt[3]{3}$

33. $\sqrt[3]{4}(\sqrt[3]{2} - 3)$
$= \sqrt[3]{4}(\sqrt[3]{2}) + \sqrt[3]{4}(-3)$ Distributive property
$= \sqrt[3]{8} - 3\sqrt[3]{4}$ Product rule
$= 2 - 3\sqrt[3]{4}$ $\sqrt[3]{8} = 2$

35. $\sqrt{\dfrac{5}{8}} = \dfrac{\sqrt{5}}{\sqrt{8}} = \dfrac{\sqrt{5}\cdot\sqrt{2}}{\sqrt{4}\cdot\sqrt{2}\cdot\sqrt{2}}$
$= \dfrac{\sqrt{10}}{2\cdot 2}$
$= \dfrac{\sqrt{10}}{4}$

37. $S = 28.6\sqrt[3]{A}$

(a) Substitute 8 for A.
$S = 28.6\sqrt[3]{8} = 28.6(2) = 57.2$
There would be 57 species (rounded).

(b) Substitute 27,000 for A.
$S = 28.6\sqrt[3]{27{,}000} = 28.6(30) = 858$
There would be 858 species.

8.6 Solving Equations with Radicals

8.6 Margin Exercises

1. (a) $\sqrt{k} = 3$
   $\left(\sqrt{k}\right)^2 = 3^2$ *Square each side.*
   $k = 9$

   Check $k = 9$: $\sqrt{k} = 3$
   $\sqrt{9} = 3$?   *Let k = 9.*
   $3 = 3$ *True*

   The solution is 9.

   (b) $\sqrt{x - 2} = 4$
   $\left(\sqrt{x-2}\right)^2 = 4^2$ *Square each side.*
   $x - 2 = 16$
   $x = 18$ *Add 2.*

   Check $x = 18$: $\sqrt{x - 2} = 4$
   $\sqrt{18 - 2} = 4$?   *Let x = 18.*
   $\sqrt{16} = 4$?
   $4 = 4$ *True*

   The solution is 18.

   (c) $\sqrt{9 - t} = 4$
   $\left(\sqrt{9-t}\right)^2 = 4^2$ *Square each side.*
   $9 - t = 16$
   $-t = 7$ *Subtract 9.*
   $t = -7$ *Multiply by −1.*

   Check $t = -7$: $\sqrt{9 - t} = 4$
   $\sqrt{9 - (-7)} = 4$?   *Let t = −7.*
   $\sqrt{16} = 4$?
   $4 = 4$ *True*

   The solution is −7.

2. (a) $\sqrt{3x + 9} = 2\sqrt{x}$
   $\left(\sqrt{3x+9}\right)^2 = \left(2\sqrt{x}\right)^2$ *Square each side.*
   $3x + 9 = 4x$
   $9 = x$ *Subtract 3x.*

   Check $x = 9$:
   $\sqrt{3x + 9} = 2\sqrt{x}$
   $\sqrt{3(9) + 9} = 2\sqrt{9}$?   *Let x = 9.*
   $\sqrt{27 + 9} = 2(3)$?
   $\sqrt{36} = 6$?
   $6 = 6$ *True*

   The solution is 9.

   (b) $5\sqrt{x} = \sqrt{20x + 5}$
   $\left(5\sqrt{x}\right)^2 = \left(\sqrt{20x+5}\right)^2$ *Square each side.*
   $25x = 20x + 5$
   $5x = 5$ *Subtract 20x.*
   $x = 1$ *Divide by 5.*

   Check $x = 1$:
   $5\sqrt{x} = \sqrt{20x + 5}$
   $5\sqrt{1} = \sqrt{20(1) + 5}$?   *Let x = 1.*
   $5(1) = \sqrt{25}$?
   $5 = 5$ *True*

   The solution is 1.

3. (a) $\sqrt{x} + 4 = 0$
   $\sqrt{x} = -4$ *Subtract 4.*
   $\left(\sqrt{x}\right)^2 = (-4)^2$ *Square each side.*
   $x = 16$

   Check $x = 16$: $\sqrt{x} + 4 = 0$
   $\sqrt{16} + 4 = 0$?   *Let x = 16.*
   $4 + 4 = 0$?
   $8 = 0$ *False*

   Because the statement $8 = 0$ is false, the number 16 is *not* a solution. The equation has no solution.

   (b) $x = \sqrt{x^2 - 4x - 16}$
   $x^2 = \left(\sqrt{x^2 - 4x - 16}\right)^2$ *Square each side.*
   $x^2 = x^2 - 4x - 16$
   $0 = -4x - 16$ *Subtract x^2.*
   $4x = -16$ *Add 4x.*
   $x = -4$ *Divide by 4.*

   Check $x = -4$:
   $x = \sqrt{x^2 - 4x - 16}$
   $-4 = \sqrt{(-4)^2 - 4(-4) - 16}$?   *Let x = −4.*
   $-4 = \sqrt{16 + 16 - 16}$?
   $-4 = \sqrt{16}$?
   $-4 = 4$ *False*

   The only potential solution does not check, so −4 is an extraneous solution, and the equation has no solution.

4. **(a)** Use the pattern
$$(a-b)^2 = a^2 - 2ab + b^2$$
with $a = w$ and $b = 5$.
$$(w-5)^2 = w^2 - 2(w)(5) + 5^2$$
$$= w^2 - 10w + 25$$

(b) $(2k-5)^2 = (2k)^2 - 2(2k)(5) + 5^2$
$$= 4k^2 - 20k + 25$$

(c) $(3m - 2p)^2 = (3m)^2 - 2(3m)(2p) + (2p)^2$
$$= 9m^2 - 12mp + 4p^2$$

5. **(a)** $\sqrt{6w+6} = w + 1$

$\left(\sqrt{6w+6}\right)^2 = (w+1)^2$ Square each side.

$6w + 6 = w^2 + 2w + 1$

$0 = w^2 - 4w - 5$ Subtract $6w + 6$.

$0 = (w-5)(w+1)$ Factor.

$w - 5 = 0 \quad \text{or} \quad w + 1 = 0$
$w = 5 \quad \text{or} \quad w = -1$

Check both of these potential solutions in the original equation.

Check $w = 5$:
$$\sqrt{6w+6} = w + 1$$
$$\sqrt{6(5)+6} = 5 + 1 \text{ ?} \quad \text{Let } w = 5.$$
$$\sqrt{30+6} = 6 \quad \text{?}$$
$$\sqrt{36} = 6 \quad \text{?}$$
$$6 = 6 \quad \text{True}$$

Check $w = -1$:
$$\sqrt{6w+6} = w + 1$$
$$\sqrt{6(-1)+6} = -1 + 1 \text{ ?} \quad \text{Let } w = -1.$$
$$\sqrt{-6+6} = 0 \quad \text{?}$$
$$\sqrt{0} = 0 \quad \text{?}$$
$$0 = 0 \quad \text{True}$$

The solutions are 5 and -1.

(b) $2u - 1 = \sqrt{10u + 9}$

$(2u-1)^2 = \left(\sqrt{10u+9}\right)^2$ Square each side.

$4u^2 - 4u + 1 = 10u + 9$

$4u^2 - 14u - 8 = 0$ Subtract $10u + 9$.

$2(2u^2 - 7u - 4) = 0$ Factor.

$2(2u+1)(u-4) = 0$ Factor.

$2u + 1 = 0 \quad \text{or} \quad u - 4 = 0$
$2u = -1$
$u = -\tfrac{1}{2} \quad \text{or} \quad u = 4$

Check $u = -\tfrac{1}{2}$:
$$2u - 1 = \sqrt{10u + 9}$$
$$2\left(-\tfrac{1}{2}\right) - 1 = \sqrt{10\left(-\tfrac{1}{2}\right) + 9} \text{ ?} \quad \text{Let } u = -\tfrac{1}{2}.$$
$$-1 - 1 = \sqrt{-5 + 9} \quad \text{?}$$
$$-2 = \sqrt{4} \quad \text{?}$$
$$-2 = 2 \quad \text{False}$$

Check $u = 4$:
$$2u - 1 = \sqrt{10u + 9}$$
$$2(4) - 1 = \sqrt{10(4) + 9} \text{ ?} \quad \text{Let } u = 4.$$
$$8 - 1 = \sqrt{49} \quad \text{?}$$
$$7 = 7 \quad \text{True}$$

The number $-\tfrac{1}{2}$ does not satisfy the original equation, so it is extraneous. The only solution is 4.

6. **(a)** $\sqrt{x} - 3 = x - 15$

$\sqrt{x} = x - 12$ Add 3 to get $\sqrt{x}$ alone.

$\left(\sqrt{x}\right)^2 = (x-12)^2$ Square each side.

$x = x^2 - 24x + 144$

$0 = x^2 - 25x + 144$ Subtract x.

$0 = (x-16)(x-9)$ Factor.

$x - 16 = 0 \quad \text{or} \quad x - 9 = 0$
$x = 16 \quad \text{or} \quad x = 9$

Check $x = 16$:
$$\sqrt{x} - 3 = x - 15$$
$$\sqrt{16} - 3 = 16 - 15 \text{ ?} \quad \text{Let } x = 16.$$
$$4 - 3 = 1 \quad \text{?}$$
$$1 = 1 \quad \text{True}$$

Check $x = 9$:
$$\sqrt{x} - 3 = x - 15$$
$$\sqrt{9} - 3 = 9 - 15 \text{ ?} \quad \text{Let } x = 9.$$
$$3 - 3 = -6 \quad \text{?}$$
$$0 = -6 \quad \text{False}$$

The number 9 does not satisfy the original equation, so it is extraneous. The only solution is 16.

(b) $\sqrt{z+5} + 2 = z + 5$

$\sqrt{z+5} = z + 3$ *Subtract 2.*

$\left(\sqrt{z+5}\right)^2 = (z+3)^2$ *Square each side.*

$z + 5 = z^2 + 6z + 9$

$0 = z^2 + 5z + 4$ *Subtract $z + 5$.*

$0 = (z+4)(z+1)$ *Factor.*

$z + 4 = 0$ or $z + 1 = 0$

$z = -4$ or $z = -1$

Check $z = -4$:

$\sqrt{z+5} + 2 = z + 5$

$\sqrt{-4+5} + 2 = -4 + 5$? *Let $z = -4$.*

$\sqrt{1} + 2 = 1$?

$1 + 2 = 1$?

$3 = 1$ *False*

Check $z = -1$:

$\sqrt{z+5} + 2 = z + 5$

$\sqrt{-1+5} + 2 = -1 + 5$? *Let $z = -1$.*

$\sqrt{4} + 2 = 4$?

$2 + 2 = 4$?

$4 = 4$ *True*

The number -4 does not satisfy the original equation, so it is extraneous. The only solution is -1.

7. (a) $\sqrt{p+1} - \sqrt{p-4} = 1$

$\sqrt{p+1} = \sqrt{p-4} + 1$

$\left(\sqrt{p+1}\right)^2 = \left(\sqrt{p-4} + 1\right)^2$

$p + 1 = p - 4 + 2\sqrt{p-4} + 1$

$4 = 2\sqrt{p-4}$

$2 = \sqrt{p-4}$

$2^2 = \left(\sqrt{p-4}\right)^2$

$4 = p - 4$

$8 = p$

Check $p = 8$:

$\sqrt{p+1} - \sqrt{p-4} = 1$

$\sqrt{8+1} - \sqrt{8-4} = 1$? *Let $p = 8$.*

$\sqrt{9} - \sqrt{4} = 1$?

$3 - 2 = 1$?

$1 = 1$ *True*

The solution is 8.

(b) $\sqrt{2x+1} + \sqrt{x+4} = 3$

$\sqrt{2x+1} = 3 - \sqrt{x+4}$

$\left(\sqrt{2x+1}\right)^2 = \left(3 - \sqrt{x+4}\right)^2$

$2x + 1 = 9 - 2 \cdot 3 \cdot \sqrt{x+4} + x + 4$

$x - 12 = -6\sqrt{x+4}$

$(x - 12)^2 = \left(-6\sqrt{x+4}\right)^2$

$x^2 - 24x + 144 = 36(x+4)$

$x^2 - 24x + 144 = 36x + 144$

$x^2 - 60x = 0$

$x(x - 60) = 0$

$x = 0$ or $x = 60$

Check $x = 0$:

$\sqrt{2x+1} + \sqrt{x+4} = 3$

$\sqrt{2(0)+1} + \sqrt{0+4} = 3$? *Let $x = 0$.*

$\sqrt{1} + \sqrt{4} = 3$?

$1 + 2 = 3$?

$3 = 3$ *True*

Check $x = 60$:

$\sqrt{2x+1} + \sqrt{x+4} = 3$

$\sqrt{2(60)+1} + \sqrt{60+4} = 3$? *Let $x = 60$.*

$\sqrt{121} + \sqrt{64} = 3$?

$11 + 8 = 3$?

$19 = 3$ *False*

The only solution is 0.

8.6 Section Exercises

1. $\sqrt{x} = 7$

Use the *squaring property of equality* to square each side of the equation.

$\left(\sqrt{x}\right)^2 = 7^2$

$x = 49$

Now check this proposed solution in the original equation.

Check $x = 49$: $\sqrt{x} = 7$

$\sqrt{49} = 7$?   *Let $x = 49$.*

$7 = 7$ *True*

Since this statement is true, the solution of the original equation is 49.

3.
$$\sqrt{t+2} = 3$$
$$\left(\sqrt{t+2}\right)^2 = 3^2 \quad \text{Square each side.}$$
$$t+2 = 9$$
$$t = 7$$

Check $t = 7$:
$$\sqrt{t+2} = 3$$
$$\sqrt{7+2} = 3 \text{ ?} \quad \text{Let } t = 7.$$
$$\sqrt{9} = 3 \text{ ?}$$
$$3 = 3 \quad \text{True}$$

Since this statement is true, the solution of the original equation is 7.

5.
$$\sqrt{r-4} = 9$$
$$\left(\sqrt{r-4}\right)^2 = 9^2 \quad \text{Square each side.}$$
$$r - 4 = 81$$
$$r = 85$$

Check $r = 85$:
$$\sqrt{r-4} = 9$$
$$\sqrt{85-4} = 9 \text{ ?} \quad \text{Let } r = 85.$$
$$\sqrt{81} = 9 \text{ ?}$$
$$9 = 9 \quad \text{True}$$

Since this statement is true, the solution of the original equation is 85.

7.
$$\sqrt{4-t} = 7$$
$$\left(\sqrt{4-t}\right)^2 = 7^2 \quad \text{Square each side.}$$
$$4 - t = 49$$
$$-t = 45$$
$$t = -45$$

Check $t = -45$:
$$\sqrt{4-t} = 7$$
$$\sqrt{4-(-45)} = 7 \text{ ?} \quad \text{Let } t = -45.$$
$$\sqrt{49} = 7 \text{ ?}$$
$$7 = 7 \quad \text{True}$$

Since this statement is true, the solution of the original equation is -45.

9.
$$\sqrt{2t+3} = 0$$
$$\left(\sqrt{2t+3}\right)^2 = 0^2 \quad \text{Square each side.}$$
$$2t + 3 = 0$$
$$2t = -3$$
$$t = -\tfrac{3}{2}$$

Check $t = -\tfrac{3}{2}$:
$$\sqrt{2t+3} = 0$$
$$\sqrt{2\left(-\tfrac{3}{2}\right)+3} = 0 \text{ ?} \quad \text{Let } t = -\tfrac{3}{2}.$$
$$\sqrt{-3+3} = 0 \text{ ?}$$
$$\sqrt{0} = 0 \text{ ?}$$
$$0 = 0 \quad \text{True}$$

Since this statement is true, the solution of the original equation is $-\tfrac{3}{2}$.

11.
$$\sqrt{3x-8} = -2$$
$$\left(\sqrt{3x-8}\right)^2 = (-2)^2 \quad \text{Square each side.}$$
$$3x - 8 = 4$$
$$3x = 12$$
$$x = 4$$

Check $x = 4$:
$$\sqrt{3x-8} = -2$$
$$\sqrt{3(4)-8} = -2 \text{ ?} \quad \text{Let } x = 4.$$
$$\sqrt{12-8} = -2 \text{ ?}$$
$$\sqrt{4} = -2 \text{ ?}$$
$$2 = -2 \quad \text{False}$$

Since this statement is false, there is no solution to the original equation.

13. $\sqrt{w} - 4 = 7$

Add 4 to both sides of the equation before squaring.
$$\sqrt{w} = 11$$
$$\left(\sqrt{w}\right)^2 = (11)^2$$
$$w = 121$$

Check $w = 121$:
$$\sqrt{w} - 4 = 7$$
$$\sqrt{121} - 4 = 7 \text{ ?} \quad \text{Let } w = 121.$$
$$11 - 4 = 7 \text{ ?}$$
$$7 = 7 \quad \text{True}$$

Since this statement is true, the solution of the original equation is 121.

8.6 Solving Equations with Radicals

15. $\sqrt{10x-8} = 3\sqrt{x}$
$\left(\sqrt{10x-8}\right)^2 = \left(3\sqrt{x}\right)^2$ Square sides.
$10x - 8 = (3)^2\left(\sqrt{x}\right)^2$ $(ab)^2 = a^2b^2$
$10x - 8 = 9x$
$x = 8$

Check $x = 8$:

$\sqrt{10x-8} = 3\sqrt{x}$
$\sqrt{10(8)-8} = 3\sqrt{8}$? Let x = 8.
$\sqrt{72} = 3\sqrt{8}$?
$\sqrt{36 \cdot 2} = 3 \cdot 2\sqrt{2}$?
$6\sqrt{2} = 6\sqrt{2}$ True

Since this statement is true, the solution of the original equation is 8.

17. $5\sqrt{x} = \sqrt{10x+15}$
$\left(5\sqrt{x}\right)^2 = \left(\sqrt{10x+15}\right)^2$
$25x = 10x + 15$
$15x = 15$
$x = 1$

Check $x = 1$:

$5\sqrt{x} = \sqrt{10x+15}$
$5\sqrt{1} = \sqrt{10 \cdot 1 + 15}$? Let x = 1.
$5 \cdot 1 = \sqrt{25}$?
$5 = 5$ True

Since this statement is true, the solution of the original equation is 1.

19. $\sqrt{3x-5} = \sqrt{2x+1}$
$\left(\sqrt{3x-5}\right)^2 = \left(\sqrt{2x+1}\right)^2$
$3x - 5 = 2x + 1$
$x = 6$

Check $x = 6$:

$\sqrt{3x-5} = \sqrt{2x+1}$
$\sqrt{3(6)-5} = \sqrt{2(6)+1}$? Let x = 6.
$\sqrt{13} = \sqrt{13}$ True

Since this statement is true, the solution of the original equation is 6.

21. $k = \sqrt{k^2 - 5k - 15}$
$(k)^2 = \left(\sqrt{k^2 - 5k - 15}\right)^2$
$k^2 = k^2 - 5k - 15$
$0 = -5k - 15$
$5k = -15$
$k = -3$

Check $k = -3$:

$k = \sqrt{k^2 - 5k - 15}$
$-3 = \sqrt{(-3)^2 - 5(-3) - 15}$? Let k = -3.
$-3 = \sqrt{9 + 15 - 15}$?
$-3 = \sqrt{9}$?
$-3 = 3$ False

Since this statement is false, there is no solution to the original equation.

23. $7x = \sqrt{49x^2 + 2x - 10}$
$(7x)^2 = \left(\sqrt{49x^2 + 2x - 10}\right)^2$
$49x^2 = 49x^2 + 2x - 10$
$0 = 2x - 10$
$10 = 2x$
$5 = x$

Check $x = 5$:

$7x = \sqrt{49x^2 + 2x - 10}$
$7(5) = \sqrt{49(5)^2 + 2(5) - 10}$? Let x = 5.
$35 = \sqrt{1225 + 10 - 10}$?
$35 = \sqrt{1225}$?
$35 = 35$ True

Since this statement is true, the solution of the original equation is 5.

25. $\sqrt{2x+1} = x - 7$

The first step in solving this equation is to square both sides of the equation. The right side is a binomial which must be squared as a quantity, not term by term. The correct square of the right side is

$(x-7)^2 = x^2 - 2(x)(7) + (7)^2$
$= x^2 - 14x + 49.$

27. $\sqrt{2x+1} = x - 7$
$\left(\sqrt{2x+1}\right)^2 = (x-7)^2$
$2x + 1 = x^2 - 14x + 49$
$0 = x^2 - 16x + 48$
$0 = (x-4)(x-12)$
$x = 4$ or $x = 12$

Check $x = 4$: $3 = -3$ False
Check $x = 12$: $5 = 5$ True

The solution is 12.

29.
$\sqrt{3k+10} + 5 = 2k$
$\sqrt{3k+10} = 2k - 5$
$\left(\sqrt{3k+10}\right)^2 = (2k-5)^2$
$3k + 10 = 4k^2 - 20k + 25$
$0 = 4k^2 - 23k + 15$
$0 = (4k-3)(k-5)$
$k = \frac{3}{4}$ or $k = 5$

Check $k = \frac{3}{4}$: $\frac{17}{2} = \frac{3}{2}$ False
Check $x = 5$: $10 = 10$ True

The solution is 5.

31.
$\sqrt{5x+1} - 1 = x$
$\sqrt{5x+1} = x + 1$
$\left(\sqrt{5x+1}\right)^2 = (x+1)^2$
$5x + 1 = x^2 + 2x + 1$
$0 = x^2 - 3x$
$0 = x(x-3)$
$x = 0$ or $x = 3$

Check $x = 0$: $0 = 0$ True
Check $x = 3$: $3 = 3$ True

The solutions are 0 and 3.

33.
$\sqrt{6t+7} + 3 = t + 5$
$\sqrt{6t+7} = t + 2$
$\left(\sqrt{6t+7}\right)^2 = (t+2)^2$
$6t + 7 = t^2 + 4t + 4$
$0 = t^2 - 2t - 3$
$0 = (t+1)(t-3)$
$t = -1$ or $t = 3$

Check $t = -1$: $4 = 4$ True
Check $t = 3$: $8 = 8$ True

The solutions are -1 and 3.

35.
$x - 4 - \sqrt{2x} = 0$
$x - 4 = \sqrt{2x}$
$(x-4)^2 = \left(\sqrt{2x}\right)^2$
$x^2 - 8x + 16 = 2x$
$x^2 - 10x + 16 = 0$
$(x-8)(x-2) = 0$
$x = 8$ or $x = 2$

Check $x = 8$: $0 = 0$ True
Check $x = 2$: $-4 = 0$ False

The solution is 8.

37.
$\sqrt{x} + 6 = 2x$
$\sqrt{x} = 2x - 6$
$\left(\sqrt{x}\right)^2 = (2x-6)^2$
$x = 4x^2 - 24x + 36$
$0 = 4x^2 - 25x + 36$
$0 = (4x-9)(x-4)$
$x = \frac{9}{4}$ or $x = 4$

Check $x = \frac{9}{4}$: $\frac{15}{2} = \frac{9}{2}$ False
Check $x = 4$: $8 = 8$ True

The solution is 4.

39.
$\sqrt{x+1} - \sqrt{x-4} = 1$
$\sqrt{x+1} = \sqrt{x-4} + 1$
$\left(\sqrt{x+1}\right)^2 = \left(\sqrt{x-4}+1\right)^2$
$x + 1 = x - 4 + 2\sqrt{x-4} + 1$
$4 = 2\sqrt{x-4}$
$2 = \sqrt{x-4}$
$2^2 = \left(\sqrt{x-4}\right)^2$
$4 = x - 4$
$8 = x$

Check $x = 8$:
$\sqrt{x+1} - \sqrt{x-4} = 1$
$\sqrt{8+1} - \sqrt{8-4} = 1$? Let $x = 8$.
$\sqrt{9} - \sqrt{4} = 1$?
$3 - 2 = 1$?
$1 = 1$ True

The solution is 8.

41.
$\sqrt{x} = \sqrt{x-5} + 1$
$\left(\sqrt{x}\right)^2 = \left(\sqrt{x-5}+1\right)^2$
$x = x - 5 + 2\sqrt{x-5} + 1$
$4 = 2\sqrt{x-5}$
$2 = \sqrt{x-5}$
$2^2 = \left(\sqrt{x-5}\right)^2$
$4 = x - 5$
$9 = x$

Check $x = 9$: $3 = 3$ True

The solution is 9.

8.6 Solving Equations with Radicals

43.
$$\sqrt{3x+4} - \sqrt{2x-4} = 2$$
$$\sqrt{3x+4} = 2 + \sqrt{2x-4}$$
$$\left(\sqrt{3x+4}\right)^2 = \left(2+\sqrt{2x-4}\right)^2$$
$$3x+4 = 4 + 4\sqrt{2x-4} + 2x - 4$$
$$x+4 = 4\sqrt{2x-4}$$
$$(x+4)^2 = \left(4\sqrt{2x-4}\right)^2$$
$$x^2 + 8x + 16 = 16(2x-4)$$
$$x^2 + 8x + 16 = 32x - 64$$
$$x^2 - 24x + 80 = 0$$
$$(x-4)(x-20) = 0$$
$$x = 4 \text{ or } x = 20$$

Check $x = 4$: $4 - 2 = 2$ True
Check $x = 20$: $8 - 6 = 2$ True

The solutions are 4 and 20.

45.
$$\sqrt{2x+11} + \sqrt{x+6} = 2$$
$$\sqrt{2x+11} = 2 - \sqrt{x+6}$$
$$\left(\sqrt{2x+11}\right)^2 = \left(2-\sqrt{x+6}\right)^2$$
$$2x+11 = 4 - 4\sqrt{x+6} + x + 6$$
$$x+1 = -4\sqrt{x+6}$$
$$(x+1)^2 = \left(-4\sqrt{x+6}\right)^2$$
$$x^2 + 2x + 1 = 16(x+6)$$
$$x^2 + 2x + 1 = 16x + 96$$
$$x^2 - 14x - 95 = 0$$
$$(x+5)(x-19) = 0$$
$$x = -5 \text{ or } x = 19$$

Check $x = -5$: $1 + 1 = 2$ True
Check $x = 19$: $7 + 5 = 2$ False

The solution is -5.

47. Let $x =$ the number.
"The square root of the sum of a number and 4 is 5" translates to
$$\sqrt{x+4} = 5.$$
$$\left(\sqrt{x+4}\right)^2 = 5^2$$
$$x + 4 = 25$$
$$x = 21$$

Check $x = 21$: $5 = 5$ True

The number is 21.

49. Let $x =$ the number.
"Three times the square root of 2 equals the square root of the sum of some number and 10" translates to
$$3\sqrt{2} = \sqrt{x+10}.$$
$$\left(3\sqrt{2}\right)^2 = \left(\sqrt{x+10}\right)^2$$
$$9 \cdot 2 = x + 10$$
$$18 = x + 10$$
$$8 = x$$

Check $x = 8$: $3\sqrt{2} = \sqrt{18}$ True

The number is 8.

51. $s = 30\sqrt{\dfrac{a}{p}}$

Use a calculator and round answers to the nearest tenth.

(a) $s = 30\sqrt{\dfrac{862}{156}}$ Let $a = 862$ and $p = 156$.
≈ 70.5 miles per hour

(b) $s = 30\sqrt{\dfrac{382}{96}}$ Let $a = 382$ and $p = 96$.
≈ 59.8 miles per hour

(c) $s = 30\sqrt{\dfrac{84}{26}}$ Let $a = 84$ and $p = 26$.
≈ 53.9 miles per hour

53. Refer to the right triangle shown in the figure in the textbook. Note that the given distances are the lengths of the hypotenuse (193.0 feet) and one of the legs (110.0 feet) of the triangle. Use the Pythagorean formula with $a = 110.0$, $c = 193.0$, and $b =$ the height of the building.
$$a^2 + b^2 = c^2$$
$$(110.0)^2 + b^2 = (193.0)^2$$
$$12,100 + b^2 = 37,249$$
$$b^2 = 25,149$$
$$b = \sqrt{25,149} \approx 158.6$$

The height of the building, to the nearest tenth, is 158.6 feet.

55. $s = \dfrac{1}{2}(a+b+c)$
$= \dfrac{1}{2}(7+7+12)$
$= \dfrac{1}{2}(26) = 13$ units

56. $A = \sqrt{s(s-a)(s-b)(s-c)}$
$= \sqrt{13(13-7)(13-7)(13-12)}$
$= \sqrt{13(6)(6)(1)}$
$= 6\sqrt{13}$ square units

334 Chapter 8 Roots and Radicals

57. $c^2 = a^2 + b^2$
$7^2 = 6^2 + h^2$
$49 = 36 + h^2$
$h^2 = 13$
$h = \sqrt{13}$ units

58. $A = \dfrac{1}{2}bh$
$= \tfrac{1}{2}(6)\left(\sqrt{13}\right)$
$= 3\sqrt{13}$ square units

59. $2\left(3\sqrt{13}\right) = 6\sqrt{13}$ square units

60. They are both $6\sqrt{13}$.

Chapter 8 Review Exercises

1. The square roots of 49 are -7 and 7 because $(-7)^2 = 49$ and $7^2 = 49$.

2. The square roots of 81 are -9 and 9 because $(-9)^2 = 81$ and $9^2 = 81$.

3. The square roots of 196 are -14 and 14 because $(-14)^2 = 196$ and $14^2 = 196$.

4. The square roots of 121 are -11 and 11 because $(-11)^2 = 121$ and $11^2 = 121$.

5. The square roots of 225 are -15 and 15 because $(-15)^2 = 225$ and $15^2 = 225$.

6. The square roots of 729 are -27 and 27 because $(-27)^2 = 729$ and $27^2 = 729$.

7. $\sqrt{16} = 4$ because $4^2 = 16$.

8. $\sqrt{.36} = .6$ because $(.6)^2 = .36$. Thus,
$$-\sqrt{.36} = .6.$$

9. $\sqrt[3]{1000} = 10$ because $10^3 = 1000$.

10. $\sqrt[4]{81} = 3$ because 3 is positive and $3^4 = 81$.

11. $\sqrt{-8100}$ is not a real number.

12. $-\sqrt{4225}$ represents the negative square root of 4225. Since $65 \cdot 65 = 4225$,
$$-\sqrt{4225} = -65.$$

13. $\sqrt{\dfrac{49}{36}} = \dfrac{\sqrt{49}}{\sqrt{36}} = \dfrac{7}{6}$

14. $\sqrt{\dfrac{100}{81}} = \dfrac{\sqrt{100}}{\sqrt{81}} = \dfrac{10}{9}$

15. $\sqrt{64} = 8$; **B**

16. $-\sqrt{64} = -8$; **F**

17. $\sqrt{-64}$ is not a real number; **D**

18. $\sqrt[3]{64} = 4$; **A**

19. $\sqrt[3]{-64} = -4$; **C**

20. $\sqrt[3]{64} = (4)$; **A**

21. Use the Pythagorean formula with $a = 15$, $b = x$, and $c = 17$.
$$c^2 = a^2 + b^2$$
$$17^2 = 15^2 + x^2$$
$$289 = 225 + x^2$$
$$64 = x^2$$
$$x = \sqrt{64} = 8$$

22. Use the Pythagorean formula with $a = 24.4$ cm and $b = 32.5$ cm.
$$c^2 = a^2 + b^2$$
$$= (24.4)^2 + (32.5)^2$$
$$= 595.36 + 1056.25$$
$$= 1651.61$$
$$c = \sqrt{1651.61} \approx 40.6 \text{ cm}$$

23. This number is *irrational* because 23 is not a perfect square.
$$\sqrt{23} \approx 4.796$$

24. This number is *rational* because 169 is a perfect square.
$$\sqrt{169} = 13$$

25. $-\sqrt{25}$

This number is *rational* because 25 is a perfect square.
$$-\sqrt{25} = -5$$

26. $\sqrt{-4}$

This is not a real number.

27. $\sqrt{2} \cdot \sqrt{7} = \sqrt{2 \cdot 7} = \sqrt{14}$

28. $\sqrt{5} \cdot \sqrt{15} = \sqrt{5} \cdot \sqrt{5} \cdot \sqrt{3}$
$= \sqrt{25} \cdot \sqrt{3} = 5\sqrt{3}$

29. $-\sqrt{27} = -\sqrt{9 \cdot 3} = -\sqrt{9} \cdot \sqrt{3} = -3\sqrt{3}$

30. $\sqrt{48} = \sqrt{16 \cdot 3} = \sqrt{16} \cdot \sqrt{3} = 4\sqrt{3}$

31. $\sqrt{160} = \sqrt{16 \cdot 10} = \sqrt{16} \cdot \sqrt{10} = 4\sqrt{10}$

32. $\sqrt{12} \cdot \sqrt{27} = \sqrt{4 \cdot 3} \cdot \sqrt{9 \cdot 3}$
$= 2\sqrt{3} \cdot 3\sqrt{3}$
$= 2 \cdot 3 \cdot \left(\sqrt{3}\right)^2$
$= 2 \cdot 3 \cdot 3 = 18$

33. $\sqrt{32} \cdot \sqrt{48} = \sqrt{16 \cdot 2} \cdot \sqrt{16 \cdot 3}$
$= 4\sqrt{2} \cdot 4\sqrt{3}$
$= 4 \cdot 4 \cdot \sqrt{2 \cdot 3}$
$= 16\sqrt{6}$

34. $\sqrt{50} \cdot \sqrt{125} = \sqrt{25 \cdot 2} \cdot \sqrt{25 \cdot 5}$
$= 5\sqrt{2} \cdot 5\sqrt{5}$
$= 5 \cdot 5 \cdot \sqrt{2 \cdot 5}$
$= 25\sqrt{10}$

35. $\sqrt{\dfrac{9}{4}} = \dfrac{\sqrt{9}}{\sqrt{4}} = \dfrac{3}{2}$

36. $-\sqrt{\dfrac{121}{400}} = -\dfrac{\sqrt{121}}{\sqrt{400}} = -\dfrac{11}{20}$

37. $\sqrt{\dfrac{7}{169}} = \dfrac{\sqrt{7}}{\sqrt{169}} = \dfrac{\sqrt{7}}{13}$

38. $\sqrt{\dfrac{1}{6}} \cdot \sqrt{\dfrac{5}{6}} = \sqrt{\dfrac{1}{6} \cdot \dfrac{5}{6}}$
$= \sqrt{\dfrac{5}{36}}$
$= \dfrac{\sqrt{5}}{\sqrt{36}} = \dfrac{\sqrt{5}}{6}$

39. $\sqrt{\dfrac{2}{5}} \cdot \sqrt{\dfrac{2}{45}} = \sqrt{\dfrac{2}{5} \cdot \dfrac{2}{45}}$
$= \sqrt{\dfrac{4}{225}}$
$= \dfrac{\sqrt{4}}{\sqrt{225}} = \dfrac{2}{15}$

40. $\dfrac{3\sqrt{10}}{\sqrt{5}} = \dfrac{3 \cdot \sqrt{5} \cdot \sqrt{2}}{\sqrt{5}}$
$= 3\sqrt{2}$

41. $\dfrac{24\sqrt{12}}{6\sqrt{3}} = \dfrac{24 \cdot \sqrt{4} \cdot \sqrt{3}}{6\sqrt{3}}$
$= 4\sqrt{4} = 4 \cdot 2 = 8$

42. $\dfrac{8\sqrt{150}}{4\sqrt{75}} = \dfrac{8 \cdot \sqrt{75} \cdot \sqrt{2}}{4\sqrt{75}}$
$= 2\sqrt{2}$

43. $\sqrt{p} \cdot \sqrt{p} = p$

44. $\sqrt{k} \cdot \sqrt{m} = \sqrt{km}$

45. $\sqrt{r^{18}} = r^9$ because $(r^9)^2 = r^{18}$.

46. $\sqrt{x^{10}y^{16}} = x^5 y^8$ because $(x^5 y^8)^2 = x^{10}y^{16}$.

47. $\sqrt{x^9} = \sqrt{x^8 \cdot x} = \sqrt{x^8} \cdot \sqrt{x} = x^4\sqrt{x}$

48. $\sqrt{\dfrac{36}{p^2}} = \dfrac{\sqrt{36}}{\sqrt{p^2}} = \dfrac{6}{p}$

49. $\sqrt{a^{15}b^{21}} = \sqrt{a^{14}b^{20} \cdot ab}$
$= \sqrt{a^{14}b^{20}} \cdot \sqrt{ab}$
$= a^7 b^{10} \sqrt{ab}$

50. $\sqrt{121x^6 y^{10}} = 11x^3 y^5$ because
$(11x^3 y^5)^2 = 121x^6 y^{10}$.

51. $\sqrt[3]{y^6} = y^2$ because $(y^2)^3 = y^6$.

52. $\sqrt[3]{216x^{15}} = 6x^5$ because $(6x^5)^3 = 216x^{15}$.

53. Using a calculator,
$$\sqrt{.5} \approx .7071067812$$
and
$$\dfrac{\sqrt{2}}{2} \approx \dfrac{1.4142135624}{2}$$
$$= .7071067812$$

It looks like these two expressions represent the same number. In fact, they do represent the same number because

$$\sqrt{.5} = \sqrt{\dfrac{5}{10}} = \sqrt{\dfrac{1}{2}} = \dfrac{\sqrt{1}}{\sqrt{2}}$$
$$= \dfrac{1 \cdot \sqrt{2}}{\sqrt{2} \cdot \sqrt{2}} = \dfrac{\sqrt{2}}{2}.$$

54. $\sqrt{11} + \sqrt{11} = 1\sqrt{11} + 1\sqrt{11} = 2\sqrt{11}$

55. $3\sqrt{2} + 6\sqrt{2} = (3+6)\sqrt{2} = 9\sqrt{2}$

56. $3\sqrt{75} + 2\sqrt{27}$
$= 3\left(\sqrt{25} \cdot \sqrt{3}\right) + 2\left(\sqrt{9} \cdot \sqrt{3}\right)$
$= 3\left(5\sqrt{3}\right) + 2\left(3\sqrt{3}\right)$
$= 15\sqrt{3} + 6\sqrt{3} = 21\sqrt{3}$

57. $4\sqrt{12} + \sqrt{48}$
$= 4\left(\sqrt{4} \cdot \sqrt{3}\right) + \sqrt{16} \cdot \sqrt{3}$
$= 4\left(2\sqrt{3}\right) + 4\sqrt{3}$
$= 8\sqrt{3} + 4\sqrt{3} = 12\sqrt{3}$

58. $4\sqrt{24} - 3\sqrt{54} + \sqrt{6}$
$= 4(\sqrt{4} \cdot \sqrt{6}) - 3(\sqrt{9} \cdot \sqrt{6}) + \sqrt{6}$
$= 4(2\sqrt{6}) - 3(3\sqrt{6}) + \sqrt{6}$
$= 8\sqrt{6} - 9\sqrt{6} + 1\sqrt{6}$
$= 0\sqrt{6} = 0$

59. $2\sqrt{7} - 4\sqrt{28} + 3\sqrt{63}$
$= 2\sqrt{7} - 4(\sqrt{4} \cdot \sqrt{7}) + 3(\sqrt{9} \cdot \sqrt{7})$
$= 2\sqrt{7} - 4(2\sqrt{7}) + 3(3\sqrt{7})$
$= 2\sqrt{7} - 8\sqrt{7} + 9\sqrt{7} = 3\sqrt{7}$

60. $\frac{2}{5}\sqrt{75} + \frac{3}{4}\sqrt{160}$
$= \frac{2}{5}(\sqrt{25} \cdot \sqrt{3}) + \frac{3}{4}(\sqrt{16} \cdot \sqrt{10})$
$= \frac{2}{5}(5\sqrt{3}) + \frac{3}{4}(4\sqrt{10})$
$= 2\sqrt{3} + 3\sqrt{10}$

61. $\frac{1}{3}\sqrt{18} + \frac{1}{4}\sqrt{32}$
$= \frac{1}{3}(\sqrt{9} \cdot \sqrt{2}) + \frac{1}{4}(\sqrt{16} \cdot \sqrt{2})$
$= \frac{1}{3}(3\sqrt{2}) + \frac{1}{4}(4\sqrt{2})$
$= 1\sqrt{2} + 1\sqrt{2} = 2\sqrt{2}$

62. $\sqrt{15} \cdot \sqrt{2} + 5\sqrt{30} = \sqrt{30} + 5\sqrt{30}$
$= 1\sqrt{30} + 5\sqrt{30}$
$= 6\sqrt{30}$

63. $\sqrt{4x} + \sqrt{36x} - \sqrt{9x}$
$= \sqrt{4}\sqrt{x} + \sqrt{36}\sqrt{x} - \sqrt{9}\sqrt{x}$
$= 2\sqrt{x} + 6\sqrt{x} - 3\sqrt{x} = 5\sqrt{x}$

64. $\sqrt{16p} + 3\sqrt{p} - \sqrt{49p}$
$= \sqrt{16}\sqrt{p} + 3\sqrt{p} - \sqrt{49}\sqrt{p}$
$= 4\sqrt{p} + 3\sqrt{p} - 7\sqrt{p}$
$= 0\sqrt{p} = 0$

65. $\sqrt{20m^2} - m\sqrt{45}$
$= \sqrt{4m^2 \cdot 5} - m(\sqrt{9} \cdot \sqrt{5})$
$= \sqrt{4m^2} \cdot \sqrt{5} - m(3\sqrt{5})$
$= 2m\sqrt{5} - 3m\sqrt{5} = -m\sqrt{5}$

66. $3k\sqrt{8k^2n} + 5k^2\sqrt{2n}$
$= 3k(\sqrt{4k^2} \cdot \sqrt{2n}) + 5k^2\sqrt{2n}$
$= 3k(2k\sqrt{2n}) + 5k^2\sqrt{2n}$
$= 6k^2\sqrt{2n} + 5k^2\sqrt{2n}$
$= (6k^2 + 5k^2)\sqrt{2n}$
$= 11k^2\sqrt{2n}$

67. $\frac{10}{\sqrt{3}} = \frac{10 \cdot \sqrt{3}}{\sqrt{3} \cdot \sqrt{3}} = \frac{10\sqrt{3}}{3}$

68. $\frac{8\sqrt{2}}{\sqrt{5}} = \frac{8\sqrt{2} \cdot \sqrt{5}}{\sqrt{5} \cdot \sqrt{5}} = \frac{8\sqrt{10}}{5}$

69. $\frac{12}{\sqrt{24}} = \frac{12}{\sqrt{4 \cdot 6}} = \frac{12}{2\sqrt{6}}$
$= \frac{12 \cdot \sqrt{6}}{2\sqrt{6} \cdot \sqrt{6}} = \frac{12\sqrt{6}}{2 \cdot 6}$
$= \frac{12\sqrt{6}}{12} = \sqrt{6}$

70. $\sqrt{\frac{2}{5}} = \frac{\sqrt{2}}{\sqrt{5}} = \frac{\sqrt{2} \cdot \sqrt{5}}{\sqrt{5} \cdot \sqrt{5}} = \frac{\sqrt{10}}{5}$

71. $\sqrt{\frac{5}{14}} \cdot \sqrt{28} = \sqrt{\frac{5}{14} \cdot 28}$
$= \sqrt{5 \cdot 2} = \sqrt{10}$

72. $\sqrt{\frac{2}{7}} \cdot \sqrt{\frac{1}{3}} = \sqrt{\frac{2}{7} \cdot \frac{1}{3}}$
$= \sqrt{\frac{2}{21}} = \frac{\sqrt{2}}{\sqrt{21}}$
$= \frac{\sqrt{2} \cdot \sqrt{21}}{\sqrt{21} \cdot \sqrt{21}} = \frac{\sqrt{42}}{21}$

73. $\sqrt{\frac{r^2}{16x}} = \frac{\sqrt{r^2}}{\sqrt{16x}}$
$= \frac{r \cdot \sqrt{x}}{\sqrt{16x} \cdot \sqrt{x}}$
$= \frac{r\sqrt{x}}{\sqrt{16x^2}} = \frac{r\sqrt{x}}{4x}$

74. $\sqrt[3]{\frac{1}{3}} = \frac{\sqrt[3]{1}}{\sqrt[3]{3}} = \frac{1 \cdot \sqrt[3]{3^2}}{\sqrt[3]{3} \cdot \sqrt[3]{3^2}}$
$= \frac{\sqrt[3]{3^2}}{\sqrt[3]{3^3}} = \frac{\sqrt[3]{9}}{3}$

75. $r = \sqrt{\dfrac{3V}{\pi h}} = \dfrac{\sqrt{3V}}{\sqrt{\pi h}}$

$= \dfrac{\sqrt{3V}}{\sqrt{\pi h}} \cdot \dfrac{\sqrt{\pi h}}{\sqrt{\pi h}}$

$= \dfrac{\sqrt{3V\pi h}}{\pi h}$

76. $r = \sqrt{\dfrac{S}{4\pi}} = \dfrac{\sqrt{S}}{\sqrt{4\pi}}$

$= \dfrac{\sqrt{S}}{\sqrt{4\pi}} \cdot \dfrac{\sqrt{\pi}}{\sqrt{\pi}}$

$= \dfrac{\sqrt{\pi S}}{\sqrt{4\pi^2}}$

$= \dfrac{\sqrt{\pi S}}{2\pi}$

77. $-\sqrt{3}\left(\sqrt{5} + \sqrt{27}\right)$

$= -\sqrt{3}\left(\sqrt{5}\right) + \left(-\sqrt{3}\right)\left(\sqrt{27}\right)$

$= -\sqrt{3 \cdot 5} - \sqrt{3 \cdot 27}$

$= -\sqrt{15} - \sqrt{81}$

$= -\sqrt{15} - 9$

78. $3\sqrt{2}\left(\sqrt{3} + 2\sqrt{2}\right)$

$= 3\sqrt{2}\left(\sqrt{3}\right) + 3\sqrt{2}\left(2\sqrt{2}\right)$

$= 3\sqrt{6} + 6 \cdot 2$

$= 3\sqrt{6} + 12$

79. $\left(2\sqrt{3} - 4\right)\left(5\sqrt{3} + 2\right)$

$= 2\sqrt{3}\left(5\sqrt{3}\right) + \left(2\sqrt{3}\right)(2) - 4\left(5\sqrt{3}\right)$
$\quad - 4(2)$ FOIL

$= 10 \cdot 3 + 4\sqrt{3} - 20\sqrt{3} - 8$

$= 30 - 16\sqrt{3} - 8$

$= 22 - 16\sqrt{3}$

80. $\left(\sqrt{7} + 2\sqrt{6}\right)\left(\sqrt{12} - \sqrt{2}\right)$

$= \sqrt{7}\left(\sqrt{12}\right) - \sqrt{7}\left(\sqrt{2}\right) + 2\sqrt{6}\left(\sqrt{12}\right)$
$\quad - 2\sqrt{6}\left(\sqrt{2}\right)$

$= \sqrt{84} - \sqrt{14} + 2\sqrt{72} - 2\sqrt{12}$

$= \sqrt{4 \cdot 21} - \sqrt{14} + 2\sqrt{36 \cdot 2} - 2\sqrt{4 \cdot 3}$

$= 2\sqrt{21} - \sqrt{14} + 2 \cdot 6\sqrt{2} - 2 \cdot 2\sqrt{3}$

$= 2\sqrt{21} - \sqrt{14} + 12\sqrt{2} - 4\sqrt{3}$

81. $\left(2\sqrt{3} + 5\right)\left(2\sqrt{3} - 5\right)$

$= \left(2\sqrt{3}\right)^2 - (5)^2$

$= 4 \cdot 3 - 25$

$= 12 - 25 = -13$

82. $\left(\sqrt{x} + 2\right)^2 = \left(\sqrt{x}\right)^2 + 2\left(\sqrt{x}\right)(2) + 2^2$

$= x + 4\sqrt{x} + 4$

83. $\dfrac{1}{2 + \sqrt{5}}$

$= \dfrac{1\left(2 - \sqrt{5}\right)}{\left(2 + \sqrt{5}\right)\left(2 - \sqrt{5}\right)}$ Multiply by the conjugate.

$= \dfrac{2 - \sqrt{5}}{(2)^2 - \left(\sqrt{5}\right)^2} = \dfrac{2 - \sqrt{5}}{4 - 5}$

$= \dfrac{2 - \sqrt{5}}{-1} = -2 + \sqrt{5}$

84. $\dfrac{2}{\sqrt{2} - 3}$

$= \dfrac{2\left(\sqrt{2} + 3\right)}{\left(\sqrt{2} - 3\right)\left(\sqrt{2} + 3\right)}$ Multiply by the conjugate.

$= \dfrac{2\left(\sqrt{2} + 3\right)}{\left(\sqrt{2}\right)^2 - (3)^2} = \dfrac{2\left(\sqrt{2} + 3\right)}{2 - 9}$

$= \dfrac{2\left(\sqrt{2} + 3\right)}{-7} = \dfrac{\left(2\sqrt{2} + 6\right)(-1)}{-7(-1)}$

$= \dfrac{-2\sqrt{2} - 6}{7}$

85. $\dfrac{3}{1 + \sqrt{x}} = \dfrac{3\left(1 - \sqrt{x}\right)}{\left(1 + \sqrt{x}\right)\left(1 - \sqrt{x}\right)}$

$= \dfrac{3\left(1 - \sqrt{x}\right)}{1 - x}$

86. $\dfrac{\sqrt{8}}{\sqrt{2} + 6}$

$= \dfrac{\sqrt{8}\left(\sqrt{2} - 6\right)}{\left(\sqrt{2} + 6\right)\left(\sqrt{2} - 6\right)}$ Multiply by the conjugate.

$= \dfrac{\sqrt{16} - 6\sqrt{8}}{\left(\sqrt{2}\right)^2 - 6^2}$

$= \dfrac{4 - 6 \cdot \sqrt{4 \cdot 2}}{2 - 36}$

continued

$$= \frac{4 - 6 \cdot 2\sqrt{2}}{-34}$$

$$= \frac{4 - 12\sqrt{2}}{-34}$$

$$= \frac{-2(-2 + 6\sqrt{2})}{-2(17)} \quad \text{Factor numerator and denominator}$$

$$= \frac{-2 + 6\sqrt{2}}{17} \quad \text{Lowest terms}$$

87. $\dfrac{\sqrt{5} - 1}{\sqrt{2} + 3} = \dfrac{(\sqrt{5} - 1)(\sqrt{2} - 3)}{(\sqrt{2} + 3)(\sqrt{2} - 3)}$

$$= \frac{\sqrt{10} - 3\sqrt{5} - \sqrt{2} + 3}{2 - 9}$$

$$= \frac{\sqrt{10} - 3\sqrt{5} - \sqrt{2} + 3}{-7}$$

$$= \frac{-\sqrt{10} + 3\sqrt{5} + \sqrt{2} - 3}{7}$$

88. $\dfrac{2 + \sqrt{6}}{\sqrt{3} - 1} = \dfrac{(2 + \sqrt{6})(\sqrt{3} + 1)}{(\sqrt{3} - 1)(\sqrt{3} + 1)}$

$$= \frac{2\sqrt{3} + 2 + \sqrt{18} + \sqrt{6}}{3 - 1}$$

$$= \frac{2\sqrt{3} + 2 + \sqrt{9 \cdot 2} + \sqrt{6}}{2}$$

$$= \frac{2\sqrt{3} + 2 + 3\sqrt{2} + \sqrt{6}}{2}$$

89. $\dfrac{15 + 10\sqrt{6}}{15} = \dfrac{5(3 + 2\sqrt{6})}{5(3)} \quad \text{Factor.}$

$$= \frac{3 + 2\sqrt{6}}{3} \quad \text{Lowest terms}$$

90. $\dfrac{3 + 9\sqrt{7}}{12} = \dfrac{3(1 + 3\sqrt{7})}{3(4)} \quad \text{Factor.}$

$$= \frac{1 + 3\sqrt{7}}{4} \quad \text{Lowest terms}$$

91. $\dfrac{6 + \sqrt{192}}{2} = \dfrac{6 + \sqrt{64 \cdot 3}}{2}$

$$= \frac{6 + 8\sqrt{3}}{2}$$

$$= \frac{2(3 + 4\sqrt{3})}{2}$$

$$= 3 + 4\sqrt{3}$$

92. $\sqrt{x} + 5 = 0$

$\sqrt{x} = -5$

Since a square root cannot equal a negative number, there is no solution.

93. $\sqrt{k + 1} = 7$

$(\sqrt{k + 1})^2 = 7^2$

$k + 1 = 49$

$k = 48$

Check $k = 48$: $\sqrt{49} = 7$ True

The solution is 48.

94. $\sqrt{5t + 4} = 3\sqrt{t}$

$(\sqrt{5t + 4})^2 = (3\sqrt{t})^2$

$5t + 4 = 9t$

$4 = 4t$

$1 = t$

Check $t = 1$: $\sqrt{9} = 3\sqrt{1}$ True

The solution is 1.

95. $\sqrt{2p + 3} = \sqrt{5p - 3}$

$(\sqrt{2p + 3})^2 = (\sqrt{5p - 3})^2$

$2p + 3 = 5p - 3$

$6 = 3p$

$2 = p$

Check $p = 2$: $\sqrt{7} = \sqrt{7}$ True

The solution is 2.

96. $\sqrt{4x + 1} = x - 1$

$(\sqrt{4x + 1})^2 = (x - 1)^2$

$4x + 1 = x^2 - 2x + 1$

$0 = x^2 - 6x$

$0 = x(x - 6)$

$x = 0 \quad \text{or} \quad x = 6$

Check $x = 0$: $\quad 1 = -1$ False
Check $x = 6$: $\sqrt{25} = 5 \quad$ True

Of the two potential solutions, 6 checks in the original equation, but 0 does not. Thus, the solution is 6.

97. $\sqrt{13 + 4t} = t + 4$

$(\sqrt{13 + 4t})^2 = (t + 4)^2$

$13 + 4t = t^2 + 8t + 16$

$0 = t^2 + 4t + 3$

$0 = (t + 3)(t + 1)$

$t = -3 \quad \text{or} \quad t = -1$

Check $t = -3$: $1 = 1 \quad$ True
Check $t = -1$: $3 = 3 \quad$ True

The solutions are -3 and -1.

98. $\sqrt{2-x} + 3 = x + 7$
$\sqrt{2-x} = x + 4$ *Isolate the radical.*
$\left(\sqrt{2-x}\right)^2 = (x+4)^2$
$2 - x = x^2 + 8x + 16$
$0 = x^2 + 9x + 14$
$0 = (x+2)(x+7)$
$x = -2 \text{ or } x = -7$

Check $x = -2$: $2 + 3 = 5$ *True*
Check $x = -7$: $3 + 3 = 0$ *False*

Of the two potential solutions, -2 checks in the original equation, but -7 does not. Thus, the solution is -2.

99. $\sqrt{x} - x + 2 = 0$
$\sqrt{x} = x - 2$ *Isolate the radical.*
$\left(\sqrt{x}\right)^2 = (x-2)^2$
$x = x^2 - 4x + 4$
$0 = x^2 - 5x + 4$
$0 = (x-4)(x-1)$
$x = 4 \text{ or } x = 1$

Check $x = 4$: $0 = 0$ *True*
Check $x = 1$: $2 = 0$ *False*

Of the two potential solutions, 4 checks in the original equation, but 1 does not. Thus, the solution is 4.

100. $\sqrt{x+2} - \sqrt{x-3} = 1$
$\sqrt{x+2} = 1 + \sqrt{x-3}$
$\left(\sqrt{x+2}\right)^2 = \left(1 + \sqrt{x-3}\right)^2$
$x + 2 = 1 + 2\sqrt{x-3} + x - 3$
$4 = 2\sqrt{x-3}$
$2 = \sqrt{x-3}$
$2^2 = \left(\sqrt{x-3}\right)^2$
$4 = x - 3$
$7 = x$

Check $x = 7$: $1 = 1$ *True*

The solution is 7.

101. $\sqrt{3} \cdot \sqrt{27} = \sqrt{3 \cdot 27} = \sqrt{81} = 9$

102. $2\sqrt{27} + 3\sqrt{75} - \sqrt{300}$
$= 2\sqrt{9 \cdot 3} + 3\sqrt{25 \cdot 3} - \sqrt{100 \cdot 3}$
$= 2 \cdot 3\sqrt{3} + 3 \cdot 5\sqrt{3} - 10\sqrt{3}$
$= 6\sqrt{3} + 15\sqrt{3} - 10\sqrt{3}$
$= 11\sqrt{3}$

103. $\sqrt{\dfrac{121}{t^2}} = \dfrac{\sqrt{121}}{\sqrt{t^2}} = \dfrac{11}{t}$

104. $\dfrac{1}{5 + \sqrt{2}} = \dfrac{1\left(5 - \sqrt{2}\right)}{\left(5 + \sqrt{2}\right)\left(5 - \sqrt{2}\right)}$
$= \dfrac{5 - \sqrt{2}}{(5)^2 - \left(\sqrt{2}\right)^2}$
$= \dfrac{5 - \sqrt{2}}{25 - 2}$
$= \dfrac{5 - \sqrt{2}}{23}$

105. $\sqrt{\dfrac{1}{3}} \cdot \sqrt{\dfrac{24}{5}} = \sqrt{\dfrac{1}{3} \cdot \dfrac{24}{5}} = \sqrt{\dfrac{8}{5}} = \dfrac{\sqrt{8}}{\sqrt{5}}$
$= \dfrac{\sqrt{8} \cdot \sqrt{5}}{\sqrt{5} \cdot \sqrt{5}} = \dfrac{\sqrt{40}}{5}$
$= \dfrac{\sqrt{4 \cdot 10}}{5} = \dfrac{2\sqrt{10}}{5}$

106. $\sqrt{50y^2} = \sqrt{25y^2 \cdot 2}$
$= \sqrt{25y^2} \cdot \sqrt{2}$
$= 5y\sqrt{2}$

107. $\sqrt[3]{-125} = -5$ because $(-5)^3 = -125$.

108. $-\sqrt{5}\left(\sqrt{2} + \sqrt{75}\right)$
$= -\sqrt{5}\left(\sqrt{2}\right) + \left(-\sqrt{5}\right)\left(\sqrt{75}\right)$
$= -\sqrt{10} - \sqrt{375}$
$= -\sqrt{10} - \sqrt{25 \cdot 15}$
$= -\sqrt{10} - 5\sqrt{15}$

109. $\sqrt{\dfrac{16r^3}{3s}} = \dfrac{\sqrt{16r^3}}{\sqrt{3s}} = \dfrac{\sqrt{16r^2} \cdot \sqrt{r}}{\sqrt{3s}}$
$= \dfrac{4r\sqrt{r}}{\sqrt{3s}} = \dfrac{4r\sqrt{r} \cdot \sqrt{3s}}{\sqrt{3s} \cdot \sqrt{3s}}$
$= \dfrac{4r\sqrt{3rs}}{3s}$

110. $\dfrac{12 + 6\sqrt{13}}{12} = \dfrac{6\left(2 + \sqrt{13}\right)}{6(2)}$
$= \dfrac{2 + \sqrt{13}}{2}$

111. $-\sqrt{162} + \sqrt{8} = -\sqrt{81 \cdot 2} + \sqrt{4 \cdot 2}$
$= -9\sqrt{2} + 2\sqrt{2}$
$= -7\sqrt{2}$

340 Chapter 8 Roots and Radicals

112. $\left(\sqrt{5} - \sqrt{2}\right)^2$
$= \left(\sqrt{5}\right)^2 - 2\sqrt{5}\sqrt{2} + \left(\sqrt{2}\right)^2$
    Square of a binomial
$= 5 - 2\sqrt{10} + 2$
$= 7 - 2\sqrt{10}$

113. $\left(6\sqrt{7} + 2\right)\left(4\sqrt{7} - 1\right)$
$= 6\sqrt{7}\left(4\sqrt{7}\right) - 1\left(6\sqrt{7}\right)$
$\quad + 2\left(4\sqrt{7}\right) + 2(-1)$ FOIL
$= 24 \cdot 7 - 6\sqrt{7} + 8\sqrt{7} - 2$
$= 168 - 2 + 2\sqrt{7}$
$= 166 + 2\sqrt{7}$

114. $-\sqrt{121} = -11$

115. $\sqrt{98} = \sqrt{49 \cdot 2} = \sqrt{49} \cdot \sqrt{2} = 7\sqrt{2}$

116. $\sqrt{x+2} = x - 4$
$\left(\sqrt{x+2}\right)^2 = (x-4)^2$
$x + 2 = x^2 - 8x + 16$
$0 = x^2 - 9x + 14$
$0 = (x-2)(x-7)$
$x = 2 \quad \text{or} \quad x = 7$

Check $x = 2$: $\sqrt{4} = -2$ *False*
Check $x = 7$: $\sqrt{9} = 3$ *True*

The solution is 7.

117. $\sqrt{k} + 3 = 0$
$\sqrt{k} = -3$

Since a square root cannot equal a negative number, there is no solution.

118. $\sqrt{1+3t} - t = -3$
$\sqrt{1+3t} = t - 3$
$\left(\sqrt{1+3t}\right)^2 = (t-3)^2$
$1 + 3t = t^2 - 6t + 9$
$0 = t^2 - 9t + 8$
$0 = (t-1)(t-8)$
$t = 1 \quad \text{or} \quad t = 8$

Check $t = 1$: $2 - 1 = -3$ *False*
Check $t = 8$: $5 - 8 = -3$ *True*

The solution is 8.

119. $S = \dfrac{2.74D}{\sqrt{h}}$
$= \dfrac{2.74(32)}{\sqrt{5}}$ Let $D = 32$, $h = 5$.
≈ 39.2

The fall speed is about 39.2 mph.

In Exercises 120–124, consider the points
$A\left(2\sqrt{14}, 5\sqrt{7}\right)$ and $B\left(-3\sqrt{14}, 10\sqrt{7}\right)$.

120. $m = \dfrac{y_2 - y_1}{x_2 - x_1} = \dfrac{10\sqrt{7} - 5\sqrt{7}}{-3\sqrt{14} - 2\sqrt{14}}$

An equivalent expression for the slope, obtained by using the two points in the reverse order is

$\dfrac{5\sqrt{7} - 10\sqrt{7}}{2\sqrt{14} - \left(-3\sqrt{14}\right)} = \dfrac{5\sqrt{7} - 10\sqrt{7}}{2\sqrt{14} + 3\sqrt{14}}$.

121. $\dfrac{10\sqrt{7} - 5\sqrt{7}}{-3\sqrt{14} - 2\sqrt{14}} = \dfrac{5\sqrt{7}}{-5\sqrt{14}}$ or $-\dfrac{\sqrt{7}}{\sqrt{14}}$

122. $-\dfrac{\sqrt{7}}{\sqrt{14}} = -\sqrt{\dfrac{7}{14}} = -\sqrt{\dfrac{1}{2}}$

123. $-\sqrt{\dfrac{1}{2}} = -\dfrac{\sqrt{1}}{\sqrt{2}} = -\dfrac{1}{\sqrt{2}}$
$= -\dfrac{1}{\sqrt{2}} \cdot \dfrac{\sqrt{2}}{\sqrt{2}} = -\dfrac{\sqrt{2}}{2}$

124. The slope is negative, so the line AB falls from left to right.

Chapter 8 Test

1. The square roots of 196 are -14 and 14 because $(-14)^2 = 196$ and $14^2 = 196$.

2. (a) $\sqrt{142}$ is *irrational* because 142 is not a perfect square.

   (b) $\sqrt{142} \approx 11.916$

3. If $\sqrt{a}$ is not a real number, then a must be a negative number.

4. $\sqrt[3]{216} = 6$ because $6^3 = 216$.

5. $-\sqrt{27} = -\sqrt{9 \cdot 3} = -\sqrt{9} \cdot \sqrt{3} = -3\sqrt{3}$

6. $\sqrt{\dfrac{128}{25}} = \dfrac{\sqrt{128}}{\sqrt{25}} = \dfrac{\sqrt{64 \cdot 2}}{5} = \dfrac{8\sqrt{2}}{5}$

7. $\sqrt[3]{32} = \sqrt[3]{8 \cdot 4} = \sqrt[3]{8} \cdot \sqrt[3]{4} = 2\sqrt[3]{4}$

8. $\dfrac{20\sqrt{18}}{5\sqrt{3}} = \dfrac{4\sqrt{9 \cdot 2}}{\sqrt{3}}$

$= \dfrac{4 \cdot 3\sqrt{2}}{\sqrt{3}}$

$= \dfrac{12\sqrt{2} \cdot \sqrt{3}}{\sqrt{3} \cdot \sqrt{3}}$

$= \dfrac{12\sqrt{6}}{3} = 4\sqrt{6}$

9. $3\sqrt{28} + \sqrt{63} = 3\left(\sqrt{4 \cdot 7}\right) + \sqrt{9 \cdot 7}$

$= 3\left(2\sqrt{7}\right) + 3\sqrt{7}$

$= 6\sqrt{7} + 3\sqrt{7} = 9\sqrt{7}$

10. $3\sqrt{27x} - 4\sqrt{48x} + 2\sqrt{3x}$

$= 3\left(\sqrt{9 \cdot 3x}\right) - 4\left(\sqrt{16 \cdot 3x}\right) + 2\sqrt{3x}$

$= 3\left(3\sqrt{3x}\right) - 4\left(4\sqrt{3x}\right) + 2\sqrt{3x}$

$= 9\sqrt{3x} - 16\sqrt{3x} + 2\sqrt{3x} = -5\sqrt{3x}$

11. $\sqrt{32x^2y^3} = \sqrt{16x^2y^2 \cdot 2y}$

$= \sqrt{16x^2y^2} \cdot \sqrt{2y}$

$= 4xy\sqrt{2y}$

12. $\left(6 - \sqrt{5}\right)\left(6 + \sqrt{5}\right)$

$= (6)^2 - \left(\sqrt{5}\right)^2$

$= 36 - 5 = 31$

13. $\left(2 - \sqrt{7}\right)\left(3\sqrt{2} + 1\right)$

$= 2\left(3\sqrt{2}\right) + 2(1) - \sqrt{7}\left(3\sqrt{2}\right) - \sqrt{7}(1)$

$= 6\sqrt{2} + 2 - 3\sqrt{14} - \sqrt{7}$

14. $\left(\sqrt{5} + \sqrt{6}\right)^2$

$= \left(\sqrt{5}\right)^2 + 2\left(\sqrt{5}\right)\left(\sqrt{6}\right) + \left(\sqrt{6}\right)^2$

$= 5 + 2\sqrt{30} + 6$

$= 11 + 2\sqrt{30}$

15. Use the Pythagorean formula with $c = 9$ and $b = 3$.

$c^2 = a^2 + b^2$
$9^2 = a^2 + 3^2$
$81 = a^2 + 9$
$72 = a^2$
$\sqrt{72} = a$

(a) $a = \sqrt{72} = \sqrt{36 \cdot 2} = 6\sqrt{2}$ inches

(b) $a = \sqrt{72} \approx 8.485$ inches

16. $Z = \sqrt{R^2 + X^2}$

$= \sqrt{40^2 + 30^2}$ Let $R = 40$, $X = 30$.

$= \sqrt{1600 + 900}$

$= \sqrt{2500} = 50$ ohms

17. $\dfrac{5\sqrt{2}}{\sqrt{7}} = \dfrac{5\sqrt{2} \cdot \sqrt{7}}{\sqrt{7} \cdot \sqrt{7}} = \dfrac{5\sqrt{14}}{7}$

18. $\sqrt{\dfrac{2}{3x}} = \dfrac{\sqrt{2}}{\sqrt{3x}} = \dfrac{\sqrt{2} \cdot \sqrt{3x}}{\sqrt{3x} \cdot \sqrt{3x}} = \dfrac{\sqrt{6x}}{3x}$

19. $\dfrac{-2}{\sqrt[3]{4}} = \dfrac{-2 \cdot \sqrt[3]{2}}{\sqrt[3]{4} \cdot \sqrt[3]{2}} = \dfrac{-2\sqrt[3]{2}}{\sqrt[3]{8}}$

$= \dfrac{-2\sqrt[3]{2}}{2} = -\sqrt[3]{2}$

20. $\dfrac{-3}{4 - \sqrt{3}} = \dfrac{-3\left(4 + \sqrt{3}\right)}{\left(4 - \sqrt{3}\right)\left(4 + \sqrt{3}\right)}$

$= \dfrac{-12 - 3\sqrt{3}}{(4)^2 - \left(\sqrt{3}\right)^2}$

$= \dfrac{-12 - 3\sqrt{3}}{16 - 3}$

$= \dfrac{-12 - 3\sqrt{3}}{13}$

21. $\dfrac{\sqrt{12} + 3\sqrt{128}}{6} = \dfrac{\sqrt{4} \cdot \sqrt{3} + 3 \cdot \sqrt{64} \cdot \sqrt{2}}{6}$

$= \dfrac{2\sqrt{3} + 24\sqrt{2}}{6}$

$= \dfrac{2\left(\sqrt{3} + 12\sqrt{2}\right)}{2(3)}$

$= \dfrac{\sqrt{3} + 12\sqrt{2}}{3}$

22. $\sqrt{p} + 4 = 0$

$\sqrt{p} = -4$

Since a square root cannot equal a negative number, there is no solution.

23. $\sqrt{x + 1} = 5 - x$

$\left(\sqrt{x + 1}\right)^2 = (5 - x)^2$

$x + 1 = 25 - 10x + x^2$

$0 = x^2 - 11x + 24$

$0 = (x - 3)(x - 8)$

$x = 3$ or $x = 8$

Check $x = 3$: $\sqrt{4} = 2$ *True*
Check $x = 8$: $\sqrt{9} = -3$ *False*

The solution is 3.

342 Chapter 8 Roots and Radicals

24. $3\sqrt{x} - 2 = x$
$3\sqrt{x} = x + 2$
$(3\sqrt{x})^2 = (x+2)^2$
$9x = x^2 + 4x + 4$
$0 = x^2 - 5x + 4$
$0 = (x-4)(x-1)$
$x - 4 = 0$ or $x - 1 = 0$
$x = 4$ or $x = 1$

Check $x = 4$: $4 = 4$ True
Check $x = 1$: $1 = 1$ True

The solutions are 1 and 4.

25. Nothing is wrong with the steps taken so far, but the potential solution must be checked.

Let $x = 12$ in the original equation.

$\sqrt{2x+1} + 5 = 0$
$\sqrt{2(12)+1} + 5 = 0$? *Let $x = 12$.*
$\sqrt{25} + 5 = 0$?
$5 + 5 = 0$?
$10 = 0$ False

12 is not a solution because it does not satisfy the original equation. The equation has no solution.

Cumulative Review Exercises
(Chapters R–8)

1. $3(6+7) + 6 \cdot 4 - 3^2$
$= 3(13) + 24 - 9$
$= 39 + 24 - 9$
$= 63 - 9 = 54$

2. $\dfrac{3(6+7) + 3}{2(4) - 1} = \dfrac{3(13) + 3}{8 - 1}$
$= \dfrac{39 + 3}{7}$
$= \dfrac{42}{7} = 6$

3. $|-6| - |-3| = 6 - 3 = 3$

4. $5(k-4) - k = k - 11$
$5k - 20 - k = k - 11$
$4k - 20 = k - 11$
$3k = 9$
$k = 3$

The solution is 3.

5. $-\dfrac{3}{4}x \leq 12$
$-\dfrac{4}{3}\left(-\dfrac{3}{4}x\right) \geq -\dfrac{4}{3}(12)$
$x \geq -16$

6. $5z + 3 - 4 > 2z + 9 + z$
$5z - 1 > 3z + 9$
$2z > 10$
$z > 5$

7. Let $x =$ the amount earned by Brazile.
Then $x + \$22{,}422 =$ the amount earned by Ohl.
Together they earned \$570,416, so

$x + (x + 22{,}422) = 570{,}416$
$2x + 22{,}422 = 570{,}416$
$2x = 547{,}994$
$x = 273{,}997$

Brazile earned \$273,997 and Ohl earned
$\$273{,}997 + \$22{,}422 = \$296{,}419$.

8. $-4x + 5y = -20$

Find the intercepts.

If $y = 0$, $x = 5$, so the x-intercept is $(5, 0)$.
If $x = 0$, $y = -4$, so the y-intercept is $(0, -4)$.

Draw the line that passes through the points $(5, 0)$ and $(0, -4)$.

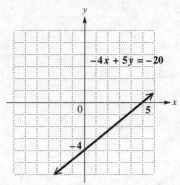

9. $x = 2$

For any value of y, the value of x is 2, so this is a vertical line through $(2, 0)$.

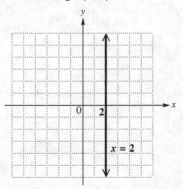

10. $2x - 5y > 10$

The boundary, $2x - 5y = 10$, is the line that passes through $(5, 0)$ and $(0, -2)$; draw it as a dashed line because of the $>$ symbol. Use $(0, 0)$ as a test point. Because

$$2(0) - 5(0) > 10$$

is a false statement, shade the side of the dashed boundary that does not include the origin, $(0, 0)$.

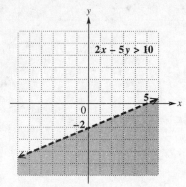

11. The slope of the line through the points $(1976, 2.2)$ and $(2000, 13.5)$ is

$$m = \frac{y_2 - y_1}{x_2 - x_1}$$
$$= \frac{13.5 - 2.2}{2000 - 1976}$$
$$= \frac{11.3}{24} \approx .47$$

An interpretation of the slope is that convention spending increased $.47 million per year.

12. We will use the point-slope form of a line with $m = .47$ and $(x_1, y_1) = (2000, 13.5)$.

$$y - y_1 = m(x - x_1)$$
$$y - 13.5 = .47(x - 2000)$$
$$y - 13.5 = .47x - 940$$
$$y = .47x - 926.5$$

13. $y = .47(2004) - 926.5$ Let x = 2004.
    $= 941.88 - 926.5$
    $= 15.38$

The projected convention spending for 2004 is about $15.4 million.

14. $4x - y = 19$ (1)
    $3x + 2y = -5$ (2)

We will solve this system by the elimination method. Multiply both sides of equation (1) by 2, and then add the result to equation (2).

$$\begin{array}{rcrcr} 8x & - & 2y & = & 38 \\ 3x & + & 2y & = & -5 \\ \hline 11x & & & = & 33 \\ & & x & = & 3 \end{array}$$

Let $x = 3$ in equation (1).

$$4(3) - y = 19$$
$$12 - y = 19$$
$$-y = 7$$
$$y = -7$$

The solution is $(3, -7)$.

15. $2x - y = 6$ (1)
    $3y = 6x - 18$ (2)

We will solve this system by the substitution method. Solve equation (2) for y by dividing both sides by 3.

$$y = 2x - 6$$

Substitute $2x - 6$ for y in (1).

$$2x - (2x - 6) = 6$$
$$2x - 2x + 6 = 6$$
$$6 = 6$$

This true statement indicates that the two original equations both describe the same line. This system has an infinite number of solutions.

16. Let $x =$ the average speed of the slower car (departing from Des Moines).
    Then $x + 7 =$ the average speed of the faster car (departing from Chicago).

In 3 hours, the slower car travels $3x$ miles and the faster car travels $3(x + 7)$ miles. The total distance traveled is 345 miles, so

$$3x + 3(x + 7) = 345$$
$$3x + 3x + 21 = 345$$
$$6x = 324$$
$$x = 54.$$

The car departing from Des Moines averaged 54 miles per hour and traveled $3(54) = 162$ miles. The car departing from Chicago averaged 61 miles per hour and traveled $3(61) = 183$ miles.

17. $(3x^6)(2x^2y)^2$
   $= (3x^6)(2)^2(x^2)^2(y)^2$
   $= (3x^6) \cdot 4x^4y^2$
   $= 12x^{10}y^2$

18. $\left(\dfrac{3^2y^{-2}}{2^{-1}y^3}\right)^{-3} = \left(\dfrac{2^{-1}y^3}{3^2y^{-2}}\right)^3$
   $= \left(\dfrac{y^3 \cdot y^2}{2^1 \cdot 3^2}\right)^3$
   $= \dfrac{(y^5)^3}{(18)^3} = \dfrac{y^{15}}{5832}$

19. $(10x^3 + 3x^2 - 9) - (7x^3 - 8x^2 + 4)$
   $= 10x^3 + 3x^2 - 9 - 7x^3 + 8x^2 - 4$
   $= 3x^3 + 11x^2 - 13$

20.
```
              4t² -  8t +  5
      ┌─────────────────────
2t + 3│8t³ -  4t² - 14t + 15
       8t³ + 12t²
      ─────────────
            -16t² - 14t
            -16t² - 24t
            ─────────────
                    10t + 15
                    10t + 15
                    ─────────
                           0
```
   The remainder is 0, so the answer is the quotient, $4t^2 - 8t + 5$.

21. $\$6.9 \times 10^{12} = \$6,900,000,000,000$
   Move the decimal point 12 places to the right.

22. $m^2 + 12m + 32 = (m + 8)(m + 4)$

23. $25t^4 - 36 = (5t^2)^2 - (6)^2$
   $= (5t^2 + 6)(5t^2 - 6)$

24. $12a^2 + 4ab - 5b^2 = (6a + 5b)(2a - b)$

25. $81z^2 + 72z + 16$
   $= (9z)^2 + 2(9z)(4) + 4^2$
   $= (9z + 4)^2$

26. $x^2 - 7x = -12$
   $x^2 - 7x + 12 = 0$
   $(x - 3)(x - 4) = 0$
   $x = 3 \text{ or } x = 4$
   The solutions are 3 and 4.

27. $(x + 4)(x - 1) = -6$
   $x^2 + 3x - 4 = -6$
   $x^2 + 3x + 2 = 0$
   $(x + 2)(x + 1) = 0$
   $x = -2 \text{ or } x = -1$
   The solutions are -2 and -1.

28. $\dfrac{x^2 - 3x - 4}{x^2 + 3x} \cdot \dfrac{x^2 + 2x - 3}{x^2 - 5x + 4}$
   $= \dfrac{(x-4)(x+1)}{x(x+3)} \cdot \dfrac{(x-1)(x+3)}{(x-4)(x-1)}$ Factor.
   $= \dfrac{x+1}{x}$ Lowest terms

29. $\dfrac{t^2 + 4t - 5}{t + 5} \div \dfrac{t - 1}{t^2 + 8t + 15}$
   $= \dfrac{t^2 + 4t - 5}{t + 5} \cdot \dfrac{t^2 + 8t + 15}{t - 1}$
   Multiply by the reciprocal.
   $= \dfrac{(t+5)(t-1)}{t+5} \cdot \dfrac{(t+5)(t+3)}{t-1}$ Factor.
   $= (t+5)(t+3)$ Lowest terms

30. $\dfrac{2}{x+3} - \dfrac{4}{x-1}$
   $= \dfrac{2(x-1)}{(x+3)(x-1)} - \dfrac{4(x+3)}{(x-1)(x+3)}$
   $= \dfrac{2(x-1) - 4(x+3)}{(x+3)(x-1)}$
   $= \dfrac{2x - 2 - 4x - 12}{(x+3)(x-1)}$
   $= \dfrac{-2x - 14}{(x+3)(x-1)}$

31. $\sqrt{27} - 2\sqrt{12} + 6\sqrt{75}$
   $= \sqrt{9} \cdot \sqrt{3} - 2\sqrt{4} \cdot \sqrt{3} + 6\sqrt{25} \cdot \sqrt{3}$
   $= 3\sqrt{3} - 2(2\sqrt{3}) + 6(5\sqrt{3})$
   $= 3\sqrt{3} - 4\sqrt{3} + 30\sqrt{3} = 29\sqrt{3}$

32. $\dfrac{2}{\sqrt{3} + \sqrt{5}} = \dfrac{2(\sqrt{3} - \sqrt{5})}{(\sqrt{3} + \sqrt{5})(\sqrt{3} - \sqrt{5})}$
   $= \dfrac{2(\sqrt{3} - \sqrt{5})}{3 - 5}$
   $= \dfrac{2(\sqrt{3} - \sqrt{5})}{-2}$
   $= \dfrac{\sqrt{3} - \sqrt{5}}{-1} = -\sqrt{3} + \sqrt{5}$

33. $\sqrt{200x^2y^5} = \sqrt{100x^2y^4 \cdot 2y}$
   $= \sqrt{100x^2y^4} \cdot \sqrt{2y}$
   $= 10xy^2\sqrt{2y}$

34. $\left(3\sqrt{2}+1\right)\left(4\sqrt{2}-3\right)$
$= 3\sqrt{2}\left(4\sqrt{2}\right) - 3\sqrt{2}(3) + 1\left(4\sqrt{2}\right)$
$\quad + 1(-3)$
$= 12 \cdot 2 - 9\sqrt{2} + 4\sqrt{2} - 3$
$= 24 - 3 - 5\sqrt{2}$
$= 21 - 5\sqrt{2}$

35. $\sqrt{x} + 2 = x - 10$
$\quad \sqrt{x} = x - 12$
$\quad \left(\sqrt{x}\right)^2 = (x-12)^2$
$\quad x = x^2 - 24x + 144$
$\quad 0 = x^2 - 25x + 144$
$\quad 0 = (x-16)(x-9)$
$x = 16 \quad \text{or} \quad x = 9$

Check $x = 16$: $6 = 6$ *True*
Check $x = 9$: $5 = -1$ *False*

The solution is 16.

CHAPTER 9 QUADRATIC EQUATIONS

9.1 Solving Quadratic Equations by the Square Root Property

9.1 Margin Exercises

1. **(a)** $x^2 = 49$

   Solve by the square root property.

   $x = \sqrt{49}$ or $x = -\sqrt{49}$
   $x = 7$ or $x = -7$

   The solutions are 7 and -7, which may be written as ± 7.

   (b) $x^2 = 11$

   By the square root property,

   $x = \sqrt{11}$ or $x = -\sqrt{11}$.

   The solutions are $\sqrt{11}$ and $-\sqrt{11}$.

   (c) $2x^2 + 8 = 32$
   $2x^2 = 24$
   $x^2 = 12$

   Use the square root property.

   $x = \sqrt{12}$ or $x = -\sqrt{12}$
   $x = \sqrt{4} \cdot \sqrt{3}$ or $x = -\sqrt{4} \cdot \sqrt{3}$
   $x = 2\sqrt{3}$ or $x = -2\sqrt{3}$

   The solutions are $2\sqrt{3}$ and $-2\sqrt{3}$.

   (d) $x^2 = -9$

   The square of a real number cannot be negative. (The square root property cannot be used because k must be positive.) Thus, there is *no real number solution* for this equation.

2. **(a)** $(x+2)^2 = 36$

   Use the square root property.

   $x + 2 = 6$ or $x + 2 = -6$
   $x = 4$ or $x = -8$

   The solutions are 4 and -8.

   (b) $(x-4)^2 = 3$

   Use the square root property.

   $x - 4 = \sqrt{3}$ or $x - 4 = -\sqrt{3}$
   $x = 4 + \sqrt{3}$ or $x = 4 - \sqrt{3}$

   The solutions are $4 + \sqrt{3}$ and $4 - \sqrt{3}$.

3. $(2x-5)^2 = 18$

   $2x - 5 = \sqrt{18}$ or $2x - 5 = -\sqrt{18}$
   $2x = 5 + \sqrt{18}$ or $2x = 5 - \sqrt{18}$
   $x = \dfrac{5 + \sqrt{18}}{2}$ or $x = \dfrac{5 - \sqrt{18}}{2}$
   $x = \dfrac{5 + \sqrt{9} \cdot \sqrt{2}}{2}$ or $x = \dfrac{5 - \sqrt{9} \cdot \sqrt{2}}{2}$
   $x = \dfrac{5 + 3\sqrt{2}}{2}$ or $x = \dfrac{5 - 3\sqrt{2}}{2}$

   The solutions are $\dfrac{5 + 3\sqrt{2}}{2}$ and $\dfrac{5 - 3\sqrt{2}}{2}$.

4. **(a)** $(5x+1)^2 = 7$

   $5x + 1 = \sqrt{7}$ or $5x + 1 = -\sqrt{7}$
   $5x = -1 + \sqrt{7}$ or $5x = -1 - \sqrt{7}$
   $x = \dfrac{-1 + \sqrt{7}}{5}$ or $x = \dfrac{-1 - \sqrt{7}}{5}$

   The solutions are $\dfrac{-1 + \sqrt{7}}{5}$ and $\dfrac{-1 - \sqrt{7}}{5}$.

   (b) $(7x-1)^2 = -1$

   Since the square root of -1 is not a real number, there is *no real number solution*.

5. $w = \dfrac{L^2 g}{1200}$ *Given formula*

   $2.80 = \dfrac{L^2 \cdot 11}{1200}$ *Let $w = 2.80$, $g = 11$.*

   $3360 = 11L^2$ *Multiply by 1200.*

   $L^2 \approx 305.45$ *Divide by 11.*

   $L \approx 17.48$ *Approximate $L > 0$.*

   The length of the bass is approximately 17.48 in.

9.1 Section Exercises

1. This statement is *true*.

3. This statement is *false*.

   If k is a positive integer that is not a perfect square, the solutions will be irrational.

5. This statement is *false*.

   For values of k that satisfy $0 \leq k < 10$; there are real solutions.

7. $x^2 = 81$

   Use the square root property to get

   $x = \sqrt{81} = 9$ or $x = -\sqrt{81} = -9$.

   The solutions are 9 and -9.

348 Chapter 9 Quadratic Equations

9. $k^2 = 14$

Use the square root property to get
$$k = \sqrt{14} \quad \text{or} \quad k = -\sqrt{14}.$$
The solutions are $\sqrt{14}$ and $-\sqrt{14}$.

11. $t^2 = 48$
$$t = \sqrt{48} \quad \text{or} \quad t = -\sqrt{48}$$
Write $\sqrt{48}$ in simplest form.
$$\sqrt{48} = \sqrt{16} \cdot \sqrt{3} = 4\sqrt{3}$$
The solutions are $4\sqrt{3}$ and $-4\sqrt{3}$.

13. $x^2 = \dfrac{25}{4}$
$$x = \sqrt{\dfrac{25}{4}} \quad \text{or} \quad x = -\sqrt{\dfrac{25}{4}}$$
$$x = \tfrac{5}{2} \quad \text{or} \quad x = -\tfrac{5}{2}$$
The solutions are $\tfrac{5}{2}$ and $-\tfrac{5}{2}$.

15. $x^2 = -100$

This equation has *no real number solution* because the square of a real number cannot be negative. The square root property cannot be used because it requires that k be positive.

17. $z^2 = 2.25$
$$z = \sqrt{2.25} \quad \text{or} \quad z = -\sqrt{2.25}$$
$$z = 1.5 \quad \text{or} \quad z = -1.5$$
The solutions are 1.5 and -1.5.

19. $r^2 - 3 = 0$
$$r^2 = 3$$
$$r = \sqrt{3} \quad \text{or} \quad r = -\sqrt{3}$$
The solutions are $\sqrt{3}$ and $-\sqrt{3}$.

21. $7x^2 = 4$
$$x^2 = \dfrac{4}{7} \quad \text{Divide by 7.}$$
$$x = \sqrt{\dfrac{4}{7}} \quad \text{or} \quad x = -\sqrt{\dfrac{4}{7}}$$
$$= \dfrac{\sqrt{4}}{\sqrt{7}} \cdot \dfrac{\sqrt{7}}{\sqrt{7}} \quad\quad = -\dfrac{\sqrt{4}}{\sqrt{7}} \cdot \dfrac{\sqrt{7}}{\sqrt{7}}$$
$$= \dfrac{2\sqrt{7}}{7} \quad\quad = -\dfrac{2\sqrt{7}}{7}$$
The solutions are $\dfrac{2\sqrt{7}}{7}$ and $-\dfrac{2\sqrt{7}}{7}$.

23. $5x^2 + 4 = 8$
$$5x^2 = 4 \quad \text{Subtract 4.}$$
$$x^2 = \dfrac{4}{5} \quad \text{Divide by 5.}$$
$$x = \sqrt{\dfrac{4}{5}} \quad \text{or} \quad x = -\sqrt{\dfrac{4}{5}}$$
$$= \dfrac{\sqrt{4}}{\sqrt{5}} \cdot \dfrac{\sqrt{5}}{\sqrt{5}} \quad\quad = -\dfrac{\sqrt{4}}{\sqrt{5}} \cdot \dfrac{\sqrt{5}}{\sqrt{5}}$$
$$= \dfrac{2\sqrt{5}}{5} \quad\quad = -\dfrac{2\sqrt{5}}{5}$$
The solutions are $\dfrac{2\sqrt{5}}{5}$ and $-\dfrac{2\sqrt{5}}{5}$.

25. $3x^2 - 8 = 64$
$$3x^2 = 72$$
$$x^2 = 24$$
Now use the square root property.
$$x = \pm\sqrt{24} = \pm\sqrt{4 \cdot 6} = \pm 2\sqrt{6}$$
The solutions are $2\sqrt{6}$ and $-2\sqrt{6}$.

27. $(x - 3)^2 = 25$

Use the square root property.
$$x - 3 = \sqrt{25} \quad \text{or} \quad x - 3 = -\sqrt{25}$$
$$x - 3 = 5 \quad \text{or} \quad x - 3 = -5$$
$$x = 8 \quad \text{or} \quad x = -2$$
The solutions are -2 and 8.

29. $(x + 5)^2 = -13$

The square root of -13 is not a real number, so there is *no real number solution* for this equation.

31. $(x - 8)^2 = 27$

Begin by using the square root property.
$$x - 8 = \sqrt{27} \quad \text{or} \quad x - 8 = -\sqrt{27}$$
Now simplify the radical.
$$\sqrt{27} = \sqrt{9} \cdot \sqrt{3} = 3\sqrt{3}$$
$$x - 8 = 3\sqrt{3} \quad \text{or} \quad x - 8 = -3\sqrt{3}$$
$$x = 8 + 3\sqrt{3} \quad \text{or} \quad x = 8 - 3\sqrt{3}$$
The solutions are $8 + 3\sqrt{3}$ and $8 - 3\sqrt{3}$.

33. $(3x + 2)^2 = 49$
$$3x + 2 = \sqrt{49} \quad \text{or} \quad 3x + 2 = -\sqrt{49}$$
$$3x + 2 = 7 \quad \text{or} \quad 3x + 2 = -7$$
$$3x = 5 \quad \text{or} \quad 3x = -9$$
$$x = \tfrac{5}{3} \quad \text{or} \quad x = -3$$
The solutions are $\tfrac{5}{3}$ and -3.

9.2 Solving Quadratic Equations by Completing the Square

35. $(4x - 3)^2 = 9$

$4x - 3 = \sqrt{9}$ or $4x - 3 = -\sqrt{9}$
$4x - 3 = 3$ or $4x - 3 = -3$
$4x = 6$ or $4x = 0$
$x = \frac{6}{4} = \frac{3}{2}$ or $x = 0$

The solutions are 0 and $\frac{3}{2}$.

37. $(5 - 2x)^2 = 30$

$5 - 2x = \sqrt{30}$ or $5 - 2x = -\sqrt{30}$
$-2x = -5 + \sqrt{30}$ or $-2x = -5 - \sqrt{30}$
$x = \frac{-5 + \sqrt{30}}{-2}$ or $x = \frac{-5 - \sqrt{30}}{-2}$
$x = \frac{-5 + \sqrt{30}}{-2} \cdot \frac{-1}{-1}$ or $x = \frac{-5 - \sqrt{30}}{-2} \cdot \frac{-1}{-1}$
$x = \frac{5 - \sqrt{30}}{2}$ or $x = \frac{5 + \sqrt{30}}{2}$

The solutions are $\frac{5 + \sqrt{30}}{2}$ and $\frac{5 - \sqrt{30}}{2}$.

39. $(3x + 1)^2 = 18$

$3x + 1 = \sqrt{18}$ or $3x + 1 = -\sqrt{18}$
$3x = -1 + 3\sqrt{2}$ or $3x = -1 - 3\sqrt{2}$

Note that $\sqrt{18} = \sqrt{9 \cdot 2} = 3\sqrt{2}$.

$x = \frac{-1 + 3\sqrt{2}}{3}$ or $x = \frac{-1 - 3\sqrt{2}}{3}$

The solutions are $\frac{-1 + 3\sqrt{2}}{3}$ and $\frac{-1 - 3\sqrt{2}}{3}$.

41. $\left(\frac{1}{2}x + 5\right)^2 = 12$

$\frac{1}{2}x + 5 = \sqrt{12}$ or $\frac{1}{2}x + 5 = -\sqrt{12}$
$\frac{1}{2}x = -5 + 2\sqrt{3}$ or $\frac{1}{2}x = -5 - 2\sqrt{3}$

Note that $\sqrt{12} = \sqrt{4 \cdot 3} = 2\sqrt{3}$.

$x = 2\left(-5 + 2\sqrt{3}\right)$ or $x = 2\left(-5 - 2\sqrt{3}\right)$
$x = -10 + 4\sqrt{3}$ or $x = -10 - 4\sqrt{3}$

The solutions are $-10 + 4\sqrt{3}$ and $-10 - 4\sqrt{3}$.

43. $(4x - 1)^2 - 48 = 0$

$(4x - 1)^2 = 48$

$4x - 1 = \sqrt{48}$ or $4x - 1 = -\sqrt{48}$
$4x - 1 = 4\sqrt{3}$ or $4x - 1 = -4\sqrt{3}$
$4x = 1 + 4\sqrt{3}$ or $4x = 1 - 4\sqrt{3}$
$x = \frac{1 + 4\sqrt{3}}{4}$ or $x = \frac{1 - 4\sqrt{3}}{4}$

The solutions are $\frac{1 + 4\sqrt{3}}{4}$ and $\frac{1 - 4\sqrt{3}}{4}$.

45. Michael's first solution, $\frac{5 + \sqrt{30}}{2}$, is equivalent to Lindsay's second solution, $\frac{-5 - \sqrt{30}}{-2}$. This can be verified by multiplying $\frac{5 + \sqrt{30}}{2}$ by 1 in the form $\frac{-1}{-1}$. Similarly, Michael's second solution is equivalent to Lindsay's first one.

47. $d = 16t^2$
$4 = 16t^2$ Let $d = 4$.
$t^2 = \frac{4}{16} = \frac{1}{4}$
$t = \pm\sqrt{\frac{1}{4}} = \pm\frac{1}{2}$

Reject $-\frac{1}{2}$ as a solution, since negative time does not make sense. About $\frac{1}{2}$ second elapses between the dropping of the coin and the shot.

49. $A = \pi r^2$
$81\pi = \pi r^2$ Let $A = 81\pi$.
$81 = r^2$ Divide by π.
$r = 9$ or $r = -9$

Discard -9 since the radius cannot be negative. The radius is 9 inches.

51. Let $A = 110.25$ and $P = 100$.

$A = P(1 + r)^2$
$110.25 = 100(1 + r)^2$
$(1 + r)^2 = \frac{110.25}{100} = 1.1025$
$1 + r = \pm\sqrt{1.1025}$
$1 + r = \pm 1.05$
$r = -1 \pm 1.05$

So $r = -1 + 1.05 = .05$ or $r = -1 - 1.05 = -2.05$. Reject the solution -2.05. The rate is $r = .05$ or 5%.

9.2 Solving Quadratic Equations by Completing the Square

9.2 Margin Exercises

1. (a) $x^2 + \underline{} + 36$

Here $k^2 = 36$, so $k = 6$ and $2kx = 2(6)(x) = 12x$. The perfect square trinomial is $x^2 + \underline{12x} + 36$.

(b) $25x^2 - \underline{} + 4$

Replace x^2 by $25x^2 = (5x)^2$ and k^2 by $4 = 2^2$, to get the middle term

$$2(2)(5x) = 20x.$$

The perfect square trinomial is $25x^2 - \underline{20x} + 4$.

(c) $x^2 + 14x + \underline{}$

Here the middle term $14x$ must equal $2kx$.

$$2kx = 14x$$
$$k = 7 \quad \text{Divide by } 2x.$$

Thus, $k = 7$ and $k^2 = 49$. The required trinomial is $x^2 + 14x + \underline{49}$.

2. $x^2 - 4x - 1 = 0$

$$x^2 - 4x = 1$$
$$x^2 - 4x + 4 = 1 + 4 \quad \begin{array}{l} 2kx = -4x, \text{ so} \\ k = -2 \text{ and } k^2 = 4. \end{array}$$
$$(x - 2)^2 = 5$$
$$x - 2 = \sqrt{5} \quad \text{or} \quad x - 2 = -\sqrt{5}$$
$$x = 2 + \sqrt{5} \quad \text{or} \quad x = 2 - \sqrt{5}$$

The solutions are $2 + \sqrt{5}$ and $2 - \sqrt{5}$.

3. (a) $x^2 + 4x = 1$

Take half of the coefficient of x and square the result.

$$\tfrac{1}{2}(4) = 2, \text{ and } 2^2 = 4.$$

Add 4 to each side of the equation, and write the left side as a perfect square.

$$x^2 + 4x + 4 = 1 + 4$$
$$(x + 2)^2 = 5 \quad \text{Factor.}$$

Use the square root property to solve for x.

$$x + 2 = \sqrt{5} \quad \text{or} \quad x + 2 = -\sqrt{5}$$
$$x = -2 + \sqrt{5} \quad \text{or} \quad x = -2 - \sqrt{5}$$

The solutions are $-2 + \sqrt{5}$ and $-2 - \sqrt{5}$.

(b) $z^2 + 6z - 3 = 0$

First add 3 to each side of the equation.

$$x^2 + 6z = 3$$

Now take half the coefficient of z and square the result.

$$\tfrac{1}{2}(6) = 3, \text{ and } 3^2 = 9.$$

Add 9 to each side of the equation, and then write the left side as a perfect square.

$$z^2 + 6z + 9 = 3 + 9$$
$$(z + 3)^2 = 12 \quad \text{Factor.}$$

Use the square root property to solve for z.

$$z + 3 = \sqrt{12} \quad \text{or} \quad z + 3 = -\sqrt{12}$$
$$z + 3 = 2\sqrt{3} \quad \text{or} \quad z + 3 = -2\sqrt{3}$$
$$z = -3 + 2\sqrt{3} \quad \text{or} \quad z = -3 - 2\sqrt{3}$$

The solutions are $-3 + 2\sqrt{3}$ and $-3 - 2\sqrt{3}$.

4. (a) $9x^2 + 18x = -5$

Divide each side by 9 to get a coefficient of 1 for the x^2-term.

$$x^2 + 2x = -\tfrac{5}{9} \quad \text{Divide by 9.}$$

Take half the coefficient of x, or $\left(\tfrac{1}{2}\right)(2) = 1$, and square the result: $1^2 = 1$. Then add 1 to each side.

$$x^2 + 2x + 1 = -\tfrac{5}{9} + 1 \quad \text{Add 1.}$$
$$x^2 + 2x + 1 = \tfrac{4}{9} \quad 1 = \tfrac{9}{9}$$
$$(x + 1)^2 = \tfrac{4}{9} \quad \text{Factor.}$$

Apply the square root property, and solve for x.

$$x + 1 = \sqrt{\tfrac{4}{9}} \quad \text{or} \quad x + 1 = -\sqrt{\tfrac{4}{9}}$$
$$x + 1 = \tfrac{2}{3} \quad \text{or} \quad x + 1 = -\tfrac{2}{3}$$
$$x = -1 + \tfrac{2}{3} \quad \text{or} \quad x = -1 - \tfrac{2}{3}$$
$$x = -\tfrac{1}{3} \quad \text{or} \quad x = -\tfrac{5}{3}$$

The solutions are $-\tfrac{1}{3}$ and $-\tfrac{5}{3}$.

(b) $4t^2 - 24t + 11 = 0$

Divide each side of the equation by 4 to get 1 as the coefficient of the t^2-term.

$$t^2 - 6t + \tfrac{11}{4} = 0 \quad \text{Divide by 4.}$$
$$t^2 - 6t = -\tfrac{11}{4} \quad \text{Subtract } \tfrac{11}{4}.$$

Square half the coefficient of t.

$$\left[\tfrac{1}{2}(-6)\right]^2 = (-3)^2 = 9$$

Then add 9 to each side.

$$t^2 - 6t + 9 = -\tfrac{11}{4} + 9 \quad \text{Add 9.}$$
$$t^2 - 6t + 9 = \tfrac{25}{4} \quad 9 = \tfrac{36}{4}$$
$$(t - 3)^2 = \tfrac{25}{4} \quad \text{Factor.}$$

Use the square root property.

$$t - 3 = \sqrt{\frac{25}{4}} \quad \text{or} \quad t - 3 = -\sqrt{\frac{25}{4}}$$
$$t - 3 = \tfrac{5}{2} \quad \text{or} \quad t - 3 = -\tfrac{5}{2}$$
$$t = 3 + \tfrac{5}{2} \quad \text{or} \quad t = 3 - \tfrac{5}{2}$$
$$t = \tfrac{11}{2} \quad \text{or} \quad t = \tfrac{1}{2}$$

The solutions are $\tfrac{11}{2}$ and $\tfrac{1}{2}$.

5. **(a)**
$$3x^2 + 5x = 2$$
$$x^2 + \tfrac{5}{3}x = \tfrac{2}{3} \qquad \text{Divide by 3.}$$
$$x^2 + \tfrac{5}{3}x + \tfrac{25}{36} = \tfrac{2}{3} + \tfrac{25}{36} \qquad \text{Add} \left(\tfrac{1}{2} \cdot \tfrac{5}{3}\right)^2 = \tfrac{25}{36}.$$
$$\left(x + \tfrac{5}{6}\right)^2 = \tfrac{49}{36} \qquad \tfrac{2}{3} = \tfrac{24}{36}$$

Use the square root property.

$$x + \tfrac{5}{6} = \sqrt{\tfrac{49}{36}} \quad \text{or} \quad x + \tfrac{5}{6} = -\sqrt{\tfrac{49}{36}}$$
$$x + \tfrac{5}{6} = \tfrac{7}{6} \quad \text{or} \quad x + \tfrac{5}{6} = -\tfrac{7}{6}$$
$$x = \tfrac{2}{6} \quad \text{or} \quad x = -\tfrac{12}{6}$$
$$x = \tfrac{1}{3} \quad \text{or} \quad x = -2$$

The solutions are $\tfrac{1}{3}$ and -2.

(b) $2x^2 - 4x - 1 = 0$

Divide each side of the equation by 2 to get the coefficient of 1 for the x^2-term.

$$x^2 - 2x - \tfrac{1}{2} = 0$$

Get all the terms with variables on one side of the equals sign.

$$x^2 - 2x = \tfrac{1}{2}$$

Square half the coefficient of x.

$$\tfrac{1}{2}(-2) = -1, \text{ and } (-1)^2 = 1.$$

Add 1 to each side of the equation, and then write the left side as a perfect square.

$$x^2 - 2x + 1 = \tfrac{1}{2} + 1$$
$$(x - 1)^2 = \tfrac{3}{2}$$

Use the square root property.

$$x - 1 = \sqrt{\tfrac{3}{2}} \quad \text{or} \quad x - 1 = -\sqrt{\tfrac{3}{2}}$$

Simplify the radical.

$$\sqrt{\tfrac{3}{2}} = \tfrac{\sqrt{3}}{\sqrt{2}} = \tfrac{\sqrt{3} \cdot \sqrt{2}}{\sqrt{2} \cdot \sqrt{2}} = \tfrac{\sqrt{6}}{2}$$

Thus,

$$x - 1 = \tfrac{\sqrt{6}}{2} \quad \text{or} \quad x - 1 = -\tfrac{\sqrt{6}}{2}$$
$$x = 1 + \tfrac{\sqrt{6}}{2} \quad \text{or} \quad x = 1 - \tfrac{\sqrt{6}}{2}$$
$$x = \tfrac{2}{2} + \tfrac{\sqrt{6}}{2} \quad \text{or} \quad x = \tfrac{2}{2} - \tfrac{\sqrt{6}}{2}$$
$$x = \tfrac{2 + \sqrt{6}}{2} \quad \text{or} \quad x = \tfrac{2 - \sqrt{6}}{2}.$$

The solutions are $\tfrac{2 + \sqrt{6}}{2}$ and $\tfrac{2 - \sqrt{6}}{2}$.

6.
$$5x^2 + 3x + 1 = 0$$
$$x^2 + \tfrac{3}{5}x + \tfrac{1}{5} = 0 \qquad \text{Divide by 5.}$$
$$x^2 + \tfrac{3}{5}x = -\tfrac{1}{5} \qquad \text{Subtract } \tfrac{1}{5}.$$
$$x^2 + \tfrac{3}{5}x + \tfrac{9}{100} = -\tfrac{1}{5} + \tfrac{9}{100} \qquad \text{Add} \left(\tfrac{1}{2} \cdot \tfrac{3}{5}\right)^2 = \tfrac{9}{100}.$$
$$\left(x + \tfrac{3}{10}\right)^2 = -\tfrac{11}{100} \qquad -\tfrac{1}{5} = -\tfrac{20}{100}$$

The square root of $-\tfrac{11}{100}$ is not a real number, so the square root property does not apply. This equation has *no real number solution*.

7. **(a)**
$$r(r - 3) = -1$$
$$r^2 - 3r = -1$$
$$r^2 - 3r + \tfrac{9}{4} = -1 + \tfrac{9}{4} \qquad \text{Add} \left[\tfrac{1}{2}(-3)\right]^2 = \tfrac{9}{4}.$$
$$\left(r - \tfrac{3}{2}\right)^2 = \tfrac{5}{4}$$
$$r - \tfrac{3}{2} = \tfrac{\sqrt{5}}{2} \quad \text{or} \quad r - \tfrac{3}{2} = -\tfrac{\sqrt{5}}{2}$$
$$r = \tfrac{3 + \sqrt{5}}{2} \quad \text{or} \quad r = \tfrac{3 - \sqrt{5}}{2}$$

The solutions are $\tfrac{3 + \sqrt{5}}{2}$ and $\tfrac{3 - \sqrt{5}}{2}$.

(b) $(x+2)(x+1) = 5$

$$x^2 + 3x + 2 = 5$$

$$x^2 + 3x = 3$$

$$x^2 + 3x + \frac{9}{4} = 3 + \frac{9}{4} \quad \text{Add } \left[\frac{1}{2}(3)\right]^2 = \frac{9}{4}.$$

$$\left(x + \frac{3}{2}\right)^2 = \frac{21}{4}$$

$$x + \frac{3}{2} = \sqrt{\frac{21}{4}} \quad \text{or} \quad x + \frac{3}{2} = -\sqrt{\frac{21}{4}}$$

$$x + \frac{3}{2} = \frac{\sqrt{21}}{2} \quad \text{or} \quad x + \frac{3}{2} = -\frac{\sqrt{21}}{2}$$

$$x = -\frac{3}{2} + \frac{\sqrt{21}}{2} \quad \text{or} \quad x = -\frac{3}{2} - \frac{\sqrt{21}}{2}$$

The solutions are $\dfrac{-3 + \sqrt{21}}{2}$ and $\dfrac{-3 - \sqrt{21}}{2}$.

8. $s = -16t^2 + 128t$

Let $s = 48$ and solve for t.

$$-16t^2 + 128t = 48$$

$$t^2 - 8t = -3 \quad \text{Divide by } -16.$$

$$t^2 - 8t + 16 = -3 + 16 \quad \text{Add } \left[\frac{1}{2}(-8)\right]^2 = 16.$$

$$(t - 4)^2 = 13 \quad \text{Factor.}$$

$$t - 4 = \sqrt{13} \quad \text{or} \quad t - 4 = -\sqrt{13}$$

$$t = 4 + \sqrt{13} \quad \text{or} \quad t = 4 - \sqrt{13}$$

$$t \approx 7.6 \quad \text{or} \quad t \approx .4$$

The ball will be 48 feet above the ground after about .4 second and again after about 7.6 seconds.

9.2 Section Exercises

1. $x^2 + \underline{} + 25$

If this expression is a perfect square trinomial, it must be $(x + 5)^2$. Expanding $(x + 5)^2$ gives us

$$(x + 5)^2 = x^2 + 2(x)(5) + 5^2$$

$$= x^2 + 10x + 25.$$

Thus, the missing term is $10x$.

3. $z^2 + 14z + \underline{}$

Here, the middle term, $14z$, must equal $2kz$. So $14 = 2k$ and $k = 7$. The third term is $k^2 = 49$.

5. $2x^2 - 4x = 9$

Before completing the square, the coefficient of x^2 must be 1. Dividing each side of the equation by 2 is the correct way to begin solving the equation, and this corresponds to choice **D**.

7. $x^2 + 20x$

Take half of the coefficient of x and square it.

$$\frac{1}{2}(20) = 10, \quad \text{and} \quad 10^2 = 100.$$

Adding 100 to the expression $x^2 + 20x$ will make it a perfect square.

9. $x^2 - 5x$

Take half of the coefficient of x and square it.

$$\frac{1}{2}(-5) = -\frac{5}{2}, \quad \text{and} \quad \left(-\frac{5}{2}\right)^2 = \frac{25}{4}.$$

Adding $\frac{25}{4}$ to the expression $x^2 - 5x$ will make it a perfect square.

11. $r^2 + \frac{1}{2}r$

Take half of the coefficient of r and square it.

$$\frac{1}{2}\left(\frac{1}{2}\right) = \frac{1}{4}, \quad \text{and} \quad \left(\frac{1}{4}\right)^2 = \frac{1}{16}.$$

Adding $\frac{1}{16}$ to the expression $r^2 + \frac{1}{2}r$ will make it a perfect square.

13. $x^2 - 4x = -3$

Take half of the coefficient of x and square it. Half of -4 is -2, and $(-2)^2 = 4$. Add 4 to each side of the equation, and write the left side as a perfect square.

$$x^2 - 4x + 4 = -3 + 4$$

$$(x - 2)^2 = 1$$

Use the square root property.

$$x - 2 = \sqrt{1} \quad \text{or} \quad x - 2 = -\sqrt{1}$$

$$x - 2 = 1 \quad \text{or} \quad x - 2 = -1$$

$$x = 3 \quad \text{or} \quad x = 1$$

A check verifies that the solutions are 1 and 3.

15. $x^2 + 5x + 6 = 0$

Subtract 6 from each side.

$$x^2 + 5x = -6$$

Take half the coefficient of x and square it, and add this to each side.

$$x^2 + 5x + \frac{25}{4} = -6 + \frac{25}{4} \quad \text{Add } \left[\frac{1}{2}(5)\right]^2 = \frac{25}{4}.$$

$$\left(x + \frac{5}{2}\right)^2 = \frac{1}{4} \quad \text{Factor.}$$

9.2 Solving Quadratic Equations by Completing the Square

$x + \frac{5}{2} = \sqrt{\frac{1}{4}}$ or $x + \frac{5}{2} = -\sqrt{\frac{1}{4}}$

$x + \frac{5}{2} = \frac{1}{2}$ or $x + \frac{5}{2} = -\frac{1}{2}$

$x = -\frac{5}{2} + \frac{1}{2}$ or $x = -\frac{5}{2} - \frac{1}{2}$

$x = -\frac{4}{2}$ or $x = -\frac{6}{2}$

$x = -2$ or $x = -3$

A check verifies that the solutions are -2 and -3.

17. $x^2 + 2x - 5 = 0$

Add 5 to each side.

$$x^2 + 2x = 5$$

Take half the coefficient of x and square it.

$$\frac{1}{2}(2) = 1, \text{ and } 1^2 = 1.$$

Add 1 to each side of the equation, and write the left side as a perfect square.

$$x^2 + 2x + 1 = 5 + 1$$
$$(x+1)^2 = 6$$

Use the square root property.

$x + 1 = \sqrt{6}$ or $x + 1 = -\sqrt{6}$

$x = -1 + \sqrt{6}$ or $x = -1 - \sqrt{6}$

A check verifies that the solutions are $-1 + \sqrt{6}$ and $-1 - \sqrt{6}$. Using a calculator for your check is highly recommended.

19. $t^2 + 6t + 9 = 0$

The left-hand side of this equation is already a perfect square.

$$(t+3)^2 = 0$$
$$t + 3 = 0$$
$$t = -3$$

A check verifies that the solution is -3.

21. $x^2 + x - 1 = 0$

Add 1 to each side.

$$x^2 + x = 1$$

Take half of 1, square it, and add it to each side.

$x^2 + x + \frac{1}{4} = 1 + \frac{1}{4}$ Add $\left[\frac{1}{2}(1)\right]^2 = \frac{1}{4}$.

$\left(x + \frac{1}{2}\right)^2 = \frac{5}{4}$ Factor.

$x + \frac{1}{2} = \sqrt{\frac{5}{4}}$ or $x + \frac{1}{2} = -\sqrt{\frac{5}{4}}$

$x + \frac{1}{2} = \frac{\sqrt{5}}{2}$ or $x + \frac{1}{2} = -\frac{\sqrt{5}}{2}$

$x = \frac{-1 + \sqrt{5}}{2}$ or $x = \frac{-1 - \sqrt{5}}{2}$

A check verifies that the solutions are $\frac{-1 \pm \sqrt{5}}{2}$.

23. $4x^2 + 4x - 3 = 0$

Add 3 to each side.

$$4x^2 + 4x = 3$$

Divide each side by 4 so that the coefficient of x^2 is 1.

$$x^2 + x = \frac{3}{4}$$

The coefficient of x is 1. Take half of 1, square the result, and add this square to each side.

$$\frac{1}{2}(1) = \frac{1}{2} \text{ and } \left(\frac{1}{2}\right)^2 = \frac{1}{4}$$

$$x^2 + x + \frac{1}{4} = \frac{3}{4} + \frac{1}{4}$$

The left-hand side can then be written as a perfect square.

$$\left(x + \frac{1}{2}\right)^2 = 1$$

Use the square root property.

$x + \frac{1}{2} = 1$ or $x + \frac{1}{2} = -1$

$x = -\frac{1}{2} + 1$ or $x = -\frac{1}{2} - 1$

$x = \frac{1}{2}$ or $x = -\frac{3}{2}$

A check verifies that the solutions are $-\frac{3}{2}$ and $\frac{1}{2}$.

25. $2x^2 - 4x = 5$

$x^2 - 2x = \frac{5}{2}$ *Divide by 2.*

$x^2 - 2x + 1 = \frac{5}{2} + 1$ Add $\left[\frac{1}{2}(-2)\right]^2 = 1.$

$(x-1)^2 = \frac{7}{2}$ *Factor.*

$x - 1 = \sqrt{\frac{7}{2}}$ or $x - 1 = -\sqrt{\frac{7}{2}}$

Simplify the radical:

$$\sqrt{\frac{7}{2}} = \frac{\sqrt{7}}{\sqrt{2}} = \frac{\sqrt{7}}{\sqrt{2}} \cdot \frac{\sqrt{2}}{\sqrt{2}} = \frac{\sqrt{14}}{2}$$

continued

$$x - 1 = \frac{\sqrt{14}}{2} \quad \text{or} \quad x - 1 = -\frac{\sqrt{14}}{2}$$

$$x = 1 + \frac{\sqrt{14}}{2} \quad \text{or} \quad x = 1 - \frac{\sqrt{14}}{2}$$

$$x = \frac{2}{2} + \frac{\sqrt{14}}{2} \quad \text{or} \quad x = \frac{2}{2} - \frac{\sqrt{14}}{2}$$

$$x = \frac{2 + \sqrt{14}}{2} \quad \text{or} \quad x = \frac{2 - \sqrt{14}}{2}$$

A check verifies that the solutions are $\frac{2 \pm \sqrt{14}}{2}$.

27. $2p^2 - 2p + 3 = 0$

Divide each side by 2.

$$p^2 - p + \frac{3}{2} = 0$$

Subtract $\frac{3}{2}$ from each side.

$$p^2 - p = -\frac{3}{2}$$

Take half the coefficient of p and square it.

$$\frac{1}{2}(-1) = -\frac{1}{2}, \text{ and } \left(-\frac{1}{2}\right)^2 = \frac{1}{4}.$$

Add $\frac{1}{4}$ to each side of the equation.

$$p^2 - p + \frac{1}{4} = -\frac{3}{2} + \frac{1}{4}$$

Factor on the left side and add on the right.

$$\left(p - \frac{1}{2}\right)^2 = -\frac{5}{4}$$

The square root of $-\frac{5}{4}$ is not a real number, so there is *no real number solution*.

29. $3k^2 + 7k = 4$

Divide each side by 3.

$$k^2 + \frac{7}{3}k = \frac{4}{3}$$

Take half of the coefficient of k and square it.

$$\frac{1}{2}\left(\frac{7}{3}\right) = \frac{7}{6} \text{ and } \left(\frac{7}{6}\right)^2 = \frac{49}{36}.$$

Add $\frac{49}{36}$ to each side of the equation.

$$k^2 + \frac{7}{3}k + \frac{49}{36} = \frac{4}{3} + \frac{49}{36}$$

$$\left(k + \frac{7}{6}\right)^2 = \frac{97}{36}$$

Use the square root property.

$$k + \frac{7}{6} = \sqrt{\frac{97}{36}} \quad \text{or} \quad k + \frac{7}{6} = -\sqrt{\frac{97}{36}}$$

$$k + \frac{7}{6} = \frac{\sqrt{97}}{6} \quad \text{or} \quad k + \frac{7}{6} = -\frac{\sqrt{97}}{6}$$

$$k = -\frac{7}{6} + \frac{\sqrt{97}}{6} \quad \text{or} \quad k = -\frac{7}{6} - \frac{\sqrt{97}}{6}$$

$$k = \frac{-7 + \sqrt{97}}{6} \quad \text{or} \quad k = \frac{-7 - \sqrt{97}}{6}$$

A check verifies that the solutions are $\frac{-7 \pm \sqrt{97}}{6}$.

31. $(x + 3)(x - 1) = 5$

$$x^2 + 2x - 3 = 5$$
$$x^2 + 2x = 8$$
$$x^2 + 2x + 1 = 8 + 1$$
$$(x + 1)^2 = 9$$

$$x + 1 = 3 \quad \text{or} \quad x + 1 = -3$$
$$x = 2 \quad \text{or} \quad x = -4$$

A check verifies that the solutions are -4 and 2.

33. $-x^2 + 2x = -5$

Divide each side by -1.

$$x^2 - 2x = 5$$

Take half of the coefficient of x and square it. Half of -2 is -1, and $(-1)^2 = 1$. Add 1 to each side of the equation, and write the left side as a perfect square.

$$x^2 - 2x + 1 = 5 + 1$$
$$(x - 1)^2 = 6$$

Use the square root property.

$$x - 1 = \sqrt{6} \quad \text{or} \quad x - 1 = -\sqrt{6}$$
$$x = 1 + \sqrt{6} \quad \text{or} \quad x = 1 - \sqrt{6}$$

A check verifies that the solutions are $1 \pm \sqrt{6}$.

35. $s = -16t^2 + 96t$

Find the value of t when $s = 80$.

$$80 = -16t^2 + 96t \quad \text{Let } s = 80.$$
$$-16t^2 + 96t = 80$$
$$t^2 - 6t = -5 \quad \text{Divide by } -16.$$
$$t^2 - 6t + 9 = -5 + 9 \quad \text{Add 9.}$$
$$(t - 3)^2 = 4 \quad \text{Factor; add.}$$
$$t - 3 = \pm\sqrt{4} = \pm 2$$
$$t = 3 + 2 = 5 \quad \text{or} \quad t = 3 - 2 = 1$$

The object will reach a height of 80 feet at 1 second (on the way up) and at 5 seconds (on the way down).

37.
$$s = -13t^2 + 104t$$
$$195 = -13t^2 + 104t \quad \text{Let } s = 195.$$
$$-15 = t^2 - 8t \quad \text{Divide by } -13.$$
$$t^2 - 8t + 16 = -15 + 16$$
$$\text{Add } \left[\frac{1}{2}(-8)\right]^2 = 16.$$
$$(t - 4)^2 = 1$$
$$t - 4 = \pm\sqrt{1} = \pm 1$$
$$t = 4 \pm 1$$
$$= 3 \text{ or } 5$$

The object will be at a height of 195 feet at 3 seconds (on the way up) and 5 seconds (on the way down).

39. Let $x = $ the width of the pen.
Then $175 - x = $ the length of the pen.

Use the formula for the area of a rectangle.
$$A = LW$$
$$7500 = (175 - x)x$$
$$7500 = 175x - x^2$$
$$x^2 - 175x + 7500 = 0$$

Solve this quadratic equation by completing the square.
$$x^2 - 175x = -7500$$
$$x^2 - 175x + \frac{30{,}625}{4} = -\frac{30{,}000}{4} + \frac{30{,}625}{4}$$
$$\text{Add } \left(\frac{175}{2}\right)^2 = \frac{30{,}625}{4}.$$
$$\left(x - \frac{175}{2}\right)^2 = \frac{625}{4}$$
$$x - \frac{175}{2} = \pm\sqrt{\frac{625}{4}} = \pm\frac{25}{2}$$
$$x = \frac{175}{2} \pm \frac{25}{2}$$
$$x = \frac{175}{2} + \frac{25}{2} \quad \text{or} \quad x = \frac{175}{2} - \frac{25}{2}$$
$$x = \frac{200}{2} \quad \text{or} \quad x = \frac{150}{2}$$
$$x = 100 \quad \text{or} \quad x = 75$$

If $x = 100$, $175 - x = 175 - 100 = 75$.

If $x = 75$, $175 - x = 175 - 75 = 100$.

The dimensions of the pen are 75 feet by 100 feet.

9.3 Solving Quadratic Equations by the Quadratic Formula

9.3 Margin Exercises

1. (a) $5x^2 + 2x - 1 = 0$ has the form of the standard quadratic equation $ax^2 + bx + c = 0$.

Thus, $a = 5$, $b = 2$, and $c = -1$.

(b) $3x^2 = x - 2$

Rewrite in $ax^2 + bx + c = 0$ form.
$$3x^2 - x + 2 = 0$$

Then $a = 3$, $b = -1$, and $c = 2$.

(c) $9x^2 - 13 = 0$ has the form of the standard quadratic equation $ax^2 + bx + c = 0$.

Thus, $a = 9$, $b = 0$, and $c = -13$.

(d) $(3x + 2)(x - 1) = 8$
$$3x^2 - x - 2 = 8 \quad \text{FOIL}$$
$$3x^2 - x - 10 = 0 \quad \text{Subtract 8.}$$

Thus, $a = 3$, $b = -1$, and $c = -10$.

2. (a) $2x^2 + 3x - 5 = 0$

The quadratic equation is in standard form, so $a = 2$, $b = 3$, and $c = -5$. Substitute these values into the quadratic formula.

$$x = \frac{-b \pm \sqrt{b^2 - 4ac}}{2a}$$
$$x = \frac{-3 \pm \sqrt{3^2 - 4(2)(-5)}}{2(2)}$$
$$x = \frac{-3 \pm \sqrt{9 + 40}}{4}$$
$$x = \frac{-3 \pm \sqrt{49}}{4}$$
$$x = \frac{-3 \pm 7}{4}$$
$$x = \frac{-3 + 7}{4} \quad \text{or} \quad x = \frac{-3 - 7}{4}$$
$$x = \frac{4}{4} \quad \text{or} \quad x = \frac{-10}{4}$$
$$x = 1 \quad \text{or} \quad x = -\frac{5}{2}$$

The solutions are 1 and $-\frac{5}{2}$.

Chapter 9 Quadratic Equations

(b) $6x^2 + x - 1 = 0$

Substitute $a = 6$, $b = 1$, and $c = -1$ into the quadratic formula.

$$x = \frac{-b \pm \sqrt{b^2 - 4ac}}{2a}$$

$$x = \frac{-1 \pm \sqrt{1^2 - 4(6)(-1)}}{2(6)}$$

$$x = \frac{-1 \pm \sqrt{1 + 24}}{12}$$

$$x = \frac{-1 \pm 5}{12}$$

$$x = \frac{-1 + 5}{12} \quad \text{or} \quad x = \frac{-1 - 5}{12}$$

$$x = \frac{4}{12} \quad \text{or} \quad x = \frac{-6}{12}$$

$$x = \frac{1}{3} \quad \text{or} \quad x = -\frac{1}{2}$$

The solutions are $\frac{1}{3}$ and $-\frac{1}{2}$.

3. $x^2 + 1 = -8x$

Write the equation in standard form.

$$x^2 + 8x + 1 = 0$$

Substitute $a = 1$, $b = 8$, and $c = 1$ into the quadratic formula.

$$x = \frac{-b \pm \sqrt{b^2 - 4ac}}{2a}$$

$$x = \frac{-8 \pm \sqrt{8^2 - 4(1)(1)}}{2(1)}$$

$$x = \frac{-8 \sqrt{64 - 4}}{2} = \frac{-8 \pm \sqrt{60}}{2}$$

$$= \frac{-8 \pm \sqrt{4} \cdot \sqrt{15}}{2} = \frac{-8 \pm 2\sqrt{15}}{2}$$

$$= \frac{2(-4 \pm \sqrt{15})}{2} = -4 \pm \sqrt{15}$$

The solutions are $-4 + \sqrt{15}$ and $-4 - \sqrt{15}$.

4. $9x^2 - 12x + 4 = 0$

Substitute $a = 9$, $b = -12$, and $c = 4$ into the quadratic formula.

$$x = \frac{-b \pm \sqrt{b^2 - 4ac}}{2a}$$

$$x = \frac{-(-12) \pm \sqrt{(-12)^2 - 4(9)(4)}}{2(9)}$$

$$x = \frac{12 \pm \sqrt{144 - 144}}{18}$$

$$x = \frac{12}{18} = \frac{2}{3}$$

Since there is just one solution, $\frac{2}{3}$, the trinomial $9x^2 - 12x + 4$ is a perfect square.

5. (a) $x^2 - \frac{4}{3}x + \frac{2}{3} = 0$

$3x^2 - 4x + 2 = 0$ Multiply by 3.

Substitute $a = 3$, $b = -4$, and $c = 2$ into the quadratic formula.

$$x = \frac{-b \pm \sqrt{b^2 - 4ac}}{2a}$$

$$x = \frac{-(-4) \pm \sqrt{(-4)^2 - 4(3)(2)}}{2(3)}$$

$$x = \frac{4 \pm \sqrt{16 - 24}}{6}$$

$$x = \frac{4 \pm \sqrt{-8}}{6}$$

Because $\sqrt{-8}$ does not represent a real number, there is *no real number solution*.

(b) $x^2 - \frac{9}{5}x = \frac{2}{5}$

$5x^2 - 9x = 2$ Multiply by 5.

$5x^2 - 9x - 2 = 0$ Subtract 2.

Substitute $a = 5$, $b = -9$, and $c = -2$ into the quadratic formula.

$$x = \frac{-b \pm \sqrt{b^2 - 4ac}}{2a}$$

$$x = \frac{-(-9) \pm \sqrt{(-9)^2 - 4(5)(-2)}}{2(5)}$$

$$x = \frac{9 \pm \sqrt{81 + 40}}{10}$$

$$x = \frac{9 \pm \sqrt{121}}{10}$$

$$x = \frac{9 \pm 11}{10}$$

$$x = \frac{9 + 11}{10} = 2 \quad \text{or} \quad x = \frac{9 - 11}{10} = -\frac{1}{5}$$

The solutions are 2 and $-\frac{1}{5}$.

9.3 Section Exercises

1. In $4x^2 + 5x - 9 = 0$, the coefficient of the x^2-term is 4, so $a = 4$. The coefficient of the x-term is 5, so $b = 5$. The constant is -9, so $c = -9$. (Note that one side of the equation was equal to 0 before we started.)

3. $3x^2 = 4x + 2$

First, write the equation in standard form, $ax^2 + bx + c = 0$.

$$3x^2 - 4x - 2 = 0$$

Now, identify the values: $a = 3$, $b = -4$, and $c = -2$.

5. $3x^2 = -7x$

Write the equation in standard form.
$$3x^2 + 7x = 0$$

Now, identify the values: $a = 3$, $b = 7$, and $c = 0$.

7. $k^2 = -12k + 13$

Write the equation in standard form.
$$k^2 + 12k - 13 = 0$$

Substitute $a = 1$, $b = 12$, and $c = -13$ into the quadratic formula.

$$k = \frac{-b \pm \sqrt{b^2 - 4ac}}{2a}$$

$$k = \frac{-12 \pm \sqrt{12^2 - 4(1)(-13)}}{2(1)}$$

$$= \frac{-12 \pm \sqrt{144 + 52}}{2}$$

$$= \frac{-12 \pm \sqrt{196}}{2}$$

$$= \frac{-12 \pm 14}{2}$$

$$k = \frac{-12 + 14}{2} = \frac{2}{2} = 1$$

or $\quad k = \frac{-12 - 14}{2} = \frac{-26}{2} = -13$

A check verifies that the solutions are -13 and 1.

9. $p^2 - 4p + 4 = 0$

Substitute $a = 1$, $b = -4$, and $c = 4$ into the quadratic formula.

$$p = \frac{-b \pm \sqrt{b^2 - 4ac}}{2a}$$

$$p = \frac{-(-4) \pm \sqrt{(-4)^2 - 4(1)(4)}}{2(1)}$$

$$= \frac{4 \pm \sqrt{16 - 16}}{2}$$

$$= \frac{4 \pm 0}{2} = \frac{4}{2} = 2.$$

The solution is 2. Note that the discriminant is 0.

11. $2x^2 + 12x = -5$

Write the equation in standard form.
$$2x^2 + 12x + 5 = 0$$

Substitute $a = 2$, $b = 12$, and $c = 5$ into the quadratic formula.

$$x = \frac{-b \pm \sqrt{b^2 - 4ac}}{2a}$$

$$x = \frac{-12 \pm \sqrt{12^2 - 4(2)(5)}}{2(2)}$$

$$= \frac{-12 \pm \sqrt{144 - 40}}{4}$$

$$= \frac{-12 \pm \sqrt{104}}{4} = \frac{-12 \pm \sqrt{4} \cdot \sqrt{26}}{4}$$

$$= \frac{-12 \pm 2\sqrt{26}}{4} = \frac{2(-6 \pm \sqrt{26})}{2 \cdot 2}$$

$$= \frac{-6 \pm \sqrt{26}}{2}$$

A check verifies that the solutions are $\frac{-6 \pm \sqrt{26}}{2}$.

13. $2x^2 = 5 + 3x$

Write the equation in standard form.
$$2x^2 - 3x - 5 = 0$$

Substitute $a = 2$, $b = -3$, and $c = -5$ into the quadratic formula.

$$x = \frac{-b \pm \sqrt{b^2 - 4ac}}{2a}$$

$$x = \frac{-(-3) \pm \sqrt{(-3)^2 - 4(2)(-5)}}{2(2)}$$

$$= \frac{3 \pm \sqrt{9 + 40}}{4}$$

$$= \frac{3 \pm \sqrt{49}}{4} = \frac{3 \pm 7}{4}$$

$$x = \frac{3 + 7}{4} = \frac{10}{4} = \frac{5}{2}$$

or $\quad x = \frac{3 - 7}{4} = \frac{-4}{4} = -1$

A check verifies that the solutions are -1 and $\frac{5}{2}$.

15. $6x^2 + 6x = 0$

Substitute $a = 6$, $b = 6$, and $c = 0$ into the quadratic formula.

$$x = \frac{-b \pm \sqrt{b^2 - 4ac}}{2a}$$

$$x = \frac{-6 \pm \sqrt{6^2 - 4(6)(0)}}{2(6)}$$

$$= \frac{-6 \pm \sqrt{36 - 0}}{12}$$

$$= \frac{-6 \pm 6}{12}$$

$$x = \frac{-6 + 6}{12} = \frac{0}{12} = 0$$

or $\quad x = \frac{-6 - 6}{12} = \frac{-12}{12} = -1$

A check verifies that the solutions are -1 and 0.

17. $-2x^2 = -3x + 2$

Write the equation in standard form.
$$-2x^2 + 3x - 2 = 0$$

Substitute $a = -2$, $b = 3$, and $c = -2$ into the quadratic formula.

$$x = \frac{-b \pm \sqrt{b^2 - 4ac}}{2a}$$

$$x = \frac{-3 \pm \sqrt{3^2 - 4(-2)(-2)}}{2(-2)}$$

$$= \frac{-3 \pm \sqrt{9 - 16}}{-4}$$

$$= \frac{-3 \pm \sqrt{-7}}{-4}$$

Because $\sqrt{-7}$ does not represent a real number, there is *no real number solution*.

19. $3x^2 + 5x + 1 = 0$

Substitute $a = 3$, $b = 5$, and $c = 1$ into the quadratic formula.

$$x = \frac{-5 \pm \sqrt{5^2 - 4(3)(1)}}{2(3)}$$

$$x = \frac{-5 \pm \sqrt{25 - 12}}{6}$$

$$x = \frac{-5 \pm \sqrt{13}}{6}$$

A check verifies that the solutions are $\frac{-5 \pm \sqrt{13}}{6}$.

21. $7x^2 = 12x$

Write the equation in standard form.
$$7x^2 - 12x = 0$$

Substitute $a = 7$, $b = -12$, and $c = 0$ into the quadratic formula.

$$x = \frac{-b \pm \sqrt{b^2 - 4ac}}{2a}$$

$$x = \frac{-(-12) \pm \sqrt{(-12)^2 - 4(7)(0)}}{2(7)}$$

$$= \frac{12 \pm \sqrt{144 - 0}}{14}$$

$$= \frac{12 \pm 12}{14}$$

$$x = \frac{12 + 12}{14} = \frac{24}{14} = \frac{12}{7}$$

or $x = \frac{12 - 12}{14} = \frac{0}{14} = 0$

A check verifies that the solutions are 0 and $\frac{12}{7}$.

23. $x^2 - 24 = 0$

Substitute $a = 1$, $b = 0$, and $c = -24$ into the quadratic formula.

$$x = \frac{-b \pm \sqrt{b^2 - 4ac}}{2a}$$

$$x = \frac{0 \pm \sqrt{0^2 - 4(1)(-24)}}{2(1)}$$

$$= \frac{\pm \sqrt{96}}{2} = \frac{\pm \sqrt{16} \cdot \sqrt{6}}{2}$$

$$= \frac{\pm 4\sqrt{6}}{2} = \pm 2\sqrt{6}$$

A check verifies that the solutions are $\pm 2\sqrt{6}$.

25. $25x^2 - 4 = 0$

Substitute $a = 25$, $b = 0$, and $c = -4$ into the quadratic formula.

$$x = \frac{-b \pm \sqrt{b^2 - 4ac}}{2a}$$

$$x = \frac{-0 \pm \sqrt{0^2 - 4(25)(-4)}}{2(25)}$$

$$= \frac{\pm \sqrt{400}}{50}$$

$$= \frac{\pm 20}{50} = \pm \frac{2}{5}$$

A check verifies that the solutions are $-\frac{2}{5}$ and $\frac{2}{5}$.

27. $3x^2 - 2x + 5 = 10x + 1$

Write the equation in standard form.
$$3x^2 - 12x + 4 = 0$$

Substitute $a = 3$, $b = -12$, and $c = 4$ into the quadratic formula.

$$x = \frac{-b \pm \sqrt{b^2 - 4ac}}{2a}$$

$$x = \frac{-(-12) \pm \sqrt{(-12)^2 - 4(3)(4)}}{2(3)}$$

$$= \frac{12 \pm \sqrt{144 - 48}}{6}$$

$$= \frac{12 \pm \sqrt{96}}{6} = \frac{12 \pm \sqrt{16} \cdot \sqrt{6}}{6}$$

$$= \frac{12 \pm 4\sqrt{6}}{6} = \frac{2(6 \pm 2\sqrt{6})}{2 \cdot 3}$$

$$= \frac{6 \pm 2\sqrt{6}}{3}$$

A check verifies that the solutions are $\frac{6 \pm 2\sqrt{6}}{3}$.

29. $2x^2 + x + 5 = 0$

Substitute $a = 2$, $b = 1$, and $c = 5$ into the quadratic formula.

$$x = \frac{-1 \pm \sqrt{1^2 - 4(2)(5)}}{2(2)}$$

$$= \frac{-1 \pm \sqrt{1 - 40}}{4}$$

$$= \frac{-1 \pm \sqrt{-39}}{4}$$

Because $\sqrt{-39}$ does not represent a real number, there is *no real number solution*.

31. If the radicand in the quadratic formula, $b^2 - 4ac$, is negative, then the original equation has *no real number solutions*. (The square root of a negative number is not a real number.)

33. $\frac{3}{2}k^2 - k - \frac{4}{3} = 0$

Eliminate the denominators by multiplying each side by the least common denominator, 6.

$$9k^2 - 6k - 8 = 0$$

Substitute $a = 9$, $b = -6$, and $c = -8$ into the quadratic formula.

$$k = \frac{-(-6) \pm \sqrt{(-6)^2 - 4(9)(-8)}}{2(9)}$$

$$= \frac{6 \pm \sqrt{36 + 288}}{18} = \frac{6 \pm \sqrt{324}}{18}$$

$$= \frac{6 \pm 18}{18}$$

$$k = \frac{6 + 18}{18} = \frac{24}{18} = \frac{4}{3}$$

or $k = \frac{6 - 18}{18} = \frac{-12}{18} = -\frac{2}{3}$

A check verifies that the solutions are $-\frac{2}{3}$ and $\frac{4}{3}$.

35. $\frac{1}{2}x^2 + \frac{1}{6}x = 1$

Eliminate the denominators by multiplying each side by the least common denominator, 6.

$$3x^2 + x = 6$$
$$3x^2 + x - 6 = 0$$

Here, $a = 3$, $b = 1$, and $c = -6$.

$$x = \frac{-1 \pm \sqrt{1^2 - 4(3)(-6)}}{2(3)}$$

$$= \frac{-1 \pm \sqrt{1 + 72}}{6}$$

$$= \frac{-1 \pm \sqrt{73}}{6}$$

A check verifies that the solutions are $\frac{-1 \pm \sqrt{73}}{6}$.

37. $.5x^2 = x + .5$

To eliminate the decimals, multiply each side by 10. Write this equation in standard form.

$$5x^2 = 10x + 5$$
$$5x^2 - 10x - 5 = 0$$

Divide each side by 5 so that we can work with smaller coefficients in the quadratic formula.

$$x^2 - 2x - 1 = 0$$

Use the quadratic formula with $a = 1$, $b = -2$, and $c = -1$.

$$x = \frac{-(-2) \pm \sqrt{(-2)^2 - 4(1)(-1)}}{2(1)}$$

$$= \frac{2 \pm \sqrt{4 + 4}}{2}$$

$$= \frac{2 \pm \sqrt{8}}{2} = \frac{2 \pm \sqrt{4} \cdot \sqrt{2}}{2}$$

$$= \frac{2 \pm 2\sqrt{2}}{2} = \frac{2(1 \pm \sqrt{2})}{2}$$

$$= 1 \pm \sqrt{2}$$

A check verifies that the solutions are $1 \pm \sqrt{2}$.

39. $\frac{3}{8}x^2 - x + \frac{17}{24} = 0$

Multiply each side by the least common denominator, 24.

$$9x^2 - 24x + 17 = 0$$

Use the quadratic formula with $a = 9$, $b = -24$, and $c = 17$.

$$x = \frac{-(-24) \pm \sqrt{(-24)^2 - 4(9)(17)}}{2(9)}$$

$$= \frac{24 \pm \sqrt{576 - 612}}{18}$$

$$= \frac{24 \pm \sqrt{-36}}{18}$$

Because $\sqrt{-36}$ does not represent a real number, there is *no real number solution*.

41. $h = -.5x^2 + 1.25x + 3$
$1.25 = -.5x^2 + 1.25x + 3$ *Let h = 1.25.*
$.5x^2 - 1.25x - 1.75 = 0$
$2x^2 - 5x - 7 = 0$ *Multiply by 4.*

Use the quadratic formula with $a = 2$, $b = -5$, and $c = -7$.

$$x = \frac{-(-5) \pm \sqrt{(-5)^2 - 4(2)(-7)}}{2(2)}$$

$$= \frac{5 \pm \sqrt{25 + 56}}{4}$$

$$= \frac{5 \pm \sqrt{81}}{4} = \frac{5 \pm 9}{4}$$

$$x = \frac{5 + 9}{4} = \frac{14}{4} = 3.5$$

or $x = \frac{5 - 9}{4} = \frac{-4}{4} = -1$

x must be positive, so the frog was 3.5 feet from the base of the stump when he was 1.25 feet above the ground.

43. $\left(\dfrac{d-4}{4}\right)^2 = 9$

$\dfrac{d-4}{4} = \pm\sqrt{9} = \pm 3$

$d - 4 = \pm 12$
$d = 4 \pm 12$
$= 16$ or -8

The solutions for the equation are -8 and 16. Only 16 board feet is a reasonable answer.

Summary Exercises on Quadratic Equations

1. $x^2 = 36$

Use the square root property.

$x = \sqrt{36}$ or $x = -\sqrt{36}$
$x = 6$ or $x = -6$

A check verifies that the solutions are ± 6.

3. $x^2 - \dfrac{100}{81} = 0$

$x^2 = \dfrac{100}{81}$

Use the square root property.

$x = \sqrt{\dfrac{100}{81}}$ or $x = -\sqrt{\dfrac{100}{81}}$

$x = \dfrac{10}{9}$ or $x = -\dfrac{10}{9}$

A check verifies that the solutions are $\pm \dfrac{10}{9}$.

5. $z^2 - 4z + 3 = 0$

Solve this equation by factoring.

$(z - 3)(z - 1) = 0$

$z - 3 = 0$ or $z - 1 = 0$
$z = 3$ or $z = 1$

A check verifies that the solutions are 1 and 3.

7. $z(z - 9) = -20$
$z^2 - 9z = -20$
$z^2 - 9z + 20 = 0$

Solve this equation by factoring.

$(z - 4)(z - 5) = 0$

$z - 4 = 0$ or $z - 5 = 0$
$z = 4$ or $z = 5$

A check verifies that the solutions are 4 and 5.

9. $(3k - 2)^2 = 9$

Use the square root property.

$3k - 2 = \sqrt{9}$ or $3k - 2 = -\sqrt{9}$
$3k - 2 = 3$ or $3k - 2 = -3$
$3k = 5$ or $3k = -1$
$k = \dfrac{5}{3}$ or $k = -\dfrac{1}{3}$

A check verifies that the solutions are $-\dfrac{1}{3}$ and $\dfrac{5}{3}$.

11. $(x + 6)^2 = 121$

Use the square root property.

$x + 6 = \sqrt{121}$ or $x + 6 = -\sqrt{121}$
$x + 6 = 11$ or $x + 6 = -11$
$x = 5$ or $x = -17$

A check verifies that the solutions are -17 and 5.

13. $(3r - 7)^2 = 24$

Use the square root property.

$3r - 7 = \sqrt{24}$ or $3r - 7 = -\sqrt{24}$

Now simplify the radical.

$\sqrt{24} = \sqrt{4} \cdot \sqrt{6} = 2\sqrt{6}$

$3r - 7 = 2\sqrt{6}$ or $3r - 7 = -2\sqrt{6}$
$3r = 7 + 2\sqrt{6}$ or $3r = 7 - 2\sqrt{6}$
$r = \dfrac{7 + 2\sqrt{6}}{3}$ or $r = \dfrac{7 - 2\sqrt{6}}{3}$

A check verifies that the solutions are $\dfrac{7 \pm 2\sqrt{6}}{3}$.

15. $(5x - 8)^2 = -6$

The square root of -6 is not a real number, so the square root property does not apply. This equation has *no real number solution*.

17. $\quad -2x^2 = -3x - 2$
$\quad 2x^2 - 3x - 2 = 0$

Solve this equation by factoring.
$$(2x + 1)(x - 2) = 0$$
$2x + 1 = 0 \quad$ or $\quad x - 2 = 0$
$\quad x = -\frac{1}{2} \quad$ or $\quad x = 2$

A check verifies that the solutions are $-\frac{1}{2}$ and 2.

19. $\quad 8z^2 = 15 + 2z$
$8z^2 - 2z - 15 = 0$

Solve this equation by factoring.
$$(4z + 5)(2z - 3) = 0$$
$4z + 5 = 0 \quad$ or $\quad 2z - 3 = 0$
$\quad z = -\frac{5}{4} \quad$ or $\quad z = \frac{3}{2}$

A check verifies that the solutions are $-\frac{5}{4}$ and $\frac{3}{2}$.

21. $\quad 0 = -x^2 + 2x + 1$
$\quad x^2 - 2x - 1 = 0$

Use the quadratic formula with $a = 1$, $b = -2$, and $c = -1$.
$$x = \frac{-b \pm \sqrt{b^2 - 4ac}}{2a}$$
$$x = \frac{-(-2) \pm \sqrt{(-2)^2 - 4(1)(-1)}}{2(1)}$$
$$= \frac{2 \pm \sqrt{4 + 4}}{2} = \frac{2 \pm \sqrt{8}}{2}$$
$$= \frac{2 \pm 2\sqrt{2}}{2} = \frac{2(1 \pm \sqrt{2})}{2}$$
$$= 1 \pm \sqrt{2}$$

A check verifies that the solutions are $1 \pm \sqrt{2}$.

23. $\quad 5x^2 - 22x = -8$
$5x^2 - 22x + 8 = 0$

Solve this equation by factoring.
$$(5x - 2)(x - 4) = 0$$
$5x - 2 = 0 \quad$ or $\quad x - 4 = 0$
$\quad x = \frac{2}{5} \quad$ or $\quad x = 4$

A check verifies that the solutions are $\frac{2}{5}$ and 4.

25. $\quad (x + 2)(x + 1) = 10$
$\quad x^2 + 3x + 2 = 10$
$\quad x^2 + 3x - 8 = 0$

Use the quadratic formula with $a = 1$, $b = 3$, and $c = -8$.
$$x = \frac{-b \pm \sqrt{b^2 - 4ac}}{2a}$$
$$x = \frac{-3 \pm \sqrt{3^2 - 4(1)(-8)}}{2(1)}$$
$$= \frac{-3 \pm \sqrt{9 + 32}}{2}$$
$$= \frac{-3 \pm \sqrt{41}}{2}$$

A check verifies that the solutions are $\dfrac{-3 \pm \sqrt{41}}{2}$.

27. $\quad 4x^2 = -1 + 5x$
$4x^2 - 5x + 1 = 0$

Solve this equation by factoring.
$$(x - 1)(4x - 1) = 0$$
$x - 1 = 0 \quad$ or $\quad 4x - 1 = 0$
$\quad x = 1 \quad$ or $\quad x = \frac{1}{4}$

A check verifies that the solutions are $\frac{1}{4}$ and 1.

29. $\quad 3x(3x + 4) = 7$
$\quad 9x^2 + 12x = 7$
$9x^2 + 12x - 7 = 0$

Use the quadratic formula with $a = 9$, $b = 12$, and $c = -7$.
$$x = \frac{-b \pm \sqrt{b^2 - 4ac}}{2a}$$
$$x = \frac{-12 \pm \sqrt{12^2 - 4(9)(-7)}}{2(9)}$$
$$= \frac{-12 \pm \sqrt{144 + 252}}{18}$$
$$= \frac{-12 \pm \sqrt{396}}{18} = \frac{-12 \pm \sqrt{36} \cdot \sqrt{11}}{18}$$
$$= \frac{-12 \pm 6\sqrt{11}}{18} = \frac{6(-2 \pm \sqrt{11})}{6 \cdot 3}$$
$$= \frac{-2 \pm \sqrt{11}}{3}$$

A check verifies that the solutions are $\dfrac{-2 \pm \sqrt{11}}{3}$.

31. $\dfrac{x^2}{2} + \dfrac{7x}{4} + \dfrac{11}{8} = 0$

Multiply each side by the least common denominator, 8.

$$8\left(\dfrac{x^2}{2} + \dfrac{7x}{4} + \dfrac{11}{8}\right) = 8(0)$$
$$4x^2 + 14x + 11 = 0$$

Use the quadratic formula with $a = 4$, $b = 14$, and $c = 11$.

$$x = \dfrac{-b \pm \sqrt{b^2 - 4ac}}{2a}$$

$$x = \dfrac{-14 \pm \sqrt{14^2 - 4(4)(11)}}{2(4)}$$

$$= \dfrac{-14 \pm \sqrt{196 - 176}}{8}$$

$$= \dfrac{-14 \pm \sqrt{20}}{8} = \dfrac{-14 \pm 2\sqrt{5}}{8}$$

$$= \dfrac{2(-7 \pm \sqrt{5})}{2(4)} = \dfrac{-7 \pm \sqrt{5}}{4}$$

A check verifies that the solutions are $\dfrac{-7 \pm \sqrt{5}}{4}$.

33.
$$9k^2 = 16(3k + 4)$$
$$9k^2 = 48k + 64$$
$$9k^2 - 48k - 64 = 0$$

Use the quadratic formula with $a = 9$, $b = -48$, and $c = -64$.

$$k = \dfrac{-b \pm \sqrt{b^2 - 4ac}}{2a}$$

$$k = \dfrac{-(-48) \pm \sqrt{(-48)^2 - 4(9)(-64)}}{2(9)}$$

$$= \dfrac{48 \pm \sqrt{2304 + 2304}}{18}$$

$$= \dfrac{48 \pm \sqrt{4608}}{18} = \dfrac{48 \pm \sqrt{2304} \cdot \sqrt{2}}{18}$$

$$= \dfrac{48 \pm 48\sqrt{2}}{18} = \dfrac{6(8 \pm 8\sqrt{2})}{6 \cdot 3}$$

$$= \dfrac{8 \pm 8\sqrt{2}}{3}$$

A check verifies that the solutions are $\dfrac{8 \pm 8\sqrt{2}}{3}$.

35. $x^2 - x + 3 = 0$

Use the quadratic formula with $a = 1$, $b = -1$, and $c = 3$.

$$x = \dfrac{-b \pm \sqrt{b^2 - 4ac}}{2a}$$

$$x = \dfrac{-(-1) \pm \sqrt{(-1)^2 - 4(1)(3)}}{2(1)}$$

$$= \dfrac{1 \pm \sqrt{1 - 12}}{2} = \dfrac{1 \pm \sqrt{-11}}{2}$$

Because $\sqrt{-11}$ does not represent a real number, there is *no real number solution*.

37.
$$-3x^2 + 4x = -4$$
$$3x^2 - 4x - 4 = 0$$

Solve this equation by factoring.

$$(3x + 2)(x - 2) = 0$$

$3x + 2 = 0$ or $x - 2 = 0$
$x = -\tfrac{2}{3}$ or $x = 2$

A check verifies that the solutions are $-\tfrac{2}{3}$ and 2.

39.
$$5k^2 + 19k = 2k + 12$$
$$5k^2 + 17k - 12 = 0$$

Solve this equation by factoring.

$$(5k - 3)(k + 4) = 0$$

$5k - 3 = 0$ or $k + 4 = 0$
$k = \tfrac{3}{5}$ or $k = -4$

A check verifies that the solutions are -4 and $\tfrac{3}{5}$.

41. $x^2 - \dfrac{4}{15} = -\dfrac{4}{15}x$

Multiply each side by 15 to clear fractions.

$$15x^2 - 4 = -4x$$

Write this equation in standard form.

$$15x^2 + 4x - 4 = 0$$

Solve by factoring.

$$(3x + 2)(5x - 2) = 0$$

$3x + 2 = 0$ or $5x - 2 = 0$
$x = -\tfrac{2}{3}$ or $x = \tfrac{2}{5}$

A check verifies that the solutions are $-\tfrac{2}{3}$ and $\tfrac{2}{5}$.

9.4 Graphing Quadratic Equations

9.4 Margin Exercises

1. $y = x^2$

   To find y-values, substitute each x-value into $y = x^2$.

x	y
3	9
2	4
1	1
0	0
-1	1
-2	4
-3	9

2. $y = \dfrac{1}{2}x^2$

   To find y-values, substitute each x-value into $y = \frac{1}{2}x^2$.

x	y
-2	2
-1	$\frac{1}{2}$
0	0
1	$\frac{1}{2}$
2	2

   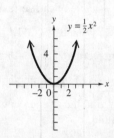

3. $y = -x^2 - 1$

   To find the y-values for the ordered pairs, substitute each x-value into $y = -x^2 - 1$.

   If $x = -2$, $y = -(-2)^2 - 1 = -4 - 1 = -5$.
   If $x = -1$, $y = -(-1)^2 - 1 = -1 - 1 = -2$.
   If $x = 1$, $y = -1^2 - 1 = -1 - 1 = -2$.
   If $x = 2$, $y = -2^2 - 1 = -4 - 1 = -5$.

   The ordered pairs are $(-2, -5)$, $(-1, -2)$, $(1, -2)$, and $(2, -5)$.

4. **(a)** $y = -x^2 - 3$

   Find several ordered pairs. To begin, check for intercepts.

   If $x = 0$,
   $$y = -0^2 - 3 = -3,$$
   so the y-intercept is $(0, -3)$.

   If $y = 0$,
   $$0 = -x^2 - 3,$$
   and hence, $x^2 = -3$.

   This equation has no real number solution, so there are no x-intercepts.

   We can choose additional x-values near $x = 0$ to find other ordered pairs. We obtain the following table for ordered pairs.

x	y
-2	-7
-1	-4
0	-3
1	-4
2	-7

   Plot these points and connect them with a smooth curve. From the graph, we see that the vertex, $(0, -3)$, is the highest point of this graph.

   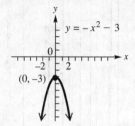

   (b) $y = x^2 + 3$

   Find several ordered pairs. To begin, check for intercepts.

   If $x = 0$,
   $$y = 0^2 + 3 = 3,$$
   so the y-intercept is $(0, 3)$.

   If $y = 0$,
   $$0 = x^2 + 3,$$
   and hence, $x^2 = -3$.

   This equation has no real number solution, so there are no x-intercepts.

   Choose additional x-values near $x = 0$. We obtain the following table for ordered pairs.

x	y
-2	7
-1	4
0	3
1	4
2	7

   Plot these points and connect them with a smooth curve. From the graph, we see that the vertex, $(0, 3)$, is the lowest point of this graph.

   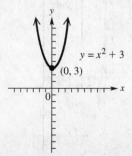

364 Chapter 9 Quadratic Equations

5. $y = x^2 + 2x - 8$

Find any x-intercepts by substituting 0 for y in the equation.

$$y = x^2 + 2x - 8$$
$$0 = x^2 + 2x - 8 \quad \text{Let } y = 0.$$
$$0 = (x+4)(x-2) \quad \text{Factor.}$$
$$x + 4 = 0 \quad \text{or} \quad x - 2 = 0$$
$$x = -4 \quad \text{or} \quad x = 2$$

The x-intercepts are $(-4, 0)$ and $(2, 0)$.

Now find any y-intercepts by substituting 0 for x.

$$y = x^2 + 2x - 8$$
$$y = 0^2 + 2(0) - 8 \quad \text{Let } x = 0.$$
$$y = -8$$

The y-intercept is $(0, -8)$. The x-value of the vertex is halfway between the x-intercepts, $(-4, 0)$ and $(2, 0)$.

$$x = \frac{1}{2}(-4 + 2) = \frac{1}{2}(-2) = -1$$

Find the y-value of the vertex by substituting -1 for x in the given equation.

$$y = x^2 + 2x - 8$$
$$y = (-1)^2 + 2(-1) - 8 \quad \text{Let } x = -1.$$
$$y = 1 - 2 - 8$$
$$y = -9$$

The vertex is at $(-1, -9)$.

The axis of the parabola is the line $x = -1$.

x	y
-4	0
-3	-5
-2	-8
-1	-9
0	-8
1	-5
2	0

Plot the intercepts, vertex, and additional points shown in the table of values and connect them with a smooth curve.

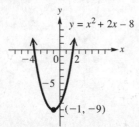

6. $y = x^2 - 4x + 1$

If $x = 5$, $y = 5^2 - 4(5) + 1 = 6$, giving the ordered pair $(5, 6)$.

If $x = 1$, $y = 1^2 - 4(1) + 1 = -2$, giving the ordered pair $(1, -2)$.

If $x = 4$, $y = 4^2 - 4(4) + 1 = 1$, giving the ordered pair $(4, 1)$.

If $x = 3$, $y = 3^2 - 4(3) + 1 = -2$, giving the ordered pair $(3, -2)$.

If $x = -1$, $y = (-1)^2 - 4(-1) + 1 = 6$, giving the ordered pair $(-1, 6)$.

7. (a) $y = x^2 - 3x - 3$

Here, $a = 1$, $b = -3$, and $c = -3$. The x-value of the vertex is

$$x = -\frac{b}{2a} = -\frac{-3}{2(1)} = \frac{3}{2} = 1.5.$$

The y-value of the vertex is

$$y = (1.5)^2 - 3(1.5) - 3 = -5.25,$$

so the vertex is $(1.5, -5.25)$.

The axis of the parabola is the line $x = 1.5$.

Now find the x-intercepts by using the quadratic formula.

$$x = \frac{-(-3) \pm \sqrt{(-3)^2 - 4(1)(-3)}}{2(1)}$$

$$x = \frac{3 \pm \sqrt{21}}{2}$$

Using a calculator, $x \approx -.8$ and $x \approx 3.8$.

The x-intercepts are $(-.8, 0)$ and $(3.8, 0)$.

Find the y-intercept by letting $x = 0$.

$$y = 0^2 - 3(0) - 3 = -3$$

The y-intercept is $(0, -3)$.

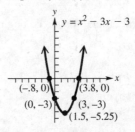

(b) $y = -x^2 + 2x + 4$

Here, $a = -1$, $b = 2$, and $c = 4$.

The x-value of the vertex is

$$x = -\frac{b}{2a} = -\frac{2}{2(-1)} = 1.$$

The y-value of the vertex is
$$y = -1^2 + 2(1) + 4 = 5,$$
so the vertex is $(1, 5)$.

The axis of the parabola is the line $x = 1$.

Now find the x-intercepts by using the quadratic formula.
$$x = \frac{-2 \pm \sqrt{2^2 - 4(-1)(4)}}{2(-1)}$$
$$x = \frac{-2 \pm \sqrt{20}}{-2}$$

Using a calculator, $x \approx 3.2$ and $x \approx -1.2$.

The x-intercepts are $(-1.2, 0)$ and $(3.2, 0)$.

Find the y-intercept by letting $x = 0$.
$$y = -0^2 + 2(0) + 4 = 4$$

The y-intercept is $(0, 4)$.

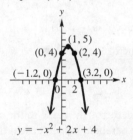

8. As in Example 5, one point on the graph is $\left(\frac{1}{2}(350), 48\right) = (175, 48)$.

$$y = ax^2$$
$$48 = a(175)^2 \quad \text{Let } x = 175, y = 48.$$
$$48 = 30{,}625a$$
$$a = \frac{48}{30{,}625}$$

The equation is $y = \frac{48}{30{,}625}x^2$.

9.4 Section Exercises

1. The vertex of a parabola is the lowest or highest point on the graph.

3. $y = 2x^2$

   In $y = 2x^2$, $a = 2$, $b = 0$, and $c = 0$.

   The x-value of the vertex is
   $$x = -\frac{b}{2a} = -\frac{0}{2(2)} = 0.$$

   The y-value of the vertex is
   $$y = 2(0)^2 = 0,$$
   so the vertex is $(0, 0)$.

   The axis of the parabola is the line $x = 0$, which is the y-axis.

   Now find the intercepts.

   Let $x = 0$ to get $y = 2 \cdot 0^2 = 0$. The y-intercept is $(0, 0)$.

   Let $y = 0$ to get $0 = 2x^2$, which implies $x = 0$. The only x-intercept is $(0, 0)$.

   Make a table of ordered pairs.

x	y
-2	8
-1	2
0	0
1	2
2	8

   Plot these five ordered pairs and connect them with a smooth curve.

   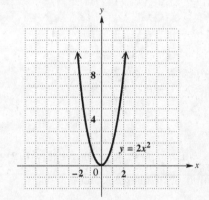

5. $y = x^2 - 4$

   In $y = x^2 - 4$, $a = 1$, $b = 0$, and $c = -4$.

   The x-value of the vertex is
   $$x = -\frac{b}{2a} = -\frac{0}{2(1)} = 0.$$

   The y-value of the vertex is
   $$y = 0^2 - 4 = -4,$$
   so the vertex is $(0, -4)$.

   The axis of the parabola is the line $x = 0$, which is the y-axis.

   Now find the intercepts.

   Let $x = 0$ to get $y = 0^2 - 4 = -4$; the y-intercept is $(0, -4)$. Let $y = 0$ and solve for x.
   $$0 = x^2 - 4$$
   $$0 = (x + 2)(x - 2)$$

   continued

$x + 2 = 0$ or $x - 2 = 0$
$x = -2$ or $x = 2$

The x-intercepts are $(-2, 0)$ and $(2, 0)$.

Make a table of ordered pairs.

x	y
0	-4
± 1	-3
± 2	0
± 3	5

Plot these seven ordered pairs and connect them with a smooth curve.

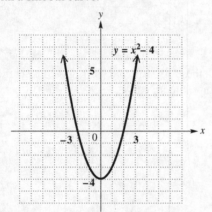

7. $y = -x^2 + 2$

If $x = 0$, $y = 2$, so the y-intercept is $(0, 2)$.

To find any x-intercepts, let $y = 0$.

$$0 = -x^2 + 2$$
$$x^2 = 2$$
$$x = \pm\sqrt{2} \approx \pm 1.41$$

The x-intercepts are $\left(\pm\sqrt{2}, 0\right)$.

The x-value of the vertex is

$$x = -\frac{b}{2a} = -\frac{0}{2(-1)} = 0.$$

Thus, the vertex is the same as the y-intercept (since $x = 0$). The axis of the parabola is the vertical line $x = 0$.

Make a table of ordered pairs whose x-values are on either side of the vertex's x-value of $x = 0$.

x	y
0	2
± 1	1
± 2	-2
± 3	-7

Plot these seven ordered pairs and connect them with a smooth curve.

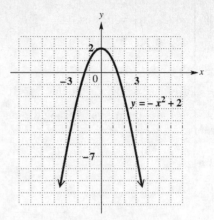

9. $y = (x + 3)^2 = x^2 + 6x + 9$

If $x = 0$, $y = 9$, so the y-intercept is $(0, 9)$.

To find any x-intercepts, let $y = 0$.

$$0 = (x + 3)^2$$
$$0 = x + 3$$
$$-3 = x$$

The x-intercept is $(-3, 0)$.

The x-value of the vertex is

$$x = -\frac{b}{2a} = -\frac{6}{2(1)} = -3.$$

Thus, the vertex is the same as the x-intercept. The axis of the parabola is the vertical line $x = -3$.

Make a table of ordered pairs whose x-values are on either side of the vertex's x-value of $x = -3$.

x	y
-6	9
-5	4
-4	1
-3	0
-2	1
-1	4
0	9

Plot these seven ordered pairs and connect them with a smooth curve.

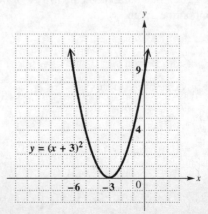

11. $y = x^2 + 2x + 3$

Let $x = 0$ to get
$$y = 0^2 + 2(0) + 3 = 3;$$
the y-intercept is $(0, 3)$.

To find any x-intercepts, let $y = 0$.
$$0 = x^2 + 2x + 3$$

The trinomial on the right cannot be factored. Because the discriminant $b^2 - 4ac = 2^2 - 4(1)(3) = -8$ is negative, this equation has no real number solutions. Thus, the parabola has no x-intercepts.

The x-value of the vertex is
$$x = -\frac{b}{2a} = -\frac{2}{2(1)} = -1.$$

The y-value of the vertex is
$$y = (-1)^2 + 2(-1) + 3$$
$$= 1 - 2 + 3 = 2,$$

so the vertex is $(-1, 2)$. The axis of the parabola is the vertical line $x = -1$.

Make a table of ordered pairs whose x-values are on either side of the vertex's x-value of $x = -1$.

x	y
-4	11
-3	6
-2	3
-1	2
0	3
1	6
2	11

Plot these seven ordered pairs and connect them with a smooth curve.

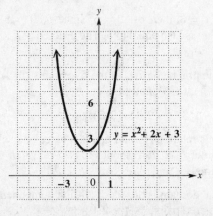

13. $y = -x^2 + 6x - 5$

Let $x = 0$ to get
$$y = -(0)^2 + 6(0) - 5 = -5;$$
the y-intercept is $(0, -5)$.

Let $y = 0$ and solve for x.
$$0 = -x^2 + 6x - 5$$
$$x^2 - 6x + 5 = 0$$
$$(x - 1)(x - 5) = 0$$
$$x - 1 = 0 \quad \text{or} \quad x - 5 = 0$$
$$x = 1 \quad \text{or} \quad x = 5$$

The x-intercepts are $(1, 0)$ and $(5, 0)$.

The x-value of the vertex is
$$x = -\frac{b}{2a} = -\frac{6}{2(-1)} = 3.$$

The y-value of the vertex is
$$y = -(3)^2 + 6(3) - 5$$
$$= -9 + 18 - 5 = 4,$$

so the vertex is $(3, 4)$.

The axis of the parabola is the line $x = 3$.

Make a table of ordered pairs whose x-values are on either side of the vertex's x-value of $x = 3$.

x	y
0	-5
1	0
2	3
3	4
4	3
5	0
6	-5

Plot these seven ordered pairs and connect them with a smooth curve.

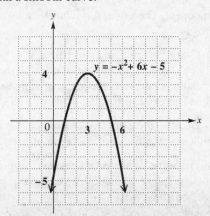

Chapter 9 Quadratic Equations

15. If $a > 0$, the parabola $y = ax^2 + bx + c$ opens upward. If $a < 0$, the parabola $y = ax^2 + bx + c$ opens downward.

17. Because the vertex is at the origin, an equation of the parabola is of the form

$$y = ax^2.$$

As shown in the figure, one point on the graph has coordinates $(150, 44)$.

$y = ax^2$ General equation
$44 = a(150)^2$ Let $x = 150, y = 44$.
$44 = 22{,}500a$
$a = \dfrac{44}{22{,}500} = \dfrac{4 \cdot 11}{4 \cdot 5625} = \dfrac{11}{5625}$

Thus, an equation of the parabola is

$$y = \dfrac{11}{5625}x^2.$$

9.5 Introduction to Functions

9.5 Margin Exercises

1. **(a)** Consider the following relation:

$$\{(5, 10), (15, 20), (25, 30), (35, 40)\}$$

The domain is the set of all first components in the ordered pairs, $\{5, 15, 25, 35\}$. The range is the set of all second components in the ordered pairs, $\{10, 20, 30, 40\}$.

(b) The relation $\{(1, 4), (2, 4), (3, 4)\}$ has domain $\{1, 2, 3\}$ and range $\{4\}$.

2. **(a)** $\{(-2, 8), (-1, 1), (0, 0), (1, 1), (2, 8)\}$
Notice that each first component appears once and only once. Because of this, the relation *is a function*.

(b) $\{(5, 2), (5, 1), (5, 0)\}$
The first component 5 appears in all three ordered pairs, and corresponds to more than one second component. Therefore, this relation *is not a function*.

3. **(a)** Every first component is paired with one and only one second component, and furthermore, no vertical line intersects the graph in more than one point. Therefore, this is the graph of a function.

(b) Because there are two ordered pairs with first component 0, this is not the graph of a function.

(c) The vertical line test shows that this graph is not the graph of a function; a vertical line could cross the graph twice.

(d) The vertical line test shows that this graph is not the graph of a function; a vertical line could cross the graph twice.

(e) The graph of $y = 3$ is a horizontal line, so the equation defines a function. A vertical line will only intersect the graph in one point for every value of x.

4. $f(x) = 6x - 2$

(a) $f(-1) = 6(-1) - 2$ Let $x = -1$.
$= -6 - 2$
$= -8$

(b) $f(0) = 6(0) - 2 = 0 - 2 = -2$

(c) $f(1) = 6(1) - 2 = 6 - 2 = 4$

5. **(a)** Choose the years as the domain elements and the numbers of U.S. children (in millions) as the range elements.

$f = \{(1997, 1.1), (1998, 1.2), (1999, 1.3),$
$(2000, 1.5), (2001, 1.7)\}.$

(b) The domain is the set of years, or of x-values:

$$\{1997, 1998, 1999, 2000, 2001\}$$

The range is the set of numbers of children, or y-values:

$$\{1.1, 1.2, 1.3, 1.5, 1.7\}$$

(c) $f(1999) = 1.3$, or 1.3 million

(d) For what x-values does $f(x) = 1.5$?

$$f(2000) = 1.5$$

In 2000, the number of children was 1.5 million.

9.5 Section Exercises

1. If $x = 1$, then $x + 2 = 3$. Since $f(x) = x + 2$, $f(x)$ is also equal to 3. The ordered pair (x, y) is $(1, 3)$.

3. If $x = 3$, then $x + 2 = 5$. Since $f(x) = x + 2$, $f(x)$ is also equal to 5. The ordered pair (x, y) is $(3, 5)$.

5. The graph consists of the five points $(0, 2), (1, 3), (2, 4), (3, 5),$ and $(4, 6)$.

7. $\{(-4, 3), (-2, 1), (0, 5), (-2, -8)\}$

This relation *is not a function* since one value of x, namely -2, corresponds to two values of y, namely 1 and -8.

The domain is the set of all first components in the ordered pairs, $\{-4, -2, 0\}$.

The range is the set of all second components in the ordered pairs, $\{3, 1, 5, -8\}$.

9. The relation *is a function* since each of the first components A, B, C, D, and E corresponds to exactly one second component.

The domain is $\{A, B, C, D, E\}$.
The range is $\{2, 3, 6, 4\}$.

11. The graph consists of the following set of six ordered pairs:

$\{(-4, 1), (-2, 0), (-2, 2), (0, -2), (2, 1), (3, 3)\}$

This relation *is not a function* since one value of x, namely -2, corresponds to two values of y, namely 0 and 2.

The domain is the set of all first components in the ordered pairs, $\{-4, -2, 0, 2, 3\}$.

The range is the set of all second components in the ordered pairs, $\{1, 0, 2, -2, 3\}$.

13. Any vertical line will intersect the graph in only one point. The graph passes the vertical line test, so this is the graph of a function.

15. A vertical line can cross the graph twice, so this is not the graph of a function.

17. $y = 5x + 3$

Every value of x will give one and only one value of y, so the equation defines a function.

19. $x = |y|$

Every positive value of x will give two values of y. For example, if $x = 4$, $4 = |y|$ and $y = +4$ or -4. Therefore, the equation does not define a function.

21. If $f(2) = 4$, one point on the line has coordinates $(2, 4)$.

22. If $f(-1) = -4$, then another point on the line has coordinates $(-1, -4)$.

23. Using the points $(2, 4)$ and $(-1, -4)$, we obtain

$$m = \frac{-4 - 4}{-1 - 2} = \frac{-8}{-3} = \frac{8}{3}.$$

24. Start with the point-slope form of a line using $m = \frac{8}{3}$ and $(x_1, y_1) = (2, 4)$.

$$y - y_1 = m(x - x_1)$$
$$y - 4 = \tfrac{8}{3}(x - 2)$$

Now solve for y to write this equation in slope-intercept form.

$$y - 4 = \tfrac{8}{3}x - \tfrac{16}{3}$$
$$y = \tfrac{8}{3}x - \tfrac{16}{3} + \tfrac{12}{3}$$
$$y = \tfrac{8}{3}x - \tfrac{4}{3}$$

Therefore,
$$f(x) = \tfrac{8}{3}x - \tfrac{4}{3}.$$

25. $f(x) = 4x + 3$

(a) $f(2) = 4(2) + 3 = 8 + 3 = 11$
(b) $f(0) = 4(0) + 3 = 0 + 3 = 3$
(c) $f(-3) = 4(-3) + 3 = -12 + 3 = -9$

27. $f(x) = x^2 - x + 2$

(a) $f(2) = (2)^2 - (2) + 2 = 4 - 2 + 2 = 4$
(b) $f(0) = (0)^2 - (0) + 2 = 0 - 0 + 2 = 2$
(c) $f(-3) = (-3)^2 - (-3) + 2$
$= 9 + 3 + 2 = 14$

29. $f(x) = |x|$

(a) $f(2) = |2| = 2$
(b) $f(0) = |0| = 0$
(c) $f(-3) = |-3| = -(-3) = 3$

31. Write the information in the graph as a set of ordered pairs of the form (year, population). The set is $\{(1970, 9.6), (1980, 14.1), (1990, 19.8), (2000, 28.4)\}$. Since each year corresponds to exactly one number, the set defines a function.

33. $g(1980) = 14.1$ (million);
$g(1990) = 19.8$ (million)

35. For the year 2002, the function indicates 30.3 million foreign-born residents in the United States.

36.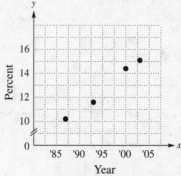

Yes, the points lie approximately on a straight line, so a linear function would give a reasonable approximation of the data.

37. Using $(1987, 10.2)$ and $(2003, 15.1)$, we have slope

$$m = \frac{15.1 - 10.2}{2003 - 1987} = \frac{4.9}{16} = .30625.$$

Now use the point $(1987, 10.2)$ for (x_1, y_1) and $m = .30625$ in the point-slope form.

$$y - y_1 = m(x - x_1)$$
$$y - 10.2 = .30625(x - 1987)$$
$$y - 10.2 = .30625x - 608.51875$$
$$y = .30625x - 598.31875$$
or $y = f(x) = .30625x - 598.32$

38. $f(1993) = .30625(1993) - 598.32$
$= 610.35625 - 598.32$
$= 12.03625 \approx \mathbf{12.0\%}$

$f(2000) = .30625(2000) - 598.32$
$= 612.5 - 598.32$
$= 14.18 \approx \mathbf{14.2\%}$

39. Using $(1993, 11.6)$ and $(2000, 14.4)$, we have slope

$$m = \frac{14.4 - 11.6}{2000 - 1993} = \frac{2.8}{7} = .4.$$

Now use the point $(1993, 11.6)$ for (x_1, y_1) and $m = .4$ in the point-slope form.

$$y - y_1 = m(x - x_1)$$
$$y - 11.6 = .4(x - 1993)$$
$$y - 11.6 = .4x - 797.2$$
$$y = .4x - 785.6$$
or $y = f(x) = .4x - 785.6$

40. $f(1987) = .4(1987) - 785.6$
$= 794.8 - 785.6$
$= \mathbf{9.2\%}$

$f(2003) = .4(2003) - 785.6$
$= 801.2 - 785.6$
$= \mathbf{15.6\%}$

The equation from Exercise 38 gives better approximations. The results in Exercise 38 vary by .4% and .2% from the data. The results here give answers that vary by 1.0% and .5%.

Chapter 9 Review Exercises

1. $y^2 = 144$

$y = \sqrt{144}$ or $y = -\sqrt{144}$
$y = 12$ or $y = -12$

A check verifies that the solutions are 12 and -12.

2. $x^2 = 37$

$x = \sqrt{37}$ or $x = -\sqrt{37}$

A check verifies that the solutions are $\pm\sqrt{37}$.

3. $m^2 = 128$

$m = \pm\sqrt{128}$
$= \pm\sqrt{64 \cdot 2}$
$= \pm 8\sqrt{2}$

A check verifies that the solutions are $\pm 8\sqrt{2}$.

4. $(k + 2)^2 = 25$

$k + 2 = \sqrt{25}$ or $k + 2 = -\sqrt{25}$
$k + 2 = 5$ or $k + 2 = -5$
$k = 3$ or $k = -7$

A check verifies that the solutions are -7 and 3.

5. $(r - 3)^2 = 10$

$r - 3 = \sqrt{10}$ or $r - 3 = -\sqrt{10}$
$r = 3 + \sqrt{10}$ or $r = 3 - \sqrt{10}$

A check verifies that the solutions are $3 \pm \sqrt{10}$.

6. $(2p + 1)^2 = 14$

$2p + 1 = \sqrt{14}$ or $2p + 1 = -\sqrt{14}$
$2p = -1 + \sqrt{14}$ or $2p = -1 - \sqrt{14}$
$p = \dfrac{-1 + \sqrt{14}}{2}$ or $p = \dfrac{-1 - \sqrt{14}}{2}$

A check verifies that the solutions are $\dfrac{-1 \pm \sqrt{14}}{2}$.

7. $(3k + 2)^2 = -3$

$3k + 2 = \sqrt{-3}$ or $3k + 2 = -\sqrt{-3}$

Because $\sqrt{-3}$ does not represent a real number, there is *no real number solution*.

8. $(3x + 5)^2 = 0$
$3x + 5 = 0$
$3x = -5$
$x = -\frac{5}{3}$

A check verifies that the solution is $-\frac{5}{3}$.

9. $m^2 + 6m + 5 = 0$

Rewrite the equation with the variable terms on one side and the constant on the other side.

$$m^2 + 6m = -5$$

Take half the coefficient of m and square it.

$$\frac{1}{2}(6) = 3, \text{ and } (3)^2 = 9.$$

Add 9 to each side of the equation.

$$m^2 + 6m + 9 = -5 + 9$$
$$m^2 + 6m + 9 = 4$$
$$(m+3)^2 = 4 \quad \textit{Factor.}$$
$$m + 3 = \sqrt{4} \quad \text{or} \quad m + 3 = -\sqrt{4}$$
$$m + 3 = 2 \quad \text{or} \quad m + 3 = -2$$
$$m = -1 \quad \text{or} \quad m = -5$$

A check verifies that the solutions are -5 and -1.

10. $p^2 + 4p = 7$

Take half the coefficient of p and square it.

$$\frac{1}{2}(4) = 2, \text{ and } (2)^2 = 4.$$

Add 4 to each side of the equation.

$$p^2 + 4p + 4 = 7 + 4$$
$$(p+2)^2 = 11$$
$$p + 2 = \sqrt{11} \quad \text{or} \quad p + 2 = -\sqrt{11}$$
$$p = -2 + \sqrt{11} \quad \text{or} \quad p = -2 - \sqrt{11}$$

A check verifies that the solutions are $-2 \pm \sqrt{11}$.

11. $-x^2 + 5 = 2x$

Divide each side of the equation by -1 to make the coefficient of the squared term equal to 1.

$$x^2 - 5 = -2x$$

Rewrite the equation with the variable terms on one side and the constant on the other side.

$$x^2 + 2x = 5$$

Take half the coefficient of x and square it.

$$\frac{1}{2}(2) = 1, \text{ and } 1^2 = 1.$$

Add 1 to each side of the equation.

$$x^2 + 2x + 1 = 5 + 1$$
$$(x+1)^2 = 6$$
$$x + 1 = \sqrt{6} \quad \text{or} \quad x + 1 = -\sqrt{6}$$
$$x = -1 + \sqrt{6} \quad \text{or} \quad x = -1 - \sqrt{6}$$

A check verifies that the solutions are $-1 \pm \sqrt{6}$.

12. $2x^2 - 3 = -8x$

Divide each side by 2 to get the x^2 coefficient equal to 1.

$$x^2 - \frac{3}{2} = -4x$$

Rewrite the equation with the variable terms on one side and the constant on the other side.

$$x^2 + 4x = \frac{3}{2}$$

Take half the coefficient of x and square it.

$$\frac{1}{2}(4) = 2, \text{ and } 2^2 = 4.$$

Add 4 to each side of the equation.

$$x^2 + 4x + 4 = \frac{3}{2} + 4$$
$$(x+2)^2 = \frac{11}{2}$$
$$x + 2 = \pm\sqrt{\frac{11}{2}}$$
$$x + 2 = \pm\frac{\sqrt{11}}{\sqrt{2}} \cdot \frac{\sqrt{2}}{\sqrt{2}}$$
$$x + 2 = \pm\frac{\sqrt{22}}{2}$$
$$x = -2 \pm \frac{\sqrt{22}}{2}$$
$$x = \frac{-4}{2} \pm \frac{\sqrt{22}}{2}$$
$$x = \frac{-4 \pm \sqrt{22}}{2}$$

A check verifies that the solutions are $\dfrac{-4 \pm \sqrt{22}}{2}$.

13. $4(x^2 + 7x) + 29 = -20$

$$\begin{aligned} 4x^2 + 28x + 49 &= 0 & &\textit{Multiply by 4; add 20.} \\ (2x+7)^2 &= 0 & &\textit{Factor.} \\ 2x + 7 &= 0 & &\textit{Take square root.} \\ 2x &= -7 & &\textit{Subtract 7.} \\ x &= -\tfrac{7}{2} & &\textit{Divide by 2.} \end{aligned}$$

The solution is $-\frac{7}{2}$.

14. $(4x+1)(x-1) = -7$

Multiply on the left side and then simplify. Get all variable terms on one side and the constant on the other side.

$$4x^2 - 4x + x - 1 = -7$$
$$4x^2 - 3x = -6$$

Divide each side by 4 so that the coefficient of x^2 will be 1.

$$x^2 - \frac{3}{4}x = -\frac{6}{4} = -\frac{3}{2}$$

Square half the coefficient of x and add it to each side.

$$x^2 - \frac{3}{4}x + \frac{9}{64} = -\frac{3}{2} + \frac{9}{64}$$
$$\left(x - \frac{3}{8}\right)^2 = -\frac{96}{64} + \frac{9}{64}$$
$$\left(x - \frac{3}{8}\right)^2 = -\frac{87}{64}$$

The square root of $-\frac{87}{64}$ is not a real number, so there is *no real number solution*.

15. $h = -16t^2 + 32t + 50$

Let $h = 30$ and solve for t (which must have a positive value since it represents a number of seconds).

$$30 = -16t^2 + 32t + 50$$
$$16t^2 - 32t - 20 = 0$$

Divide each side by 16.

$$t^2 - 2t - \frac{20}{16} = 0$$
$$t^2 - 2t = \frac{5}{4}$$

Half of -2 is -1, and $(-1)^2 = 1$.

Add 1 to each side of the equation.

$$t^2 - 2t + 1 = \frac{5}{4} + 1$$
$$(t-1)^2 = \frac{9}{4}$$

$t - 1 = \sqrt{\frac{9}{4}}$ or $t - 1 = -\sqrt{\frac{9}{4}}$

$t - 1 = \frac{3}{2}$ or $t - 1 = -\frac{3}{2}$

$t = 1 + \frac{3}{2}$ or $t = 1 - \frac{3}{2}$

$t = \frac{5}{2} = 2\frac{1}{2}$ or $t = -\frac{1}{2}$

Reject the negative value of t. The object will reach a height of 30 feet after $2\frac{1}{2}$ seconds.

16. Use the Pythagorean formula with legs x and $x + 2$ and hypotenuse $x + 4$.

$$a^2 + b^2 = c^2$$
$$(x)^2 + (x+2)^2 = (x+4)^2$$
$$x^2 + x^2 + 4x + 4 = x^2 + 8x + 16$$
$$x^2 - 4x - 12 = 0$$
$$(x-6)(x+2) = 0$$

$x - 6 = 0$ or $x + 2 = 0$
$x = 6$ or $x = -2$

Reject the negative value because x represents a length. The value of x is 6. The lengths of the three sides are 6, 8, and 10.

17. Take half the coefficient of x and square the result.

$$\frac{1}{2} \cdot k = \frac{k}{2}$$
$$\left(\frac{k}{2}\right)^2 = \frac{k^2}{4}$$

Add $\left(\frac{k}{2}\right)^2$ or $\frac{k^2}{4}$ to $x^2 + kx$ to get the perfect square $x^2 + kx + \frac{k^2}{4}$.

18. $x^2 - 9 = 0$, or $1x^2 + 0x - 9 = 0$

(a) $(x+3)(x-3) = 0$

$x + 3 = 0$ or $x - 3 = 0$
$x = -3$ or $x = 3$

A check verifies that the solutions are -3 and 3.

(b) $x^2 = 9$

$x = \pm\sqrt{9} = \pm 3$

A check verifies that the solutions are -3 and 3.

(c) Here, $a = 1$, $b = 0$, and $c = -9$.

$$x = \frac{-0 \pm \sqrt{0^2 - 4(1)(-9)}}{2(1)}$$
$$= \frac{\pm\sqrt{36}}{2}$$
$$= \frac{\pm 6}{2} = \pm 3$$

A check verifies that the solutions are -3 and 3.

(d) Because there is only one pair of solutions, we always get the same results, no matter which method of solution is used.

19. $-4x^2 - 2x + 7 = 0$
$4x^2 + 2x - 7 = 0$ Multiply by -1.

Use the quadratic formula with $a = 4$, $b = 2$, and $c = -7$.

$$x = \frac{-2 \pm \sqrt{(2)^2 - 4(4)(-7)}}{2(4)}$$

$$= \frac{-2 \pm \sqrt{4 + 112}}{8} = \frac{-2 \pm \sqrt{116}}{8}$$

$$= \frac{-2 \pm 2\sqrt{29}}{8} = \frac{2(-1 \pm \sqrt{29})}{2(4)}$$

$$= \frac{-1 \pm \sqrt{29}}{4}$$

A check verifies that the solutions are $\frac{-1 \pm \sqrt{29}}{4}$.

20. $2x^2 + 8 = 4x + 11$
$2x^2 - 4x - 3 = 0$

Use the quadratic formula with $a = 2$, $b = -4$, and $c = -3$.

$$x = \frac{-(-4) \pm \sqrt{(-4)^2 - 4(2)(-3)}}{2(2)}$$

$$= \frac{4 \pm \sqrt{16 + 24}}{4} = \frac{4 \pm \sqrt{40}}{4}$$

$$= \frac{4 \pm \sqrt{4 \cdot 10}}{4} = \frac{4 \pm 2\sqrt{10}}{4}$$

$$= \frac{2(2 \pm \sqrt{10})}{2(2)} = \frac{2 \pm \sqrt{10}}{2}$$

A check verifies that the solutions are $\frac{2 \pm \sqrt{10}}{2}$.

21. $x(5x - 1) = 1$
$5x^2 - x = 1$
$5x^2 - x - 1 = 0$

Use the quadratic formula with $a = 5$, $b = -1$, and $c = -1$.

$$x = \frac{-b \pm \sqrt{b^2 - 4ac}}{2a}$$

$$x = \frac{-(-1) \pm \sqrt{(-1)^2 - 4(5)(-1)}}{2(5)}$$

$$x = \frac{1 \pm \sqrt{1 + 20}}{10}$$

$$x = \frac{1 \pm \sqrt{21}}{10}$$

A check verifies that the solutions are $\frac{1 \pm \sqrt{21}}{10}$.

22. $\frac{1}{4}x^2 = 2 - \frac{3}{4}x$
$\frac{1}{4}x^2 + \frac{3}{4}x - 2 = 0$

Multiply each side by the LCD, 4.

$$4\left(\tfrac{1}{4}x^2 + \tfrac{3}{4}x - 2\right) = 4(0)$$
$$x^2 + 3x - 8 = 0$$

Use the quadratic formula with $a = 1$, $b = 3$, and $c = -8$.

$$x = \frac{-3 \pm \sqrt{3^2 - 4(1)(-8)}}{2(1)}$$

$$= \frac{-3 \pm \sqrt{9 + 32}}{2}$$

$$= \frac{-3 \pm \sqrt{41}}{2}$$

A check verifies that the solutions are $\frac{-3 \pm \sqrt{41}}{2}$.

23. $\frac{1}{2}x^2 + 3x = 5$
$\frac{1}{2}x^2 + 3x - 5 = 0$
$x^2 + 6x - 10 = 0$ Multiply by 2.

Use the quadratic formula with $a = 1$, $b = 6$, and $c = -10$.

$$x = \frac{-6 \pm \sqrt{6^2 - 4(1)(-10)}}{2(1)}$$

$$x = \frac{-6 \pm \sqrt{36 + 40}}{2} = \frac{-6 \pm \sqrt{76}}{2}$$

$$= \frac{-6 \pm 2\sqrt{19}}{2} = -3 \pm \sqrt{19}$$

A check verifies that the solutions are $-3 \pm \sqrt{19}$.

24. The correct statement of the quadratic formula is

$$x = \frac{-b \pm \sqrt{b^2 - 4ac}}{2a}.$$

To state that

$$x = -b \pm \frac{\sqrt{b^2 - 4ac}}{2a}$$

is not equivalent because the $-b$ term should be above the fraction bar rather than a separate term from the fraction.

374   Chapter 9   Quadratic Equations

25. $y = -3x^2$

If $x = 0$, $y = 0$, so the y- and x-intercepts are $(0, 0)$.

The x-value of the vertex is

$$x = -\frac{b}{2a} = -\frac{0}{2(-3)} = 0.$$

Thus, the vertex is the same as the y-intercept (since $x = 0$). The axis of the parabola is the vertical line $x = 0$.

Make a table of ordered pairs whose x-values are on either side of the vertex's x-value of $x = 0$.

x	y
0	0
± 1	-3
± 2	-12
± 3	-27

Plot these seven ordered pairs and connect them with a smooth curve.

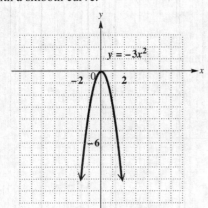

26. $y = -x^2 + 5$

If $x = 0$, $y = 5$, so the y-intercept is $(0, 5)$.

To find any x-intercepts, let $y = 0$.

$$0 = -x^2 + 5$$
$$x^2 = 5$$
$$x = \pm\sqrt{5} \approx \pm 2.24$$

The x-intercepts are $\left(\pm\sqrt{5}, 0\right)$.

The x-value of the vertex is

$$x = -\frac{b}{2a} = -\frac{0}{2(-1)} = 0.$$

Thus, the vertex is the same as the y-intercept (since $x = 0$). The axis of the parabola is the vertical line $x = 0$.

Make a table of ordered pairs whose x-values are on either side of the vertex's x-value of $x = 0$.

x	y
0	5
± 1	4
± 2	1
± 3	-4

Plot these seven ordered pairs and connect them with a smooth curve.

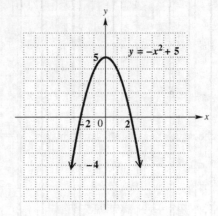

27. $y = x^2 - 2x + 1$

Let $x = 0$ to get

$$y = 0^2 - 2(0) + 1 = 1;$$

the y-intercept is $(0, 1)$.

Let $y = 0$ and solve for x.

$$0 = x^2 - 2x + 1$$
$$0 = (x - 1)^2$$
$$x - 1 = 0$$
$$x = 1$$

The x-intercept is $(1, 0)$.

The x-value of the vertex is

$$x = -\frac{b}{2a} = -\frac{-2}{2(1)} = 1.$$

The y-value of the vertex is

$$y = 1^2 - 2(1) + 1 = 0,$$

so the vertex is $(1, 0)$.

Make a table of ordered pairs whose x-values are on either side of the vertex's x-value of $x = 1$.

x	y
-1	4
0	1
1	0
2	1
3	4

Plot these five ordered pairs and connect them with a smooth curve.

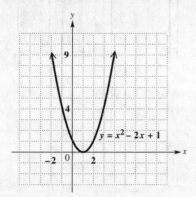

28. $y = -x^2 + 2x + 3$

Let $x = 0$ to get
$$y = -0^2 + 2(0) + 3 = 3;$$
the y-intercept is $(0, 3)$.

Let $y = 0$ and solve for x.
$$0 = -x^2 + 2x + 3$$
$$x^2 - 2x - 3 = 0$$
$$(x - 3)(x + 1) = 0$$
$$x - 3 = 0 \quad \text{or} \quad x + 1 = 0$$
$$x = 3 \quad \text{or} \quad x = -1$$

The x-intercepts are $(3, 0)$ and $(-1, 0)$.

The x-value of the vertex is
$$x = -\frac{b}{2a} = -\frac{2}{2(-1)} = 1.$$

The y-value of the vertex is
$$y = -1^2 + 2(1) + 3 = 4,$$
so the vertex is $(1, 4)$.

Make a table of ordered pairs whose x-values are on either side of the vertex's x-value of $x = 1$.

x	y
-1	0
0	3
1	4
2	3
3	0

Plot these five ordered pairs and connect them with a smooth curve.

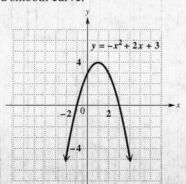

29. $y = x^2 + 4x + 2$

Let $x = 0$ to get
$$y = 0^2 + 4(0) + 2 = 2;$$
the y-intercept is $(0, 2)$.

Let $y = 0$ and solve for x.
$$x^2 + 4x + 2 = 0$$
$$x^2 + 4x = -2$$
$$x^2 + 4x + 4 = -2 + 4$$
$$(x + 2)^2 = 2$$

$x + 2 = \sqrt{2}$ or $x + 2 = -\sqrt{2}$
$x = -2 + \sqrt{2}$ or $x = -2 - \sqrt{2}$
$x \approx -.6$ or $x \approx -3.4$

The x-intercepts are approximately $(-.6, 0)$ and $(-3.4, 0)$.

The x-value of the vertex is
$$x = -\frac{b}{2a} = -\frac{4}{2(1)} = -2.$$

The y-value of the vertex is
$$y = (-2)^2 + 4(-2) + 2 = -2,$$
so the vertex is $(-2, -2)$.

Make a table of ordered pairs whose x-values are on either side of the vertex's x-value of $x = -2$.

x	y
-4	2
-3.4	0
-3	-1
-2	-2
-1	-1
-0.6	0
0	2

Plot these seven ordered pairs and connect them with a smooth curve.

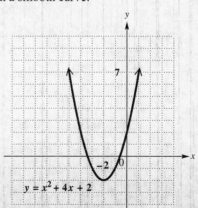

30. $y = (x+4)^2 = x^2 + 8x + 16$

If $x = 0$, $y = 16$, so the y-intercept is $(0, 16)$.

To find any x-intercepts, let $y = 0$.

$$0 = (x+4)^2$$
$$0 = x + 4$$
$$-4 = x$$

The x-intercept is $(-4, 0)$.

The x-value of the vertex is

$$x = -\frac{b}{2a} = -\frac{8}{2(1)} = -4.$$

Thus, the vertex is the same as the x-intercept. The axis of the parabola is the vertical line $x = -4$.

Make a table of ordered pairs whose x-values are on either side of the vertex's x-value of $x = -4$.

x	y
-7	9
-6	4
-5	1
-4	0
-3	1
-2	4
-1	9

Plot these seven ordered pairs and connect them with a smooth curve.

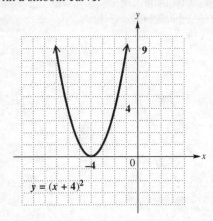

31. As in Exercise 17 in Section 9.4, one point on the graph has coordinates $(100, 30)$.

$$y = ax^2 \quad \text{General equation}$$
$$30 = a(100)^2 \quad \text{Let } x = 100, y = 30.$$
$$a = \frac{30}{10{,}000} = \frac{3}{1000}$$

Thus, an equation for a cross section of the parabolic dish is

$$y = \frac{3}{1000}x^2.$$

32. $\{(-2, 4), (0, 8), (2, 5), (2, 3)\}$

Since $x = 2$ appears in two ordered pairs, one value of x yields more than one value of y. Hence, this relation *is not a function*.

The domain is the set of first components of the ordered pairs, $\{-2, 0, 2\}$.

The range is the set of second components of the ordered pairs, $\{4, 8, 5, 3\}$.

33. $\{(8, 3), (7, 4), (6, 5), (5, 6), (4, 7)\}$

Since each first component of the ordered pairs corresponds to exactly one second component, the relation *is a function*.

The domain is the set of first components of the ordered pairs, $\{8, 7, 6, 5, 4\}$.

The range is the set of second components of the ordered pairs, $\{3, 4, 5, 6, 7\}$.

34. Since a vertical line may cross the graph twice, this is not the graph of a function.

35. Any vertical line will cut this graph in exactly one point, so it is the graph of a function.

36. $2x + 3y = 12$

Solve the equation for y.

$$3y = -2x + 12$$
$$y = -\tfrac{2}{3}x + 4$$

Since one value of x will lead to only one value of y, $2x + 3y = 12$ is a function.

37. $y = x^2$

Each value of x will lead to only one value of y, so $y = x^2$ is a function.

38. $x = 2|y|$

Notice that for $x = 4$, $y = -2$ **or** 2. Since there is not a *unique* value of y for $x = 4$, $x = 2|y|$ does not define y as a function of x.

39. $f(x) = 3x + 2$

(a) $f(2) = 3(2) + 2 = 6 + 2 = 8$

(b) $f(-1) = 3(-1) + 2 = -3 + 2 = -1$

40. $f(x) = 2x^2 - 1$

(a) $f(2) = 2(2)^2 - 1$
$= 2(4) - 1 = 8 - 1 = 7$

(b) $f(-1) = 2(-1)^2 - 1$
$= 2(1) - 1 = 2 - 1 = 1$

41. $f(x) = |x+3|$

(a) $f(2) = |2+3| = |5| = 5$

(b) $f(-1) = |-1+3| = |2| = 2$

42.
$$y = -x^2 + 12x - 26$$
$$6 = -x^2 + 12x - 26 \quad \text{Let } y = 6.$$
$$x^2 - 12x + 32 = 0$$
$$(x-4)(x-8) = 0 \quad \text{Factor.}$$
$$x = 4 \quad \text{or} \quad x = 8$$

Since the demand x is measured in hundreds, a demand of 400 or 800 cards produces a price of $6.

43. The x-value of the vertex is
$$x = -\frac{b}{2a} = -\frac{12}{2(-1)} = 6.$$

The y-value of the vertex is
$$y = -6^2 + 12(6) - 26$$
$$= -36 + 72 - 26 = 10.$$

So the vertex of the parabola is $(6, 10)$.

44. The vertex is $(6, 10)$, which indicates that a demand of 600 cards ($x = 6$) produces a price of $10 ($y = 10$).

45. $(2t-1)(t+1) = 54$

Write the equation in standard form.
$$2t^2 + t - 1 = 54$$
$$2t^2 + t - 55 = 0$$

Solve this equation by factoring.
$$(2t+11)(t-5) = 0$$
$$2t+11 = 0 \quad \text{or} \quad t-5 = 0$$
$$t = -\tfrac{11}{2} \quad \text{or} \quad t = 5$$

A check verifies that the solutions are $-\tfrac{11}{2}$ and 5.

46. $(2p+1)^2 = 100$

Use the square root property.

$2p+1 = \sqrt{100}$ or $2p+1 = -\sqrt{100}$
$2p+1 = 10$ or $2p+1 = -10$
$2p = 9$ or $2p = -11$
$p = \tfrac{9}{2}$ or $p = -\tfrac{11}{2}$

A check verifies that the solutions are $-\tfrac{11}{2}$ and $\tfrac{9}{2}$.

47. $(k+2)(k-1) = 3$

Write the equation in standard form.
$$(k+2)(k-1) = 3$$
$$k^2 + k - 2 = 3$$
$$k^2 + k - 5 = 0$$

The left side cannot be factored, so use the quadratic formula with $a = 1$, $b = 1$, and $c = -5$.

$$k = \frac{-b \pm \sqrt{b^2 - 4ac}}{2a}$$
$$k = \frac{-1 \pm \sqrt{1^2 - 4(1)(-5)}}{2(1)}$$
$$= \frac{-1 \pm \sqrt{1+20}}{2}$$
$$= \frac{-1 \pm \sqrt{21}}{2}$$

A check verifies that the solutions are $\dfrac{-1 \pm \sqrt{21}}{2}$.

48. $6t^2 + 7t - 3 = 0$

Solve by factoring.
$$(3t-1)(2t+3) = 0$$
$3t - 1 = 0$ or $2t + 3 = 0$
$t = \tfrac{1}{3}$ or $t = -\tfrac{3}{2}$

A check verifies that the solutions are $-\tfrac{3}{2}$ and $\tfrac{1}{3}$.

49. $2x^2 + 3x + 2 = x^2 - 2x$

Write the equation in standard form.
$$x^2 + 5x + 2 = 0$$

The left side cannot be factored, so use the quadratic formula with $a = 1$, $b = 5$, and $c = 2$.

$$x = \frac{-b \pm \sqrt{b^2 - 4ac}}{2a}$$
$$x = \frac{-5 \pm \sqrt{5^2 - 4(1)(2)}}{2(1)}$$
$$= \frac{-5 \pm \sqrt{25-8}}{2}$$
$$= \frac{-5 \pm \sqrt{17}}{2}$$

A check verifies that the solutions are $\dfrac{-5 \pm \sqrt{17}}{2}$.

50. $x^2 + 2x + 5 = 7$

Write the equation in standard form.
$$x^2 + 2x - 2 = 0$$

The left side cannot be factored, so use the quadratic formula with $a = 1$, $b = 2$, and $c = -2$.

continued

Chapter 9 Quadratic Equations

$$x = \frac{-2 \pm \sqrt{2^2 - 4(1)(-2)}}{2(1)}$$

$$= \frac{-2 \pm \sqrt{4+8}}{2}$$

$$= \frac{-2 \pm \sqrt{12}}{2} = \frac{-2 \pm 2\sqrt{3}}{2}$$

$$= \frac{2(-1 \pm \sqrt{3})}{2} = -1 \pm \sqrt{3}$$

A check verifies that the solutions are $-1 \pm \sqrt{3}$.

51. $m^2 - 4m + 10 = 0$

Use the quadratic formula with $a = 1, b = -4,$ and $c = 10$.

$$m = \frac{-(-4) \pm \sqrt{(-4)^2 - 4(1)(10)}}{2(1)}$$

$$= \frac{4 \pm \sqrt{16 - 40}}{2}$$

$$= \frac{4 \pm \sqrt{-24}}{2}$$

Because $\sqrt{-24}$ does not represent a real number, there is *no real number solution*.

52. $k^2 - 9k + 10 = 0$

The left side cannot be factored, so use the quadratic formula with $a = 1, b = -9,$ and $c = 10$.

$$k = \frac{-b \pm \sqrt{b^2 - 4ac}}{2a}$$

$$k = \frac{-(-9) \pm \sqrt{(-9)^2 - 4(1)(10)}}{2(1)}$$

$$= \frac{9 \pm \sqrt{81 - 40}}{2} = \frac{9 \pm \sqrt{41}}{2}$$

A check verifies that the solutions are $\frac{9 \pm \sqrt{41}}{2}$.

53. $(5x + 6)^2 = 0$

$5x + 6 = 0$

$5x = -6$

$x = -\frac{6}{5}$

A check verifies that the solution is $-\frac{6}{5}$.

54. $\frac{1}{2}r^2 = \frac{7}{2} - r$

Multiply by 2 to clear fractions; then rewrite the result in standard form.

$$r^2 = 7 - 2r$$

$$r^2 + 2r - 7 = 0$$

The left side cannot be factored, so use the quadratic formula with $a = 1, b = 2,$ and $c = -7$.

$$r = \frac{-2 \pm \sqrt{2^2 - 4(1)(-7)}}{2(1)}$$

$$= \frac{-2 \pm \sqrt{4 + 28}}{2}$$

$$= \frac{-2 \pm \sqrt{32}}{2} = \frac{-2 \pm 4\sqrt{2}}{2}$$

$$= \frac{2(-1 \pm 2\sqrt{2})}{2} = -1 \pm 2\sqrt{2}$$

A check verifies that the solutions are $-1 \pm 2\sqrt{2}$.

55. $x^2 + 4x = 1$

$x^2 + 4x - 1 = 0$

The left side cannot be factored, so use the quadratic formula with $a = 1, b = 4,$ and $c = -1$.

$$x = \frac{-4 \pm \sqrt{4^2 - 4(1)(-1)}}{2(1)}$$

$$= \frac{-4 \pm \sqrt{16 + 4}}{2}$$

$$= \frac{-4 \pm \sqrt{20}}{2} = \frac{-4 \pm 2\sqrt{5}}{2}$$

$$= \frac{2(-2 \pm \sqrt{5})}{2} = -2 \pm \sqrt{5}$$

A check verifies that the solutions are $-2 \pm \sqrt{5}$.

56. $7x^2 - 8 = 5x^2 + 8$

$2x^2 = 16$

$x^2 = 8$

$x = \pm\sqrt{8} = \pm 2\sqrt{2}$

A check verifies that the solutions are $\pm 2\sqrt{2}$.

Chapter 9 Test

1. $x^2 = 39$

Use the square root property.

$x = \sqrt{39} \quad$ or $\quad x = -\sqrt{39}$

A check verifies that the solutions are $\pm\sqrt{39}$.

2. $(x + 3)^2 = 64$

Use the square root property.

$x + 3 = \sqrt{64} \quad$ or $\quad x + 3 = -\sqrt{64}$

$x + 3 = 8 \quad$ or $\quad x + 3 = -8$

$x = 5 \quad$ or $\quad x = -11$

A check verifies that the solutions are -11 and 5.

3. $(4x+3)^2 = 24$

Use the square root property.

$$4x + 3 = \sqrt{24} \quad \text{or} \quad 4x + 3 = -\sqrt{24}$$

Note that $\sqrt{24} = \sqrt{4 \cdot 6} = 2\sqrt{6}$.

$$4x + 3 = 2\sqrt{6} \quad \text{or} \quad 4x + 3 = -2\sqrt{6}$$
$$4x = -3 + 2\sqrt{6} \quad \text{or} \quad 4x = -3 - 2\sqrt{6}$$
$$x = \frac{-3 + 2\sqrt{6}}{4} \quad \text{or} \quad x = \frac{-3 - 2\sqrt{6}}{4}$$

A check verifies that the solutions are $\dfrac{-3 \pm 2\sqrt{6}}{4}$.

4. $x^2 - 4x = 6$

Solve by completing the square.

$$x^2 - 4x + 4 = 6 + 4 \qquad \text{Add } \left[\tfrac{1}{2}(-4)\right]^2 = 4.$$
$$(x-2)^2 = 10$$
$$x - 2 = \sqrt{10} \quad \text{or} \quad x - 2 = -\sqrt{10}$$
$$x = 2 + \sqrt{10} \quad \text{or} \quad x = 2 - \sqrt{10}$$

A check verifies that the solutions are $2 \pm \sqrt{10}$.

5. $2x^2 + 12x - 3 = 0$

Solve by completing the square.

$$x^2 + 6x - \tfrac{3}{2} = 0 \qquad \text{Divide by 2.}$$
$$x^2 + 6x = \tfrac{3}{2}$$
$$x^2 + 6x + 9 = \tfrac{3}{2} + 9 \qquad \text{Add } \left[\tfrac{1}{2}(6)\right]^2 = 9.$$
$$(x+3)^2 = \tfrac{21}{2}$$
$$x + 3 = \sqrt{\tfrac{21}{2}} \quad \text{or} \quad x + 3 = -\sqrt{\tfrac{21}{2}}$$

Note that

$$\sqrt{\tfrac{21}{2}} = \frac{\sqrt{21}}{\sqrt{2}} = \frac{\sqrt{21} \cdot \sqrt{2}}{\sqrt{2} \cdot \sqrt{2}} = \frac{\sqrt{42}}{2}.$$

$$x + 3 = \frac{\sqrt{42}}{2} \quad \text{or} \quad x + 3 = -\frac{\sqrt{42}}{2}$$
$$x = -3 + \frac{\sqrt{42}}{2} \quad \text{or} \quad x = -3 - \frac{\sqrt{42}}{2}$$
$$x = \frac{-6 + \sqrt{42}}{2} \quad \text{or} \quad x = \frac{-6 - \sqrt{42}}{2}$$

A check verifies that the solutions are $\dfrac{-6 \pm \sqrt{42}}{2}$.

6. $2x^2 + 5x - 3 = 0$

Use $a = 2$, $b = 5$, and $c = -3$ in the quadratic formula.

$$x = \frac{-b \pm \sqrt{b^2 - 4ac}}{2a}$$
$$x = \frac{-5 \pm \sqrt{5^2 - 4(2)(-3)}}{2(2)}$$
$$= \frac{-5 \pm \sqrt{25 + 24}}{4}$$
$$= \frac{-5 \pm \sqrt{49}}{4} = \frac{-5 \pm 7}{4}$$
$$x = \frac{-5 + 7}{4} \quad \text{or} \quad x = \frac{-5 - 7}{4}$$
$$x = \frac{2}{4} = \frac{1}{2} \quad \text{or} \quad x = \frac{-12}{4} = -3$$

A check verifies that the solutions are -3 and $\tfrac{1}{2}$.

7. $3w^2 + 2 = 6w$

$$3w^2 - 6w + 2 = 0 \qquad \text{Rewrite.}$$

Use $a = 3$, $b = -6$, and $c = 2$ in the quadratic formula.

$$w = \frac{-(-6) \pm \sqrt{(-6)^2 - 4(3)(2)}}{2(3)}$$
$$= \frac{6 \pm \sqrt{36 - 24}}{6}$$
$$= \frac{6 \pm \sqrt{12}}{6} = \frac{6 \pm 2\sqrt{3}}{6}$$
$$= \frac{2(3 \pm \sqrt{3})}{2(3)} = \frac{3 \pm \sqrt{3}}{3}$$

A check verifies that the solutions are $\dfrac{3 \pm \sqrt{3}}{3}$.

8. $4x^2 + 8x + 11 = 0$

Use $a = 4$, $b = 8$, and $c = 11$ in the quadratic formula.

$$x = \frac{-8 \pm \sqrt{8^2 - 4(4)(11)}}{2(4)}$$
$$x = \frac{-8 \pm \sqrt{64 - 176}}{8}$$
$$x = \frac{-8 \pm \sqrt{-112}}{8}$$

The radical $\sqrt{-112}$ is not a real number, so the equation has *no real number solution*.

380   Chapter 9   Quadratic Equations

9. $t^2 - \dfrac{5}{3}t + \dfrac{1}{3} = 0$

$3\left(t^2 - \tfrac{5}{3}t + \tfrac{1}{3}\right) = 3(0)$ *Multiply by 3.*

$3t^2 - 5t + 1 = 0$

Use $a = 3$, $b = -5$, and $c = 1$ in the quadratic formula.

$t = \dfrac{-(-5) \pm \sqrt{(-5)^2 - 4(3)(1)}}{2(3)}$

$= \dfrac{5 \pm \sqrt{25 - 12}}{6} = \dfrac{5 \pm \sqrt{13}}{6}$

A check verifies that the solutions are $\dfrac{5 \pm \sqrt{13}}{6}$.

10. $p^2 - 2p - 1 = 0$

Solve by completing the square.

$p^2 - 2p = 1$

$p^2 - 2p + 1 = 1 + 1$

$(p - 1)^2 = 2$

$p - 1 = \sqrt{2}$ or $p - 1 = -\sqrt{2}$

$p = 1 + \sqrt{2}$ or $p = 1 - \sqrt{2}$

A check verifies that the solutions are $1 \pm \sqrt{2}$.

11. $(2x + 1)^2 = 18$

Use the square root property.

$2x + 1 = \pm \sqrt{18}$

$2x + 1 = \pm 3\sqrt{2}$

$2x = -1 \pm 3\sqrt{2}$

$x = \dfrac{-1 \pm 3\sqrt{2}}{2}$

A check verifies that the solutions are $\dfrac{-1 \pm 3\sqrt{2}}{2}$.

12. $(x - 5)(2x - 1) = 1$

$2x^2 - 11x + 5 = 1$ *FOIL*

$2x^2 - 11x + 4 = 0$ *Subtract 1.*

Use $a = 2$, $b = -11$, and $c = 4$ in the quadratic formula.

$x = \dfrac{-(-11) \pm \sqrt{(-11)^2 - 4(2)(4)}}{2(2)}$

$= \dfrac{11 \pm \sqrt{121 - 32}}{4} = \dfrac{11 \pm \sqrt{89}}{4}$

A check verifies that the solutions are $\dfrac{11 \pm \sqrt{89}}{4}$.

13. $t^2 + 25 = 10t$

$t^2 - 10t + 25 = 0$ *Subtract 10t.*

$(t - 5)^2 = 0$ *Factor.*

$t - 5 = 0$ *Take square root.*

$t = 5$

A check verifies that the solution is 5.

14. $s = -16t^2 + 64t$

Let $s = 64$ and solve for t.

$64 = -16t^2 + 64t$

$16t^2 - 64t + 64 = 0$

$t^2 - 4t + 4 = 0$ *Divide by 16.*

$(t - 2)^2 = 0$ *Factor.*

$t - 2 = 0$ *Take square root.*

$t = 2$

The object will reach a height of 64 feet after 2 seconds.

15. Use the Pythagorean formula.

$c^2 = a^2 + b^2$

$(x + 8)^2 = (x)^2 + (x + 4)^2$

$x^2 + 16x + 64 = x^2 + x^2 + 8x + 16$

$0 = x^2 - 8x - 48$

$0 = (x - 12)(x + 4)$

$x - 12 = 0$ or $x + 4 = 0$

$x = 12$ or $x = -4$

Disregard a negative length. The sides measure 12, $x + 4 = 12 + 4 = 16$, and $x + 8 = 12 + 8 = 20$.

16. $y = (x - 3)^2 = x^2 - 6x + 9$

The x-value of the vertex is

$x = -\dfrac{b}{2a} = -\dfrac{-6}{2(1)} = 3.$

The y-value of the vertex is

$y = (3 - 3)^2 = 0,$

so the vertex is $(3, 0)$.

Make a table of ordered pairs whose x-values are on either side of the vertex's x-value of $x = 3$.

x	y
0	9
1	4
2	1
3	0
4	1
5	4
6	9

Plot these seven ordered pairs and connect them with a smooth curve.

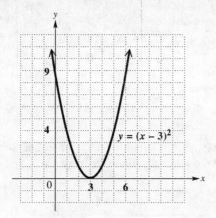

17. $y = -x^2 - 2x - 4$

The x-value of the vertex is

$$x = -\frac{b}{2a} = -\frac{-2}{2(-1)} = -1.$$

The y-value of the vertex is

$$y = -(-1)^2 - 2(-1) - 4 = -3,$$

so the vertex is $(-1, -3)$.

Make a table of ordered pairs whose x-values are on either side of the vertex's x-value of $x = -1$.

x	y
-3	-7
-2	-4
-1	-3
0	-4
1	-7

Plot these five ordered pairs and connect them with a smooth curve.

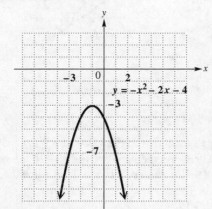

18. **(a)** $\{(2,3), (2,4), (2,5)\}$
Since $x = 2$ appears in two ordered pairs, one value of x yields more than one value of y. Hence, this relation *is not a function*.

(b) $\{(0,2), (1,2), (2,2)\}$
Since each first component of the ordered pairs corresponds to exactly one second component, the relation *is a function*. The domain is $\{0, 1, 2\}$ and the range is $\{2\}$.

19. The vertical line test shows that this graph is not the graph of a function; a vertical line could cross the graph twice.

20. $f(x) = 3x + 7$
$f(-2) = 3(-2) + 7 = -6 + 7 = 1$

Cumulative Review Exercises (Chapters R–9)

1. $\dfrac{-4 \cdot 3^2 + 2 \cdot 3}{2 - 4 \cdot 1} = \dfrac{-4 \cdot 9 + 6}{2 - 4}$

$$= \dfrac{-36 + 6}{-2} = \dfrac{-30}{-2} = 15$$

2. $-9 - (-8)(2) + 6 - (6 + 2)$
$= -9 - (-8)(2) + 6 - 8$
$= -9 - (-16) + 6 - 8$
$= -9 + 16 + 6 - 8$
$= 7 + 6 - 8$
$= 13 - 8 = 5$

3. $|-3| - |1 - 6| = |-3| - |-5|$
$= 3 - 5 = -2$

4. $-4r + 14 + 3r - 7 = -r + 7$

5. $13k - 4k + k - 14k + 2k$
$= (13 - 4 + 1 - 14 + 2)k$ *Distributive property*
$= -2k$

6. $5(4m - 2) - (m + 7)$
$= 5(4m - 2) - 1(m + 7)$
$= 20m - 10 - m - 7$
$= 19m - 17$

7. $6x - 5 = 13$
$6x = 18$
$x = 3$

The solution is 3.

8. $3k - 9k - 8k + 6 = -64$
$-14k + 6 = -64$
$-14k = -70$
$k = 5$

The solution is 5.

382 Chapter 9 Quadratic Equations

9. $2(m-1) - 6(3-m) = -4$
$2m - 2 - 18 + 6m = -4$
$8m - 20 = -4$
$8m = 16$
$m = 2$

The solution is 2.

10. Let $x =$ the length of the court.
Then $x - 44 =$ the width of the court.

Use the formula for the perimeter of a rectangle, $P = 2L + 2W$, with $P = 288$, $L = x$, and $W = x - 44$.

$288 = 2x + 2(x - 44)$
$288 = 2x + 2x - 88$
$376 = 4x$
$94 = x$

If $x = 94$, $x - 44 = 94 - 44 = 50$.
The length of the court is 94 feet and the width of the court is 50 feet.

11. Together, the two angles form a straight angle, so the sum of their measures is 180°.

$(20x - 20) + (12x + 8) = 180$
$32x - 12 = 180$
$32x = 192$
$x = \frac{192}{32} = 6$

If $x = 6$,
$20x - 20 = 20(6) - 20$
$= 120 - 20 = 100,$

and

$12x + 8 = 12(6) + 8$
$= 72 + 8 = 80.$

The measures of the angles are 100° and 80°.

12. Solve $P = 2L + 2W$ for L.

$P - 2W = 2L$
$\frac{P - 2W}{2} = L$ or $L = \frac{P}{2} - W$

13. $-8m < 16$

Divide each side by -8 and reverse the inequality symbol.

$\frac{-8m}{-8} > \frac{16}{-8}$
$m > -2$

To graph this solution on a number line, place an open circle at -2 and draw an arrow extending to the right.

$m > -2$
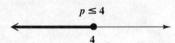

14. $-9p + 2(8 - p) - 6 \geq 4p - 50$
$-9p + 16 - 2p - 6 \geq 4p - 50$
$-11p + 10 \geq 4p - 50$
$-15p \geq -60$
$\frac{-15p}{-15} \leq \frac{-60}{-15}$ *Divide by -15; reverse symbol.*
$p \leq 4$

To graph this solution on a number line, place a solid dot at 4 and draw an arrow extending to the left.

$p \leq 4$

15. $2x + 3y = 6$

Find the intercepts.

Let $x = 0$. $2(0) + 3y = 6$
$3y = 6$
$y = 2$

The y-intercept is $(0, 2)$.

Let $y = 0$. $2x + 3(0) = 6$
$2x = 6$
$x = 3$

The x-intercept is $(3, 0)$.

The graph is the line through the points $(0, 2)$ and $(3, 0)$.

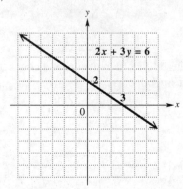

16. $y = 3$

For any value of x, the value of y will always be 3. Three ordered pairs are $(-2, 3)$, $(0, 3)$, and $(4, 3)$. Plot these points and draw a line through them. This will be a horizontal line.

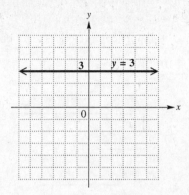

17. $2x - 5y < 10$

First, graph the boundary line $2x - 5y = 10$. If $x = 0$, then $-5y = 10$ and $y = -2$, so the y-intercept is $(0, -2)$. If $y = 0$, then $2x = 10$ and $x = 5$, so the x-intercept is $(5, 0)$. Because of the "$<$" sign, the line through $(0, -2)$ and $(5, 0)$ should be dashed.

Next, use $(0, 0)$ as a test point in $2x - 5y < 10$.

$$2(0) - 5(0) < 10$$
$$0 < 10 \quad \text{True}$$

Since $0 < 10$ is a true statement, shade the region on the side of the line that contains $(0, 0)$. This is the region above the line.

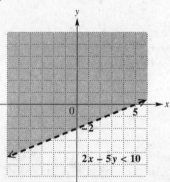

18. The slope m of the line passing through the points $(-1, 4)$ and $(5, 2)$ is

$$m = \frac{\text{change in } y}{\text{change in } x} = \frac{2 - 4}{5 - (-1)} = \frac{-2}{6} = -\frac{1}{3}.$$

19. Slope 2; y-intercept $(0, 3)$

Let $m = 2$ and $b = 3$ in slope-intercept form.

$$y = mx + b$$
$$y = 2x + 3$$

Now rewrite the equation in the form $Ax + By = C$.

$$-2x + y = 3$$
$$2x - y = -3 \quad \text{Multiply by } -1.$$

20. $2x + y = -4$ (1)
$-3x + 2y = 13$ (2)

Use the elimination method.

Multiply equation (1) by -2 and add the result to equation (2).

$$\begin{array}{rcl}
-4x - 2y &=& 8 \\
-3x + 2y &=& 13 \\ \hline
-7x &=& 21 \\
x &=& -3
\end{array}$$

To find y, substitute -3 for x in equation (1).

$$2x + y = -4$$
$$2(-3) + y = -4$$
$$-6 + y = -4$$
$$y = 2$$

The solution of the system is $(-3, 2)$.

21. $3x - 5y = 8$ (1)
$-6x + 10y = 16$ (2)

Use the elimination method.

Multiply equation (1) by 2 and add the result to equation (2).

$$\begin{array}{rcl}
6x - 10y &=& 16 \\
-6x + 10y &=& 16 \\ \hline
0 &=& 32 \quad \text{False}
\end{array}$$

The false statement indicates that there *is no solution*.

22. Let $x = $ the price of a Motorola phone and $y = $ the price of a Kyocera phone.

We have the system

$$3x + 2y = 379.95 \quad (1)$$
$$2x + 3y = 369.95 \quad (2)$$

To solve the system by the elimination method, we multiply equation (1) by 2 and equation (2) by -3, and then add the results.

$$\begin{array}{rcl}
6x + 4y &=& 759.90 \\
-6x - 9y &=& -1109.85 \\ \hline
-5y &=& -349.95 \\
y &=& 69.99
\end{array}$$

To find the value of x, substitute 69.99 for y in equation (1).

$$3x + 2(69.99) = 379.95$$
$$3x + 139.98 = 379.95$$
$$3x = 239.97$$
$$x = 79.99$$

The price of a single Motorola phone is \$79.99, and the price of a single Kyocera phone is \$69.99.

23.
$2x + y \leq 4$ (1)
$x - y > 2$ (2)

For inequality (1), draw a solid boundary line through $(2, 0)$ and $(0, 4)$, and shade the side that includes the origin (since substituting 0 for x and 0 for y results in a true statement).

For inequality (2), draw a dashed boundary line through $(2, 0)$ and $(0, -2)$, and shade the side that *does not* include the origin.

The solution of the system of inequalities is the intersection of these two shaded half-planes.

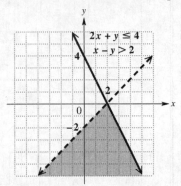

24. $(3^2 \cdot x^{-4})^{-1} = \left(\dfrac{3^2}{x^4}\right)^{-1}$

$= \left(\dfrac{x^4}{3^2}\right)^{1} = \dfrac{x^4}{3^2}$ or $\dfrac{x^4}{9}$

25. $\left(\dfrac{b^{-3}c^4}{b^5 c^3}\right)^{-2} = \left(b^{-3-5} c^{4-3}\right)^{-2}$

$= \left(b^{-8} c^1\right)^{-2}$

$= \left(b^{-8}\right)^{-2} \left(c^1\right)^{-2}$

$= b^{16} c^{-2}$

$= b^{16} \cdot \dfrac{1}{c^2} = \dfrac{b^{16}}{c^2}$

26. $\left(\dfrac{5}{3}\right)^{-3} = \left(\dfrac{3}{5}\right)^{3} = \dfrac{3^3}{5^3}$ or $\dfrac{27}{125}$

27. $(5x^5 - 9x^4 + 8x^2) - (9x^2 + 8x^4 - 3x^5)$

$= 5x^5 - 9x^4 + 8x^2 - 9x^2 - 8x^4 + 3x^5$

$= 8x^5 - 17x^4 - x^2$

28. $(2x - 5)(x^3 + 3x^2 - 2x - 4)$

Multiply vertically.

$$\begin{array}{rrrrr} & x^3 + & 3x^2 - & 2x - & 4 \\ & & & 2x - & 5 \\ \hline -5x^3 - & 15x^2 + & 10x + & 20 & \\ 2x^4 + 6x^3 - & 4x^2 - & 8x & & \\ \hline 2x^4 + x^3 - & 19x^2 + & 2x + & 20 & \end{array}$$

29. $(5t + 9)^2 = (5t)^2 + 2(5t)(9) + (9)^2$

$= 25t^2 + 90t + 81$

30. $\dfrac{3x^3 + 10x^2 - 7x + 4}{x + 4}$

$$\begin{array}{r} 3x^2 - 2x + 1 \\ x+4 \overline{\smash{\big)}\, 3x^3 + 10x^2 - 7x + 4} \\ \underline{3x^3 + 12x^2} \\ -2x^2 - 7x \\ \underline{-2x^2 - 8x} \\ x + 4 \\ \underline{x + 4} \\ 0 \end{array}$$

The remainder is 0, so the answer is the quotient, $3x^2 - 2x + 1$.

31. $16x^3 - 48x^2 y = 16x^2(x - 3y)$

$GCF = 16x^2$

32. $16x^4 - 1$

$= (4x^2 + 1)(4x^2 - 1)$ Difference of squares

$= (4x^2 + 1)(2x + 1)(2x - 1)$

Difference of squares

33. $2a^2 - 5a - 3$

Use the grouping method. Look for two integers whose product is $2(-3) = -6$ and whose sum is -5. The integers are -6 and 1.

$2a^2 - 5a - 3 = 2a^2 - 6a + 1a - 3$

$= 2a(a - 3) + 1(a - 3)$

$= (a - 3)(2a + 1)$

34. $25m^2 - 20m + 4$

Since $25m^2 = (5m)^2$, $4 = 2^2$, and $20m = 2(5m)(2)$, $25m^2 - 20m + 4$ is a perfect square trinomial.

$25m^2 - 20m + 4 = (5m)^2 - 2(5m)(2) + (2)^2$

$= (5m - 2)^2$

35. $x^2 + 3x - 54 = 0$

$(x + 9)(x - 6) = 0$ Factor.

$x + 9 = 0$ or $x - 6 = 0$

$x = -9$ or $x = 6$

A check verifies that the solutions are -9 and 6.

36. $3x^2 = x + 4$

$3x^2 - x - 4 = 0$

$(3x - 4)(x + 1) = 0$ Factor.

$$3x - 4 = 0 \quad \text{or} \quad x + 1 = 0$$
$$3x = 4 \quad \text{or}$$
$$x = \tfrac{4}{3} \quad \text{or} \quad x = -1$$

A check verifies that the solutions are $\tfrac{4}{3}$ and -1.

37. Let x represent the width of the rectangle. Then $2.5x$ represents the length.

Use the formula for the area of a rectangle.

$$A = LW$$
$$1000 = (2.5x)x \quad \text{Let } A = 1000.$$
$$1000 = 2.5x^2$$
$$x^2 = \frac{1000}{2.5} = 400$$
$$x = \pm\sqrt{400} = \pm 20$$

Reject $x = -20$ since the width cannot be negative. The width is 20 meters, so the length is $2.5(20) = 50$ meters.

38. $\dfrac{2}{a-3} \div \dfrac{5}{2a-6}$

$= \dfrac{2}{a-3} \cdot \dfrac{2a-6}{5}$ *Multiply by reciprocal of divisor.*

$= \dfrac{2}{a-3} \cdot \dfrac{2(a-3)}{5}$ *Factor.*

$= \dfrac{4(a-3)}{(a-3)5}$ *Multiply.*

$= \dfrac{4}{5}$ *Lowest terms*

39. $\dfrac{1}{k} - \dfrac{2}{k-1}$

$= \dfrac{1(k-1)}{k(k-1)} - \dfrac{2(k)}{k(k-1)}$ *LCD = $k(k-1)$*

$= \dfrac{(k-1) - 2k}{k(k-1)}$ *Subtract numerators.*

$= \dfrac{-k-1}{k(k-1)}$ *Combine terms.*

40. $\dfrac{2}{a^2-4} + \dfrac{3}{a^2-4a+4}$

$= \dfrac{2}{(a+2)(a-2)} + \dfrac{3}{(a-2)(a-2)}$

Factor denominators.

$= \dfrac{2(a-2)}{(a+2)(a-2)(a-2)}$

$\quad + \dfrac{3(a+2)}{(a+2)(a-2)(a-2)}$

LCD = $(a+2)(a-2)(a-2)$

$= \dfrac{2(a-2) + 3(a+2)}{(a+2)(a-2)(a-2)}$

Add numerators.

$= \dfrac{2a - 4 + 3a + 6}{(a+2)(a-2)(a-2)}$

Distributive property

$= \dfrac{5a+2}{(a+2)(a-2)(a-2)}$

Combine terms.

$= \dfrac{5a+2}{(a+2)(a-2)^2}$

41. $\dfrac{6 + \dfrac{1}{x}}{3 - \dfrac{1}{x}}$

Multiply the numerator and the denominator of the complex fraction by x, which is the LCD of all the denominators in the complex fraction.

$\dfrac{x\left(6 + \tfrac{1}{x}\right)}{x\left(3 - \tfrac{1}{x}\right)} = \dfrac{6x + x\left(\tfrac{1}{x}\right)}{3x - x\left(\tfrac{1}{x}\right)}$

$= \dfrac{6x+1}{3x-1}$

42. $\dfrac{1}{x+3} + \dfrac{1}{x} = \dfrac{7}{10}$

Multiply each side by the LCD, $10x(x+3)$.

$10x(x+3)\left(\dfrac{1}{x+3} + \dfrac{1}{x}\right) = 10x(x+3)\left(\dfrac{7}{10}\right)$

$10x(x+3)\left(\dfrac{1}{x+3}\right) + 10x(x+3)\left(\dfrac{1}{x}\right)$
$= 10x(x+3)\left(\dfrac{7}{10}\right)$

$10x + 10(x+3) = 7x(x+3)$
$10x + 10x + 30 = 7x^2 + 21x$
$20x + 30 = 7x^2 + 21x$
$0 = 7x^2 + x - 30$
$0 = (7x+15)(x-2)$

$7x + 15 = 0 \quad \text{or} \quad x - 2 = 0$
$x = -\tfrac{15}{7} \quad \text{or} \quad x = 2$

A check verifies that the solutions are $-\tfrac{15}{7}$ and 2.

43. $\sqrt{100} = 10$ since $10^2 = 100$ and $\sqrt{100}$ represents the positive square root.

44. $\dfrac{6\sqrt{6}}{\sqrt{5}} = \dfrac{6\sqrt{6} \cdot \sqrt{5}}{\sqrt{5} \cdot \sqrt{5}} = \dfrac{6\sqrt{30}}{5}$

45. $3\sqrt{5} - 2\sqrt{20} + \sqrt{125}$
$= 3\sqrt{5} - 2\sqrt{4 \cdot 5} + \sqrt{25 \cdot 5}$
$= 3\sqrt{5} - 2 \cdot 2\sqrt{5} + 5\sqrt{5}$
$= 3\sqrt{5} - 4\sqrt{5} + 5\sqrt{5}$
$= (3 - 4 + 5)\sqrt{5} = 4\sqrt{5}$

46. $\sqrt[3]{16a^3b^4} - \sqrt[3]{54a^3b^4}$
$= \sqrt[3]{8a^3b^3 \cdot 2b} - \sqrt[3]{27a^3b^3 \cdot 2b}$
$= \sqrt[3]{8a^3b^3} \cdot \sqrt[3]{2b} - \sqrt[3]{27a^3b^3} \cdot \sqrt[3]{2b}$
$= 2ab\sqrt[3]{2b} - 3ab\sqrt[3]{2b}$
$= (2 - 3)ab\sqrt[3]{2b} = -ab\sqrt[3]{2b}$

47. $\sqrt{x + 2} = x - 4$
$\left(\sqrt{x + 2}\right)^2 = (x - 4)^2$
$x + 2 = x^2 - 8x + 16$
$0 = x^2 - 9x + 14$
$0 = (x - 7)(x - 2)$
$x - 7 = 0 \quad \text{or} \quad x - 2 = 0$
$x = 7 \quad \text{or} \quad x = 2$

Check $x = 7$: $\sqrt{9} = 3$ True
Check $x = 2$: $\sqrt{4} = -2$ False

The solution is 7.

48. $2a^2 - 2a = 1$
$2a^2 - 2a - 1 = 0$

Use the quadratic formula with $a_1 = 2, b = -2,$ and $c = -1$.

$a = \dfrac{-b \pm \sqrt{b^2 - 4a_1c}}{2a_1}$

$a = \dfrac{-(-2) \pm \sqrt{(-2)^2 - 4(2)(-1)}}{2(2)}$

$a = \dfrac{2 \pm \sqrt{4 + 8}}{4}$

$a = \dfrac{2 \pm \sqrt{12}}{4} = \dfrac{2 \pm 2\sqrt{3}}{4}$

$= \dfrac{2\left(1 \pm \sqrt{3}\right)}{4} = \dfrac{1 \pm \sqrt{3}}{2}$

A check verifies that the solutions are $\dfrac{1 \pm \sqrt{3}}{2}$.

49. $y = x^2 - 4$

The x-value of the vertex is

$x = -\dfrac{b}{2a} = \dfrac{0}{2(1)} = 0.$

The y-value of the vertex is

$y = 0^2 - 4 = -4,$

so the vertex is $(0, -4)$.

The axis of the parabola is the line $x = 0$, which is the y-axis.

Now find the intercepts.

Find the y-intercept.

Let $x = 0$; then $y = -4$. The y-intercept is $(0, -4)$, which is also the vertex.

Find the x-intercepts.

$0 = x^2 - 4 \quad \text{Let } y = 0.$
$0 = (x - 2)(x + 2)$

$x - 2 = 0 \quad \text{or} \quad x + 2 = 0$
$x = 2 \quad \text{or} \quad x = -2$

The x-intercepts are $(2, 0)$ and $(-2, 0)$.

Find some other points on the graph.

Let $x = \pm 1$. Then

$y = (\pm 1)^2 - 4 = 1 - 4 = -3.$

Plot the points $(0, -4), (2, 0), (-2, 0), (1, -3),$ and $(-1, -3)$ and connect them with a smooth curve.

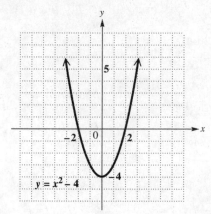

50. (a) $\{(0, 4), (1, 2), (3, 5)\}$

Notice that each first component appears once and only once. Therefore, this relation *is a function*.

(b) The domain is $\{0, 1, 3\}$ and the range is $\{4, 2, 5\}$.

(c) $f(x) = -2x + 7$
$f(-2) = -2(-2) + 7$
$= 4 + 7$
$= 11$

(d) The vertical line test shows that this graph is not the graph of a function; a vertical line could cross the graph twice.

APPENDIX A STRATEGIES FOR PROBLEM SOLVING

Appendix A Margin Exercises

1. Step 1 compares to Polya's first step, Steps 2 and 3 compare to his second step, Step 4 compares to his third step, and Step 6 compares to his fourth step.

2.

Chosen Terms	First chosen term	Fourth chosen term	Product	Second chosen term	Third chosen term	Product
4th–7th	3	13	39	5	8	40
6th–9th	8	34	272	13	21	273
8th–11th	21	89	1869	34	55	1870

The product of the two middle terms is always 1 more than the product of the first and last term.

3. **(a)** Suppose she started with x dollars. She bought a book for $10, so now she has $x - 10$ dollars. She spent half her remaining money on a train ticket, so she has $\frac{1}{2}(x - 10) = \frac{1}{2}x - 5$ dollars remaining. She bought lunch for $4, so she has $\left(\frac{1}{2}x - 5\right) - 4 = \frac{1}{2}x - 9$ dollars remaining. She spent half her remaining money at a bazaar, so she has $\frac{1}{2}\left(\frac{1}{2}x - 9\right) = \frac{1}{4}x - \frac{9}{2}$ dollars remaining. This must equal the $20 she left the bazaar with, so solve the equation

$$\tfrac{1}{4}x - \tfrac{9}{2} = 20.$$
$$\tfrac{1}{4}x = \tfrac{49}{2} \quad \text{Add } \tfrac{9}{2}.$$
$$x = 98 \quad \text{Multiply by 4.}$$

She started with $98.

(b) The given subtraction problem:
```
  7 a 2
- 4 8 b
-------
  c 7 3
```

Change to an addition problem:
```
  4 8 b
+ c 7 3
-------
  7 a 2
```

Now $b + 3 = 2$, so $b = 9$ and we carry 1 to the tens column. Thus, $1 + 8 + 7 = 16$, so $a = 6$ and we carry 1 to the hundreds column. Thus, $1 + 4 + c = 7$, so $c = 2$. Hence, $a + b + c = 6 + 9 + 2 = 17$, which is choice **D**.

4. **(a)** By inspection, $44^2 = 1936$, $45^2 = 2025$, and $46^2 = 2116$. Only 2025 makes sense for this problem, and $2025 - 76 = 1949$, which must be the year he was born.

(b) Use trial and error for this problem. Here is one possible solution.

```
    3 5
  7 1 8 2
    4 6
```

5. **(a)** Let x represent the positive number. "If I square it" gives x^2. "Double the result" gives $2x^2$. "Take half of that result" gives $\frac{1}{2}(2x^2) = x^2$ (the last two steps canceled each other). "Then add 12" gives $x^2 + 12$. "I get 21" implies that $x^2 + 12 = 21$. Solve this equation.

$$x^2 + 12 = 21$$
$$x^2 = 9 \quad \text{Subtract 12.}$$
$$x = 3 \quad \text{Positive square root}$$

The number is 3.

(b) Condition (2) leads us to a year in the 1960s. If x is the ones digit, then condition (1) gives $1 + 9 + 6 + x = 23$, so $x = 7$, and the year is 1967.

6. **(a)** Examine the units digit in powers of 7.

$$7^1 = 7 \qquad 7^5 = 16{,}807$$
$$7^2 = 49 \qquad 7^6 = 117{,}649$$
$$7^3 = 343 \qquad 7^7 = 823{,}543$$
$$7^4 = 2401 \qquad 7^8 = 5{,}764{,}801$$

We see that the units digit cycles through 7, 9, 3, and 1. Since $491 = 4 \cdot 122 + 3$ (122 full cycles), we know that 7^{491} and 7^3 have the same units digit. Thus, the units digit in 7^{491} is 3.

(b) Since $\frac{1}{11} = .0909\ldots$, we see that every odd-numbered digit in the decimal representation of $\frac{1}{11}$ is 0. Since 103 is odd, the 103rd digit in the decimal representation of $\frac{1}{11}$ is 0.

7. (a) Since there are 3 rows, and there could only be a maximum of 2 crosses per row, we know that there are a maximum of 6 crosses in the 9 squares. In the first row, put a cross in the second and third squares. In the second row, put a cross in the first square (since the first column already has one blank square), and then put a cross in the third square (if we put a cross in the second square, we wouldn't be able to put a cross in the first square of the third row—why?). In the third row, put a cross in the first and second squares. This gives us the maximum number of crosses, 6.

	X	X
X		X
X	X	

(b) The sum of the numbers 1 to 12 on the face of a clock is 78, and 78 divided by 3 is 26. Start with the number 12 and investigate what numbers can be combined with 12 to add up to 26. Moving left, $12 + 11 = 23$. We can't include 10 with 12 and 11 because the sum is greater than 26, so we'll move right and include 1 and 2 for a sum of 26. Draw a line that separates 11, 12, 1, and 2 from the rest of the numbers.

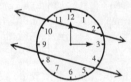

Now we can group the 10 and 9 together (but not with the 8—why?) along with the 3 and the 4 for a sum of 26. Draw another line and check that the remaining numbers add up to 26.

8. (a) Neither is correct, since $3^3 = 27$. However, "three cubed *is* twenty-seven" would be correct.

(b) If you take 7 bowling pins from 10 bowling pins, you have <u>7 bowling pins</u>.

(c) If you boil all the eggs at the same time, it still take $7\frac{1}{2}$ minutes.

Appendix A Exercises

1. You could choose a sock from the box labeled *red and green socks*. Since it is mislabeled, it contains only red socks or only green socks, determined by the sock you choose. If the sock is green, relabel this box *green socks*. Since the other two boxes were mislabeled, switch the remaining label to the other box and place the label that says *red and green socks* on the unlabeled box. No other choice guarantees a correct relabeling, since you can remove only one sock.

3. Let $b =$ Mr. Green's birth year and let $c =$ the current year. So the year of Mr. Green's tenth birthday is $(b + 10)$, the year of his fiftieth birthday is $(b + 50)$, and his current age is $(c - b)$. From the statement, write an equation.

$$b + c - (b + 10) - (b + 50) + (c - b) = 80$$
$$b + c - b - 10 - b - 50 + c - b = 80$$
$$2c - 2b = 140$$
$$c - b = 70$$

So his current age, $c - b$, is 70.

5. There are 4 choices to place your next O.

(1) Row 2, column 2: If you place an O here, then your friend will place an X in row 3, column 3, which forces you to get three Os in a row, and you lose.

(2) Row 2, column 3: If you place an O here, then your friend will place an X in row 3, column 1, which forces you to get three Os in a row, and you lose.

(3) Row 3, column 1: If you place an O here, then your friend can't use row 2, column 2 (Why?), so they must use row 2, column 3 or row 3, column 3.

(A) If they place an X in row 2, column 3, then you would use row 3, column 3 and they are forced to place an X in row 2, column 2, which gives them 3 Xs in a row (2nd column).

(B) If they place an X in row 3, column 3, then you would use row 2, column 3 and they are forced to place an X in row 2, column 2, which gives them 3 Xs in a row (2nd column).

Thus, you win in all cases by placing an O in the *bottom-left square*.

(4) Row 3, column 3: If you place an O here, then your friend would place an X in row 3, column 1, you would place an O in row 2, column 2, your friend would place an X in row 2, column 3, and there would not be a winner.

7. One strategy is to organize a table such as the one which follows. Let $x =$ Chris's current age.

	Current Age	Past Age	Elapsed No. of Years
Pat	24	x	$24 - x$
Chris	x	$x - (24 - x)$	

Since Chris's past age can be represented as

$$x - (24 - x) = -24 + 2x = 2x - 24,$$

and Pat's current age, 24, is twice that of Chris's past age, we have

$$24 = 2(2x - 24)$$
$$24 = 4x - 48$$
$$72 = 4x$$
$$18 = x.$$

Thus, Chris's current age is 18 years.

9. Use a calculator to find the square root of each number. Only 329,476 has a square root, 524, without a decimal remainder. Thus,

$$524^2 = 329{,}476.$$

11. Fill the big bucket. Pour into the small bucket. This leaves 4 gallons in the larger bucket. Empty the small bucket. Pour from the big bucket to fill up the small bucket. This leaves 1 gallon in the big bucket. Empty the small bucket. Pour 1 gallon from the big bucket to the small bucket. Fill up the big bucket. Pour into the small bucket. This leaves 5 gallons in the big bucket. Pour out the small bucket. This leaves exactly 5 gallons in the big bucket to take home. The above sequence is indicated by the following table.

Big	7	4	4	1	1	0	7	5	5
Small	0	3	0	3	0	1	1	3	0

13. Similar to Example 5 in the text, we might examine the units place and tens place for repetitive powers of 7 in order to explore possible patterns.

$$7^1 = \mathbf{07} \qquad 7^5 = 16{,}8\mathbf{07}$$
$$7^2 = \mathbf{49} \qquad 7^6 = 117{,}6\mathbf{49}$$
$$7^3 = 3\mathbf{43} \qquad 7^7 = 823{,}5\mathbf{43}$$
$$7^4 = 24\mathbf{01} \qquad 7^8 = 5{,}764{,}8\mathbf{01}$$

Since the final two digits cycle over four values, we might consider dividing the successive exponents by 4 and examining their remainders. (Note: We are using inductive reasoning when we assume that this pattern will continue and will apply when the exponent is 1997.) Dividing the exponent 1997 by 4, we get a remainder of 1. This is the same remainder we get when dividing the exponent 1 (on 7^1) and 5 (on 7^5). Thus, we expect that the last two digits for 7^{1997} would be 07 as well.

15. At the end of 1st day, the frog has a net progression of 1 foot; day 2: 2 feet; day 3: 3 feet; ... ; day 16: 16 feet (it crawls up 4 feet from 15 to 19 feet and then falls back 3 feet to 16 feet); on the 17th day it crawls up 4 feet from the 16-foot level, which takes it to the top.

17. Set 4 opposite 12, then 8 is halfway around from 4 to 12. Opposite 8 must be 16 to allow for three equally spaced numbers (children) between each of these values. Note that children 1, 2, and 3 stand between 16 and 4. So there are 16 children in the circle.

19. Solve by drawing a sketch. The following figure satisfies the description. Only three birds are needed.

21. Add the given diagonal elements together to get 15. Each row, column, and other diagonal must also add to 15. This yields the following perfect square.

6	1	8
7	5	3
2	9	4

23. 25 pitches (The visiting team's pitcher retires 24 consecutive batters through the first eight innings, using only one pitch per batter. His team does not score either. Going into the bottom of the ninth inning tied 0–0, the first batter for the home team hits his first pitch for a homerun. The pitcher threw 25 pitches and loses the game by a score of 1–0.)

25. For three weighings, first balance four against four. Of the lighter four, balance two against the other two. Finally, of the lighter two, balance them one against the other.

    To find the bad coin in two weighings, divide the eight coins into groups of 3, 3, 2. Weigh the groups of three against each other on the scale. If the groups weigh the same, the fake is in the two left out and can be found in one additional weighing. If the two groups of three do not weigh the same, pick the lighter group. Choose any two of the coins and weigh them. If one of these is lighter, it is the fake; if they weigh the same, then the third coin is the fake.

27. Draw a sketch, visualize, or cut a piece of paper to build the cube. The cube may be folded with Z on the front.

    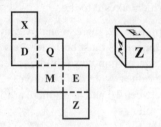

    Then, E is on top and M is on the left face. This places Q opposite the face marked Z. (D is on the bottom and X is on the right face.)

29. This may be worked algebraically or in reverse as Example 2 in the text. Multiplying 2 by 10 gives 20. Subtract 8 to give 12 and then square to get 144. Add 52 to get 196. This represents a number times itself. The number is 14, from the fact that $14 \times 14 = 196$ (or the square root of $196 = 14$). The quotient must be 21 since $21 - \frac{1}{3} \times 21 = 14$. Multiplying 21 by 7, we get 147, which represents 3 times the original number plus $\frac{3}{4}$ of that same product. The original number must be 28 since $3 \times 28 = 84$ and $\frac{3}{4}$ of 84 is 63, and $84 + 63 = 147$.

31. A solution, found by trial and error, is shown here.

    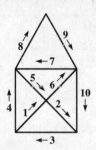

33. Jessica is married to James or Dan. Since Jessica is married to the oldest person in the group, she is not married to James, who is younger than Cathy. So Jessica is married to Dan, and Cathy is married to James. Since Jessica is married to the oldest person, we know that Dan is 36. Since James is older than Jessica but younger than Cathy, we conclude that Cathy is 31, Jame is 30, and Jessica is 29.

35. Common sense tell you that the CEO is a woman.

37. This is a problem with a "catch." Someone reading this problem might go ahead and calculate the volume of a cube 6 feet on each side, to get the answer 216 cubic feet. However, common sense tells us that since holes are by definition empty, there is no dirt in a hole.

39. One solution is
    $1 + 2 + 3 + 4 + 5 + 6 + 7 + 8 \times 9 = 100.$

41. The first digit in the answer cannot be 0, 2, 3, or 5, since these digits have already been used. It cannot be more than 3, since one of the factors is a number in the 30's, making it impossible to get a product over 45,000. Thus, the first digit of the answer must be 1. To find the first digit in the 3-digit factor, use estimation. Dividing a number between 15,000 and 16,000 by a number between 30 and 40 could give a result with a first digit of 3, 4, or 5. Since 3 and 5 have already been used, this first digit must be 4. Thus, the 3-digit factor is 402. We now have the following.

$$\begin{array}{r} \mathbf{4}\,0\,2 \\ \times \quad\ \ 3 \\ \hline \mathbf{1}\,5, \end{array}$$

To find the units digit of the 2-digit factor, use trial and error with the digits that have not yet been used: 6, 7, 8, and 9.

$36 \times 402 = 14{,}472$ (Too small and reuses 2 and 4)
$37 \times 402 = 14{,}874$ (Too small and reuses 4)
$38 \times 402 = 15{,}276$ (Reuses 2)
$39 \times 402 = 15{,}678$ (Correct)

The correct problem is as follows.

$$\begin{array}{r} \mathbf{4}\,0\,2 \\ \times \quad\ 3\,\mathbf{9} \\ \hline \mathbf{1}\,5,\mathbf{6}\,\mathbf{7}\,\mathbf{8} \end{array}$$

Notice that a combination of strategies was used to solve this problem.

43. To count the triangles, it helps to draw sketches of the figure several times. There are 5 triangles formed by two sides of the pentagon and a diagonal. There are 4 triangles formed with each side of the pentagon as a base, so there are $4 \times 5 = 20$ triangles formed in this way. Each point of the star forms a small triangle, so there are 5 of these. Finally, there are 5 triangles formed with a diagonal as a base. In each, the other two sides are inside the pentagon. (None of these triangles has a side common to the pentagon.) Thus, the total number of triangles in the figure is $5 + 20 + 5 + 5 = 35$.

45. None. To see this result, set two books beside each other and note the location of page 1 of the 1st book and the location of the last page of the 2nd book.

47. By Eve's statement, Adam must have $2 more than Eve. But according to Adam, a loss of $1 from Eve to Adam gives Adam twice the amount that Eve has. By trial and error, the counting numbers 5 and 7 are the first to satisfy both conditions. Thus, Eve has $5 and Adam has $7.

APPENDIX B SETS

Appendix B Margin Exercises

1. **(a)** The set of states in the United States whose names begin with the letter O is {Oregon, Ohio, Oklahoma}.

   (b) The set of letters of the alphabet that follow T is {U, V, W, X, Y, Z}.

   (c) The set of even natural numbers less than 10 is $\{2, 4, 6, 8\}$.

   (d) The set of odd counting numbers between 15 and 20 is $\{17, 19\}$.

2. **(a)** The set of whole numbers
   One way to list the elements is $\{0, 1, 2, 3, \ldots\}$. The set is *infinite*.

   (b) The set of odd natural numbers between 10 and 20 is $\{11, 13, 15, 17, 19\}$. The set is *finite*.

   (c) The set of integers greater than 3
   One way to list the elements is $\{4, 5, 6, \ldots\}$. The set is *infinite*.

   (d) The set of rational numbers is an *infinite* set whose elements cannot be listed.

3. **(a)** $B \subseteq A$ is *true* since all the elements of $B = \{2, 4, 8, 10\}$ are also elements of $A = \{2, 4, 6, 8, 10, 12\}$.

   (b) $C \subseteq B$ is *false* since $C = \{4, 10, 12\}$ has an element, 12, which is not in B.

   (c) $A \not\subseteq C$ is *true*. A is not a subset of C because A has elements $(2, 6, 8)$ which are not also elements of C.

   (d) $B \not\subseteq C$ is *true*. B is not a subset of C because B has elements $(2, 8)$ which are not also elements of C.

4. $U = \{0, 1, 2, 3, 4, 5, 6, 7, 8\}$

   (a) $M = \{0, 2, 4, 6, 8\}$, so $M' = \{1, 3, 5, 7\}$, which is the set of elements that are in U but not in M.

   (b) $N = \{1, 3, 5, 7\}$, so $N' = \{0, 2, 4, 6, 8\}$, which is the set of elements that are in U but not in N.

   (c) $Q = \{0, 1, 2, 3, 4\}$, so $Q' = \{5, 6, 7, 8\}$, which is the set of elements that are in U but not in Q.

5. **(a)** $M = \{0, 2, 4, 6, 8\}$ and $N = \{1, 3, 5, 7\}$, so $M \cup N = \{0, 1, 2, 3, 4, 5, 6, 7, 8\} = U$, the set of elements that are in *either M or N*.

   (b) $N = \{1, 3, 5, 7\}$ and $Q = \{0, 1, 2, 3, 4\}$, so $N \cup Q = \{0, 1, 2, 3, 4, 5, 7\}$, the set of elements that are in *either N or Q*.

6. **(a)** $M = \{0, 2, 4, 6, 8\}$ and $Q = \{0, 1, 2, 3, 4\}$, so $M \cap Q = \{0, 2, 4\}$, the set of elements that are in *both M and Q*.

   (b) $N = \{1, 3, 5, 7\}$ and $Q = \{0, 1, 2, 3, 4\}$, so $N \cap Q = \{1, 3\}$, the set of elements that are in *both N and Q*.

   (c) $M = \{0, 2, 4, 6, 8\}$ and $N = \{1, 3, 5, 7\}$, so $M \cap N = \emptyset$, because they have no elements in common.

7. $U = \{1, 2, 3, 4, 6, 8, 10\}$,
   $A = \{1, 3, 4, 6\}$,
   $B = \{2, 4, 6, 8, 10\}$, and
   $C = \{2, 8\}$.

   (a) $B \cup C = \{2, 4, 6, 8, 10\} = B$, the set of elements that are in either B or C.

   (b) $A \cap B = \{4, 6\}$, the set of elements that are in both A and B.

   (c) $A \cap C = \emptyset$, the set of elements that are in both A and C.

   (d) $A' = \{2, 8, 10\}$, the set of elements that are in U but not in A.

   (e) $A \cup B = \{1, 2, 3, 4, 6, 8, 10\} = U$, the set of elements that are in either A or B.

Appendix B Exercises

1. The set of all natural numbers less than 8 is
$$\{1, 2, 3, 4, 5, 6, 7\}.$$

3. The set of seasons is
$$\{\text{winter, spring, summer, fall}\}.$$
The seasons may be written in any order within the braces.

5. To date, there have been no women presidents, so this set is the empty set, written $\emptyset$, or $\{\ \}$.

7. The set of letters of the alphabet between K and M is $\{L\}$.

9. The set of positive even integers is
$$\{2, 4, 6, 8, 10, \ldots\}.$$

11. The sets in Exercises 9 and 10 are infinite, since each contains an unlimited number of elements.

13. $5 \in \{1, 2, 5, 8\}$

5 is an element of the set, so the statement is true.

15. $2 \in \{1, 3, 5, 7, 9\}$

2 is not an element of the set, so the statement is false.

17. $7 \notin \{2, 4, 6, 8\}$

7 is not an element of the set, so the statement is true.

19. $\{2, 4, 9, 12, 13\} = \{13, 12, 9, 4, 2\}$

The two sets have exactly the same elements, so they are equal. The statement is true. (The order in which the elements are written does not matter.)

21. $A \subseteq U$

$A = \{1, 3, 4, 5, 7, 8\}$

$U = \{1, 2, 3, 4, 5, 6, 7, 8, 9, 10\}$

Since all the elements of A are elements of U, the statement $A \subseteq U$ is true.

23. $\emptyset \subseteq A$

Since the empty set contains no elements, the empty set is a subset of every set. The statement $\emptyset \subseteq A$ is true.

25. $C \subseteq A$

$C = \{1, 3, 5, 7\}$

$A = \{1, 3, 4, 5, 7, 8\}$

Since all the elements of C are elements of A, the statement $C \subseteq A$ is true.

27. $D \subseteq B$

$D = \{1, 2, 3\}$

$B = \{2, 4, 6, 8\}$

Since 1 and 3 are elements of D but are not elements of B, the statement $D \subseteq B$ is false.

29. $D \nsubseteq E$

$D = \{1, 2, 3\}$

$E = \{3, 7\}$

Since 1 and 2 are elements of D and are not elements of E, D is not a subset of E, so the statement $D \nsubseteq E$ is true.

31. There are exactly 4 subsets of E.

$E = \{3, 7\}$

Since E has 2 elements, the number of subsets is $2^2 = 4$. The statement is true.

33. There are exactly 12 subsets of C.

$C = \{1, 3, 5, 7\}$

Since C has 4 elements, the number of subsets is $2^4 = 16$. The statement is false.

35. $\{4, 6, 8, 12\} \cap \{6, 8, 14, 17\} = \{6, 8\}$

The symbol $\cap$ means the intersection of the two sets, which is the set of elements that belong to both sets. Since 6 and 8 are the only elements belonging to both sets, the statement is true.

37. $\{3, 1, 0\} \cap \{0, 2, 4\} = \{0\}$

Only 0 belongs to both sets, so the statement is true.

39. $\{3, 9, 12\} \cap \emptyset = \{3, 9, 12\}$

Since 3, 9, and 12 are not elements of the empty set, they are not in the intersection of the two sets. The intersection of any set with the empty set is the empty set. The statement is false.

41. $\{4, 9, 11, 7, 3\} \cup \{1, 2, 3, 4, 5\}$

$= \{1, 2, 3, 4, 5, 7, 9, 11\}$

The union of the two sets is the set of all elements that belong to either one of the sets or to both sets. The statement is true.

43. $\{3, 5, 7, 9\} \cup \{4, 6, 8\} = \emptyset$

The union of the two sets is the set of all elements that belong to either one of the sets or to both sets.

$\{3, 5, 7, 9\} \cup \{4, 6, 8\}$

$= \{3, 4, 5, 6, 7, 8, 9\} \neq \emptyset$

The statement is false.

45. A'

$U = \{a, b, c, d, e, f, g, h\}$

$A = \{a, b, c, d, e, f\}$

A' contains all elements in U that are not in A, so

$$A' = \{g, h\}.$$

47. C'

$U = \{a, b, c, d, e, f, g, h\}$

$C = \{a, f\}$

C' contains all elements in U that are not in C, so

$$C' = \{b, c, d, e, g, h\}.$$

49. $A \cap B$

$A = \{a, b, c, d, e, f\}$

$B = \{a, c, e\}$

The intersection of A and B is the set of all elements belonging to both A and B, so

$$A \cap B = \{a, c, e\} = B.$$

51. $A \cap D$

$A = \{a, b, c, d, e, f\}$

$D = \{d\}$

Since d is the only element in both A and D,

$$A \cap D = \{d\} = D.$$

53. $B \cap C$

$B = \{a, c, e\}$

$C = \{a, f\}$

Since a is the only element that belongs to both sets, the intersection is the set with a as its only element, so

$$B \cap C = \{a\}.$$

55. $B \cup D$

$B = \{a, c, e\}$

$D = \{d\}$

The union of B and D is the set of elements belonging to either B or D or both, so

$$B \cup D = \{a, c, d, e\}.$$

57. $C \cup B$

$C = \{a, f\}$

$B = \{a, c, e\}$

The union of C and B is the set of elements belonging to either C or B or both, so

$$C \cup B = \{a, c, e, f\}.$$

59. $A \cap \emptyset$

Since $\emptyset$ has no elements, there is no element that belongs to both A and $\emptyset$, so the intersection is the empty set. Thus,

$$A \cap \emptyset = \emptyset.$$

61. $A = \{a, b, c, d, e, f\}$ $\quad C = \{a, f\}$

$B = \{a, c, e\}$ $\quad D = \{d\}$

Disjoint sets are sets which have no elements in common.

B and D are disjoint since they have no elements in common. Also, C and D are disjoint since they have no elements in common.

APPENDIX C   MEAN, MEDIAN, AND MODE

Appendix C Margin Exercises

1. $\text{mean} = \dfrac{\text{sum of all values}}{\text{number of values}}$

    $= \dfrac{96 + 98 + 84 + 88 + 82 + 92}{6}$

    $= \dfrac{540}{6} = 90$

    Her mean (average) score was 90.

2. $\text{mean} = \dfrac{\text{sum of all values}}{\text{number of values}}$

    $= \dfrac{(\$25.12 + \$42.58 + \$76.19 + \$32.00 + \$81.11 + \$26.41 + \$19.76 + \$59.32 + \$71.18 + \$21.03)}{10}$

    $= \dfrac{\$454.70}{10} = \$45.47$

    The mean (average) monthly long-distance phone bill was $45.47.

3.
Parking Fee	Frequency	Product
$6	2	$12
$7	6	$42
$8	3	$24
$9	4	$36
$10	6	$60
Totals	21	$174

    $\text{weighted mean} = \dfrac{\text{sum of products}}{\text{total number of days}}$

    $= \dfrac{\$174}{21} \approx \$8.29$

    The average daily parking cost was about $8.29.

4.
Course	Credits	Grade	Credits · Grade
Mathematics	5	A (= 4)	5 · 4 = 20
English	3	C (= 2)	3 · 2 = 6
Biology	4	B (= 3)	4 · 3 = 12
History	3	B (= 3)	3 · 3 = 9
Totals	15		47

    $\text{GPA} = \dfrac{\text{sum of Credits} \cdot \text{Grade}}{\text{total number of credits}}$

    $= \dfrac{47}{15} \approx 3.13$

    The student's GPA was 3.13 (rounded).

5. Arrange the numbers in numerical order from least to greatest.

    25, 27, 30, 31, 33, 35, 39, 50, 59

    The list has 9 numbers. The middle number is the 5th number, so the median is 33.

6. Arrange the numbers in numerical order from least to greatest.

    121, 126, 178, 189, 195, 261

    The list has 6 numbers. The middle numbers are the 3rd and 4th numbers, so the median is

    $\dfrac{178 + 189}{2} = 183.5$ ft.

7. **(a)** <u>28</u>, 16, 22, <u>28</u>, 34, 22, <u>28</u>

    The number 28 occurs three times, which is more often than any other number. Therefore, 28 is the mode.

    (b) <u>312</u>, <u>219</u>, 782, <u>312</u>, <u>219</u>, 426, 507, 600

    Because both 219 and 312 occur twice, which is more often than any other number, each is a mode. This list is bimodal.

    (c) $1706, $1289, $1653, $1892, $1301, $1782

    No number occurs more than once. This list has no mode.

Appendix C Exercises

1. $\text{mean} = \dfrac{\text{sum of all values}}{\text{number of values}}$

    $= \dfrac{4 + 9 + 6 + 4 + 7 + 10 + 9}{7}$

    $= \dfrac{49}{7} = 7$

    The mean (average) age of the infants at the child care center was 7 months.

3. $\text{mean} = \dfrac{\text{sum of all values}}{\text{number of values}}$

    $= \dfrac{92 + 51 + 59 + 86 + 68 + 73 + 49 + 80}{8}$

    $= \dfrac{558}{8} = 69.75$

    The mean (average) final exam score was 69.8 (rounded).

5. $\text{mean} = \dfrac{\text{sum of all values}}{\text{number of values}}$

$= \dfrac{(31{,}900 + 32{,}850 + 34{,}930 + 39{,}712 + 38{,}340 + 60{,}000)}{6}$

$= \dfrac{237{,}732}{6} = 39{,}622$

The mean (average) annual salary was $39,622.

7. $\text{mean} = \dfrac{\text{sum of all values}}{\text{number of values}}$

$= \dfrac{(75.52 + 36.15 + 58.24 + 21.86 + 47.68 + 106.57 + 82.72 + 52.14 + 28.60 + 72.92)}{10}$

$= \dfrac{582.40}{10} = 58.24$

The mean (average) shoe sales amount was $58.24.

9.

Policy Amount ($)	Number of Policies Sold	Product ($)
10,000	6	60,000
20,000	24	480,000
25,000	12	300,000
30,000	8	240,000
50,000	5	250,000
100,000	3	300,000
250,000	2	500,000
Totals	60	2,130,000

$\text{weighted mean} = \dfrac{\text{sum of products}}{\text{total number of policies}}$

$= \dfrac{2{,}130{,}000}{60} = 35{,}500$

The mean (average) amount for the policies sold was $35,500.

11.

Quiz Scores	Frequency	Product
3	4	12
5	2	10
6	5	30
8	5	40
9	2	18
Totals	18	110

$\text{weighted mean} = \dfrac{\text{sum of products}}{\text{total number of quizzes}}$

$= \dfrac{110}{18} = 6.\overline{1}$

The mean (average) quiz score was 6.1 (rounded).

13.

Hours Worked	Frequency	Product
12	4	48
13	2	26
15	5	75
19	3	57
22	1	22
23	5	115
Totals	20	343

$\text{weighted mean} = \dfrac{\text{sum of products}}{\text{total number of workers}}$

$= \dfrac{343}{20} = 17.15$

The mean (average) number of hours worked was 17.2 (rounded).

15.

Course	Credits	Grade	Credits · Grade
Biology	4	B (= 3)	4 · 3 = 12
Biology Lab	2	A (= 4)	2 · 4 = 8
Mathematics	5	C (= 2)	5 · 2 = 10
Health	1	F (= 0)	1 · 0 = 0
Psychology	3	B (= 3)	3 · 3 = 9
Totals	15		39

$\text{GPA} = \dfrac{\text{sum of Credits} \cdot \text{Grade}}{\text{total number of credits}}$

$= \dfrac{39}{15} = 2.60$

17. (a) In Exercise 15, replace 1 · 0 with 1 · 3 to get
$\text{GPA} = \dfrac{42}{15} = 2.80$.

(b) In Exercise 15, replace 5 · 2 with 5 · 3 to get
$\text{GPA} = \dfrac{44}{15} = 2.9\overline{3} = 2.93$ (rounded).

(c) Making both of those changes gives us
$\text{GPA} = \dfrac{47}{15} = 3.1\overline{3} = 3.13$ (rounded).

19. Arrange the numbers in numerical order from least to greatest (they already are).

9, 12, 14, 15, 23, 24, 28

The list has 7 numbers. The middle number is the 4th number, so the median is 15.

21. Arrange the numbers in numerical order from least to greatest.

328, 420, 483, 549, 592, 715

The list has 6 numbers. The middle numbers are the 3rd and 4th numbers, so the median is

$\dfrac{483 + 549}{2} = 516.$

23. Arrange the numbers in numerical order from least to greatest.

$$23, 34, 40, 47, 48, 48, 51, 56, 95, 96$$

The list has 10 numbers. The middle numbers are the 5th and 6th numbers, so the median is

$$\frac{48 + 48}{2} = 48.$$

25. mean = $\dfrac{\text{sum of all values}}{\text{number of values}}$

$= \dfrac{7650 + 6450 + 1100 + 5225 + 1550 + 2875}{6}$

$= \dfrac{24{,}850}{6} = 4141.\overline{6}$

The mean distance flown without refueling was 4142 miles (rounded).

27. **(a)** $1100, 1550, 2875, 5225, 6450, 7650$

There are 6 distances. The middle numbers are the 3rd and 4th numbers, so the median is

$$\frac{2875 + 5225}{2} = 4050 \text{ miles.}$$

(b) The median is somewhat different from the mean; the mean is more affected by the high and low numbers.

29. $3, \underline{8}, 5, 1, 7, 6, \underline{8}, 4, 5, \underline{8}$

The number 8 occurs three times, which is more often than any other number. Therefore, 8 is the mode.

31. $\underline{74}, \underline{68}, \underline{68}, \underline{68}, 75, 75, \underline{74}, \underline{74}, 70, 77$

Because both 68 and 74 occur three times, which is more often than any other value, each is a mode. This list is *bimodal*.

33. $5, 9, 17, 3, 2, 8, 19, 1, 4, 20, 10, 6$

No number occurs more than once. This list has *no mode*.

APPENDIX D FACTORING SUMS AND DIFFERENCES OF CUBES

Appendix D Margin Exercises

1. (a) $t^3 - 64 = t^3 - 4^3$
$= (t-4)(t^2 + 4t + 4^2)$
$= (t-4)(t^2 + 4t + 16)$

    (b) $2x^3 - 54 = 2(x^3 - 27)$
$= 2(x^3 - 3^3)$
$= 2(x-3)(x^2 + 3x + 3^2)$
$= 2(x-3)(x^2 + 3x + 9)$

    (c) $8k^3 - y^3 = (2k)^3 - y^3$
$= (2k-y)\left[(2k)^2 + (2k)(y) + y^2\right]$
$= (2k-y)(4k^2 + 2ky + y^2)$

2. (a) $x^3 + 8 = x^3 + 2^3$
$= (x+2)(x^2 - 2x + 2^2)$
$= (x+2)(x^2 - 2x + 4)$

    (b) $64y^3 + 1 = (4y)^3 + 1^3$
$= (4y+1)\left[(4y)^2 - 4y(1) + 1^2\right]$
$= (4y+1)(16y^2 - 4y + 1)$

    (c) $27m^3 + 343n^3$
$= (3m)^3 + (7n)^3$
$= (3m+7n)\left[(3m)^2 - (3m)(7n) + (7n)^2\right]$
$= (3m+7n)(9m^2 - 21mn + 49n^2)$

Appendix D Exercises

1. $1^3 = \underline{1}$ $2^3 = \underline{8}$ $3^3 = \underline{27}$
$4^3 = \underline{64}$ $5^3 = \underline{125}$ $6^3 = \underline{216}$
$7^3 = \underline{343}$ $8^3 = \underline{512}$ $9^3 = \underline{729}$
$10^3 = \underline{1000}$

3. **C.** $x^3 - 1 = x^3 - 1^3$
D. $8x^3 - 27y^3 = (2x)^3 - (3y)^3$
C and **D** are differences of cubes.

5. $a^3 + 1$
Use the pattern for the sum of cubes.
$a^3 + 1 = a^3 + 1^3$
$= (a+1)(a^2 - a \cdot 1 + 1^2)$
$= (a+1)(a^2 - a + 1)$

7. $a^3 - 1$
Use the pattern for the difference of cubes.
$a^3 - 1 = a^3 - 1^3$
$= (a-1)(a^2 + a \cdot 1 + 1^2)$
$= (a-1)(a^2 + a + 1)$

9. $p^3 + q^3 = (p+q)(p^2 - pq + q^2)$

11. $y^3 - 216 = y^3 - 6^3$
$= (y-6)(y^2 + 6y + 6^2)$
$= (y-6)(y^2 + 6y + 36)$

13. $k^3 + 1000 = k^3 + 10^3$
$= (k+10)(k^2 - 10k + 10^2)$
$= (k+10)(k^2 - 10k + 100)$

15. $27x^3 - 1 = (3x)^3 - 1^3$
$= (3x-1)\left[(3x)^2 + (3x)(1) + 1^2\right]$
$= (3x-1)(9x^2 + 3x + 1)$

17. $125a^3 + 8 = (5a)^3 + 2^3$
$= (5a+2)\left[(5a)^2 - (5a)(2) + 2^2\right]$
$= (5a+2)(25a^2 - 10a + 4)$

19. $y^3 - 8x^3 = y^3 - (2x)^3$
$= (y-2x)\left[y^2 + y(2x) + (2x)^2\right]$
$= (y-2x)(y^2 + 2xy + 4x^2)$

21. $27a^3 - 64b^3$
$= (3a)^3 - (4b)^3$
$= (3a-4b)\left[(3a)^2 + (3a)(4b) + (4b)^2\right]$
$= (3a-4b)(9a^2 + 12ab + 16b^2)$

23. $8p^3 + 729q^3$
$= (2p)^3 + (9q)^3$
$= (2p+9q)\left[(2p)^2 - (2p)(9q) + (9q)^2\right]$
$= (2p+9q)(4p^2 - 18pq + 81q^2)$

25. $16t^3 - 2 = 2(8t^3 - 1)$
$= 2\left[(2t)^3 - 1^3\right]$
$= 2(2t-1)\left[(2t)^2 + (2t)(1) + 1^2\right]$
$= 2(2t-1)(4t^2 + 2t + 1)$

27. $40w^3 + 135$
$= 5(8w^3 + 27)$
$= 5\left[(2w)^3 + 3^3\right]$
$= 5(2w+3)\left[(2w)^2 - (2w)(3) + 3^2\right]$
$= 5(2w+3)(4w^2 - 6w + 9)$

29. $x^3 + y^6 = x^3 + (y^2)^3$
$= (x+y^2)\left[x^2 - x(y^2) + (y^2)^2\right]$
$= (x+y^2)(x^2 - xy^2 + y^4)$

31. $125k^3 - 8m^9$
$= (5k)^3 - (2m^3)^3$
$= (5k - 2m^3)\left[(5k)^2 + (5k)(2m^3) + (2m^3)^2\right]$
$= (5k - 2m^3)(25k^2 + 10km^3 + 4m^6)$

33. $x^6 - 1 = (x^3)^2 - 1^2$
$= (x^3 + 1)(x^3 - 1)$
or $(x^3 - 1)(x^3 + 1)$

34. From Exercise 33, we have
$$x^6 - 1 = (x^3 - 1)(x^3 + 1).$$
Use the rules for the difference and sum of cubes to factor further.
Since
$$x^3 - 1 = (x - 1)(x^2 + x + 1)$$
and
$$x^3 + 1 = (x + 1)(x^2 - x + 1),$$
we obtain the factorization
$x^6 - 1 = (x - 1)(x^2 + x + 1)$
$\cdot (x + 1)(x^2 - x + 1).$

35. $x^6 - 1 = (x^2)^3 - 1^3$
$= (x^2 - 1)\left[(x^2)^2 + x^2 \cdot 1 + 1^2\right]$
$= (x^2 - 1)(x^4 + x^2 + 1)$

36. From Exercise 35, we have
$$x^6 - 1 = (x^2 - 1)(x^4 + x^2 + 1).$$
Use the rule for the difference of two squares to factor the binomial.
$$x^2 - 1 = (x - 1)(x + 1)$$
Thus, we obtain the factorization
$$x^6 - 1 = (x - 1)(x + 1)(x^4 + x^2 + 1).$$

37. The result in Exercise 34 is the completely factored form.

38. Multiply the trinomials from the factored form in Exercise 34 vertically.

$$
\begin{array}{rrrr}
& x^2 & +x & +1 \\
& x^2 & -x & +1 \\
\hline
& x^2 & +x & +1 \\
-x^3 & -x^2 & -x & \\
x^4 & +x^3 & +x^2 & \\
\hline
x^4 & & +x^2 & +1 \\
\end{array}
$$

39. In general, if I must choose between factoring first using the method for a difference of squares or the method for a difference of cubes, I should choose the __difference of squares__ method to eventually obtain the complete factored form.

40. $x^6 - 729 = (x^3)^2 - 27^2$
$= (x^3 - 27)(x^3 + 27)$
$= (x^3 - 3^3)(x^3 + 3^3)$
$= (x - 3)(x^2 + 3x + 9)$
$\cdot (x + 3)(x^2 - 3x + 9)$